Chemistry

EXPERIMENTAL FOUNDATIONS

ROBERT W. PARRY

Professor of Chemistry
University of Utah
Salt Lake City, Utah

HERB BASSOW

Chemistry Teacher
Germantown Friends School
Philadelphia, Pennsylvania

PHYLLIS MERRILL

Chemistry Teacher
Fountain Valley High School
Fountain Valley, California

Prentice-Hall, Inc., Englewood Cliffs, New Jersey

Chemistry

EXPERIMENTAL FOUNDATIONS

Fourth Edition

Chemistry: Experimental Foundations
  Fourth Edition

Robert W. Parry, Herb Bassow, and Phyllis Merrill

Supplementary Material

  Laboratory Manual

  Laboratory Notebook

  Teachers Guide (Includes Achievement Tests)

ISBN 0-13-129081-9

10 9 8 7 6 5 4 3 2 1

Prentice-Hall of Australia, Pty. Ltd., *Sydney*
Prentice-Hall Canada, Inc., *Toronto*
Prentice-Hall Hispanoamericana, S.A., *Mexico*
Prentice-Hall of India Private Ltd., *New Delhi*
Prentice-Hall International (U.K.) Ltd., *London*
Prentice-Hall of Japan, Inc., *Tokyo*
Prentice-Hall of Southeast Asia Pte. Ltd., *Singapore*
Editora Prentice-Hall do Brasil Ltda., *Rio de Janeiro*
Whitehall Books Limited, Wellington, *New Zealand*

PHOTO CREDITS

Cover, title page and back: Ken Karp.
  Chapter 1: Pages 15,16 Center for Disease Control, Atlanta, Ga. Page 18 (top) Ken Karp; (bottom) Larry Mulvehill. Page 19 Photo Researchers, Inc./Herman Emmet. Page 21 Grant Heilman/Barry L. Runk. Page 23 Courtesy of Hamilton Watch Co., Inc. Page 24 Charles F. Plath. Page 25 (left) Woodfin Camp and Associates/Jim Anderson; (right) DPI/George Roos. Page 30 Jerrold J. Stefl. Page 31 United Technologies Research Center.
  Chapter 2: Page 33 DPI/Wil Blanche. Pages 34,35 Jerrold J. Stefl. Page 45 C.L. Martonyi, Photographic Services, University of Michigan.
  Chapter 3: Page 61 Jack D. Griffith. Page 67 From the CHEM Study film *Molecular Motions*.
  Chapter 4: Page 83 United Press International. Page 84 DPI/Hal McKusick. Page 93 Photo Researchers/Carl Frank. Page 94 Stock Boston/Frank Siteman. Page 96 Ireco, Inc.
  Chapter 5: Page 103 DPI/John Di Jardins. Page 120 Black Star/Leo Choplin.
  Chapter 6: Page 133 DPI/John V. Dunigan. Page 150 Ken Karp. Page 151 Corning Glass Works.
  Chapter 7: Page 161 Kitt Peak National Observatory. Page 165 Grant Heilman/Barry L. Runk. Page 169 Peter Arnold, Inc./David Scharf. Page 170 California Institute of Technology.
  Chapter 8: Page 183 Brookhaven National Laboratory/Leonard Newman. Page 190 Courtesy of T. T. Tsong, Pennsylvania State University.
  Chapter 9: Page 211 Sovfoto. Page 226 Godfrey Argent. Page 227 Neil Bartlett. Page 228 E.R. Degginger. Page 231 DPI/Chris Reeberg.
  Chapter 10: Page 239 Ken Karp/courtesy of Let There Be Neon, New York, N.Y. Page 245 Gerhard Herzberg. Page 252 Ken Karp.
  Chapter 11: Andy Walsh. Page 264 C.L. Martonyi, Photographic Services, University of Michigan. Page 266 (top) Andy Walsh; (bottom) C.L. Martonyi, Photographic Services, University of Michigan. Page 269 Grant Heilman/Runk-Schoenberger. Page 275 Courtesy of Education Development Center, Newton, Mass. Pages 277,278 Courtesy of Don T. Cromer. Page 282 Luis W. Alvarez.
  Chapter 12: Page 299 Photograph by Goji Kodama of researchers Steven Snow, Sarah Seversen, and John Higashi, University of Utah. Page 315 Dorothy C. Hodgkin. Page 316 C.L. Martonyi, Photographic Services, University of Michigan. Page 325 Jerrold J. Stefl.
  Chapter 13: Page 331 DPI/W.A. Bentley Collection. Page 335 Dan Hightower. Page 338 Ken Karp. Page 345 Naval Research Laboratory. Page 346 (left) Herb Bassow; (right) Ward's Natural Science Establishment, Inc. Page 349 Jerome Karle. Page 356 Joel Gordon. Page 357 Bell Telephone Laboratories.
  Chapter 14: Page 359 Photo Researchers/Charles Cocaine. Page 370 Oak Ridge National Laboratory. Page 375 Black Star/Tom Tracy. Page 379 University of California, Lawrence Berkeley Laboratory, Photographic Services.
  Chapter 15: Page 389 Photo Researchers/William and Marcie Levy. Page 390 Goji Kodama, University of Utah. Page 398 Monkmeyer/Hays. Page 405 Courtesy of Celanese Corporation.
  Chapter 16: Page 417 Goji Kodama, University of Utah. Pages 419, 421, 425 C.L. Martonyi, Photographic Services, University of Michigan. Page 440 Peter Arnold, Inc./H. Gritscher. Page 441 (top) Charles F. Plath; (bottom) C.L. Martonyi, Photographic Services, University of Michigan. Page 448 Peter Arnold, Inc./Bjorn Bolstad.
  Chapter 17: Page 455 DPI/Ivor A. Parry. Page 459 Corning Medical, Corning Glass Works. Page 464 Herb Bassow. Page 478 Merck and Co., Inc.
  Chapter 18: Page 485 General Motors Corporation. Page 495 Henry Taube. Page 505 Office of University Relations, News Bureau of University of Pennsylvania. Page 511 Ken Karp. Page 520 Courtesy of the National Automotive History Collection, Detroit Public Library.
  Chapter 19: Page 525 Courtesy of *Seventeen* magazine. Page 538 Herbert C. Brown. Page 553 Bill Powers/Criminal Justice Publications.
  Chapter 20: Page 559 Woodfin Camp/Marc and Evelyn Bernheim. Page 567 Marshall W. Nirenberg. Page 570 Medical Media Production Services, Veterans' Administration Medical Center. Page 573 Florida Department of Commerce Division of Tourism. Page 574 (left) M.H.C. Wilkins; (right) Photo Researchers/Lawrence Livermore National Laboratory. Page 575 United Press International/L.A. Kornberg. Page 576 Ken Karp.
  Chapter 21: Page 585 British Airways. Page 590 Ernst O. Fischer. Page 596 John C. Bailer.
  Chapter 22: Page 603 DPI/George Roos. Page 608 Brookhaven National Laboratory/Leonard Newman. Page 609 Merck and Co., Inc. Page 610 Courtesy of the Metropolitan Museum of Art.
  Chapter 23: Page 619 Hale Observatories. Pages 625,626,628,630 National Aeronautics and Space Administration. Page 631 Margaret Burbidge. Pages 632,633 National Aeronautics and Space Administration.

PREFACE

Chemistry: Experimental Foundations is a direct descendent of the now classic CHEM Study program which was developed by a group of leading, practicing chemists in the decade of the 1960's. The original program was focused quite sharply on observation as the key to the development of the science of chemistry. It was hoped that even if the student did not know all about chemistry at the end of the course, he or she would at least know something about how scientific information is obtained and how models and theories are developed. A proper appreciation of these points frequently leads to the questioning of dogmatic assertions and promotes an examination of the evidence (if any) upon which such assertions are based. These lessons extend far beyond science. We live in an age of easy communication; we hear "news" many times each day. Information reaching us varies widely in reliability. A healthy, noncynical skepticism is our best hope for progress in science and for advancement in all other areas of human endeavor. This book has retained the philosophical approach of the original CHEM Study. The point is illustrated by the development of the gas laws in Chapter 2 and by emphasis, where possible, upon the experimental origins of other principles as they are introduced.

In this book, as in the Third Edition, "Extensions" are frequently used to develop: (1) experimental and historical perspective, (2) new information related to topics developed in the text, and (3) other technical topics which the student encounters in daily life. In this context the Extensions provide valuable supplemental reading, but they are not essential for the smooth development of the subject matter. In the Fourth Edition, significant parts of the text have been rewritten and many new and completely revised problems now replace the problems of the Third Edition.

The authors are particularly appreciative of help received from Richard F. Ebeling, West Aurora High School, Aurora, Illinois and George H. Stevens, Lansing High School, Lansing, New York. They reviewed all the material and gave very valuable suggestions and criticisms. Professor Goji Kodama, of the University of Utah, also provided very helpful comments and photographs.

Finally, we want to thank the teachers and students who have used the book in the past for many helpful comments and suggestions. Their help has added significantly to the value and accuracy of the text. This book results from the work of many people. We are grateful to all.

Robert W. Parry

Herb Bassow

Phyllis Merrill

CONTENTS

I wish you not only the joy of great discovery; I wish for you a world of confidence in . . . *humanity,* a world of confidence in *reason,* so that as you work you may be inspired by the hope that what you find *will make men freer and better.*

J. ROBERT OPPENHEIMER

1

EXPLAINING CHEMISTRY: OBSERVATIONS, MODELS, AND EXPERIMENTS

Objectives

After completing this chapter, you should be able to:

- Describe the activities used by scientists in solving problems.
- Use a model for a simple and well-understood situation to explain a related but less-understood situation.
- Understand the reasons for uncertainty and the limits of accuracy in making scientific measurements.

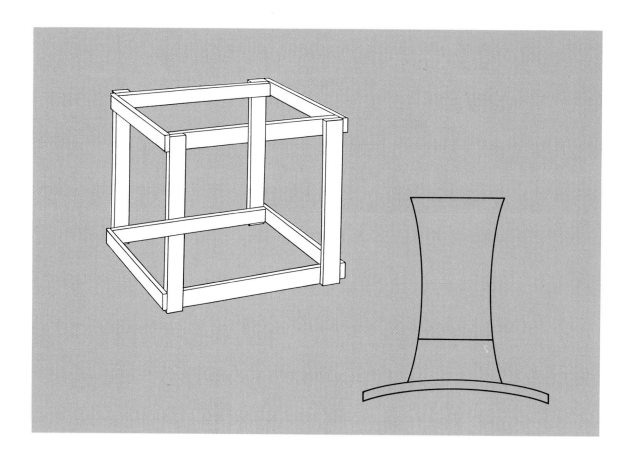

Careful observation is needed to solve these illusions. The "crazy crate" consists of two separate pieces seen from a certain angle of vision. The height of the crown and width of the brim of this top hat are actually equal.

What is chemistry? As most people know, chemistry is one branch of science. But what is science? To answer this question let us look at the activities of a group of scientists who were given the job of investigating an outbreak of disease that appeared at an American Legion Convention in 1976. The following account of their efforts gives an excellent example of how trained scientists approached a difficult and puzzling problem. The example is useful because it illustrates methods of science in action. Although scientists do not follow a set procedure for their work, all scientists engage in certain related activities such as observation and interpretation of facts.

1.1 The activities of science: A case history

In the summer of 1976, a sudden outbreak of a severe pneumonia-like disease occurred at the American Legion Convention in Philadelphia. Nearly 200 people became ill and 29 of them died. The newly discovered disease had taken a heavy toll. A team of scientists from the Center for

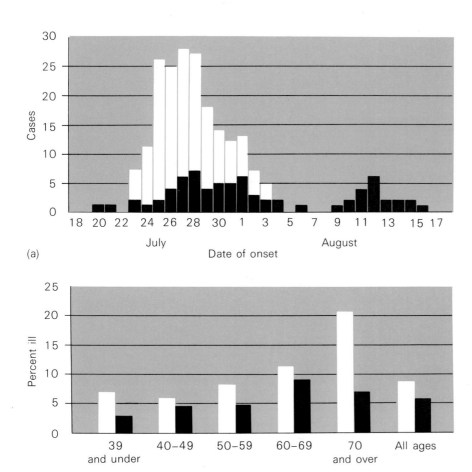

Figure 1-1

(a) Between July 22 and July 25, the number of cases of Legionnaires' disease increased rapidly among the delegates to the convention (white bars) and other people (black bars). (b) The number of cases of Legionnaires' disease varied among the delegates according to age and the hotel at which they stayed. More older delegates than younger ones and more delegates who stayed at one particular hotel (white bars) than those who stayed at other hotels in the area (black bars) contracted the disease.

Disease Control in Atlanta rushed to the scene to try to unravel the mysteries of that deadly disease.

They examined the bodies of the victims and attempted to reconstruct the movements of each one while at the convention. What did the victims have in common? Scientists observed the following:

1. The victims had symptoms similar to those caused by pneumonia.
2. The disease was not transmitted from person to person, because victims' roommates and family members were not infected.
3. All victims either had visited or were near the same hotel during the same ten-day period. Thirty-seven of the victims had not actually

been in the hotel, but they were within one block of the hotel during the ten-day period.

4. No particular room or meeting appeared to be the source of the disease.
5. The lobby was the only place in the hotel where all the victims from the hotel had been.
6. None of the hotel employees contracted the disease.

What questions did the scientists have to answer?

1. How was the disease transmitted?
2. What caused the disease?
3. What treatment could be given for the disease?
4. How could the disease be diagnosed?
5. What was the source of the disease?
6. Why did hotel employees escape the disease?

The scientists knew that to control the disease, they had to find out how it was transmitted. In the case of Legionnaires' disease, solving that mystery proved difficult. Only after persistent research did the scientists begin to observe certain patterns that helped explain how the disease was transmitted. Systematically they ruled out food, water, animals, and person-to-person contact. Finally, the scientists concluded that the disease was transmitted through the air, because that was the only explanation consistent with all the known facts.

It took many months of intensive research to isolate and identify the disease agent, or cause. At first, scientists were not sure what kind of agent produced the disease. Toxins, bacteria, fungi, parasites, viruses—the symptoms might have been caused by any one of them. At first, researchers could not even grow the disease-producing agent in tissue samples from victims. Eventually they succeeded in transmitting the disease from the lung tissue of several victims to guinea pigs. Then they injected fertilized chicken eggs with an agent from those guinea pigs. Using special staining techniques, researchers observed a new rod-shaped bacterium in the dead embryos.

Finally, scientists isolated the new bacterium, *Legionella pneumophilia*, from victims' lung tissue. They discovered antibodies of the bacterium in the blood of victims recovering from the disease. And they found high antibody counts in the hotel employees' blood, which explained their apparent immunity to the disease.

By the time scientists detected the bacterium, Legionnaires' disease had struck several other parts of the country. Hundreds of samples of water, air, dust, soil, and other materials were collected from disease sites. Scientists analyzed the samples and found *Legionella* widely distributed in the soil and, unexpectedly, in the filters and evaporation pans of large air conditioners. Further investigations showed that the air conditioners provided ideal temperatures and humidity for rapid growth of the bacteria. The bacteria became airborne when the air conditioner fans were turned on. Air conditioner cooling towers also transmitted the bacteria by means of a fine water vapor mist.*

*An excellent review of how scientists solved the mysteries of Legionnaire's disease may be found in an article by David W. Fraser and Joseph E. McDade: "Legionellosis," *Scientific American*, vol. 241 (October 1979) pp. 82–99.

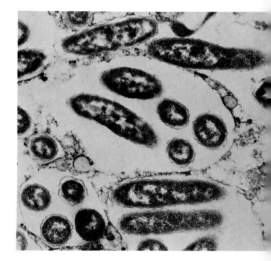

Figure 1-2

Electron micrograph of *Legionella pneumophila*, the rodlike bacterium that causes Legionnaires' disease.

In May 1980, researchers at Rockefeller University in New York announced that *Legionella pneumophila,* like the bacteria that cause tuberculosis and leprosy, can avoid the body's immune system. *Legionella* invades white blood cells called monocytes, which normally kill bacteria, and uses their cellular material to reproduce in large numbers. In one experiment, *Legionella* increased 100,000-fold within days of being added to a culture of monocytes.

In their search, scientists discovered

1. the method by which the disease was transmitted
2. the bacterium that caused the disease
3. antibody tests to diagnose the disease
4. an antibiotic to combat the disease
5. the source of the disease-causing bacteria
6. the method by which the bacterium reproduces in the body

However, questions remained:

1. What procedures were necessary to prevent future outbreaks of the disease?
2. Could different strains or varieties of this bacterium be diagnosed?
3. Could the disease be transmitted by other methods?

Figure 1-3

Laboratory researcher working on a specimen suspected of containing *Legionella pneumophila.*

The attempt to unravel the mysteries of Legionnaires' disease illustrates four basic activities of science:

1. assembling important facts and observations
2. organizing the information and seeking **regularities,** or patterns, in it
3. trying to explain why the regularities exist
4. communicating the findings and their explanations to others

There is no fixed order in which these activities must be carried out. The "scientific method" does not require that the steps be done in the order given above. In fact, the third step usually suggests the need for more carefully controlled observations—a return to steps 1 and 2. A carefully controlled sequence of observations is frequently called an **experiment.** We usually think of scientists conducting experiments only in a laboratory. But the investigation of Legionnaires' disease shows that the work of science takes place all around us.

1.2 Chemistry: A search for regularities

What does the word *chemistry* mean to you? Perhaps it means medicines, fuels, detergents, no-iron shirts, or any of thousands of useful products. These products are the results of chemistry. Chemistry involves the activities that produce them.

Chemistry, like baseball, is not easily defined. We must participate as players or fans to really appreciate the excitement of baseball. In the same way, chemistry is a series of activities in which many people participate. Like the attempt to unravel Legionnaire's disease, chemistry is a search for regularities in the nature of substances and their interactions.

For example, let us consider some observed regularities in the melting of solids. In the laboratory, we observe that certain solid substances melt when heated sufficiently. Those liquids form solids as they cool. From these observations we can make three general statements: (1) certain solid substances melt to form liquids when their temperatures are raised high enough; (2) each of these solid substances melts at a definite temperature; (3) as the resulting liquid cools, the original solid is reformed at the same temperature at which it originally melted.

These regularities hold true for hundreds of thousands of substances. Chemists use the term *substance* to mean a definite kind of material. The solid form of a substance is called its *solid phase.* The liquid form of a substance is called its *liquid phase.* The common melting-freezing temperature of a substance is often called its **melting point.** For example, the melting point of ice is 0.00 °Celsius (°C). The freezing temperature of water in its liquid phase is also 0.00 °C. The change that occurs when a solid melts or a liquid freezes is called a **phase change.** In later chapters, we shall consider other phase changes, such as the change of a liquid to a gas or a gas to a liquid.

The range of melting points is very great. Mercury freezes to its solid phase at −38.9 °C. The temperature range of a mercury thermometer is therefore limited. Tungsten, on the other hand, melts at 3380 °C. This is a very high temperature.

Figure 1-4

Dry Ice changing from a solid directly to a gas is an example of the process of sublimation.

Figure 1-5

The specific melting point of pure ice is characteristic of the substance and does not vary. But the melting point varies when ice is part of a mixture.

Many solid materials do not melt. For example, some substances decompose when heated. Wood chars and loses material until charcoal remains. Wood does not have a liquid phase. Limestone, when heated, loses a gas, carbon dioxide. The remaining solid is quicklime. Some solids soften gradually over a range of temperatures. Glass and wax are examples of this. A few solid substances, such as Dry Ice, which is frozen carbon dioxide, evaporate when warmed. They turn directly into gases without ever melting. This is called **sublimation.** A solid such as iodine is another example of sublimation. Iodine vaporizes directly without passing through the liquid phase.

The specific melting temperature of those solids that do melt is a regularity that can be used to identify them. For example, pure ice always melts at 32.00 °F or 0.00 °C. In contrast, the melting point of a mixture varies over a fairly wide range. For example, when we put salt on ice, the ice melts because the ice-salt mixture has a lower melting point than that of pure ice. Furthermore, we know that as more salt is added, the melting point goes to a lower temperature. In short, a *pure* substance has a fixed and reproducible melting point, if it melts at all, while a mixture of two substances has a melting point that is dependent on the composition of the mixture.

This generalization is widely used in the laboratory to identify a pure substance. If a very pure sample of solid silver is heated steadily, no change in the sample is observed until the melting point is reached. For pure silver, this temperature is 960.8 °C. At the melting point, the solid begins to change to a liquid and the temperature remains constant at 960.8 °C until all the solid has disappeared. After all of the solid has gone, the temperature will again rise as heat is applied. This readily observed sharp melting point at 960.8 °C is characteristic of pure silver. In contrast, butter softens as we heat it and gradually melts over a wide temperature range. It is not possible to identify a sharp and characteristic melting temperature for butter. Butter is identified as a mixture. Thus, the melting behavior of a solid tells whether the material is made of one substance or several substances. The characteristic melting behavior of a solid helps to determine its identity.

EXERCISE 1-1

You are given a solid material that may be any one of four pure substances: camphene, naphthalene, phenol, or quinol. How could you tell which one of the four solid substances had been given to you? Use the following information:

Substance	Melting point (°C)	Color
phenol	40.9	white
camphene	51.0	white
napthalene	80.2	white
quinol	172.3	white

1.3 Explaining regularities

We have considered some of the activities of science: careful observation under controlled conditions, organization of the information obtained, and the search for regularities in the information. We now take a closer look at an activity that might be called wondering why. When we wonder why something happens, we have a desire to know more than merely what happens. Seeking such deeper explanations is probably the most creative and rewarding activity of science.

THE USE OF MODELS

Let us consider what happens when you blow up a balloon. Why does the balloon expand? Why are the sides of the balloon more difficult to push in?

There are two ways to proceed in an attempt to explain what happens. One is to make careful observations and seek regularities. Another is to look for similar behavior in another situation that we understand better. Sometimes a well-understood situation, or **model**, can be very useful in helping us think about a problem. Let us try to build a model to represent the gas in a balloon by using a system that is familiar to us all.

Most of us have played the game of ping-pong. When ping-pong balls drop to the floor, they rebound *almost* to the height from which they were dropped. When hit by a paddle, a ball rebounds. It continues bouncing until it gradually slows down and finally stops. Could there be a connection between the motion of ping-pong balls and the behavior of gas in a balloon?

A collection of bouncing balls is a relatively simple system for experimental study. This system can be described in quantitative terms. Thus, it might be a good model for representing the gas in a balloon. Suppose we picture air, or any other gas, as a collection of miniature ping-pong balls bouncing around inside a container and colliding with its walls. When a ball strikes the container, it pushes against the wall. But the wall pushes back with equal force, and the ball rebounds in a different direction. Remember that the wall does not move. It forces the ball to change direction. If there was a very

Figure 1-6

Adding gas to a balloon makes its surface harder. A model of gas behavior can be used to help explain this observation.

Figure 1-7

(a) Ping-pong balls in motion and (b) ping-pong balls after many collisions.

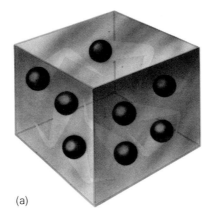

(a)

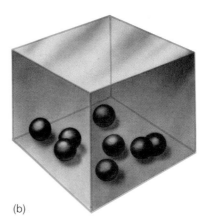

(b)

large number of balls, there would be many collisions and rebounds per second. Such a model could account for the "push" of air on the inner surface of a balloon. The push acting on a given area of surface is called **gas pressure.**

We could say that the collisions of the gas "balls" with the balloon surface account for the observed pressure of the gas. Adding gas to the balloon by blowing air into it increases the number of gas "balls." Thus, there are more surface collisions per second. This means more **pressure** (*push per unit area*) on the balloon surface. The rubber balloon stretches outward, exposing *more surface area*. This process continues until the push per unit of area is only slightly larger than it was before to compensate for the slightly higher pressure exerted by the more tightly stretched surface of the balloon. In short, the balloon grows bigger when gas is added.

Our ping-pong ball model has passed its first test. It reproduces several of the behaviors we can observe when we blow up a balloon. In that sense, it is a good model and deserves more careful examination. It leads to many predictions about the nature of gases. We will explore these predictions more carefully in Chapter 5.

THE LIMITATIONS OF MODELS

The ping-pong ball model for the behavior of particles in a gas explains the observations made so far. But the model cannot explain all gaseous behavior.

For example, when a ping-pong ball is dropped from a certain height, it rebounds, but not quite to the same height from which it was dropped. In a perfectly elastic bounce, the ball would return to its original height. A perfectly elastic ball would bounce forever. Real balls slow down after each collision and finally stop. Do particles in gases behave this way?

To answer the question, consider the following experiment. Air is forced into a scuba-diving tank to the desired pressure. The ping-pong model predicts that the pressure in the tank will fall as time passes. All the particles will eventually drop to the bottom of the tank. But this does not happen. As long as the tank is at the same temperature and does not leak, the air pressure in the tank stays the same. So the particles in a gas must make perfectly elastic collisions with the tank walls. Hence, there is at least one difference between a gas and the ping-pong ball model.

EXERCISE 1-2

A balloon filled with air gradually shrinks on standing for some time. Suggest an explanation.

The ping-pong ball model is good for explaining the behavior of gas in a balloon if we understand that the collisions of particles in a real gas are more elastic than those in the model. We have seen how a model can be used to answer a question about something that is not well understood. The

Figure 1-8
The pressure of gas in a tank remains unchanged at a given temperature as long as the gas is not used or does not leak from the tank.

answer is given in terms of something that is well understood. The model may suggest new experiments or observations. The results may show that the unknown phenomenon is similar to the model or that it differs. If the differences are not great, the model can be retained and used.

The search for explanation, then, is the search for likenesses that connect the system under study with the model. The explanation is good when

1. the model is well understood
2. there are close similarities between the model and the system being explained

1.4   Conclusions and the accuracy of observations

The regularities you read about earlier can help solve practical problems. Suppose that in the laboratory, two chemicals are mixed and a reaction occurs. One of the isolated products is a white crystalline substance. Can the product be identified by its melting temperature (see Table 1-1)?

In an experiment, the melting temperature of the white solid is found to be 27 °C. According to Table 1-1, material C melts at 27.5 °C. Since that is closest to the experimental melting temperature, the product can be tentatively identified as material C.

But wait! Can material D, with a melting temperature of 26.0 °C, be ruled out? That depends on the accuracy of the table values and the experimental

Table 1-1

Material	Melting point (°C)
A	32.0
B	10.1
C	27.5
D	26.0
E	12.0
F	18.0
G	24.5
H	40.0
I	0.0
J	−25.0

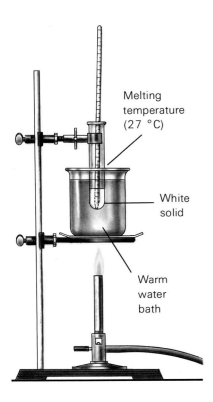

Figure 1-9

Apparatus for observing the behavior of the white solid at its melting point.

value, 27 °C. Suppose that the table values can vary by 0.5 °C either way. Then the table value for material C could be any value between 27.0 °C and 28.0 °C (27.5 °C ± 0.5 °C). And the value for material D could range from 25.5 °C to 26.5 °C.

What about the experimental observation? If the temperature reading could be off by as much as one degree in either direction, the melting point could be any value between 26 °C and 28 °C (27 °C ± 1 °C), falling within the range of possible values for material C and material D. The measured values and the table values overlap. Therefore, it can be concluded that the white solid could be either C or D, within the limits of the specified uncertainties, as shown in Figure 1-10.

EXERCISE 1-3

Suppose the laboratory measurement could be in error by as much as 2 °C either way. What melting point temperatures for the unknown solid now exist?

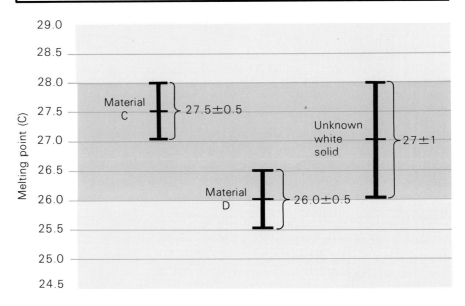

Figure 1-10

Within the limits defined by uncertainty, the unknown solid could be either material C or material D.

UNCERTAINTY

As illustrated, the ability to use melting points to identify substances depends upon how accurately the melting points are known. You might wonder why the exact melting points were not determined in the first place.

Unfortunately, every measurement involves unavoidable **uncertainty.** For example, in reading the melting temperature of a substance as shown in Figure 1-11, most would agree that the temperature is at least 30.9 °C. However, the top of the mercury column extends slightly higher than 30.9 °C.

Different observers would not agree on how much higher. Estimates might range from 30.91 °C to 30.93 °C, with an average of 30.92 °C. In other words, the first three figures would be certain, and the fourth would not.

All figures that are certain in a measurement plus one uncertain figure are called the **significant figures.** Remember, the last figure is used to show where uncertainty begins. Scientists use significant figures to indicate the reproducibility, or precision, of their measurements. In our example, the significant figures indicate that all observers would agree on the value 30.9 but that the last number might vary from one observer to another. We shall use the concept of significant figures frequently to indicate the precision of our measurements.

The size of the uncertainty in the last place can be recorded so as to cover the several possible readings by different observers. In the case of the thermometer in Figure 1-11, we might reasonably indicate the uncertainty in the last figure by writing 30.92 °C ± 0.01. Then, in addition to showing that the last numeral is uncertain, we are showing *how* uncertain. Try Exercise 1-4.

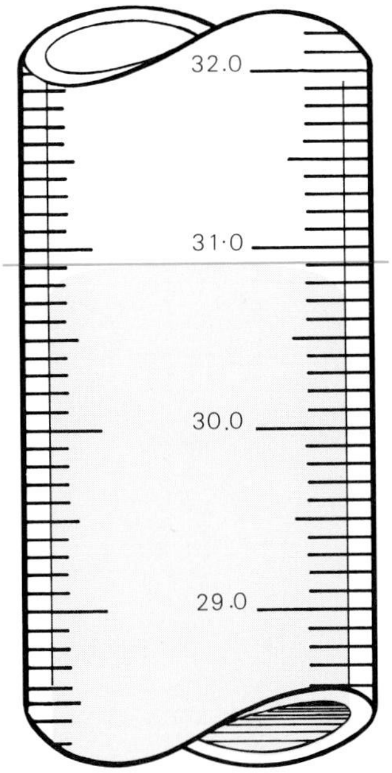

EXERCISE 1-4

You are given a thermometer reading of 30.92 °C ± 0.02. How many certain figures are there in the number? Considering the uncertainty of the last figure, ± 0.02, what are the highest and lowest possible readings in this case?

Figure 1-11
Some uncertainty will result when reading this thermometer.

ACCURACY AND PRECISION

So far, our discussion of uncertainty seems to have centered on the question, how closely can different observers read the same instrument? Answers to this question tell us something about the precision of the instrument. A measurement of melting point by a given thermometer may be extremely reproducible, giving the same result to about ±0.02 °C each time. Such a result is *precise,* but it is not necessarily accurate.

Why not? Different observers can use this thermometer and get the same result each time. Why is the result not necessarily accurate?

The answer to this question may lie in the nature of the measuring instruments used. The point can be illustrated by answering the question, what time is it? Suppose that your classroom has a large clock in the front. At a given instant all members of the class are asked to read the time. Some of the readings are exactly 10:25.0, exactly 10:27.0, and exactly 10:26.5.

The readings seem to be precise (10:26 ± 1 minute), but it would be foolish to believe that we know the exact time to within one minute. Why? Because the clock in front of the room may be slow or fast or even stopped. Suppose it was four minutes slow and the time for the region was 10:30.0 by an astronomical observatory. The reading by the class would be accurate to no better than ±4 minutes. In short, **accuracy** is the closeness of a measurement to a best or accepted value. **Precision** describes the reproducibility of the measurement.

Figure 1-12
Modern watches that use batteries and solid state circuitry are more accurate than watches that must be wound. However, all watches must be corrected periodically to be accurate.

EXERCISE 1-5

Three targets with shot marks are shown below. Indicate which target represents (a) good precision but poor accuracy, (b) poor precision and poor accuracy, and (c) good precision and good accuracy.

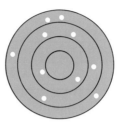 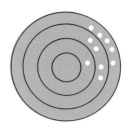 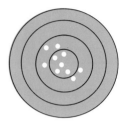

All measurements are uncertain, due to the limitations of measuring devices and of human abilities to use them. Estimates of uncertainty in the use of various pieces of your laboratory equipment are given in the Laboratory

OPPORTUNITIES

Writing and editing

Career opportunities for people with scientific training and writing talent are considerable. The need for current information about new and exciting scientific topics is great. Someone must present the best up-to-date information in a simple, accurate, and lively manner. Therein lies the challenge of the scientific writer.

Scientific writers have the important responsibility of keeping the public informed about developments that could have profound effects upon future generations. An informed public is better able to make decisions about supporting scientific research. In addition to writing books, writers with scientific training work for newspapers and magazines. They report firsthand such major scientific events as rocket launchings and breakthroughs in energy research. They also cover important developments in medical research. They are, in fact, concerned with every aspect of scientific research and development.

Some writers are employed by industries engaged in researching and manufacturing drugs, plastics, cosmetics, and other products. These writers produce promotional literature and press releases about discoveries made by scientists working in industry. Publishing companies also employ science writers and editors. Chemistry textbooks need constant revision because of new developments in content and new methods of presentation. Writers and editors are required to update textbooks in order to keep them useful.

A college degree is basic to success in science writing and editing. Highly technical writing and editing require graduate training in chemistry and related areas. Another prerequisite, of course, is the ability to write about complex ideas in a simple but interesting manner.

The demand for people with such a rare combination of abilities and interests is likely to grow in the fu-

ture. The public is showing increased interest in what is going on in the scientific laboratories of the world. Additional information about scientific writing can be obtained from the American Chemical Society, Department of Educational Activities, 1155 Sixteenth Street NW, Washington, D.C. 20036.

(a)

(b)

Manual. The estimates indicate how much confidence you can have in any measurement you take with any given instrument.

It is important to indicate the uncertainty of your measurements. You can do this most easily by using the proper number of significant figures. Your results will be even more meaningful if you also indicate uncertainty by including plus-or-minus values after the result. Uncertainty in measurements produces uncertainty in results calculated from those measurements. In your laboratory work, you will explore how the value of calculated results depends on the uncertainty of measurements. (See Appendix 1.)

1.5 Communicating scientific information

One of the underlying reasons for human progress is the ability to communicate information to others. It is not necessary for each of us to invent the atomic description of matter. That was done by others and passed on to us through lectures, books, and papers. But a scientific advance is useful only if it is explained to others in a way they can understand. The importance of communication is vividly illustrated by the following episode from the early history of the atomic theory.

In 1789, William Higgins published a book in Britain entitled *A Comparative View of the Phlogiston and Antiphlogiston Theories*. In that book, Higgins introduced many of the ideas that were to form the basis of the atomic theory. However, Higgins's writing was difficult to follow, and his views were not widely read or accepted. Nineteen years later, John Dalton, a Manchester schoolteacher, published a book entitled *A New System of Chemical Philosophy*. Dalton, though a man of limited education and background, wrote clearly. He used pictures and symbols to make the concept of atoms easily understood. Dalton's book attracted much attention and led to the acceptance of the atomic theory.

In 1814, Higgins claimed that his book, published in 1789, had contained a description of the atomic theory. Higgins's claim to priority was supported by the prominent chemist Humphry Davy, who wrote in 1811: "It is not a little curious that the first views of the Atomic Chemistry, which has been so expanded by Dalton, are to be found in a work published in 1789 by William Higgins." Many historians of science have since examined Higgins's book, but few have recognized the atomic theory in its pages. For that reason, Dalton is usually given sole credit for the atomic theory and Higgins is seldom mentioned. This is unfortunate, since it appears that Higgins did indeed foresee some of the concepts of the atomic theory. But because his writing failed to present his ideas in a way that could be understood and used by most scientists of the day, his contribution attracted little attention and was not considered to be of great value. Even today, he does not usually get proper credit for his great insight.

A scientific advance is useful *only* if it is told to others in a manner that they can understand. In this book, you will encounter many different ways of presenting an idea. In the end, it is the reader, not the author, who decides which method is best.

Today, scientific ideas come at us from all directions. Besides actual laboratory records of experiments and observations, we have books, review and research journals, newspapers, magazines, and television. Scientists frequently communicate with one another through journals. Review journals contain articles that summarize all the available information on a given topic, but such articles go out of date rapidly. Research journals record the results of original experiments and describe new ideas. The American Chemical Society publishes a review journal entitled *Chemical Reviews.* It also publishes such research journals as the *Journal of the American Chemical Society,* the *Journal of Organic Chemistry,* the *Journal of Physical Chemistry, Inorganic Chemistry* and *Biochemistry.* A magazine for high school students called *Chem Matters* is also published. The National Research Council of Canada publishes the *Canadian Journal of Chemistry,* and the Chemical Institute of Canada publishes *Chemistry in Canada.* Throughout the world, chemical societies publish journals recording ideas and experiments on the frontiers of science.

Key Terms

accuracy
experiment
gas pressure
melting point
model
phase change
precision
pressure
regularity
significant figures
sublimation
uncertainty

SUMMARY

The basic activities of science include

1. assembling important facts and observations
2. organizing the information and seeking regularities, or patterns, in it
3. trying to explain why the regularities exist
4. communicating the findings and their explanations to others

Regularities in chemistry can be determined through observation and experiment. Such activities enable us to observe, for example, that the temperature at which a solid melts is characteristic of that solid.

A model is a well-understood system that can be used to help explain a system that is poorly understood. A model is good when there are close similarities between the model and the system being explained.

There is some uncertainty in every measurement we make. In recorded values of measurements, the certain figures and one uncertain figure are called significant figures. Precision is the degree to which various observers make the same reading using the same measuring instrument or experimental setup. Accuracy is the degree to which the observed reading agrees with a best or accepted value.

The ability to communicate scientific information is an underlying reason for human progress. Today, we have more methods of receiving such information than at any time in the past.

QUESTIONS AND PROBLEMS

A

1. What caused Legionnaires' disease? How was it spread in Philadelphia?
2. Why were some people immune to the disease?
3. Why was the source of the disease agent so difficult to trace?
4. What basic activities of science are illustrated by the story of Legionnaires' disease?
5. A substance melts sharply and freezes sharply at 6.25 °C. Is this material a mixture or a pure substance?
6. What is meant by a phase change for a substance?
7. The *Handbook of Chemistry and Physics* lists a melting point for ordinary table salt but it does not give a melting point for wax, wood, or limestone. Why?
8. What is the term for the push of a gas on a unit area of the surface of a balloon?
9. What important factors must be considered when comparing an observed melting point to a handbook value?
10. What are significant figures?
11. Different students read a thermometer. The following values are recorded: 66.5, 65.5, 66, 67.5, 65. The temperature reading should be correctly expressed to how many significant figures?
12. What is the difference between accuracy and precision?
13. A digital watch can be read to the closest second. Will readings on this watch be precise? Will they be accurate? Explain.

B

14. How was it determined that Legionnaires' disease was not transmitted from person to person? What was the most important regularity that scientists discovered about Legionnaires' disease?
15. What would happen to the pressure in a scuba tank of oxygen if the "balls" making up the gas behaved just like real ping pong balls? What would happen to the pressure in the tank if all of the "balls" making up the gas began to move more rapidly?
16. (a) What is a good model? (b) How does the ping-pong ball model of a gas explain the increasing pressure that expands a balloon? (c) Explain how the ping-pong ball model fails to explain gas behavior. (d) Is a collision between automobiles an elastic collision? Explain your answer.
17. Another model for air is a group of bees flying around in a closed jar. (a) In what ways is this model similar to the ping-pong ball model for air? (b) In what ways is the bee model inadequate for explaining the behavior of air?

18. Given the following table of melting and boiling points of various substances, list as many regularities as you can for these data.

Substance	Melting point (°C)	Boiling point (°C)
1. copper metal	1083	2583
2. silver metal	961	2193
3. iron metal	1535	2800
4. moth balls*	53	175
5. sugar	112	—
6. water	0	100
7. wax	53–55	282

* Substances 4–7 are nonmetals.

19. How would the regularities from question 18 be affected if the table were to include the following substances as well?

mercury metal	−39	357
salt (a nonmetal)	800	1413

20. Assume that you mixed two solid substances each of which has a definite melting point. Would you expect the mixture to melt at a definite temperature? Explain.
21. Read the four measurements at the left to the correct number of significant figures and indicate the uncertainty you feel there is in the reading.
22. The mass of an empty sugar bowl was found to be 585 ± 1 g. When filled with sugar, the mass was 943 ± 1 g. Which of the following statements about this operation is *not* true? (a) The mass of the sugar added to the bowl can be correctly expressed as 358 ± 1 g. (b) The balance has an uncertainty of ±1 g. (c) Each of the masses was determined to three significant figures. (d) The largest mass that the empty sugar bowl could have had is 586 g. (e) The sugar could have had a mass of anywhere between 356 and 360 g.
23. Determine the number of significant figures in each of the following numbers: (a) 10.4 (b) 46.32 (c) 98.40 (d) 10.020 (e) 100.4620.

C

24. A dieter weighed himself one week and found his weight to be 175 pounds. Two weeks later his weight was 170 pounds. Assume that his scale has an uncertainty of ±1 pound. (a) What is the maximum possible weight loss? (b) What is the minimum possible weight loss? (c) Express the weight loss using the ± notation.
25. The mass of a piece of metal may be as much as 264 grams (g) or as little as 260 g. How many significant figures are used to express the mass of the metal? How would the mass be given using the ± designation?
26. The piece of metal used in question 25 is cut in two. One piece has a mass of 230 ± 2 g. What is the *largest* mass that the other piece could

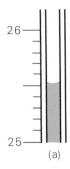

(a)

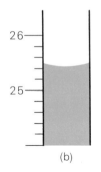

(b)

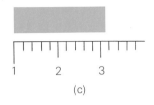

(c)

(d)

have? What is the *smallest* mass that the other piece could have? Express the mass of the other piece using a plus-or-minus estimate to indicate the uncertainty of your measurement.

27. The accepted value for the melting point of a substance is 35.0 ± 0.5 °C. (a) Which of the following melting points correspond to the accepted value: substance A = 34 ± 1 °C, substance B = 34.0 ± 0.5 °C, substance C = 34.0 ± 0.2 °C, substance D = 36.0 ± 0.5 °C? (b) If the accepted value had been 35.0 ± 0.2 °C, which of the substances would correspond?

28. Suppose substances A and B both have a melting point of 27 ± 2 °C and your melting point measurement on an unknown substance is 26 ± 2 °C. Can these numbers be used to identify the unknown as either A or B? Explain.

29. Suppose you were attempting to identify a substance by its melting point and you found that it started melting at 26 °C and finished melting at 30 °C. How would you record the melting point? What would your observations indicate?

30. How many significant figures are present in each of the following numbers: (a) 0.0255 (b) 0.00205 (c) 10200 (d) 0.0020200?

31. Round off each of the numbers in question 30 to two significant figures.

32. Solve each of the following problems and express the result to the correct number of significant figures using exponential notation where necessary: (a) 4.0 × 0.2 (b) 4.00 × 0.20 (c) 4.0 × 0.002 (d) 40 × 2.0 (e) 400 × 20 (f) 40 × 0.0200 (g) 4.0 / 0.2 (h) 4.00 / 0.20 (i) 4.0 / 0.002 (j) 40 / 2.0 (k) 400 / 0.2 (l) 40 / 0.0200.

E X T E N S I O N

Scientific observation: Studying flames

One of the basic activities of science is the assembly of facts and observations. In this chapter you have read how scientists investigating Legionnaires' disease began their study by making a number of observations. Usually, scientists make both qualitative and quantitative observations. To make quantitative observations, they use various instruments to measure the phenomenon under investigation.

To illustrate this point, let us observe and describe a burning candle. How many observations of this "simple" phenomenon would you be likely to make? A half dozen? A dozen? As the discussion below reveals, there are at least 53 separate and distinct observations that can be made. Each observa-

tion is indicated by a superscript. Be sure to note the difference between observation and interpretation.

Observations of a burning candle Figure 1-14 (see page 30) is a photograph of a burning candle.[1] The candle is cylindrical in shape[2] and has a diameter[3] of about $1\frac{1}{2}$ cm. The length[4] of the candle is initially about 8 cm. Its length changes slowly[5] during observation, decreasing about 1 cm per hour.[6] The candle is made of a translucent,[7] white[8] solid[9] with a slight odor[10] and no taste.[11] It is soft enough to be scratched with your fingernail.[12] There is a wick[13] that extends from the top to the bottom[14] of the candle along its central axis.[15] The wick protrudes[16] above the top of the candle about 1 cm. It is made of three strands of string braided together.[17]

The candle is lighted by holding a source of flame close to the wick for a few seconds. The source of the flame can be removed and the flame will sustain

Figure 1-14
A burning candle.

the flame has a blue tint.[27] Immediately around the wick in an area about $\frac{1}{2}$ cm wide and extending about 1 cm above the top of the wick,[28] the flame is dark.[29] This dark area is roughly conical in shape.[30] Around the dark area and extending about 1 cm above it is a section that emits yellow light[31] that is bright but not blinding.[32] The flame has rather sharply defined sides[33] but a ragged top.[34]

The wick is white where it emerges from the candle.[35] It is black from the base of the flame to the end of the wick, appearing burned,[36] except for the last 0.1 cm, which glows red.[37] The wick curls over about $\frac{1}{2}$ cm from its end.[38] As the candle becomes shorter, the wick becomes shorter too, so as to extend roughly a constant length above the top of the candle.[39]

Heat is emitted by the flame.[40] There is enough heat so that your finger becomes uncomfortable in 10 or 20 seconds if you hold it $\frac{1}{2}$ cm to the side of the flame[41] or 6 to 8 cm above it.[42]

The top of the quietly burning candle becomes bowl shaped[43] and wet with a colorless liquid.[44] If the flame is blown, one side of the bowl-shaped top may liquefy, and the liquid trapped in the bowl may drain down the side of the candle.[45] As the liquid runs down, it cools[46] and becomes translucent.[47] The liquid gradually solidifies from the outside[48] and attaches itself to the side of the candle.[49] When there is no draft, the candle can burn for hours without such drippings.[50] Under such conditions, a stable pool of clear liquid remains in the bowl-shaped top of the candle.[51] The liquid rises slightly around the wick,[52] wetting the base of the wick up to the base of the flame.[53]

Several aspects of this description deserve specific mention:

1. The description is comprehensive in *qualitative* terms. It includes reference to appearance, smell, taste, feel, and sound. (Note: *A chemist quickly becomes reluctant to taste or smell an unknown chemical. A chemical should be considered poisonous unless it is known not to be.*)
2. Whenever possible, the description is stated *quantitatively*. This means that the question, how much? is answered. The quantity

itself at the wick.[18] The burning candle makes no sound.[19] While burning, the body of the candle remains cool to the touch[20] except near the top. Within about 1 cm from the top, the candle is warm (but not hot)[21] and sufficiently soft to mold easily.[22]

The flame flickers in response to air currents[23] and tends to give off smoke while flickering.[24] In the absence of air currents, the flame is of the form shown in Figure 1-14, although it exhibits some movement at all times.[25] The flame begins about $\frac{1}{4}$ cm above the top of the candle,[26] and at its base

is specified. The observation that the flame emits yellow light is made more meaningful by the phrase "bright but not blinding," which indicates how much light is emitted. Any statement to the effect that heat is emitted might lead a cautious investigator who is lighting a candle for the first time to stand some distance away. A few words telling how much heat would save the investigator the trouble of taking this unnecessary precaution.

3. The description does not make assumptions regarding the relative importance of observations. Thus, the observation that a burning candle does not emit sound deserves to be mentioned just as much as the observation that it does emit light.

4. The description does not confuse observation and interpretation. To say that the top of the burning candle is wet with a colorless liquid is to make an observation. To suggest a possible composition for this liquid is to offer an interpretation.

New techniques for studying flames What goes on inside a flame has become the focus of recent scientific research. The reason is that most of our energy comes from the combustion, or burning, of fuels. New techniques have been developed to study the effects of different temperatures, pressures, and rates of fuel flow on a flame. By 1981, these new techniques and the research underlying them were developing so rapidly that the journal *Combustion and Flame* increased its frequency of publication.

Today, researchers are able to probe the complex processes inside a flame using instruments such as lasers, spectrometers, and computers. Precise temperatures and atomic and molecular components at any point within a flame can be determined by the new technology. Temperature ranges and composition can be determined for flames produced by different fuels under differing pressures and with varying amounts of oxygen.

A powerful method of analyzing the chemistry of flames involves the use of a well-known instrument, the mass spectrometer. Through ingenious new sampling processes, the masses of individual

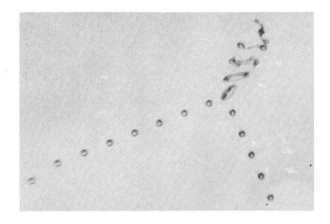

Figure 1-15

Scientists study the size of and spacing between fuel droplets, and the amount of oxygen present, in order to determine their effects on the burning efficiency of the fuel.

atomic or molecular ions at many different locations throughout the flame can be determined by this instrument. The flame chemistry of natural gas, or methane, is now known almost completely. This knowledge should help scientists increase the burning efficiency of methane.

The new technologies are also helping to determine the effects of fuel additives on combustion reactions and to find the optimum burning temperatures of many different fuels. Computerized equipment has been developed to observe liquid fuel injected into a combustion chamber as a stream of individual droplets (3,600 droplets per second). Studies show that the size and spacing between droplets, and the amount of oxygen present, affect combustion efficiency. Changes in spacing between droplets also affect the amount of incompletely burned substances, such as the carbon in soot.

Using the new techniques, automobile and jet engine manufacturers can uncover the "history" of a flame in a combustion chamber. Automobile builders can track flames inside operating cylinders to see whether fuel is being burned evenly in all parts of the chamber. If the fuel is not being burned evenly or completely, the engine will run less efficiently and poorer gas mileage will result. If this is the case, the design of combustion chambers can be modified to improve fuel-burning efficiency.

> Hypotheses ought to be fitted merely to explain the properties of things and not attempt to predetermine them except insofar as they can be an aid to experiments.
>
> SIR ISAAC NEWTON

OUR MODEL GROWS: MOLECULES, MOLES, AND MOLECULAR WEIGHTS

Objectives

After completing this chapter, you should be able to:

- Explain the relative molecular weight scale based on Avogadro's hypothesis.
- Explain why chemists invented the mole.
- Explain how the mole is used as a unit of quantity of substance measured, either in mass or number of particles.
- Solve problems using the number of moles or the molar mass of a substance.

Inflating a balloon is a simple demonstration of the particle model for gases.

In Chapter 1, the behavior of gases became less mysterious. We related gas behavior to the behavior of ping-pong balls in constant motion. Building on this model, we will be able to compare the number of particles in a standard quantity of different gases. We will also be able to find the relative mass of each kind of particle.

Gas particles are much too small to be counted and weighed individually by balances or scales. Instead, chemists use a quantity of a substance called a *mole*. This quantity is like a chemist's "dozen." Just as a dozen eggs, a dozen cupcakes, and a dozen watermelons have different weights, a mole of one substance usually weighs more or less than a mole of another substance. You will learn more about the mole later.

In this chapter, you will be working with gases. You may wonder why you should start with gases. After all, gases are difficult to handle. Nearly all of them are invisible. They must be carefully contained, and they are more difficult to measure than solids and liquids. In spite of these problems, your study begins with gases because regularities in gas behavior are easier to discover in the laboratory and to understand in model building.

One of the activities of science is to search for regularities. Scientists search because they have learned that a true regularity predicts the results of many experiments. When an apparent regularity is false, it soon runs into conflict with experimental results. For example, a statement of regularity such as "all balloons are red" must be either rejected or modified. But when an apparent regularity is true, it will predict with reasonable accuracy the results of many experiments. The predictions may be either quantitative or qualitative.

A true regularity should express a relationship between different measurable quantities or observations in the laboratory. A quantitative relationship may take the form of an equation. An equation can eliminate the need for lengthy observations because, if you know one quantity, you can compute the other. You will soon see many examples of such regularities.

As a regularity predicts more and more laboratory results, its worth grows and it becomes a **law.** A law indicates a proven relationship between observations. In explaining a law or a regularity, we usually use a model.

In Chapter 1, for example, we explained the behavior of a gas by imagining that it was composed of particles that moved like ping-pong balls. That mental picture was the model for the gas. Useful models are at the heart of scientific thought. The ping-pong ball model helps us explain the kinetic theory describing gas behavior. The kinetic theory will be discussed in Chapter 5.

2.1 Growth of a model

The explanation of the behavior of a gas in terms of particles is called a **theory.** But how did this theory develop? First, review the model used to describe a gas confined in a balloon. We pictured the gas as a collection of small, rapidly moving particles. These particles rebound from the walls of the balloon and do not lose energy. Their speed is unchanged. As the particles rebound from the balloon wall, they push on it. These pushes are responsible for the pressure exerted on the balloon wall. As more gas particles are forced into the balloon, the pressure on the walls increases and the balloon becomes larger. The volume of the gas in the balloon increases as the result of the increased pressure generated by the gas particles *inside* the balloon. If we squeeze the balloon between our hands—that is, exert pressure on the *outside* of the balloon—the volume becomes smaller. There is apparently a relationship between the pressure exerted on a gas and the volume it occupies. When the pressure is increased, the volume decreases.

Regularities in science are most useful when they are **quantitative.** Then they answer the question, how much? as well as the question, what happens? For example, in explaining the behavior of gases, we would like to know "how much" pressure and "how much" volume. A numerical relationship between the pressure exerted by the gas and the measured volume of the gas is called a quantitative relationship. The simple experiment described below helps us develop a quantitative relationship between the pressure exerted by a gas and its volume.

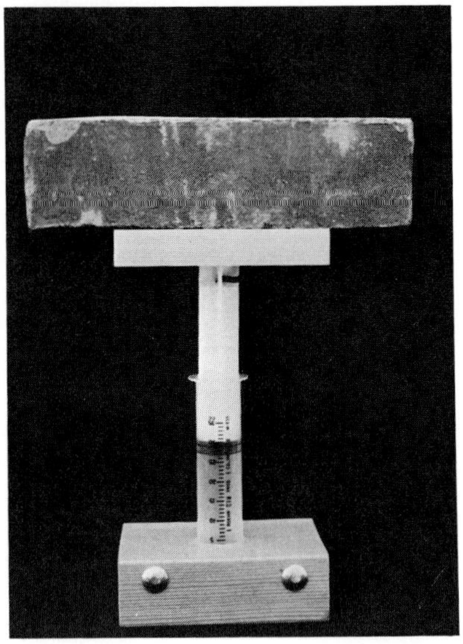

Figure 2-1

Apparatus for studying the change in volume as pressure on a gas is increased.

PRESSURE-VOLUME MEASUREMENTS FOR AIR

You can use the following equipment to study pressure-volume relationships of air: a plastic syringe sealed at the end, a thin piece of paper, a wooden block to hold the syringe, and eight ordinary bricks.

First, fill the syringe with air to the 35.0 millilitre (ml) mark. Insert the paper beside the plunger to let air escape as you adjust the plunger to read

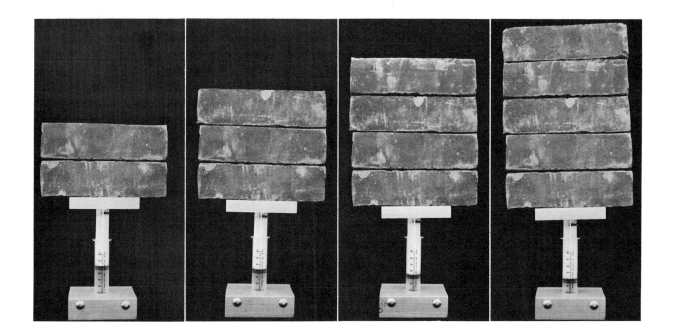

Figure 2-2

As pressure increases, the volume of air in the syringe decreases.

35.0 millilitres.* Then remove the paper. Balance a brick on top of the plunger, as shown in Figure 2-1. Now read the volume of air trapped in the syringe under the load of one brick. Record this value as the pressure of one brick. Add a second brick and record the new volume as the pressure of two bricks. Figure 2-2 shows the apparatus and the steps in the experiment. Actual values for air volume in five trials are shown in Table 2-1.

*Although the customary American spellings are *milliliters, liters, kilometers,* and *meters,* we will use the spellings *millilitres, litres, kilometres,* and *metres,* as established by the Nomenclature Committee of the International Union of Pure and Applied Chemistry (IUPAC).

Table 2-1

Preliminary pressure-volume measurements for air

Pressure* (bricks)	Volume (ml)†					Average volume (ml)	$\frac{1}{volume}$ (1/ml)
	Trial 1	Trial 2	Trial 3	Trial 4	Trial 5		
1	27.0	29.0	28.0	28.0	29.5	28.5	3.50×10^{-2}
2	23.0	22.0	23.0	22.0	22.0	22.5	4.45×10^{-2}
3	18.0	17.0	17.0	17.5	18.0	17.5	5.70×10^{-2}
4	14.0	14.5	13.5	14.0	13.5	14.0	7.15×10^{-2}
5	12.0	12.0	12.0	12.5	11.5	12.0	8.30×10^{-2}
6	10.5	11.0	9.5	10.0	10.5	10.0	1.0×10^{-1}
7	9.0	9.5	9.5	8.5	9.0	9.0	1.1×10^{-1}
8	8.5	8.0	9.0	8.5	8.0	8.5	1.2×10^{-1}

*Only whole bricks were used.   †Read to closest 0.5.

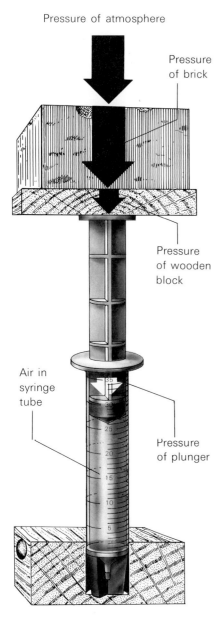

Pressure of atmosphere

Pressure of brick

Pressure of wooden block

Air in syringe tube

Pressure of plunger

Figure 2-3

An invisible column of air pushes down (increases pressure) and confines the gas (decreases volume) in the cylinder.

PRESSURE DUE TO THE ATMOSPHERE: TOTAL PRESSURE

The values in Table 2-1 show that as brick pressure increases, air volume decreases. This is a simple **qualitative** relationship. To answer the question, "how much?" we need numbers to express gas volume and gas pressure. We must determine the total pressure exerted by the gas.

What is pushing down on the air in the syringe? Because of their weight, the wooden block and one or more bricks are exerting pressure. *In addition, there is an invisible column of air, with a diameter equal to that of the plunger, extending from the top of the plunger up into the atmosphere.* The air in this column has weight and must be considered in determining the total pressure on the air in the syringe. (See Figure 2-3). You can sense such pressure, called atmospheric pressure, in a simple experiment.

Expel all the air from a plastic syringe by pushing down the plunger completely. Then put your finger firmly over the hole in the tip so that no air can enter. Now try to pull the plunger up again. Atmospheric pressure prevents you from pulling the plunger out. Now remove your finger from the tip and try to pull the plunger up again. You can do it because the atmospheric pressure is counteracted by the pressure of the air you let into the syringe when you removed your finger. Thus, the total pressure is equal to the pressure of bricks plus the pressure of the atmosphere and block.

PRESENTATION OF DATA

We can now prepare a new table, Table 2-2. In the first column we show the pressure due to the bricks. The second column gives the pressure due to the atmosphere and block. The third column gives the total pressure. The fourth column shows average volumes. These numbers show quantitatively how much the volume of air decreased when the pressure on it increased.

Table 2-2

Corrected pressure-volume measurements for air

(1) Pressure due to bricks*	(2) Pressure due to atmosphere and block (bricks)†	(3) Total pressure (bricks) (1) + (2)	(4) Average volume (ml)	(5) Pressure × volume (bricks × ml)
1	1.7	2.7	28.5	77
2	1.7	3.7	22.5	83
3	1.7	4.7	17.5	82
4	1.7	5.7	14.0	80
5	1.7	6.7	12.0	80
6	1.7	7.7	10.0	77
7	1.7	8.7	9.0	78
8	1.7	9.7	8.0	78

*Consider only whole bricks here.
†By methods shown in Experiment 4 in the Laboratory Manual, we can measure the amount of atmospheric pressure plus the block. It is equal to about 1.7 bricks.

Another way to present the data is in the form of a graph. In Figure 2-4 we plot total pressure against volume. A *curved* line is obtained. In Figure 2-5 we plot total pressure against 1/volume. This time a *straight* line can be drawn through the plotted points. The straight line suggests that total pressure is directly proportional to 1/volume. This relationship can be expressed mathematically as

$$P = \text{proportionality constant} \times \left(\frac{1}{V}\right)$$

If both sides of this equation are multiplied by V, we obtain the mathematical expression

$$PV = \text{a constant}$$

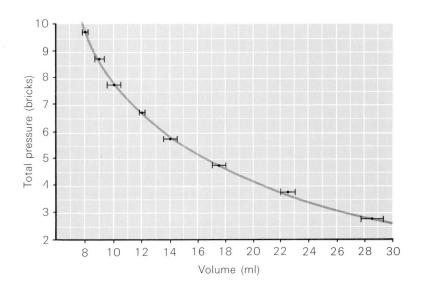

Figure 2-4

Pressure versus volume of air.

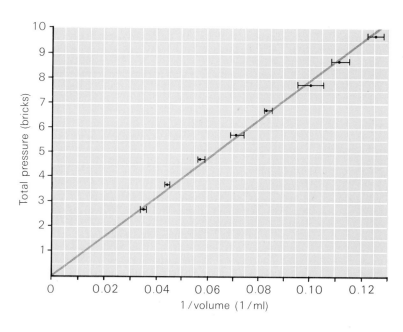

Figure 2-5

Pressure versus 1/volume of air.

This expression leads us to multiply each value of pressure by volume. The results are shown in the last column of Table 2-2. The $P \times V$ values shown in Table 2-2 are nearly constant. They are as constant as one could expect from our relatively crude measurements. More precise measurements demonstrate clearly that, for air, $P \times V$ is indeed a constant under the conditions used here.

Up to now we have been discussing the pressure *on* our gas sample, air. But what pressure was exerted *by* the gas itself? The plunger could move up or down until the internal pressure became equal to the external pressure. Therefore, the pressure (P) also represents the pressure exerted by the gas at that volume.

In summary, we have presented our data on pressure and volume of an air sample in four ways:

1. in a qualitative statement indicating that volume decreases as pressure increases
2. in a table that summarizes quantitatively how much the volume of air decreases as the pressure increases
3. in two graphs, one showing pressure plotted against volume, which gives a curved line; the other showing pressure plotted against $1/V$, which gives a straight line
4. in a mathematical statement, $P \times V =$ a constant, suggested by the straight line found when we plotted P against $1/V^*$

EXERCISE 2-1

Are the values of PV really as constant as you would expect them to be? Look at the numbers in Table 2-2. (a) Suppose the value of 14.0 ml shown as the volume under a pressure of 5.7 bricks was really 14.5 ml. What would the value of $P \times V$ be? (b) Suppose it was 13.5 ml. What would the value of $P \times V$ be? (c) What range in values for $P \times V$ is indicated by the precision with which the volume was read? Express $P \times V$ uncertainty using $\pm$ notation. (d) Repeat steps (a), (b), and (c) for the volume at a pressure of 3.7 bricks. (e) Is the variation in PV reasonable?

QUANTITIVE DATA AND THE GROWTH OF THE MODEL

We have discovered a regularity. We now have a mathematical relationship between the pressure on an air sample and the volume of the air sample. Does the particle model agree with the data that have been accumulated? Let us examine the model in more detail.

---

*The relationship between pressure and volume at constant temperature was first established experimentally by the Irish chemist and physicist Robert Boyle in 1662. Boyle's law states that the volume of a gas varies inversely with its pressure, provided the temperature and quantity of the gas remain constant.

Picture the small particles that make up air as rebounding back and forth between the container walls. In such a model the pressure is determined by

1. the amount of push each colliding particle gives to the walls of the container
2. the number of collisions made with a unit of wall area in any one second

Imagine that the volume of air is reduced to one half while the number of gas particles remains the same. Then each unit of volume will have twice as many particles. This means twice as many collisions per second between particles and a unit of wall area. Doubling the number of collisions with the wall doubles the push per unit area. This doubles the pressure exerted by the gas. Therefore, the experiment and the model agree.

PRESSURE-VOLUME MEASUREMENTS FOR OTHER GASES

The ping-pong ball model can explain all the observations of air discussed so far. Do gases other than air behave in the same way? Many gases with sharply contrasting properties, such as odor and color, are available for our study. Oxygen, hydrogen, carbon dioxide, carbon tetrafluoride, ammonia, hydrogen chloride, and hydrogen bromide are a few of the gases that can be tested. Each of these pure gases can be put into the pressure-volume syringe and studied. The results of several such studies are shown in Table 2-3 (see page 40). The same volumes of gases were used at the atmospheric pressure in the room. Bricks were added to increase the pressure. The volume was recorded for each pressure reading. The data in Table 2-3 indicate a surprising regularity. This regularity is the value for "PV = a constant." The relationship "PV = a constant" seems to apply to all gases, including air. The numerical value for the constant seems to be the *same* for all gases.

Why should this be? What properties are common to all gases? For one thing, the behavior of all gases is consistent with the particle model we used for air. Each gas behaves as air does. This fact suggests that all gases are made up of particles.

The particles that make up gases are called **molecules.** The more molecules in a given volume of any gas, the higher the pressure. But are all gas molecules the same? No. Since gases have different properties, all gas molecules are *not* the same. Different gases must be made up of *different* kinds of molecules. If this were not so, the differences between gases would be very difficult to understand.

EXERCISE 2-2

On a sheet of graph paper, plot pressure against $1/V$ for carbon dioxide, using the data in Table 2-3. How would the shape of the curve differ if a larger syringe was used and there was a greater initial volume of gas?

How would the numerical value for $P \times V$ differ if a mixture of carbon dioxide and nitrogen was used in the experiment? What does your answer suggest about the ability of different gas particles to produce pressure?

Table 2-3

Pressure-volume measurements for selected gases at room temperature (25 °C)

Selected gases	Pressure due to bricks	Total pressure (bricks)	Volume (ml)	Pressure × volume (bricks × ml)
carbon dioxide	1	2.7	29.2	79
	2	3.7	21.5	80
	3	4.7	17.0	80
	4	5.7	13.5	77
	5	6.7	11.7	78
ammonia	1	2.7	27.0	73
	2	3.7	22.1	82
	3	4.7	18.0	84
	4	5.7	13.6	78
	5	6.7	12.0	80
nitrogen (trial 1)	1	2.7	29.2	79
	2	3.7	22.8	84
	3	4.7	17.8	84
	4	5.7	15.0	85
	5	6.7	12.5	84
nitrogen (trial 2)	1	2.7	29.2	79
	2	3.7	22.5	83
	3	4.7	18.0	85
	4	5.7	14.7	84
	5	6.7	12.1	81
hydrogen bromide	1	2.7	29.2	79
	2	3.7	21.6	80
	3	4.7	17.0	80
	4	5.7	13.8	79
	5	6.7	11.9	80

COMBINING VOLUMES OF GASES

What happens when volumes of different gases combine? We have already established that when more gas molecules are forced into a given volume, the pressure due to the gas rises. Thus, if the volume occupied by a given quantity of gas at constant temperature is cut in half, the pressure doubles. Forcing one gas into another should bring about an increase in the number of molecules per unit volume, hence an increase in pressure.

To test this hypothesis, a syringeful of nitrogen is forced into a syringeful of hydrogen bromide. We see that the pressure doubles. In a similar way a syringeful of nitrogen is forced into a syringeful of ammonia. Again the pressure doubles, since all the gas is now in one syringe. The particle model explains the behavior of these gas mixtures.

Let us now check this result by forcing a syringeful of ammonia into a syringeful of hydrogen bromide. We see a surprising result. No force is necessary! A white smoke and powder form in the hydrogen bromide syringe. The smoke condenses to a powder. *At the same time* both plungers move to the bottoms of the syringes. *Both* gases disappear and a white solid appears.

$$1 \left\{ \begin{array}{l} \text{syringeful of} \\ \text{hydrogen} \\ \text{bromide} \end{array} \right\} \text{plus } 1 \left\{ \begin{array}{l} \text{syringeful} \\ \text{of} \\ \text{ammonia} \end{array} \right\} \text{gives} \left\{ \begin{array}{l} \text{white solid of very} \\ \text{small volume} \\ \text{compared to gases} \end{array} \right\}$$

What happens if we start with one syringeful of ammonia and *half* a syringeful of hydrogen bromide? White smoke and white powder again appear, but less gas disappears. Half a syringeful of gas remains. This remaining gas has the properties of ammonia. The experiment shows that only half the ammonia is used.

$$\tfrac{1}{2} \left\{ \begin{array}{l} \text{syringeful} \\ \text{of} \\ \text{hydrogen} \\ \text{bromide} \end{array} \right\} \text{plus } 1 \left\{ \begin{array}{l} \text{syringeful} \\ \text{of} \\ \text{ammonia} \end{array} \right\} \text{gives} \left\{ \begin{array}{l} \text{white} \\ \text{solid} \end{array} \right\} \text{plus } \tfrac{1}{2} \left\{ \begin{array}{l} \text{syringeful} \\ \text{of} \\ \text{ammonia} \end{array} \right\}$$

If we repeat the experiment using one syringeful of hydrogen bromide and only half a syringeful of ammonia, the white solid again appears. This time only *half* the hydrogen bromide gas disappears.

$$1 \left\{ \begin{array}{l} \text{syringeful} \\ \text{of} \\ \text{hydrogen} \\ \text{bromide} \end{array} \right\} \text{plus } \tfrac{1}{2} \left\{ \begin{array}{l} \text{syringeful} \\ \text{of} \\ \text{ammonia} \end{array} \right\} \text{gives} \left\{ \begin{array}{l} \text{white} \\ \text{solid} \end{array} \right\} \text{plus } \tfrac{1}{2} \left\{ \begin{array}{l} \text{syringeful} \\ \text{of} \\ \text{hydrogen} \\ \text{bromide} \end{array} \right\}$$

The result of these three experiments can be summarized by the statement

$$1 \left\{ \begin{array}{l} \text{volume of} \\ \text{hydrogen bromide} \end{array} \right\} \text{plus } 1 \left\{ \begin{array}{l} \text{volume of} \\ \text{ammonia} \end{array} \right\} \text{gives} \left\{ \begin{array}{l} \text{white solid} \\ \text{only} \end{array} \right\}$$

If one volume of hydrogen bromide contains the same number of molecules as one volume of ammonia, we could say that one molecule of hydrogen bromide combines with one molecule of ammonia to give the white solid.

$$1 \left\{ \begin{array}{l} \text{molecule of} \\ \text{hydrogen bromide} \end{array} \right\} \text{plus } 1 \left\{ \begin{array}{l} \text{molecule of} \\ \text{ammonia} \end{array} \right\} \text{gives} \left\{ \begin{array}{l} \text{substance that} \\ \text{appears as a} \\ \text{white solid} \end{array} \right\}$$

Then, if there are more molecules of one gas than of the other gas, the excess molecules remain. This explanation would agree with the observed combining volumes. The key assumption is that one volume of hydrogen bromide contains the same number of molecules as one volume of ammonia if both are measured at the same temperature and pressure. This is a new hypothesis. We have already learned that the two gases have the same pressure-volume constant. If our hypothesis is true, one molecule of ammonia must be as effective as one molecule of hydrogen bromide in exerting pressure.

EXERCISE 2-4

What products would be formed or left over if the following gases at equal conditions of temperature and pressure were combined: (a) 1 litre of hydrogen bromide gas and $\frac{1}{2}$ litre of ammonia, (b) $\frac{1}{2}$ litre of hydrogen bromide gas and $\frac{1}{2}$ litre of ammonia, (c) $\frac{1}{2}$ litre of hydrogen bromide gas and 1 litre of ammonia?

AVOGADRO'S HYPOTHESIS

Around 1810, the pressure-volume relationship for gases led the French physicist André Ampère (1775–1836) to suggest that equal volumes of different gases at the same temperature and pressure contain an equal number of molecules. Working independently, the Italian chemist Amedeo Avogadro (1776–1856) also studied the simple combining volumes of gases. In 1811, from both lines of evidence, Avogadro proposed what is usually called **Avogadro's hypothesis:** equal volumes of gases, measured at the same temperature and pressure, contain an equal number of molecules.

You may object, saying that the evidence is consistent with Avogadro's hypothesis but that it certainly does not prove the hypothesis. Your position would be reasonable. In fact, you would be in rather distinguished company. The generally recognized founder of the atomic theory, John Dalton, never believed Avogadro's hypothesis. Neither did Jöns Jakob Berzelius, one of the great chemists of all time. For almost 50 years, Avogadro's hypothesis was not widely accepted. *No single experiment or series of experiments ever established the validity of Avogadro's hypothesis.* Rather, the slow accumulation of facts, all of which could be explained by Avogadro's hypothesis, finally brought about its general acceptance. It stands today as one of the cornerstones of modern chemistry. The controversy over Avogadro's views between 1800 and 1860 make that period in the history of science fascinating reading.[*]

Using Avogadro's hypothesis, we shall see how it can explain many of the observations we shall make throughout this course. That, in fact, is precisely how the hypothesis was incorporated into the structure of modern chemistry.

[*] See, for example, Thomas Thomson, *History of Chemistry,* 2nd ed., (New York: Arno Press, Inc., 1975).

2.2 Molecular weight, the mole, and molar volume

You already know that gases differ from each other. Some, like ammonia, have unforgettable odors. Some, like chlorine, are colored. Some dissolve in water readily, and others do not. Molecules of different gases must differ in some way. Do they have different masses? The molecules are invisible and too small to be weighed directly. However, Avogadro's hypothesis enables us to calculate their masses from experimental data.

THE RELATIVE MASSES OF GASES

In the laboratory you found the masses of several gases by weighing bagfuls of them. The bagful of carbon dioxide had a mass about 1.4 times greater than the mass of the bagful of oxygen. The bagfuls were at the same temperature and pressure and had the same volume. Therefore, according to Avogadro's hypothesis, the two bags contained the same number of molecules. If the hypothesis is true, we can reach a *qualitative* conclusion: the average carbon dioxide molecule has a greater mass than the average oxygen molecule. A *quantitative* conclusion is that all the carbon dioxide molecules have a mass 1.4 times greater than all the oxygen molecules. The *average* masses of the two kinds of molecules will also be in the same ratio. Notice that we do not need to know how many molecules were present as long as both samples have the same number of molecules.

More exact masses can be obtained with a glass flask and a more precise balance. Some masses, to three significant figures, are given in the second column of Table 2-4 (see page 44). Using the masses in grams for carbon dioxide and oxygen, we can write

$$\frac{\text{mass of carbon dioxide sample}}{\text{mass of oxygen sample}} = \frac{1.93 \, \cancel{g}}{1.40 \, \cancel{g}} = \frac{1.38}{1}$$

$$\frac{\text{mass of carbon dioxide molecule}}{\text{mass of oxygen molecule}} = \frac{1.38}{1}$$

Notice that the mass units in grams cancel out. The ratio 1.38/1 is the same for grams, kilograms, or tons of the gases. Notice also that the molecular mass for a gas is for an *average* molecule of that gas. Other experiments would be needed if we were to try to prove that all the molecules of a substance have the *same* mass.

MOLECULAR WEIGHT

The values in the third column of Table 2-4 give a set of relative masses of molecules assuming that the average mass of an oxygen molecule is 1.00. Note that the hydrogen bromide molecule becomes 2.53. Carbon dioxide becomes 1.38. Many years ago chemists found that if oxygen molecules are *assigned* a value of 32.00 instead of 1.00, then most gases have molecular masses that are whole numbers. Thus, if we multiply all the values in the third column of Table 2-4 by 32.00, we obtain new molecular masses for other gases. Notice that most of the values are whole numbers or nearly whole numbers.

Table 2-4

Some molecular weights and relative masses of molecules

(1) Name of gas	(2) Mass of gas (g) in a flask	(3) Mass of a gas molecule relative to mass of an oxygen molecule*	(4) Column 3 multiplied by the molecular weight of oxygen (32.00) = molecular weight of gas†
hydrogen bromide	3.54	2.53	80.90
carbon dioxide	1.93	1.38	44.00
oxygen	1.40	1.00	32.00
nitrogen	1.23	0.879	28.10
ammonia	0.743	0.531	17.00
hydrogen	.0883	0.0631	2.02

*This ratio can also be stated as molecular weight of a gas relative to the molecular weight of oxygen. The values are obtained by dividing each of the numbers in column 2 by the mass of oxygen in the flask (1.40 g).
†These and many other molecular weights have been recorded in the International Molecular Scale, a valuable reference tool for chemists.

Notice also that all the values are greater than 1. Is there some hidden regularity that results in whole numbers for some molecular masses? Some answers will emerge in later chapters.

By tradition, the values in the last column of Table 2-4 are called **molecular weights.***

Let us now summarize the procedure we can use to determine the molecular weights of gaseous substances:

1. Find the mass of a given volume of a standard gas at a given temperature and pressure.
2. Find the recorded molecular weight of the standard gas on the International Molecular Scale. Chemists use oxygen and the value 32.00 as its molecular weight.†
3. Find the mass of an equal volume of an unknown gas using the same temperature and pressure as was used for the standard gas.
4. Calculate the molecular weight of the unknown gas by means of the relationship

$$\frac{\text{molecular weight of unknown}}{\text{molecular weight of standard}} = \frac{\text{mass of unknown gas sample}}{\text{mass of standard gas sample}}$$

*Mass and weight are frequently used as equivalent terms. However, they do have different meanings. *Weight* refers to the gravitational attraction between an object and the earth. *Mass* refers to the actual quantity of material present. We shall use the more scientifically precise term, *mass*, instead of *weight* wherever possible. But we still use the archaic terms *molecular weight* and *atomic weight*, and IUPAC still recognizes them.
† More precise methods for determining molecular weight utilize the mass spectrometer and use carbon = 12.00 as the primary standard.

Neither the temperature nor the pressure need be known as long as the unknown and the standard are weighed under the same conditions. Here is an example to help you calculate the molecular weight of a gas:

Example: The mass of a sample of gaseous methane is determined in the laboratory at 25 °C in a 1-liter bulb, shown in Figure 2-6. It has a mass of 0.654 grams. The mass of an equal volume of oxygen under the same conditions is 1.308 grams. What is the molecular weight of methane?

Answer: Because the measurements for the two gases were taken under the same conditions, the temperature and the volume of the bulb do not affect the result. We can use the relationship

$$\frac{\text{molecular weight of methane}}{\text{molecular weight of standard}} = \frac{\text{mass of unknown sample}}{\text{mass of standard gas sample}}$$

$$\frac{\text{molecular weight of methane}}{32.00 \ (\text{oxygen})} = \frac{0.654 \text{ g}}{1.308 \text{ g}}$$

$$\text{molecular weight of methane} = \frac{0.654 \text{ g}}{1.308 \text{ g}} \times 32.00$$

$$\text{molecular weight of methane} = 16.0$$

Note that molecular weights as obtained here are *relative* and therefore do not have units. Grams cancel out.

EXERCISE 2-5

The mass of a bulb containing a sample of a gas named diborane is determined. The gas is found to have a mass of 0.600 g. The diborane is removed from the bulb and replaced with oxygen gas measured under the same conditions of temperature and pressure. The sample of oxygen has a mass of 0.696 g. What is the molecular weight of the diborane?

Figure 2-6
A one-litre bulb used to weigh and handle gases in the laboratory.

THE MOLE

The molecular weights in Table 2-4 came from measurements on real gas samples. The values 32.00 for oxygen, 17.00 for ammonia, and so on, represent the relative masses of individual molecules. We now reverse the procedure, moving from individual molecules to laboratory quantities of the gases. Chemists have taken the 32.00 associated with oxygen and the gram as a convenient unit of mass and combined them as in 32.00 grams. They then defined a standard quantity of oxygen as one whose mass is 32.0 grams.

How many molecules are present? There are enough so that the sum of their masses equals 32.0 grams. For example, ammonia has a molecular weight of 17.00. Hence, the standard quantity weighs 17.0 grams. The number of molecules in 17.00 grams of ammonia will be the same as the number of molecules in 32.00 grams of oxygen. This conclusion is true if Avogadro's hypothesis is true: equal volumes of gases at the same temperature and pressure have the same number of molecules. Equal volumes of ammonia and oxygen have relative masses of 17.00 and 32.00.

The name given to a 32.00 gram sample of oxygen or a 17.00 gram sample of ammonia is the **mole.** The name for the *number of molecules per mole* is **Avogadro's number.** It is the same for all gases. Avogadro's number is very large and difficult to visualize. The value has been established as 6.02×10^{23}, which is approximately 600,000 billion billion molecules. The following example, given by the American chemist Glenn Crosby, may help you better understand the large amount represented by this number. Crosby has suggested that if *1 mole* of particles the size of sand grains were released by the eruption of Mount Saint Helens, they would cover the entire state of Washington to a height greater than that of a ten-story building!

The mass per mole of gas, the **molar mass*,** differs from gas to gas because the mass per molecule differs. The numerical value for the molar mass of a gas is the number of grams per mole of that gas. Some molar mass values are: oxygen, 32.0 grams per mole; carbon dioxide, 44.0 grams per mole; hydrogen, 2.02 grams per mole.

EXERCISE 2-6

For carbon dioxide gas, what is (a) the molecular weight, (b) the mass of one mole, and (c) the number of molecules in one mole?

EXERCISE 2-7

Given a 32.0 kilogram (kg) sample of oxygen, (a) how many moles of oxygen are present and (b) how many molecules are present?

DETERMINING THE NUMBER OF MOLES

On pages 41–42 we considered the volumes of hydrogen bromide and ammonia gases that combine to form a white solid. Our interpretation was that each molecule of hydrogen bromide combines with an ammonia molecule. Since a mole of each gas has Avogadro's number of molecules, we can say that 1 mole of hydrogen bromide combines with 1 mole of ammonia to form the white solid. From Table 2-4 we obtain the molecular weights 80.90 for hydrogen bromide and 17.00 for ammonia. The molar masses are, there-

---

*Sometimes the mole is called the *gram molecule,* and the molar mass is called the *gram molecular weight* or the *molecular weight in grams.*

fore, 80.90 grams per mole for hydrogen bromide and 17.00 grams per mole for ammonia. These masses are large enough for experiments in the laboratory, whereas the masses of single molecules are not. In the laboratory, we usually weigh a sample of a substance and express the mass as a number of grams. By comparing the mass with the mass of 1 mole, we find out how many moles are present. Knowing the number of moles, we can find out how many molecules are present.

We can use 6.02×10^{23} units in a mole just as we use 12 eggs in a dozen. Let us solve some problems using moles and dozens as units.

We will use the **unit factor method,** or **dimensional analysis,** because it can be used with many kinds of calculations. Let us start with a very simple example.

Example 1a: How many dozen (doz) eggs are there in 48 eggs? You know that there are 12 eggs in a dozen eggs. This can be stated formally by writing

$$1 \text{ doz eggs} = 12 \text{ eggs}$$

Divide both sides of this equation by 12 eggs:

$$\frac{1 \text{ doz eggs}}{12 \text{ eggs}} = \frac{12 \text{ eggs}}{12 \text{ eggs}} = 1$$

The identity 1 doz eggs/12 eggs is called a unit factor because it is 1.

Answer: To find out how many dozen eggs are in 48 eggs we carry out the following calculations. Note that you start with the given: 48 eggs. If we multiply our number of eggs by the unit factor and treat the units as numbers we get

$$\text{given quantity} \times \text{unit factor*} = \text{number of doz}$$

$$48 \text{ eggs} \times \frac{1 \text{ doz eggs}}{12 \text{ eggs}} = 4 \text{ doz eggs}$$

The units can be canceled just as numbers can. The unit left over for the answer is the unit we wanted: dozens of eggs. We use the unit factor method to help detect or avoid mistakes. If the units on the two sides of the equation are *not* equal, the answer is incorrect.

Example 1b: Let us now try a related chemical problem. How many moles of carbon dioxide are contained in 18.06×10^{23} *molecules* of carbon dioxide?

Answer: From the definition of a mole given on page 46 we can write

1 mole of carbon dioxide $= 6.02 \times 10^{23}$ molecules of carbon dioxide

---

Useful Unit Factors

$$\frac{1 \text{ doz eggs}}{12 \text{ eggs}} = 1$$

or

$$\frac{12 \text{ eggs}}{1 \text{ doz eggs}} = 1$$

$$\frac{1 \text{ mole of molecules}}{6.02 \times 10^{23} \text{ molecules}} = 1$$

or

$$\frac{6.02 \times 10^{23} \text{ molecules}}{1 \text{ mole of molecules}} = 1$$

---

*Since the unit factor is equal to 1, we do *not* change the *quantity* when we multiply by the unit factor. Only the units of measurement change. For example, 12 eggs and one dozen eggs are identical.

To obtain the unit factor, divide both sides of this expression by 6.02×10^{23} molecules of carbon dioxide. This unit must be canceled by the unit factor.

given number $\qquad \times \qquad$ unit factor $=$ number of moles

$$18.06 \times 10^{23} \text{ molecules of carbon dioxide} \times \frac{1 \text{ mole of carbon dioxide}}{6.02 \times 10^{23} \text{ molecules of carbon dioxide}} = 3 \text{ moles of carbon dioxide}$$

Note that 6.02×10^{23} units in a mole are just like 12 units in a dozen.

Example 2a: How many dozen eggs are there in a crate containing 7.5 kilograms of eggs? Assume that 1 dozen eggs has a mass of 0.75 kilograms. The defining relationship is

$$1 \text{ doz eggs} = 0.75 \text{ kg eggs}$$

Answer:

given number $\times$ unit factor $=$ number of doz

$$7.5 \text{ kg eggs} \times \frac{1 \text{ doz eggs}}{0.75 \text{ kg eggs}} = 10 \text{ doz eggs}$$

Note that the unit factor method again indicates the correct units for the answer.

Example 2b: How many moles of hydrogen gas are there in a tank containing 40 grams of hydrogen?

Answer: We know that the defining relationship is that each mole of hydrogen gas has a mass of 2.0 grams.* Thus, we can write

$$40 \text{ g hydrogen} \times \frac{1 \text{ mole of hydrogen}}{2.0 \text{ g hydrogen}} = 20 \text{ moles of hydrogen}$$

Example 3a: If a dozen eggs has a mass of 0.75 kilograms, what is the mass of 6 dozen eggs? Again, the defining relationship is

$$1 \text{ doz eggs} = 0.75 \text{ kg eggs}$$

Answer:

$$6 \text{ doz eggs} \times \frac{0.75 \text{ kg eggs}}{1 \text{ doz eggs}} = 4.5 \text{ kg eggs}$$

Example 3b: If 1 mole of oxygen has a mass of 32.0 grams, what is the mass of 3.0 moles of oxygen gas? The defining relationship is

$$1 \text{ mole of oxygen} = 32.0 \text{ g oxygen*}$$

---

*The molar mass to only 2 or 3 significant figures may be used if answers containing only 2 or 3 significant figures are acceptable.

Useful Unit Factors

$$\frac{1 \text{ doz eggs}}{0.75 \text{ kg eggs}} = 1$$

or

$$\frac{0.75 \text{ kg eggs}}{1 \text{ doz eggs}} = 1$$

$$\frac{1.0 \text{ mole of hydrogen}}{2.0 \text{ g of hydrogen}} = 1$$

or

$$\frac{2.0 \text{ g of hydrogen}}{1.0 \text{ mole of hydrogen}} = 1$$

$$\frac{1.0 \text{ mole of oxygen}}{32.0 \text{ g of oxygen}} = 1$$

or

$$\frac{32.0 \text{ g of oxygen}}{1.0 \text{ mole of oxygen}} = 1$$

Answer:

given number $\qquad \times \qquad$ unit factor $\qquad$ = number of g

$$3.0 \text{ \sout{moles of oxygen gas}} \times \frac{32.0 \text{ g oxygen gas}}{1 \text{ \sout{mole of oxygen gas}}} = \begin{array}{l} 96 \text{ g} \\ \text{oxygen gas} \end{array}$$

EXERCISE 2-8

How many moles of carbon dioxide are in a tank containing 88 g of carbon dioxide? Remember that the molar mass (or molecular weight in grams) of carbon dioxide is 44.0 g per mole. Thus, the defining relationship is, 1 mole of carbon dioxide = 44.0 g of carbon dioxide.

EXERCISE 2-9

How many moles of oxygen molecules are contained in a 16.0-g sample of oxygen? How many individual molecules of oxygen are contained in this sample?

EXERCISE 2-10

How many grams of hydrogen gas are contained in a 5.0-mole sample of hydrogen? Remember that the molar mass of hydrogen is 2.0 g per mole.

EXERCISE 2-11

The molar mass of carbon tetrafluoride gas is 88.0 g per mole. How many grams of carbon tetrafluoride gas are contained in a 0.10-mole sample of that gas? How many individual molecules of gas are contained in this sample?

MOLAR VOLUME

There is another way of determining the number of moles present in a sample of gas. First, you measure the volume of the gas under well-defined conditions of temperature and pressure. Then you can calculate the number

Table 2-5

Laboratory data for determining the volume occupied by 1 mole* of a gas at 0 °C and
1 atmosphere pressure

Gas	Mass (grams) of				Molar volume† (litres per mole) $E = \dfrac{D}{C}$
	1.00 litre flask		1.00 litre of gas $C = B - A$	1 mole of gas (molar mass) D	
	empty A	plus gas B			
oxygen	157.35	158.78	1.43	32.0	22.4
nitrogen	157.35	158.60	1.25	28.0	22.4
carbon monoxide	157.35	158.59	1.24	28.0	22.5
carbon dioxide	157.35	159.32	1.97	44.0	22.3

*Remember that 1 mole is equivalent to 1 molecular weight in grams or the molar mass in grams.
†The value for the volume of 1 mole of a gas can be obtained from laboratory data by dividing the mass of 1 mole of gas by the mass of 1 litre of the gas, measured in the laboratory at 0 °C and 1 atmosphere pressure. We shall use this relationship later. A unit check shows

$$\frac{32.0 \text{ g oxygen}}{1 \text{ mole of oxygen}} \times \frac{1 \text{ litre oxygen}}{1.43 \text{ g oxygen}} = 22.4 \text{ litres per mole of oxygen}$$

This is the molar volume of oxygen gas at STP.

of moles of gas present if you know the volume occupied by 1 mole of gas under the specified conditions.

What volume does 1 mole of a gas such as oxygen occupy? Our previous study of pressure-volume relationships on pages 34–40 showed that the volume of a gas depends upon the pressure on that gas. In order to determine the volume occupied by 1 mole of gas, the pressure must be specified. We shall use a fixed pressure called standard atmospheric pressure. (Air pressure changes from day to day and from place to place. Air pressure also decreases as altitude increases.) Standard atmospheric pressure will be defined later.

The temperature of the gas also must be specified. The Celsius temperature scale will be the standard used. For example, 0 on that scale is the temperature of an ice-water slush under 1 atmosphere pressure in an insulated container such as a Thermos bottle. These conditions are relatively easy to duplicate in the laboratory. Zero Celsius temperature and 1 standard atmosphere pressure are called **standard temperature and pressure,** abbreviated **STP.** Under STP conditions (1 atmosphere, 0 °C), 1 mole of oxygen (32.0 grams) has a volume of 22.4 litres. Values for other gases are shown in Table 2-5. Notice that these values also are close to 22.4 litres per mole. Similar measurements made on many other gases show that one mole of any gas occupies approximately 22.4 litres at 0 °C and one atmosphere of pressure. The molar volume and molar mass of oxygen and ammonia are compared in Figure 2-7.

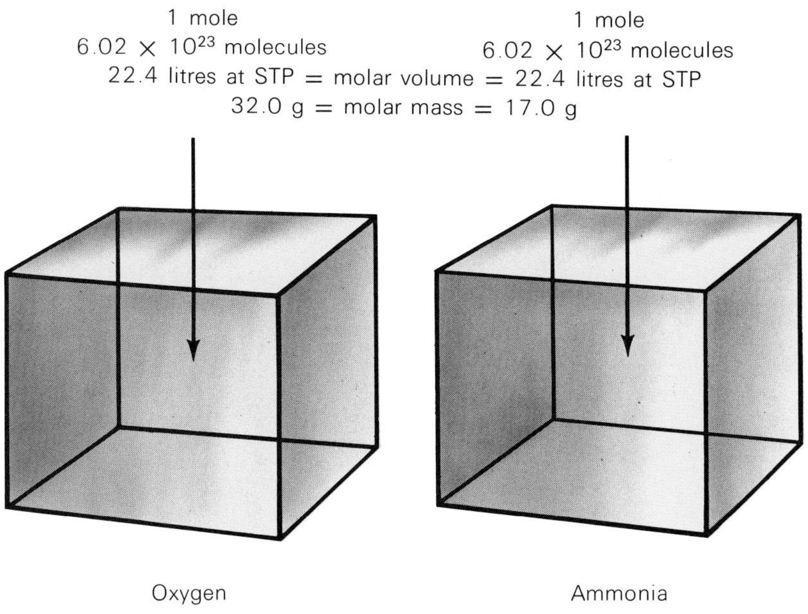

1 mole   1 mole
6.02×10^{23} molecules 6.02×10^{23} molecules
22.4 litres at STP = molar volume = 22.4 litres at STP
32.0 g = molar mass = 17.0 g

Oxygen   Ammonia

Figure 2-7
One mole of any gas occupies approximately 22.4 litres at 0 °C and 1 atmosphere pressure.

You have seen on pages 37–38 that for a given quantity of gas at constant temperature, the pressure multiplied by the volume is a constant. One mole of a gas occupies 22.4 litres at 0 °C and 1 atmosphere of pressure. The volume occupied by one mole of a gas is called the **molar volume.**

$$\text{molar volume} = \frac{22.4 \text{ litres}}{\text{mole}}$$

Written another way we have

$$22.4 \text{ litres of gas} = 1 \text{ mole of gas}$$

This expression can be converted to a unit factor for calculating the number of moles present in a given volume of gas.

Example: We have 44.8 litres of gas at STP (0 °C and 1 atmosphere). How many moles of gas are present in the sample?

Answer:

given quantity $\times$ unit factor   = number of moles

$$44.8 \text{ litres of gas} \times \frac{1 \text{ mole of gas}}{22.4 \text{ litres of gas}} = 2.00 \text{ moles of gas}$$

Useful Unit Factors

$$\frac{1 \text{ mole of gas}}{22.4 \text{ litres of gas*}} = 1$$

or

$$\frac{22.4 \text{ litres of gas*}}{1 \text{ mole of gas}}$$

*The gas volume is measured at 0 °C and 1 atmosphere pressure.

Example: How many moles of hydrogen gas are contained in 11.2 litres of hydrogen gas measured at STP?

Answer:

$$11.2 \text{ litres of hydrogen} \times \frac{1 \text{ mole of hydrogen}}{22.4 \text{ litres of hydrogen}} = \frac{0.500 \text{ mole of}}{\text{hydrogen}}$$

Example: What volume will 0.85 moles of oxygen occupy at STP? This example permits us to use the other unit factor.

Answer:

$$0.85 \text{ mole} \times 22.4 \frac{\text{litres}}{\text{mole}} = 19 \text{ litres}$$

EXERCISE 2-12

How many moles of hydrogen molecules are contained in 22.4 litres of hydrogen gas measured at STP?

EXERCISE 2-13

How many moles of carbon dioxide molecules are contained in 10.0 litres of carbon dioxide gas at STP?

EXERCISE 2-14

What volume will 3.0 moles of gaseous oxygen molecules occupy at 0 °C and 1 atmosphere (atm) pressure?

SUMMARY

The model discussed in Chapter 1 to explain the behavior of a gas in a balloon has grown rapidly. We now know that (1) pressure × volume = a constant, (2) the pressure of a gas is directly proportional to $1/V$, and (3) the volume decreases as the pressure on a fixed quantity of gas is increased. This increase in pressure is explained by the increasing numbers of gas particles colliding with a given area of the container wall in a given unit of time.

Gas particles are called molecules. Equal volumes of gases at the same conditions of temperature and pressure contain the same number of molecules. This statement is known as Avogadro's hypothesis.

Using Avogadro's hypothesis as a starting point, we obtained relative masses of gaseous molecules. A standard gas was needed to establish a molecular weight scale. Oxygen was selected as the standard gas and assigned a value of 32.00 for its molecular weight. The relative mass of any other gas on this scale is called the molecular weight of that gas. A mole of gas is a quantity whose mass equals its molecular weight in grams. There are 6.02×10^{23} molecules in a mole. The value 6.02×10^{23} is known as Avogadro's number. The mass in grams of 1 mole of any gas is called its molar mass. For example, the molar mass of oxygen is 32.0 grams per mole. One mole of gas occupies 22.4 liters at 0 °C and 1 atmosphere pressure. These conditions are known as standard temperature and pressure and abbreviated STP.

Key Terms
Avogadro's hypothesis
Avogadro's number
law
mass
molar mass
molar volume
mole
molecular weight
molecule
qualitative
quantitative
standard temperature and pressure (STP)
theory
unit factor method (dimensional analysis)
weight

QUESTIONS AND PROBLEMS

A

1. What name is given to a regularity as it predicts more and more laboratory facts?
2. (a) Describe a very simple experiment that would show *qualitatively* what happens to the volume of a fixed quantity of gas at room temperature when the pressure is increased. (b) Describe a somewhat more complicated experiment to determine *quantitatively* how volume changes as the pressure is increased. (c) What two variables were held constant in this study?
3. At a total pressure of 5.5 bricks the value of $1/V$ is 0.070. (a) What will the pressure be at a value of $1/V = 0.140$? (b) At $1/V = 0.110$? (c) At $1/V = 0.020$? First try to calculate these answers, then see Figure 2-5 on page 37.
4. If a gas has a volume of 600 ml at a pressure of 2 atmospheres, what will the volume be at 4 atmospheres? The amount of gas and the temperature are constant.
5. Assuming constant temperature and volume for a container, what will be the effect on pressure if the number of molecules in the container is doubled?
6. (a) What is Avogadro's hypothesis? (b) Why was it important in getting relative weights for different kinds of molecules?
7. (a) What is the mass of one mole of oxygen molecules? In other words, what is the molar mass of gaseous oxygen? (b) What is the mass of one mole of hydrogen molecules? (c) Of one mole of methane molecules, which is a natural gas burned as fuel? (d) What is the molar mass of carbon dioxide?
8. (a) How many molecules does one mole of nitrogen gas contain? One mole of hydrogen bromide? (b) The numerical answer of part (a) has a name. What is that name?
9. You are given the defining relationship: 10 doz eggs = 1 crate of eggs. How many crates of eggs could a farmer sell if he had 120 doz eggs? Set this problem up using the unit factor procedure and show that your answer *must have* units of "crates of eggs."

10. Remembering that 12 eggs = 1 doz, calculate *by the unit factor method* the number of eggs in a crate. Show all units and cancel them as appropriate.

11. Use the unit factor method to calculate the number of moles of carbon dioxide in 154 g of carbon dioxide. The appropriate defining relationship is one mole of carbon dioxide = 44.0 g of carbon dioxide.

12. (a) What volume does one mole of a gas occupy at standard temperature and pressure (STP = 0 °C and 1 atm)? What is the name of this quantity? (b) How many molecules are there in the volume described in part (a)?

B

13. How does a law differ from a model?

14. (a) Describe three ways in which the quantitative relationship between the pressure and volume of a gas can be presented. (b) What two important quantities were held constant in this study?

15. A beaker of water at room temperature is heated, with temperature readings taken each minute. All readings are ±0.2 °C. They are as follows: 26.0, 32.4, 39.4, 46.0, 52.2, 58.8, 65.6, 72.2, 79.0, 85.6, and 91.8. (a) Make a qualitative statement about the observed data. (b) Express the results quantitatively in a table. (c) Express the results graphically. (d) Develop a mathematical relationship between the time of heating, t, and the temperature, T.

16. What simple experiment demonstrates that the column of air above the plunger pushes down on the plunger? Refer to Figure 2-3 on page 36.

17. (a) What volume in millilitres is indicated by a value of 0.070 for $1/V$? Refer to Figure 2-5 on page 37. (b) What total pressure corresponds to a $1/V$ value of 0.070 1/ml? (c) What volume in millilitres is indicated by a value of 0.115 1/ml for $1/V$? (d) What total pressure in Figure 2-5 corresponds to a $1/V$ value of 0.115 1/ml. Locate these two points in Figure 2-4 on page 37. Is pressure directly proportional to volume? If not, what is the relationship?

18. (a) We have three syringes. The first contains 50 ml of nitrogen gas at room temperature and one atmosphere pressure. The second contains 50 ml of oxygen gas under the same conditions. The third contains 25 ml of carbon dioxide under the same conditions. The oxygen and the carbon dioxide are both forced into the syringe containing the nitrogen. If the volume in the syringe is held constant at 50 ml, what will be the final pressure in the nitrogen syringe? (b) Suppose the gases in the syringes were 50 ml of nitrogen, 50 ml of hydrogen bromide, and 25 ml of ammonia. What would the final pressure in the syringe be?

19. (a) If one mole of oxygen molecules had been assigned a mass of 1.00, what would the mass of a hydrogen molecule be on this scale? (b) What is the major advantage in making oxygen 32.00 instead of 1.00?

20. You have one mole of gas at 0 °C and 1 atmosphere. The pressure is increased to 2 atmospheres. (a) What will the new volume be? (b) How has the *number of particles per unit volume* changed as a result of the pressure increase?

21. If two volumes of hydrogen bromide were allowed to react with one volume of ammonia, what would remain after the reaction?

22. Data obtained rapidly in a demonstration show that a bagful of carbon dioxide has a mass that is 1.4 ± 0.1 times the mass of a bagful of oxygen. (a) What molar mass could we write for carbon dioxide based on

this measurement? Put ± error limits on your value. (b) How did this problem use Avogadro's hypothesis?

23. How many molecules make up 17.00 g of ammonia? How many molecules make up 44.00 g of carbon dioxide?

24. How many moles of nitrogen gas are there in a tank containing 84 g of nitrogen? We know that each mole of nitrogen gas has a mass of 28.00 g.

25. Which represents more molecules: 80.90 g of hydrogen bromide or 44.00 g of carbon dioxide? Show your calculations.

26. (a) How many moles of carbon dioxide are in 12.04×10^{23} molecules of carbon dioxide? (b) How many moles of oxygen gas are in 16.00 g of oxygen? Defining relationship: one mole oxygen = 32.00 g oxygen.

27. If the molecular weight of nitrogen is 28.00, what would be the mass of (a) $\frac{1}{2}$ mole of nitrogen gas and (b) $\frac{1}{10}$ mole of nitrogen gas?

28. How many moles of each substance are present in the following: (a) 22 g carbon dioxide (molecular weight = 44.00), (b) 85.0 g ammonia (molecular weight = 17.00), (c) 0.342 g sugar (molecular weight = 180.00)?

29. An unknown gas is weighed in a bag. The sample weighed 1.38 grams. Under the same experimental conditions, oxygen in the bag weighed 0.62 grams. What is the *molar mass* of the gas? Could the gas be carbon dioxide, ammonia, or hydrogen? Explain.

30. (a) What volume is occupied by an 84.0 g sample of nitrogen gas? Assume STP conditions. (b) How many nitrogen molecules are in the sample? Hint: set up (a) and (b) as a unit factor problem using the following defining relations: one mole nitrogen = 28.0 grams; one mole gas (STP) = 22.4 litres; and one mole gas = 6.02×10^{23} molecules.

C

31. On page 45 it is stated that molecular weights do not have units. (a) Why is this so? (b) On the other hand a molar mass has units of grams per mole. What is the difference? Sometimes we give units to "molecular weights." What are we really considering when we do this?

32. Assume that one can count 100 molecules per minute. How many years would be required to count all the molecules in a mole of a substance? Compare your answer with the estimated age of the earth, 5 billion years (1 billion = 1×10^9).

33. In an experiment similar to Experiment 5 in your Laboratory Manual, the air is pumped out of a glass container. The evacuated flask has a mass of 126.44 ± 0.01 g. When filled with oxygen at room temperature and pressure, the flask now has a mass of 127.79 ± 0.01 g. The flask is then emptied and refilled with an unknown gas, again at room temperature and pressure. The flask now has a mass of 127.15 ± 0.01 g.

(a) What is the ratio

$$\frac{\text{mass of unknown gas contained in the flask}}{\text{mass of oxygen contained in the flask}}$$

(b) What is the ratio

$$\frac{\text{mass of one molecule of unknown gas}}{\text{mass of one molecule of oxygen}}$$

(c) What is the ratio

$$\frac{\text{molecular weight of unknown gas}}{\text{molecular weight of oxygen}}$$

(d) Since the molecular weight of oxygen has been established at 32.00, what is the molecular weight of the unknown gas?

34. A laboratory tank of oxygen contains 8.50 pounds of oxygen. (1.00 pound = 454 g). (a) How many *moles* of oxygen are in the tank? How many oxygen molecules? (b) If this gas were at STP (0 °C and 1 atm), what volume would it occupy? (c) If the volume of the tank was 10.0 litres, what would be the pressure in the tank? Give pressure first in atmospheres then in pounds per square inch. The defining relationship is: 1.00 atmosphere = 14.7 pounds per square inch. Most of our laboratory tank gauges still read in pounds per square inch.

35. Suppose you were told that you could have one trillionth (1/1,000,000,000,000) of a mole of silver dollars if you could carry them away in a 10-ton truck. Could you do it? Show your calculations. The following defining relationships will be useful: 1 silver dollar = 26.7 g (These silver dollars were minted from 1837–1935); 454 g = 1.00 pound; 1.00 mole = 6.02×10^{23} units; 2,000 pounds = 1 ton.

36. The present population of the United States is approximately 238,420,000. A presidential candidate seeking office promises that he would distribute one mole of dollars *equally* to every man, woman, and child in the United States. If he were elected, what would he owe you?

37. A 10.00 g sample of a gas occupies 2.00 litres at STP. What is the molar mass of the gas? (a) 5.5 g/mole, (b) 55 (no units), (c) 112 g/mole, (d) 220 (no units), (e) 44.8 g/mole; (f) None of these.

EXTENSION

Phlogiston and the "chemical revolution"

Probably the most common chemical reaction is ordinary burning—of candles, wood, paper, coal, or oil. It is a reaction that has been observed since the discovery of fire. What happens when a substance burns is a question that has intrigued the curious for many centuries. Today we know that air contains oxygen. When anything burns, heat and light are released by a chemical reaction involving oxygen. This is combustion.

The nature of burning To understand how combustion was first satisfactorily—although incorrectly—explained, we must know something of the ideas of Aristotle (384–322 B.C.). According to Aristotle, everything was made of just four basic elements: earth, air, fire, and water. Each of these had a "natural" place in the universe. The characteristic properties of any substance would depend on the type and quantity of each element it contained. For example, water was a pure element whose natural place was between earth and air. Heating it over a flame, however, added a considerable quantity of the element fire, whose natural place was "up." The water rose as vapor because fire pulled it up. Then the liquid water fell back to earth as rain and returned to its natural place between earth and air.

It is not surprising that the first reasonably complete theory of burning was greatly influenced by Aristotle's widely accepted ideas. George Ernst Stahl (1660–1734), a German professor of medicine, postulated the existence of an invisible substance in all combustible substances. He called that substance *phlogiston.* Substances burned, he said, because they contained phlogiston, which was released into the air as burning proceeded. The more phlogiston a substance contained, the more completely it burned. A substance crumbled to ash when burned because phlogiston had left the substance. A *metallic oxide,* or *calx,* an ashy powder, became metallic again when heated with charcoal because it "took" phlogiston from the charcoal. Phlogiston was thought to be usually invisible, although it often gave rise to flames and must therefore have been closely related to Aristotle's element fire—though no one seemed to know just how.

This so-called *phlogiston theory* was successful in explaining almost all of the then-known facts about burning. Stahl's hypothesis offered a plausible explanation of combustion. For example, a candle burns continually, if fresh air is available, gradually diminishing in size until only the wax drippings and charred wick remain. According to the theory, a candle was largely phlogiston, which was continually released into the air as burning proceeded. That explanation accounted for the virtual disappearance of the original candle. If a candle is enclosed in a limited amount of air, as shown in Figure 2-8, it burns for a few moments and then goes out. The enclosed air no longer supports burning. According to the theory, the trapped air could hold just so much phlogiston before it became saturated and could hold no more. A similar explanation accounts for the well-known fact that when humid air becomes saturated with water vapor, evaporation can no longer take place. Since the candle's phlogiston

Figure 2-8

Burning a candle in an enclosed amount of air.

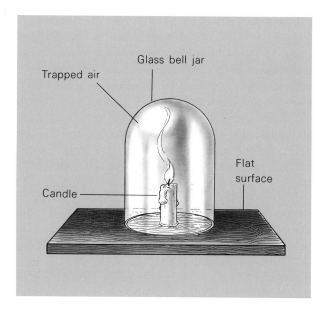

Glass bell jar

Trapped air

Flat surface

Candle

could no longer "evaporate," burning stopped. More phlogiston could leave the candle when fresh, "unsaturated" air was admitted. Then the candle would burn when ignited.

Another striking success of the phlogiston theory was the way it explained how metals could be obtained by heating the metallic oxide, or calx, of a metal with charcoal, which was believed to be rich in phlogiston. You can do this in the laboratory by combining equal volumes of powdered red lead, which is a metallic oxide, and powdered charcoal in a porcelain crucible. Heat the mixture with a Bunsen burner. Then empty the contents of the crucible onto a fireproof surface. A pool of molten lead will appear.

To explain all this, assume that powdered charcoal is almost pure phlogiston. This is reasonable, because charcoal burns so well and leaves no ash— all of it supposedly going up as phlogiston. What could be more logical than to suppose that phlogiston from the charcoal must in each case be responsible for the transformation of the calxes to their corresponding metals? This process may be summarized as a word equation:

metallic oxide (calx) + phlogiston (from charcoal)

$$\longrightarrow \text{metal}$$

The consistency of the phlogiston theory is apparent when you consider the reverse process: the burning of a metal. For example, when shiny magnesium metal burns, a brilliant white light is given off. According to the theory, phlogiston is released. Replacing the shiny metal is a dull white powder, which is the metallic oxide. This is the reaction that occurs in a photographer's flash bulb. Thus, when phlogiston is *added* to a calx you would have a metal. When phlogiston is *removed* from a metal you would again have a calx. In word equation form, the reverse process would be expressed:

metal $\longrightarrow$

metallic oxide (calx) + phlogiston (into the air)

Note that the phlogiston theory considered the calxes, *not* the metals, to be the *simpler* materials. Phlogiston must be *added* to the calxes to make them metals.

The phlogiston theory was widely accepted at the time of the American Revolution and formed the basis of the chemistry taught to students then. But there was a flaw in the phlogiston argument. The flaw was recognized as early as 1630, but the time was not yet appropriate for it to be considered scientifically significant. What was that flaw, and how did it ultimately overthrow the phlogiston theory?

Problems with the phlogiston theory As we know, the phlogiston theory maintained that when a metal burned, the phlogiston in it escaped into the air, leaving the simpler calx. Thus, the calx would be lighter than the metal and of less mass. This certainly appears to be the case when a candle, log, or piece of paper burns. Actually, appearances are misleading in these cases because the products of burning are invisible gases, which escape into the air. But that is jumping ahead of our discussion.

When a metal burns, the calx formed *gains* weight instead of losing it. This fact was known as early as 1630, over 150 years before the phlogiston theory was seriously questioned. The weight gain seems to contradict the theory, because phlogiston is supposedly lost during burning. But the phlogiston theory satisfactorily explained almost everything else known about burning at that time.

The contradiction was explained by phlogistonists in an ingenious way. Phlogiston was closely related to Aristotle's element fire, whose natural place was above the air. It is reasonable to suppose that the phlogiston in a metal acts like helium in a partially filled balloon. Even though a small amount of helium might not have enough lifting power to raise a balloon, the balloon would certainly weight less on a scale than it would if it contained no helium.

The phlogistonists did not know about helium because it was not discovered until 1868, but they imagined that phlogiston acted in much the same way. They said that phlogiston had levity, which meant it tried to pull a metal up—thereby canceling some of the metal's weight. Naturally, when the phlogiston escaped during burning, the upward pull disappeared, and the full weight of the metal, which was now transformed into its calx, would register on a scale. Hence, the gain in weight could be explained without giving up the theory—an example of rationalizing a seemingly contradictory fact.

It was also known that when a metal burns in an enclosed space, some of the enclosed air disappears. You can illustrate this effect by allowing steel wool without oil or soap on it to burn slowly in some trapped air, as shown in Figure 2-9. Simply wedge some steel wool into the bottom of a drinking glass, turn the glass over, and place it upside down in a saucer of water. The steel wool will burn slowly overnight, without a flame, turning into its calx, which you will recognize as rust. Next day, you will see the water level has risen inside the glass, evidently taking the place of some of the air formerly inside.

If phlogiston was given off during the steel wool's burning, it should have been added to the trapped air in the glass, and thus the volume should have *increased* instead of getting smaller. This was never satisfactorily explained by the phlogistonists and is an example of a disturbing fact for which the theory could not account. However, the theory clearly explained all the other known facts. Besides, no one had anything as good or better with which to replace it—and a good theory is better than no theory at all!

The "chemical revolution": Lavoisier and the phlogiston theory The phlogiston theory prevailed until the French scientist Antoine Laurent Lavoisier (1743–1794) gathered enough evidence to overthrow it. As he did so, Lavoisier presented a new theory that was better than the old one.

Figure 2-9
Water level and air volume changes when rust (calx) forms on the steel wool.

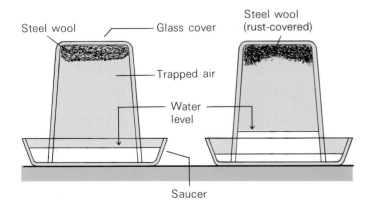

Steel wool — Glass cover

Steel wool (rust-covered)

Trapped air

Water level

Saucer

Lavoisier's experiments were similar to the experiment with the steel wool just described. He burned a metal and therefore obtained a calx. If you think about your steel wool experiment, you will probably arrive at the same conclusion Lavoisier did: if some of the air disappears during burning, then "something" in that air must have been *consumed.* Although Lavoisier was not the originator of this idea, it was he who emphasized its importance.

Lavoisier was convinced that a portion of the air was combining with the metal as it burned and was thereby removed from the air. He wanted to prove this hypothesis by an actual experiment—an experiment whose results could be explained only by the new idea and not by the phlogiston theory. Lavoisier was not sure what had been removed from the air when *calcination,* or calx formation, and combustion occurred during his experiments. He described that unknown "something" as an "atmospheric principle."

By a fortunate coincidence, an English scientist and Presbyterian minister, Joseph Priestley (1733–1804), had just discovered that the metal mercury burns to form a red powdery calx that is easy to break down. In August 1774, Priestley heated the red calx of mercury, or mercuric oxide, by focusing sunlight through a magnifying lens. Little metallic globules of mercury were formed and a gas was given off. The gas was colorless and quite similar to ordinary air. But a candle burned much brighter in this gas than in ordinary air. As a follower of Stahl and a confirmed phlogistonist, Priestley concluded that the new gas did not contain phlogiston.* He therefore named the gas *dephlogisticated air.* Several months later Lavoisier learned of this experiment and its results when he met Priestley in Paris.

Using balances more accurate than any used before, Lavoisier recovered some dephlogisticated air in a new experiment and found it to be his atmospheric principle. This was the gas responsible for both calcination and combustion. Because this gas had acid-forming properties, Lavoisier named it *oxygène* in 1786, from the Greek *oxys,* for acid, and the French suffix *-gène,* for forming.

----

* Recall that the phlogistonists believed that phlogiston combined with a calx to form its metal. Therefore, the resulting air did not contain phlogiston.

The beginnings of modern chemistry Let us summarize the impact of Lavoisier's work. First, it gave chemists a new and better explanation of burning, different from that provided by the phlogiston theory. This difference can best be seen by comparing word equations summarizing the two ideas. Burning was pictured by the phlogistists as

metal $\longrightarrow$

metallic oxide (calx) + phlogiston (into the air)

and by Lavoisier as

metal + oxygen (from the air) $\longrightarrow$

metallic oxide (calx)

The new theory pictured the metal as the simpler substance, because the metallic oxide was considered a compound of the metal and oxygen. Lavoisier proved that combustion involves the combining of oxygen with a substance and not the giving off of phlogiston.

Second, from studying chemical reactions in closed systems, Lavoisier established an important principle: the total weight of the products of a reaction exactly equals the total weight of the reactants. In other words, *chemical change does not result in a loss or gain of mass.* This generalization is known as the *Law of Conservation of Mass.** For example, a candle loses weight when it burns because the products of its burning are gases. If these gases are collected instead of being allowed to escape into the air, they are found to weigh more than that part of the candle that has disappeared. Finally, if the oxygen used up during burning is also considered, it is found that the weight of the oxygen plus the burned-away portion of the candle *does* equal the combined weight of the gases produced. Thus, the use of a closed experimental system is crucial to the accurate study of chemical reactions.

Lavoisier's quantitative examination of the combustion process initiated the decline of the phlogiston theory. Numerous other experiments by Lavoisier and others resulted in the final rejection of the phlogiston theory by most scientists. The modern study of chemistry had begun.

----

* Priestley coined the term *conservation,* and he and Lavoisier *assumed* the conservation of mass in their experiments.

Matter . . . is not infinitely divisible. The existence of . . . ultimate particles of matter can scarcely be doubted, though they are probably much too small ever to be exhibited by microscopic improvements.

JOHN DALTON

THE ATOMIC THEORY: ONE OF OUR BEST SCIENTIFIC MODELS

Objectives

After completing this chapter, you should be able to:

- Explain why different molecules have different masses.
- Write and understand chemical equations.
- Explain the differences between elements and compounds.
- Use atomic weights to solve molecular weight and molar mass problems.

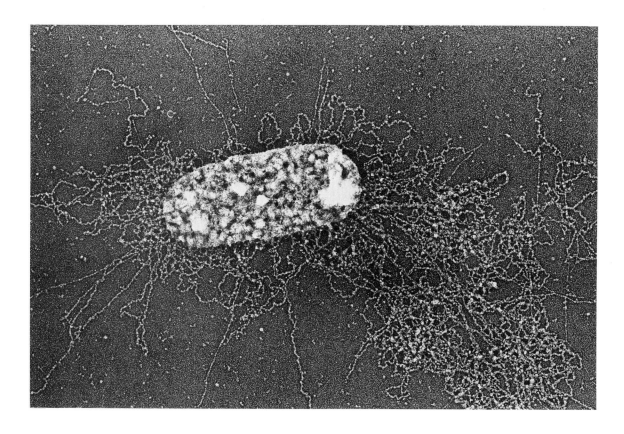

Complex strands of DNA molecules are ejected from this *E. coli* cell. The DNA substructure provides the key to heredity.

In the previous chapter, with the help of Avogadro's hypothesis, we were able to explain the similarities shown by all gases. Our ping-pong ball model of gas molecules showed that one kind of gas molecule is just as effective as another in exerting pressure on the walls of a container. We also saw that gases have differences. For example, gases vary in color, odor, and mass.

In this chapter we will focus on the differences among gases and, hence, on the differences among their molecules. Do molecules have a substructure that accounts for their different masses? As we search for an explanation, we shall find it necessary to expand our ping-pong ball model of gases.

3.1 Atoms and molecules

The molecule is the unit particle found in a gas. But most molecules can be broken into simpler particles we call atoms.* Indeed, the word *atom* is firmly established in our culture. How often have you heard such expressions as *atomic energy, atomic power plant, atomic age* and *atomic theory*? What is the scientific basis for our belief in atoms? To answer that question, let us look again at some experiments.

*In a few cases, gas molecules have only one atom (such as He, Ne, and Ar).

CHEMICAL CHANGE AND THE COMBINING VOLUMES OF GASES

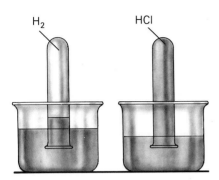

Figure 3-1

Hydrogen gas does not dissolve appreciably in water, whereas hydrogen chloride gas dissolves completely in water.

Suppose we mix equal volumes of gaseous hydrogen and gaseous chlorine in a darkened room. One litre of each of these two gases is placed in a container. When an ultraviolet sunlamp shines into the container, we see a dangerous explosion! A sizable amount of heat and light energy has been released. Examination of the container after the explosion reveals that no hydrogen or chlorine gas remains.* A new colorless gas, hydrogen chloride, now fills the container. Its volume, when measured at the same temperature and pressure as the original hydrogen and chlorine, is found to be 2 litres. This is twice the volume of either of the original gases.

Hydrogen chloride also has properties completely different from those of hydrogen or chlorine. Chlorine is a green gas. Both hydrogen and hydrogen chloride are colorless. But hydrogen chloride behaves differently with water than does hydrogen, as is shown in Figure 3-1. Hydrogen chloride dissolves.

The change resulting from a reaction such as the one between hydrogen gas and chlorine gas is known as a **chemical change**. *A chemical change produces new substances with new properties. An energy change is usually observed also.* In our example, the flash, bang, and heat indicated that energy was released. The original substances, hydrogen and chlorine, are called *reactants*. The new substances made in the reaction are known as *products*. The reaction between gaseous hydrogen and chlorine has only one product, hydrogen chloride.

EXERCISE 3-1

> When a spark is passed through a mixture of hydrogen gas and oxygen gas, an explosion occurs, and a small amount of liquid water is formed. Is this a chemical change? Defend your answer by comparing properties of reactants and product, such as their melting points, role in combustion, and role in supporting life.

What kinds of molecular changes accompany the reaction between hydrogen and chlorine? We look first at the volumes of the gases involved and use Avogadro's hypothesis to "count" molecules. Experimental measurements of the gas volumes, all made at the same temperature and pressure, show that

$$1 \left\{ \begin{array}{l} \text{litre of} \\ \text{hydrogen} \end{array} \right\} \text{plus } 1 \left\{ \begin{array}{l} \text{litre of} \\ \text{chlorine} \end{array} \right\} \text{gives } 2 \left\{ \begin{array}{l} \text{litres of} \\ \text{hydrogen chloride} \end{array} \right\}$$

It is not necessary for us to know exactly how many molecules are present in 1 litre of hydrogen and 1 litre of chlorine. But we can arbitrarily represent this number by the letter n. Because we began with one litre of each gas, Avogadro's hypothesis tells us that we started with n molecules of each. The product, hydrogen chloride, occupied twice this volume, so it must have

*The container might be shattered by the explosion.

contained $2n$ molecules. (This is true because equal volumes of gases at the same temperature and pressure contain equal numbers of molecules.) We can now write

$$n \left\{ \begin{array}{l} \text{hydrogen} \\ \text{molecules} \end{array} \right\} + n \left\{ \begin{array}{l} \text{chlorine} \\ \text{molecules} \end{array} \right\} \longrightarrow 2n \left\{ \begin{array}{l} \text{hydrogen chloride} \\ \text{molecules} \end{array} \right\}$$

Note that the symbol $(+)$ has replaced the word "plus" and the arrow $(\rightarrow)$ is used in place of the word "gives" in the earlier expression. If we let n equal 1, our second expression becomes the word equation

$$1 \left\{ \begin{array}{l} \text{hydrogen} \\ \text{molecule} \end{array} \right\} + 1 \left\{ \begin{array}{l} \text{chlorine} \\ \text{molecule} \end{array} \right\} \longrightarrow 2 \left\{ \begin{array}{l} \text{hydrogen chloride} \\ \text{molecules} \end{array} \right\}$$

But how can two molecules of hydrogen chloride be formed from only one molecule of hydrogen and one molecule of chlorine? Clearly, both the hydrogen and chlorine molecules must split into two identical halves in the reaction, as represented in Figure 3-2. Such splitting must occur no matter how big n is because the number of hydrogen chloride molecules is always twice the number of hydrogen or chlorine molecules.

One hydrogen molecule   One chlorine molecule   Two hydrogen chloride molecules

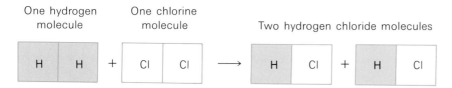

Figure 3-2
One diatomic hydrogen molecule reacts with one diatomic chlorine molecule to form two hydrogen chloride molecules.

Because hydrogen and chlorine molecules are broken into two parts in this reaction, it is convenient to identify those parts. The smaller parts, or building blocks, of molecules are **atoms.** A molecule containing two, four, six, or any other *even* number of atoms, can split into two equal halves. To explain the reaction between hydrogen and chlorine, the simplest possible assumption is that both the hydrogen molecule and the chlorine molecule must be made of *two* atoms. Molecules made of two atoms are said to be **diatomic.** Notice that it is not the atoms that break apart. According to the definitions and models we are using, atoms are assumed to be indivisible and *cannot* be split in a normal chemical reaction. But molecules, made of two or more atoms, can and do split in such reactions.

A similar volume study of the reaction between nitric oxide and oxygen shows that each oxygen molecule must also contain at least two atoms:

$$2n \left\{ \begin{array}{l} \text{nitric oxide} \\ \text{molecules} \end{array} \right\} + n \left\{ \begin{array}{l} \text{oxygen} \\ \text{molecules} \end{array} \right\} \longrightarrow 2n \left\{ \begin{array}{l} \text{nitrogen dioxide} \\ \text{molecules} \end{array} \right\}$$

Thus, oxygen molecules are diatomic. In addition to hydrogen, chlorine, and oxygen, the elements nitrogen, fluorine, bromine, and iodine also exist as diatomic molecules. Some gaseous molecules consist of only one atom. They

Table 3-1

Formulas for diatomic molecules of some common elements

Diatomic molecules	Formulas	Phase
bromine	Br_2	liquid
chlorine	Cl_2	gas
fluorine	F_2	gas
hydrogen	H_2	gas
iodine	I_2	solid
nitrogen	N_2	gas
oxygen	O_2	gas

are called **monatomic** molecules. Helium, neon, and argon are examples of monatomic gases.

Some molecules are found to contain more than one or two atoms. For example, the following volume relationships describe the formation of ammonia gas from the gases nitrogen and hydrogen:

$$1\begin{Bmatrix}\text{litre of}\\\text{nitrogen}\end{Bmatrix} + 3\begin{Bmatrix}\text{litres of}\\\text{hydrogen}\end{Bmatrix} \longrightarrow 2\begin{Bmatrix}\text{litres of}\\\text{ammonia}\end{Bmatrix}$$

which is pictured by assuming that

$$1\begin{Bmatrix}\text{nitrogen}\\\text{molecule}\end{Bmatrix} + 3\begin{Bmatrix}\text{hydrogen}\\\text{molecules}\end{Bmatrix} \longrightarrow 2\begin{Bmatrix}\text{ammonia}\\\text{molecules}\end{Bmatrix}$$

As shown in Figure 3-3, each ammonia molecule must contain half a nitrogen molecule and three halves of the three hydrogen molecules. Since half of a diatomic nitrogen molecule (N_2) is one nitrogen atom, and half of a diatomic hydrogen molecule (H_2) is one hydrogen atom, each ammonia molecule must contain one nitrogen atom and three hydrogen atoms. Thus the ammonia molecule contains four atoms.

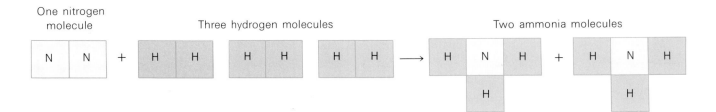

Figure 3-3

One diatomic nitrogen molecule reacts with three diatomic hydrogen molecules to form two ammonia molecules. Each ammonia molecule is made of one nitrogen atom and three hydrogen atoms.

Why do molecules differ? That question can be answered using the concept of atoms. We can explain the differences between various molecules in terms of the kinds, numbers, and arrangements of atoms in each molecule. For example, hydrogen chloride is different from ammonia because its molecules contain a different combination of atoms arranged in a different way. Throughout the course we shall investigate properties of substances. We shall seek explanations in terms of the numbers, types, and arrangements of the atoms present.

MOLECULAR FORMULAS AND CHEMICAL EQUATIONS

The concept of *molecules* enabled us to explain the pressure-volume relationship of gases. The concept of *atoms* enables us to explain the chemical changes these gases undergo.

In trying to describe the chemical changes that occur when hydrogen gas and chlorine gas react, we wrote

$$n \left\{ \begin{array}{c} \text{hydrogen} \\ \text{molecules} \end{array} \right\} + n \left\{ \begin{array}{c} \text{chlorine} \\ \text{molecules} \end{array} \right\} \longrightarrow 2n \left\{ \begin{array}{c} \text{hydrogen chloride} \\ \text{molecules} \end{array} \right\}$$

While this statement as written is accurate and contains much information, it can be condensed into a kind of scientific shorthand.

Suppose we represent one hydrogen atom by the symbol H. Since a hydrogen molecule contains two hydrogen atoms, it can be represented as H—H. The line between the H's suggests some kind of bond holding the two atoms together. If a chlorine atom is represented as Cl, a chlorine molecule would be written as Cl—Cl, and a hydrogen chloride molecule would be represented as H—Cl. The chemical reaction can now be summarized by writing a shorter equation:

$$\text{H—H} + \text{Cl—Cl} \longrightarrow 2\,\text{H—Cl}$$

The equation clearly shows that one diatomic molecule of hydrogen combines with one diatomic molecule of chlorine. In the process, two molecules of hydrogen chloride are formed. Bonds between the hydrogen atoms in the hydrogen molecule are broken during the reaction. Bonds between the chlorine atoms in the chlorine molecule are also broken during the reaction. (As you will learn later, the ultraviolet light used to start the reaction breaks some of the bonds between the chlorine atoms. That is what initiates the explosion.) Finally, new bonds between hydrogen and chlorine atoms must form to produce hydrogen chloride molecules. The breaking of existing bonds and formation of new bonds is one of the important characteristics of a chemical reaction.

EXERCISE 3-2

(a) Write an *equation,* using the notations H—H for hydrogen and N—N for nitrogen, to represent the reaction between nitrogen and hydrogen to form ammonia. (b) Two litres of hydrogen gas combine with 1 litre of oxygen gas to form 2 litres of water vapor. Write an equation using the notation described in (a) to represent this reaction.

While we have simplified the notation by using H—H instead of the words "hydrogen molecule," we can simplify it even further. A hydrogen molecule can be represented as H_2. The subscript $_2$ indicates that there are two hydrogen atoms in each molecule. Similarly, the chlorine molecule can be represented as Cl_2. Since the hydrogen chloride molecule contains one hydrogen atom and one chlorine atom, it is written as HCl. The bond between atoms is implied. The **chemical equation** representing the reaction between hydrogen and chlorine is then

$$H_2 + Cl_2 \longrightarrow 2\,\text{HCl}$$

A chemical equation is simply a representation, in an abbreviated form, of laboratory observations. The notations H_2, Cl_2, and HCl are called the **molecular formulas** of the molecules they represent.

Write the following formulas and equations using these symbols: carbon = C, oxygen = O, and nitrogen = N.

(a) Two litres of carbon monoxide combine with 1 litre of oxygen to produce 2 litres of carbon dioxide. (All volumes are measured at the same temperature and pressure.) If carbon dioxide has the formula CO_2, what is the formula for carbon monoxide? The burning of carbon monoxide is the process by which that gas combines with molecular oxygen to produce carbon dioxide. Write an equation to represent the burning of carbon monoxide.

(b) Two litres of nitric oxide break down, or decompose, to form 1 litre of nitrogen and 1 litre of oxygen gas. (Again, all volumes are measured at the same temperature and pressure.) Write the word equation for the decomposition of nitric oxide. Write formulas for each product formed and indicate the relative numbers of molecules of each product. Then write a formula for nitric oxide that is consistent with the above. Finally, write a chemical equation to represent the decomposition of nitric oxide.

STRUCTURAL FORMULAS AND MOLECULAR MODELS

You have learned that the sequence in which atoms are bonded together in a molecule is very important in determining the properties of a substance. Also the position of each atom within the molecule and the shape of the molecule are important in explaining the properties of that molecule. Thus, it is helpful to represent the *arrangement* of atoms in the molecule as well as the number and kinds of atoms present. For this purpose, **structural formulas** and *molecular models* are used. The structural formula for steam, the gaseous form of water, is

$$\underset{\text{H}\qquad\text{H}}{\overset{\text{O}}{\diagup\diagdown}}$$

As with the H—H and H—Cl molecules, the dashes indicate the chemical bonds, or connections, between atoms. The angular arrangement indicates that the two hydrogen atoms and the oxygen atom do not lie in a straight line. The two hydrogen atoms join the oxygen atom at an angle of 104 °. A structural formula is a representation on a piece of paper of a three-dimensional molecule. For example, the molecule methane (CH_4) is represented by the planar structural formula

$$\begin{array}{c} \text{H} \\ | \\ \text{H—C—H} \\ | \\ \text{H} \end{array}$$

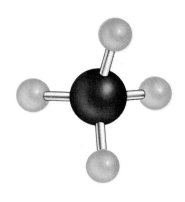

Figure 3-4

Model of a methane (CH_4) molecule.

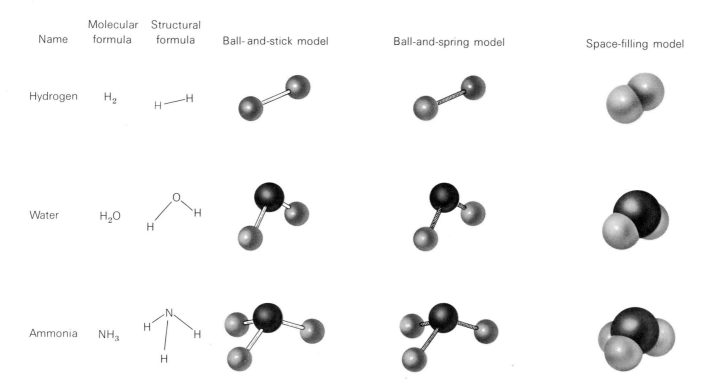

Name	Molecular formula	Structural formula	Ball-and-stick model	Ball-and-spring model	Space-filling model
Hydrogen	H_2	H——H			
Water	H_2O				
Ammonia	NH_3				

Figure 3-5

Different representations of hydrogen (H_2), water (H_2O), and ammonia (NH_3) molecules.

even though the actual molecule has the three-dimensional shape shown by the model in Figure 3-4 on page 66.

No written formula is quite as effective as a molecular model in helping us visualize molecular shapes. A number of physical models can be used to represent molecules. Some of the more commonly used models are shown in Figure 3-5. The *ball-and-stick model* and *ball-and-spring model* indicate bond arrangements. Ball-and-spring models are useful to indicate molecular flexibility and molecular vibration. The so-called *space-filling model* provides a more realistic view of the spatial relationships and the packing of nonbonded atoms and molecules. Figure 3-6 shows a model representing such packing of nonbonded molecules.

Figure 3-6

A model showing the crowding of nonbonded molecules.

ATOMS AND MOLECULES IN LIQUIDS AND SOLIDS

So far, we have examined evidence indicating that gases are made up of molecules. Gas molecules, in turn, are made up of atoms. But what of liquids and solids? Are they also made up of atoms and molecules?

A burning candle is a common sight. The products formed when a candle burns are found to be carbon dioxide and water. This prompts us to ask: what is a solid candle made of? Since a candle burns to produce two new gaseous products, we can assume the candle is also made of molecules that are made of atoms. As the candle burns, the "candle molecules" must combine with diatomic oxygen molecules to form gaseous products. Liquid candle material collects below the wick as burning proceeds. Thus, both liquid and solid candle material also contains atoms, and such atoms combine to give molecules of the liquid or solid. This simple assumption is consistent with all we know about both liquids and solids, which can be thought of as condensed forms of matter. The assumption that all matter is made of atoms is the fundamental postulate of the atomic theory. The atomic theory is amazingly powerful and can well be considered one of the cornerstones of modern science.

3.2 Elements and compounds

In giving an equation for the reaction between gaseous chlorine and hydrogen, we wrote

$$\text{H--H} + \text{Cl--Cl} \longrightarrow 2 \text{ H--Cl}$$

It is obvious that there are two classes of molecules in this reaction. In the first class, *identical atoms* are bound together to form molecules, such as H—H and Cl—Cl. Of course, the atoms in H—H are different from those in Cl—Cl. A substance made up of one kind of atom is called an **element.** The second class of molecule observed in our equation is hydrogen chloride. Pure hydrogen chloride has *identical molecules,* each of which contains one hydrogen *atom* and one chlorine *atom* in chemical combination. Hydrogen chloride, a substance that contains more than one kind of atom in chemical combination, is a **compound.**

THE ELEMENTS

As we have said, an element is a substance that contains only one kind of atom. All the materials on earth, as well as throughout the known universe, are made of one or more elements. A few of the elements, such as gold, silver, copper, and iron, were known to the ancient peoples of China, the Middle East, and the Mediterranean. By the beginning of the nineteenth century only about 26 elements had been discovered. One hundred years later, 81 elements were known. Figure 3-7 shows the number of known elements as a function of time. The graph on the right shows that the number of new elements discovered is declining. At present, we know of 109

Figure 3-7

At left, the discovery of the elements: total number of known elements as a function of time. At right, the discovery of the elements: number of elements discovered in each half century since 1700.

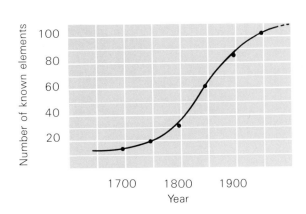

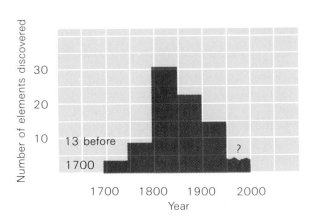

different elements. An important fact not shown by the graphs is that all elements discovered since 1940 were produced by nuclear reactions. Those newer elements do not occur naturally on earth. In fact, only one atom of element 109 and three atoms of element 108 have ever been made. For this reason, one can question their inclusion in the normal list of elements.

SYMBOLS OF THE ELEMENTS

In Section 3.1, we wrote word equations and equations using symbols. The most obvious choice for a symbol to represent an element is the first letter of the name of that element. In our equations we used O for oxygen, H for hydrogen, N for nitrogen, and C for carbon. Eight other elements are commonly represented by this simple convention. They are boron, fluorine, sulfur, phosphorus, iodine, vanadium, uranium, and yttrium.

Because there are 109 elements to name but only 26 letters in our alphabet, a second letter must be used to identify some elements. What choices for the second letter are most reasonable? Since chemistry is an international subject, international agreement is desirable. The symbols listed in Table 3-2 (see page 70) have been adopted by the International Union of Pure and Applied Chemistry (IUPAC), an international organization of chemists. After using the first letter of an element's name as its symbol, the next most obvious choice is the first *two* letters of the name. Many elements are represented by this convention, including helium, lithium, beryllium, neon, aluminum, silicon, argon, calcium, and titanium.

In other cases it is convenient to use the first letter of an element's name followed by a letter *other than* the second. Frequently the choice of the second letter is related to the sound of the name. For example, magnesium is Mg, chlorine is Cl, chromium is Cr, manganese is Mn, and zinc is Zn. Many additional elements have been given symbols in this way. The newer synthetic elements 104 to 109 are best represented by their atomic numbers at this time.[*]

Finally, ten common elements have symbols derived from their Latin names rather than their English names. For example, gold is Au, from the Latin *aurum*, silver is Ag, from the Latin *argentum* and iron is Fe, from the Latin *ferrum*. Because the elements in this group are common and their symbols are frequently used, they are listed in Table 3-3 (see page 71). Take a good look at these symbols.

NAMES OF COMPOUNDS

We have mentioned the names and symbols of many *elements*. What about the names of *compounds?* Is there logic to the system for naming them? The answer is a qualified yes. There are some rules, but—as with the rules of grammar—there are also exceptions.

---

[*]Considerable confusion and uncertainty is currently associated with names and symbols for elements 104 and 105. When these elements were reported by Albert Ghiorso and coworkers at The University of California at Berkeley, they suggested that 104 be called rutherfordium, Rf, in honor of Ernest Rutherford and 105 be called hahnium, Ha, in honor of Otto Hahn. Because some scientific and priority questions arose about these elements, IUPAC has not yet decided on permanent names.

Table 3-2

Atomic weights accepted by the International Union of Pure and Applied Chemistry

Name	Symbol	Atomic number	Atomic weight	Name	Symbol	Atomic number	Atomic weight
actinium	Ac	89	(227)*	neon	Ne	10	20.179
aluminum	Al	13	26.9815	neptunium	Np	93	(237)
americium	Am	95	(243)	nickel	Ni	28	58.70
antimony	Sb	51	121.75	niobium	Nb	41	92.906
argon	Ar	18	39.948	nitrogen	N	7	14.0067
arsenic	As	33	74.9216	nobelium	No	102	(259)
astatine	At	85	(210)	osmium	Os	76	190.2
barium	Ba	56	137.34	oxygen	O	8	15.9994
berkelium	Bk	97	(247)	palladium	Pd	46	106.4
beryllium	Be	4	9.0122	phosphorus	P	15	30.9738
bismuth	Bi	83	208.980	platinum	Pt	78	195.09
boron	B	5	10.811	plutonium	Pu	94	(244)
bromine	Br	35	79.904	polonium	Po	84	(209)
cadmium	Cd	48	112.40	potassium	K	19	39.098
calcium	Ca	20	40.08	praseodymium	Pr	59	140.908
californium	Cf	98	(251)	promethium	Pm	61	(145)
carbon	C	6	12.01115	protactinium	Pa	91	(231)
cerium	Ce	58	140.12	radium	Ra	88	(226)
cesium	Cs	55	132.905	radon	Rn	86	(222)
chlorine	Cl	17	35.453	rhenium	Re	75	186.2
chromium	Cr	24	51.996	rhodium	Rh	45	102.906
cobalt	Co	27	58.9332	rubidium	Rb	37	85.47
copper	Cu	29	63.546	ruthenium	Ru	44	101.07
curium	Cm	96	(247)	samarium	Sm	62	150.35
dysprosium	Dy	66	162.50	scandium	Sc	21	44.956
einsteinium	Es	99	(254)	selenium	Se	34	78.96
erbium	Er	68	167.26	silicon	Si	14	28.086
europium	Eu	63	151.96	silver	Ag	47	107.868
fermium	Fm	100	(257)	sodium	Na	11	22.9898
fluorine	F	9	18.9984	strontium	Sr	38	87.62
francium	Fr	87	(223)	sulfur	S	16	32.064
gadolinium	Gd	64	157.25	tantalum	Ta	73	180.948
gallium	Ga	31	69.72	technetium	Tc	43	(97)
germanium	Ge	32	72.59	tellurium	Te	52	127.60
gold	Au	79	196.967	terbium	Tb	65	158.925
hafnium	Hf	72	178.49	thallium	Tl	81	204.37
helium	He	2	4.0026	thorium	Th	90	232.038
holmium	Ho	67	164.930	thulium	Tm	69	168.934
hydrogen	H	1	1.00797	tin	Sn	50	118.69
indium	In	49	114.82	titanium	Ti	22	47.90
iodine	I	53	126.9045	tungsten	W	74	183.85
iridium	Ir	77	192.2	uranium	U	92	238.03
iron	Fe	26	55.847	vanadium	V	23	50.941
krypton	Kr	36	83.80	xenon	Xe	54	131.30
lanthanum	La	57	138.91	ytterbium	Yb	70	173.04
lawrencium	Lr	103	(260)	yttrium	Y	39	88.906
lead	Pb	82	207.2	zinc	Zn	30	65.38
lithium	Li	3	6.941	zirconium	Zr	40	91.22
lutetium	Lu	71	174.97	—	—	104	(261)
magnesium	Mg	12	24.305	—	—	105	(262)
manganese	Mn	25	54.9380	—	—	106	(263)
mendelevium	Md	101	(258)	—	—	107	(262)
mercury	Hg	80	200.59	—	—	108	(265)
molybdenum	Mo	42	95.94	—	—	109	(266?)
neodymium	Nd	60	144.24				

* For those elements all of whose isotopes are radioactive, the parentheses indicate the isotope with the longest half-life. (For a definition of *half-life,* see page 198.)

Table 3-3

Chemical symbols that are not derived from the common English name of the element

Common name	Symbol	Symbol source*
antimony	Sb	*stibnum*
copper	Cu	*cuprum*
gold	Au	*aurum*
iron	Fe	*ferrum*
lead	Pb	*plumbum*
mercury	Hg	*hydrargyrum*
potassium	K	*kalium*
silver	Ag	*argentum*
sodium	Na	*natrium*
tin	Sn	*stannum*
tungsten	W	*wolfram*

*All entries in this column are Latin, except *wolfram* and *kalium*, which are German.

The simplest kind of compound is one containing only two kinds of atoms. This is called a **binary compound.** The name of a binary compound is formed by taking the name of the first element in the formula, joining it to the stem of the name of the second element, and adding *-ide.* Let us see how this works. As you know, common table salt is represented as NaCl and named sodium chloride. *Chlor-* is the stem of chlorine. We add *-ide* to the second part of the name of the substance. Similarly, the compound $CaCl_2$ is named calcium chloride. We have already called the compound HCl hydrogen chloride. The gas H_2S is named hydrogen sulfide.

In some cases two elements combine to form more than one chemical compound. For example, carbon combines with oxygen to form different compounds, such as CO and CO_2. The name carbon oxide would not differentiate between these two molecules. Thus, we name the compound with the formula CO carbon *mon*oxide. The other compound, with the formula CO_2, is known as carbon *di*oxide. The prefix *mono-* means one and the prefix *di-* means two. Other prefixes are listed in Table 3-4. Carbon monoxide contains one oxygen atom per molecule, and carbon dioxide contains two oxygen atoms per molecule. Usually, but not always, the absence of a prefix before the name of an element means only one of its atoms is present in the molecule. The absence of a prefix before carbon tells us that only one atom of carbon is present in each molecule. Carbon tetrafluoride is CF_4. Sulfur trioxide is SO_3. Nitrogen trifluoride is NF_3, while N_2F_2 is dinitrogen difluoride.

Table 3-4

Chemical prefixes

Prefix	Meaning
mono-	one
di-	two
tri-	three
tetra-	four
penta-	five
hexa-	six
hepta-	seven
octa-	eight
nona-	nine
deca-	ten

> Name each of the following compounds, using appropriate prefixes: SO_2, SO_3, CF_4, NF_3, NO_2, SF_6, Na_2S, $ScCl_3$. The stem for oxygen is *ox-*, for fluorine *fluor-*, and for sulfur *sulf-*.

ATOMIC AND MOLECULAR WEIGHTS

We have defined *molecular weight* as the *relative mass of a molecule* of a given substance. The molecular weights in the fourth column of Table 2-4 on page 44 are based on a scale in which the oxygen molecule is assigned a value of 32.00 units. One hydrogen molecule must then be assigned a molecular weight of 2.02 of these units. Both the oxygen molecule and the hydrogen molecule are diatomic. Each contains two atoms. Thus, it is reasonable to define **atomic weight** as the *relative mass of an atom* on this same scale. Each oxygen atom must have an atomic weight that is half 32.00, or 16.00, since the diatomic oxygen molecule is assigned a molecular weight of 32.00. Each hydrogen atom must have an atomic weight that is half 2.02, or 1.01.

Such atomic and molecular weights give only the relative masses of atoms and molecules. They tell us, for example, that one oxygen atom (O) has about 16 times the mass of one hydrogen atom (H). One oxygen molecule (O_2) has about 32 times the mass of one hydrogen atom. Chemists today compare the mass of all atoms and molecules to the mass of one atom of a form of carbon known as carbon-12.* Modern atomic and molecular weights are defined in terms of this specific carbon atom and expressed as *unified atomic mass units,* abbreviated *u*.† This new unit is so small that 32.00 units are present in the mass of a *single* oxygen molecule.

The actual mass in grams of *one* oxygen *molecule* (O_2) can be calculated, knowing the mass of 1 mole of oxygen molecules, 32 grams, and knowing the number of molecules in 1 mole. As you read in Chapter 2, the number of molecules in 1 mole has been measured by various experiments as 6.02×10^{23}. This is Avogadro's number.

Just as you know that the mass of one dime is $\frac{1}{12}$ the mass of a dozen dimes, so the mass of one oxygen molecule is $1/6.02 \times 10^{23}$ the mass of 1 mole of oxygen molecules. Thus the mass of one O_2 molecule is

$$\frac{32.00 \text{ g}}{1 \text{ mole } O_2} \times \frac{1 \text{ mole } O_2}{6.02 \times 10^{23} \text{ molecules}} = 5.31 \times 10^{-23} \text{ g/molecule}$$

Written without the negative exponent, 10^{-23}, this number would be written 0.000,000,000,000,000,000,000,0531 g/molecule!

The relative mass of *atoms* is of more practical value to chemists. One mole of atoms of an element is 6.02×10^{23} atoms. It has a mass equal to the atomic weight of the element expressed in grams. For oxygen, the atomic weight is 16.00 units. One mole of oxygen *atoms* has a mass of 16.00 g, which

---

*The experiments that led to the choice of carbon-12 as the standard are difficult to explain at this point in our discussion. The change from oxygen to a specific carbon atom (a change of -0.0045 percent) is important in nuclear processes but unimportant in most chemistry.
†This is the symbol approved by IUPAC.

is half the mass of 1 mole of oxygen *molecules.** For hydrogen the atomic weight is 1.008, so 1 mole of hydrogen atoms has a mass of 1.008 g. For carbon the atomic weight is 12.01, and 1 mole of carbon atoms has a mass of 12.01 g. Atomic weight values are used frequently. A table of atomic weight values is given at the back of this text. Table 3-2, on page 70, lists more precise atomic weight values.

Chemists use atomic weights to find the molecular weights of compounds. Common sense tells us that the mass of an object equals the sum of the masses of its parts. If we want to know the *molecular weight* of HCl, we simply add the atomic weight of one hydrogen atom and the atomic weight of one chlorine atom:

$$\text{mol wt HCl} = \text{at. wt H} + \text{at. wt Cl}$$
$$\text{mol wt HCl} = 1.0 + 35.5$$
$$\text{mol wt HCl} = 36.5$$

By definition, 1 *mole* of HCl molecules has a mass of 36.5 g.

EXERCISE 3-5

(a) Show that the mass of 1 mole of CO_2 molecules is 44.0 g. (b) Show that the mass of 1 mole of SO_2 molecules is 64.1 g. (c) Calculate the mass in grams of 6.02×10^{23} molecules of CO. Calculate the mass in grams of 3.01×10^{23} molecules of fluorine (F_2).

EXERCISE 3-6

What is the molecular weight of hydrogen cyanide (HCN)? (Use the table of atomic weights inside the back cover.) How many moles of HCN are present in a 2.7-g sample?

EXERCISE 3-7

What is the molecular weight of carbon tetrachloride (CCl_4)? How many moles are contained in a 7.7-g sample of CCl_4?

EXERCISE 3-8

What is the mass, in grams, of 0.20 mole of HCl? (Use atomic weights.)

*The mass per mole, or *molar mass*, of a substance in grams will always be the same number as that substance's atomic or molecular weight expressed in unified atomic mass units (u). The calculation of molar mass was discussed in Chapter 2.

More complicated molecules work the same way. The very corrosive compound sulfuric acid (H_2SO_4) is a colorless, oily liquid. It is important not only because of its own properties but also because of its widespread use in the chemical industry. What is the molar mass of this important chemical? The molecular weight of one H_2SO_4 molecule is the sum of the masses of its atoms based on their atomic weights:

$$\text{mass of 2 H atoms} = 2\,\text{H atoms} \times 1\,u/\text{H atom} = 2\,u$$
$$\text{mass of 1 S atom} = 1\,\text{S atom} \times 32\,u/\text{S atom} = 32\,u$$
$$\text{mass of 4 O atoms} = 4\,\text{O atoms} \times 16\,u/\text{O atom} = \underline{64\,u}$$
$$\text{mol wt of } H_2SO_4 = 98\,u$$

Hence, the *molar mass* of H_2SO_4 is 98 grams. The mass of 2 moles of H_2SO_4 is twice that figure, or 196 grams. The mass of $\frac{1}{10}$ mole (0.1 mole) of H_2SO_4 is $\frac{1}{10}$ of 98 grams or 9.8 grams. These relationships are reviewed in the margin.

Some of the advantages of the atomic theory and its symbols and formulas are obvious, but some very important advantages remain hidden. For example, by using the atomic theory we can usually tell how much of a given product will be obtained from known amounts of reactants. Often, this is of great practical as well as theoretical value. In Chapter 4 we will examine these numerical, or quantitative, relationships using the mole as the measuring unit.

Reviewing Unit Factors

The molar mass relationships of H_2SO_4 can be clearly shown by using the unit factor method. We can write:

98.1 g H_2SO_4 = 1 mole H_2SO_4

from the definition of a mole. The unit factors are:

$$\frac{1 \text{ mole } H_2SO_4}{98.1 \text{ g } H_2SO_4} = 1 \quad \text{or}$$

$$\frac{98.1 \text{ g } H_2SO_4}{1 \text{ mole } H_2SO_4} = 1$$

We then write:

$$2.00 \text{ moles } H_2SO_4 \times \frac{98.1 \text{ g } H_2SO_4}{1 \text{ mole } H_2SO_4} = 196 \text{ g } H_2SO_4$$

$$0.10 \text{ mole } H_2SO_4 \times \frac{98.1 \text{ g } H_2SO_4}{1 \text{ mole } H_2SO_4} = 9.8 \text{ g } H_2SO_4$$

EXERCISE 3-9

Find the molar mass of each of the following compounds: lye, formula NaOH; baking soda, formula $NaHCO_3$; iron rust, formula Fe_2O_3; acetic acid from vinegar, formula CH_3COOH.

SUMMARY

A chemical change produces new substances with new properties and is usually accompanied by an observable energy change. Such changes can be represented by chemical equations.

The unit particles of any substance are called molecules. Molecules are made of simpler particles called atoms. A molecule can be broken down into its atoms, but the atoms themselves cannot be split in normal chemical reactions.

All matter is made of substances containing one or more different kinds of atoms. Substances containing only one kind of atom are known as elements and are represented by symbols consisting of letters or of numbers and letters.

Substances containing more than one *kind* of atom are known as compounds. Their formulas indicate both the kind and number of each atom in each of their molecules.

Compounds containing only two different kinds of atoms are called binary compounds. When the same elements form more than one compound, a prefix is used to distinguish them. The prefix *mono-* in carbon monoxide indicates one oxygen atom per molecule. The prefix *di-* in carbon dioxide indicates two oxygen atoms per molecule.

The atomic weight of an atom is the mass of that atom relative to the mass of an atom of carbon-12 that has an assigned mass of 12.000. The molecular weight of a molecule is equal to the sum of the atomic masses of all atoms in the molecule.

One mole of atoms of an element, or 6.02×10^{23} atoms, has a mass equal to the element's atomic weight expressed in grams. For example, 1 mole of oxygen atoms with atomic weight 16.00 has a mass of 16.00 grams.

The molecular weight of a substance is the sum of the atomic weights of *all* the atoms in the molecule. One mole of molecules has a mass equal to the molecular weight in grams. One oxygen molecule (O_2) has a molecular weight of 2×16.00, or 32.00, units. One mole of oxygen molecules has a mass of 32.00 grams.

Key Terms

atomic weight

atom

binary compound

chemical change

chemical equation

compound

diatomic

element

molecular formula

monatomic

structural formula

QUESTIONS AND PROBLEMS

A

1. Explain the difference between a chemical change and a physical change?
2. Which of the following processes would be classified as a physical change and which ones as a chemical change: (a) the boiling of water; (b) the burning of a candle; (c) the formation of a shell by an oyster using materials dissolved in sea water; (d) the inflation of a bicycle tire; (e) the conversion of NO to NO_2 (NO and NO_2 are pollutants in air that are present in automobile exhaust gases)?
3. Consider the reaction of gaseous nitric oxide with gaseous fluorine. The following observations are made: (1) a one litre sample of gaseous nitric oxide reacts with a 0.50 litre sample of gaseous fluorine to give one litre of gaseous product; (2) experiment shows that only one product forms. (a) Write the word equation for this process. (b) Write the word equation converting volumes to relative numbers of molecules. Use n as shown on page 63. (c) Write the word equation using the smallest set of whole numbers as coefficients in the word equation. (Hint: let $n = 1$ or the smallest possible whole number value.) Refer to page 63. (d) Do these observations indicate that fluorine has an even number of atoms in each molecule? Explain. (e) Do these observations indicate that the nitric oxide molecule must be broken in the reaction? Explain. (f) Write the chemical equation for the process using the formula NO and the symbol F. The final product is named nitrosyl fluoride.
4. One molecule of solid white phosphorus (symbol for phosphorus is P) will give 4 molecules of H_3PO_4. What is the simplest formula for a *molecule* of white phosphorus?
5. Using oxygen, nitrogen, and chlorine as examples, explain the difference between atomic weight and molecular weight.

6. Write the structural formula for: (a) water, (b) ammonia, (c) carbon tetrachloride, (d) hydrogen fluoride.

7. (a) How does the molecule of an element differ from the molecule of a compound? (b) Which of the following are elements and which are compounds: (1) He (2) SO_2 (3) P_4 (4) S_8 (5) KCl (6) CH_4 (7) CO (8) MgO (9) CH_3OH (10) NF_3 (11) $Al_2(SO_4)_3$ (12) carbon as diamond (13) O_3 (14) I_2? (c) Name the following: (1) SO_2 (2) KCl (3) CF_4 (4) S_8 (5) NF_3 (6) N_2F_4 (7) CO (8) MgO (9) CS_2 (10) N_2O_3 (11) Cu_2O (12) Na_2S.

8. (a) Give the symbols and names of four elements that are represented by the first letter in the name of the element. See Table 3-2 on page 70. (b) Give the symbols and names of four elements that are represented by the first two letters in the name of the element. (c) Give the symbols and names of two elements that are represented by one or two letters derived from the Latin name of the element.

9. (a) What is meant by one mole of a substance? (b) How many grams of each of the following substances would have to be weighed out in order to get one mole of each: (1) oxygen gas, (2) water, (3) H_2SO_4, (4) P_4.

10. (a) If one mole of water weighs 18 grams, how many moles do you have in 36 g? (b) What would $\frac{1}{2}$ mole of water weigh? What would 0.10 of a mole of water weigh? (c) How many moles of CBr_4 are contained in a 100.0 g sample of CBr_4? (d) You need 0.1 mole of NaOH. How many grams of NaOH do you weigh out?

B

11. One litre of laughing gas will react with one litre of carbon monoxide to give one litre of carbon dioxide and one litre of gaseous nitrogen. (a) Write the word equation for this process. (b) Replace the words carbon monoxide, carbon dioxide, and nitrogen gas in the word equation with chemical formulas. (c) Based on the information in the equation of part (b), what is the formula and scientific name of laughing gas?

12. What is the mass in grams of white phosphorus that must be weighed out to get one mole of phosphorus *atoms*?

13. (a) Write a balanced equation for the reaction of hydrogen gas with oxygen gas to give water vapor. (b) On the basis of the equation in part (a), what gases would result if 2 litres of hydrogen gas were allowed to react with a half litre of oxygen gas? with 2 litres of oxygen gas? (c) Which reactant is limiting the process of each reaction in part (b)?

14. (a) Write a balanced equation for the reaction between nitrogen gas and hydrogen gas to give ammonia (NH_3). (b) On the basis of the equation in part (a), how many litres of ammonia can be formed by a reaction mixture containing 6 litres of nitrogen gas and 6 litres of hydrogen gas? Which reactant is limiting the process of this reaction?

15. If the molecular weight of sulfur is 256.8 and the atomic weight of sulfur is 32.1, what is the molecular formula for elemental sulfur?

16. When an orange powder is heated strongly over a Bunsen burner flame in a test tube, a colorless gas is released and a silvery deposit appears on the side of the tube. (a) Is the orange powder an element or a compound? Explain. (b) Is there enough information to decide whether the silvery substance and the colorless gas are elements or compounds?

17. For each of the following substances: (a) List the kinds of atoms present

by writing the names and symbols of the atoms. (b) State how many atoms of each kind are in the molecule. (c) Indicate which of the substances are elements and which are compounds. (d) Give the molecular weight for each substance: (1) CO_2 (2) $AgNO_3$ (3) S_4 (4) Cu (5) $K_2Cr_2O_7$ (6) C_2H_5OH (ethyl alcohol) (7) $CaH_4(PO_4)_2$.

18. You are given 64.0 grams of CH_4 molecules: (a) What is the molecular weight of methane? (b) How many *moles* of CH_4 are present? (c) How many *molecules* of CH_4 are present? (d) What is the total number of atoms in each molecule? (e) What is the total number of *moles of atoms* in the sample? (f) What is the total number of *atoms* in the sample?

19. How many moles are in each of the following: (a) 36.0 g H_2O (b) 8.50 g NH_3 (c) 6.35 g Cu (d) 2.16 g Ag (e) 0.85 g $AgNO_3$ (f) 7.10 g Cl_2?

C

20. Two litres of hydrogen fluoride combine completely with one litre of the gas dinitrogen difluoride to form two litres of a new gaseous product that we will call "G." (a) Write formulas for hydrogen fluoride and dinitrogen difluoride based on the rules of nomenclature. (b) If the number of moles of dinitrogen difluoride is n, how many moles of G are formed? (c) What hypothesis have you assumed to be true in obtaining your answer? (d) Write the equation for the process described, including the formula of G. (e) Do the data require the formula N_2F_2 or would they be equally consistent with the formula N_4F_4 or NF? Explain.

21. You are given 58.0 grams of butane, C_4H_{10}: (a) What is the molecular weight of butane, C_4H_{10}? (b) How many *moles* of butane molecules are present? (c) How many *molecules* of C_4H_{10} are present? (d) What is the total number of *atoms* in each individual molecule of C_4H_{10}? (e) What is the total number of *moles of atoms* in the sample? (f) What is the total number of atoms in the sample?

22. Element A has an atomic weight of 12.0 and element B has an atomic weight of 35.5. A 12.0 gram sample of element A combines with 142.0 grams of element B to give one mole of a new compound, C. Substance B is a *diatomic* gas. Substance A is a solid. (a) How many moles of *atoms* of element A are taken? (b) How many moles of *atoms* of element B are taken? How many moles of *molecules* of B are taken? (c) Write the balanced equation for the process, including the formula of compound C in terms of A and B. (d) What is the molecular weight of C? (e) What is the actual formula for compound C?

23. When sodium metal is dropped into water a vigorous reaction occurs. Hydrogen gas is released and a solution of lye, NaOH, remains in the reaction vessel. (a) Write the balanced equation for this process. (b) If 23.0 g of sodium react with water, how many moles of hydrogen will be produced? (c) What will be the mass of this hydrogen?

24. (a) Determine the mass in grams of one gold atom. Gold has an atomic weight of 197. (b) A person has a gold filling in a tooth that weighs 0.296 g. How many *moles* of gold are in the filling? (c) How many *atoms* of gold are in the filling? (d) Suppose you were to divide the gold atoms in this filling equally among all of the people in the world. The world's population is estimated at 4.7 billion, which can be written as 4.7×10^9. How many gold atoms would each person get?

The birth of the atomic theory

The French chemist Antoine Lavoisier has been called the father of modern chemistry. In 1789 he proposed that no mass is lost during a chemical reaction. This generalization is known as the *Law of Conservation of Mass*. About 1800, another important mass relationship was revealed by Joseph Proust, a French chemist who found that pure water is *always* composed of one part hydrogen and eight parts oxygen by weight. An 18-gram sample would have 2 grams of hydrogen and 16 grams of oxygen, and so on. Other pure compounds also had constant compositions. This generalization is known as the *Law of Constant or Definite Composition*. Formally, it states that every sample of any pure compound is always composed of the same elements in the same proportions by mass.

Dalton's atomic theory Most scientists at the start of the nineteenth century believed in the atomistic nature of matter. The English schoolteacher John Dalton (1766–1844) was able to see that by taking the idea of atoms and giving to them a set of consistent masses, he could explain both conservation of mass and constant composition. Specifically, Dalton assumed in 1808 that each element was made of its own kind of atoms, with their own sizes, shapes, and masses. All atoms of a given element were seen as identical in these properties, but *different* from the atoms of all other elements. Further, he believed that atoms were *indivisible* and could not be split apart—even during the most violent chemical change.

To appreciate how neatly this new picture explained the observable, we use it to picture the reaction between hydrogen and oxygen to form water. Dalton *assumed* that *one* hydrogen atom combined with *one* oxygen atom to form one molecule of water.* Dalton's water molecule would be represented by modern symbols as HO. Using this incorrect assumption and the experimental fact that water contains 8.0 g of oxygen for every gram of hydrogen, Dalton concluded that one oxygen atom

*Dalton's model does not include the term *molecule* but states that all substances are composed of combinations of atoms.

must be eight times as massive as one hydrogen atom. Extending arguments of this type to other compounds, he was able to construct the first atomic weight table. Hydrogen, the lightest element, was assigned a mass of one atomic weight unit (abbreviated *u*). Oxygen was assigned a mass of 8 *u* on the Dalton scale. So far so good, but a quick look at the modern table of atomic weights shows that oxygen has a mass of 16.0 *u*, not 8 and hydrogen is 1 *u* on both scales. Dalton's error arose when he assumed a formula of HO for water. In his formula, the molecule contained one hydrogen atom and one oxygen atom rather than two hydrogen atoms and one oxygen atom. His incorrect assumptions about formulas gave erroneous weights, yet he did a lot with this oversimplified concept.

Advantages of Dalton's atomic theory Any chemical reaction was pictured by Dalton as a *rearrangement* of atoms, which involved no splitting of atoms or change of mass. This meant the total mass of all atoms involved in the reaction must remain constant throughout. A useful analogy is the rearranging of a child's alphabet blocks into different words. Literally and figuratively the letters are the building blocks from which many different words can be constructed. So were atoms, according to Dalton. The same few atoms could form many different chemical compounds. Whether several letter blocks are considered separately, brought together to form one word, or rearranged into a different word, their masses do not change. Their *total* mass remains constant because the same number of blocks are only moved into different positions. By analogy, this must also be the case with atoms. Thus, Dalton explained the Law of Conservation of Mass.

How did Dalton explain the formation of a weighable amount of a compound such as water? If the process in which one hydrogen atom unites with one oxygen atom could be repeated billions and billions of times, a detectable amount of water would result. No matter how many HO molecules were formed, 8 mass units of oxygen would always combine with one mass unit of hydrogen. Dalton's atomic weight scale used this experimental fact. His reasoning was applied to many pure compounds other than water, and it provided a reasonable explanation for the Law of Constant Composition.

Modification and extension of the atomic theory
So far, we have examined the atomic theory in essentially its original form, as proposed by Dalton in 1808. As such, we have seen that it is both convincing and useful in explaining laws such as conservation of mass and constant composition. But all theories, even those we accept today, have their limitations, and Dalton's was no exception. We have already spotted the most severe limitation of Dalton's atomic theory. His assumption of a formula of HO for water gave erroneous atomic weights. But notice that while his atomic weights were wrong, he could still explain other observed mass relationships.

Nevertheless, chemists became increasingly troubled by the arbitrariness of Dalton's rules. They wanted additional evidence to help them decide just how atoms do combine to form molecules. To illustrate the nature of their dilemma, consider the example of just how hydrogen and oxygen do combine to form water. Obviously, there are an infinite number of possibilities, so we will limit our discussion to the three simplest ones. One possibility is that one atom of each forms the molecule HO, as Dalton thought. Another is that one H atom and two O atoms form an HO_2 molecule. Still another possibility is that two H atoms and one O atom form H_2O.

These three different assumptions give vastly different atomic weights for oxygen. On page 78 we found that an assumed formula of HO gave oxygen an atomic mass of 8, if hydrogen were assigned a value of 1: $\frac{O}{H} = \frac{8}{1}$ or $O = 8$. If a formula of HO_2 were the true formula for water we would find, if hydrogen were assigned a value of 1: $\frac{2O}{1H} = \frac{8}{1}$ or $O = \frac{8 \times 1H}{2}$ or $O = 4$. With this assumption, the atomic mass of oxygen would be 4 when $H = 1$. Finally, if H_2O were the true formula for water we would find, if hydrogen were assigned a value of 1: $\frac{O}{2H} = \frac{8}{1}$ or $O = \frac{2 \times H \times 8}{1}$ or $O = 16$. With this assumption, the atomic mass of oxygen would be 16 when $H = 1$. The key question involves the number of atoms of oxygen and hydrogen in a single molecule of water. To answer this question, people again appealed to experiments.

The point is that atomic weights are just as arbitrary as formulas because they are derived from formulas. Of the three so-called cornerstones of the atomic theory—combining weights, formulas, and atomic weights—only the first was based on actual experimentation. The other two were pure guesswork. But once one guess was made, the other followed logically from it. What was badly needed was additional evidence to help eliminate the guesswork, or at least to make it less arbitrary. In 1811, a modification of the atomic theory was suggested, and that modification is essentially what is taught in our schools today. We now turn to this new evidence and how it led to the modern atomic theory.

Gay-Lussac's discovery The new evidence came from studies of gases. Since gases can be liquefied and even solidified if they are cooled and/or compressed enough, they too must have atoms. Because many substances can exist either as solids, liquids, or gases, depending on their temperatures, it seems reasonable to assume that they are all made of the same small particles.

Keeping this concept in mind, let us consider the discovery, in 1808, by the French chemist, Joseph Louis Gay-Lussac (1778–1850), of the *Law of Combining Gas Volumes*. This law states that volumes of gases that react with each other can be expressed in ratios of small whole numbers.

You have probably seen at least one example of what Gay-Lussac found when you observed a demonstration of the decomposition of water. With an apparatus like the one in Figure 4-7, water can be broken down into its elements: the gases hydrogen and oxygen. The amazing thing about the experiment is that no matter how much water decomposes, the volume of hydrogen formed is exactly *twice* the volume of oxygen formed. Conversely, if two volumes of hydrogen are ignited in one volume of oxygen, all of both gases combine explosively to form water. If this is done above 100 °C, the water forms as steam, and it occupies two volumes. Figure 3-8 illustrates Gay-Lussac's Law of Combining Gas Volumes. The two-to-one volume relationship is no accident. If 2 litres of hydrogen are ignited with 2 litres of oxygen, 2 litres of steam forms, but 1 litre of oxygen remains uncombined at the end of the reaction. Evidently, the chemical combination involves hydrogen and oxygen in a two-to-one ratio by volume.

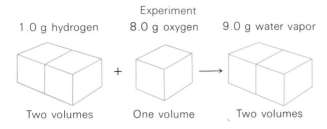

Experiment

1.0 g hydrogen   8.0 g oxygen   9.0 g water vapor

Two volumes   One volume   Two volumes

Figure 3-8

Gay-Lussac's Law of Combining Gas Volumes. Numbers above the boxes give the combining weights of the gaseous elements in these reactions. The boxes show the combining volumes as determined by experiment and expressed in their simplest whole number ratios. All gases are at the same conditions of temperature and pressure.

The Law of Combining Gas Volumes works for every reaction involving gases. Combining volumes are always found to be simple whole-number ratios. Gay-Lussac's discovery provides us with valuable additional information about how the gaseous elements combine. If this information could be interpreted in terms of atoms and molecules, we might be less arbitrary than Dalton was in assigning molecular formula and atomic weights. The two-to-one volume relation of the gases in the formation of water suggests that its formula might be H_2O instead of Dalton's HO. Yet Dalton rejected the Law of Combining Gas Volumes. As we trace the logical argument behind the acceptance of H_2O, you will see why he rejected this concept.

Avogadro's hypothesis The formula H_2O means twice as many H atoms as O atoms would be needed, regardless of how many water molecules formed. This follows from the volume ratios *only* if two volumes of hydrogen contain twice as many atoms as one volume of oxygen. This means that equal volumes of the two gases contain the *same* number of particles. All this may seem perfectly obvious to you, but it is only an *assumption* that could not be proven with the knowledge and techniques of Dalton's time. This was the assumption made in 1811 by the Italian physicist Amedeo Avogadro (1776–1856). Although he could not

prove it, he saw this assumption as a way to clear up the formula–atomic weight stalemate. In 1811, Avogadro's assumption that equal gas volumes contain equal numbers of particles was only a shrewd guess. It was no more startling than Dalton's rule of greatest simplicity regarding the formation of a compound from the atoms of elements.

But why did Dalton find this concept impossible to accept? The main reason is a logical consequence of Avogadro's assumption. Consider the reaction between hydrogen and chlorine gas to form gaseous hydrogen chloride, as shown in Figure 3-9. The volume ratios are one to one, forming two volumes of hydrogen chloride. But if you try to picture this reaction using atoms and Avogadro's assumption, you run into trouble.

For simplicity, let us assume that one volume contains only one particle that we will call a *molecule*. Since "molecules" of elements were Dalton's

Figure 3-9

Avogadro's explanation of Gay-Lussac's experimental observation of combining gas volumes. Any number of n molecules will result in the same 1-to-1-to-2 ratio when hydrogen chloride forms from its elements. Dividing through by n gives the simplest balanced equation for this reaction.

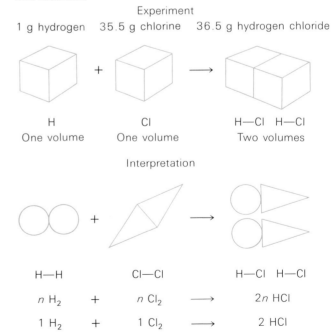

Experiment

1 g hydrogen   35.5 g chlorine   36.5 g hydrogen chloride

H   Cl   H—Cl H—Cl
One volume   One volume   Two volumes

Interpretation

H—H   Cl—Cl   H—Cl H—Cl

$n\ H_2$ + $n\ Cl_2$ ⟶ $2n\ HCl$

$1\ H_2$ + $1\ Cl_2$ ⟶ $2\ HCl$

atoms, *one* atom of hydrogen plus *one* atom of chlorine becomes *two* molecules of hydrogen chloride. The two molecules of hydrogen chloride follow from Avogadro: equal volumes contain equal numbers of particles, in this case, *one* particle or molecule per volume. Since there are *two* volumes of hydrogen chloride there must be *two* particles or molecules of it. Each of these particles or molecules must contain at least one atom of each element to be a compound of them. But count the atoms. Notice that there is an extra atom of each element on the right of our equation. (See the first equation in Figure 3-9.) This would seem to violate the Law of Conservation of Mass and was the problem that troubled Dalton. He could not accept either of the two ways to overcome this difficulty. Increasing the number of particles per volume would not help, because the two volumes of product will always contain twice as many of its molecules as atoms of hydrogen and chlorine.

One way out of the difficulty would be to imagine that each hydrogen and each chlorine molecule *split* during the reaction to become part of the two hydrogen chloride molecules. But this is contrary to Dalton's entire concept of indivisible atoms*, which so neatly accounted for the conservation of mass and constant composition laws.

Avogadro proposed another alternative. An element's atoms do not split, he said, but each particle of such gaseous elements as hydrogen and chlorine is itself made of two atoms joined together. During a reaction, the *atoms* are indivisible, but the *molecules* may split. As you know, *molecule* is the name now given to the existing particle of an element or compound under the conditions of the experiment. Thus, Avogadro's picture of the formation of hydrogen chloride has each molecule of hydrogen and chlorine splitting to become part of the two HCl molecules. Avogadro was able to overcome this difficulty by introducing the idea of *diatomic* molecules of certain gaseous elements, which could split during a reaction.

Now, argued Avogadro, if both hydrogen and oxygen are pictured as diatomic gases, then any sample of either element must be made of these diatomic molecules. Thus, the formation of two vol-

*Dalton believed that the molecule and atom of an element were the same.

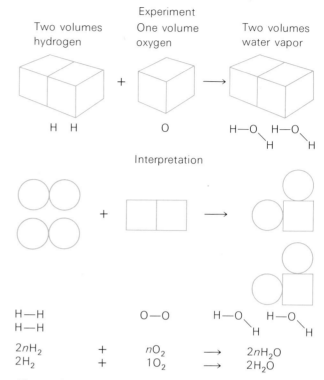

Figure 3-10

Avogadro's explanation of the formation of water vapor from its elements at constant temperature (above 100 °C) and pressure. The ratio in this example is 2-to-1-to-2.

umes of steam from two volumes of hydrogen and one volume of oxygen would be pictured as shown in Figure 3-10. This is the way we picture the process today.

Dalton could not accept Avogadro's diatomic molecules because he could not imagine what force held two identical atoms together in such molecules. The discovery that water could be pulled apart into its elements by electricity suggested to Dalton that opposite electric charges held atoms together in molecules. If so, all oxygen atoms must have the same electric charge, and should repel each other. How could two such atoms stay together in an oxygen molecule? Avogadro had no explanation for this. Nor did anyone else until nearly 50 years later. Thus Avogadro's ideas were not generally accepted in 1811, and chemistry remained about where Dalton had left it. It was not until 1860 that another Italian scientist, Stanislao Cannizzaro, was able to present enough evidence, in such an organized way, that Avogadro's ideas were finally accepted.

Because of its practical use, and for its own intrinsic interest, the principle of the conservation of energy may be regarded as one of the great achievements of the human mind.

SIR WILLIAM CECIL DAMPIER

THE CONSERVATION OF MASS AND ENERGY IN CHEMICAL REACTIONS

Objectives

After completing this chapter, you should be able to:

● Write, and understand the meaning of, a balanced chemical equation.

● Use the unit factor method to solve numerical problems.

● Calculate the relative amounts of each substance needed in a chemical reaction.

● Predict the effect of a limiting reagent on equation calculations.

● Support the statement that energy is conserved when water is decomposed and then reformed from its elements.

Tremendous amounts of energy put the space shuttle into orbit.

We live in an age marked by high speed travel. Some commercial planes now fly faster than the speed of sound and space vehicles orbit Earth in a very few hours. "Around the World in Eighty Days" is now just a song, a movie, and an 1872 novel by Jules Verne. It is no longer a technical goal. Cars and trucks make ground travel easy and heavy work is usually done by very large machines. We can live comfortably in very cold or very warm climates because we know how to heat or cool large amounts of air. All of these things are now possible because human beings have learned to focus needed amounts of energy on a single goal. Our ability to use such large amounts of energy effectively is a direct result of knowledge and experience. The subject of thermodynamics or heat transfer stands today as one of our most powerful tools. In this chapter, we will start to examine energy changes and chemical processes. Each year we burn millions of tons of coal and millions of barrels of oil to get the energy stored in them. Carbon dioxide, carbon monoxide, water vapor, sulfur dioxide, smoke, ash, and a variety of other trace products are an inevitable consequence of our need for the energy of fossil fuels. One of the most important problems of the day is to obtain the most energy with the least environmental damage. Chemistry dictates the rules of the game. Let us examine these rules.

4.1 Balanced chemical equations and the conservation of mass

When a substance burns, it combines chemically with oxygen in the air, forming a binary compound known as an *oxide*. The process of burning coal is represented approximately by the following equation:

$$C + O_2 \longrightarrow CO_2$$

Because coal consists mostly of carbon atoms, it is represented here as C.

Note that the equation "conserves" atoms, as do all correctly written equations. One carbon and two oxygen atoms are shown on either side of the arrow. In other words, *atoms are neither created nor destroyed in chemical reactions*. The atoms we start with may be rearranged into new molecules, but they are still present. The carbon atoms in the coal we burn usually become part of carbon dioxide molecules in the atmosphere, although some CO and other carbon compounds may also be produced. Because atoms have mass, conserving atoms means conserving mass. This principle, known as the **Law of Conservation of Mass,** is the reason why *equations must balance*. This law allows us to determine and predict the quantities of reactants and products involved in a chemical change.

In order to write a correctly balanced equation, you must know the correct formula for each reactant and product. The needed information comes from laboratory experiments. The formulas obtained are fixed *by nature* and can not be changed in the balancing process. Still, the Law of Conservation of Mass must be obeyed. *All the atoms shown on one side of an equation must also appear on the other side.* To conserve atoms, the **coefficients** in front of the formulas of the equation must be adjusted.

As an example, consider the burning of methane, one of the major components of natural gas used for heating. Chemical analysis shows that methane has the formula CH_4. Laboratory work also shows that two products are formed. They are carbon dioxide, CO_2, and water, H_2O. No formula can be changed during the balancing process. To balance the equation we first write the unbalanced or "skeleton" equation using the experimentally correct formulas to represent all the reactants and products:

$$? CH_4 + ? O_2 \longrightarrow ? CO_2 + ? H_2O$$

Figure 4-1

Burning of natural gas in a kitchen range.

The question marks remind us that we have not yet adjusted the coefficients for reactants and products to conserve atoms. In this equation there is one carbon atom on each side of the equation so carbon atoms are conserved if CH_4 and CO_2 are given coefficients of one. In contrast, we have four hydrogen atoms in the reactants and only two in the products. In order to conserve hydrogen atoms we need two water molecules. To show that these two water molecules are produced, a coefficient of *two* is placed in front of the formula for water. The equation then becomes:

$$1 CH_4 + ? O_2 \longrightarrow 1 CO_2 + 2 H_2O$$

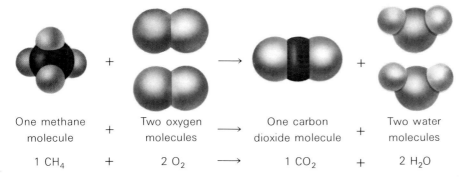

One methane + Two oxygen ⟶ One carbon + Two water
molecule   molecules   dioxide molecule   molecules

$1\ CH_4$ + $2\ O_2$ ⟶ $1\ CO_2$ + $2\ H_2O$

Figure 4-2
The balanced chemical equation for the combustion of CH_4 shown with molecular models.

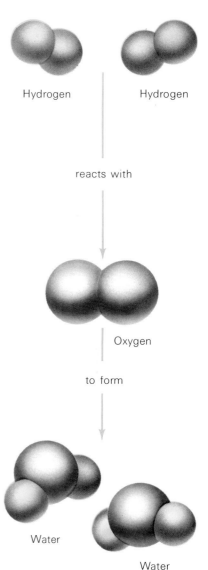

Hydrogen   Hydrogen

reacts with

Oxygen

to form

Water

Water

Figure 4-3
The formation of water shown with molecular models.

We now have four oxygen atoms on the product side. Oxygen atoms can be conserved by using $2\ O_2$ molecules on the reactant side:

$$1\ CH_4 + 2\ O_2 \longrightarrow 1\ CO_2 + 2\ H_2O$$

A quick check now shows that we have one carbon atom on each side, four hydrogen atoms on each side, and four oxygen atoms on each side. The equation is now balanced. The coefficient in front of each molecular formula indicates the number of molecules of each substance taking part in the process. Note that only coefficients in front of formulas were changed during balancing. Formulas were never changed. A balanced equation can be interpreted in terms of moles as well as in terms of molecules or atoms. In all equations, the coefficient 1 may be dropped but it is not wrong to retain it.

EXERCISE 4-1

Correctly balance each of the following unbalanced, or "skeleton," equations. Assume all molecular formulas are correct as written. They *cannot* be changed. Adjust only coefficients.

(a) $Mg + O_2 \rightarrow MgO$
(b) $Na + Cl_2 \rightarrow NaCl$
(c) $N_2 + O_2 \rightarrow NO$
(d) $Fe + O_2 \rightarrow Fe_2O_3$
(e) $CuS + O_2 \rightarrow CuO + SO_2$
(f) $N_2H_4 + O_2 \rightarrow N_2 + H_2O$

THE MEANING OF A CHEMICAL EQUATION

The equation

$$2\ H_2 + 1\ O_2 \longrightarrow 2\ H_2O$$

represents the reaction in which water is formed from hydrogen and oxygen. Regardless of how much or how little water is formed, the same equation represents the process. The equation does not tell us the actual number, but

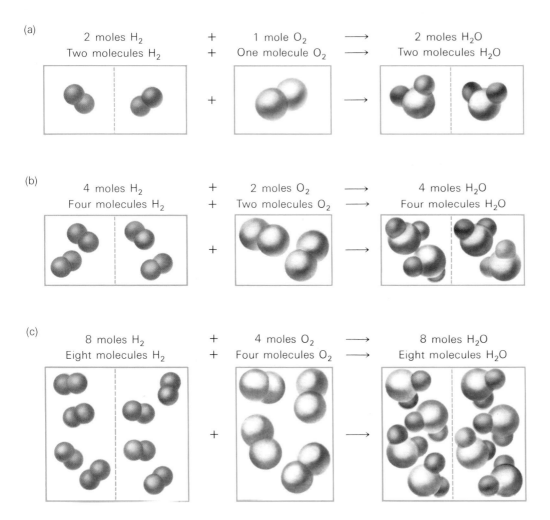

(a)

2 moles H$_2$ +   1 mole O$_2$ $\longrightarrow$ 2 moles H$_2$O
Two molecules H$_2$ +   One molecule O$_2$ $\longrightarrow$ Two molecules H$_2$O

+

(b)

4 moles H$_2$ +   2 moles O$_2$ $\longrightarrow$ 4 moles H$_2$O
Four molecules H$_2$ +   Two molecules O$_2$ $\longrightarrow$ Four molecules H$_2$O

+

(c)

8 moles H$_2$ +   4 moles O$_2$ $\longrightarrow$ 8 moles H$_2$O
Eight molecules H$_2$ +   Four molecules O$_2$ $\longrightarrow$ Eight molecules H$_2$O

+

Figure 4-4
The ratio of H$_2$ to O$_2$ to H$_2$O molecules or moles is 2 to 1 to 2 in each case.

only the *relative* number, of molecules involved. The actual number might be different each time the reaction occurs. But the equation does tell us the simplest ratio between the numbers of molecules reacting and forming. It says that if 2 hydrogen molecules (H$_2$) react with 1 oxygen molecule (O$_2$), then 2 water molecules (H$_2$O) form. It also indicates that 20 H$_2$ molecules would react with 10 O$_2$ molecules to form 20 H$_2$O molecules. In addition, the equation tells us that 2 million H$_2$ molecules would react with 1 million O$_2$ molecules to form 2 million H$_2$O molecules.

Suppose we started with 2 million molecules of H$_2$ and 2 million molecules of O$_2$. According to the equation, 2 million molecules of H$_2$ would react with only 1 million molecules of O$_2$. Then 1 million molecules of O$_2$ would remain after the reaction. Examples of such a reaction are shown in Figure 4-5. Regardless of the number of molecules of each gas present before a reaction, a chemical equation gives only the ratio of molecules that react. The presence of leftover, unreacted O$_2$ molecules in the last example is readily explained; not enough H_2 molecules were present initially to react with the O_2 molecules. On the other hand, an excessive number of O_2 molecules were present, more than could react. We therefore speak of an *excess* of oxygen.

The amount of oxygen that could react was determined by the amount of hydrogen initially present, *all* of which was consumed. The amount of the substance that is completely consumed in any reaction determines how much product is formed. This substance is called the **limiting reagent.** We use the word **reagent** here to describe *a substance that takes part in a particular chemical reaction.* In our example, H_2 was the limiting reagent that determined how much H_2O was formed.

EXERCISE 4-2

> One million oxygen molecules react with sufficient hydrogen molecules to form water molecules. How many water molecules are formed? How many hydrogen molecules are consumed?

BALANCED EQUATIONS AND THE MOLE

Recall from Chapter 2 that the mole is often referred to as the chemist's dozen and that 1 mole of molecules is equal to 6.02×10^{23} molecules. Thus, 2 moles of H_2 molecules would react with 1 mole of O_2 molecules to form 2 moles of H_2O molecules. In this example, $2 \times (6.02 \times 10^{23})$ H_2 molecules react with $1 \times (6.02 \times 10^{23})$ O_2 molecules to form $2 \times (6.02 \times 10^{23})$ H_2O molecules. Note that the ratio of H_2 to O_2 to H_2O molecules is 2 to 1 to 2 in each case. The ratio 2 to 1 to 2 contains the *same* numbers in the same order as those used to balance the equation:

$$2\,H_2 + 1\,O_2 \longrightarrow 2\,H_2O$$

Because a chemical equation gives the numerical ratio of *molecules* reacting and forming, it also gives that ratio of *moles* reacting and forming. To illustrate why this is so, let us compare several of the examples previously given for the hydrogen-oxygen reaction:

hydrogen (H_2) + oxygen (O_2) $\longrightarrow$ water (H_2O)

2 molecules + 1 molecule $\longrightarrow$ 2 molecules	*(1)*
$2 \times (6.02 \times 10^{23}$ molecules$) + 1 \times (6.02 \times 10^{23}$ molecules$) \longrightarrow$ $2 \times (6.02 \times 10^{23}$ molecules$)$	*(2)*
2 moles + 1 mole $\longrightarrow$ 2 moles	*(3)*

Note that lines 1 and 2 each show the same 2-to-1-to-2 ratio of molecules. By substituting in line 2 the word *mole* for its equivalent, *6.02×10^{23} molecules,* we derive line 3. Thus, an equation can be read in terms of molecules, as in line 1, or moles, as in line 3.

It is sometimes convenient to write an equation that shows 1 mole of some substance. This may necessitate the use of a fraction of a mole of some other

Figure 4-5

The effect of excess O_2 on the reaction $2 H_2(g) + 1 O_2(g) \longrightarrow 2 H_2O(g)$

when (a) two moles of O_2 are present and (b) four moles of O_2 are present.

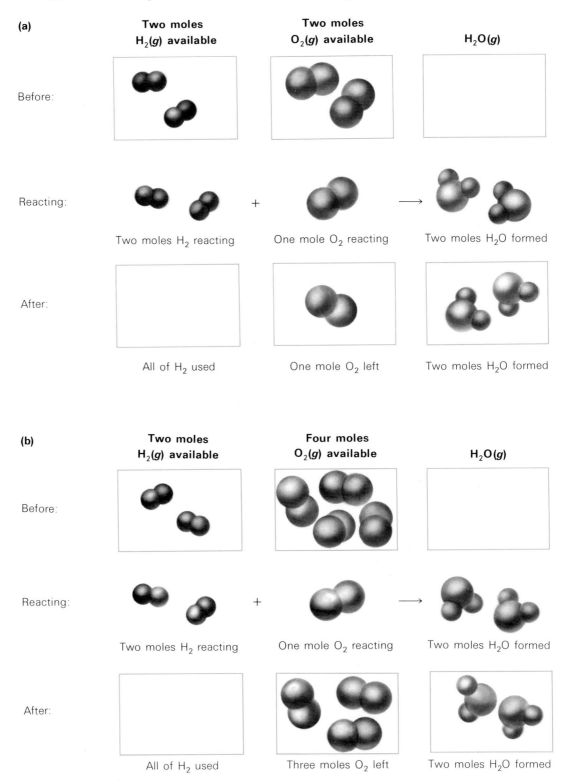

substance. The equation for the hydrogen-oxygen-water reaction, for example, may be written either

$$2\,H_2 + 1\,O_2 \longrightarrow 2\,H_2O$$

or

$$1\,H_2 + \tfrac{1}{2}\,O_2 \longrightarrow 1\,H_2O$$

When looking at the second equation, remember that it is as reasonable to think of half a mole of O_2 as it is to think of half a dozen eggs.

EXERCISE 4-3

The balanced equation for the burning of methane (CH_4) is

$$1\,CH_4 + 2\,O_2 \longrightarrow 1\,CO_2 + 2\,H_2O$$

Read this equation in terms of (a) molecules and (b) moles. (c) Rewrite the equation starting with 1 mole of O_2.

EXERCISE 4-4

Ammonia gas (NH_3) can be burned with oxygen gas (O_2) to form nitrogen gas (N_2) and water (H_2O). Try to follow the logic of the following steps in balancing this reaction:

$$NH_3 + O_2 \longrightarrow N_2 + H_2O$$
$$NH_3 + O_2 \longrightarrow \underline{1}\,N_2 + H_2O$$
$$\underline{2}\,NH_3 + O_2 \longrightarrow 1\,N_2 + H_2O$$
$$2\,NH_3 + O_2 \longrightarrow 1\,N_2 + \underline{3}\,H_2O$$
$$2\,NH_3 + \underline{\tfrac{3}{2}}\,O_2 \longrightarrow 1\,N_2 + 3\,H_2O$$

State briefly what occurs in each step.

CALCULATIONS OF MASS AND VOLUME

In Section 2.2 of Chapter 2, you learned to express grams of a pure substance as moles of substance, and litres of gas as moles of gas. In this chapter, you have learned to balance chemical equations. These two skills can be combined to help you determine the quantities of reactants and products involved in a chemical process. Let us see how this works.

A balanced chemical equation tells us that a definite number of *moles* of reactants will give a definite number of *moles* of products. The mole is the unit indicated in the chemical equation. On the other hand, a laboratory chemist measures materials in units such as grams or litres. If laboratory observations are to be used in chemical equations, masses or volumes must be converted to moles. When this has been done, the balanced equation can be used to determine how many moles of product are produced from the reactants used. Once the number of moles of products has been established, it is an easy matter to convert this quantity back to appropriate laboratory units such as grams or litres. We shall start with a simple example to illustrate a general procedure that is applicable to many chemical problems.

Coal contains mostly carbon, although it also contains some oxygen, hydrogen, nitrogen, and sulfur. Hard coal contains 85 to 95 percent carbon. It is not pure carbon. Each kilogram of coal we burn completely forms, among other things, a predictable mass of carbon dioxide. The burning of carbon in coal has already been represented by the equation on page 84. The information contained in this equation is summarized below:

$$1\,C + 1\,O_2 \longrightarrow 1\,CO_2$$

$$1 \text{ molecule} + 1 \text{ molecule} \longrightarrow 1 \text{ molecule} \qquad (4)$$

$$1 \text{ mole} + 1 \text{ mole} \longrightarrow 1 \text{ mole} \qquad (5)$$

$$12.0 \text{ g} + 32.0 \text{ g} \longrightarrow 44.0 \text{ g} \qquad (6)$$

Lines 4 and 5 are derived directly from the equation. Line 6 represents the mass of 1 mole of each substance in the equation. The number of grams per mole is calculated from atomic and molecular weights, as described on pages 72–74.

EXERCISE 4-5

Find the mass, in grams, of 1 mole of *each* substance in the following equation:

$$2\,Mg + O_2 \longrightarrow 2\,MgO$$

Is mass conserved?

Let us now determine how many kilograms of carbon dioxide are formed when 1 kilogram of carbon is *completely* burned. The word *completely* indicates that plenty of oxygen is available. Carbon is the limiting reagent. The equation

$$1\,C + 1\,O_2 \longrightarrow 1\,CO_2$$

tells us that 1 mole of carbon forms 1 mole of carbon dioxide. The moles of carbon burned must equal the moles of carbon dioxide formed. Since we are given the mass of carbon, we must first change the mass in grams to moles of carbon. The needed equivalents of mass are

$$1 \text{ mole C} = 12.0 \text{ g C}$$

and

$$1 \text{ kg} = 1,000 \text{ g}$$

We can now apply the unit conversion factors:*

$$1 \text{ kg C} \times \frac{1,000 \text{ g}}{1 \text{ kg}} \times \frac{1 \text{ mole C}}{12.0 \text{ g C}} = 83.3 \text{ moles C}$$

Because the ratio of moles of C to moles of CO_2 is 1 to 1, 83.3 moles of C generate 83.3 moles of CO_2. We complete our calculations by converting 83.3 moles of CO_2 to the desired unit, kilograms, of CO_2:

$$83.3 \text{ moles CO}_2 \times \frac{44.0 \text{ g CO}_2}{1 \text{ mole CO}_2} \times \frac{1 \text{ kg}}{1,000 \text{ g}} = 3.67 \text{ kg CO}_2$$

The final answer is in the desired unit: kilograms. An estimate of the approximate size of your answer may help you check your calculations. The calculation, $83 \times 44/1,000$, is about equal to $100 \times 40/1,000$. Thus, we can estimate that our answer will be slightly less than 4.0. Our actual answer, 3.67, is therefore quite reasonable.

Sometimes, it is useful to calculate the *volume* of a gaseous substance. Volume depends not only on the amount of gas but also on its temperature and pressure. To simplify matters, chemists usually express gas volumes at the same conditions of temperature and pressure, usually 0 °C and 760 millimetres of mercury. As you recall, these conditions are known as Standard Temperature and Pressure, abbreviated STP. Thus, the volume of any gas at STP depends only on how much of the gas is present. We know from laboratory studies that 1 mole of any gas at STP occupies a volume of approximately 22.4 litres. One mole of one gas contains the *same* number of molecules, 6.02×10^{23}, as 1 mole of any other gas. (See Figure 2-7.) We now have a new equivalent of volume:

1 mole of any gas at STP = 22.4 litres

Let us find the volume at STP occupied by the 83.3 moles of CO_2 in the previous problem. Using the proper equivalent of volume, we write

$$83.3 \text{ moles CO}_2 \times \frac{22.4 \text{ litres}}{1 \text{ mole}} = 1,870 \text{ litres CO}_2$$

The answer is in the desired unit: litres. The numerical calculation rounds off to about 100×20, or 2,000. Thus, an answer of 1,870 is reasonable.

EXERCISE 4-6

Find the number of moles in each of the following and the volume that each occupies at STP: (a) 20 g hydrogen (H_2), (b) 16 g oxygen (O_2), (c) 64 g methane (CH_4), (d) 1 kg propane (C_3H_8).

In problems involving reactions between gases, it is often useful to know the volumes of the gases involved. For example, what volume of oxygen gas

*You may want to review pages 47–51 for information about unit conversion factors.

(O_2) is needed to burn 50 litres of methane (CH_4), both measured at standard temperature and pressure?

The equation for the burning of methane has already been balanced on page 89. The information it contains is written beneath the equation:

$$1\ CH_4 + \qquad 2\ O_2 \longrightarrow 1\ CO_2 + 2\ H_2O$$

$$1\ \text{molecule} + 2\ \text{molecules} \longrightarrow 1\ \text{molecule} + 2\ \text{molecules}$$

$$1\ \text{mole} + \quad 2\ \text{moles} \longrightarrow 1\ \text{mole} + 2\ \text{moles}$$

$$22.4\ \text{litres} + \quad 44.8\ \text{litres} \longrightarrow 22.4\ \text{litres} + 44.8\ \text{litres}$$

Because the molecule and mole ratios of CH_4 to O_2 are 1 to 2, their *volume* ratios must be 1 to 2. Therefore, the 50 litres of methane need 100 litres of O_2 in order to burn completely, if they are measured at the same conditions of temperature and pressure.

EXERCISE 4-7

How many litres of oxygen gas are needed to burn 20 litres of hydrogen, if both gases are at STP? How many grams of water are formed in the reaction?

EXERCISE 4-8

Propane gas (C_3H_8) is often burned as a fuel. When burned in excess oxygen, CO_2 and water vapor are produced. Write the balanced equation for this reaction. If 100 litres of propane are burned, find the volume, in litres, of oxygen consumed and of CO_2 produced. Assume all gases are at the same temperature and pressure.

The methods used to solve the problems already considered are widely applicable. Some equations, as well as calculations made from them, are more complicated than those studied so far. The following is an example.

The gasoline burned in the internal combustion engine of an automobile makes that engine a major source of carbon dioxide in the atmosphere. One component of gasoline, octane, has the formula C_8H_{18}. Like methane (CH_4), octane is a hydrocarbon. When excess oxygen is available, octane burns completely to give carbon dioxide and water vapor. If less oxygen is present, products include carbon monoxide (CO) or unburned carbon and water. Carbon monoxide is a deadly poison and can be a serious hazard when *any* carbon-containing compound is burned in poorly ventilated areas that lack sufficient oxygen.

The approximate amount of CO_2 formed from burning the octane in 10 gallons of gasoline in an automobile engine can be determined by solving the following problem: One gallon of gasoline contains about 25 moles of

octane (C_8H_{18}). Find the number of moles of O_2 needed to burn this amount completely to CO_2 and H_2O. Also find the number of moles, and the mass, in kilograms, of CO_2 produced.

We start, as always, with the "skeleton," or unbalanced, equation:

$$1 \, C_8H_{18} + O_2 \longrightarrow CO_2 + H_2O$$

In deciding what coefficients to use in balancing an equation, it is often convenient to focus attention on the most complicated molecule, in this case C_8H_{18}. We will balance the equation in terms of 1 mole of C_8H_{18}. The C_8H_{18} molecule contains 8 carbon atoms, all of which must become part of CO_2 molecules. We place the coefficient 8 in front of CO_2. Similarly, the 18 H atoms must, if they are to be conserved, become part of 9 H_2O:

$$1 \, C_8H_{18} + O_2 \longrightarrow 8 \, CO_2 + 9 \, H_2O$$

The C and H atoms are now conserved, but the O atoms are not. The 8 CO_2 and 9 H_2O molecules contain 25 oxygen atoms. We need $12\frac{1}{2}$ O_2 molecules as reactants to conserve O atoms. We can write

$$1 \, C_8H_{18} + 12\tfrac{1}{2} \, O_2 \longrightarrow 8 \, CO_2 + 9 \, H_2O$$
$$1 \text{ mole} + 12\tfrac{1}{2} \text{ moles} \longrightarrow 8 \text{ moles} + 9 \text{ moles}$$

The completed equation indicates that $12\frac{1}{2}$ moles of O_2 are needed for *each* mole of C_8H_{18}. The 25 moles of C_8H_{18} contained in 1 gallon of gasoline needs $12\frac{1}{2} \times 25$ moles, or 310 moles of O_2 to burn it. This equation also tells us that 1 mole of C_8H_{18} forms 8 moles of CO_2. Thus, 25 moles of C_8H_{18} must form 8×25, or 200, moles of CO_2.

The *mass* of CO_2, in kilograms, involves the now-familiar conversion:

$$\frac{200 \text{ mole } CO_2}{1 \text{ gal}} \times \frac{44.0 \text{ g } CO_2}{1 \text{ mole } CO_2} \times \frac{1 \text{ kg}}{1,000 \text{ g}} = 8.80 \text{ kg } CO_2/\text{gal}$$

This means that every time your automobile burns one gallon of gasoline, about 9 kilograms of CO_2 are released into the atmosphere from this process.

EXERCISE 4-9

Acetylene gas (C_2H_2) burns in air to produce carbon dioxide (CO_2) and water (H_2O). (a) Write the balanced equation for this reaction. (b) How many grams of CO_2 and H_2O are produced when 65.0 g of C_2H_2 are burned? (c) What volume at STP is occupied by the CO_2 produced from 65.0 g of C_2H_2?

EXERCISE 4-10

Table sugar has the formula $C_{12}H_{22}O_{11}$. When we eat sugar, our bodies burn it up to produce CO_2 and water. (a) Write the balanced equation for this reaction. (b) How many grams of CO_2 and water does your body produce when you burn 34.2 g sugar? (c) What volume, at STP, does the CO_2 produced in (b) occupy?

OPPORTUNITIES

Teaching chemistry

Teaching chemistry can be a very rewarding career. There is great satisfaction in explaining new ideas to students and in watching what could be the development of a lifelong interest in chemistry.

High school chemistry teachers plan lessons, conduct classes, correct homework and notebooks, and select laboratory experiences for their students. In addition, chemistry teachers often participate in extracurricular activities related to science. Secondary school teachers work in the classroom with students 80 percent of the school day and usually have no research facilities.

Many chemistry teachers are also employed by colleges and universities. Teaching at that level involves research and writing about chemistry as well as the normal duties of conducting lectures and seminars and directing student research. Teaching chemistry at any level requires the abilities to use words effectively and to understand mathematical concepts and experimental procedures.

A bachelor's degree is essential for employment at the secondary school level. Many schools prefer candidates with master's degrees. Public schools require the completion of a prescribed course of study and licensing by the state department of education. Many two-year colleges require their teachers to have only a master's degree in chemistry. Teachers at four-year colleges and universities must have a doctorate.

Further information about this

career can be obtained from your chemistry teacher and from your state department of education. You may also want to contact the American Chemical Society, Department of Education, Dr. Moses Passer, 1155 Sixteenth Street NW, Washington, D.C. 20036.

CHEMICAL CALCULATIONS BASED
ON A LIMITING REAGENT

In all of our past problems involving combustion, a fuel gas such as methane or hydrogen was burned in air. Oxygen gas, which makes up about 20 percent of air is available without cost. On the other hand, fuel gases such as methane, acetylene, and hydrogen are expensive. Their cost dictates that they will always be the limiting reagent in a combustion process. Hydrogen gas has many attractive characteristics as a fuel. It gives off lots of energy when it burns and the only product of combustion is water. Unfortunately, hydrogen gas is not available in natural deposits. It must be produced by chemical means, which is expensive. The combustion equation for hydrogen is a very familiar one that was reviewed in some detail in Figure 4-5:

$$2 H_2 + 1 O_2 \longrightarrow 2 H_2O$$

When working with a chemical system involving a limiting reagent, such as hydrogen, it is helpful to consider the system at three different times: (1) before the reaction begins, (2) during the reaction, and (3) after the reaction is completed. This is done by "taking inventory" on three separate lines below the equation itself. For example, suppose we begin with 2 moles of H_2 and 4 moles of O_2. The "inventory" in Figure 4-5b shows:

	$2 H_2 +$	$1 O_2 \longrightarrow$	$2 H_2O$
BEFORE	2 moles	4 moles	0 moles
REACTING	2 moles	1 mole $\longrightarrow$	2 moles
AFTER	0 moles	3 moles	2 moles

The hydrogen is used up, producing 2 moles of water, and there are 3 moles of oxygen left over. You can see that the "inventory" is useful in helping to record the amounts of reactants and products involved. Except for this "inventory" procedure, we solve limiting reagent problems in the same way as we handled previous problems. The amount of limiting reagent determines the amount of product generated.

EXERCISE 4-11

The gas nitrogen monoxide (NO) combines with oxygen gas to form nitrogen dioxide (NO_2). (a) Write the balanced equation for this reaction. (b) Find the number of grams and the volume at STP in litres of NO_2 produced when 45 g of NO combines with 16 g, 24 g, and 32 g of O_2. What is the limiting reagent in each case?

4.2 Energy changes in chemical reactions and the conservation of energy

The principal source of hydrogen is liquid water. But liquid water is difficult to decompose chemically. Heating warms water and eventually brings it to a boil. However, this procedure merely changes the state of water from a liquid to a gas. The water does not change chemically. Even at temperatures

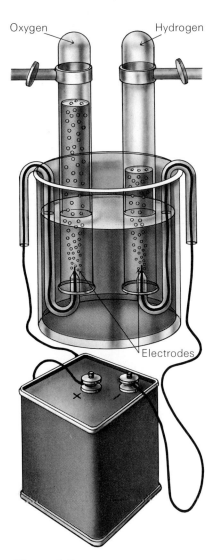

Oxygen   Hydrogen

Electrodes

Figure 4-7

Apparatus for the decomposition of water by direct electrical current.

as high as 1000 °C, water molecules are still dominant. This change-of-state process can be represented by the equation

$$\text{heat energy} + H_2O(l) = H_2O(g)$$

Molecules of liquid water are represented as $H_2O(l)$, while molecules of gaseous water, or steam, are represented as $H_2O(g)$.

When electrical energy is applied to liquid water, the water can be changed into different chemical substances. Figure 4-7 shows an apparatus that uses direct electrical current to decompose water. Such a device changes water into hydrogen gas and oxygen gas, the elements from which water is made. In the apparatus shown, two electical conductors, or electrodes, are immersed in water. When the electrodes are connected to a source of electrical energy, hydrogen gas appears at one electrode and oxygen gas appears at the other. The equation representing this process is written

$$\text{electrical energy} + 2\,H_2O(l) \longrightarrow 2\,H_2(g) + 1\,O_2(g)$$

Chemists have measured the amount of energy required to decompose 1 mole, or 18 grams, of liquid water into hydrogen gas and oxygen gas. The energy required is 68,300 calories, or 68.3 kilocalories. To decompose 2 moles, or 36.0 grams, of water requires *twice* as much energy as is needed to decompose 1 mole. It therefore requires 136,600 calories, or 136.6 kilocalories. The **calorie** is the unit that we use to measure heat energy. To understand this unit, think of heating a small mass of water with a gas flame or an electric heating device of the kind used in electric coffee makers. Imagine placing an accurate thermometer into a 100-gram mass of water. If you added enough heat to the 100 grams of water to raise its temperature only 1 °C, you would have added 100 calories of energy. One calorie is the *quantity of heat energy required to raise the temperature of 1 gram of water 1 °C.** Since the calorie is a very small unit, chemists often find it convenient to use a larger unit: the kilocalorie. One kilocalorie (kcal) is equal to 1,000 calories. The kilocalorie is used in defining the energy content of foods even though we mistakenly use the term calorie.

We can add the above information to the equation representing the decomposition of water:

$$136.6\ \text{kcal} + 2\,H_2O(l) \longrightarrow 2\,H_2(g) + 1\,O_2(g)$$

*The joule (J) is the metric unit of energy. One calorie has the same amount of energy as 4.184 joules. We can write this as 1 cal = 4.184 J.

Figure 4-8

Energy is released in this carefully controlled exothermic reaction.

Sometimes it is more convenient to write this equation in terms of 1 mole of water. To do this, we divide each term of the equation by 2:

$$68.3 \text{ kcal} + 1 \text{ H}_2\text{O}(l) \longrightarrow 1 \text{ H}_2(g) + \tfrac{1}{2}\text{O}_2(g)$$

The 68.3 kilocalories of energy discussed above are stored in the hydrogen and oxygen and are released when the gases are burned to give 1 mole of water. Stored energy is called *potential energy*. As an example, potential energy is released when H_2 and $\tfrac{1}{2}\text{O}_2$ react slowly and continually in a fuel cell. Potential molecular energy is converted to direct current electrical energy. Fuel cells using hydrogen and oxygen have supplied life-supporting electrical energy for the astronauts aboard spacecrafts.

Hydrogen gas and oxygen gas can also react rapidly. For example, if two volumes of hydrogen gas and one volume of oxygen gas are mixed and ignited, a loud explosion results. Water is formed during this very rapid reaction, and there is a sudden release of energy. When we measure the amount of energy released, the measurement agrees with our expectations. This is true whether the energy is released slowly in a fuel cell or rapidly in an explosion. Exactly 68.3 kilocalories of energy are released when 1 mole of water is formed from hydrogen gas and oxygen gas.

We can now rewrite the equation for the formation of liquid water given on page 89 as

$$1 \text{ H}_2 + \tfrac{1}{2}\text{O}_2 \longrightarrow 1 \text{ H}_2\text{O} + 68.3 \text{ kcal}$$

Multiplying each term in this equation by 2, we obtain

$$2 \text{ H}_2 + 1 \text{ O}_2 \longrightarrow 2 \text{ H}_2\text{O} + 136.6 \text{ kcal}$$

Chemical reactions in which energy is released are known as **exothermic reactions.** Chemical reactions that absorb energy are called **endothermic reactions.** Thus, the formation of water is exothermic: energy is *released*. The decomposition of water is endothermic: energy is *absorbed*.

If we look at both processes together, we see that energy, as well as mass, is conserved. Neither energy nor mass can be created or destroyed. A very large collection of experimental evidence indicates that this statement is true for all chemical reactions. This principle is referred to as the **Law of Conservation of Energy.** Like the Law of Conservation of Mass, the Law of Conservation of Energy is one of the cornerstones of modern science. A more detailed discussion of energy changes that occur in chemical and nuclear reactions is found in Chapter 14.

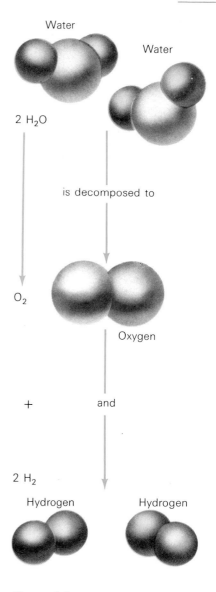

Figure 4-9

The decomposition of water shown with molecular models.

SUMMARY

Chemical equations represent chemical reactions. In an equation, the starting materials are known as reactants and are placed to the left of the arrow. Substances produced are known as products and are placed to the right of the arrow. To conserve atoms in an equation, we place the proper coefficient in front of each formula to ensure that equal numbers of each type of atom

Key Terms

Key Terms

calorie

coefficient

endothermic reaction

exothermic reaction

Law of Conservation
of Energy

Law of Conservation
of Mass

limiting reagent

reagent

appear both to the left and to the right of the arrow. This is known as balancing the equation.

The sum of reactant masses always equals the sum of product masses in a chemical reaction. This regularity is known as the Law of Conservation of Mass and is explained by the conservation of atoms in the equation representing the process.

A chemical equation gives the relative number of molecules or moles of substances that react and form in a chemical reaction. The substance completely consumed in a reaction is called the limiting reagent. The limiting reagent determines how much of the other substances can react and form. If we know the amount of any one limiting substance involved in a chemical reaction, we can calculate the amounts of the other substances involved.

The energy accompanying chemical reactions is measured in calories. One calorie of heat energy is the amount necessary to raise the temperature of 1 gram of water 1 °C.

Energy is conserved in any chemical process. This regularity is known as the Law of Conservation of Energy. Exothermic reactions release energy, while endothermic reactions absorb energy. Energy absorbed in an endothermic reaction is stored in its products as potential energy. This stored energy is released when the endothermic reaction is reversed.

QUESTIONS AND PROBLEMS

A

1. (a) What is the name of the product formed when carbon burns in plenty of oxygen? (b) Write the balanced equation for the process. (c) What is the name of the substance formed when hydrogen gas burns in air? (d) Write the equation for the process.

2. (a) When you balance a chemical equation, can you ever change the *formulas* of reactants or products? What is changed in the balancing process? (b) Balance the following equations by starting with one molecule of the underlined substance in each case. (1) $Na + \underline{Cl_2} \longrightarrow NaCl$ (2) $\underline{CH_4} + O_2 \longrightarrow CO_2 + H_2O$ (3) $\underline{N_2} + H_2 \longrightarrow NH_3$ (4) $\underline{Mg} + HCl \longrightarrow MgCl_2 + H_2$

3. Write balanced chemical equations for the following processes: (a) One atom of elemental sulfur combines with one molecule of oxygen to give one molecule of sulfur dioxide. (b) Two molecules of sulfur dioxide gas combine with one molecule of oxygen gas to give two molecules of sulfur trioxide. (c) One atom of uranium from uranium metal combines with three diatomic molecules of gaseous fluorine to give one molecule of uranium hexafluoride. (d) One mole of nitrogen gas combines with three molecules of hydrogen gas to give two molecules of ammonia gas.

4. Normal-propyl alcohol is burned. The "skeleton" equation for the process is: $? C_3H_7OH + ? O_2 \longrightarrow ? CO_2 + ? H_2O$. Give the logic for each of the steps listed below. In other words, what is balanced in each step?
   (a) $\underline{1} C_3H_7OH + ? O_2 \longrightarrow \underline{3} CO_2 + ? H_2O$
   (b) $1 C_3H_7OH + ? O_2 \longrightarrow 3 CO_2 + \underline{4} H_2O$
   (c) $1 C_3H_7OH + 4\frac{1}{2} O_2 \longrightarrow 3 CO_2 + 4 H_2O$
   (d) $2 C_3H_7OH + \underline{9} O_2 \longrightarrow 6 CO_2 + 8 H_2O$

Count carbons, hydrogens, and oxygens on each side to see if equation (d) is properly balanced.

5. The gas hydrogen chloride is made by the following process:

$$NaCl + H_2SO_4 \xrightarrow{\text{heat}} HCl + Na_2SO_4$$

| TABLE SALT | SULFURIC ACID | HYDROGEN CHLORIDE | SODIUM SULFATE |

(a) Balance this equation. (b) Two moles of table salt are mixed with four moles of H_2SO_4. How many moles of HCl will be formed? (c) Two moles of H_2SO_4 are mixed with 10 moles of NaCl. How many moles of HCl will be formed?

6. Laughing gas (N_2O) will combine with carbon monoxide (CO) to give carbon dioxide (CO_2) and dinitrogen (N_2). (a) Write the equation for the process. (b) What will be the composition of the final mixture if 6 litres of CO and 3 litres of N_2O are mixed?

7. Magnesium metal burns in oxygen gas to produce magnesium oxide (MgO). (a) Write the balanced equation for the process. (b) If a 6.1 g sample of Mg is burned in 8.0 g of O_2 gas, which element is the limiting reagent? (c) How many *grams* of MgO will be formed?

8. (a) What is the "standard temperature and pressure" as it applies to a gas? (b) What is the volume of 1 mole of gas at STP? Give units.

9. Metallic chromium burns in chlorine gas to give beautiful purple crystals of Cr_2Cl_6. (a) Write the equation for the process. (b) How many *moles* of chromium are contained in 104 g of Cr? (c) How many moles of chlorine would be required to react completely with this chromium? (d) What volume would this chlorine occupy at STP? (e) What would be the mass of the $CrCl_3$ produced?

10. (a) How much energy in kilocalories is required to convert *one mole* of liquid H_2O to its elements by direct electrical current? Write the equation. Is the process exothermic or endothermic? (b) Where does the energy go that is put into the system? (c) How could the energy be recovered? Is the process exothermic or endothermic?

11. (a) How much energy would be released if a mixture containing 32 g of oxygen and 10 g of hydrogen were exploded to give water? (Hint: how many moles of each gas were present in the mixture?) (b) What was the limiting reagent? (c) How many moles of water were formed? (d) How much energy was liberated?

B

12. The commercial heating gas methane (CH_4) burns in a limited amount of oxygen. Under this condition, water and carbon monoxide are the products. (a) Write the balanced equation for the process. (b) What volume of oxygen would be required to burn 4 litres of methane?

13. Hydrogen gas (H_2) combines with fluorine (F_2) to give hydrogen fluoride gas (HF). (a) Write the balanced equation for the reaction. (b) If 12 g of hydrogen gas and 50.0 g of fluorine gas are allowed to react, how many grams of hydrogen fluoride gas will be produced? What is the limiting reagent?

14. Balance the following equations by starting with one mole of the underlined reagent:
(a) $\underline{KClO_3} \longrightarrow KCl + O_2$
(b) $\underline{Al} + O_2 \longrightarrow Al_2O_3$

(c) $Na_2O_2 + H_2O \longrightarrow NaOH + O_2$

(d) $\overline{C_3H_8} + O_2 \longrightarrow CO_2 + H_2O(g)$

(e) $\overline{CaC_2} + H_2O \longrightarrow C_2H_2(g) + Ca(OH)_2$

15. State the logic of each step in the balancing of the equation for the decomposition of hydrazine: $? N_2H_4 \longrightarrow ? N_2 + ? H_2 + ? NH_3$.
    (a) $2 N_2H_4 \longrightarrow \underline{1} N_2 + ? H_2 + \underline{2} NH_3$
    (b) $2 N_2H_4 \longrightarrow 1 N_2 + \underline{1} H_2 + 2 NH_3$

16. (a) When propane gas (C_3H_8), which is used for heating, is burned with insufficient air, the flame is yellow and smokey because the products are elemental carbon and water vapor. (b) Write the balanced equation for the process. Write balanced equations for each of the following:
    (1) Potassium chloride is synthesized from its elements.
    (2) Butane (C_4H_{10}) burns completely in air.
    (3) Nitrogen monoxide plus oxygen gas gives nitrogen dioxide gas.
    (4) Ammonia gas burns in air to give water vapor.

17. Supply the coefficients needed to properly balance each of the following "skeleton" equations:
    (a) $Ca + O_2 \longrightarrow CaO$
    (b) $P_4 + O_2 \longrightarrow P_4O_{10}$
    (c) $NO_2 + H_2O \longrightarrow HNO_3 + NO$
    (d) $FeS_2 + O_2 \longrightarrow Fe_2O_3 + SO_2$
    (e) $Fe_2O_3 + CO \longrightarrow Fe + CO_2$

18. A 112-litre container is filled with neon gas at STP. (a) How many moles of neon molecules are in the container? (b) How many molecules of neon are in the container? (c) What is the mass of neon in the container?

19. How many moles are present in each of the following:
    (a) 42.0 g H_2O_2 (b) 445.6 g $Ba(NO_3)_2$ (c) 0.40 g C_6H_6 (d) 100.0 g Zn
    (e) 218.3 g $Cu(OH)_2$ (f) 94.5 g Cl_2.

20. What is the mass, in grams of each of the following: (a) 1 mole of sulfur dioxide molecules (b) 2.50 moles of carbon monoxide molecules (c) 0.10 mole of hydrogen fluoride molecules (d) 0.500 mole of oxygen molecules (e) 0.50 mole of aluminum atoms (f) 0.0400 mole of phosphorus atoms.

21. When sodium metal is dropped into water, a vigorous reaction occurs. Hydrogen gas (H_2) is given off and a solution of sodium hydroxide (NaOH) is formed. (a) Write the balanced equation for this reaction. (b) If 2.30 g of sodium are put into water, how many moles of hydrogen are produced? (c) What volume does this hydrogen occupy at STP? (d) What is the mass of this hydrogen?

22. When mercuric oxide (HgO) is heated strongly, it decomposes into its elements. (a) Write the balanced equation for this reaction. (b) How many moles of HgO are necessary to produce 11.2 litres of oxygen at STP? (c) What is the mass of the HgO in (b)? (d) How many grams of mercury are produced by the reaction?

23. In a blast furnace, iron oxide is reduced to metallic iron by CO through the following process: $Fe_2O_3 + CO \longrightarrow Fe + CO_2$.
    (a) Balance the equation for the process.
    (b) A 320 g sample of iron is reduced. How many moles of CO at STP would be needed?
    (c) What volume of CO at STP would be required?
    (d) What mass in grams of iron would be obtained? How many moles of iron atoms is this?

24. The equation for the burning of hydrogen to produce water is: $H_2 + \frac{1}{2}O_2 \longrightarrow H_2O + 68.3$ kcal. (a) If 255 g of H_2O are produced by this process, how many kilocalories of energy will be liberated? Is the reaction exothermic or endothermic? (b) If 75.0 g of oxygen are to be produced by the electrolysis of water, how many kilocalories of electrical energy must be supplied? Is the process exothermic or endothermic?

25. A form of sulfur (S_8) burns to form sulfur dioxide. (a) Write the balanced equation for the process. (b) Find the number of grams and the volume at STP in litres of the sulfur dioxide formed when 512 g of sulfur are burned in (1) 320 g (2) 512 g (3) 640 g of oxygen gas.

26. Calcium metal can be produced by passing an electric current through molten $CaCl_2$. Chlorine gas is the other product of this reaction. (a) Write the balanced equation for this reaction. (b) How many grams of calcium would be obtained from the electrolysis of 1.50 kg (1 kg = 1,000 g) of $CaCl_2$? (c) What volume of chlorine, measured at STP, would be produced?

27. When copper (II) oxide (CuO) is heated in the presence of hydrogen gas, the products are copper metal and water vapor. (a) Write the balanced equation for this reaction. (b) How many grams of copper metal are obtained from heating 10.0 g of CuO with excess H_2?

28. (a) Write the balanced equation for the synthesis of liquid water from its elements. (b) What mass in grams of water forms when 160 g of oxygen gas reacts with each of the following amounts of hydrogen gas: (1) 4 g (2) 10 g (3) 20.0 g (4) 50 g? (c) How many kcal of heat are released for each part of (b)? (d) Name the limiting reagent for each part of (b).

C

29. The mass of 1.00 litre of a certain gas at STP is 6.88 ± 0.01 g. (a) What is the mass of 22.4 litres of the gas? (b) How many molecules of gas are present in this volume? (c) What is the molecular weight of the gas? (d) Which of the following gases could it be: Cl_2, SO_3, CCl_4, or UF_6?

30. Five litres of a gas have a mass of 9.80 ± 0.01 g at STP. (a) What is the molecular weight of the gas? (b) Which, if any, of the following could it be: N_2, O_2, Cl_2, CO_2, or SO_2?

31. A 10.0 g sample of metal M burns in oxygen to produce 14.0 g of a metal oxide with formula MO. What is the atomic weight of M?

32. Methane (CH_4) has a molar heat of combustion of 18,000 calories per mole. The equation is: $CH_4 + 2O_2 \longrightarrow CO_2 + 2H_2O + 18,000$ cal. How many grams of methane would have to be burned to warm 100 g of water from 10 °C to 100 °C?

33. (a) What mass in grams of CaO could be obtained from the heating of 2.00 moles of $CaCO_3$ to give CaO and CO_2? (b) What volume at STP of CO_2 would be obtained? Hint: You should begin to solve this problem by first writing the balanced equation.

34. What mass in grams of $Ca(NO_3)_2$ can be prepared by the reaction of 18.9 g of HNO_3 with 7.4 g of $Ca(OH)_2$? What is the limiting reagent?

35. Beautiful blue $CuSO_4 \cdot 5H_2O$ will lose water when heated to give white, anhydrous $CuSO_4$. (a) Write the balanced equation for the process. (b) How much of the $CuSO_4 \cdot 5H_2O$ must be dehydrated (water driven off) to get 80.0 g of anhydrous $CuSO_4$?

I am never content until I have constructed a mechanical model of the object that I am studying. If I succeed in making one, I understand; otherwise, I do not.

WILLIAM THOMSON, LORD KELVIN

5

MORE ABOUT GASES: THE KINETIC-MOLECULAR THEORY

Objectives

After completing this chapter, you should be able to:

- Use the kinetic-molecular theory to explain the properties and behavior of gases.

- Explain how a mercury barometer measures atmospheric pressure and be able to get the pressure of a gas from a manometer.

- Use Dalton's Law of Partial Pressures to calculate the pressure of each gas in a mixture of several gases.

- Understand and use the absolute, or kelvin, temperature scale.

- Explain the relationship between temperature and the average molecular-kinetic energy of a gas sample.

- Differentiate between the behavior of ideal and real gases.

- Use the ideal gas law, $PV = nRT$, to determine $P, V, n,$ or T for a gas if the other three variables and values of R are given.

Understanding the behavior of gases can help explain the rising of hot air balloons.

The model we suggested in Chapter 1 to explain the behavior of air in a balloon has grown rapidly. In Chapter 2, we made observations on the way in which the volume of a given quantity of gas changes as pressure on the gas is increased. The model explained these observations in both a qualitative and a quantitative way. Next, we *assumed* that Avogadro's hypothesis was true. We used this hypothesis, together with our model for gases and some chemical observations, to develop the concepts of molecules and molecular weights. In Chapters 3 and 4, it was necessary to break molecules into atoms in order to understand some chemical reactions. These concepts then provided a quantitative procedure for examining masses of chemically reacting materials and thus established a substantial basis for further development of the atomic theory. The study of gases, therefore, is fundamental to our understanding of chemical concepts.

In some ways, gas is the simplest form of matter. We found in Chapter 2 that the mathematical statement "pressure $\times$ volume = a constant" is true for gases in general. The similar behavior of gases helped scientists develop the particle model of gases we have already found so useful. This regularity, along with other regularities discussed in this chapter, led to the development of the *kinetic theory* and an understanding of temperature on the molecular level.

Gases react with each other in simple volume ratios, provided that they are at the same temperature and pressure. For example, one volume of hydrogen gas reacts with one volume of chlorine gas to form two volumes of hydrogen chloride gas, as shown in Figure 3-2, on page 63. In the reaction between hydrogen and oxygen, two volumes of hydrogen gas react with one volume of oxygen gas to form liquid water. If this reaction occurs at a temperature slightly above 100 °C, the water formed is in a gaseous state. In this case, we observe that *two* volumes of hydrogen gas react with *one* volume of oxygen gas to form *two* volumes of water vapor, as shown in Figure 3-8 (see page 80). These simple combining volumes were first noticed by the French scientist Joseph Louis Gay-Lussac. In 1808 he formulated the **Law of Combining Gas Volumes,** which stated that if gases are measured at constant temperature and pressure, the volumes of gases used in a chemical reaction and the volumes of gases produced in that chemical reaction can be expressed as a ratio of small, whole numbers.

5.1 Molar volumes and distance between molecules

Our model of a gas proposed the existence of particles that have a rapid, random motion and that collide with one another and with the walls of a container. One of the interesting questions that we have not yet answered is: how far apart are these particles? Is the distance between molecules much larger than the size of the molecules themselves?

The answer can be found by comparing the volumes occupied by 1 mole (6.02×10^{23} molecules) of a substance when it exists as a solid, a liquid, and a gas. In Chapter 2, we calculated the volume of 1 mole of a *gas* at standard temperature and pressure. Recall that STP is 0 °C and 1 atmosphere pressure. This quantity of gas was called the **molar volume** of the gas. We found that *all gases* have a molar volume fairly close to 22.4 litres at 0 °C and 1 atmosphere pressure.

One mole of a pure material in the solid state has a definite and reproducible volume, too, if we specify temperature and pressure. In contrast to gases, however, volumes for *different* solids are *not* the same, even at the same temperature and pressure. Similarly, 1 mole of a pure material in the liquid state has a definite volume at a given temperature and pressure, but *different* liquids have different molar volumes. Let us examine molar volumes for a relatively simple material and see what this study tells us about the distance between molecules in the gas phase.

Nitrogen gas (N_2), which makes up almost 80 percent of the air we breathe, changes from a liquid to a solid at −210 °C. Solid nitrogen has a density of 1.03 grams per millilitre. Liquid nitrogen has a density of 0.81 gram per millilitre. But gaseous nitrogen at STP has a density of 0.00125 gram per millilitre. What do these densities tell us about the molar volume of nitrogen in its solid, liquid, and gas phases?

Density is the mass per unit volume of a substance. Using the units given above, the formula

$$\text{density} = \frac{\text{mass (g)}}{\text{volume (ml)}}$$

allows us to calculate the molar volume of nitrogen in each of its three phases. Solved for volume, the formula becomes

$$\text{volume} = \frac{\text{mass}}{\text{density}}$$

Dividing the molar mass of nitrogen by its density gives its molar volume.

EXERCISE 5-1

Nitrogen gas has the molecular formula N_2. Using the table of atomic weights on page 70, find the *molecular weight* of nitrogen in unified atomic mass units (u). Then find *molar mass* in grams.

EXERCISE 5-2

Using the molar mass of nitrogen, found in Exercise 5-1, and the densities given above for solid, liquid, and gaseous N_2, calculate the *molar volume* of (a) solid N_2, (b) liquid N_2, and (c) gaseous N_2 at STP. Compare your answers with those given below.

The molar mass of N_2 is 28.0 grams. Using the densities given above for the three states of nitrogen, the molar volume of N_2 is found to be 27.2 millilitres per mole as a white solid (below -210 °C), 34.6 millilitres per mole as a liquid (just above -210 °C), and 22,400 millilitres per mole as a gas (at STP). The volume of N_2 as a gas is about 750 times as great as the volume of the *same mass* of solid or liquid N_2! Experiments with other substances measured in the solid, liquid, and gas phases show similar results.

Experiments with nitrogen reveal a regularity that holds not only for nitrogen but for other substances as well. The molar volume of the gas phase is much greater than the molar volume of the liquid or solid phase. We also observe that nitrogen is invisible and easy to pass through. Each day, we walk through the invisible air that surrounds us. In marked contrast, both liquid and solid N_2 are clearly visible. Objects cannot easily pass through them. Liquids and solids have definite volume. However, gases spread out to fill the entire volume of any closed container in which they are placed.

5.2   A model for the kinetic-molecular theory

How can we explain the differences shown by N_2 gas on one hand, and solid or liquid N_2 on the other? We use the kinetic-molecular theory to help us interpret these differences. By combining the above information with the model provided by the theory, we can answer some interesting questions. How far apart are the molecules of a gas? How far apart are the molecules when the gas becomes a liquid or a solid?

The kinetic-molecular theory states that all substances, regardless of their state, are made up of tiny particles. One mole of any substance contains the *same* number, 6.02×10^{23}, of particles. The dramatic increase in volume occupied by this number of particles when the substance becomes a gas is observed for all substances. We have already noted that for nitrogen, this increase is about 750-fold. *If the size of a single molecule is assumed to be the same in each phase, then the molecules must be widely separated in the gas phase.* If we assume that molecules are touching each other in the solid and liquid states, then the volume in these states is very close to the molecular volume. Since gases have a molar volume, which is 750 times the molar volume of the liquid, the space between gas molecules must be 750 times the actual molecular volume.

Imagine that a swarm of bees represents the molecules in a given mass of a substance such as nitrogen. When crowded together in the beehive, the bees are in contact, with little or no space between them. You could not pass an object through them without pushing some bees aside. Similarly, solids and liquids cannot be passed through easily without being pushed aside. If you imagine the bees "frozen" in place, making the kind of regular arrangement seen in their honeycombs, you have a rather good picture of the arrangements of molecules in most solids. The way bees are able to maneuver past each other, keeping in contact as they move around the hive, accurately represents the way molecules behave in the liquid state. Indeed, the way the bees seem to "flow" past each other suggests liquid flowing into a container.

When bees leave the hive, they spread out and cover a large volume in their search for pollen and nectar. Each bee is then widely separated from the other bees. You have often been in places where individual bees are searching for pollen yet, at the time, you may not have seen them. In their constantly moving, widely separated state, the bees can represent the molecules of a gas. The same number of bees, once crowded together in the hive, is now spread out over a much greater volume. The 750-fold increase in volume observed when liquid nitrogen becomes a gas is directly comparable to the above model.

The model of a gas as a collection of particles in endless motion requires that each particle posses *energy of motion,* called **kinetic energy.** We should not be surprised to learn, then, that the model for gases is called the **kinetic-molecular theory of gases.** An application of the quantitative mathematical concepts of the kinetic theory permits us to be a little more specific in our description of molecular motion.

Considerable evidence, direct and indirect, indicates that gas molecules travel in straight lines until they meet other gas molecules or the walls of a container. Then they bounce off. Some hit head-on and some at an angle. The net result is a helter-skelter movement of molecules in all directions and at all speeds. Since an individual molecule will change both speed and direction of motion even in a single second, thus traveling in a zig-zag path, it is pointless to speak of the velocity of a given molecule. We can, however, speak of the *average speed* of a collection of molecules at any instant. The average speed of a collection of gas molecules does not change with time unless the temperature is changed.

At room temperature, the average speed of a nitrogen molecule is about 400 metres per second, or about 900 miles per hour. Although the average distance between molecules is small in an absolute sense, it is large in comparison to the size of the molecules themselves. This means that molecules in a gas can travel relatively long distances without colliding. On the average, at room temperature and 1 atmosphere pressure, a molecule will travel about 15 times the average distance between molecules before colliding with another molecule. The molecules making up solids and liquids are also in motion, but such motion is harder to describe because the particles are closer together and bump into each other constantly. They slide past each other while still in contact.

5.3 The pressure of gases

Gases exert pressure. The pressure of a gas and the pressure-volume behavior of different gases were explained with the aid of the kinetic-molecular theory in Chapter 2. Understanding the pressure behavior of gases makes it easier to understand the nature of gases.

Pressure is defined as the *force,* or push, *per unit area* of surface. The equation for this relationship is

$$\text{pressure} = \frac{\text{force}}{\text{area}}$$

The air we breathe exerts considerable pressure all the time. Because air is present all around us, it pushes on us from all directions at the same time. These pushes usually equalize each other, and we are often unaware of them. When outside air pressure falls, as it does in a rapidly climbing airplane, we can feel the pressure within our bodies pushing on our eardrums. Discomfort persists until the pressure insides us becomes equal to outside air pressure.

The result of removing air from inside a metal can is frequently dramatic. The can collapses as if hit by a sledgehammer! The air is still pushing each square centimetre of surface area on the outside of the can. But the equalizing push from inside disappears as the air is removed by a vacuum pump. The air pressure on its outside surfaces makes the can collapse.

MEASURING PRESSURE: THE BAROMETER

The effects of air pressure may be explained by a concept first suggested in 1643 by the Italian scientist Evangelista Torricelli (1608–1647). Torricelli pictured people as living at the bottom of a "sea of air," which is our atmosphere. We now know that the atmosphere completely surrounds the earth. It becomes less dense with altitude. Even though molecules of air can be found at higher altitudes, most air is within 160 kilometres of the earth's surface.

Torricelli reasoned that the air exerts pressure because it pushes from above. Air pressure is pushing down on each square centimetre of this page. Pressure is caused by the weight of the column of air resting on each square centimetre of the page. Each column must be pictured as rising straight up to the outer limit of the atmosphere, hundreds of kilometres above us. Each column weighs enough to exert considerable pressure at its base. Pressure on a high mountain top would be less than at sea level because there is less air above the mountain.

To measure atmospheric pressure, Torricelli invented the **barometer** (from the Greek *baros,* pressure, and *metron,* to measure). A barometer can be made by filling a long tube, closed at one end, with mercury and then placing the open end under the surface of a pool of mercury in a dish, as shown in Figure 5-1. If the tube is long enough, mercury will flow from the tube until the column of mercury exerts a downward pressure that is exactly balanced by the pressure of the air. In Figure 5-1, the pressure of the air is 760 millimetres. The air pressure pushes down on the surface of the mercury in the dish, thus holding up the mercury in the tube.

As weather reports remind us, atmospheric pressure changes almost daily. Therefore, the height of the mercury column fluctuates. However, its average height at sea level is observed to be 760 millimetres. Recall that standard atmospheric pressure corresponds to a pressure that will support a column of mercury 760 millimetres in height. This amount of pressure is known as 1.0 atmosphere. Half the amount of pressure is 0.5 atmosphere. Twice the amount of pressure is 2.0 atmospheres. *A pressure corresponding to a height of 1 millimetre of mercury* is referred to as 1 torr, after Torricelli. Thus, a pressure of 1.0 atmosphere is equivalent to 760 torr or 760 millimetres of mercury (Hg). To avoid confusion, we will use mm Hg as the unit for atmospheric pressure in further discussions.

Pressure, by definition, is force per unit area. This concept helps explain how a person who would normally sink in soft snow can walk on it with ease when wearing snowshoes. The total force exerted on the snow is probably a bit more with snowshoes because the shoes add a small amount to the total weight of the person. However, the snowshoes distribute a person's total force on the snow over a larger area. Thus, the pressure, or force per unit area, on the snow becomes small enough so that the snow may support even a heavy person.

MEASURING PRESSURE: THE OPEN-END MANOMETER

An instrument commonly used to measure gas pressure is the *manometer*. The **open-end manometer** is a U-shaped tube that is about half-filled with mercury. This instrument permits us to measure the *difference* in pressure exerted on the surfaces of the two arms of the manometer. One end of the tube, arm A, is connected to a container that can be filled with the gas whose pressure is to be measured. The other end of the tube, arm B, is open to the atmosphere, as shown in Figure 5-2. When the mercury levels are at the *same* height in both arms, the pressure exerted on the mercury surfaces in each side of the tube will be the same. Thus, the pressure of the gas sample being measured in Figure 5-2a is equal to atmospheric pressure, 760 mm Hg, at the time of measurement.

If some of the container gas is now pumped out through the valve, the mercury level drops in the open tube as it *rises* in the tube connected to the gas container. The mercury comes to rest in the position shown in Figure 5-2b. The *difference* between mercury levels measures 400 millimetres. The appearance of the manometer in Figure 5-2b tells us two things. It tells us that the pressure on the container gas is now *less* than atmospheric pressure, and it tells us how much less the pressure is. The combined use of a barometer to measure atmospheric pressure and a manometer to measure the difference between atmospheric and container-gas pressure can tell us the pressure of the container gas.

To read the manometer in Figure 5-2b, we focus on the level in arm B. Total pressure on the mercury surface in the open tube is still 760 mm Hg, or

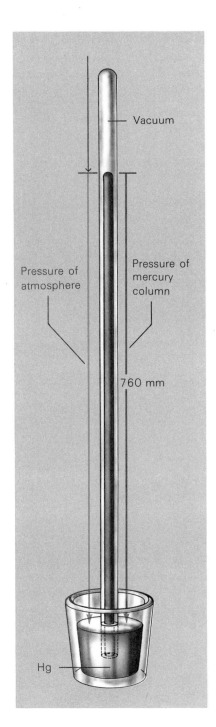

Figure 5-1

A barometer.

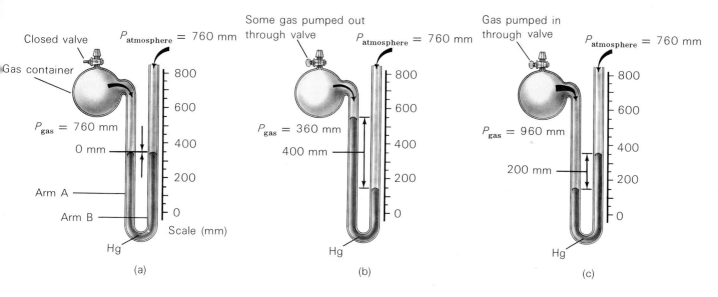

Closed valve

Gas container

$P_{gas} = 760$ mm

0 mm

Arm A

Arm B

Hg

(a)

$P_{atmosphere} = 760$ mm

800

600

400

200

0

Scale (mm)

Some gas pumped out through valve

$P_{atmosphere} = 760$ mm

$P_{gas} = 360$ mm

400 mm

Hg

(b)

800

600

400

200

0

Gas pumped in through valve

$P_{atmosphere} = 760$ mm

$P_{gas} = 960$ mm

200 mm

Hg

(c)

800

600

400

200

0

Figure 5-2

Open-end manometer. (a) Atmospheric pressure of 760 mm is present on both sides. (b) Manometer after some gas has been removed from the container. (c) Manometer after additional gas has entered the container.

atmospheric pressure. Total pressure on the mercury at the same level in arm A, the gas-container side of the tube, must also be 760 mm Hg. But the total pressure at this level in arm A is the *sum* of two pressures. Added to the pressure of the container gas, P_{gas}, is the pressure due to a height of 400 mm Hg. Thus,

$$P_{gas} + 400 \text{ mm Hg} = 760 \text{ mm Hg}$$
$$P_{gas} = 360 \text{ mm Hg}$$

Let us now use our open-end manometer to make an additional measurement. We open the valve and pump gas into the container. Now, the mercury level falls in the tube directly attached to the gas container and, at the same time, rises in the tube's open end. The mercury comes to rest in the position shown in Figure 5-2c. Now the manometer tells us that the pressure in the container is more than atmospheric pressure, or more than the 760 mm Hg indicated by a barometer. Since the *difference* in mercury levels now measures 200 millimetres, we know that the container pressure is 200 mm Hg *above* atmospheric pressure. Thus, the gas-container pressure indicated by the manometer shown in Figure 5-2c is 960 mm Hg.

EXERCISE 5-3

A flask of gas is attached to an open-end manometer. In the following three cases, decide whether the gas pressure is equal to, greater than, or less than atmospheric pressure: (a) the mercury level on the flask side is 20 mm lower than the level on the atmosphere side; (b) the mercury levels are equal; (c) the mercury level on the flask side is 20 mm higher than the level on the atmosphere side. Explain your reasoning.

> If the atmospheric pressure is 748 mm, what is the pressure of the gas sample for the three cases in Exercise 5-3?

MEASURING PRESSURE: THE CLOSED-END MANOMETER

Another type of manometer is also used to measure gas pressure. This type can be made by connecting the open-end of the manometer in Figure 5-2 to a vacuum pump. The pump removes virtually all gases from the formerly open end of the manometer. When the gases have been removed as well as possible, the tube is sealed shut with a torch. No vacuum pump is perfect, but a good one is able to remove virtually all traces of gas from a closed container. The modified manometer shown in Figure 5-3a is obtained. If the gas container is now attached to the vacuum pump, *zero pressure,* a so-called laboratory **vacuum,** will exist above the mercury in both sides of the tubing. The instrument we have just made is known as a **closed-end manometer.** The *levels* of mercury shown in Figure 5-3a are equal, indicating equal pressure on the mercury surfaces. But, unlike the equal-level condition in the open-end manometer, this equal-level condition indicates *zero* pressure in the closed-end type of manometer. Zero pressure in the gas container equals zero pressure above the mercury in the closed-end manometer.

To get an idea of how the closed-end manometer measures pressure, we allow a small amount of air to enter the container. The mercury level falls in the tube connected to the container and rises in the closed end. It stops at the positions shown in Figure 5-3b. The difference in levels is 150 mm Hg, indicating a pressure *difference* of 150 mm Hg. Since zero pressure exists above the mercury in the closed end, the pressure of the gas sample in the container must be 150 mm Hg.

We now open the container valve completely and allow full atmospheric pressure to enter the container. This is accompanied by a dramatic change

Figure 5-3

Closed-end manometer. (a) Virtually all gas has been removed by a vacuum pump, so that zero pressure is above the mercury on either side. (b) Manometer after some gas enters shows a small rise in arm B. (c) Manometer after valve is completely opened shows a large rise in arm B. Manometer is now a barometer.

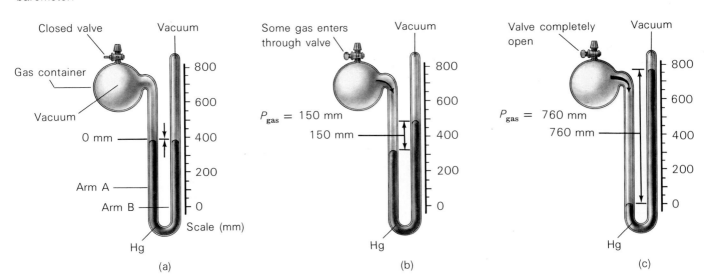

in mercury levels: a rapid fall in the container end, a rapid rise in the closed-tube end. The final positions this time are shown in Figure 5-3c. The *difference* in levels is 760 millimetres. The barometer now also shows a mercury column height of 760 millimetres. As we said earlier, atmospheric pressure changes from day to day. The closed-end manometer shown in Figure 5-3c has now become a barometer. Zero pressure, a laboratory vacuum, exists above the mercury in the sealed tube in both manometer and barometer. Atmospheric pressure exists in the container above the mercury in the manometer tube and above the mercury in the dish of the barometer. Both instruments would indicate the pressure of the atmosphere at the time of this measurement as equivalent to 760 mm Hg.

PARTIAL PRESSURE

The closed-end manometer shown in Figure 5-3c contains a sample of the atmosphere. This sample is the air admitted to the gas container just before our last reading was taken. When a 1 mole air sample is analyzed, it is found to contain *two* principle gases, plus smaller amounts of several other gases. The principal components are 0.781 mole of nitrogen (N_2) and 0.210 mole of oxygen (O_2). Small quantities of water vapor (H_2O), argon (Ar), and carbon dioxide (CO_2) are also present. Acting together, these gases exert a total pressure of 760 mm Hg.

Suppose the air sample is now removed from the gas container of Figure 5-3 with a vacuum pump. The mercury levels return to the positions shown in Figure 5-3a. A few mercury atoms escaping from the liquid mercury surfaces generate a very, very small pressure above the mercury surfaces. Except for this slight pressure, a condition of zero pressure exists. Let us experiment separately with the different gases contained in the air sample to find the pressure each would exert when acting alone. This can be done by placing only the N_2 from the sample in the gas container and reading the pressure in the container. The N_2 sample is then removed with a vacuum pump and the O_2 from the sample is then introduced. The results of these experiments are summarized in Table 5-1.

As indicated in Table 5-1, the pressure of *each* gas alone is known as its **partial pressure.** Examination of the data shows that the sum of the partial

Table 5-1

An approximate analysis of laboratory air

Component	Mole fraction		Total pressure (mm Hg)		Partial pressure (mm Hg)
nitrogen (N_2)	0.781	×	760	=	593
oxygen (O_2)	0.210	×	760	=	160
argon (Ar)	0.009	×	760	=	7
totals	1.000				760

pressures equals the total pressure of the air sample. Whenever several gases are analyzed separately and when mixed together, we get the same result. The **total pressure** of a gaseous system is the sum of the partial pressures of each gas present.

Let us use the kinetic-molecular model to explain the observations summarized in Table 5-1. Pressure, according to that model, is caused by molecular collisions with container walls. *There is so much space between molecules that each molecule behaves almost independently.* Total pressure, due to *all* collisions, is the sum of all the collisions made by all the different molecules. As long as each set of molecules is able to act independently of the other sets and make its own collisions, we expect each gas to exert its own partial pressure. A gas should exert its own pressure when it is alone in a container, as well as when it shares a container with other gases. The kinetic-molecular theory we have discussed could thus be used to *predict* Dalton's Law of Partial Pressures.*

5.4 Temperature and gas laws

When a given mass of any gas is placed in a container of fixed volume, the pressure exerted by the gas depends on its temperature. The higher the temperature, the more pressure the gas exerts. But what is the exact nature of this regularity? Can we formulate a law to describe it? For example, if we double the temperature, holding the mass and volume constant, does the gas pressure double?

THE RELATIONSHIP BETWEEN PRESSURE AND TEMPERATURE

In order to discover the answers to the questions posed above, we can use a fixed volume of air that is trapped in a container fitted with a pressure gauge. The gauge is read with the container at room temperature, 25 °C,

*This relationship states that the total pressure of a gaseous system is the sum of the partial pressures of each gas present.

Figure 5-4

Apparatus for determining the pressure of a given mass of gas at constant volume and different temperatures. The apparatus is shown immersed in an ice-water mixture at 0 °C.

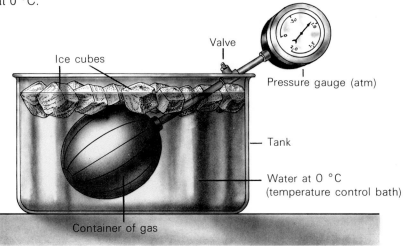

Table 5-2

Pressure exerted by a given mass of air at constant volume as temperature varies from $+100\ °C$ to $-79\ °C$ (data collected from apparatus shown in Figure 5-4)

Temperature control bath	Temperature of air (°C)	Pressure exerted by air (atm)
boiling water	+100	1.26
room temperature air	+25	1.00
ice-water mixture	0	0.92
Dry Ice–acetone mixture	−79	0.65

and again at the temperature of boiling water, 100 °C. The gauge is then read with the container placed in ice water, at a temperature of 0 °C, as shown in Figure 5-4. Finally, it is read after the container has been placed in a mixture of Dry Ice and acetone, where the temperature is −79 °C. The data collected are summarized in Table 5-2.

To begin our search for regularities, we might consider the following. A temperature of 100 °C is four times as great as a temperature of 25 °C. Is the pressure of our air sample at 100 °C four times as great as it is at 25 °C? Table 5-2 clearly indicates that this is not the case. A pressure of 1.26 atmospheres is not four times 1.00 atmosphere. However, if we graph our data as shown in Figure 5-5, we find that the pressure points do lie on the same

Figure 5-5

A plot of pressure versus temperature for a fixed mass of air at constant volume.

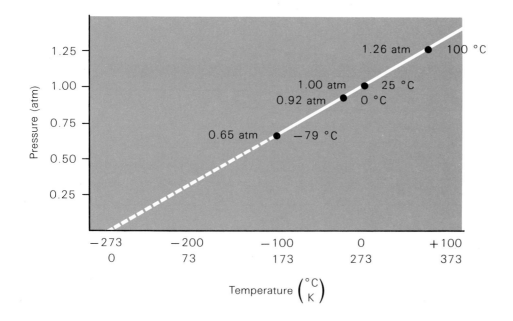

straight line. Thus, there is a regularity. If we extend the line to the left, as represented by the dashed line in the figure, it *crosses* the horizontal axis at a temperature of −273 °C. The horizontal axis represents zero pressure.

When we repeat the experiment suggested in Figure 5-4 using equimolar amounts of other gases in place of our air sample, similar results are obtained. When connected, the temperature-pressure points lie on a straight line. The highest pressure of other gases is 1.26 atmospheres, the pressure of our air sample. The absolute pressure reading at each temperature will depend on the *amount* of gas in the sample: But when the straight lines are extended, they *all* cross the zero-pressure axis at −273 °C!

What is the meaning of zero pressure? According to the kinetic-molecular model, at zero pressure there are no collisions and no molecular motion. Presumably, all the air molecules are lying still at the bottom of the gas container. However, this is only speculation because all gases liquefy before reaching −273 °C. Since they become liquids, they no longer have the properties of gases. New models are needed to describe liquids. When we extend the straight line on the graph for gases, we are assuming that the gas remains a gas and continues to behave as a gas down to −273 °C. Even though we know this is not true, the exercise does suggest some interesting ideas.

Clearly, 0 °C is not the lowest possible temperature. Refrigerator freezers are colder. Many winter days are colder. *Is there such a thing as a lowest possible temperature?* Is there a temperature minimum below which matter cannot be cooled? The meaning of zero pressure—the complete *absence* of molecular motion—suggests that −273 °C *is* that minimum temperature. Historically, the idea of an absolute zero temperature came from experimental results such as those shown in Figure 5-5. Absolute zero represents the lowest possible temperature that can exist. In 1848, the English physicist Lord Kelvin proposed a new temperature scale based on this idea. This **absolute temperature scale** is also referred to as the **kelvin scale**. On this scale, absolute zero, 0 K, corresponds to −273 °C. Values on the kelvin scale are expressed in kelvins (K).* Both kelvin and Celsius temperatures are shown in Figure 5-5. Notice that all numerical values on the kelvin scale are 273 degrees higher than the corresponding temperatures on the Celsius scale.

EXERCISE 5-5

The relationship K = °C + 273 can be derived from the above discussion. Using this formula and what you know about significant figures, convert each of the following temperatures to K:

(a) 0 °C (b) 20 °C (c) 25 °C (d) 100 °C (e) −79 °C (f) −210 °C (g) −273 °C.

Any absolute measurement scale must have a zero point representing the complete absence of the property being measured. Absolute scales cannot

* By international agreement, the degree sign (°) is omitted when referring to absolute temperatures. The word *kelvin* is not capitalized because it is one of the internationally defined base units of science.

have negative values. The metric scale of mass is an absolute scale. Zero grams represents the complete absence of mass. Four grams of mass are twice as many as 2 grams because they are both absolute measurements.

The Celsius temperature scale is not an absolute scale. Its zero point, the freezing point of water, is not the lowest possible temperature. Negative Celcius temperature exist, and, as we discovered, quadrupling the Celsius temperature of our gas sample did *not* quadruple its pressure. Doubling its Celsius temperature would not have doubled its pressure. But the kelvin scale *is* absolute. Zero degrees kelvin is the lowest possible temperature, for negative values on the scale are as impossible as negative mass. *Doubling the kelvin temperature of a gas with fixed mass and volume does double its pressure.*

There is a simple relationship between gas pressure and *absolute* temperature. For that reason, chemists prefer the kelvin scale to the Celsius scale when working with gases. We can use the symbol T for kelvin, or absolute, temperatures and the symbol t for Celsius temperatures. The data shown in the graph in Figure 5-5 reveal that the ratio of pressure (P) to absolute temperature (T) is constant for a fixed mass of gas at constant volume. This gives us a second regularity, or law, of gas behavior:

$$\frac{P}{T} = K_2$$

The symbol K_2 represents a constant for a given amount of gas in a given volume.* Pressure (P) must increase with absolute temperature (T), and in the same ratio. For example, doubling the absolute temperature must double the pressure.

The first regularity, or law, of gas behavior was described in Chapter 2. This relationship, known as Boyle's law, states that pressure (P) times volume (V) is constant for a fixed mass of gas at constant temperature. Using the constant symbol (K) for a gas of constant mass and temperature, the relationship is represented as

$$P \times V = K_1$$

Both of these laws,

$$\frac{P}{T} = K_2 \quad \text{and} \quad PV = K_1$$

hold for a fixed mass of gas. But note the additional restrictions.

$$PV = K_1$$

holds only at constant *temperature,* as well as for a constant amount, and

$$\frac{P}{T} = K_2$$

holds only at constant *volume,* as well as for a constant amount.

*Other K values in this chapter, with different subscripts, represent different constants.

THE RELATIONSHIP BETWEEN VOLUME AND TEMPERATURE

Let us further investigate gas behavior, again using a fixed mass of gas. This time, we will look at how *volume* changes with temperature, keeping pressure constant. We seal one end of a glass tube having a uniformly narrow inside diameter. A drop of mercury placed in the open end of the tube traps a small, fixed mass of air. This setup is shown in Figure 5-6. The drop of mercury has normal atmospheric pressure pushing on it. Since it is free to move inside the tube, it moves until the pressure of the trapped air is equal to the outside air pressure. When the pressure exerted by the trapped air on the drop of mercury is equal to the outside air pressure, the mercury stops moving. The length of the column of trapped air is a measure of the volume of the gas.

Since the outside air pressure is very close to being constant during the time of the experiment, we can assume that the pressure on our sample is very close to being constant. Any increase in the pressure of our sample is compensated by movement of the mercury plug. As volume increases, the

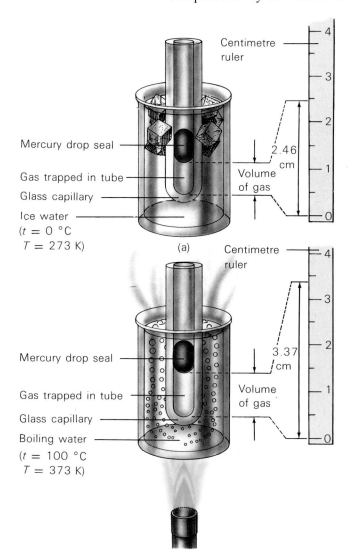

(a)

(b)

Figure 5-6

(a) Set-up for determining the relationship between temperature and volume of a fixed amount of air at constant pressure. The relative volumes are determined by measuring the distance between the glass tube's sealed end and the bottom of the mercury drop. At ice-water temperature (0 °C), this distance is 2.46 cm. (b) At boiling-water temperature (100 °C), the distance has increased to 3.37 cm.

Table 5-3

Relative volume occupied by a given mass of air at constant pressure as temperature varies, from +100 °C to −79 °C (data collected from apparatus shown in Figure 5-6)

Temperature control bath	Temperature of air (°C)	Volume of air (length of tube in cm occupied by air)
boiling water	+100	3.37
room temperature air	+25	2.70
ice-water mixture	0	2.46
Dry Ice–acetone mixture	−79	1.76

Figure 5-7

A plot of relative volume versus temperature for a fixed mass of air at constant pressure.

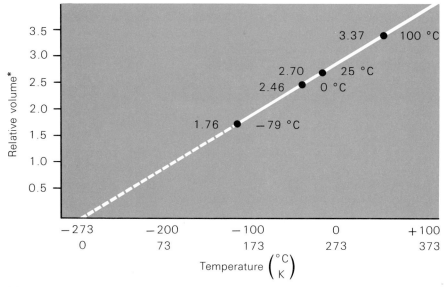

*The length of the column is proportional to the volume of the gas.

mercury moves farther up the tube. An increase in temperature is observed as an increase in volume of the air sample. In the same way, a decrease in temperature would be observed as a decrease in the sample's volume. Since the diameter of the opening in the glass tube is uniform, the *volume* of the trapped air sample depends on the *distance* between the drop of mercury and the sealed end of the tube. If one distance is twice as long as another, the air sample's volume becomes twice as large as it was at the shorter distance.

Let us now use the same four experimental temperatures we used before; boiling water at 100 °C, room air at 25 °C, water and ice mixture at 0 °C, and Dry Ice and acetone mixture at −79 °C. In each case, we wait until the drop of mercury stops moving in the experimental setup before measuring the lengths that represent volume. The data are recorded in Table 5-3 and are graphed in Figure 5-7. Volume (V) on the vertical axis is plotted against temperature (t) on the horizontal axis. We again obtain a straight line by connecting the data points. This line, when extended downward, crosses the zero *volume* axis at -273 °C. Since this corresponds to 0 K, absolute zero, we again see a direct ratio, this time between *volume* and absolute temperature. Thus, 0 K corresponds to zero volume. Doubling the absolute temperature (T) doubles the volume (V).

To summarize, we find that the volume of a fixed amount of gas at constant pressure varies directly with temperature on an absolute temperature scale. Using symbols to represent this relationship, we have

$$V = KT$$

Thus, for a fixed mass of gas at constant pressure, we have a third regularity, or constant (K):

$$\frac{V}{T} = K_3$$

The direct relationship between volume and temperature at constant pressure is known as **Charles's law.** It is named for Jacques Charles (1746–1823), the French physicist who first observed the principle in 1787.

Another regularity, or constant (K), that involves a fixed mass of gas at constant pressure is Avogadro's hypothesis: equal volumes of gases measured at the same temperature and pressure contain equal numbers of molecules. Because the number of molecules is directly related to the number of moles (n), we can write

$$\frac{V}{n} = K_4$$

for any gas at constant temperature and pressure.

TEMPERATURE AND KINETIC ENERGY

Gases are invisible unless they are colored or they visibly interact. Two familiar gases, ammonia and hydrogen chloride, are both invisible. However, they combine with each other to form a white smoke that is clearly visible. We can use this fact to learn something about how fast the molecules of each gas can travel at the same temperature and pressure. We place some ammonia (NH_3) at one end of a narrow, metre-long glass tube and some hydrogen chloride (HCl) at the other end. A few minutes later, the appearance of a white smoke tells us where the molecules of the two gases met. The smoke was closer to the HCl side than to the NH_3 side. Apparently, NH_3 molecules traveled farther than did HCl molecules in the same time period.

The results of the NH_3-HCl interaction, and data from other experiments, show that molecules of different gases travel at different speeds, even at the same pressure and temperature. These experiments show that lighter molecules travel faster, on the average, than heavier molecules. HCl molecules, for example, each have a mass more than twice that of NH_3 molecules. The difference in mass can be shown by comparing the atomic weight of each compound:

$$hydrogen + chlorine \longrightarrow HCl$$
$$1 + 35.5 = 36.5$$
$$hydrogen + nitrogen \longrightarrow NH_3$$
$$1 \times 3 + 14 = 17$$

The experiment described above shows us that the lighter NH_3 molecules do, on the average, move faster.

Gases may be separated by taking advantage of differences in molecular velocities. The average speed of nitrogen molecules, which comprise almost 80 percent of normal air, is 334 metres per second at 0 °C. Lighter helium molecules, at the same temperature, have an average speed of 965 metres per second, almost three times that of nitrogen. Table 5-4 gives the average speeds of various gas molecules at 0 °C. Helium may be separated from nitrogen by diffusion.

Our use of the word *average* is deliberate. Additional experiments with gases indicate that every group of gas molecules contains a few very fast-moving members, some very slow-moving ones, and many with speeds in between. The additional experiments reveal the average speed of the molecules tested. The experiments also reveal that average speed depends on

Table 5-4

Average velocities of various gases at 0 °C

Gas	Velocity	
	(metres/sec)	(miles/hr)
air (dry)	331	740
ammonia (NH_3)	415	928
argon (Ar)	319	714
carbon dioxide (CO_2)	259	579
chlorine (Cl_2)	206	461
helium (He)	965	2,760
hydrogen (H_2)	1,284	2,870
hydrogen chloride (HCl)	296	662
methane (CH_4)	430	962
nitrogen (N_2)	334	747
oxygen (O_2)	316	707
sulfur dioxide (SO_2)	213	476

temperature. The higher the temperature, the higher the average speed. Kinetic-molecular theory, as we shall see, enables us to predict just such results.

The kinetic-molecular model we have used to describe gas behavior tells us that pressure is the result of molecular collisions. It also tells us that gas molecules move faster at high temperatures than at low ones. This corresponds with the observed rise in pressure with rising temperature, as summarized in Table 5-2. We have also explained the differences in chemical properties of different gases by assuming that molecules of one gas differ from those of another. For example, molecules differ in their mass. An O_2 molecule has a molecular weight of 32.00. An H_2 molecule has a molecular weight of 2.00. Yet, equal volumes of 1 mole of each gas exert the same pressure at the same temperature. How can light H_2 molecules exert the same pressure as heavier O_2 molecules?

The answer is revealed by experiments such as the NH_3-HCl reaction: lighter molecules move faster, on the average, than heavier molecules. Because lighter molecules move faster, they can exert the same pressure as slower-moving, heavier molecules at the same temperature. An analogy may help explain why. A heavy bowling ball moving slowly can push your hand just as hard as can a light baseball moving rapidly. Thus, the pressure

OPPORTUNITIES

Quality control

Quality control chemists are responsible for maintaining proper standards of manufacturing in chemical companies. Such chemists coordinate all activities necessary to evaluate the efficacy of gases used for anesthesia, other pharmaceuticals, fertilizers, cosmetics, paints, adhesives, or any product made by chemical companies. A quality control chemist is a specialist in sampling techniques, standard testing procedures, and, frequently, statistical analyses.

Quality control chemists should have a thorough knowledge of federal and state regulations controlling the production and marketing of chemicals. They must constantly review guidelines regarding the use of specific chemicals and provide analytical data to help determine how chemicals may affect the environment. They may collect samples of gases from smokestacks, auto-

mobile exhausts, and other potential sources of atmospheric pollution. They must test water samples to ensure that no pollutants are released by the company into waste water. Another important area of concern is the testing of specific quantities of materials to see if the materials meet company standards. Most companies are careful to see that their products meet high quality standards.

Some quality control chemists specialize in computer analysis of data. They supervise the preparation of information for computer storage. Such information is very useful in the solution of specific problems. The retrieval of stored information helps chemists better evaluate the products and processes to be tested.

Employment opportunities in this area are very good. The need is increasing for chemists who are capable, thorough, and well-trained in sampling and testing techniques. Such chemists must also have the

ability to develop tests for new products. The success or failure of a new chemical product often depends on the work of quality control chemists.

A quality control chemist must have a bachelor's degree in chemistry. Graduate work, particularly in analytical procedures, is essentil for advancement and increased responsibility. For further information, contact the Chemical Manufacturers Association, 2501 M Street NW, Washington, D.C. 20037.

an object can produce is related to the property known as **momentum.** This property depends both on the object's mass and on its velocity. Momentum is equal to mass × velocity. Heavy molecules can exert the same pressure as light molecules if they have *different* velocities.

Moving particles possess energy. The energy of motion is called *kinetic energy* (*KE*), and is defined by the formula

$$KE = \tfrac{1}{2}mv^2$$

in terms of both mass (*m*) and velocity (*v*). According to kinetic-molecular theory, two gases at the *same* temperature have the *same* average kinetic energy. This means that heavier molecules (large *m*) have slower average velocities (smaller *v*) at a given temperature than do lighter molecules. Since lighter molecules have smaller masses, they must have greater average velocity. As we have seen, this result is in accord with the findings of experiments. Thus, hydrogen molecules at room temperature move much faster, on the average, than do oxygen molecules at the same temperature, as shown in Table 5-4.

Let us look finally at the kinetic-molecular view of two samples of the *same* gas initially at *different* temperatures. Suppose we start with two different samples of ammonia gas, each in its own container. One is hot, the other is cold. The kinetic energy of the hot ammonia molecules is greater, on the average, than the kinetic energy of the cold ammonia molecules. This statement is easily verified. If the ammonia and hydrogen chloride used in the experiment described on page 118 had both been *heated* before use, the white smoke would have formed *sooner* than it did. Therefore, both NH_3 and HCl molecules must move faster, on the average, after being heated.

According to the kinetic-molecular theory, gases have greater average kinetic energy, and hence greater average velocity, when they are hot than when they are cold. This is in agreement with the experimental results described above.

Finally, suppose we have two samples of ammonia gas, one hot, and one cold. We allow the hot and cold gas samples to mix in one large container. The *same* temperature will eventually exist throughout the container. According to the kinetic-molecular model, fast-moving, hot NH_3 molecules collide with slower-moving, cold NH_3 molecules. Such collisions tend, on the average, to slow down the fast-moving molecules and speed up the slow-moving ones. Eventually, the formerly separate samples become one sample at some intermediate temperature. Each portion of this larger sample has the same average kinetic energy, and thus the same temperature, as any other portion of it.

A GENERAL GAS-LAW EQUATION

In considering gases, we have identified four general variables that are interrelated. These variables—pressure, volume, temperature, and number of moles of gas—have been considered in pairs until now. We have held two variables constant and then studied variations in the remaining two. For example, in our very first investigation of the relationship between pressure and volume, in Chapter 2, we considered a fixed quantity or number of moles of gas at constant temperature. Number of moles and temperature were therefore held constant while pressure and volume were varied. In studying the temperature-volume relationship in section 5.4, we have held the number of moles of gas and the pressure constant while varying volume and temperature. Is this pattern always necessary or desirable? Let us see if a more general equation can be developed.

In studying the general relationship between pressure and volume, we found that pressure $\times$ volume = a constant. The values of $P \times V$ for different amounts of gas at two temperatures are given in Tables 5-5 and 5-6. An examination of the data shows that the size of the constant depends on (1) the quantity of gas and (2) the temperature. It is convenient to measure the quantity of gas in moles. We remember that for 1 mole of gas at 0 °C and 1 atmosphere, $V = 22.4$ litres. Then for 1 mole of gas,

$$P \times V = 1 \text{ atm} \times 22.4 \text{ litres/mole}$$

Since the constant is twice as large for 2 moles and three times as large for 3 moles, we can write (for 0 °C)

$$P \times V = \underbrace{1 \text{ atm}}_{P} \times \underbrace{n \times 22.4 \text{ litres/mole}}_{\text{VOLUME AT 273 K}}$$

where n is the number of moles of gas taken.

We now have the three variables P, V, and n related through a single expression, but the equation is valid only at 0 °C. How can the effects of temperature be included? We said earlier in section 5.4 that an increase in temperature increases the volume of the gas. For example, at 25 °C and 1 atmosphere the volume of 1 mole of gas is 24.4 litres. The general expression showing how the volume of 1 mole of gas at 1 atmosphere varies with temperature is

$$V = 22.4 \text{ litres/mole} \times \frac{T}{273 \text{ K}}$$

where T is the temperature on the kelvin scale. Substituting this expression for the volume of 1 mole of gas, we obtain the expression

$$P \times V = \underbrace{1 \text{ atm}}_{P} \times \underbrace{n \times 22.4 \text{ litres/mole} \times \frac{T}{273 \text{ K}}}_{\text{VOLUME OF } n \text{ MOLES AT } T}$$

or

$$P \cdot V = nT \left(\frac{22.4}{273} \text{ litres} \times \text{atm/mole} \times \text{K} \right)$$

The quantity in parentheses does not change as we change P, V, n, or T. It simply reflects our earlier arbitrary choices for molar volume, standard temperature, and standard pressure. Since this number does not change, it is worthwhile doing the division needed to evaluate the constant.

$$\left(\frac{22.4 \times 1}{1 \times 273} \right)\left(\frac{\text{litres} \times \text{atmospheres}}{\text{number moles} \times \text{K}} \right) = 0.0821 \left(\frac{\text{litres} \times \text{atm}}{\text{mole} \times \text{K}} \right)$$

Table 5-5

Pressure-volume products for any gas at 0 °C and 1 atm

Number of molecules $(\times 10^{23})$	Volume (litres)	$P \times V$ (atm $\times$ litres)
0.268	1.00	1.00
0.536	2.00	2.00
1.07	4.00	4.00
3.01	11.2	11.2
6.02	22.4	22.4
12.0	44.8	44.8
18.1	67.2	67.2

Table 5-6

Pressure-volume products for any gas at 25 °C and 1 atm

Number of molecules $(\times 10^{23})$	Volume (litres)	$P \times V$ (atm $\times$ litres)
0.268	1.09	1.09
0.536	2.18	2.18
1.07	4.37	4.37
3.01	12.2	12.2
6.02	24.4	24.4
12.0	48.8	48.8
18.1	73.2	73.2

Table 5-7

Values of the ideal gas constant

Pressure in	Volume in	Value of R
atm	litres	$0.0821 \left(\dfrac{\text{litres} \times \text{atm}}{\text{mole} \times \text{K}} \right)$
mm Hg	litres	$62.4 \left(\dfrac{\text{litres} \times \text{mm Hg}}{\text{mole} \times \text{K}} \right)$
mm Hg	ml	$62{,}400 \left(\dfrac{\text{ml} \times \text{mm Hg}}{\text{mole} \times \text{K}} \right)$
atm	ml	$82.1 \left(\dfrac{\text{ml} \times \text{atm}}{\text{mole} \times \text{K}} \right)$

The unit "mole" must be included in the denominator because the volume of 22.4 litres is for *1 mole* of gas at 1 atmosphere and 273 K.

This quantity is of great importance in physical chemistry. It is assigned the symbol R and is known as the **ideal gas constant.** The general gas equation can then be written as

$$PV = nRT$$

If P is given in millimetres of Hg and V is given in millilitres, R has a different numerical value but the same overall significance.

$$\frac{22{,}400 \text{ ml} \times 760 \text{ mm Hg}}{\text{mole} \times 273 \text{ K}} = \frac{62{,}400 \text{ ml} \times \text{mm Hg}}{\text{mole} \times \text{K}}$$

The expression $PV = nRT$ is known as the **ideal gas law.** * Chemists find it a most convenient and powerful equation. The numerical value selected for R depends on the units being used in the problem. This fact is summarized in Table 5-7.

We first described Avogadro's hypothesis in section 2.1. In section 5.4, we have combined separate observations of gas behavior algebraically to obtain the ideal gas law, $PV = nRT$. One of the triumphs of the kinetic theory is our ability to derive both Avogadro's hypothesis and the ideal gas law from a mathematical analysis of the motion of gas particles. Such an analysis is not particularly hard to follow. Ask your teacher, if you are interested. Our ability to use the kinetic theory to duplicate several sets of independent observations gives us confidence in the theoretical model upon which the kinetic theory rests.

*Using Avogadro's hypothesis and any two of the other three regularities discussed in this chapter, you can derive the ideal gas law. For example, $PV = k_1$; $P = k_2 \times T$; $V = k_4 \times n$. Then we can write

$$PV = (k_2 \times T)(k_4 \times n) = \underbrace{k_2 \times k_4}_{R} n \times T$$

$$PV = nRT$$

SOME EXAMPLES OF GAS-LAW PROBLEMS

Some sample problems are included here to indicate how gas-law calculations can be made.

Example 1: A gas occupies 0.300 litre and exerts a pressure of 0.800 atmosphere. What must the volume become if the pressure is 1.200 atmospheres? The temperature and number of moles of gas are constant.

Answer: Qualitative common sense arguments tell us that if the *pressure* is to *go up*, the *volume must go down*. Remember that

$$PV = K \quad \text{and} \quad P_1V_1 = P_2V_2$$

What is the factor that we must multiply the original volume (V_1) by in order to get the final volume (V_2)? The answer is the ratio of the two pressures that will give a smaller volume.

$$V_2 = V_1 \times \frac{P_1}{P_2} \quad \text{or} \quad V_2 = 0.300 \text{ litre} \times \frac{0.800 \text{ atm}}{1.200 \text{ atm}}$$

$$V_2 = 0.200 \text{ litre}$$

Example 2: What pressure is required to compress a 2.00-mole sample of gas into 10.0 litres of 0 °C?

Answer: Use the general gas equation, $PV = nRT$. Since pressure is given in atmospheres and the volume in litres, the value of R, as listed in Table 5-7, is 0.0821 (litres × atm/mole × K) The temperature, 0 °C, is 273 K. The required pressure is

$$P = \frac{nRT}{V} = \frac{2.00 \text{ moles} \times 0.0821 \left(\frac{\text{litres} \times \text{atm}}{\text{mole} \times \text{K}} \right) \times 273 \text{ K}}{10.0 \text{ litres}}$$

$$P = 4.48 \text{ atm}$$

Example 3: Barometer A (a barometer is used to measure atmospheric pressure) has some gas on top of the column. It reads 742 mm Hg. A new barometer, B, from the storeroom reads 757 mm Hg. What is the gas pressure above the mercury in barometer A, as shown in Figure 5-8?

Answer: Barometer A is now serving as a manometer, which measures the *difference* between the pressure above the mercury column and the pressure of the atmosphere outside. Pressure reading by barometer A = atmospheric pressure − pressure inside. Thus,

$$742 \text{ mm Hg} = 757 \text{ mm Hg} - \text{pressure inside}$$
$$\text{pressure inside} = 15 \text{ mm Hg}$$

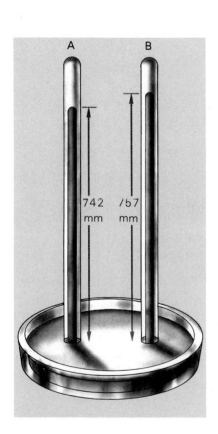

Figure 5-8

Barometers A and B.

Example 4a: What is the partial pressure of each gas in a mixture containing by volume 20 percent He, 30 percent CO, 10 percent H_2, and 40 percent CH_4 if the total pressure is 800 mm Hg?

Answer: Twenty percent He by volume means that 20 percent of the molecules in the container are He. Thus,

$$800 \text{ mm Hg} \times 0.20 = 160 \text{ mm He}$$
$$800 \text{ mm Hg} \times 0.30 = 240 \text{ mm CO}$$
$$800 \text{ mm Hg} \times 0.10 = 80 \text{ mm H}_2$$
$$800 \text{ mm Hg} \times 0.40 = \underline{320 \text{ mm CH}_4}$$
$$800 \text{ mm total}$$

Example 4b: What would be the *total pressure* if water vapor at a pressure of 18 mm were added to the dry gases with no change in total volume?

Answer:

$$\text{total pressure} = 800 + 18 = 818 \text{ mm Hg}$$

Example 5: What volume is occupied by 3 moles of nitrogen under a pressure of 120 atmospheres and a temperature of 50 °C? (Use the ideal gas law.)

Answer: Given $PV = nRT$,

$$P = 1.20 \times 10^2 \text{ atm}$$
$$V = \text{volume (litres)}$$
$$T = 50 \text{ °C} + 273 = 323 \text{ K}$$
$$n = 3.00 \text{ moles}$$

Since P is given in atmospheres and V in litres, use R in these units:

$$R = 0.0821 \left(\frac{\text{litres} \times \text{atm}}{\text{mole} \times \text{K}} \right)$$

$$120 \text{ atm} \times V = 3.00 \text{ \cancel{moles}} \times 0.0821 \left(\frac{\text{litres} \times \text{atm}}{\cancel{\text{mole}} \times \cancel{\text{K}}} \right) \times 323 \, \cancel{K}$$

$$V = \frac{3.00 \times 0.0821 \times 323}{120 \, \cancel{\text{atm}}} \text{ litres} \times \cancel{\text{atm}} = 0.663 \text{ litre}$$

EXERCISE 5-6

How many moles of a gas will occupy 900 ml at a pressure of 4,500 mm Hg and a temperature of −73 °C?

Table 5-8

Accurate pressure-volume measurement of 1 mole of ammonia gas (NH_3) at 25 °C

Pressure (atm)	Volume (litres)	$P \times V$ (atm × litres)
0.1000	244.5	24.45
0.2000	122.2	24.44
0.4000	61.02	24.41
0.8000	30.44	24.35
2.000	12.17	24.34
4.000	5.975	23.90
8.000	2.925	23.40

The gas laws we have studied accurately describe the behavior of many different gases. But we have looked at such behavior under rather limited conditions. The gases studied have all been at low pressures: 1 to 5 atmospheres or less. They have also been at relatively high temperatures: hundreds of degrees above absolute zero. The lowest temperature we considered, that of Dry Ice and acetone, was only 79 degrees below zero on the Celsius scale. The temperature is equivalent to 194 kelvin!

The reason we have used these rather limited conditions is illustrated in Table 5-8. Note that ammonia gas accurately obeys Boyle's law—$PV = $ a constant—until pressure climbs to about 3 atmospheres. At higher pressures, ammonia gas starts to deviate from $PV = K$. Other gases show similar deviations from expected gas-law behavior as pressure rises, and/or as temperature falls.

The term *ideal gas* has been used to describe a gas that shows the exact behavior described by the gas laws. Real gases, such as ammonia, clearly do *not* exhibit this so-called ideal behavior, except at low pressures and relatively high temperatures. Recall the sample of air we used in the fixed volume experiment described in Figure 5-4. We knew we were only speculating when, in Figure 5-5, we extended the straight line summarizing the air's behavior. We knew that the air would not continue to show this behavior at very low temperatures because it would liquefy. It would then no longer be a gas.

How can we explain the non-ideal behavior shown by real gases? The experiments just referred to give us two hints. One is that ideal behavior is seen at high temperatures and low pressures. The other is that real gases do liquefy. There is a connection between these two facts: gases liquefy when their temperatures are low enough or when the pressure on them is high enough. According to the kinetic-molecular theory, molecules of a liquid are in clusters. The fact that real gases *condense* to form liquids at low temperatures suggests the existence of forces of attraction between molecules. The kinetic-molecular theory assumes that ideal gas molecules are not attracted to each other. Recall from Chapter 2 that when gas molecules collide, they are assumed to rebound like colliding ping-pong balls, without the loss of energy.

Real gas molecules attract each other, but at high molecular velocities (high temperatures) and at large molecular separations (large gas volumes) the kinetic energy of the molecules overcomes the forces of attraction. As molecules are pushed closer together, the forces of attraction between molecules increase and the actual non-compressible volume of the molecules occupies more of the total gas volume. Further, at low temperatures, molecules slow down and have less kinetic energy. Thus, at low temperature and small volume, forces of attraction begin to overcome forces associated with molecular motion. The molecules begin to form clusters or potential clusters. Consequently, deviations from the ideal gas laws are seen. As gases are cooled and compressed, they will turn into a liquid. The volume of the liquid is a fairly good approximation to the actual volume of the gas molecules themselves. If the forces acting between molecules are very weak, as we see in H_2 and He, the gas is very hard to liquefy. As the forces get stronger, as we see in H_2O, the gas is easier to liquefy.

SUMMARY

Molar volume is the volume occupied by 1 mole of any substance. It can be calculated from density measurements. The molar volume of a gas is almost 1,000 times larger than that of a liquid or a solid. The molar volume of *any* gas is 22.4 litres at 0 °C and 1 atmosphere.

Since liquids and solids can not be easily compressed, we assume that there is little space between molecules in liquids and solids. On the other hand, molecules in the gas phase are so very widely separated that the actual volume occupied by the molecules themselves is negligible in comparison to the total volume of the gas.

Gas pressure is measured with an instrument known as a manometer. The open-end manometer compares pressure of a gas to atmospheric pressure. The closed-end manometer gives a direct reading of the absolute pressure of a gas.

Four variables, important in considering a gas, are: (1) pressure, (2) temperature, (3) number of moles of gas, and (4) volume. If the number of moles of gas and the temperature are held constant, we find $PV = K$. If the number of moles and volume are held constant, we find $P/T = K_2$, where T is the temperature on the Kelvin scale (0 K = −273.1 °C). If pressure and the number of moles of gas are held constant, we find $V = K_3T$. Finally, Avogadro's hypothesis tells us that $P = K_4n$, if volume and temperature are held constant.

By combining these four relationships involving variables taken two at a time, we can get the ideal gas law, $PV = nRT$, where R is the gas constant that we evaluated. An ideal gas is one whose behavior is accurately described by the ideal gas law under all conditions. There are no true ideal gases. Real gases deviate from ideal gas behavior at small gas volumes (high pressure) and low temperatures.

Key Terms

absolute temperature scale

barometer

Charles's law

closed-end manometer

ideal gas constant

ideal gas law

kelvin scale

kinetic energy

kinetic-molecular theory of gases

Law of Combining Gas Volumes

molar volume

momentum

open-end manometer

partial pressure

pressure

total pressure

vacuum

QUESTIONS AND PROBLEMS

A

1. The volume occupied by *one mole* of N_2 molecules in the liquid and solid states is about 31 ± 4 ml. The volume occupied by one mole of gaseous N_2 molecules is about 22,400 ml. (a) On the basis of these numbers, can we say that the *actual volume* of the molecules is not significant compared to the *total volume* of the gas? Explain. (b) How much bigger is the volume occupied by the gas than is the volume occupied by the liquid and solid? What meaning can be attached to this number in terms of a kinetic-molecular model?

2. At sea level, the barometer reads 760 mm Hg (760 torr). What reading might you expect (a) on top of one of the peaks in the Sierra-Nevada mountains? (b) at Furnace Creek, which is below sea level in Death Valley, California? General qualitative estimates are acceptable for both answers.

3. Explain in terms of a kinetic-molecular model why a metal can collapses when air is pumped out of it with a vacuum pump.

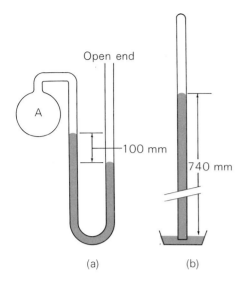

Open end

—100 mm

740 mm

(a)   (b)

Open end

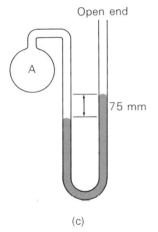

A

75 mm

(c)

Closed end

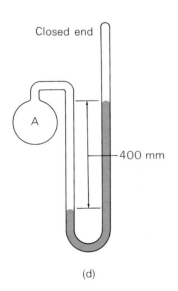

A

—400 mm

(d)

4. A man weighing 60 kg steps into powdery snow and sinks to his waist. The area of each of the man's feet is about 250 cm². He puts on a pair of skis, each of which has an area of about 1,800 cm². (a) What pressure, expressed in kg/cm², does each of the man's feet exert on the snow? (b) What pressure, again in kg/cm², is exerted on the bottom of each ski? (c) Why can he walk on top of the snow wearing skis without sinking up to his waist? Assume in each case that the man puts all of his weight on one foot as the pressure is measured.

5. If the mercury level in the open arm of a manometer is 100 mm higher than the mercury surface in the other arm, is the gas pressure higher or lower than atmospheric pressure?

6. What are the advantages and disadvantages of an open-end manometer compared to a closed-end manometer?

7. Give the pressure in bulb A for each example shown in the margin.

8. Use a kinetic-molecular model to explain why each gas in a mixture exerts its own partial pressure quite independently of the other gases.

9. (a) In order to decrease the pressure exerted by a gas, how must the volume be changed if all other conditions remain constant? (b) If the volume of a container holding a gas is decreased, what will be the effect on the pressure if all other conditions remain constant? (c) If the number of gas molecules is decreased, what will be the effect on the pressure if all other conditions remain constant? (d) If the number of gas molecules is decreased, what must happen to the volume if all other conditions are to remain constant?

10. A gas occupies 0.400 litre and exerts a pressure of 0.900 atm. What must its volume become in order for the pressure to be 1.80 atm?

11. What pressure would be needed to force a 5.00-litre sample of gaseous N_2 at 1 atm into a 500 ml tank? (Assume that n and T are constant.)

12. What volume does 1 mole of NH_3 occupy at 1 atm pressure and 25 °C?

13. A sample of O_2 under 2 atm pressure occupies 500 ml at 25 °C. (a) What volume will the sample occupy at 0 °C? (Assume that P and n are constant.) (b) What volume will the sample occupy at 200 °C and 3 atm? (Assume that n is constant.)

14. (a) What volume will the sample in problem 11 occupy at 1 atm pressure and 0 °C if the temperature in problem 11 was 25 °C? (b) If the gas sample has a mass of 1.80 g, what is the density of the gas in grams per litre at STP? (c) What is the molecular weight of the gas?

15. A gas sample is originally at 0 °C. At what temperature will the volume of this sample be twice as large as it was at 0 °C if P is constant?

16. A gas at 20 °C increases in temperature by 10 °C. By what fraction does the volume increase? (Assume that P and n are constant.)

17. What volume would 8.00 g of methane (CH_4) occupy at 25 °C?

18. Make the temperature conversions indicated here: (a) 100 °C to Kelvin (b) 325 K to °C (c) −259 °C to K (d) −117 °C to K (e) 298 K to °C (f) 78 K to °C (g) −40 °C to K (h) −182 °C to K (i) 78.5 °C to K

B

19. Hydrogen gas at one atmosphere liquifies at −252 °C to give a liquid of density 0.0700 g/ml. (a) What is the *molar volume* for liquid H_2? (b) What is the ratio of the molar volume of liquid H_2 at −252 °C and one atm to that of gaseous hydrogen at STP?

20. Clean dry air has a molecular composition as follows: $N_2 = 78.1\%$,

$O_2 = 21.0\%$, and $Ar = 0.9\%$. The barometer in the room reads 640 mm. What is the partial pressure of N_2 in the room?

21. A 2.00-mole sample of oxygen is put into an 11.2-litre tank. Then a 3.00-mole sample of N_2 is pumped into the same tank. Finally, a 4.00-mole sample of He is put in the tank. (a) What is the total pressure in the tank if the temperature is 25 °C? (The ideal gas law is useful. Hint: Count the number of moles of gas.) (b) What is the partial pressure of each gas in the tank? (c) What is the total *mass* of all gases in the tank? What is the gas density in the tank? (d) Using the ideal gas law, what volume would this total gas mixture occupy at STP? (e) What density does the gas mixture have at STP? (f) What is the "apparent molecular weight" of this gas mixture? (g) The addition of a gas of low molecular weight, such as He, will reduce the density of the gas as soon as it is allowed to expand to one atmosphere pressure. To show that this statement is true, calculate the density of the gas sample at STP without the helium. Since air has an apparent molecular weight of about 29.0 and water vapor has a molecular weight of 18.0, it should be clear why adding water vapor to air makes the air lighter as long as the air is allowed to expand to a final pressure of one atm.

22. Metallic silver can be reclaimed from silver chloride by the following reaction: $AgCl + H_2 \longrightarrow Ag + HCl$. (a) Balance the equation. (b) What volume of hydrogen, measured at 25 °C and 1 atm, would be used in changing 28.7 g of AgCl back to metallic silver?

23. How many moles of gas are present in each of the following samples: (a) 120 g of a gas occupying 89.6 litres measured at STP, (b) 245 g of a gas occupying 122 litres measured at 25 °C and 1 atm? (c) What is the molecular weight of each gas? Note that the molar volume of a gas at 25 °C and 1 atm is 24.4 litres.

24. Ammonia (NH_3) and sulfur dioxide (SO_2) are both gases with readily distinguishable odors. If a cylinder of each was opened in a draftless room, which odor would you expect to smell first? Why?

25. (a) What happens to the average kinetic energy of a gas if its absolute temperature is doubled? (b) What happens to the absolute temperature of a gas if its average kinetic energy is cut in half?

26. For each of the following, calculate the volume of the gas sample under the conditions specified: (a) 0.200 mole gas at STP, (b) 0.75 mole gas at 25 °C and 1.00 atm, (c) 2.3 moles gas at 50 °C and 3.00 atm, (d) If sample (a) weighs 5.6 g, what is the molecular weight of (a)? (e) Sample weights for (b) and (c) are 33.0 g and 147.2 g, respectively. What are the molecular weights of each?

27. In order to determine the value of *absolute zero* on the Fahrenheit scale, a tank of helium gas is placed in four different temperature control baths. The temperature and pressure of the gas in each bath are read. (Assume that n and V are constant.) The following data are obtained:

Temperature control bath	Temperature (°F)	Pressure (atm)
boiling water	212	1.26
ice-water mixture	32	0.93
dry ice-acetone mixture	−110	0.67
liquid nitrogen	−321	0.27

Plot these data on a graph. From the graph, determine the value of absolute zero on the Fahrenheit scale.

28. A CO_2 fire extinguisher contains about 4.4 kg of CO_2. What volume of gas could this extinguisher deliver at 25°C and 1 atm?

29. The density of liquid carbon dioxide at room temperature is 0.80 g/ml. How large a cartridge of liquid CO_2 must be provided to inflate a life-jacket of 4.0-litre capacity at STP? (Hint: Calculate the number of moles of CO_2 needed to fill the jacket, then calculate the mass of CO_2 and the volume of the liquid.)

30. Why does a blowout of an automobile tire occur most frequently when the tire is moving over asphalt roads in the middle of the summer?

31. Compressed oxygen is sold in 40.0-litre steel cylinders. The pressure at 25 °C is 130 atm. (a) How many moles does such a filled cylinder contain? (b) What is the mass of the O_2 in the cylinder? (Hint: The ideal gas law is useful.)

32. Calculate R from the following data. Give units in both cases. (a) $P = 760$ mm, $V = 12.2$ litres, $T = 298$ K, $n = 0.500$, (b) $P = 108$ cm, $V = 3.36$ litres, mass of gas $= 3.03$ g, molecular weight of gas $= 2.02$ g/mole, T $= 38.6$ K

C

33. The acetylene (C_2H_2) in acetylene lanterns is generated by the reaction of water and calcium carbide (CaC_2) as follows:

$$CaC_2 + H_2O \longrightarrow Ca(OH)_2 + C_2H_2$$

(a) Balance the equation. (b) Calculate the volume of C_2H_2 (measured at 25 °C and 1 atm) released if 5.00 moles of CaC_2 are used in the above reaction.

34. When hydrogen is needed in the laboratory, it is commonly made by the reaction of zinc metal with hydrochloric acid (HCl). The products of the reaction are hydrogen gas and zinc dichloride. (a) Write the balanced equation for the reaction. (b) How many moles of zinc must be used to produce 10.0 litres of hydrogen gas measured at room conditions? (c) How many grams of zinc would be necessary to produce the 10.0 litres of hydrogen gas?

35. Oxygen is often generated in the laboratory by the reaction between sodium peroxide (Na_2O_2) and water, *under appropriate conditions*. Sodium hydroxide (NaOH) is the other product. (a) Write the balanced equation for the reaction. (b) How many grams of Na_2O_2 must be used to generate 5.0 litres of oxygen at 25 °C and 1 atm? Assume all Na_2O_2 reacts.

36. Given the following unbalanced equation:

$$Cu_2O + H_2 \longrightarrow Cu + H_2O$$

(a) Write the balanced equation for the reaction. (b) Calculate the volume of hydrogen, measured at STP, needed to react with 8.30 moles of Cu_2O.

37. Hydrogen can be made from iron and steam at high pressure using the following process:

$$Fe + H_2O \xrightarrow[\text{heat}]{} Fe_3O_4 + H_2$$

(a) Write the balanced equation for the reaction. (b) Calculate the number of moles of iron needed to produce 8.50 litres of hydrogen by this reaction if the volume of hydrogen is measured at STP.

38. Liquid vegetable oils, such as cottonseed oil, can be converted into solids of the margarine type by the process of hydrogenation. Although cottonseed oil is not a pure chemical substance, but a mixture of esters of fatty acids, the reaction can be typified by the following *unbalanced* equation:

$$(C_{17}H_{33}COO)_3C_3H_5 + H_2 \longrightarrow (C_{17}H_{35}COO)_3C_3H_5$$

OLEIN   STEARIN

(a) Write the balanced equation for the reaction. (b) If 425 litres of hydrogen gas (measured at 25 °C and 1 atm) enter the reaction, how many moles of olein are hydrogenated?

39. A mixture of gases contains 0.60 g H_2, 4.4 g CO_2, and 0.80 g He. The total pressure is 1.3 atm. If the CO_2 is solidified and then removed and the other gases are warmed to their original temperature, (a) what is the new total pressure of the remaining gases and (b) what is the partial pressure of each remaining gas? (Assume that there is no interaction among the gases.)

40. A gas occupies 10.0 litres at a measured pressure of 1.4 atm. Consider each of the following problems independently of each other: (a) if the volume is changed to 25.0 litres, what is the pressure (assuming no change in the amount of gas or the temperature); (b) if one fourth of the molecules are removed, what must the volume be in order that there be no change in pressure or temperature; (c) what must the volume be if the pressure registers 0.80 atm (no change in temperature or amount of gas); (d) if gas is added so that the number of moles of gas is quadrupled, what is the pressure if there is no change in volume and no change in temperature?

41. A sample of nitrogen gas in a syringe occupies 15.0 ml when under a pressure of 6 bricks. Given the further information that the sample has a mass of 0.040 g and the temperature is 25 °C, calculate the value of R in the general gas equation. Express your answer in the proper units.

42. Suppose standard conditions had been selected as 25 °C and 2 atm instead of 0 °C and 1 atm. What would the new value of R have been?

43. A glass bulb has a mass of 108.11 g after all of the gas has been removed from it. When filled with O_2 gas at 1 atm and 25 °C, the mass of the bulb is 109.56 g. (a) What is the volume of the glass bulb? (The ideal gas law would be useful.) (b) This same empty bulb is filled with a gas taken from the mouth of a volcano (some geologists are fearless). The temperature is still 25 °C and the pressure 1 atm. The bulb now has a mass of 111.01 g. Which one of the following gases could be coming from the volcano on the basis of these data? (a) CO_2 (b) CH_4 (c) Si_2H_6 (d) NF_3 (e) SO_2 (f) S_8 (g) a mixture containing half CO_2 and half krypton?

Almost all the chemical processes which occur in nature, whether in animal or vegetable organisms, or in the nonliving surface of the earth, . . . take place between substances in solution.

WILHELM OSTWALD

LIQUIDS AND SOLIDS: CONDENSED PHASES OF MATTER

Objectives

After completing this chapter, you should be able to:

● Explain why real gases depart from ideal gas behavior at high pressure.

● Describe the difference between solids, liquids, and gases at the molecular level.

● Use the kinetic-molecular theory to explain evaporation and vapor pressure.

● Understand the meaning of the terms *molar heat of vaporization* and *molar heat of melting*.

● Express the concentration of a solution in terms of molarity (*M*) and calculate *M* for various solutions.

The amount of attraction between particles of a substance determines whether it will be a solid or liquid at a given temperature.

So far, we have thought of an ideal gas as a collection of rebounding particles. The particles of such a gas are far enough apart so that the attraction between them can be ignored. However, no gas is ideal. The attraction between gas particles *is* important, and the amount of attraction varies in different gases. It is as a result of this attraction that liquids and solids exist.

Gas particles can come together, or condense, to form various different substances. The amount of attraction between the particles accounts for many of the properties of the substances formed. Such properties include boiling point, freezing point, the energy required to produce a phase change, and mixing behavior with other substances. Solids and liquids are not as "simple" to study as free, unattached gas particles. But at least we can see, feel, measure, and weigh liquids and solids.

6.1 Phase changes

Our model for gas behavior has grown from a qualitative description to one that is quantitative and mathematical. The expression $PV = $ a constant describes the behavior of fixed quantities of gases at the same temperature. A number of separate quantitative relationships are combined in the general expression $PV = nRT$. How useful is this relationship? What happens when gases are subjected to very high pressures and low temperatures? Experiment provides the answers.

GAS BEHAVIOR AT HIGH PRESSURE

The results of compressing 1 mole of ammonia gas (NH_3) at 25 °C are summarized in Table 6-1. Gas volume was measured as pressure was gradually increased. The relationship $PV = 24.4 \pm 0.1$ held true at pressures from 0.1 to 2.0 atmospheres. However, there was a gradual decrease of the "constant" from 24.45 at 0.1 atmosphere to 24.34 at 2.0 atmospheres. What happens when we go beyond 2.0 atmospheres? Is the relationship $PV = 24.4 \pm 0.1$ still true?

The data shown in Table 6-1 indicate that the relationship $PV =$ a constant is not valid at pressures greater than 2.0 atmospheres. The value of the "constant," originally 24.4, decreases noticeably above 2.0 atmospheres until, at a pressure of 9.8 atmospheres, liquid droplets begin to appear in the gas container. The value of the product obtained by multiplying pressure and volume drops rapidly from 23.10. As we try to increase the gas pressure above 9.8 atmospheres by pushing a piston into the container, *we find that more gas is converted to liquid and the pressure returns to 9.8 atmospheres*. We are surprised to learn that *the pressure on the ammonia cannot be raised above 9.8 atmospheres at 25 °C until all the gas has been converted to liquid*. These results are illustrated in Figure 6-1 and in the close-up in Figure 6-2.

Table 6-1

Accurate pressure-volume measurements for 17.00 grams of ammonia gas at 25 °C

Pressure (atm)	Volume (litres)	$P \times V$ (atm $\times$ litres)
0.1000	244.5	24.45
0.2000	122.2	24.44
0.4000	61.02	24.41
0.8000	30.44	24.35
2.000	12.17	24.34
4.000	5.975	23.90
8.000	2.925	23.40
9.800	2.360	23.10 (condensation beginning)
9.800	0.020	0.20 (no gas left, liquid only)
20.00	0.020	0.40 (only liquid present)
50.00	0.020	1.0 (only liquid present)

Figure 6-1

Behavior of ammonia as pressure is increased while temperature remains at 25 °C.

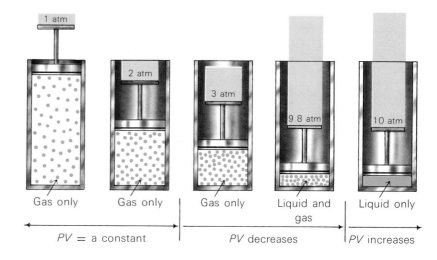

Figure 6-2

Enlarged view of ammonia becoming a liquid at 25 °C.

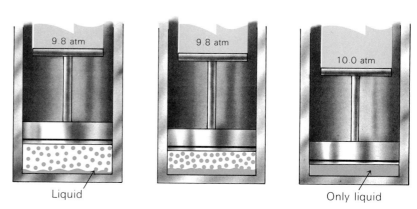

As soon as all the gas has disappeared, pressure on the liquid can be increased without any obvious volume change. In a liquid, the PV product increases as pressure increases but without significant change in volume. It is clear that the simple relationship $PV = $ a constant is not a good description of the liquid phase. Our simple gas law does not apply during the change from gas to liquid.

These observations give us much to think about. For example,

1. Why does the value of the constant 24.4 begin to decrease at about 2 atmospheres of pressure?

2. What is so special about a pressure of 9.8 atmospheres for ammonia vapor at 25 °C?

3. What is the relationship between the liquid and gaseous phases of a substance?

4. Why does the expression $PV = $ a constant not apply to liquids?

5. What change in the gas model is needed to explain the properties of liquids?

Question 5 is the most important. How must the model for gases be altered to account for the nature of liquids? Let us suppose that a liquid is also composed of ping-pong balls that collect in the lower part of a container. The balls have the same average kinetic energy as they did in the gas phase because they are still at the same temperature. Now, however, the balls (or molecules) are in contact with each other. They move back and forth, pushing the other balls (or molecules) about. Some *forces* must be acting among the balls (or molecules) as a group to hold them together.

The existence of forces between molecules is an important consideration. In developing the kinetic theory for gases, we assumed that gas particles were not attracted to each other. Yet, we cannot understand the existence of the liquid phase unless we assume that real molecules *are* attracted to each other. In fact, such intermolecular forces do exist. But they are very small when the molecules are far apart, as in a gas at low pressure. But these forces are significant when the molecules are close together. At intermediate pressures, molecules are attracted to each other just a little. The volume is a little smaller than it would be if there was no attraction between molecules. The PV product drops *below* 24.4 for NH_3 at 25 °C and pressures above 2 atmospheres. This is a reasonable answer to question 1.

In question 2, what is so special about a pressure of 9.8 atmospheres for ammonia vapor at 25 °C? The answer lies in the experiment. In order to raise the pressure of the NH_3 gas above 9.8 atmospheres, the volume of the system was reduced. That is, the molecules were pushed closer together. When the molecules were close together, the forces of attraction between them were great enough to pull them together into liquid droplets. As droplets of liquid formed, molecules were removed from the vapor phase until the pressure fell back to the stable value of 9.8 atmospheres at 25 °C.

This explanation leads us to the answer to question 3: what is the relationship between the liquid and gaseous phases of a substance? In liquids, molecules are close together. Forces between molecules are strong enough to hold them together in a droplet. In gases, molecules are far apart, and forces acting between them are relatively small. A gas has no visible boundary.

For question 4 (why does the expression $PV =$ constant not apply to liquids?), the model again provides an answer. In gases, an increase in pressure pushes molecules together, but there is still a lot of space between them. Thus, $PV =$ constant. In liquids, molecules are touching and cannot be pushed much closer together as pressure is raised. As pressure is raised the volume remains constant, so PV rises. We describe this behavior by saying that liquids are not compressible.

A summary of our discussion answers question 5. The kinetic theory for gases is based on an ideal gas: one in which the molecules exert no force on each other, and in which the actual volume of the molecules is negligible in comparison to the gas volume.* Every gas approaches such ideal behavior at sufficiently low pressures and high temperatures. At very low pressures, molecules are, on the average, so far apart that their forces of attraction and molecular volume can be disregarded. An *ideal gas* obeys the ideal gas law: $PV = nRT$. Attractions between molecules and actual volumes of molecules cause deviations from the ideal gas law and lead to the formation of liquids. Most gases show some deviation from the ideal condition.

*Actually, molecules of ideal gases are assumed to have zero volume.

The table below gives the boiling points and molar volumes (at 0 °C and 1 atm) of some common gases:

Gas	Formula	Boiling point (°C)	Molar volume (litres)
helium	He	−269	22.426
nitrogen	N_2	−196	22.402
carbon monoxide	CO	−190	22.402
oxygen	O_2	−183	22.393
methane	CH_4	−161	22.360
hydrogen chloride	HCl	−84.0	22.248
ammonia	NH_3	−33.3	22.094
chlorine	Cl_2	−34.6	22.063
sulfur dioxide	SO_2	−10.0	21.888

(a) What regularity is suggested by the relationship between boiling points and molar volumes? Plot the values. Set up the graph so that the range 21.800–22.500 occupies a full page of graph paper. (b) Account for this regularity. (*Hint:* what does a high boiling point suggest about intermolecular forces?) (c) What relationship between molar volume and ease of liquefying a gas is indicated by these data?

MELTING POINT

From the foregoing investigation, we see that it is possible to neglect intermolecular forces when we treat gases, but it is *not* possible to do so for liquids. Unfortunately, only a handful of substances are gases under normal conditions of temperature and pressure. Of the 100 or so elements, 11 are gases and 2 are liquids at STP (0 °C and 1 atmosphere).* The rest are solids. Almost 3 million compounds have been prepared. Yet more than 99 percent of those are liquids or solids at STP. Clearly, gases are in the minority.†

Look again at the data for ammonia in Table 6-1. Ammonia is a gas or a liquid, depending on its temperature and pressure. Ammonia is a gas at 25 °C and 1 atmosphere and a liquid at 25 °C and 10 atmospheres. The process by which a gas changes to a liquid or a liquid changes to a solid is called a *phase change*. For example, ice melts to form water and water boils to become water vapor (steam). Solid lead melts to form liquid lead. Gaseous ammonia condenses to form liquid ammonia. Candle wax melts to form liquid wax. These are a few familiar phase changes.

Every pure substance melts at a different temperature. This temperature, known as the melting point, is always the same for a given substance. We

*The gaseous elements at STP are hydrogen (H_2), helium (He), nitrogen (N_2), oxygen (O_2), fluorine (F_2), neon (Ne), chlorine (Cl_2), argon (Ar), krypton (Kr), xenon (Xe), and radon (Rn). The liquid elements at STP are bromine (Br) and mercury (Hg).
†Even so, gases have been very important in the development of chemical theory. See Chapters 3, 4, and 5.

Table 6-2

Melting points of some pure substances

Substance	Melting point (°C)	Molecular weight
oxygen	−218	32
nitrogen	−210	28
mercury	−39	201
oil of winter-green	−8.6	152
bromine	−7.2	160
p-dichloro-benzene	+53	147
trinitro-toluene (TNT)	+80.7	227
lead	+327	207
iron	+1535	56
potassium	+63.6	39
platinum	+1769	195

138

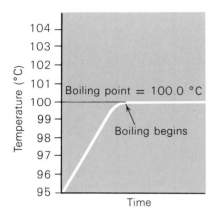

Figure 6-3

Temperature of water above a flame as a function of time (pressure of 1 atmosphere = 760 mm Hg).

stated in Chapter 1 that the melting point is helpful in identifying a substance. For example, if we know that a material melts at −8.6 °C, then it cannot be trinitrotoluene (TNT) because TNT melts at 80.7 °C. However, it could be oil of wintergreen, which is known to melt at −8.6 °C. A wide range of melting points for various substances are listed in Table 6-2. Pressure has a very small effect on melting temperature. For this reason, **melting point** is *precisely* defined as the temperature at which a pure solid changes to a liquid under a given pressure, usually 1 atmosphere.

ENERGY REQUIREMENTS: HEATS OF VAPORIZATION

When a pan of water is warmed over a gas flame, the input of energy causes the water temperature to rise. If the pressure is 1 atmosphere, the water begins to boil when the thermometer reads 100 °C. Bubbles of water vapor form in the liquid. The bubbles rise to the surface and break. Liquid water changes to gaseous water. The temperature stays at 100 °C as long as the liquid water is boiling, as shown in Figure 6-3. See Experiment 11 (Part 2) on page 44 in the Laboratory Manual.

The energy added to the water by the gas flame is used to change liquid water to gaseous water at the same temperature (100 °C). When the gas flame is turned off, boiling stops immediately. In summary, we can say, *when water changes from the liquid phase to the gaseous phase at constant pressure, energy must be supplied, but the boiling temperature remains constant.* Because of the necessary energy input, water molecules in the gas phase must have more energy than water molecules in the liquid phase. The equation representing the change is

$$H_2O(liquid) + \text{thermal energy}^* \longrightarrow H_2O(gas) \text{ at 100 °C and 1 atm}$$

How much energy is involved in the process? Using a device known as a *calorimeter,* (which you may have used in the laboratory), we discover that it takes 9.70 kilocalories (kcal) of energy to vaporize 1 mole of water (6.02 ×

*This term is usually referred to as heat energy.

Table 6-3

Normal boiling points and molar heats of vaporization of some pure substances

Substance	Phase change (*liquid*) $\longrightarrow$ (*gas*)	Boiling point (K)	(°C)	Molar heat of vaporization (kcal/mole)
neon	Ne(*l*) $\longrightarrow$ Ne(*g*)	27.2	−245.8	0.405
chlorine	Cl$_2$(*l*) $\longrightarrow$ Cl$_2$(*g*)	238.9	−34.1	4.88
water	H$_2$O(*l*) $\longrightarrow$ H$_2$O(*g*)	373	100	9.7
sodium	Na(*l*) $\longrightarrow$ Na(*g*)	1162	889	24.1
sodium chloride	NaCl(*l*) $\longrightarrow$ NaCl(*g*)	1738	1465	40.8
copper	Cu(*l*) $\longrightarrow$ Cu(*g*)	2855	2582	72.8

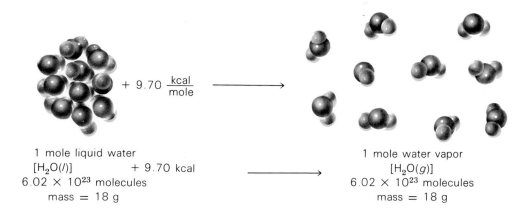

1 mole liquid water
$[H_2O(l)]$ + 9.70 kcal
6.02×10^{23} molecules
mass = 18 g

1 mole water vapor
$[H_2O(g)]$
6.02×10^{23} molecules
mass = 18 g

Figure 6-4
Schematic representation of the vaporization of 1 mole of liquid water at 100 °C to form 1 mole of water vapor at 100 °C.

10^{23} molecules, or 18.0 grams). The equation can be written in somewhat abbreviated, but more precise, form as

$$H_2O(l) + 9.70 \text{ kcal} \longrightarrow H_2O(g) \text{ at } 100\ ^\circ C \text{ and } 1 \text{ atm}$$

The value 9.70 kilocalories is called the **molar heat of vaporization** of water. This is the energy required to separate 6.02×10^{23} molecules of water from each other without changing their temperature, as pictured in Figure 6-4.

EXERCISE 6-2

How much energy is required to evaporate (a) 2 moles of water at 100°C and (b) 0.5 mole of water at 100°C?

When water vapor condenses to form liquid water, energy is conserved. The amount of energy released is the same as the amount used to vaporize the water. Table 6-3 shows the boiling points and heats of vaporization for a variety of liquids. Note the widely varying values. In each case, energy is absorbed by the liquid as the molecules separate and enter the gaseous phase. The same amount of energy is released when gaseous molecules condense into the liquid phase.

EXERCISE 6-3

Is the molecular weight of a substance the most important factor in explaining its melting and boiling points? Explain your answer. (Refer to Tables 6-2 and 6-3.)

LIQUID-VAPOR EQUILIBRIUM: VAPOR PRESSURE

We have been considering vaporization of a liquid at its normal boiling point. But liquids can vaporize at other temperatures. Let us consider this process again, beginning with liquid water.

A 50-millilitre sample of liquid water is placed in the flask shown in Figure 6-5. The liquid is frozen with a very cold refrigerant such as liquid nitrogen (−196 °C). The air in the flask is pumped out with a vacuum pump. When the air has been removed as completely as possible, the glass tube is melted shut at the point indicated in the figure. We can now place the sealed flask in a series of constant temperature baths maintained at 0 °C, 25 °C, 50 °C, and 100 °C. The experimental setup and results are shown in Figure 6-6. Our observations indicate two conclusions:

1. There is pressure above the liquid even after the air has been removed. The mercury rises in the manometer.
2. Pressure above the liquid increases as the temperature is raised.

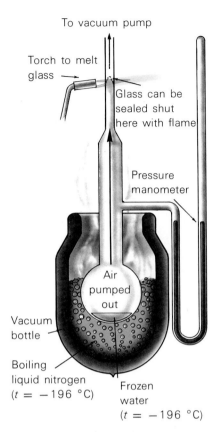

Figure 6-5
Preparing a flask to study the vapor pressure of water.

Figure 6-6
Vapor pressure of water at four different temperatures.

The pressure reading in the flask will change when the flask is first put into the bath, but experiments show that, ultimately, *many different sealed flasks of pure water will have the same reading at the same temperature*. When the pressure reading does not change with time, we say that the liquid and its vapor are in **equilibrium.** The pressure reading at that point is the *equilibrium pressure*. The equilibrium pressure reading in the flask gives the **equilibrium vapor pressure** of water at the temperature of the bath. In a more formal sense, we can say that *the pressure exerted by a vapor in equilibrium with its pure liquid phase is the vapor pressure of the liquid at the temperature used*. At equilibrium, no *net* evaporation or condensation is taking place. We are unable to detect measurable changes.*

From the experiment just described, we can tabulate the vapor pressure of water at four different temperatures, as listed in Table 6-4. The experiment can be repeated using many liquids other than water. In this way we learn another fact: *different liquids at the same temperature have different vapor pressures*. For example, if liquid benzene is used in the experiment instead of water, the vapor pressure values at 0 °C, 25 °C, 50 °C, and 100 °C are those listed in Table 6-5. A repetition of this type of experiment for many liquids indicates another generalization: *the vapor pressure of every liquid increases as the temperature is raised*. The vapor pressure of a liquid at its *normal* boiling point equals atmospheric pressure. Vapor bubbles can now form everywhere within the liquid.

Our knowledge of vapor pressure can now be summarized:

1. The pressure exerted by a vapor in equilibrium with its liquid phase at constant temperature is the vapor pressure of the liquid at that temperature.
2. The vapor pressure of a liquid is dependent only upon the nature of the liquid and the temperature.
3. Different liquids at any one temperature have different vapor pressures.
4. The vapor pressure of every liquid increases as the temperature is raised.

*At equilibrium, the rate at which molecules leave the surface is equal to the rate at which they return to the surface. No *net* change is seen, but there is lots of activity!

Table 6-4

Vapor pressure of water

Temperature (°C)	Vapor pressure (mm Hg)
0	4.6
25	23.8
50	92.5
80.1	356.5
100*	760.0

*The *normal* boiling point of water is 100 °C (373 K).

Table 6-5

Vapor pressure of benzene

Temperature (°C)	Vapor pressure (mm Hg)
0	27
25	94
50	271
80.1*	760.0
100	1,360

*The *normal* boiling point of benzene is 80.1 °C (353.1 K).

EXERCISE 6-4

Observe water as it boils. What is inside the bubbles?

EXERCISE 6-5

Compare the boiling points of water and benzene. Give the vapor pressure of each at 50 °C. Is water or benzene closer to its boiling point at 50 °C? Explain. See Tables 6-4 and 6-5.

THE KINETIC-MOLECULAR THEORY AND VAPOR PRESSURE

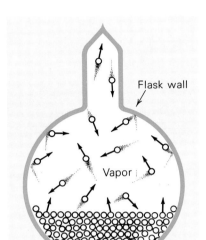

Does the *kinetic-molecular theory* of liquids and gases explain vapor pressure? Let us put it to the test. We suggested earlier that molecules at high pressure are pushed together. Molecules attract each other more readily when they are close together. If the pressure is high enough, the molecules come together to form droplets, or aggregates, that we call liquids.

The molecules in liquids are in motion. They have kinetic energy. Some of the molecules move more rapidly than others. Occasionally, energetic molecules at the surface have enough kinetic energy to break away from the liquid surface, as shown in Figure 6-7. Such molecules overcome the forces of attraction between themselves and other molecules in the liquid. The molecules that break away behave like a gas. They collide with container walls and thus exert pressure. According to our model, we must apply a large amount of energy to liquids with strong intermolecular forces before any molecules can escape from the liquid to the gas phase. Thus, liquids with strong intermolecular forces should have a lower vapor pressure at a given temperature than liquids in which there are weak intermolecular forces. Similarly, different liquids at the same temperature should have different vapor pressures. Look at Table 6-6. Our model agrees with the observations recorded in the table.

Figure 6-7
Kinetic theory view of liquid-vapor equilibrium.

EXERCISE 6-6

Which liquid has stronger forces of attraction between molecules, water or benzene?

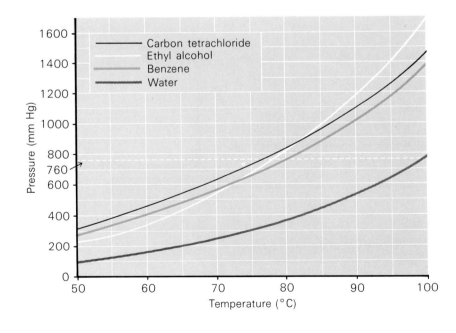

Figure 6-8
Vapor pressure versus temperature for some common liquids.

What effect does temperature have on vapor pressure? An increase in the temperature of a liquid or gas brings about an increase in the average kinetic energy of its molecules. As average kinetic energy increases, a larger number of molecules obtain enough kinetic energy to overcome the intermolecular forces in the liquid and escape into the vapor. More molecules escape into the vapor phase at higher temperatures. Thus, the increases in vapor pressure with temperature, as shown in the data in Figure 6-8, could be predicted on the basis of the kinetic theory.

Table 6-6

Vapor pressures of liquids

Temperature (°C)	Water (mm Hg)	Ethyl alcohol (mm Hg)	Carbon tetrachloride (mm Hg)	Methyl salicylate (mm Hg)	Benzene (mm Hg)	Methyl alcohol (mm Hg)
−10	2.1	5.6	19		15	16
−5	3.2	8.3	25		20	22
0	4.6	12.2	33		27	29
5	6.5	17.3	43		35	40
10	9.2	23.6	56		45	54
15	12.8	32.2	71		58	71
20	17.5	43.9	91		74	93
25	23.8	59.0	114		94	108
30	31.8	78.8	143		118	157
35	42.2	103.7	176		147	201
40	55.3	135.3	216		182	255
45	71.9	174.0	263		225	321
50	92.5	222.2	317		271	402
55	118.0	280.6	379		325	500
60	149.4	352.7	451	1.41	389	619
65	190.0	448.8	531	1.90	462	760
70	233.7	542.5	622	2.52	547	925
75	289.1	666.1	720	3.40	643	1,161
80	355.1	812.6	843	4.41	753	1,352
85	433.6	986.7	968	5.90	877	1,623
90	525.8	1187	1122	7.63	1020	1,938
95	633.9	1420	1270	9.93	1180	2,303
100	760.0	1693.3	1463	12.8	1360	2,724

THE KINETIC-MOLECULAR THEORY AND HEAT OF VAPORIZATION

Your wet skin feels cool after a shower and remains cool until you dry off. Why? Evaporating water has a "cooling effect." How does the kinetic theory explain this phenomenon?

According to the kinetic-molecular theory, those molecules on the liquid surface with the highest kinetic energy have the greatest *probability* of overcoming the forces of attraction between molecules and breaking away. In short, the more energetic, more rapidly moving molecules will escape. The less energetic, more slowly moving molecules will remain. The remaining molecules have a lower average kinetic energy than did molecules of the original liquid. Since temperature measures the average kinetic energy of molecules, the temperature of the remaining liquid is lower. In a simpler but less accurate way, we can say that the faster-moving, "hot" molecules go to the vapor, and the slower-moving, "cold" molecules remain in the liquid. The liquid cools. If we want the temperature to remain constant, external energy—the *heat of vaporization*—must be added.

BOILING POINT

It is well known that water boils at 100 °C and 1 atmosphere pressure. Bubbles form in the liquid. Experiments show that when liquid water boils, the vapor pressure of the liquid is equal to the pressure of the atmosphere. Why is the pressure of the atmosphere important to boiling? The kinetic theory provides the answer. Consider the bubble just below the liquid surface in Figure 6-9. Vapor molecules from vaporization of the liquid are present inside the bubble. These vapor molecules bombard the bubble walls

Figure 6-9

Bubble in a boiling liquid.

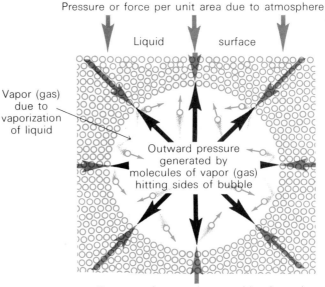

Pressure or force per unit area due to atmosphere

Liquid   surface

Vapor (gas) due to vaporization of liquid

Outward pressure generated by molecules of vapor (gas) hitting sides of bubble

Pressure of atmosphere pushing inward

and tend to push them outward. At the same time, the atmosphere pushes downward on the surface of the liquid. The push of the atmosphere is transmitted equally in all directions throughout the liquid and tends to collapse the bubble. If the atmospheric pressure is greater than the vapor pressure of the liquid, the bubble collapses. If the pressure inside the bubble (due to vapor pressure of liquid at that temperature) is very slightly higher than atmospheric pressure, the bubble grows and rises to the surface, where it breaks. Bubble formation is characteristic of the boiling process.

The **boiling point** of a liquid is the temperature at which the vapor pressure of the liquid is equal to the pressure of the atmosphere above the liquid. If atmospheric pressure is 760 mm Hg, water boils at 100 °C. If the atmospheric pressure is 750 mm Hg, water boils at 99.6 °C because the vapor pressure of water is 750 mm Hg at 99.6 °C. The *normal* boiling point of a liquid is defined as the temperature at which the vapor pressure of the liquid is exactly 1 standard atmosphere, or 760 mm Hg.

EXERCISE 6-7

Suppose a closed flask containing liquid water is connected to a vacuum pump and the pressure over the liquid is gradually lowered. If the water temperature is kept at 30 °C, at what pressure will the water boil?

EXERCISE 6-8

Answer Exercise 6-7, substituting ethyl alcohol for water. Repeat for carbon tetrachloride at 50 °C.

EXERCISE 6-9

A thermometer in a pot of boiling water on a mountain reads 95 °C. What is the atmospheric pressure on the mountain? The rule of thumb for altitude-pressure correlations is that the pressure falls about 25 mm for every 305 m (1,000 ft) of elevation. Estimate the height of this mountain.

SOLID-LIQUID PHASE CHANGES

According to the model just developed, molecules in solid and liquid phases—the *condensed phases*—are held together by the forces of attraction between them. In liquids, the molecules are irregularly spaced, randomly

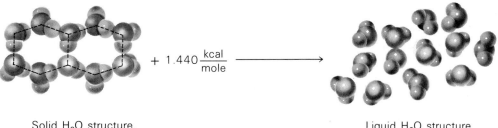

+ 1.440 $\frac{kcal}{mole}$ ⟶

Solid H_2O structure,
strongly ordered

Liquid H_2O structure,
strongly disordered

Figure 6-10
Schematic representation of the melting of 1 mole of ice to liquid water at
0 °C.

oriented, and reasonably free to move over each other. In crystalline solids, the molecules occupy regular positions and are held together more firmly, as represented in Figure 6-10. One mole of a pure solid substance has lower energy than 1 mole of the same substance in the liquid phase at the same temperature.

The difference between the energy of a substance in liquid form and in solid form is usually much smaller than the difference between the energy of the substance in liquid form and in gaseous form. For ice, experiments show that

$$H_2O(solid) + 1.440 \text{ kcal} \longrightarrow H_2O(liquid) \text{ at } 0 \text{ °C}$$

or, 1.440 kilocalories of energy would be required to convert 1 mole of ice (18.0 g) to 1 mole of liquid water at 0 °C. This is much less than the 9.7 kilocalories required to convert 1 mole of water to vapor at 100 °C. The energy required to melt 1 mole is known as the **molar heat of melting,** or the **molar heat of fusion.** Table 6-7 compares melting points of various substances with their molar heats of melting.

Table 6-7
Melting points and heats of melting of some pure substances

Substance	Phase change (*solid*) ⟶ (*liquid*)	Melting point (K)	(°C)	Molar heat of melting (kcal/mole)
neon	$Ne(s) \longrightarrow Ne(l)$	24.6	−248.4	0.080
chlorine	$Cl_2(s) \longrightarrow Cl_2(l)$	172	−101	1.53
water	$H_2O(s) \longrightarrow H_2O(l)$	273	0	1.44
sodium	$Na(s) \longrightarrow Na(l)$	371	98	0.63
sodium chloride	$NaCl(s) \longrightarrow NaCl(l)$	1081	808	6.8
copper	$Cu(s) \longrightarrow Cu(l)$	1356	1083	3.11

Heats of melting for different substances vary greatly. Molar heats of melting may range from 0.080 kilocalorie per mole for neon to 6.8 kilocalories per mole for sodium chloride. The value for sodium chloride is 85 times that for neon. As you may suspect, there are great differences in the forces that bind solid neon and solid NaCl together. These forces influence many properties other than melting point and heat of melting. We shall consider such properties later in more detail.

6.2 Solutions

Sodium chloride, ethyl alcohol, and water are three **pure substances.** Each has definite properties such as vapor pressure, melting point, molar heat of melting, boiling point at a given pressure, and density at a given temperature. Suppose we mix some of these pure substances. Sodium chloride dissolves when placed in contact with water. The solid disappears, becoming part of the liquid. Similarly, ethyl alcohol and water mix to form a liquid similar in appearance to the original liquids.

The two mixtures described above are *homogeneous,* or alike throughout. Each mixture has only a single liquid phase. Such homogeneous mixtures are called **solutions.**

There are many solutions all around us. The air we breathe is a solution of oxygen dissolved in nitrogen. There are also other gases in the clear air mixture. Soft drinks are solutions of carbon dioxide, sugar, and flavoring dissolved in water. Perfume is a solution of a fragrance dissolved in a carrier liquid. Food coloring is a solution of an organic dye in a mixture (solution) of water and some other liquids. Hair tonic is a solution as is shaving lotion.

Solutions differ from pure substances. The properties of solutions depend on the relative amounts of the pure substances in the mixture. A pure substance has its own characteristic melting point and boiling point. The boiling point and melting point of a solution depend on its composition. For example, suppose we compare salt water and distilled water. Both are homogeneous. How can we tell which sample is a pure substance and which is a mixture? More sophisticated observation is needed. Let us look further.

SOLUTIONS AND PHASE CHANGES

A sample of water freezes at 0 °C (at 760 mm Hg). If we freeze 10 grams of a 100-gram sample of water, a 10-gram ice cube and 90 grams of pure water will remain. If we continue this process, separating ice and water each time, ten ice cubes will result—all of the same size and composition, all pure water. In a similar way, we can boil the original sample of water at 100 °C until 10 millilitres have vaporized and condensed, using the apparatus shown in Figure 6-11. A 90-millilitre sample of hot water and a 10-millilitre sample of freshly distilled water result. If the boiling process is continued, nine more 10-millilitre quantities of freshly distilled water will be obtained. The properties of the different samples of a pure substance, separated through freezing or boiling, remain unchanged. A pure substance has a definite and reproducible melting point and boiling point at a given pressure. In fact, chemists use both melting and boiling points of pure substances to identify compounds.

Solution behavior on boiling or freezing depends upon composition. Suppose we start with a 100-millilitre sample of a salt solution. The solution is made by dissolving a convenient amount, such as 5 grams, of sodium chloride in water. The resulting solution is not very different from pure water, but the difference is important. We find that the solution begins to boil close to, but not exactly at, 100 °C. If we collect the steam from the boiling salt solution and condense it in a separate container, we find that the resulting liquid behaves like pure water rather than like the solution from which it came. The apparatus for carrying out this experiment is shown in Figure 6-11. Furthermore, after a 10-millilitre sample of water has been vaporized, we are left with 90 millilitres of a solution whose composition is not the same as it was originally. It is saltier than the original solution. That is because the same amount of salt that was present initially is now present in a solution containing less water. The first solution had 5 grams of salt dissolved in 100 millilitres of solution. The second solution has 5 grams of salt dissolved in only 90 millilitres of solution.

The difference between solutions noted above results in different properties for the two solutions. Figure 6-12 contrasts the behavior of pure water and a concentrated saltwater solution during boiling. The salt solution starts to boil at a temperature just above the boiling point of pure water. But as the amount of water per gram of salt decreases, the boiling point of the salt solution increases. As the boiling continues, the temperature of the pure water remains constant while the temperature of the salt solution keeps rising. If we boil off all the water in the solution, solid salt remains.

Thus, by the process of **distillation**—that is, by *evaporating* and *condensing the resulting vapor in a separate vessel*—we can frequently separate a pure *liquid* from a solution. By the process of **crystallization**—that is, by *forming a crystalline solid*—we can frequently obtain a pure *solid* from a solution. We usually call the pure substances in a solution, (such as salt and water in the above example) the *components* of the solution. Neither crystallization nor distillation separates a pure substance into components. *A pure substance consists of only one component.* The more closely alike the components of a solution, the harder it is to separate them.

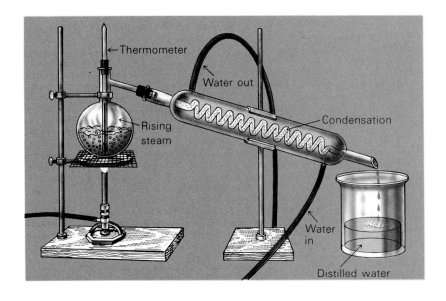

Figure 6-11
A distillation apparatus.

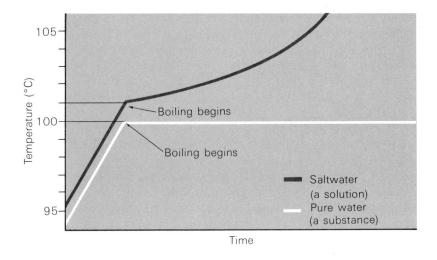

Figure 6-12
Behavior on boiling.

In nature, solutions are much more common than pure substances. To obtain pure substances, we often must prepare them from solutions through successive phase changes. *Heterogeneous systems,* in turn, are much more common than homogeneous systems such as solutions. Heterogeneous systems, such as muddy water, contain more than one phase.

Although we are primarily interested in liquid solutions, gaseous and solid solutions are also important. We shall consider them briefly before renewing our study of liquid solutions. From a chemist's point of view, liquid solutions are the most important of the three types.

GASEOUS SOLUTIONS

All gas mixtures are homogeneous. Thus, all gas mixtures are solutions. Air is an example of such a mixture, if dust is removed. There is only one phase—the gas phase—and all the molecules behave as gas molecules. Air is a single, homogeneous phase. As with other solutions, the components of air can be separated by phase changes. Through a special, low-temperature distillation process, liquid air can be separated into liquid oxygen (boiling point −182 °C), liquid nitrogen (boiling point −196 °C), and other components.

SOLID SOLUTIONS

Solid solutions are not often considered in an elementary study of chemistry, but they are very common in nature. Many common minerals and gem stones, such as emeralds, sapphires, amethysts, and rubies, are solid solutions. Rubies are solid solutions of Cr_2O_3 in Al_2O_3. The more Cr_2O_3 in the solution, the deeper the color. Steel is a solid solution with many different solid phases. Similarly, zinc atoms can replace some of the copper atoms in a copper crystal to give the zinc-copper alloy called brass. Brass is a solid solution. Copper atoms can also replace tin atoms in a tin crystal to give another solid solution, the tin-copper alloy known as bronze. Solid solutions are extremely important today. They are used in transistors for solid-state radios, amplifiers, record players, and television units.

LIQUID SOLUTIONS

In your laboratory work, you will deal mostly with liquid solutions. Liquid solutions can be made by mixing two liquids (such as alcohol and water). They can also be made by dissolving a gas in a liquid (such as carbon dioxide dissolved in water to give soda water) or by dissolving a solid in a liquid (such as sugar dissolved in water). The result is a homogeneous, transparent system that contains more than one substance. By definition, the result is a solution.

In a liquid solution, each component is diluted by the other. The effects of this dilution process are usually fairly easy to see. Consider the vapor pressure of a salt solution. Since the solution contains salt, which does not vapor-

OPPORTUNITIES

Chemical Sales

The chemical salesperson provides the link between the chemist working in the laboratory and the consumer in the field. Frequently, a chemical salesperson may deal with companies that need solvents or substances that will undergo phase changes at specific temperatures. Such companies include paint manufacturers, textile producers, and dye makers. A cosmetics house may require thickeners or fragrances. Some food processing plants purchase flavors and preservatives. Sometimes, a material has a well-recognized use. But often, a creative salesperson will find new ways in which a specific product can be used.

People who work in pharmaceutical sales positions spend the major part of their working day dealing directly with physicians, dentists, and hospitals, as well as with retail and wholesale buyers. They call on customers, inform them of new products, and provide them with samples. They may also assist customers in determining dosage requirements and possible aftereffects of the many products they sell. Pharmaceutical salespeople should have extensive knowledge of medical practice with respect to drugs and medicines.

Salespeople spend a great deal of time traveling. It is important for successful sales representatives to meet customers directly in order to present their company's products in the most favorable and effective way. Salespeople must have the ability to work well with others in a setting where evaluation and decision making are major responsibilities. Salespeople often inform chemists in industry of consumer needs for new products. Many products that are currently available were developed as a result of information furnished to a company by a sales representative.

In addition to a bachelor's degree in chemistry or chemical engineering, courses in marketing and business methods are helpful to the prospective chemical salesperson.

Many sales representatives also have master's degrees in business administration. For further information about this career, contact the Chemical Manufacturers Association, 2501 M Street NW, Washington, D.C. 20037.

ize, and water, which does, there are fewer vaporizable water molecules per unit of volume than there are in pure water. Furthermore, water molecules that are present in solution are likely to be restricted in movement by their attraction for the dissolved salt. The net effect is that the "escaping tendency" of the water in the solution is *lower* than the "escaping tendency" of the pure water. As a result the *vapor pressure* due to gaseous water above the solution is lower than the vapor pressure of pure water at the same temperature. The dissolved particles lower the vapor pressure of the liquid. We know that pure water has a vapor pressure of 760 mm Hg at 100 °C. It is necessary to heat a salt solution *above* 100 °C to reach the same vapor pressure. For this reason, the boiling point of salt water at a given pressure is higher than the boiling point of pure water at the same pressure. The amount by which the boiling point is raised depends on the relative amounts of water and salt present in the solution. The more salt in a given volume of solution, the less pure water and the higher the boiling point.

Correspondingly, a lower temperature is required to crystallize ice from salt water than is required to crystallize ice from pure water. A lower temperature is also needed to crystallize ice from an alcohol-water solution. "Antifreeze" substances added to an automobile radiator act on this principle. They dilute the water in the radiator and lower the temperature at which ice can crystallize from the solution. Again, the amount by which the freezing point is lowered depends on the relative amounts of water and antifreeze in the solution.

THE COMPOSITION OF SOLUTIONS

In general, the properties of a solution depend on the relative amounts of its components. Thus, it is important to specify **solution composition:** the kinds and relative amounts of the components, or pure substances, that were mixed to form the solution. If there are two components, one is called the **solvent** and the other is called the **solute.** *These are merely terms of convenience.* Since both components must intermingle to form the final solution, we cannot make any important distinction between them. When a liquid solution is made from a pure liquid and a solid, it is usually convenient to call the liquid component the solvent.

When we specify the relative amounts of each component present in a solution, we have specified the **concentration** of the solution. Chemists have many different ways of expressing concentration. The choice of method depends on the way in which the solution will be used. We shall need only one way to express solution concentration in this course. That method is described below.

In laboratory work, it is convenient to measure out a given quantity of a reagent by using a *measured volume of a solution of known concentration.* Suppose we weigh out carefully 1 mole of table salt. (Using the formula, this is 23.0 grams of sodium plus 35.5 grams of chloride, or 58.5 grams of sodium chloride per mole.) The salt is dissolved in about 500 millilitres of water. Then the resulting solution is transferred completely to the 1,000-millilitre volumetric flask shown in Figure 6-13. Water is added and the flask is shaken. More water is added until the volume of the solution is 1 litre (at the mark on the flask) at 25 °C. The solution is now shaken to be sure that liquid and solid are thoroughly mixed. A 1-molar solution results. This solution is

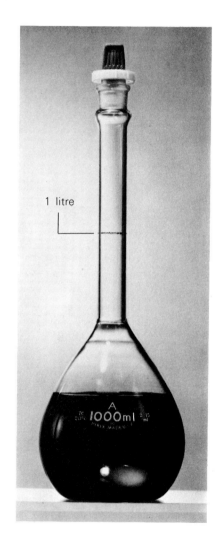

1 litre

Figure 6-13
Volumetric flask.

prepared by dissolving 1 mole of a solute in enough solvent to make a litre of solution.*

If we now pour a one-half-litre quantity (500 millilitres) of this solution into a beaker, the beaker will contain one half mole of salt. If we measure out 0.1 litre of solution into a beaker, the beaker will contain 0.1 mole of salt. In short, the number of moles of reagent is given by the expression

$$\left\{\begin{array}{c}\text{number of moles}\\ \text{of solute}\end{array}\right\} = \left\{\begin{array}{c}\text{concentration}\\ \text{of solution in}\\ \text{moles per litre}\end{array}\right\} \times \left\{\begin{array}{c}\text{volume of}\\ \text{solution}\\ \text{in litres}\end{array}\right\}$$

or

$$\begin{array}{c}\text{number of moles}\\ \text{of solute}\end{array} = \frac{1.000 \text{ mole}}{\text{litre}} \times 0.500 \text{ litre} = 0.500 \text{ mole}$$

It is frequently convenient to indicate the number of moles per litre by using a number followed by a capital M. The capital M refers to the term **molarity.** The molarity of a solution is the number of moles of solute per litre of that solution. Thus, a solution containing 2 moles per litre is designated as 2 M. Several examples of problems involving molarity will be helpful here.

Example 1: How many moles of potassium bromide (KBr) would you need to prepare 2.00 litres of 1.50 M potassium bromide solution? How many grams of KBr would be needed?

Answer: First, write the molarity relationship as

$$M = \frac{\text{moles of solute}}{\text{volume (litres) of solution}}$$

Then, substitute the given information and solve for the unknown term:

$$1.50 \ M = \frac{\text{moles of solute (KBr)}}{2.00 \text{ litres of solution}}$$

$$\text{moles of solute} = \frac{1.50 \text{ moles}}{1.00 \text{ litres of solution}} \times 2.00 \text{ litres of solution}$$

$$= 3.00 \text{ moles of solute}$$

By definition, molarity (M) is expressed in moles per litre.

---

* A solution containing 1.000 mole of solute in a litre of solution has long been called a 1-molar solution by chemists. By international agreement, concentration is now defined as moles per cubic metre, which is more useful when very precise measurements are being made. However, concentration in moles per litre is a very useful means of expressing concentration for chemists and will be used in this text.

Now, change moles of solute to grams:

$$\text{mol wt KBr} = 39.1 \text{ g} + 79.9 \text{ g} = 119.0 \text{ g KBr/mole of KBr}$$

$$3.00 \text{ moles KBr} \times \frac{119.0 \text{ g of solute (KBr)}}{1 \text{ mole of KBr}}$$

$$3.00 \text{ moles of solute} = 357 \text{ g of solute}$$

Example 2: How many litres of a 1.50 *M* solution of KBr can be prepared from 59.5 g of KBr?

Answer: Write the molarity relationship, as in the previous example. Then determine the moles of KBr:

$$59.5 \text{ g of KBr} \times \frac{1 \text{ mole of KBr}}{119.0 \text{ g of KBr}}$$

$$= 0.500 \text{ mole of KBr}$$

Substitute the appropriate figures in the molarity relationship and solve for the unknown term

$$1.50 \ M = \frac{0.500 \text{ mole of KBr}}{\text{volume (litres) of solution}}$$

$$\text{volume (litres) of solution} = \frac{0.500 \text{ mole of KBr}}{1.50 \ \frac{\text{moles}}{\text{litre}} \text{ of solution}}$$

$$= 0.500 \text{ mole of KBr} \times \frac{\text{litres of solution}}{1.50 \text{ moles of KBr}}$$

$$\text{volume} = 0.333 \text{ litres of solution}$$

Example 3: What is the molarity of a solution prepared by dissolving 29.75 g of KBr in enough water to prepare 1.00 litre of the solution?

Answer: Write the molarity relationship as in the previous examples. Then change grams of KBr to moles:

$$29.75 \text{ g of KBr} \times \frac{1 \text{ mole of KBr}}{119.0 \text{ g of KBr}} = 0.250 \text{ mole KBr}$$

Substitute the appropriate figures in the molarity relationship and solve for the unknown term:

$$M = \frac{0.250 \text{ mole of KBr}}{1.00 \text{ litre of solution}} = 0.250 \ M$$

EXERCISE 6-10

What is the molarity of a $NaNO_3$ solution that contains 8.50 g of $NaNO_3$ in enough water to make 1.00 litre of solution?

EXERCISE 6-11

How many grams of $NaNO_3$ are required to prepare 50.0 ml of 0.100 M solution?

EXERCISE 6-12

What volume of 0.100 M $NaNO_3$ solution can be prepared from 170.0 g of $NaNO_3$?

SOLUBILITY

When sugar is added to water in a beaker, the sugar dissolves. As more sugar is added, the concentration increases until the sugar begins to accumulate on the bottom of the beaker. As long as the temperature remains constant, *the concentration of sugar in the solution will not change*. This solution, containing all the sugar that it can hold at equilibrium at a given temperature, is called a **saturated solution.** A solution in equilibrium at a given temperature with excess solid is said to be saturated. The solution and solid are in equilibrium, and the solution is saturated, if a small crystal added to the solution neither increases nor decreases in volume. The rate at which the solid dissolves is equal to the rate at which the solid crystallizes out. The solid crystals can change in shape, but not in volume.

When a fixed amount of liquid has dissolved all the solid it can hold at equilibrium at a given temperature, the concentration reached is called the **solubility** of that solid in the liquid. For example, a saturated solution of sodium chloride in water at 20 °C has a concentration of about 6 moles of sodium chloride per litre of water solution (6 M). In comparison, a saturated solution of sodium chloride in ethyl alcohol at 20 °C has a concentration of only 0.009 mole sodium chloride per litre of alcohol solution (0.009 M).

The solubilities of solids in any solvent change with temperature. An increase in temperature sometimes increases solubility and sometimes reduces it. Because the conditions of solubility vary widely, the word *soluble* does not have a precise meaning. There is frequently an upper limit to the solubility of even the most soluble solid in a given solvent. Further, even the least soluble solid yields a few dissolved particles per litre of solution. If a compound has a solubility of more than 10^{-1} M, most chemists usually say it is *soluble*. Compounds with solubility below about 10^{-4} M are usually described as *insoluble* or of very *low solubility*. This property has many practical consequences. For example, we use glass containers to hold pure water because glass has negligible solubility in water.

DIFFERENCES AMONG SOLUTIONS

Though many solutions are colorless and closely resemble pure water in appearance, the differences among solutions are great. The ability of apparently similar solvents to dissolve different solids also varies widely. For example, pure sugar dissolves in both pure water and pure ethyl alcohol. Sodium chloride, or table salt, dissolves in water but does not dissolve readily in ethyl alcohol. Sulfur does not dissolve in water to any visible degree, but it dissolves readily in carbon disulfide. Thus, solubility is determined by the nature of both solute and solvent.

The examples described above yield four solutions containing a substantial amount of solute: sugar in water, sugar in ethyl alcohol, sodium chloride in water, and iodine in ethyl alcohol. The fourth solution is readily distinguishable by its dark brown color. The other three are colorless. Chemists, however, have a good way of distinguishing them. These four solutions differ greatly in their ability to conduct an electric current. Sodium chloride in water conducts an electric current much more readily than pure water does. In contrast, solutions of sugar in water and sugar in ethyl alcohol do not conduct an electric current any better than pure water.

Differences in solubility and electrical conductivity are very important in chemistry. We shall investigate electrical conductivity and the electrical nature of matter in the next chapter.

SUMMARY

Gases depart from ideal behavior—in which $P \times V = K$—at low temperature and high pressure. Under such conditions, gas molecules are pushed close enough together for their forces of attraction and molecular volume to become important. At conditions of low temperature and high pressure, gases can form liquids and solids.

Matter can exist in three states: solid, liquid, or gaseous. The change from one state to another is called a phase change. Pure substances melt and vaporize at characteristic temperatures, known respectively as melting point and boiling point.

The energy required to vaporize 1 mole of a pure substance is called the molar heat of vaporization. This amount of energy is released when 1 mole of the substance condenses to form a liquid. Vapor pressure is the pressure generated by the vapor above a pure liquid. All liquids exhibit a vapor pressure that depends only on the nature of the liquid and its temperature. A liquid boils when its vapor pressure becomes equal to atmospheric pressure.

Liquids and solids—the condensed phases of matter—exist because molecules attract each other when they are close together. Molecules can move past each other in the liquid state. In solids, however, molecules occupy

Key Terms

boiling point

concentration

crystallization

distillation

equilibrium

equilibrium vapor pressure

melting point

molar heat of melting
  or fusion

molar heat of vaporization

molarity

pure substance

saturated solution

solubility

solute

solution

solution composition

solvent

regular positions and are held in place more firmly. Liquids with strong forces of attraction between molecules require the addition of more energy before their molecules can escape from the liquid to the gas phase. A liquid cools with evaporation because the average kinetic energy of the remaining molecules is lowered as the more energetic molecules leave. A pure solid requires energy to become a liquid. The amount of energy required to convert 1 mole of solid to 1 mole of liquid is called the molar heat of melting or the *molar heat of fusion*. The same amount of energy is released when 1 mole of the liquid refreezes to form the solid.

Solutions are homogeneous mixtures: they are identical in composition throughout. A solution consists of a solute material dissolved in a solvent. The properties of a solution are usually intermediate between the properties of its components. Solutions of solids, liquids, and gases are all known to exist.

Solid solutes are often dissolved in water. The concentration of such solutions is frequently given in terms of molarity (*M*). A 1-molar solution contains 1 mole of solute per litre of solution.

QUESTIONS AND PROBLEMS

A

1. What happens to the value of $P \times V$ for ammonia at room temperature and a pressure of (a) 0.1 to 2.0 atm and (b) 2.0 to 9.8 atm?

2. Why is 9.8 atm pressure and 25 °C special for the ammonia gas in problem 1? What happens to the ammonia sample above 9.8 atm at 25 °C?

3. What is meant by the term *phase change?* Give several examples. What are the three states of matter?

4. What is meant by ideal gas behavior? At what conditions do gases approach ideal behavior? Under what conditions do gases begin to liquefy? Compare and contrast your answers.

5. (a) What properties of *molecules* were neglected when we considered ideal gases, but which can *not* be neglected when we consider liquids and solids? (b) Which gas listed in Exercise 6-1 most closely resembles an ideal gas? Explain.

6. Define the term *molar heat of vaporization*.

7. When a 3.00 g sample of liquid benzene (C_6H_6) vaporizes at its boiling point, 312 calories of energy are absorbed. (a) How many moles of benzene are vaporized? (b) What is the *molar heat of vaporization* of benzene?

8. Is the heat of vaporization for a liquid stored in the vapor as kinetic or potential energy? Explain.

9. In Exercise 6-3, it was seen that the boiling point is not directly related to the molecular weight of the substance. What property is the boiling point related to?

10. In terms of the kinetic-molecular theory and the answer to question 8, why does steam give off heat when it condenses to water?

11. In view of the answer to question 10, why should steam at 100 °C give a more severe burn than water at 100 °C?

12. How does the rate of evaporation compare with the rate of condensation

of molecules in a system at equilibrium? Do we see a *net* or observable change in a system at equilibrium? Explain.

13. (a) What two factors determine the equilibrium vapor pressure above a liquid? (b) What generalization between temperature and vapor pressure is true for all liquids?

14. Diamond has very strong bonds between carbon atoms. (a) Would you expect diamond to have a high or a low vapor pressure? Explain. (b) Should diamond have a high or low melting point? (Hint: have you ever seen melted carbon?)

15. How does the kinetic theory explain the cooling effect of evaporation?

16. (a) How is boiling point defined in terms of experimental data? (b) How do you recognize boiling water? (c) How is boiling water defined, in terms of vapor pressure? (d) Give the *normal* boiling point of water.

17. How will a solid and a liquid differ at the molecular level?

18. (a) Which is larger, the molar heat of vaporization or the molar heat of fusion of a pure substance? (b) Does the kinetic theory provide an explanation for your answer? (Hint: what changes in molecular positions are involved in each case?)

19. A solution is a homogeneous mixture. What is meant by *homogeneous?*

20. Suggest two kinds of physical processes that might be useful in separating a solution into its components?

21. Give several examples of solid solutions and gaseous solutions.

22. Name the solute and the solvent in a solution that is made by dissolving salt in water.

23. A pure substance is characterized by a fixed and reproducible melting point and boiling point. Is this also true for a "pure solution," such as sugar in water? Explain the boiling process in terms of the kinetic-molecular theory.

24. Describe how you would prepare a 1.00 *molar* solution of sugar (sucrose) in water? The molecular weight of sucrose is 342.

25. How could you use your solution from problem 24 to measure out 2 one-quarter mole samples of sucrose?

26. How many grams of K_2CrO_4 should be dissolved in enough water to make one litre of a 0.250 molar solution?

27. Suppose you take 50.0 ml of the solution used in problem 26. How many moles of K_2CrO_4 are contained in the 50.0 ml sample?

28. How many litres of 1.5 M solution can be prepared from 388.4 g of K_2CrO_4?

29. A solution of sugar in water stands above crystals of sugar for days and is shaken frequently. Finally it is warmed, shaken, then allowed to cool back to room temperature. Crystals come out during cooling. Is the solution saturated or unsaturated at room temperature?

30. Why is the term *soluble* ambiguous?

B

31. (a) Plot molar heat of melting against melting point for the substances listed in Table 6-7. (b) Do the data indicate a direct relation between melting point and heat of melting? Explain.

32. Diethyl ether boils at 35 °C. The temperature of the human body is 37.0 °C. Why does the skin feel cold when a patch of skin is cleaned by rubbing it with a piece of cotton soaked in diethyl ether?

33. Dibutyl ether boils at 142 °C at 1 atmosphere. (a) Which has the higher vapor pressure at 25 °C, dibutyl ether or diethyl ether? Explain. See problem 32. (b) Which one will evaporate faster if it is spilled?

34. Gasoline used in the winter has a higher vapor pressure at 25 °C than does gasoline used in the summer. This is achieved by adjusting the mixture of hydrocarbons used in gasoline for different seasons. Why is a higher vapor pressure for gasoline necessary during the winter?

35. A liquid is in a flask attached to a vacuum pump and a closed-end manometer. The flask is put into a thermostat at 40 °C. The vacuum pump is started and air above the liquid is gradually pumped out. When the barometer reads 216 mm Hg, bubbles begin to appear in the liquid. They rise to the surface vigorously as long as the pumping continues. On the basis of this information and the data in Table 6-6, we can conclude that the liquid is probably: (a) water, (b) ethyl alcohol, (c) carbon tetrachloride, (d) methyl salicylate, (e) benzene, (f) sodium chloride, (g) diamond, (h) nitrogen.

36. The efficiency of a car engine is improved if it can run at a higher temperature. That is the major reason why diesel engines are efficient. Why, then, do we pressurize the cooling systems of cars?

37. You need to make an estimate of vapor pressure at −50 °C. (a) Which *one* of the substances in Table 6-6 would have the highest vapor pressure at −50 °C? (b) Which liquid has the strongest intermolecular forces?

38. (a) How many kilocalories of energy are required to change 14.2 g of liquid Cl_2 to gaseous Cl_2? Is the reaction exothermic or endothermic? (b) How many moles of sodium can be vaporized by the expenditure of 65 kcal of energy? See Table 6-3 for essential information.

39. Evaporation of water also takes place at temperatures below 100 °C. If you emerge from a swimming pool with 90 g of water clinging to you, (a) why do you feel cold, even though the day may be hot, and (b) how much heat energy is necessary to evaporate the 90 g of water? (Assume 9.7 kcal/mole to be the molar heat of vaporization of water.)

40. Mercury vapor is a substance that is a cumulative poison. Repeated small doses of mercury vapor seriously disable a person. Explain why, even though the boiling point of mercury is 357 °C, it is important to clean up mercury spills thoroughly and promptly.

41. Carbon tetrachloride is another poisonous vapor. Its boiling point is 76.8 °C. Explain the following: (a) carbon tetrachloride poisoning can result from relatively short exposure to its vapor, and (b) spilled carbon tetrachloride can be removed from a room by thorough airing, while spilled mercury must be cleaned up.

42. Fog is composed of tiny droplets of *liquid* water suspended in air. Explain, in terms of the relationship between temperature and the vapor pressure of water, why fog is commonly present during early mornings and late afternoons rather than during the middle of the day.

43. How much heat must be removed from an ice cube tray full of water at 0 °C to freeze it if the tray holds 450 g of water?

44. Consider a stoppered flask partially filled with salt water, in which there is undissolved salt at the bottom of the flask even after several days of shaking and swirling. (a) How many phases are present in the flask? (b) Describe the components of each phase. (c) Which phases are pure substances and which are solutions? (d) How could you separate the solutions into their component pure substances? (e) Would you expect

the liquid phase to boil at 100 °C, at a temperature higher than 100 °C, or at a temperature lower than 100 °C? (f) Would you expect the liquid phase to freeze at 0 °C, at a temperature higher than 0 °C, or at a temperature lower than 0 °C?

45. Atmospheric pressure falls about 25 mm Hg for every 300 m above sea level. What would be the boiling temperature of water near the top of Pike's Peak, which has an altitude of about 4,300 m (over 14,000 ft)?

46. (a) Explain why food cooks more rapidly in a pressure cooker than in an open pan. (b) Why is a pressure cooker more effective in sterilizing surgical instruments than is boiling water in an unpressurized container.

47. How many millilitres of 2.00 M HCl must be added to appropriate amounts of water to make the following solutions? (a) 500 ml of 0.500 M HCl (b) 200 ml of 0.10 M HCl (c) 300 ml of 0.20 M HCl (d) 250 ml of 0.40 M HCl.

48. What is the concentration of the solution obtained when 50.0 ml of 0.20 M HCl is diluted to each of the following total volumes? (a) 100 ml (b) 250 ml (c) 500 ml (d) 750 ml.

49. How many grams of solute are necessary to make each of the following: (a) 250 ml of 0.010 M Na_2CrO_4 (b) 100 ml of 0.050 M $CoCl_2$ (c) 500 ml of 0.250 M NaBr (d) 100 ml of 0.50 M KCl?

50. How many millilitres of a 0.100 M solution contain each of the following: (a) 0.100 mole of NaCl (b) 0.010 mole of HCl (c) 0.050 mole of $Zn(NO_3)_2$ (d) 0.075 mole of NaI?

C

51. What is the concentration, in moles per litre, of solutions made in the following manner: (a) 5.85 g of NaCl dissolved in H_2O to make 250 ml total volume, (b) 4.25 g of $NaNO_3$ dissolved in H_2O to make 100 ml total volume, (c) 4.17 g of KBr dissolved in H_2O to make 500 ml total volume, (d) 1.90 g $SnCl_2$ dissolved in H_2O to make 1,000 ml total volume?

52. What is the concentration, in moles per litre, of solutions made in the following manner: (a) 4.00 g of NaOH dissolved in water to make 500 ml total volume, (b) 1.06 g of LiCl dissolved in water to make 1.00 litre total volume, (c) 1.31 g of $Ba(NO_3)_2$ dissolved in water to make 250 ml total volume, (d) 9.20 g of CH_3CH_2OH dissolved in water to make 500 ml total volume?

53. A researcher finds that 100.0 ml of 0.200 molar NaOH will react with 0.010 mole of an acid, H_xY. Write the equation for the process. (Hint: how many moles of NaOH are involved?) (b) What must x be?

54. How many millilitres of 0.300 molar $Ba(NO_3)_2$ would be needed to prepare 20.0 g of $BaSO_4$?

55. In studying the reaction between sodium hydroxide and hydrochloric acid, it is found that 55.0 ml of 0.200 molar NaOH will just react with the HCl in an aqueous (aq) or water solution in a beaker. The equation for the reaction is $NaOH(aq) + HCl(aq) \longrightarrow NaCl(aq) + HOH$. (a) How many moles of NaOH are involved? (b) From the equation, how many moles of HCl are involved? (c) How many millilitres of 0.100 molar HCl solution were in the beaker?

It is the glory of a good bit of work that it opens the way for better things and thus rapidly leads to its own eclipse.

SIR ALEXANDER FLEMING

7

WHY WE BELIEVE IN ATOMS

Objectives

After completing this chapter, you should be able to:

- Explain the Laws of Conservation of Mass, Definite Composition, and Combining Gas Volumes in terms of the atomic theory.

- Explain the electrical properties of matter and ion formation using the ionic model suggested by the atomic theory.

Lightning is a spectacular display of static electricity in the atmosphere.

Atoms are so small that it is impossible to see them by direct observation. Yet modern chemists are as convinced of the existence of atoms as you are of rain or snow. Why do chemists believe so strongly in atoms? The most direct answer is that atoms offer the simplest, most consistent explanation for *many* different chemical observations. Belief in atoms has been confirmed by a great many experiments rather than by any one conclusive experiment.

7.1 Observations and conclusions

Let us begin with an analogy. An analogy is a similarity between things. In this case, we explore the similarity between the way scientists reason that atoms exist and the way a national park ranger might make observations and arrive at conclusions.

While patrolling a wooded area, a national park ranger hears several sharp cracks coming from deep within the woods. After walking in the direction of the sounds for several minutes, he notices two sets of heavy shoe

Figure 7-1

Indirect evidence can lead to valid conclusions. The evidence here suggests that a deer has been shot by hunters.

prints in the soft ground. These are intermingled with light hoof-prints (Figure 7-1), which the shoe prints often partially obliterate. Following these tracks, the ranger comes upon a thicket that is partly flattened. Here he notices several drops of a dark reddish-brown, sticky substance on the ground.

Heading off in a different direction from the original tracks, the ranger sees a wide, flat path in the dirt. He sees flattened shrubbery and then a wide, flat area of recently disturbed dirt. More spots of the reddish-brown substance are found along this path. The flattened path of dirt and shrubbery come to an abrupt end in a clearing near the park road. Here the ranger notices two large, widely separated tire tracks leading in the direction of the paved road and ending there.

The ranger drew the following conclusions from the evidence described above. The sharp cracks were rifle shots that killed a deer. The reddish-brown substance was probably blood. Two people wearing heavy shoes or boots dragged the deer back to their vehicle, which was parked in the clearing. From the size of the tire treads and the distance between them in the soft ground, the ranger concluded that the two people were driving a small pickup truck. Because it was illegal to hunt at this time, the ranger phoned state police and asked them to find two people in a pickup truck who were wearing work shoes or boots. He indicated that they would have one or two rifles, and would be carrying the body of a deer on the truck.

The ranger's conclusions were made on the basis of indirect evidence. He did not see what took place in the woods, but still he was able to draw conclusions about what had occurred.

7.2 Observation and belief in atoms

In studying chemistry, you are like the park ranger. You will look at evidence and clues that suggest the existence of atoms. Just as the ranger found additional evidence about the people he sought, you will find additional evidence about the nature of atoms.

You read earlier in this chapter that chemists believe in atoms because atoms offer the simplest explanation for many different observations. One category of observations comes from experiments involving mass relationships in chemical reactions and the combining ratios of gas volumes. Let us consider experiments from this category.

THE LAWS OF CONSERVATION OF MASS, DEFINITE COMPOSITION, AND MULTIPLE PROPORTIONS

As we saw earlier, mass is always conserved during a chemical reaction. The atomic theory offers a reasonable explanation for this regularity. According to that model, atoms have mass and may be rearranged during chemical reactions. But the atoms themselves are *conserved*. This regularity is called the **Law of Conservation of Mass.** Conserving atoms may be compared to rearranging the same letters into different words. Thus, the phrases *A CAR TAKEN* and *NEAR A TACK* have different meanings, but they contain the same nine letters. The total mass of nine alphabet blocks containing those letters stays the same, regardless of how they are grouped. The total mass of a fixed number of atoms also stays the same, regardless of how they are grouped.

Another regularity involving the relationship of masses is known as the **Law of Definite, or Constant, Composition.** It states that *every sample of a given pure compound is always made of the same elements, in the same proportions by mass*. For example, no matter where it is found, or how it was prepared, pure water always contains 1.008 grams of hydrogen for every 8.000 grams of oxygen. When we purify any compound found in nature, we find that it is identical to another sample of the same compound in composition. This is true even though one sample may have been synthesized in the laboratory.

The Law of Definite Composition is explained by the atomic model, which assumes that:

1. all atoms of a particular element have a characteristic mass
2. every sample of a given compound contains the same relative number of atoms of each of the elements from which it is made

Figure 7-2

Alphabet blocks, like atoms, can be rearranged to form new combinations.

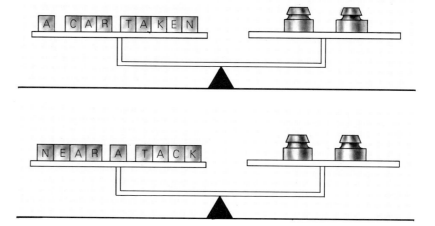

For example, all water molecules are assumed to have two hydrogen atoms for every oxygen atom. The ratio of atomic masses is 2×1.008 to 16.00, and can be represented by the fraction $2.016/16.00$.

EXERCISE 7-1

The elements hydrogen and oxygen also form another compound with which you may be acquainted. It has the formula H_2O_2 and the name hydrogen peroxide. It is a useful bleaching agent and a mild disinfectant. What is the mass ratio of hydrogen to oxygen in H_2O_2? How does this ratio compare to the ratio found in water? Calculate the numerical value of the ratio

$$\frac{\text{mass of oxygen in } H_2O_2 \text{ that combines with 1 g hydrogen}}{\text{mass of oxygen in } H_2O \text{ that combines with 1 g hydrogen}}$$

The principle illustrated in Exercise 7-1 was recognized by John Dalton as a normal and natural consequence of the atomic theory. It is stated as follows: whenever the same two elements combine to form more than one chemical compound (as with H_2O and H_2O_2), the masses of one element (in this case, oxygen) that combine with a fixed mass of the other (hydrogen) are in the ratio of small, whole numbers. This is known as the **Law of Multiple Proportions.**

THE LAW OF COMBINING GAS VOLUMES

The appearance of small-, whole-number ratios in chemical experiments is not limited to relationships between masses. When gases react at the same conditions of temperature and pressure, they combine in small-, whole-number volume ratios. This regularity is called the **Law of Combining Gas Volumes** and was discussed in Chapter 3. Remember that one volume of hydrogen gas will combine with one volume of chlorine gas to give two volumes of the gas hydrogen chloride. Hydrogen gas and oxygen gas, at the same conditions, combine in a 2-to-1 *volume* ratio to form water, as noted in Chapter 3.

Avogadro's hypothesis provides the key to these relationships. This hypothesis states that equal volumes of gases at the same temperature and pressure contain the same number of molecules. Thus, one molecule of hydrogen must combine with one molecule of chlorine to give two molecules of hydrogen chloride. Two molecules of hydrogen combine with one molecule of oxygen to give two molecules of water. Note that *one* molecule of hydrogen chloride contains *one half* of the hydrogen molecule and *one half* of the chlorine molecule. One atom from *each* molecule makes up the hydrogen chloride molecule. Hydrogen and chlorine *molecules* can be split in two. Therefore, they must contain more than one atom. From the results of this and other experiments, we now know that each molecule of hydrogen contains two hydrogen atoms. Each molecule of chlorine contains two chlorine atoms. Similar observations have been made for many other reactions involving gases. Such observations on the combining volumes of gases suggest the existence of many different kinds of atoms.

7.3 The electrical nature of matter

In Chapter 6 (page 155), we noted briefly that some solutions conduct electricity. For example, a salt solution (aqueous NaCl) conducts electricity. What is there about a salt solution that allows it to conduct electricity? A more fundamental question to ask is, how does a salt solution conduct electricity? To answer these questions, we need to know more about electrical phenomena in general. We shall then be able to explore the connection between electrical conductivity and the nature of atoms. We also shall consider how the electrical nature of matter confirms the existence of atoms.

ELECTRICAL PHENOMENA

Electrical phenomena are so common that we often take them for granted. Some, like the electric charge we generate when we comb our hair, have been known for centuries. Others are comparatively new to us. Before you continue reading, try to name several electrical phenomena you have seen recently. Does your list include any of the following?

1. the attraction of a comb for your hair on a dry day
2. the flash of a bolt of lightning
3. the shock you get if you touch a bare wire in a radio or TV set
4. the heat generated by an electric current passing through the heating element of a toaster
5. the light emitted by the filament of a light bulb as an electric current passes through it
6. the emission of "radio waves" by the antenna of a radio or television station transmitter

How are the things on this list related to each other? What does it mean when we say an electric current "passes through" a heating element or a light bulb filament? What is an electric current?

DETECTION OF ELECTRIC CHARGE

In order to answer the questions posed above, we first introduce the **electrometer,** a device that detects and measures electric charges. Figure 7-4 shows a simple electrometer. It consists of two very lightweight spheres, each coated with a thin film of metal. The spheres are hung near each other by thin metal threads, as shown in Figure 7-4. The entire arrangement is enclosed by glass sides to exclude air drafts. The top of each metal thread is connected to a brass terminal. A **battery,** or source of electrical voltage, is connected to the electrometer. The battery terminal marked − is

Figure 7-3

Opposite electric charges are produced when a brush or comb is passed through hair. The charge on the brush is negative (−), while that on the hair is positive (+).

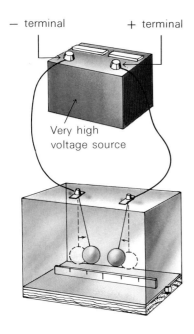

Figure 7-4

A simple electrometer.

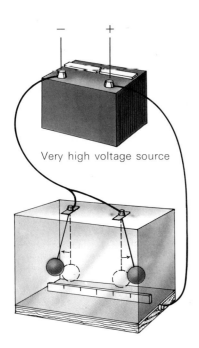

Figure 7-5

An electrometer with both spheres connected to the negative (−) terminal

connected to one electrometer terminal. The battery terminal marked + is connected to the other electrometer terminal.

When the battery and electrometer are connected in this fashion, we observe that the two spheres move *toward* each other. Evidently, the wires have transmitted to the spheres the property of exerting a *force of attraction* on each other. When air is removed from the electrometer with a vacuum pump, the spheres remain as shown in Figure 7-4. The force of attraction must still be present. The force is apparently "felt" by each sphere across the small space that still separates them.

If the battery wires are disconnected, the force of attraction remains. But the moment the two electrometer terminals are connected by a copper wire, the spheres return to their original positions and once again hang vertically. The attraction is lost.

The battery has apparently transferred to the electrometer spheres something that makes the spheres attract each other. This "something" is called **electric charge.** The *movement* of electric charge from the battery through the metal threads to the spheres is called **electric current.** The attraction due to the electric charge is lost when the two electrometer terminals are connected by a copper wire.

To learn more about electric charge, we can connect the battery terminals to the electrometer in a different way. The + terminal is connected to the base of the electrometer. The − terminal is connected to *both* electrometer terminals, as shown in Figure 7-5. This time, the two spheres move *apart:* they *repel* each other. We note that when both spheres are given electric charge from the battery terminal labeled −, they repel instead of attract each other. Such behavior suggests that there are at least *two* kinds of electric charge: one from the − battery terminal and one from the + battery terminal. We see that + and − charges attract, while two − charges repel. Therefore, we can write

1. − attracts +: unlike charges attract
   + attracts −: unlike charges attract
2. − repels −: like charges repel

It is reasonable to predict that + repels +. Experiment verifies this prediction, as shown in Figure 7-6. We now can write

3. + repels +: like charges repel

One other significant observation deserves consideration. When the terminal of a sphere holding a (+) charge is connected by a copper wire to the terminal of a sphere holding a (−) charge, the two spheres lose all attraction for each other. It is reasonable to suggest that either a (−) charge, a (+) charge, or both, moved through the wire to destroy the force of attraction between the spheres. Symbolically, the observation can be written as

− charge plus + charge = no charge

Are there more than two kinds of charge? So far, all observations indicate the existence of only two kinds. The charge on a dry comb used to comb dry hair is found to be −. The charge left on a hard rubber rod rubbed with cat's fur is −. The cat's fur is positive. The charge left on a glass rod rubbed with silk is positive. The charge on the silk is negative. No matter how an electric charge is produced, only + and − charges are found. Furthermore, the charges are always produced in equal numbers. The same number of positive charges and negative charges are produced.

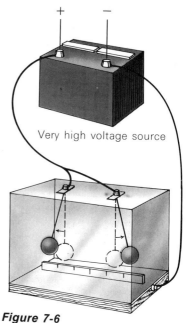

Figure 7-6
An electrometer with both spheres connected to the positive (+) terminal.

THE EFFECT OF DISTANCE

Figure 7-7 shows a battery attached to electrometers with different distances between the two spheres. The charges on the spheres are identical. *When the spheres are closer together, the deflection is larger.* As the spheres are moved apart, the deflection decreases. We conclude that the force of attraction is stronger when the charges are closer together. Careful studies show that the electric force (F) and the distance (d) between the centers of the two spheres are related by the expression: F is proportional to $1/d^2$, or

$$\text{force} = \frac{\text{constant}}{\text{distance}^2}$$

Here are some examples of this relationship. If we double the distance, the force is only $\frac{1}{4}$ as large as the original force. When we triple the distance, the force is $\frac{1}{9}$ of the original value. If the distance is 10 times as large as the original, the force will be only $\frac{1}{100}$ as large as the original. If the distance is 100 times as large as the original, the force will be only $\frac{1}{5,000}$ as large as the original. If the force is $\frac{1}{16}$ as large as the original force, the distance is four times as great as the original distance.

THE ELECTRON-PROTON MODEL OF ATOMIC STRUCTURE

The information we now have about electrical phenomena can be helpful in developing a picture of atoms. Many experiments not considered here suggest that *atoms themselves are made of still simpler particles that carry electric charge.* Let us summarize these ideas:

1. Atoms contain two kinds of electrically charged particles: electrons and protons.
2. Each electron carries one portion, or unit, of negative (−) charge.
3. Each proton carries one portion, or unit, of positive (+) charge.
4. Electrons and protons exert force on each other, according to the electrical behavior already observed. This force is related to the distance between the charged particles. It gets much smaller as the distance increases.

Since *like charges repel,* electrons repel electrons and protons repel protons. Because *unlike charges attract,* electrons attract protons. Finally, because a negative (−) charge plus a positive (+) charge equals no charge, one electron plus one proton equals no detectable charge.

This expanded version of the atomic model helps us better understand electrical phenomena. Any object (such as an electrometer sphere or an atom) that has the *same* number of electrons and protons has no charge. It is **electrically neutral.** If we *remove* some of the electrons from the object, it will then have an excess of protons (+). This object will have a net +

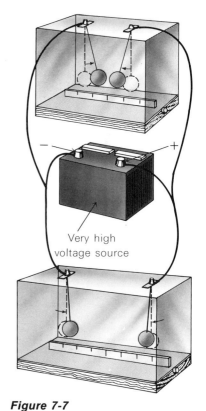

Figure 7-7
Contrast of deflections in two electrometers with different distances between spheres.

charge. If extra electrons are added to the object, it will have a net — charge. The amount of net charge is the *difference* between the amount of — charge and the amount of + charge.

EXERCISE 7-2

Determine the net charge of an object with (a) 10 protons and 10 electrons, (b) 11 protons and 10 electrons, and (c) 17 protons and 18 electrons.

ELECTRIC FORCE: A FUNDAMENTAL PROPERTY OF MATTER

We have learned that a battery can transfer a property called electric charge to the spheres of an electrometer. When this happens, the spheres exert force on each other. This observation raises several questions. First, how does a battery generate and transfer electric charge to the spheres? We shall postpone this important question until Chapter 18, where it will be examined in detail. For the present, we recognize that the operation of a battery generates electric charge.

The second question probes deeper: what is electric charge? Why do the two charged electrometer spheres attract or repel each other? The spheres have acquired an excess of electrons or protons. These electrons or protons exert force on each other. But does this really explain electric force at a distance? Why do like charges repel each other? Why do opposite charges attract each other? There is no satisfactory answer to such questions at present. We simply note that attraction between unlike charges is a **fundamental property** of matter. A property that is generally observed, but for which diligent searching has not produced a useful model, is called a fundamental property. (Matter and charge are discussed in Appendix 8.)

7.4 The electrical conductivity of water solutions

Pure water is a very poor conductor of electricity. Yet when ordinary table salt, sodium chloride (NaCl), dissolves in water, the solution conducts electricity readily. When sugar dissolves in water, the solution is as poor a conductor of electricity as is pure water. How does the presence of sodium chloride in a solution permit the movement of electric charge through that solution? The following picture has been developed. When sodium chloride dissolves in water, particles with electric charge are found to be present. The movement of these charged particles through the solution accounts for its conductivity. However, such an explanation raises some questions. Sodium chloride (NaCl) is electrically neutral. Touching it does not give you an electric shock. How, then, can neutral sodium chloride produce charged particles when dissolved?

Scientists now know that solid sodium chloride consists of sodium atoms, each of which has *lost one electron,* and an equal number of chlorine atoms, each of which has *gained one electron.* Each sodium particle carries a posi-

Na+ ions

Cl⁻ ions

Figure 7-8

The photograph at left, taken through an electron microscope, shows the crystalline structure of common table salt (NaCl). Solid NaCl forms hard, regularly shaped crystals. The crystals consist of equal numbers of Na+ and Cl⁻ ions. Larger crystals of salt with similar appearance are easily grown.

tive charge and can be represented symbolically as Na+. Each chlorine particle carries a negative charge and can be represented as Cl⁻. Atoms or groups of atoms that carry an electric charge are called **ions.** Thus, the *sodium ion* is Na+. The *chloride ion* is Cl⁻. Solid sodium chloride consists of sodium ions and an equal number of chloride ions. It is electrically neutral. The Na+ and Cl⁻ ions attract each other and the sodium chloride forms cubic crystals. This crystalline solid is the familiar white table salt.

When solid NaCl crystals are dropped into water, water molecules surround each Na+ ion and each Cl⁻ ion. The solid structure is broken up and charged Na+ ions and Cl⁻ ions float around among the water molecules in the solution. This situation can be expressed symbolically as follows:

$$\text{NaCl } (solid) + \text{water} \longrightarrow \text{Na}^+ \ (in \ water) + \text{Cl}^- \ (in \ water)$$

This equation can be shortened, while still retaining all essential information. The term *water* to the left of the arrow can be omitted because its presence is implied by the symbols on the right. We can show that ions are in water by the notation *aq*, which means aqueous. Thus, the accepted, abbreviated form of the equation is

$$\text{NaCl}(s) \longrightarrow \text{Na}^+(aq) + \text{Cl}^-(aq)$$

IONIC EQUATIONS

The equation above provides a reasonable explanation of the electrical conductivity of a salt solution. Solid NaCl dissolves, forming charged Na+(*aq*) and Cl⁻(*aq*) ions. An electric current can pass through the solution

due to the movement of these ions. The $Cl^-(aq)$ ions move in one direction, carrying negative charge. The $Na^+(aq)$ ions move in the opposite direction, carrying positive charge. As these movements carry charge through the solution, current flows.

Recall that sugar dissolves in water, but that the resulting solution is as poor a conductor of electricity as pure water. When sugar dissolves, no charged particles result. No ions form. We can assume that sugar is very different from sodium chloride.

Magnesium chloride ($MgCl_2$) is a white solid, similar in appearance to sodium chloride. A one-molar magnesium chloride solution yields more ions to carry an electric current than does a one-molar sodium chloride solution. The equation for dissolving $MgCl_2$ in water is

$$MgCl_2(s) \longrightarrow Mg^{++}(aq) + 2Cl^-(aq)$$

The equation shows that when magnesium chloride dissolves, $Mg^{++}(aq)$ and $Cl^-(aq)$ ions appear in the solution. Note that each magnesium ion carries two positive charges. Two aqueous chloride ions also are present. Comparison of the above equation with that for dissolving sodium chloride reveals that one $MgCl_2(s)$ crystal releases *twice* as many $Cl^-(aq)$ ions in solution as does one $NaCl(s)$ crystal. A one-molar $MgCl_2$ solution is a much better conductor of electricity than is a one-molar $NaCl$ solution. Both contrast sharply with sugar, which is a nonconductor.

Silver nitrate ($AgNO_3$) is another solid substance that dissolves in water to give a solution that conducts electricity. The equation for dissolving $AgNO_3$ in water is

$$AgNO_3(s) \longrightarrow Ag^+(aq) + NO_3^-(aq)$$

In this equation, the charged particles formed are silver ions, $Ag^+(aq)$, and nitrate ions, $NO_3^-(aq)$. The aqueous silver and sodium ions carry a positive charge. The aqueous nitrate and chloride ions carry a negative charge. Notice, however, that in the case of the nitrate ion, the negative charge is carried by four atoms—one nitrogen atom and three oxygen atoms—that remain together with one extra electron overall. Since the NO_3^- group remains together and acts as a unit, it is given the distinctive name *nitrate ion*.

The solids sodium chloride, magnesium chloride, and silver nitrate all dissolve in water to form aqueous ions and, thus, conducting solutions. Such solids are called **ionic solids.** The tables in Appendix 4 list the names, formulas, and charges of some common ions. The ions described in the tables are found when $NaCl(s)$, $MgCl_2(s)$, $AgNO_3(s)$, and other ionic solids dissolve in water. The solids are all electrically neutral. Thus, *the sum of their electric charges is zero.* For example, one Na^+ plus one Cl^- gives zero charge. Or, one Mg^{++} plus two Cl^- gives zero charge. A force of attraction between ions of opposite charge holds the solid together.

The formula for any ionic solid can be predicted. We know that the compound such a solid represents is electrically neutral. Magnesium nitrate, for example, needs *two* nitrate ions, each with a -1 charge, to balance each magnesium ion, with a charge of $+2$. Therefore, its formula is $Mg(NO_3)_2$.

We assume that water is able to overcome the forces of attraction between ions and to break ionic bonds. Thus, for the dissolving of magnesium nitrate, we write

$$Mg(NO_3)_2(s) \longrightarrow Mg^{++}(aq) + 2NO_3^-(aq)$$

It is helpful to review the writing of formulas of ionic compounds here. For aluminum sulfate, the table on page 645 of Appendix 4 indicates that we are combining an Al^{+++} ion with an SO_4^{--} ion. The lowest common multiple of the $+3$ and -2 charges is 6. Thus, we need 2 Al^{+++} to give 6 positive charges and 3 O^{--} to give 6 negative charges. The formula then becomes $Al_2(SO_4)_3$ and the charge on the molecule is zero.

Table 7-1

Some common positive ions (cations) and negative ions (anions)

Cations	Anions
Na^+	I^-
K^+	Br^-
Li^+	Cl^-
Mg^{++}	NO_3^-
Ca^{++}	OH^-
Ba^{++}	S^{--}
Sr^{++}	SO_4^{--}
Pb^{++}	CO_3^{--}
Fe^{++} (ous)	
Cu^{++}	
Al^{+++}	
Fe^{+++}(ic)	

EXERCISE 7-3

Use the table in the margin to predict and write formulas for the following solid, ionic compounds:

(a) sodium bromide
(b) potassium iodide
(c) sodium sulfide
(d) lithium chloride
(e) potassium nitrate
(f) sodium sulfate
(g) magnesium sulfate
(h) calcium nitrate
(i) barium hydroxide
(j) strontium carbonate
(k) aluminum hydroxide
(l) aluminum sulfide
(m) aluminum sulfate
(n) lead nitrate
(o) lead sulfide
(p) ferric chloride, or iron(III) chloride
(q) ferrous sulfate, or iron(II) sulfate
(r) copper(II) nitrate

EXERCISE 7-4

Write properly balanced ionic equations for dissolving in water each solid whose formula you wrote in Exercise 7-3.

EFFECT OF CONCENTRATION

A solution containing 0.1 mole per litre of sodium chloride is a better conductor than a solution containing 0.01 mole per litre of sodium chloride. Conductivity depends not only on the presence of ions, but also on the number of ions per litre of solution. That is, conductivity also depends on **concentration.**

Silver chloride (AgCl) is a solid that shows the effect of concentration in a dramatic way. There is only a *very slight* increase in the conductivity of a silver chloride solution over that of pure water. Very little of the substance dissolves. Thus, it is said to have very low solubility. However, there is a very slight but measurable increase in conductivity when silver chloride is added to water. This indicates that ions are formed. Clearly, silver chloride is much less soluble than sodium chloride, but it is similar to sodium chloride because the solid that does dissolve forms ions in aqueous solution.

PRECIPITATION REACTIONS IN AQUEOUS SOLUTION

We know that both silver nitrate and sodium chloride have high solubility in water and give ions in solutions. Silver chloride has low solubility. What will happen if we mix solutions of silver nitrate and sodium chloride? The $AgNO_3$ solution contains $Ag^+(aq)$ ions, while the NaCl solution contains $Cl^-(aq)$ ions. The solution obtained immediately after mixing will contain both ions in high concentration. When the concentrations of $Ag^+(aq)$ and $Cl^-(aq)$ far exceed the solubility of AgCl in a saturated solution, solid AgCl comes out of solution.

The formation of a solid from a solution is called **precipitation.** The solid itself is known as a *precipitate*. The reaction is represented by the following equation:

$$Ag^+(aq) + NO_3^-(aq) + Na^+(aq) + Cl^-(aq) \longrightarrow$$
$$AgCl(s) + Na^+(aq) + NO_3^-(aq)$$

The two ions $NO_3^-(aq)$ and $Na^+(aq)$ do not play an active role in the reaction, nor do they prevent the reaction from occurring. They are present in an uncombined state in solution both at the beginning and at the end of the process. Such ions are called **spectator ions.** They play only the role of spectators, or onlookers.

A chemical equation supposedly represents the chemical changes that occur in a reaction. Therefore spectator ions may be omitted from the equation representing a precipitation process. Thus, the above equation could be rewritten

$$Ag^+(aq) + Cl^-(aq) \longrightarrow AgCl(s)$$

The simplified equation is called the **net ionic equation** for that reaction. It shows *only* the ions participating in the chemical changes.

BALANCING EQUATIONS INVOLVING IONS

Atoms must be conserved in processes involving ions as well as in those involving neutral molecules. Furthermore, *the charges must be balanced. A chemical reaction does not change the total electric charge on the substances involved.* The sum of the electric charges on the reactants must be the same as the sum of the electric charges on the products. Calcium chloride dissolves to give aqueous Ca^{++} and two aqueous Cl^- ions. We see from the calculations that follow that electric charges before and after dissolving add up to zero:

$$\text{charge on } CaCl_2 \text{ solid} = (\text{charge on } Ca^{++}) + 2(\text{charge on } Cl^- \text{ ion})$$
$$0 = \underbrace{(+2) + 2(-1)}$$
$$0 = \qquad 0$$

In a balanced equation for a chemical reaction, charge is conserved.

The identity of the precipitate in any precipitation reaction can be revealed by experiment. The results of such experiments are summarized in the rules of solubility in Chapter 16 (pages 438–439 and 448–450).

Write ionic and net ionic equations for each of the following reactions: (a) formation of the precipitate lead iodide from solutions of lead nitrate and sodium iodide, (b) formation of the precipitate ferric hydroxide from solutions of ferric chloride and cesium hydroxide, (c) formation of the precipitate lead chromate from solutions of lead acetate and potassium chromate, (d) formation of solid silver from the reaction between solid copper and silver nitrate solution, (e) formation of $H_2(g)$ and magnesium chloride solution from the reaction between solid magnesium and $HCl(aq)$. Write formulas for reactants and products and then balance. Table 7-2 will be helpful.

Table 7-2

Some common positive ions (cations) and negative ions (anions)

Cations	Anions
Na^+	I^-
Cs^+	OH^-
Ag^+	Cl^-
H^+	NO_3^-
Pb^{++}	CrO_4^{--}
Mg^{++}	SO_4^{--}
Fe^{+++}	

7.5 Why we believe in ions: Evidence of their existence

The ionic model offers a convincing explanation of conductivity of solutions. The explanation is hardly surprising, since we specifically designed it to account for the phenomenon in question. However, a successful model should do more than explain a single observation. Other predictable consequences should follow from it—consequences whose results can then be observed.

The ionic model *does* suggest that only $Ag^+(aq)$ and $Cl^-(aq)$ ions are required to form $AgCl(s)$ precipitate. Any solution containing $Cl^-(aq)$ ions should precipitate $AgCl(s)$ when aqueous $AgNO_3$ is added to it. This is indeed found to be the case. Chemists add aqueous $AgNO_3$ to a solution as a simple test for the presence of $Cl^-(aq)$ ions. The ability to precipitate $AgCl(s)$ in the presence of aqueous $AgNO_3$ is thus a property of $Cl^-(aq)$ ions.

In the laboratory, we see that aqueous solutions of $CuSO_4$, $Cu(NO_3)_2$, and $CuCl_2$ are all light blue in color. Apparently, $Cu^{++}(aq)$ ions are responsible for the light blue color because other solutions containing $SO_4^{--}(aq)$, $NO_3^-(aq)$, and $Cl^-(aq)$ ions are colorless. The blue color must be a property of $Cu^{++}(aq)$ ions.

In reviewing how to recognize aqueous solutions of ionic compounds, you should include the following:

1. by their ability to conduct an electric current
2. by properties such as color or the ability to form a precipitate, which are known to be characteristic of aqueous ions

These characteristics are both predicted and explained by the two independent kinds of ions these solutions contain. If a solid is *not* composed of ions, its solution will conduct electricity very poorly. Such solids form *one* kind of electrically neutral solute particle, the individual molecule, when they dissolve. Sugar is an example of such a solid.

Experiments with the electrometer and battery (pages 165–168) reveal

1. the existence of two different kinds of electric charge, labeled + and −
2. that charges of opposite sign *attract* each other

(*continued on page 176*)

Improvements in the making of glass

Glass has been known and used for many centuries. It is one of the oldest manufactured chemical products in existence. Its beauty, versatility, and relative strength have been prized for thousands of years. The earliest glass objects were made in Egypt and Mesopotamia about 3500 years ago. There have been many improvements in techniques for making glass since that time. Civilizations throughout history have added to our knowledge. However, modern chemists have made the most dramatic contributions. They have succeeded in making glass that is as light as a feather, clear as air, strong as steel, and flexible as silk, as well as glass that is heatproof, photosensitive, and photochromic. Photochromic glass darkens in light and is used in special sunglasses.

Most glass is made by heating silicon dioxide (sand), sodium carbonate (washing soda), and calcium carbonate (limestone) to very high temperatures. The sodium carbonate is needed to lower the melting temperature of the silicon dioxide. The calcium carbonate makes the resulting glass insoluble in water. Carbon dioxide is released during the heating process and the final mixture consists of about 75 percent silicon dioxide, 12 percent sodium oxide, 12 percent calcium oxide, and 1 to 2 percent aluminum oxide. Thermally resistant Pyrex glass is made by substituting boric oxide for calcium or sodium oxide. Borosilicate glass has low thermal expansion and high softening temperatures. It is widely used in making cooking utensils and laboratory glassware.

Glass has no regular crystalline structure. It is formed when a material in the liquid state is cooled to a rigid condition without the formation of crystals. The molecular structure that results can be described as resembling a frozen liquid. The figure on page 175 illustrates the differences between the regular crystalline pattern of a hypothetical substance and the glassy state of the same substance.

Silicon dioxide (quartz sand) forms the molecular basis of most glass manufactured today. Each silicon atom is bonded to four oxygen atoms. In silicate crystals, such as quartz, there is a regular, repeated atomic pattern that centers around the tetrahedral silicon atom. Although bonding is similar in glass, the latter has no particular order in its molecular arrangement. The number of bonds per atom remains constant, but the pattern and repetition of set angles is missing. Thus, glassy materials have no cleavage planes and shatter along haphazard lines.

Glass is frequently described as a supercooled liquid. The atoms are bonded together according to bonding rules, but the overall picture indicates no regular pattern in the way that these groups of atoms are arranged. Although glass appears to be very rigid, it actually behaves like elastic solids at room temperature and can even be bent or otherwise deformed. We see this in glass wool and glass fibers. Glass fibers are very important today in transmission of telephone messages. Glass fibers replace copper wires. This field of science is called "fiber optics."

Colored glass is produced by the introduction of certain metallic oxides into the glass. Transition elements have been especially effective in producing vivid colors. Small quantities of these elements are sufficient to produce very deep colors. The presence of iron can be a problem when making clear glass because it is often a contaminant of the raw materials used in glass production. Iron contaminants produce a greenish-yellow glass. Very small copper and gold particles cause a ruby red color, and cobalt ions produce a beautiful blue color.

Foam glass is made by heating a mixture of ground glass and carbon in a mold. The mixture expands to many times its original volume when heated. It sets as a very light and rigid material that consists of many thousands of minute glass bubbles. Foam glass is buoyant and strong enough to be sawed. It is used as insect-proof insulating material.

Glass bricks may be used wherever strength and translucence are desired. Entire walls have been made of these bricks. Glass bricks are made by fusing molded halves together. The contraction of the residual air that is trapped within the brick as it cools creates a partial vacuum. The material that results is a highly efficient sound insulator that still permits the passage of light.

Patterned glass is produced by a photosensitive process. It can be made by exposing the glass to ultraviolet light before reheating it with metallic oxides used to produce color. Only the areas of the glass that are irradiated develop deep colors. Beautiful patterns of alternately clear and colored glass can be produced in this way. Modern methods of producing patterned glass include the use of cerium atoms. Electrons are easily removed from these atoms by ultraviolet radiation. The electrons can then be captured by gold ions that were added to the mixture while it was still in the molten state. The electrons reduce the gold ions to form gold atoms that coalesce and remain trapped in the glass. Beautiful patterns can be produced by this process.

Glass can be temporarily darkened by radiation from the visible as well as the ultraviolet regions of the electromagnetic spectrum. It is used in preparing self-darkening sunglasses. The darkening is completely reversible and relies on the fact that molten glass can dissolve a large number of materials. Even silver chloride is soluble in molten glass. The ions of silver chloride crystals separate when dissolved and spread evenly throughout the molten glass. The mixture is cooled very slowly. During the slow cooling process, minute crystals of silver chloride form throughout the mixture. The minute crystals do not interfere with the transmission of light through the glass within which they are trapped.

Silver and chloride ions in crystals are held together by electrostatic attractions known as ionic bonds. When photons pass through glass made in this manner, the photons interfere with the ionic bonds between the silver and the chloride ions. The silver ions regain the electrons they once lost, and form silver atoms. This causes darkening of the glass. When the number of photons streaming from the light source is reduced, the silver atoms again lose their electrons and become silver ions. This causes the glass to clear. Photochromic sunglasses work in this way. In bright sunlight, the silver ions that have been trapped in the glass become silver atoms and consequently retard the effects of bright light. When bright light is no longer a problem, the

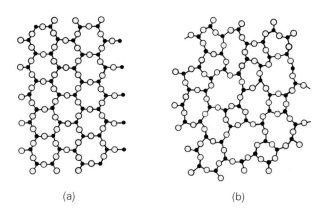

This illustration shows the differences between (a) a regular, repeated crystalline structure of a hypothetical substance and (b) an unordered, amorphous structure of the same substance.

silver atoms lose their electrons again and become silver ions. This process continues throughout the life of the sunglasses.

Although glass is tough, it is still breakable. Chemists have made considerable progress in increasing the strength of certain glasses so that they even can be used as hammers. Tempered glass is made by exchanging the sodium ions that are present in ordinary glass with larger potassium ions. This is accomplished by heating potassium nitrate to its melting point in a stainless-steel tank. Relatively large distances exist between the potassium and the nitrate ions in the molten state. The glass that is to be treated is then lowered into the tank and left there for about 16 hours. During this period, some of the sodium ions present at the surfaces of the immersed glass exchange places with the potassium ions in the molten potassium nitrate. The exchange of ions results in a glass with somewhat crowded surface atoms because the potassium ions are larger than the sodium ions they replaced. Due to the crowding at the surface, the molecules in the core are under tension and the resulting glass has considerable strength. Objects that hit glass produced in this manner will bounce off without causing damage. Tempered glass is used in building materials, automobiles, refrigerator shelving, and in any other product where both strength and beauty of material is required.

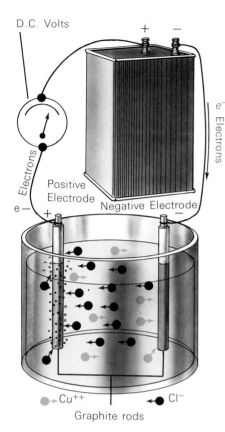

D.C. Volts

+   −

e^- Electrons

Electrons

Positive
Electrode

Negative Electrode

e −
+

$e^- \rightarrow$ Cu^{++} $\rightarrow\bullet$ Cl$^-$

Graphite rods

Figure 7-9

Schematic representation of the passage of electric current through a copper chloride solution.

What happens in a dilute aqueous solution of copper(II) chloride, $CuCl_2(aq)$, when an electric current is passed through the liquid. To find out, the setup shown in Figure 7-9 is used. In this setup, graphite rods are placed in a solution of copper(II) chloride.* The rods can be attached to a battery.

According to the ionic model, $CuCl_2(s)$ reacts in water as shown by the following equation:

$$CuCl_2(s) \longrightarrow Cu^{++}(aq) + 2Cl^-(aq)$$

This process is called dissociation. Since unlike charges attract, positive $Cu^{++}(aq)$ ions are attracted by, and move toward, the negatively charged graphite rod. The − rod is known as the **negative electrode.** At the same time, the negative $Cl^-(aq)$ ions are attracted by, and move toward, the positively charged graphite rod. The + rod is known as the **positive electrode.** See Figure 7-9. On the − rod you see an orange deposit of neutral, metallic copper developing as current flows. From the + rod we detect the choking odor of neutral chlorine gas, $Cl_2(g)$.

Positive $Cu^{++}(aq)$ ions can change to electrically neutral Cu atoms by *combining with* two electrons. The symbol e^- represents a single electron. Thus, $2e^-$ represents two electrons. Let us use the equation

$$2e^- + Cu^{++}(aq) \longrightarrow Cu \text{ (solid metal)}$$

to explain the appearance of the copper deposit on the − rod. We must now account for the presence of the chlorine gas. Negative $Cl^-(aq)$ ions can be changed to electrically neutral Cl atoms by releasing electrons. Each pair of neutral Cl atoms then joins to form a $Cl_2(g)$ molecule. The equations for this reaction are

$$2Cl^-(aq) \longrightarrow 2e^- + 2Cl$$
$$2Cl \longrightarrow Cl_2(g)$$

The equations above explain the odor of $Cl_2(g)$ detected at the + rod.

Electrode equations assume that *electrons* are gained and lost by aqueous ions. This supposition in turn leads to an additional assumption: electrons given up by the $Cl^-(aq)$ ions at the positive electrode flow through the wire to the battery. The battery serves as an electron "pump." It pushes the electrons to the negative electrode. The electrons combine with $Cu^{++}(aq)$ ions and change them to atoms of metallic copper. Our second assumption has several consequences. One consequence is that electric current moves in the wire by a net flow of electrons. Another consequence is that electric current moves in solution by migration, or movement, of ions. Remember that conductivity in a wire differs from conductivity in a solution.

The view of electrical conductivity stated above has other advantages. We can now rewrite the equations for the neutralization of both the positive and negative aqueous ions in the previous example using copper(II) chloride, $CuCl_2$. The new equations can be added algebraically:

*The symbol (II) following the word copper indicates $CuCl_2$. Copper(I) chloride is CuCl.

at − electrode: $Cu^{++}(aq) + 2e^- \longrightarrow Cu(s)^*$

at + electrode: $2Cl^-(aq) \longrightarrow 2e^{-\dagger} + Cl_2(g)$

sum: $Cu^{++}(aq) + 2Cl^-(aq) \longrightarrow Cu(s) + Cl_2(g)$

The $2e^-$ terms cancel each other out because they are on opposite sides of the two equations. *For every electron lost, there must be one gained.* The equation that results represents the sum of the two equations. It shows that *two* $Cl^-(aq)$ ions are removed from the solution every time *one* $Cu^{++}(aq)$ ion is removed. This conclusion explains the continuing electrical neutrality of the solution during the process we have described. *The process of separating a compound* (in this case, $CuCl_2$) *into simpler substances* (here, Cu and Cl_2) *by an electric current* is known as **electrolysis.** Electrolysis will be studied in more detail in Chapter 18. Next, we shall look at additional advantages of our ionic model as we explore atomic structure in Chapter 8.

*Loss of two positive charges would also give a neutral system; but this possibility is tentatively rejected since later arguments (Chapter 8) will show that gain of the electrons is the only acceptable process. Since the positively charged protons are present *in the nucleus,* they are held firmly and are not easily lost. Protons also have a large mass compared to the electron, so the loss of two protons from the copper would give a nucleus of lower mass and different properties.

†Again, a gain of two positive charges would meet all algebraic requirements for giving us neutral chlorine atoms, but this possibility is tentatively rejected. Evidence considered in Chapter 8 will justify the choice used here.

Key Terms

battery
concentration
electric charge
electric current
electrically neutral
electrolysis
electrometer
fundamental property
ion
ionic solid
Law of Combining
   Gas Volumes
Law of Conservation
   of Mass
Law of Definite, or
   Constant, Composition
Law of Multiple
   Proportions
negative electrode
net ionic equation
positive electrode
precipitation
spectator ions

SUMMARY

Atomic theory furnishes the simplest, most logical explanation for the Laws of Conservation of Mass, Definite Composition, and Combining Gas Volumes. These laws support the atomic theory.

The *Law of Conservation of Mass* assumes that atoms do not change their mass and that they are conserved. The mass of a given quantity of matter cannot be increased or decreased. Atoms may regroup, but the total mass remains unchanged.

The *Law of Definite Composition* states that every sample of a given pure compound is always made up of the same elements, in the same proportions by mass. This law assumes that each sample of a pure compound contains the same ratio of atoms. Each atom of a particular element has a characteristic mass.

The *Law of Combining Gas Volumes* states that when gases at the same temperature and pressure react, they combine in small-, whole-number volume ratios.

Observations made with an electrometer suggest the existence of two kinds of electric charge: positive (+) and negative (−). *Unlike charges attract. Like charges repel.* Substances with equal quantities of positive charges and negative charges are electrically neutral.

Electrical properties of matter may be explained by assuming that atoms contain charged particles called protons (+) and electrons (−). All atoms are electrically neutral because they contain equal numbers of protons and electrons.

An ion is an atom or group of atoms that carries an electric charge. Electrical conductivity involves the movement of positive (+) and negative (−) ions. Ionic solids dissolve in water to form aqueous ions. The solutions that result conduct an electric current. Non-ionic solutions are poor conductors of electricity.

When positive (+) and negative (−) electrodes are placed in ionic solutions, negative ions move toward the positive electrode, and positive ions move toward the negative electrode. At the electrodes, positive ions *combine* with electrons, and negative ions *release* electrons. Electrically neutral elements are produced.

The ionic model extends the atomic theory. It provides a simple and convincing explanation of electrolysis experiments.

QUESTIONS AND PROBLEMS

A

1. The Santa Claus Theory of Appearing Christmas Presents is another theory, familiar to everyone, that is built on indirect evidence. What observations (and reports from adults) lead a child to believe in the existence of Santa Claus? What theory does the child construct (with help from many adults)? How is the theory modified in the light of more mature observations?

2. Suppose all elements were continuous, as fluids appear to be, and that all elements mixed with each other in all proportions like alcohol and water. If this were so, which of the following laws would *not* be true? Explain.
   (a) Law of Definite, or Constant, Composition
   (b) Law of Multiple Proportions
   (c) Law of Combining Volumes
   (d) Law of Conservation of Mass
   (e) Law of Conservation of Energy

3. Carbon forms two oxides. Analysis shows that one has 57.1 percent oxygen and 42.9 percent carbon. (a) What is the number of grams of oxygen combined with *one gram* of carbon? (Hint: remember that 57.1 g oxygen combines with 42.9 g carbon.) The second compound has 72.7 percent oxygen and 27.3 percent carbon. (b) What is the number of grams of oxygen combined with one gram of carbon? (c) What is the relationship between the two ratios that you calculated in parts (a) and (b)? (d) What law from Question 2 is illustrated in this problem?

4. Consider the following examples of charged particles separated by a given distance. (a) Which of the examples would show attraction of the spheres? (b) In which example is the attractive force the largest? (c) How many times greater is the largest force than the smallest force?

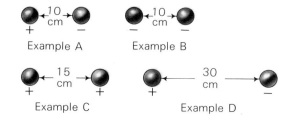

Example A   Example B

Example C   Example D

5. Name two kinds of forces other than electrostatic forces that are felt at a distance. Are they fundamental properties of matter? Explain. (Hint: Why does a rock fall to earth? Why does the old-fashioned compass work?)

6. If a hard rubber rod is rubbed vigorously with cat's fur, it becomes negatively charged and the cat's fur becomes positively charged. What kind of process can you suggest that occurs during rubbing which generates the charges? (Remember that atoms are made up of charged particles.)

7. Table salt, $NaCl(s)$, is pictured as being made up of two different kinds of charged particles called *ions*. Indicate the *name, formula,* and *electric charge* of each kind of ion in NaCl.

8. How can $NaCl(s)$ be made of electrically charged ions, and still be electrically neutral?

9. What is the net charge of a system containing each of the following: (a) 9 protons and 10 electrons, (b) 8 protons and 8 electrons, (c) 12 protons and 10 electrons?

10. Using either Table 7-1 on page 171 or Appendix 4 on page 645, write the formulas for the following compounds: (a) magnesium nitrate, (b) chromous sulfate, (c) lead sulfate, (d) silver iodide, (e) zinc sulfate, (f) copper (II) bromide, (g) barium chloride, (h) chromium (III) nitrate.

11. Write balanced ionic equations for the dissolving of the following ionic solids in water: (a) $NaCl(s)$, (b) $MgCl_2(s)$, (c) $AgNO_3(s)$, (d) $Al_2(SO_4)_3(s)$.

12. (a) What is observed to occur when solutions of $AgNO_3(aq)$ and $NaCl(aq)$ are mixed? (b) Write an ionic equation for the process described in part (a). Then write a *net* ionic equation for this process. (c) Describe a simple test that detects the presence in a water sample of $Cl^-(aq)$ ions.

13. You have three aqueous solutions for which the labels have fallen off. The first is a one-molar sugar solution; the second is a one-molar solution of NaCl; and the third is a one-molar solution of $AgNO_3$. How could you identify these solutions so that the labels could be replaced?

14. Electrodes of opposite charge are placed in a water solution of copper(II) bromide, $CuBr_2(aq)$. (a) Write the formula and charge of each kind of ion present in the solution. (b) After the electrodes are connected to a battery for a short time, what is observed at the positive $(+)$ electrode? What is observed at the negative $(-)$ electrode? (c) Write equations to represent the process seen at each electrode.

15. Use our model to describe the movement of electric current in (a) a metal wire and (b) an aqueous solution of a dissolved ionic solid.

16. Solutions of sodium chromate, potassium chromate, and ammonium chromate are all yellow in color. Indicate the formulas and charges of the ions present in each solution. Suggest a reason for their similar color.

B

17. A 4.5-g sample of water is found to contain 0.5 g hydrogen and 4.0 g oxygen. Another water sample, of 13.5-g mass, contains 1.5 g hydrogen and 12.0 g oxygen. Show how the data illustrate the Law of Constant Composition.

18. Consider the electrometer set-up shown at the top of page 180 after disconnecting the battery by opening switch B:

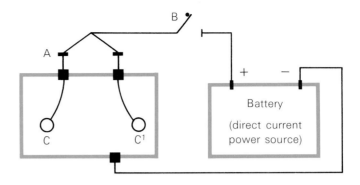

In attempting to study the electric charge associated with combing of one's hair, a researcher combs her hair vigorously for several minutes and then touches the electrode at A with the comb. The spheres C and C^1 tend to move together when the comb touches A. (a) What can we conclude about the charge on the comb? Explain. (b) What can we conclude about the charge on the hair? Explain. (c) What would happen if electrode A were touched with samples of human hair?

19. Predict combining volume ratios for the reaction between $N_2(g)$ and $H_2(g)$ to form $NH_3(g)$, where all gases are at the same temperature and pressure.

20. Two charged spheres attract each other with a force of 100 dynes when they are 1 cm apart. The spheres are then moved 5 cm apart. What is the force of attraction between the spheres at a distance of 5 cm?

21. (a) Suppose that 0.1 mole of $MgSO_4(s)$ dissolves in enough water to make 1.0 litre of solution. Give the formula, number of moles, and molar concentration of each ionic species present in the solution. (b) Repeat for 0.1 mole $MgCl_2(s)$.

22. Three electrolysis cells are connected so that the same amount of electricity passes through each. The cells contain solutions of $Cu^{++}(aq)$, $Ag^+(aq)$, and $Pb^{++}(aq)$ ions, respectively. Write equations for the process occurring at the negative electrode in each cell.

23. Explain the following observations on conductivity: (a) distilled water conducts an electric current very poorly; (b) tap water conducts an electric current only slightly; (c) salt water conducts an electric current very well; (d) a solution of sugar in distilled water conducts an electric current no better than does distilled water alone. (e) Solid silver nitrate is a very poor conductor of an electric current, but silver nitrate in solution is a good conductor; (f) solid salt (NaCl) is a poor conductor, but molten salt is a very good conductor.

24. When $Ba(OH)_2$ solution is added to H_2SO_4 solution, a white precipitate forms. (a) What is the formula of the white precipitate that forms? (b) What is the other product of the reaction? (c) As a result of this addition, we would expect the electrical conductivity of the solution to do which of the following: increase, decrease, remain unchanged? (d) Explain your answer by writing appropriate ionic equations.

25. Suggest a test for the presence of the sulfate ion in water.

26. On a separate piece of paper, reproduce the table on the following page and fill in the blanks.

Number of protons	Number of electrons	Net charge
6	6	
8	10	
20		+2
9		−1
	18	−1
	18	+1

C

27. A research worker is studying the conductivity of solutions. First he measures the conductivity of a solution of 1 molar KCl. He then begins to add a $3M$ solution of $NaNO_3$. He measures the conductivity after each addition of $NaNO_3$ solution and then plots the data. The plot is shown in the margin as curve A. In the second experiment, he starts with the same volume of an identical KCl solution but he now adds a $3M$ solution of $AgNO_3$ instead of $NaNO_3$. Again, he measures the conductivity of the solution after each addition of $AgNO_3$. Explain on an ion-molecule level why curve A differs from curve B. (Neglect the dilution factor.)

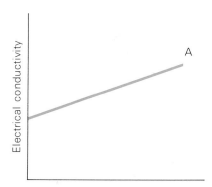

Millilitres of $3M$ $NaNO_3$ added

28. One litre of solution contains 0.100 mole of ferric nitrate, $Fe(NO_3)_3$, and 0.200 mole of calcium nitrate, $Ca(NO_3)_2$. Calculate the concentrations of the Fe^{+++}, Ca^{++}, and NO_3^- ions.

29. Write ionic and net ionic equations for each of the following reactions: (a) formation of the precipitate aluminum hydroxide from solutions of aluminum nitrate and sodium hydroxide, (b) formation of the precipitate silver sulfide from solutions of silver nitrate and potassium sulfide, (c) formation of the precipitate lead chloride from solutions of lead acetate and calcium chloride.

30. When silver nitrate ($AgNO_3$) and sodium chloride (NaCl) are mixed in solution, silver chloride (AgCl) is precipitated. (a) Write the overall and net ionic equations for this reaction. (b) If 1.0 litre each of 1.0-molar solutions of $AgNO_3$ and NaCl are mixed, what is the final concentration of each ion after mixing? (For your calculations, assume that the solubility of AgCl is negligible.)

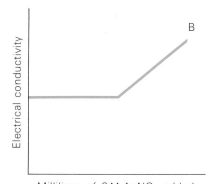

Millilitres of $3M$ $AgNO_3$ added

31. When solutions of copper(II) chloride and lead nitrate are mixed, a precipitate of lead chloride ($PbCl_2$) forms. (a) Write the overall ionic and the net ionic equations for this reaction. (b) If 0.200 litre of the $CuCl_2$ solution contains 0.300 mole of the compound, what is the concentration of each ion? (c) If 0.300 litre of the $Pb(NO_3)_2$ solution contains 0.150 mole of the compound, what is the concentration of each ion? (d) If the 0.200 litre of $CuCl_2$ solution is added to the 0.300 litre of $Pb(NO_3)_2$ solution, what is the concentration of *each ion* after the precipitate forms? (Assume that the solubility of $PbCl_2$ is negligible.)

Here was a substance emitting spontaneously and continually radiations similar to Röntgen rays, whereas ordinarily Röntgen rays can be produced only in a vacuum tube with the expenditure of electrical energy.

MARIE CURIE

ATOMIC STRUCTURE AND RADIOACTIVITY

Objectives

After completing this chapter, you should be able to:

- Explain the nuclear atom model.
- List the properties of protons, neutrons, and electrons and describe their location within the atom.
- Understand the meaning of the terms *atomic number* and *mass number.*
- Explain radioactive decay and the concept of half-life.
- Describe the nature of alpha, beta, and gamma radiation.
- Balance nuclear equations.

Tracks of atomic particles provide evidence of their existence and properties.

John Dalton viewed atoms as tiny, indivisible particles. Such a view convincingly explains many observed regularities. Laws concerning conservation of mass, definite composition, and other relationships become predictable and understandable when atoms are presumed to exist.

But Dalton's successful atomic model cannot, in its original form, account for the electrical effects described in Chapter 7. Those effects suggest that matter itself—and thus the atoms composing matter—must be electrical in nature. Electrical conductivity can be explained only by assuming a more detailed substructure for the atom. Atoms must be made up of smaller, simpler, electrically charged particles. These particles are positively charged protons and negatively charged electrons, introduced in Chapter 7. Protons and electrons were first incorporated in the current atomic model around the beginning of the present century. Changing a model, or modifying it to account for new observations, is an exciting part of science.

Changes and refinements in a model occur as scientists try to explain the results of new experiments. The experiments responsible for the model of atomic structure that we will develop in this chapter use sophisticated equipment and require considerable study in order to be well understood. Therefore, we shall present the new model without giving much experimental evidence at this point. Realize, however, that the model *is* solidly based on experiment. Some of the experimental evidence is summarized in the Extensions of Chapters 8 and 11.

8.1 Atomic architecture: Components of an atom

Experiments on the electrical conductivity of solutions (described in Section 7.4) suggest that some molecules and atoms can be changed into charged particles, or ions. Ions are viewed as being composed of protons and electrons, which have opposite electric charges. An electrically neutral atom is viewed as a combination of equal numbers of protons and electrons. Let us investigate this model.

By means of an instrument known as a **mass spectrometer** (discussed in the Extension of this chapter), scientists can determine the mass of one proton on the atomic weight scale. That mass, to two significant figures, is 1.0. The mass of an electron, determined on the same scale, is 0.00055, less than $\frac{1}{1,840}$ the mass of the proton.

Protons and electrons may seem to complicate the atomic model. Actually, they simplify it. Instead of picturing many different kinds of atoms for each of the more than 100 different known elements, we have introduced only two particles. Each atom is now viewed as containing the same particles: protons and electrons. Thus, the differences between atoms must be due to the *number* of protons and electrons that each contains.

The element hydrogen has the lightest atoms. According to our model, each hydrogen atom contains a single proton and a single electron. This agrees with the mass of hydrogen, given as 1.0 on the atomic weight scale. Experiments indicate that an atom of helium, the second lightest element, is four times as massive as an atom of hydrogen. However, experiments also demonstrate that the helium atom contains only two protons. But two protons account for only half the atomic mass, since the two electrons needed to make the helium atom neutral have virtually no mass.

Similar discrepancies in mass are found for all elements except hydrogen. Carbon has an atomic mass of 12 units, but the carbon atom contains only 6 protons. The gold atom contains 79 protons and 79 electrons but has a relative mass of 197. Clearly, these atoms must contain something *in addition* to protons and electrons.

Experiments indicate that the additional mass is due to a third particle, known as a **neutron.** A neutron has *no* electrical charge and a mass approximately equal to that of a proton: 1.0 on the atomic weight scale. Thus, helium atoms, with a mass of 4, contain two neutrons, two protons, and two electrons.

EXERCISE 8-1

Find the approximate atomic mass of chromium, fluorine, and tungsten, using the following information: (a) a chromium atom has 24 protons, 24 electrons, and 28 neutrons; (b) a fluorine atom has 9 protons, 9 electrons, and 10 neutrons; (c) a tungsten atom has 74 protons, 74 electrons, and 110 neutrons.

Table 8-1

Charge and approximate mass of some fundamental particles

Particle	Charge	Approximate mass (relative to the mass of a proton)
electron	−1	0.000 55
proton	+1	1
neutron	0	1

THE NUCLEAR ATOM MODEL

The properties of the three so-called fundamental particles that compose all atoms are summarized in Table 8-1. How are these particles arranged within the atom? This question was answered in 1911 by the British physicist Ernest Rutherford (1871–1937) and his coworkers. Their experiments indicated that atoms contain an extremely small, positively charged nucleus. Virtually all the mass of an atom is concentrated in the nucleus.*

The nucleus occupies a tiny fraction (about one trillionth, or 10^{-12}) of the total volume of an atom. The relatively large space surrounding the nucleus contains the electrons, with their negligible mass. Strange as it may seem, atoms are mostly empty space, not hard, little spheres.

A hydrogen atom, the lightest atom known, contains one proton in the nucleus and one electron outside the nucleus. The one proton accounts for the positive charge of the hydrogen nucleus. According to experiment, all other nuclei have positive charges that are whole-number multiples of the single proton's charge. Each nucleus contains a definite number of protons, and the charge on the nucleus is fixed by this number.

All atoms of a particular element have the same nuclear charge; that is, they have the same number of nuclear protons. All hydrogen atoms have a nuclear charge of +1 and one extranuclear electron. All helium atoms have a nuclear charge of +2 and two extranuclear electrons, and so on. We shall see that *the extranuclear electrons determine the chemistry of an atom.*

The nucleus, containing protons and neutrons, is located at the center of the atom and cannot be changed by normal chemical processes. Nuclear protons and/or neutrons cannot be added to or removed from atoms except by the nuclear processes discussed later in this chapter. This does not apply to electrons, however, which are located at or near the outer surface of an atom. Electrons *can* be removed from or added to a neutral atom, giving it a net charge. Thus, the atomic model accounts for the formation of ions.

ION FORMATION AND THE NUCLEAR ATOM MODEL

The nuclear atom model was developed because the view of atoms as hard spheres could not account for the electrical effects observed in experiments. To account for electrical conductivity, for example, we introduced

*See page 208 of the Extension at the end of this chapter.

(continued on page 188)

Electronic numbers

Several types of number display devices were developed during the past decade. Perhaps the most familiar is the bright-line display commonly used in calculators and digital clocks. Bright-line displays make use of the principles governing the operation of cathode-ray tubes. Other kinds of displays include those produced by plasma panels and liquid crystals.

Displays employing cathode-ray tube principles dominate the market today. Such displays commonly employ two glass plates separated by an insulated space. The space between the plates is filled with mercury vapor and neon gas under low pressure. Metallic bars are arranged on the rear glass plate in a stroke array that is familiar to all who use calculators. The bars in the common seven-stroke pattern are shown in the illustration below. The bars can be illuminated selectively to produce all the digits from 0 through 9. This is done by supplying electrical energy to the correct bars. The energized bars then cause the gases to glow along their lengths. The glow is similar to that produced in neon signs. In calculator displays, the glow is limited to the regions that receive the electric current.

A typical two-digit display is illustrated below. A seven-stroke metallic bar arrangement on the back plate is required for each digit displayed. Above each of the bar arrangements is a tin oxide electrode that acts as a source of electrons when stimulated. This illustration shows that bars in the same numer-

A metallic bar arrangement showing separate circuits for the display of a two-digit number.

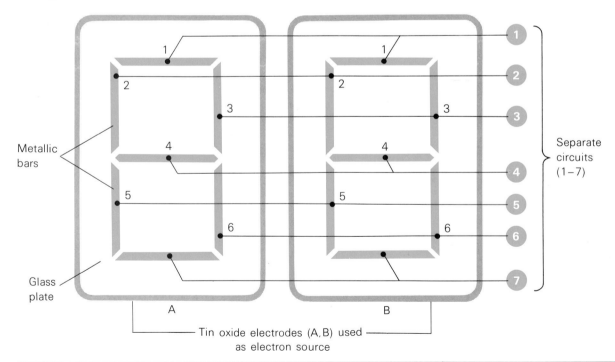

Bars	Activated tin oxide electrodes	Gas above these bars will glow
		A B
1	A, B	
2	B	
3	A	
4	A, B	
5	A, B	
6	B	
7	A, B	
Total effect =		

A two-digit display showing the numeral 26.

ical position are connected to the same circuit in a display.

The illustration on this page shows the sequence of steps in which the numeral 26 can be formed. When the tin oxide electrodes are stimulated, they release electrons that cause the gases to glow over the bar or bars that are connected to particular circuits on an *activated* electrode. The electrical stimuli are repeated very quickly and many times per second. The total effect of many individual stimuli is the appearance of the desired number. There are many kinds of cathode-ray tube displays, but they all operate on the same principles.

The plasma display panel is made up of a matrix of gas discharge cells that have insulated electrodes deposited on two glass plates separated and filled with a mixture of neon and argon gas. The whole assembly is then sealed so that the gas is trapped between the glass plates. This type of display is still in the exploratory stage and might result in flat panel television screens that could be hung from walls like pictures.

Liquid crystals do not emit light. They can only be seen when they are externally illuminated. Such crystals are organic substances that can exist only over a limited temperature range. Ordinary crystalline solids lose their crystal lattice upon heating. The molecules become free to move at random throughout the molten material. In liquid crystals, however, the molecular pattern remains. This condition can almost be considered a new state of matter and has been called the *mesophase,* an intermediate phase between a solid and a liquid.

The molecules of a liquid crystal are long and rod-shaped. They are about three times as long as they are wide, and they arrange themselves with their long axes parallel to each other. The alignment of molecules can be caused by rubbing the substance in one direction or by the application of magnetic or electrical force. There are three types of liquid crystals. The molecules in some crystals are free to rotate about their long axes. They can slip and slide along each other like crayons in a box. Another type of crystal arrangement consists of molecules that have a slight helical twist. Still other crystals have layered molecular structures. Their molecules are shorter and fit neatly over one another. So far, crystals of the first type have proven to be the most useful for numerical displays.

Research in the field of electronic numbers is ongoing, and the future possibilities for application of bright-line and liquid-crystal displays are considerable. The commercial production of flat-panel television screens is now a reality. Electronic displays will also be used in instrument panels for aircraft and other vehicles where the operator must have access to reliable, easily read information.

the idea of positive (+) and negative (−) ions. To appreciate how the model accounts for ions, let us use it to construct models of atoms of those elements we assumed formed + and − ions.

Consider first the sodium atom, with a nuclear charge of +11 and a mass of 23 atomic mass units. The nuclear charge indicates that there are 11 protons, or $11p^+$. The positive charge must be balanced in the neutral sodium atom by 11 electrons, or $11e^-$. The 11 protons account for 11 of the 23 mass units. Subtracting 11 from 23 leaves 12 mass units, indicating the presence of 12 neutrons, or $12n$.

The sodium *ion*, with a charge of +1, must have one more positive or one less negative charge. Since the nucleus is viewed as fixed, we accept the second possibility and picture the removal of one electron from the atom to form the ion. We can diagram the structure of the sodium atom and ion as follows, using e^- for electrons, p^+ for protons, and n for neutrons:

$$\left[\begin{array}{cc} \begin{array}{c} 11p^{+\,*} \\ 12n \end{array} & 11e^- \\ \text{nucleus} & \text{electrons} \end{array}\right]^0 \longrightarrow 1e^- + \left[\begin{array}{cc} \begin{array}{c} 11p^+ \\ 12n \end{array} & 10e^- \\ \text{nucleus} & \text{electrons} \end{array}\right]^+$$

SODIUM *ATOM*	electron removed	SODIUM *ION*

Thus, a sodium atom that lacks one electron is a sodium ion with a charge of +1. Because of their chemistry, magnesium and calcium atoms were represented in Chapter 7 as forming ions with charges of +2. We may thus assume that each magnesium and calcium atom loses *two* electrons in forming an ion.

EXERCISE 8-2

Use diagrams like those representing the sodium atom and ion above to represent each of the following processes: (a) formation of Mg^{++} ion from a magnesium atom, given a nuclear charge of +12 and a mass of 24 for the Mg atom; (b) formation of Ca^{++} ion from a calcium atom, given a nuclear charge of +20 and a mass of 40 for the Ca atom.

Negative ions, such as F^- and Cl^-, are formed when neutral atoms *gain* electrons. A fluorine atom has a nuclear charge of +9 and a mass of 19. The equation for forming a fluoride ion can be written as:

$$\left[\begin{array}{cc} \begin{array}{c} 9p^+ \\ 10n \end{array} & 9e^- \\ \text{nucleus} & \text{electrons} \end{array}\right]^0 + 1e^- \longrightarrow \left[\begin{array}{cc} \begin{array}{c} 9p^+ \\ 10n \end{array} & 10e^- \\ \text{nucleus} & \text{electrons} \end{array}\right]^-$$

FLUORINE *ATOM*	electron added	FLUORIDE *ION*

*The protons and neutrons are circled as a convenient way of separating them from the extranuclear electrons. Remember that these atomic structure diagrams are not scale drawings. As we mentioned earlier, a nucleus is but a tiny fraction of the entire size of an atom.

Thus, a fluorine atom containing one extra electron is a fluoride ion with a charge of -1. *Positively charged ions are formed by removing electrons from neutral atoms. Negatively charged ions are formed by adding electrons to neutral atoms. Groups of atoms* can also lose electrons to become positively charged ions, such as the ammonium ion (NH_4^+). Similarly, *groups of atoms* can pick up extra electrons to become negatively charged ions, such as nitrate (NO_3^-) and sulfate (SO_4^{--}).

EXERCISE 8-3

Use diagrams to represent each of the following: (a) formation of Cl^- ion from a chlorine atom, given a nuclear charge of $+17$ and a mass of 35 for the Cl atom; (b) formation of Br^- ion from the bromine atom, given a nuclear charge of $+35$ and a mass of 79 for the Br atom.

ENERGY RELATIONS IN ION FORMATION

Under what conditions are ions produced? In general, positive ions are formed when large amounts of energy are available. For example, atoms in a flame will lose electrons, as will atoms in the path of an electrical discharge. Thus, a lightning flash or beam of high-energy electrons will produce positive ions. Equations representing this process for gaseous sodium (Na) and magnesium (Mg) atoms can be written as

$$Na(g) + energy \longrightarrow Na^+(g) + e^-$$

$$Mg(g) + energy \longrightarrow Mg^+(g) + e^-$$
$$Mg^+(g) + energy \longrightarrow Mg^{++}(g) + e^-$$

Note the energy terms in these equations. Why are they necessary? Since the positive protons attract negative electrons, energy is needed to remove each electron. This energy is known as *ionization energy* and will be discussed in detail in Chapter 10. Energy is used to *pull* the electron away from the atom. A positive ion and a free electron result.

When neutral atoms gain electrons, negative ions are formed. A neutral fluorine atom can *gain* an electron to form a negative ion, F^-, as shown in the diagram on page 188. In the case of fluorine atoms, this change does not require energy but actually *releases* energy:

$$F(g) + e^- \longrightarrow F^-(g) + energy$$

This is not a very common situation but is observed when certain negative ions are formed. On the other hand, energy is *always* absorbed when a neutral atom is converted to a *positive ion* and an *electron*.

THE SIZE OF ATOMS AND NUCLEI

How large is an atom? We cannot answer this question for a single, isolated atom. The reason is that atoms are cloudlike structures that gradually fade away to nothing on their outer edges. We cannot really decide where

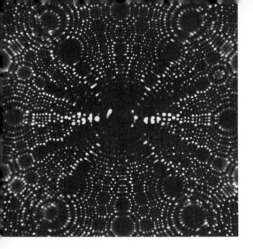

Figure 8-1

The arrangement of atomic sites in the tip of a microscopically thin tungsten crystal, as viewed through a field-ion microscope. The bright spots in the circular and straight "chains" are images of single tungsten atoms, not pictures of the actual atoms.

the outer edge of an atom is located. Still, we can devise experiments to indicate how close the nucleus of one atom can come to the nucleus of another atom. In this way we can get a working, or operational, answer. As atoms approach, they are held apart by the repulsion between positively charged nuclei. The electrons of the two atoms also repel each other, but they are attracted by both nuclei. How close two nuclei approach depends, in part, on a balance between the forces of repulsion and attraction. Experiments suggest that the diameters of atoms vary from 0.000 000 01 to 0.000 000 05 centimetres (from 1×10^{-8} to 5×10^{-8} centimetres). Nuclei are much smaller. A typical nuclear diameter is about 1×10^{-12} centimetres, or about $\frac{1}{10,000}$ the diameter of an atom.

To give you a better sense of these relative sizes, we could choose a large major league baseball stadium as the model for an atom. Assume that one atom occupies the entire stadium. To keep the proper scale, the nucleus would be about the size of a small marble in the middle of the playing field. For the hydrogen atom, the marble would represent one proton and would be located at the center of the stadium, somewhere behind second base. The one electron present in the neutral atom would move around, occupying at one time or another *all* the remaining area of the stadium. For the helium atom, the nucleus would be represented by a somewhat larger marble containing two protons and two neutrons. The two electrons of the neutral helium atom would now share the ample space of the huge stadium. In

Figure 8-2

The atomic nucleus bears the same relationship to an atom as a marble does to a sports stadium.

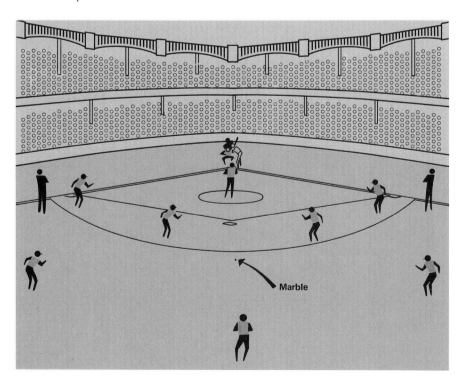

Marble

order to represent an accurate model of an atom, the marble must provide almost *all* the mass present in the entire stadium—even though it takes up almost none of the space.

EXERCISE 8-4

Imagine bees flying around their beehive as the model of an atom. The beehive represents the nucleus. The bees, flying around the countryside in search of pollen, represent the electrons. (a) If the beehive has a 20-cm radius, how far from the hive could the bees fly to maintain the proper scale of nucleus to entire atom in the nuclear atom model? Recall that the nucleus is only about $\frac{1}{10,000}$ (10^{-4}) the *diameter* of the entire atom. Express your answer in kilometres, and assume that the bees that are farthest away define the outer surface of the atom. (b) The nucleus is assumed to be located in the center of the atom. In order to make the beehive model accurate, should the roving bees spread out in all directions as they leave the hive, or should they spread out in just one or two directions?

ATOMIC NUMBER

We have seen that just three subatomic particles—electrons, protons, and neutrons—can form many different kinds of atoms. This is because the *number* of particles present in different atoms can differ. Every neutral hydrogen atom has one proton and one electron. Every helium atom has two protons in the nucleus surrounded by two electrons. Every lithium atom has three protons in the nucleus surrounded by three electrons. The lithium atom is heavier than the hydrogen or helium atom.

Thus, each chemical element consists of atoms whose nuclei contain a particular number of protons and, therefore, a particular nuclear charge. The number of protons in the nucleus is called the **atomic number.** It must be a *whole* number. Fractional protons or neutrons are *not* available. Oxygen, with atomic number 8, has eight protons in the nucleus and a nuclear charge of $+8$. The nucleus of a neutral oxygen atom must be surrounded by eight electrons. The atomic numbers of some elements are listed in Table 8-2. A complete list of atomic numbers appears inside the back cover of this book. Each element listed there has a distinctive name, symbol, atomic number, and atomic weight. A given element can be identified by its name, symbol, or atomic number. Helium is thus identified by its name, its symbol (He), or as the element with atomic number 2.

MASS NUMBERS AND ISOTOPES

All atoms of a given element have the same nuclear charge. Do all atoms of a given element have the same mass? Almost all hydrogen atoms do have the same mass: the sum of the mass of one proton and the mass of one electron. For hydrogen atoms, the nucleus consists of a single proton. But a small fraction of hydrogen atoms (0.015 percent, or 15 out of every 100,000)

Table 8-2

Atomic numbers of some elements

Element	Atomic number*
carbon (C)	6
chlorine (Cl)	17
copper (Cu)	29
fluorine (F)	9
gold (Au)	79
helium (He)	2
hydrogen (H)	1
iron (Fe)	26
lead (Pb)	82
mercury (Hg)	80
nickel (Ni)	28
nitrogen (N)	7
oxygen (O)	8
phosphorus (P)	15
potassium (K)	19
silver (Ag)	47
sodium (Na)	11
sulfur (S)	16
uranium (U)	92
zinc (Zn)	30

*Atomic number = number of protons.

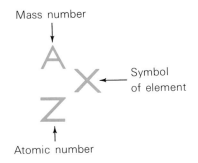

Figure 8-3

Superscripts and subscripts are used to indicate the mass number and atomic number of an element.

have nuclei whose mass is approximately twice as great as the mass of the proton. To account for the mass of these "heavy" hydrogen atoms, we add one neutron to each of their nuclei. This kind of hydrogen atom is called hydrogen-2 or **deuterium.** The two kinds of hydrogen atoms, having the *same* atomic number but *different* masses, are called **isotopes.** An isotope is identified by specifying two factors. First, the isotope is identified by the symbol or name of the element. Second, it is identified by a number that represents the sum of the number of protons and neutrons in the nucleus. The number of protons plus the number of neutrons in a nucleus is called the **mass number** of the nucleus.

Thus, two numbers—the atomic number and the mass number—define the major parts of any nucleus. The *atomic number,* often symbolized by the letter Z, is the number of protons in the nucleus. For the complete, neutral atom, it is also the number of electrons around the nucleus. The *mass number,* sometimes symbolized by the letter A, is the number of protons plus the number of neutrons in the nucleus. Since protons and neutrons are *both* nuclear particles, they are often referred to as **nucleons.** Thus, the mass number is equal to the number of nucleons. In summary,

atomic number (Z) = number of *protons* in nucleus and determines nuclear *charge*

mass number (A) = number of *nucleons* (protons + neutrons) in nucleus and determines nuclear *mass*

EXERCISE 8-5

Draw atomic structure diagrams for three elements listed, given their atomic numbers (Z) and their mass numbers (A): (a) $Z = 3$, $A = 6$; (b) $Z = 6$, $A = 14$; (c) $Z = 8$, $A = 16$. (d) What element does each atom represent? Use A and Z notation to represent these elements.

Most chemical elements consist of mixtures of isotopes. Oxygen, atomic number 8, has three stable isotopes. The use of the word *stable* in this way means *nonradioactive,* a condition discussed later in this chapter. The oxygen isotope of mass number 16 is the most abundant oxygen isotope. About 99.76 percent of all oxygen atoms (9,976 of every 10,000) are in the form of this isotope. Only 0.04 percent of oxygen atoms (4 of every 10,000) have a mass number of 17, and about 0.20 percent (20 of every 10,000) have a mass number of 18. The atomic structure of each oxygen isotope follows:

$8e^-$	$8e^-$	$8e^-$
$8p^+$ $8n$	$8p^+$ $9n$	$8p^+$ $10n$
16 nucleons	17 nucleons	18 nucleons
OXYGEN-16	OXYGEN-17	OXYGEN-18

Remember that these atomic structure diagrams are *not* to scale. If they were, the circled nucleus in each would be an invisibly small dot.

EXERCISE 8-6

Table 8-3 gives the relative amounts of the two principal isotopes of boron as found in nature. (a) How many boron-10 atoms are there in 10,000 atoms of boron as found in nature? (b) What is the total mass of these atoms in atomic mass units (*u*)? (c) How many boron-11 atoms are there in 10,000 atoms of boron, and what is their total mass? (d) What is the *total* mass of 10,000 boron atoms? (e) What is the *average* atomic mass for a boron atom as found in nature? Check your answer against the atomic mass of boron, as given in the table of atomic weights on page 70.

Table 8-3

Composition of some common isotopes

Name of isotope	Abundance in nature (%)	Atomic number	Mass number	Nucleus Composition*		Mass	Charge	Number of electrons in neutral atom	Precise isotopic mass (*u*)
hydrogen-1	99.984	1	1	1p		1	+1	1	1.008123
hydrogen-2	0.016	1	2	1p	1n	2	+1	1	2.014708
helium-3	1.34×10^{-4}	2	3	2p	1n	3	+2	2	3.01700
helium-4	100	2	4	2p	2n	4	+2	2	4.00390
lithium-6	7.30	3	6	3p	3n	6	+3	3	6.01697
lithium-7	92.70	3	7	3p	4n	7	+3	3	7.01822
beryllium-9	100	4	9	4p	5n	9	+4	4	9.01503
boron-10	18.83	5	10	5p	5n	10	+5	5	10.01618
boron-11	81.17	5	11	5p	6n	11	+5	5	11.01284
carbon-12	98.892	6	12	6p	6n	12	+6	6	12.00382
carbon-13	1.108	6	13	6p	7n	13	+6	6	13.00758
nitrogen-14	99.64	7	14	7p	7n	14	+7	7	14.00751
nitrogen-15	0.36	7	15	7p	8n	15	+7	7	15.00493
oxygen-16	99.76	8	16	8p	8n	16	+8	8	16.00000
oxygen-17	0.04	8	17	8p	9n	17	+8	8	17.00450
oxygen-18	0.20	8	18	8p	10n	18	+8	8	18.0049
fluorine-19	100	9	19	9p	10n	19	+9	9	19.00450
chlorine-35	75.4	17	35	17p	18n	35	+17	17	34.97867
chlorine-37	24.6	17	37	17p	20n	37	+17	17	36.97750
gold-197	100	79	197	79p	118n	197	+79	79	197.04
uranium-235	0.72	92	235	92p	143n	235	+92	92	———
uranium-238	99.28	92	238	92p	146n	238	+92	92	238.12

*p = proton, n = neutron.

8.2 Properties of atomic nuclei: Radioactivity and nuclear chemistry

Different isotopes of the same element have almost identical chemical properties. We believe that this is because the neutral atoms of each isotope have the same number of protons and thus the same number of electrons. Isotopes differ only in the number of neutrons their nuclei contain. Thus, the nuclei have different masses. This difference can cause striking variations in the behavior of the nuclei. For example, uranium-235 is an isotope of uranium used as the active fuel in conventional nuclear reactors. Another isotope, uranium-238, cannot be used in that way. On the other hand, the chemistry of uranium-235 and uranium-238 is almost identical because both isotopes have 92 extranuclear electrons.

NATURAL RADIOACTIVITY

The uranium isotopes mentioned above do share the property of being naturally radioactive. Both spontaneously undergo a process known as *radioactive decay,* which produces very penetrating radiation. This radiation results from a very small explosion in the uranium nucleus. Evidently, uranium nuclei are *unstable,* as shown by their explosive tendencies. Therefore, we define natural **radioactivity** as the spontaneous breakdown of unstable atomic nuclei, with the release of particles and energy. Various experiments indicate that an atom undergoing radioactive decay will emit radiation. Three types of radiation are important to us here: alpha (α), beta (β), and gamma (γ) radiation. **Alpha** "rays" are really *particles:* helium nuclei, with a charge of +2 and a mass of 4. **Beta** "rays" are also particles. They are very fast-moving electrons. **Gamma** rays are truly radiation—electromagnetic radiation—like visible light and X-rays but of even higher energy. Table 8-4 lists some of the properties of these three types of radiation.

The existence of radioactivity is striking proof that the nuclei of atoms are made up of simpler parts. For example, uranium nuclei should be composed of the simpler particles such as protons and neutrons. The heavy isotope of uranium, uranium-238, is an alpha emitter. When a uranium-238 nucleus disintegrates, it emits one alpha particle, abbreviated $^4_2$He. This abbreviation shows the atomic number (Z) at the lower left and the mass number (A) at the upper left of the element's symbol, as shown in Figure 8-3.

As might be expected, when a uranium nucleus decays by emitting an alpha particle, it ends up with two protons and two neutrons *less* than it had before. A nucleus with only 90 protons is *not* uranium, but element 90, which is thorium. We can represent this nuclear event by the following diagram, showing only the nuclei of the elements involved:

URANIUM NUCLEUS		ALPHA PARTICLE		THORIUM NUCLEUS
$92p^+$ $146n$	$\longrightarrow$	$2p^+$ $2n$	$+$	$90p^+$ $144n$
238 nucleons		4 nucleons		234 nucleons
$^{238}_{92}$U	$\longrightarrow$	$^4_2$He	$+$	$^{234}_{90}$Th

Table 8-4

Properties of three types of radiation

Radiation	Charge	Approximate mass (amu)	Composition	Penetrating power
alpha (α)	+2	4	helium nuclei, $_2^4He^{++}$	short range, stopped by a piece of paper
beta (β)	−1	$\frac{1}{1,837}$	electron, $_{-1}^0e$	intermediate range, stopped by a few centimetres of water
gamma (γ)	0	0	high-energy radiation shorter in wavelength than X-rays	long range, stopped by a few centimetres of lead

The equation below the diagrams of the nuclei is known as a **nuclear equation.** The nuclear event it describes is known as a **transmutation reaction.** A *transmutation* is a change in the identity of a nucleus because of a change in the number of its protons and neutrons.

In nuclear processes, two elementary laws must be obeyed:

1. *The total number of electrical charges represented by the atomic numbers must be the same on both sides of the equation.*
2. The total mass numbers must be the same on both sides of the equation.

These rules provide a straightforward way of balancing nuclear equations. Reexamination of the above equation shows that the two conditions have been met:

$$_{92}^{238}U \longrightarrow {}_2^4He + {}_{90}^{234}Th \quad \begin{array}{l}(238 = 4 + 234)\\(92 = 2 + 90)\end{array}$$

The original uranium is called the parent element, and the thorium-234 isotope is called the daughter element. If the parent is an alpha emitter, it decays to produce a daughter whose nuclei each contain two protons and two neutrons *less* than each parent nucleus.

EXERCISE 8-7

Plutonium-238 ($_{94}^{238}Pu$) decays to give an α-particle and a new atom. What is the atomic number of the new atom? What is its mass number?

In the case of uranium-238, the daughter isotope, thorium-234, is also radioactive. However, unlike uranium-238, thorium-234 is a beta emitter. It decays to produce element 91, a protactinium isotope with the *same* mass number as its parent, thorium-234. The symbol for protactinium is Pa. The process may be represented as follows:

THORIUM NUCLEUS		ELECTRON (BETA PARTICLE)		PROTACTINIUM NUCLEUS

$$\begin{array}{ccccc} \underset{144n}{90p^+} & \longrightarrow & e^- & + & \underset{143n}{91p^+} \\ 234 \text{ nucleons} & & & & 234 \text{ nucleons} \\ {}^{234}_{90}\text{Th} & \longrightarrow & {}^{0}_{-1}e & + & {}^{234}_{91}\text{Pa} \end{array}$$

Notice that the mass numbers total 234 on each side of the equation and that the atomic numbers total 90 on each side ($-1 + 91 = 90$). Recall that the beta particle is an *electron* and, as such, has an atomic number of -1 charge and essentially zero mass. The daughter nuclei of beta emitters always have one more proton and one less neutron than the parent nuclei. They thus have the *same* number of nucleons as the parent. These changes can be accounted for by assuming that the neutron can decay to give a proton and an electron. According to this assumption, a neutron can change to a proton by emitting its electron. The ejected *electron* is the observed *beta particle*. This process can be represented as follows:

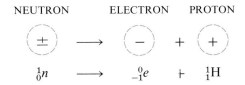

$$\begin{array}{ccccc} \text{NEUTRON} & & \text{ELECTRON} & & \text{PROTON} \\ (\pm) & \longrightarrow & (-) & + & (+) \\ {}^{1}_{0}n & \longrightarrow & {}^{0}_{-1}e & + & {}^{1}_{1}\text{H} \end{array}$$

This representation accounts for all changes associated with beta decay. The beta particles themselves are the electrons emitted by decaying neutrons. Beta-particle electrons come from the nucleus and are not to be confused with the electrons surrounding positive nuclei in neutral atoms. The change that results—one neutron into one proton—explains why daughter nuclei contain one *more* proton and one *less* neutron than their beta-decaying parent nuclei.

EXERCISE 8-8

Potassium-43(${}^{43}_{19}\text{K}$) decays by beta-particle loss. Write the nuclear equation for the process.

The third kind of radiation, gamma radiation, is often present with alpha decay and beta decay. The release of high-energy gamma radiation results in decay products of lower energy but otherwise identical nucleon makeup.

Of the three types of radiation, gamma rays are the most penetrating. Beta particles are intermediate in penetrating power, while alpha particles

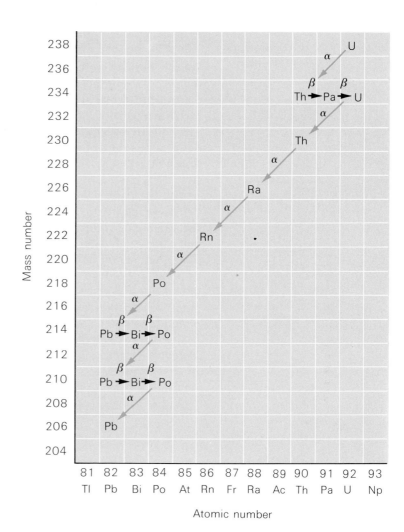

Figure 8-4

Stages in the radioactive decay of uranium-238 to lead-206.

are the least penetrating, as shown in Table 8-4. Alpha particles are easily stopped by metal foil, paper, or skin. Inside the body, however, alpha emitters are dangerous to cells because the cells stop the particles. These particles can cause a great deal of damage to cellular tissues.

The radioactive decay of uranium-238 to thorium-234 and of thorium-234 to protactinium-234 comprise a small portion of a series of other, similar, spontaneous transmutations. For example, protactinium-234 is also radioactive, emits beta particles, and forms another radioactive isotope of uranium called uranium-234. Uranium-234 is an alpha emitter that decays to thorium-230. The entire radioactive decay series is summarized in Figure 8-4. The end product is lead-206, a stable, nonradioactive isotope.

EXERCISE 8-9

Diagram and write balanced nuclear equations for the decay of (a) Pa-234 to U-234 (b) U-234 to Th-230 (c) Th-230 to Ra-226 (d) Ra-226 to Rn-222.

RATE OF NUCLEAR DECAY: HALF-LIFE

Each radioactive element has its own, steady rate of decay. The time required for half the nuclei of any radioactive element to decay is defined as the **half-life** of that element. The half-life for a given isotope is constant and is characteristic of that isotope. Half-lives of different isotopes vary widely. The half-life of iodine-131 is 8.1 days, while iodine-133 has a half-life of 21 hours. For iodine-129, it is 17 million years. The half-life of uranium-238 is 4,500,000,000 years (4.5×10^9 years). By contrast, its daughter element (thorium-234) has a half-life of only 24 days. During each half-life period, one half of the parent element will decay. As shown in Figure 8-5, this means that if we start with 1 gram of pure parent element, such as uranium-238, then $\frac{1}{2}$ gram of uranium-238 nuclei will decay to thorium-234 during the first half-life period. After 4.5×10^9 years, only $\frac{1}{2}$ gram of uranium-238 remains. The other $\frac{1}{2}$ gram has decayed to produce thorium-234, which, in turn, has decayed to produce other elements. During the next half-life period (the next 4.5×10^9 years) half the *remaining* uranium-238 ($\frac{1}{2}$ of $\frac{1}{2}$ gram, or $\frac{1}{4}$ gram) will decay to produce thorium-234. During the next, or third, half-life period (another 4.5×10^9 years), half of the remaining $\frac{1}{4}$ gram of uranium-238, or $\frac{1}{8}$ gram, will decay, and so on. *The shorter its half-life, the less stable a radioactive element is considered to be.* Thus, thorium-234, whose half-life is 24 days, is much less stable than uranium-238, whose half-life is 4.5×10^9 years.

Figure 8-5

During each half-life period, one half of a radioactive element decays.

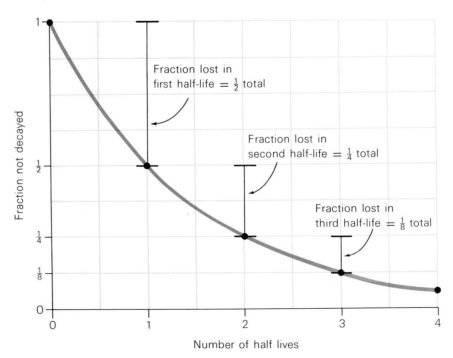

EXERCISE 8-10

One of the radioactive waste products of a nuclear power plant is iodine 131, with a half-life of about 8 days. It decays to produce an isotope of the inert gas xenon. (a) If 100 g of I-131 is present in a sample of radioactive waste, how many grams will remain after 8 days, 16 days, and 24 days? (b) If you were given a 200-g sample of I-131 now, in how many days would you have only 25 g of it remaining as I-131? (c) The iodine-131 isotope decays to give xenon-131 and another particle. Write a nuclear equation for this reaction and identify the other particle produced.

ARTIFICIAL RADIOACTIVITY

When radium-226 is placed in a lead container (such as the one shown in Figure 8-6), the emitted alpha particles are stopped by the lead walls. The particles emerge only through the container opening. The high-speed alpha particles shoot out of the opening much like a stream of tiny bullets. When directed at targets made of various nonradioactive elements, some of the alpha "bullets" are incorporated in the target nuclei. There they often cause normally stable elements to become *artificially* radioactive and thus unstable.

The problem with the use of positive alpha particles to induce radioactivity is the repulsion between the particles and the positive nuclei of the target atoms. The particles must have enough speed to overcome the repulsion of the positively charged nucleus. Although neutrons have no electric charge

Figure 8-6

An opening in the lead container directs particles of radium-226 decay toward a target of a nonradioactive element.

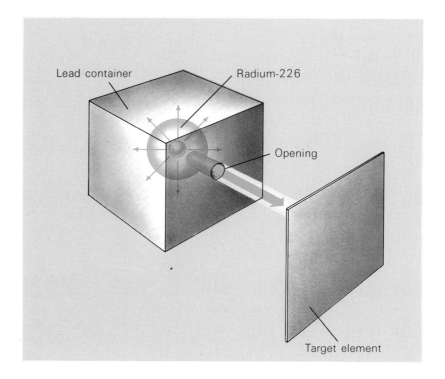

and so experience no such repulsion, their use presents other problems. Neutrons were widely used as bombarding particles in the years following their discovery. The use of neutrons in nuclear research during the 1930s and 1940s led to the development of nuclear power. Further discussions of nuclear reactions are found in Chapter 14.

Key Terms

alpha particle

atomic number

beta particle

deuterium

gamma particle

half-life

isotope

mass number

mass spectrometer

mass spectrum

neutron

nuclear equation

nucleon

radioactivity

transmutation reaction

SUMMARY

Dalton's model of the atom as a hard sphere has been modified to account for the results of more recent experiments. This modified model pictures the atom as being made up of a tiny nucleus that contains almost all the mass of the atom but that has a diameter only $\frac{1}{10,000}$ the diameter of the entire atom.

The nucleus of an atom contains positive protons and electrically neutral neutrons, each with a mass of approximately 1.0 on the atomic weight scale. Electrons are negatively charged and have a mass only 0.000 55 that of the proton on the atomic weight scale. Electrons surround the tiny, central, positively charged nucleus.

The nucleus of an atom is not changed in chemical reactions. Chemical processes are characterized by changes in extranuclear electrons. A neutral atom can *lose* electrons to form a *positive* ion or *gain* electrons to form a *negative* ion. Energy is required to remove an electron from a neutral atom because the electron is attracted to the positive nucleus. The term *ionization energy* is used to refer to the energy needed for electron removal.

The atoms of different elements contain different numbers of protons, neutrons, and electrons. All atoms of a particular element contain the *same* number of protons and thus the *same* nuclear charge. The number of electrons equals the number of protons in any neutral atom. The atomic number (Z) is the number of protons in the nucleus of an atom. The mass number (A) is the number of nucleons (protons plus neutrons) in a nucleus.

Atoms with the *same* atomic number but *different* masses are called isotopes. Most chemical elements consist of mixtures of several isotopes. Different isotopes of the same element have essentially identical chemical properties. They differ only in mass and in nuclear properties.

Certain atomic nuclei are unstable and undergo spontaneous breakdown. This process, known as natural radioactivity, is accompanied by the release of alpha, beta, and gamma radiation. Alpha radiation consists of helium nuclei: particles with a charge of $+2$ and mass of 4.0 on the atomic weight scale. Beta radiation consists of fast-moving electrons. Gamma radiation is very-short-wavelength, electromagnetic radiation similar to X-rays but with a higher level of energy.

Radioactive nuclei emit some of their nucleons and change into other elements. This process, known as transmutation, changes the identity of a nucleus since it changes the number of protons in the nucleus. Such changes are summarized by nuclear equations.

The half-life of a radioactive element is the time required for half of its nuclei to decay. Half-lives vary from a fraction of a second to millions of years, depending on the element. Normally stable, nonradioactive elements can be made artificially radioactive by bombardment with high-energy nuclear particles.

QUESTIONS AND PROBLEMS
A

1. Which of the following facts would *not* be explained by Dalton's original model in which atoms were viewed as tiny, indivisible particles?
   (a) When two atoms combine to make a molecule, the mass of the molecule is equal to the sum of the masses of the two atoms.
   (b) If energy is given off in a chemical process, an equal amount of energy must be supplied to make the process go back the other way.
   (c) A solution of sodium chloride in water conducts the electric current.
   (d) A solution of sugar in water does not conduct the electric current.
   (e) A charge builds up when hair is combed.
   (f) A charge builds up when a rubber rod is rubbed with cat's fur.
   (g) Spheres of an electrometer push apart when both are connected to the positive terminal of a battery.
   (h) Gaseous atoms in an electrical discharge tube give cathode rays.
   (i) Atoms can be converted to charged ions by a beam of x-rays.
2. Explain how Dalton's model could be modified (if needed) to explain each of the above facts.
3. An atom has 3 protons ($Z = 3$) and a mass number of 6 ($A = 6$).
   (a) Identify the element.
   (b) Write the symbol to indicate mass number and Z.
   (c) Indicate what other particles must be present in a neutral atom that fits this description.
   (d) Tell where each of the particles would be found.
   (e) If the atom is made of charged particles, why does the atom not carry a charge?
4. In only *one* case is the mass of the atom equal to its nuclear charge. In which atom is this true? Why is this true in terms of nuclear structure?
5. (a) Draw an atomic structure diagram to represent the makeup of the sodium atom of mass 23. Show the general area where each of the particles is located.
   (b) How can this atom be converted to an ion?
   (c) Is energy absorbed or released in this process? Is the process endothermic or exothermic?
   (d) What is the energy involved in the formation of an ion called?
   (e) Sodium also has an isotope of mass 24, which is *radioactive*. Draw an atomic structure diagram for one of these atoms.
   (f) Sodium-24 decays by loss of beta particles and gamma rays. Write the equation for the *transmutation reaction*.
   (g) What fraction of a Na-24 sample would be left after 30 hours? The half-life of Na-24 is 15 hours.
6. Fluorine has only a single stable isotope of mass number 19. (a) Write the symbol to show Z and A for this isotope. (b) Draw the atomic structure diagram for this isotope. (c) Write an equation to show how it can become an ion. (d) Is process exothermic or endothermic?
7. Draw an atomic structure diagram for: (a) $^{24}Mg^{++}$ ion, (b) $^{23}Na^+$ ion, (c) ^{20}Ne (neutral) atom, (d) $^{19}F^-$ ion. What is the *same* in all four cases?
8. (a) It can be shown by electron diffraction that the nuclei of the fluorine atoms in F_2 are separated by 1.44×10^{-8} cm. Suggest a reasonable

radius for a fluorine atom. (b) Why is the concept of an atomic radius rather difficult to visualize for a single isolated atom?

9. If the volume of a huge sports stadium represents the volume of an atom, describe the volume of the atom's nucleus, using the same scale.

10. Name the two rules used to help balance any nuclear equation.

11. Half-lives of different radioactive elements vary widely. Explain.

12. Explain why radioactive uranium is found in nature. Why has it not all decayed into other elements?

B

13. Identify the elements with atoms of nuclear charge $+10$, $+56$, and $+53$. Mass numbers for the most common stable isotope of each element are: $+10$, $A = 20$; $+56$, $A = 138$; $+53$, $A = 127$. (a) Draw a complete atomic structure diagram for each of these isotopes. (b) Write equations to show the conversion of each atom to a positive ion. Include "energy" on the proper side of the equation. (c) Write equations to show the conversion of each atom to a negative ion.

14. Draw complete atomic structure diagrams including diagrams of the nucleus for: (a) oxygen (nuclear charge $+8$, mass 16) neutral atom (b) oxygen (nuclear charge $+8$, mass 16) -1 ion (c) aluminum (nuclear charge $+13$, mass 27) neutral atom (d) aluminum (nuclear charge $+13$, mass 27) $+2$ ion.

15. Complete the following table:

Element	Atomic number	Number of protons	Number of electrons	Charge on nucleus	Mass number	Number of neutrons
carbon					12	
	10					10
		1			1	
			3			4
				+2	4	
calcium						20
	11				23	
		18				22
			83		209	
				+90		142

16. Draw atomic structure diagrams for the following isotopes of uranium: uranium-233, uranium-235, and uranium-238. Do the same for the isotopes chlorine-35 and chlorine-37.

17. The half-life of iodine-131 is 8 days. If you have 1 mole (6.0×10^{23}) of iodine-131 atoms today, how many will remain in your sample (a) 8 days from now? (b) 16 days from now?

18. When stable, nonradioactive nitrogen in the air is subjected to neutron bombardment during an H-bomb explosion, a single neutron is absorbed by a nitrogen nucleus and a proton is given off. Write a balanced

equation for this reaction and name the other product of the reaction.

19. The other product you named in your answer to question 18 is radioactive and undergoes beta decay. Write the balanced nuclear equation for this decay and identify the new daughter element.

20. Chlorine has two stable isotopes. One has a mass of 34.97 *u* and makes up 75.53 percent of natural chlorine. The second has a mass of 36.97 *u* and makes up 24.47 percent of natural chlorine. (a) What would be the total mass due to the isotope 34.97 in *one mole* of natural chlorine *atoms*? (b) What would be the total mass due to the isotope 36.97 in *one mole* of natural chlorine *atoms*? (c) What would be the best value for the atomic weight of natural chlorine based on the isotope information?

21. What change in mass and atomic number results when an atom emits (a) an alpha particle? (b) a beta particle? (c) a gamma ray?

C

22. The nucleus of an aluminum atom has a diameter of about 2.0×10^{-15} m. The atom has an average diameter of about 3.0×10^{-10} m. (a) What is the ratio of the diameter of the atom to the diameter of the nucleus? (b) What is the ratio of the volume of the atom to the volume of the nucleus? (The formula for the volume of a sphere is $\frac{4}{3}\pi r^3$.) (c) If 99.9 percent of the mass of the atom is contained in the nucleus, what is the ratio of the density of the extranuclear atom to the density of the nucleus?

23. Sodium has an isotope designated as $^{22}_{11}$Na which has a half-life of 2.58 years and decomposes by loss of a *positron*. (A positron is a particle just like an electron (beta particle) but it has a positive charge.) (a) Indicate the structure of an atom of Na-22. (b) Write an equation for the decomposition process. (c) How much Na-22 would be left after four half-lives, which is 10.32 years?

24. A radioactive element, A, with a mass of 214 and atomic number 84, emits an alpha particle and changes to element B. Element B emits a beta particle and is converted to element C. (a) Write nuclear reaction equations for these elements. (b) What are the atomic masses and atomic number of elements A, B, and C? (c) Name elements A, B, and C.

EXTENSION

"Seeing" parts of atoms

The nuclear atom model developed in Section 8.1 is based on laboratory evidence. Several of the classic experiments from the early 1900s supporting this model are described here. These experiments should be helpful to you in learning how a more detailed view of the atom was obtained.

"Seeing" electrons Passing electricity through aqueous solutions of salts led us to the conclusion that an electric current consists of moving ions in solutions and moving electrons in wires outside the solutions. Passing electricity through gases gives a more detailed view of electrons. Consider, for example, the familiar reddish glow of a neon sign. Closer examination of one of the segments of a neon sign reveals a glass tube fitted with electrodes so that a potential of about 10,000 volts can be applied across the space between the electrodes. The tube is filled with neon, and then the gas is gradually pumped out. When the pressure reaches 0.01 atmosphere, the familiar red glow of the neon sign appears. The color depends on the gas selected. Different gases produce different colors. If the vacuum pump continues to operate, the color will gradually disappear when the pressure reaches 10^{-6} atmosphere, and a fluorescent glow will appear on those parts of the tube wall directly in front of the negative electrode.

In order to study such a glow more carefully, we can use the experimental apparatus shown in Figure 8-7. A tube with a metal disc is placed in front of the negative electrode. A triangular hole is cut out of the metal disc, as shown in Figure 8-7a. The end of

the tube is then covered with a thin layer of a fluorescent material (such as zinc sulfide) that gives a brighter glow.

When the tube operates, a sharp triangular image (A in Figure 8-7a) appears on the tube wall opposite the disc. It appears that radiation is traveling in straight lines from the negative electrode to the opposite wall. Such behavior is characteristic of light. If we now bring an ordinary magnet near the tube, the beam of "light" can be bent and moved around.

Figure 8-7

(a) An electric discharge tube under very low pressure. Some electrons leaving the negative electrode pass through the triangular hole to produce a triangular spot on the screen. (b) Deflection of a beam of electrons by an electric charge on plates P_1 and P_2. If certain compounds such as ZnS are placed as a film over the glass wall, the glow is much brighter directly in front of the negative electrode

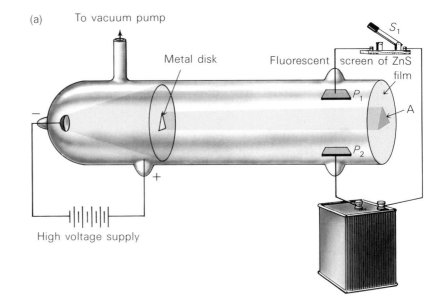

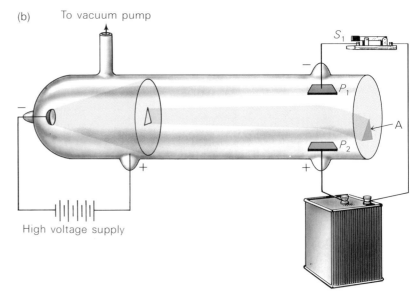

No ordinary light beam can be moved this way by any magnet. The beam is clearly *not* ordinary light. If the battery circuit shown at the right of the tube is activated by closing the switch (as shown in Figure 8-7b), Plate P_1 becomes negatively charged and P_2 becomes positively charged. The beam immediately bends toward the $+$ electrode. Further experiments will show that this behavior is independent of the kind of gas in the tube. Something in the beam is negatively charged. Many experiments have shown that the beam is a stream of the same kind of negative particles encountered in the wires of our electrolysis experiments. Thus, the beam is composed of electrons.

What have we really seen? We have *not* seen electrons directly. Rather, we have seen a burst of light on the fluorescent zinc sulfide screen. This light was caused by the collision of electrons with the zinc sulfide. The images on your TV screen are produced in the same way. Observing the *results* of electron collisions, rather than electrons themselves, typifies experiments with subatomic particles. We see their "footprints" as bursts of light on a screen, dark spots or lines on a photographic plate, or noise from a Geiger counter.

The methods used to determine the properties of subatomic particles are not unlike those used by the park ranger described in Section 7.1. The ranger was able to reconstruct the interaction between deer and hunters without ever seeing either directly.

The charge of the electron In 1909, the American physicist Robert Millikan (1868–1953) and his students determined the amount of charge of a single electron by using an apparatus similar to that shown in Figure 8-8. Tiny droplets of oil were sprayed into the space above the metal plates. Now and then, an oil droplet would fall through the tiny hole in the upper plate into the space between the plates. The rate of fall of this oil droplet was determined by watching it through a special telescope. When the rate of fall was established, the upper plate of the apparatus was connected to the positive terminal of a high-voltage battery, and the lower plate to the battery's negative terminal.

A beam of X-rays was then passed through the apparatus, ionizing some of the gas molecules pres-

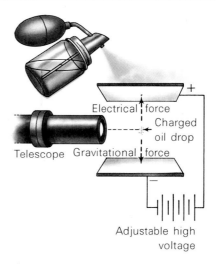

Figure 8-8

Millikan's oil-drop apparatus for determining the electron charge.

ent and providing a ready source of free electrons. After passage of the X-rays, the rate of fall of some drops was seen to change suddenly. Sometimes, the drop would fall less rapidly. Some drops even rose. The changes occurred in jumps. Millikan interpreted this as evidence for the gain of one or more electrons by the oil droplet. If the oil droplet carried an extra electron, its fall would be slowed up by the $+$ and $-$ charges on the plates. If the droplet carried several charges, it might even rise because of the action of the plates on the charges. By balancing the electrical force that was "trying" to make the droplet *rise* against the known gravitational force "trying" to make it fall, the amount of electric charge on the droplet could be estimated.

Millikan made thousands of determinations of the charges on drops of oil, glycerine, and mercury. In every case, the amount of charge was a whole number times the charge he assigned a single electron.

To understand how Millikan made his determination, suppose five measurements of oil droplet charges gave the following values: 4.83×10^{-19}, 3.24×10^{-19}, 9.62×10^{-19}, 6.44×10^{-19}, 4.80×10^{-19} coulombs.* Note that these values are obtain-

---

*The *coulomb* is a unit of electric charge (6.25×10^{18} electrons). Its magnitude can be evaluated by its relation to the *ampere*. One coulomb of charge passing a point in a wire every second is a current of 1 ampere. One mole of electrons has, then, 96,500 coulombs of charge. In a wire carrying 10 amp, it takes about $2\frac{1}{2}$ hours for 1 mole of electrons to pass any point.

able, within certain limits of uncertainty, by multi-plying a *constant* value by two or more:

$$3 \times 1.61 \times 10^{-19} = 4.83 \times 10^{-19}$$
$$2 \times 1.61 \times 10^{-19} = 3.22 \times 10^{-19}$$
$$6 \times 1.61 \times 10^{-19} = 9.66 \times 10^{-19}$$
$$4 \times 1.61 \times 10^{-19} = 6.44 \times 10^{-19}$$
$$3 \times 1.61 \times 10^{-19} = 4.83 \times 10^{-19}.$$

One could then assume that a single electron has a charge of 1.61×10^{-19} coulomb. The first drop evidently had picked up three extra electrons; the second drop, two electrons; the third, six electrons, and so on.

"Seeing" positive ions Experiments conducted in an evacuated gas discharge tube (as shown in Figure 8-7) demonstrate that electrons are present and that they are identical, regardless of the gas in the tube. Unless a small amount of some gas is present in the tube, *no* electron beam is seen. The electrons must come from the gas atoms in the tube. If the gas is neon, we can write: neon atom = neon ion$^+$ + electron$^-$. We expect an electron and a *positively charged* neon *ion* to be produced. We assume that the charge on such an ion results from the *loss* of one or more electrons from the neutral atom.

It is possible to isolate a beam of positive ions by constructing an apparatus similar to the one shown in Figure 8-9a. The beam of positive ions can be deflected in magnetic and electrostatic fields just as the electron beam could be deflected. The amount of deflection depends on (1) the *velocity,* (2) the *mass,* and (3) the *charge* of the positive ions. The instrument shown in Figure 8-9a, known as a *mass spectrometer,* is used to study such deflections. Such studies yield very precise determinations of mass. In using a mass spectrometer, we must make sure that particles entering the magnetic field all have the same velocity.

Operation of the mass spectrometer The type of mass spectrometer shown in Figure 8-9a produces positive (+) ions by bombardment of atoms in the tube with an electron beam. The rapidly moving electrons of the beam knock electrons away from atoms in the tube. The + ions produced are accelerated to a known velocity by attraction for the highly charged negative electrode, which has a slit

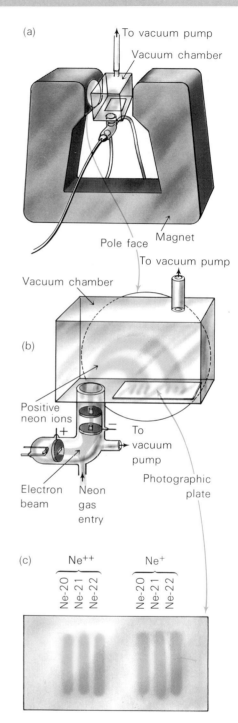

Figure 8-9

A mass spectrometer and the mass spectrum of neon: (a) a schematic representation of a mass spectrometer, (b) an enlargement of the evacuated unit between the poles of the magnet, and (c) the photographic plate from (b).

in it. This rapidly moving beam of + ions is made into a narrower beam by passage through a second slit in another charged metal disc. The beam then passes through a uniform magnetic field.

Figure 8-9b shows neon gas entering the tube at the bottom. The gas passes through the electron beam, where some atoms collide with the electrons of the beam to form neon ions. Both Ne^+ and Ne^{++} ions are formed and accelerated by moving toward the slits in the negative electrodes. After the beam of + ions is formed, it enters the magnetic field, where the charged ions move in a segment of a circular path. The piece of a circular path has a *large* radius if the ion's *mass* is high, and a *small* radius if the *charge* is high. Each + ion follows a distinctive curved path fixed by its mass and charge. In many mass spectrometers the ions hit a photographic plate after traveling through half a circle. Their impact there causes a chemical reaction that darkens the plate. A dark spot is produced as each ion strikes the plate. The place where the ion hits is determined by that ion's ratio of charge to mass. The record produced on the plate is called a **mass spectrum.**

Ions can also be detected by replacing the photographic plate with a charge detector. This, in effect, "counts" the relative number of ions of each isotope that arrives at each position. When plotted on paper, the result is a number of "peaks," as shown in Figure 8-10. For neon, the photographic plate record shows two widely separated groups of three spots each, interpreted as shown in Figure 8-9c. The "peak" record shows two separated groups of three peaks each, though only one such group is shown in Figure 8-10. Both records indicate that neon has three different isotopes. The relative abundance of each isotope can be determined by measuring the relative heights of the peaks shown in Figure 8-10. Most neon is observed to be the isotope of mass 20. As indicated in Figure 8-9c, the two distinct groups of spots were caused by the Ne^{++} ions and the Ne^+ ions.

Atomic weights today are determined by mass spectrometer measurements. The results of such measurements show the following:

1. The mass of positive ions changes if we change the gas in the instrument. The mass

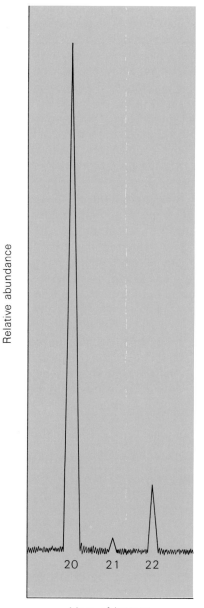

Figure 8-10

A partial mass spectrometer tracing for a sample of neon gas containing 90.5 percent $^{20}Ne^+$, 0.3 percent $^{21}Ne^+$, and 9.2 percent $^{22}Ne^+$.

of electrons, however, is always the same and independent of the type of gas present.

2. Positive ions have a much higher mass than electrons. This fact supports the model, which suggests that positive ions are fragments remaining after one or more electrons have been knocked off of gas atoms.

3. Even a pure gas such as neon will give positive ions that differ somewhat in mass. This is proof of the existence of isotopes.

"Seeing" the nucleus We have described a nucleus whose volume is but a tiny fraction of the volume of an entire atom. How did Ernest Rutherford (see page 185) and his coworkers go about "seeing" such a very small nucleus? A crude analogy would be for you to be given a rifle and asked to determine the size and shape of an object hidden in a tent. You must do this *without* entering the tent and without ever directly viewing the object inside. One way to proceed might be to fire bullets into the tent (Figure 8-11a) and then examine the holes made by the bullets that came out the far side of the tent. The pattern of holes shown in Figure 8-11b suggests that bullets that struck the object were stopped by it. Bullets that missed the object traced out a rough picture of its shape. If some bullets were deflected back, you might assume that the object was heavy and hard enough to cause bullets that hit it to bounce off.

In place of bullets, Rutherford and his colleagues used alpha particles, which are given off spontaneously by radioactive radium. Alpha particles are helium nuclei (He^{++}) shot out of radium at tremendous speeds (2×10^7 metres per second). They can be focused into a narrow beam by placing the radium in a solid lead box in which a deep hole has been drilled. Nuclear "bullets" were "fired" at a thin sheet of gold foil about 10,000 atoms thick, as shown in Figure 8-12. The alpha "bullets" were detected by the light they produced when striking zinc sulfide screens that surrounded the gold foil.

The first observation made using the apparatus (shown in Figure 8-13a) was that all the alpha particles appeared to pass through the foil undeflected. It was as if the gold atoms simply were not there. Surely, gold atoms ($197\ u$) are massive enough to deflect helium ions ($4\ u$), no matter how fast the ions are moving. Rifle bullets, by analogy, would certainly be deflected if they struck a 16-pound shot-put! Perhaps the mass of the gold atoms was distributed uniformly throughout the entire atom (Figure 8-13a). If so, a tiny rifle bullet would not encounter enough mass to hinder it in any way.

Careful examination, however, revealed that a few alpha particles *were* bounced back in the direction from which they came. Apparently, there *was* something very massive in the gold atoms. But this massive entity appeared to be so tiny that most of

Figure 8-11

Analogy to Rutherford's scattering experiment: (a) firing rifle bullets at an unseen object in a tent and (b) the bullet holes seen from the other side of the tent.

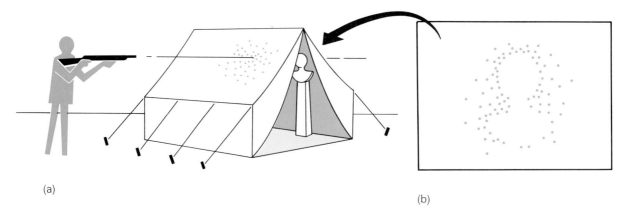

(a)

(b)

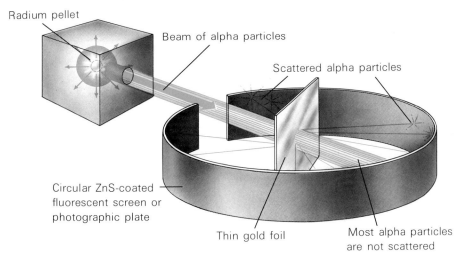

Figure 8-12
Rutherford's apparatus for observing the scattering of alpha particles by gold foil. The entire apparatus is enclosed in a vacuum chamber.

Figure 8-13
(a) Alpha particles appear to pass through Rutherford's gold foil undeflected, but (b) careful examination reveals some scattering.

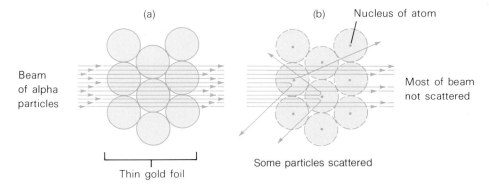

the alpha bullets *missed* it. To explain these results, Rutherford designed a new picture of the atom. His model required that most of the atom be empty space. But it also required that the atom contain a tiny, massive nucleus capable of deflecting any alpha particle that struck it directly (Figure 8-13b). Rutherford's model described most of the alpha "bullets" as missing the tiny nuclei and thus passing through the gold undeflected. The few "bullets" that did come close to a nucleus bounced off it, as a rifle bullet would bounce off a heavy shotput. In describing his experiment, Rutherford said: "It is about as incredible as if you had fired a 15-inch shell at a piece of tissue paper and it came back and hit you." The model is supported by experimental evidence, even though the evidence is indirect.

The eighth element, starting from a given one, is a kind of repetition of the first, like the eighth note of an octave in music.

JOHN A.R. NEWLANDS

ORGANIZING THE ELEMENTS: THE PERIODIC TABLE

Objectives

After completing this chapter, you should be able to:

- Understand the importance of vertical and horizontal trends in the periodic table.

- Discuss the physical and chemical properties of elements in the noble gas, alkali metal, and halogen families.

- Explain some of the consequences of the process by which closed-shell alkali metals and halogens achieve a stable electron arrangement.

- Differentiate between ionic and covalent bonding.

Таблица II.

Вторая попытка Менделѣева найти естественную систему химическихъ элементовъ. Перепечатана безъ измѣненій изъ „Журнала Русскаго Химическаго Общества", т. III, стр. 31 (1871 г.).

	Группа I.	Группа II.	Группа III.	Группа IV.	Группа V.	Группа VI.	Группа VII.	Группа VIII, переходъ къ группѣ I.
	$H=1$							
Типическіе элементы.	$Li=7$	$Be=9{,}4$	$B=11$	$C=12$	$N=14$	$O=16$	$F=19$	
1-й періодъ. Рядъ 1-й.	$Na=23$	$Mg=24$	$Al=27{,}3$	$Si=28$	$P=31$	$S=32$	$Cl=35{,}5$	
— 2-й.	$K=39$	$Ca=40$	$?=44$	$Ti=50?$	$V=51$	$Cr=52$	$Mn=55$	$Fe=56$, $Co=59$ $Ni=59$, $Cu=63$
2-й періодъ. — 3-й.	$(Cu=63)$	$Zn=65$	$?=68$	$?=72$	$As=75$	$Se=78$	$Br=80$	
— 4-й.	$Rb=85$	$Sr=87$	$Yt?=88?$	$Zr=90$	$Nb=94$	$Mo=96$	$-=100$	$Ru=104$, $Rh=104$ $Pd=104$, $Ag=108$
3-й періодъ. — 5-й.	$(Ag=108)$	$Cd=112$	$In=113$	$Sn=118$	$Sb=122$	$Te=128?$	$J=127$	
— 6-й.	$Cs=133$	$Ba=137$	$-=137$	$Ce=138?$	—	—	—	— —
4-й періодъ. — 7-й.	—	—	—	—	—	—	—	
— 8-й.	—		—	—	$Ta=182$	$W=184$	—	$Os=199?$, $Ir=198?$ $Pt=197$, $Au=197$
5-й періодъ. — 9-й.	$(Au=197)$	$Hg=200$	$Tl=204$	$Pb=207$	$Bi=208$	—		
—10-й.		—	—	$Th=232$	—	$Ur=240$	—	
Высшая соляная окись	R_2O	R_2O_2 или RO	R_2O_3	R_2O_4 или RO_2	R_2O_3	R_2O_6 или RO_3	R_2O_7	R_2O_8 или RO_4
Высшее водородное соединеніе . . .			(RH_3)	RH_4	RH_3	RH_2	RH	

The periodic table as originally published by Dmitri Mendeleev.

The world around us contains a great variety of materials—gasoline, air, rubber, sugar, wood, iron, diamond, and countless others. Some materials, such as gasoline and air, are mixtures. They can be separated into pure substances, such as pentane, octane, or oxygen. Other materials are relatively pure, such as sugar and diamond. How can we organize this vast variety of materials?

Experiments show that all pure substances that can be separated from mixtures, or that are found in nature, are really combinations of less than 100 different kinds of atoms. From these building blocks, chemists have prepared over 2 million compounds, each having its own special composition and properties. The atomic theory gives us a framework for understanding how pure substances can be described and organized.

Nevertheless, a serious problem remains. Must we memorize the properties of millions of substances? The task is frightening unless simplifications are found. Fortunately, regularities can be found to help us classify materials.

9.1 Regularities among the elements

The word *metal* is familiar to everyone. Most of us think of a metal as a shiny solid in the form of wires or sheets that can conduct heat and electricity. Our experience with copper wire, aluminum foil, cooking pans, chrome-plated toasters, and the like makes the expression *metallic properties* easy to understand in general terms. About three quarters of the elements are best described as metallic in nature. But another group of elements is completely different.

The element sulfur is a yellow solid. It is not bright and shiny like copper, but a dull yellow material. It can be melted fairly easily and is a poor conductor of heat and electricity. Furthermore, sulfur cannot be made into sturdy wires or sheets of foil. Also different is the element chlorine, a yellow gas that is highly toxic. It irritates the nose, eyes, and lungs. Substances such as sulfur and chlorine differ greatly from metals.

The obvious differences between metals and the group of elements to which sulfur and chlorine belong suggest that the elements can be sorted into two groups: metals and what we might call nonmetals. Such sorting is helpful since all metals share a set of easily recognized properties. With the exception of mercury, all metals are solids at 25 °C, which is considered room temperature. They are all shiny, flexible, and can be rolled into sheets or pulled into wires. All are good conductors of heat and electricity. By contrast, nonmetals can be gases (chlorine), liquids (bromine), or solids (sulfur and diamond) at room temperature. Some solids, such as white phosphorus, are easily vaporized. Nonmetals are not shiny. They are brittle and are poor conductors of heat and electricity.

The metal-nonmetal classification has some disadvantages however. For one thing, each group is quite large, and each group contains elements with some *different* properties. For example, sodium and gold are both metals. But sodium reacts violently with water and is so reactive with air that it is stored in kerosene. In contrast, gold does not react with water, air, or many acids. As a result, it is found free in nature as a pure element. Gold is an element of *very low reactivity*. Clearly, we need a better classification scheme.

MENDELEEV'S PERIODIC TABLE: ORDER FROM CHAOS

In 1871, after years of study, the Russian chemist Dmitri Mendeleev (1834–1907) arranged the elements in order of increasing atomic weight and found *regularly spaced, periodic recurrences of similar properties.* He emphasized the recurrence of similar properties in the table he constructed, which is known as the **periodic table.**

To get some idea of the kinds of similarities Mendeleev found, you might try listing the first 20 elements in the table, together with some of their properties as they are known today. Table 9-1 is such a list. While not all the information shown in the table was available in Mendeleev's time, you can see the periodic recurrence of similar properties that Mendeleev saw.

Notice that except for Ar (argon was not known in Mendeleev's time), the elements are ordered by increasing atomic weight. The experimentally determined formulas for hydrides and fluorides are given for each element. The elements helium (He), neon (Ne), and argon (Ar), elements 2, 10, and

Table 9-1

Some properties of the first 20 elements

Symbol of element	Atomic number	Atomic weight	Physical state at STP	Formula of hydride	Ratio of H to element	Formula of fluoride	Ratio of F to element
H	1	1.01	gas	H_2	1	HF	1
He	2	4.00	gas	—		—	
Li	3	6.94	solid	LiH	1	LiF	1
Be	4	9.01	solid	BeH_2	2	BeF_2	2
B	5	10.8	solid	$(BH_3)_2$	3	BF_3	3
C	6	12.0	solid	CH_4	4	CF_4	4
N	7	14.0	gas	NH_3	3	NF_3	3
O	8	16.0	gas	H_2O	2	F_2O	2
F	9	19.0	gas	HF	1	F_2	1
Ne	10	20.2	gas	—		—	
Na	11	23.0	solid	NaH	1	NaF	1
Mg	12	24.3	solid	MgH_2	2	MgF_2	2
Al	13	27.0	solid	$(AlH_3)_n$	3	AlF_3	3
Si	14	28.1	solid	SiH_4	4	SiF_4	4
P	15	31.0	solid	PH_3	3	PF_3	3
S	16	32.1	solid	H_2S	2	SF_2*	2
Cl	17	35.5	gas	HCl	1	ClF	1
Ar	18	39.9	gas	—		—	
K	19	39.1	solid	KH	1	KF	1
Ca	20	40.1	solid	CaH_2	2	CaF_2	2

*SF_2 is not known but its analogous compound, SCl_2, is known.

18 respectively, are similar in that they form *no* compounds with hydrogen or fluorine. These so-called noble gases were unknown to Mendeleev. Note the regularity in the combining ratios of the elements that do form hydrides and fluorides. The ratios 1, 2, 3, 4, 3, 2, 1 for elements 3 through 9 are *repeated* for elements 11 through 17. As the ratios indicate, the formulas also follow a recurring pattern. To emphasize these similarities, we list the elements in horizontal rows, placing elements with similar formulas directly *below* each other. The result, shown in Table 9-2, is a Mendeleev-like periodic grouping.

Notice also that the chemical formulas in a *horizontal* row are very different from one element to the next. On the other hand, formulas in any *vertical* column are very similar. Mendeleev could not explain why this amazingly regular order should exist. But he believed so strongly in the regularities revealed by his table that he treated the gaps he found boldly and imaginatively. He assumed that those gaps—the question marks in his original table, shown at the beginning of this chapter—represented missing elements and would be filled in when those elements were eventually discovered.*

Mendeleev predicted specific properties for the missing elements. One impressive example of such a prediction is seen below the element silicon (Si) in his original table. In 1871, there was no known element of atomic weight 72 and properties similar to those of silicon. Therefore, Mendeleev *assumed* that the element did exist but had not yet been discovered. He named it "ekasilicon" and represented this element by the symbol Es. After examining its position in his table, Mendeleev predicted the properties of ekasilicon. He suggested that it would combine with chlorine to form a compound of formula $EsCl_4$ and would have the other properties listed in Table 9-3. The table compares his predictions of the chemical and physical properties for ekasilicon (now known as germanium) and for two other, then-undiscovered elements with the properties later observed. He named these undiscovered elements ekaboron (now known as scandium) and

*Mendeleev solved the arrangement of elements by leaving gaps in his table. An earlier attempt by the English chemist John A. R. Newlands (1838–1898) to devise a periodic table failed because gaps were not left. In 1864, Newlands arranged the known elements in order of increasing atomic weight in rows of seven. He based his table on what he called the Law of Octaves. (See the quote at the beginning of this chapter.) His arrangement worked well for the elements through calcium but led to later erroneous groupings. For example, nickel and cobalt were grouped with the halogens.

Table 9-2

Periodic groups of the first 20 elements with their fluoride formulas

H 1						H 1	He 2
HF						HF	—

Li 3	Be 4	B 5	C 6	N 7	O 8	F 9	Ne 10
LiF	BeF_2	BF_3	CF_4	NF_3	OF_2*	FF	—

Na 11	Mg 12	Al 13	Si 14	P 15	S 16	Cl 17	Ar 18
NaF	MgF_2	AlF_3	SiF_4	PF_3*	(SF_2)*	ClF*	—

K 19	Ca 20
KF	CaF_2

*A number of nonmetal fluorides are now known that are not consistent with these trends, but they simply require additional refinement of the system. For example, compounds such as PF_5, SF_4, SF_6, ClF_3, ClF_5, O_4F_2, and others are now well known. Most were not known to Mendeleev. At present, the compound SF_2 is not known.

Table 9-3

Mendeleev's predictions and some observed properties of three elements

Element, year of discovery	Property	Mendeleev's predictions in 1871	Observed properties
ekaaluminum* (gallium), 1875	atomic weight	68	69.7
	density of metal	6.0 g/ml	5.96 g/ml
	melting temperature of metal	low	30 °C
	oxide formula	Ea_2O_3†	Ga_2O_3
	solubility of oxide	dissolves in ammonia solution	dissolves in ammonia solution
ekaboron* (scandium), 1877	atomic weight	44	43.7
	density of oxide	3.5 g/ml	3.86 g/ml
	oxide formula	Eb_2O_3†	Sc_2O_3
	solubility of oxide	dissolves in acids	dissolves in acids
ekasilicon* (germanium), 1886	atomic weight	72	72.6
	density of metal	5.5 g/ml	5.47 g/ml
	color of metal	dark gray	grayish white
	melting temperature of metal	high	900 °C
	density of oxide	4.7 g/ml	4.70 g/ml
	oxide formula	EsO_2†	GeO_2
	density of chloride	1.9 g/ml	1.89 g/ml
	chloride formula	$EsCl_4$†	$GeCl_4$
	boiling temperature of chloride	below 100 °C	86 °C

*Mendeleev's names for undiscovered elements.
†Symbols used to represent predicted elements: ekasilicon (Es), ekaboron (Eb), and ekaaluminum (Ea).

ekaaluminum (now known as gallium). The accuracy of his predictions should give you some idea of the importance of the periodic table and the insight of Mendeleev.

THE MODERN PERIODIC TABLE AND ATOMIC STRUCTURE

The periodic table that chemists use today closely resembles Mendeleev's original. Table 9-1 shows that the modern periodic table is organized by atomic *number*. Mendeleev used atomic weight, which resulted in almost the same order. The only exception in the first 20 elements is argon (Ar), element 18, which has a greater atomic weight than potassium (K), element 19, as shown in Table 9-1. Exceptions beyond the first 20 elements include elements 27 and 28, cobalt and nickel; and elements 52 and 53, tellurium and iodine.

Look again at Table 9-1 and at elements 2, 10, and 18. So far, no one has been able to make those elements form stable compounds. As we have said, those elements are known as *noble gases*. Because they are chemically similar, they are listed in a *single vertical column* in Table 9-2, as well as in the

Helium	**2** He
Neon	**10** Ne
Argon	**18** Ar
Krypton	**36** Kr
Xenon	**54** Xe
Radon	**86** Rn

Figure 9-1

The noble gas family.

2 He	**3** Li	Lithium
10 Ne	**11** Na	Sodium
18 Ar	**19** K	Potassium
36 Kr	**37** Rb	Rubidium
54 Xe	**55** Cs	Cesium
86 Rn	**87** Fr	Francium

Figure 9-2

The alkali metals.

complete modern periodic table on the inside back cover of this text. Elements 2, 10, and 18 make up the first three elements of the *noble gas family,* listed in Figure 9-1.

Now focus on the elements immediately following each of the noble gases in Table 9-1: elements 3, 11, and 19. They are all solids and have similar hydride and fluoride formulas. These elements, known as the *alkali metal family,* form a second group of similar elements and are listed in a second vertical column, shown in Figure 9-2.

If we look carefully at the hydride and fluoride formula columns in Table 9-1, we see repeating regularities in the formulas. For example, as we go down this table from Li (3) to Ne (10), the fluorides have the formulas LiF, BeF_2, BF_3, CF_4, NF_3, OF_2 (also written F_2O), and F_2. We have already looked at the combining ratios, defined as

$$\frac{\text{atoms of fluorine}}{\text{atoms of other element in compound}}$$

They are 1, 2, 3, 4, 3, 2, 1, as given in the last column of Table 9-1. The *same* combining ratio sequence is found for elements 11 through 17.* As you examine the formulas for fluorides and hydrides in Table 9-1, you will see that a number of vertical columns containing chemically similar elements can be identified in the periodic table.

The choice of where to begin and end each horizontal row of elements in the periodic table is somewhat arbitrary. Chemists find it convenient to end *each* horizontal row with a noble gas. This results in a periodic table of the first 20 elements as shown in Table 9-2. Looking horizontally across a row, properties differ from one element to the next. Looking vertically down a column, formulas and other properties are similar. Calling vertical groups **chemical families** emphasizes the similarities among the elements in each vertical column.

Note that the first element, hydrogen (H), appears in two places in our abbreviated periodic table in Table 9-2. It is placed above the alkali metals lithium (Li), sodium (Na), and potassium (K) because it has similar hydride and fluoride formulas. HH (H_2) and HF are similar to LiH and LiF. But HH and HF are *also* similar to the hydrides and fluorides of fluorine and chlorine. The formulas FF (F_2) and ClF indicate that hydrogen belongs to the same family as fluorine and chlorine. Chemists recognize that hydrogen is an unusual element. Some of its chemical properties indicate that it should be placed above fluorine, element 9. Other properties suggest it should be above lithium, element 3. Hydrogen is a family in itself.

Like the properties of the elements, the structure of the periodic table is related to atomic structure. According to the atomic model, the tiny, massive nucleus is inert. No chemical reactions influence or change it. Only the nuclear processes outlined in Section 8.2 (see pages 194–200) and in Chapter 14 involve changes in the nucleus. However, the nuclear *charge* does determine the chemistry of an atom because it determines the number of electrons in the neutral atom. We know that electrons are involved in the breaking and forming of bonds when chemical compounds form. We assume *it is the electrons that determine an atom's chemistry.* Since an element's chemistry determines where the element fits in the periodic table, the number of

*Except sulfur, which is not known to form SF_2.

electrons in its atom must be responsible for the position of the element in the periodic table. The key to this arrangement lies in the elements called noble gases.

9.2 Vertical trends: The noble gases

The elements helium, neon, and argon are **inert.** In other words, no stable compounds of these elements have ever been prepared. For a long time, the other three elements in the **noble gas family**—krypton, xenon, and radon— were also thought to be inert. But in 1962, Neil Bartlett (see the biography and Figure 9-7, pages 226–227) succeeded in making a relatively stable compound of xenon (Xe). Since then, compounds such as XeF_2, XeF_4, XeF_6, XeO_3, and many others have been made. Compounds of krypton and radon have also been prepared, but krypton compounds are of low stability.

While we cannot claim that the heavier three members of the noble gas family are inert, all members of that family are characterized by very low reactivity. This observation suggests that the electron arrangement of each noble gas is *stable.* Furthermore, noble gas molecules contain only *one* atom. Noble gas atoms have no tendency to combine with each other to form **diatomic** (two-atom) molecules. Everything known about the chemistry of the noble gas elements indicates that the atoms of the noble gases are nonre- active. Thus, their electron configuration must be relatively stable. Table 9-4 lists some physical properties of the noble gas elements. Let us look at some of those properties.

The noble gas elements are all gases at normal temperatures. Their boil- ing points are well below the 273 K (0 °C) freezing point of water. Helium boils at about 4 K. It cannot be solidified at any temperature unless pressure is applied. Helium becomes solid at 1.1 K (−272 °C) under a pressure of 26 atmospheres. It has the lowest freezing point, as well as the lowest boiling point, of any substance known. Note in Table 9-4 that as atomic number goes up, the boiling point of the noble gas goes up. If you plot the boiling

Table 9-4

Some properties of the noble gases

Element	Molecular formula	Atomic number	Atomic weight	Boiling point		Melting point	
				(°C)	(K)	(°C)	(K)
helium	He	2	4.00	−268.8	4.2	*	*
neon	Ne	10	20.18	−245.8	27.2	−248.4	24.6
argon	Ar	18	39.9	−185.7	87.3	−189.1	83.9
krypton	Kr	36	83.80	−153	120	−157	116
xenon	Xe	54	131.30	−108	165	−112	161
radon	Rn	86	222	−62	211	−71	202

*Helium becomes solid at 1.1 K (−272 °C) under a pressure of 26 atmospheres.

points of the noble gases against the number of electrons in the noble gas atoms, this direct relationship becomes apparent. (For neutral noble gas atoms, the atomic number gives both the number of protons *and* the number of electrons.) A higher boiling point, then, means that more energy must be added to vaporize, or pull apart, atoms in the liquid state. Thus, the weak forces of attraction that cause noble gases to liquefy increase as the number of electrons per atom increases.

NUMBER OF ELECTRONS AND STABILITY

From our study of noble gas atoms, we see that atoms with 2, 10, 18, 36, 54, and 86 electrons are relatively stable. Table 9-5 lists these electron populations, together with the increase in the number of electrons from one noble gas to the next. These numbers have special significance for the chemist and suggest several interesting questions. Why are these particular electron populations relatively stable? Why are they all even numbers? What is the significance of the regular *changes* in those numbers, from 8 to 18 to 32? We shall return to these questions in later chapters. For now, let us focus on the numbers 8 and 18.

SODIUM CHLORIDE AND NOBLE GAS ELECTRON POPULATIONS

Ordinary table salt, sodium chloride, can be made from sodium metal and chlorine gas. Sodium is so reactive that it liberates hydrogen from water. Chlorine is a poisonous green-yellow gas. Yet the salt you swallow with your food is not toxic. The element chlorine immediately precedes a noble gas in the periodic table. The element sodium immediately follows a noble gas in the table. Somehow, these extremely reactive elements, when combined, generate a compound that is quite stable and markedly different from the original substances. How can we explain such dramatic changes in properties?

Table 9-5

Stable electron arrangements of the noble gas family

Element	Total number of electrons	Change in number of electrons
helium	2	
neon	10	$10 - 2 = 8$
argon	18	$18 - 10 = 8$
krypton	36	$36 - 18 = 18$
xenon	54	$54 - 36 = 18$
radon	86	$86 - 54 = 32$

When dissolved in water, sodium chloride (NaCl) is a good conductor of electricity. As noted earlier, we picture the compound as being composed of Na^+ and Cl^- ions, and we picture ions forming when neutral atoms gain or lose electrons. Sodium, atomic number 11, has one more proton and one more electron than neon, atomic number 10. Chlorine, atomic number 17, has one less proton and one less electron than argon, atomic number 18. The sodium ion (Na^+) must have *lost* one electron to acquire its one positive charge. The chloride ion must have *gained* one electron to acquire its one negative charge. Thus, the Na^+ ion has 10 electrons, which is the same electron population as the noble gas neon. The Cl^- ion has 18 electrons, which is the same electron population as the noble gas argon. Reactive sodium and chlorine atoms form noble gas electron arrangements through compound formation. The resulting Na^+ and Cl^- ions attract each other strongly to give a solid that is of low reactivity. The noble gas configuration for ions is stable in the solid.

EXERCISE 9-1

Use atomic structure diagrams to represent the sodium-23 and the chlorine-35 atoms. Then use diagrams to show how these atoms form Na^+ and Cl^- ions.

Having looked at sodium chloride, we turn now to the chemistry of the other elements immediately preceding and immediately following the noble gases in the periodic table. These elements, listed in two vertical columns of the table, are known as the alkali metal family and the halogen family.

9.3   Vertical trends: The alkali metals

The six elements immediately following the six noble gases are *lithium, sodium, potassium, rubidium, cesium,* and *francium* (see Figure 9-2). These elements have similar chemical properties and make up the **alkali metal family,** or alkalies. In all their reactions, alkali metals form $+1$ ions by losing one electron per atom, thereby achieving the electron arrangements of the adjacent noble gas atoms (as in the case of the Na^+ ion, discussed above). Compounds of the alkalies also contain $+1$ ions.

PHYSICAL PROPERTIES

The alkalies are metallic elements. When their metal surfaces are clean, they have a bright, silvery luster. The alkali metals are excellent conductors of electricity and heat. They are soft and **malleable,** which means flexible, or able to be bent. In addition, their melting points are much lower than those of most other metals. The excellent electrical conductivity of the alkalies can be explained by assuming that some of their electrons are not attached to any single atomic core but can move freely through the array of positive ions (such as Na^+). Such electrons are called *conduction electrons.* Conduction

Table 9-6

Some properties of the alkali metals

Element	Molecular formula	Atomic number	Atomic weight	Boiling point		Melting point	
				(°C)	(K)	(°C)	(K)
lithium	Li	3	6.94	1326	1599	180	453
sodium	Na	11	23.00	889	1162	98	371
potassium	K	19	39.10	757	1030	63.4	336.4
rubidium	Rb	37	85.47	679	952	38.8	311.8
cesium	Cs	55	132.90	690	963	28.7	301.7
francium*		87					

*All the known isotopes of francium are radioactive, and it decays to form radium or astatine. The most stable isotope of francium has a mass number of 223. The French chemist Marguerite Perey first reported the discovery of this element in 1939.

electrons are able to move easily through the metal because they are *not* attached to particular atoms. The movement of such electrons would explain the electrical conductivity of metals.

Table 9-6 lists the same properties for the alkali metals that Table 9-4 listed for the noble gases. All the alkalies are solids at 25 °C (298 K), although cesium melts just above room temperature, at 28.7 °C (301.7 K). Note that both the melting and boiling points of the alkali metals generally *decrease* as atomic number increases. This behavior is opposite that of the noble gases. Note also how much greater alkali metal melting and boiling points are compared with those of the neighboring noble gases. How different are the alkali metals from the noble gases? Apparently, the addition of just one electron per atom makes a significant difference in properties. The forces that hold alkali metal atoms together as liquids and solids must be much stronger than they are in the case of the noble gases. The opposite trends in melting and boiling behavior suggest that such forces must also be different from the forces holding noble gas atoms together in the liquid and solid states. Note also the wide temperature range over which alkali metals can exist as liquids. Sodium, for example, melts at 371 K and boils at 1162 K, almost 800 degrees higher. This is in marked contrast to neighboring neon, which melts at 24.6 K and boils at 27.2 K, just 2.6 degrees higher.

CHEMICAL PROPERTIES

The alkali metals are exact opposites of the noble gases in chemical reactivity. The silvery metals all react vigorously when in contact with oxygen and chlorine, and even with a substance such as water. When chlorine gas is brought into contact with sodium metal, sodium chloride is formed with great vigor:

$$Na(s) + \tfrac{1}{2}Cl_2(g) \longrightarrow NaCl(s) + energy$$

The product, sodium chloride, is a regular arrangement, or **lattice,** of sodium ions (Na^+) and chloride ions (Cl^-), as shown in Figure 7-8, on page 169, and Figure 9-3. In this arrangement, each + ion surrounds itself with as many − ions as possible, and each − ion surrounds itself with as many + ions as possible. Each chloride ion is surrounded by six sodium ions, and each sodium ion is surrounded by six chloride ions. Distinct ions of each element are clearly present. We use the formula NaCl to represent the composition of such an array. But the formula NaCl shows only the simplest *ratio* between each kind of atom in the compound, not the total number of ions in a large or even small crystal of NaCl. Such a ratio is called an *empirical formula.* Separate and distinct NaCl molecules do not exist in such a lattice.

All of the alkali metals are very reactive. Each reacts with chlorine gas in a similar way:

$$Li(s) + \tfrac{1}{2}Cl_2(g) \longrightarrow LiCl(s) + energy$$
$$Na(s) + \tfrac{1}{2}Cl_2(g) \longrightarrow NaCl(s) + energy$$
$$K(s) + \tfrac{1}{2}Cl_2(g) \longrightarrow KCl(s) + energy$$
$$Rb(s) + \tfrac{1}{2}Cl_2(g) \longrightarrow RbCl(s) + energy$$
$$Cs(s) + \tfrac{1}{2}Cl_2(g) \longrightarrow CsCl(s) + energy$$

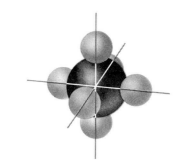

In every case, the alkali metal reacts to form a stable, ionic solid, such as the NaCl structure described above. The + and − ions in these structures are held together by *ionic bonds*—the attractions of oppositely charged ions for each other. In each case, the product is a crystalline substance with a high melting point and high solubility in water. The high melting point is due to the many ionic bonds, all tending to bind the solid tightly together. To emphasize these similarities, examine the five equations carefully. If we represent each alkali metal by the symbol M, each equation can be represented by the *single* equation

$$M(s) + \tfrac{1}{2}Cl_2(g) \longrightarrow MCl(s) + energy$$

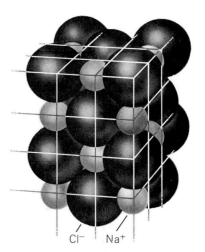

Cl^- Na^+

Figure 9-3
The packing of ions in a sodium chloride lattice.

The reaction of sodium metal with water is representative of the reactions of all the alkali metals with water. In the case of Na(s), the products are hydrogen gas and an aqueous solution of sodium hydroxide (NaOH). Because this solution conducts electricity readily, we picture it as containing mobile sodium and hydroxide ions: $Na^+(aq)$ and $OH^-(aq)$. The reaction is represented by the equation

$$Na(s) + H_2O \longrightarrow Na^+(aq) + OH^-(aq) + \tfrac{1}{2}H_2(g) + energy$$

Energy is liberated, and the reaction often takes place so rapidly that the temperature rises and the hydrogen may ignite when mixed with air. Thus, alkali metals are dangerous and must be handled with caution.

Similarities in the chemistry of alkali metals are clearly apparent in their reactions with water and chlorine. The reaction products always include metal ions of unit (+) charge. With one less electron than the alkali atoms, these ions formally resemble the inert gases in their electron arrangements.

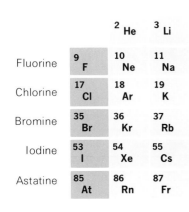

Figure 9-4
The halogens.

Figure 9-5

A covalent bond: the bonding of two chlorine atoms to form one chlorine molecule.

9.4 Vertical trends: The halogens

Next, we shall consider the column of elements that contain one *less* electron than their noble gas neighbors. These five elements—*fluorine, chlorine, bromine, iodine,* and *astatine*—make up the **halogen family.** Astatine does not occur as a stable isotope in nature and is not included in our discussion. Figure 9-4 shows the halogen family, and Table 9-7 lists some of their physical properties. The chemistry of the halogens is dominated by their tendency to form −1 ions by *gaining* one electron per atom. In this way, they display stable, noble gas electron arrangements.

PHYSICAL PROPERTIES

As indicated in Table 9-7, the halogens form stable diatomic molecules. High temperatures are required to disrupt these molecules and to form single atoms. For example, single chlorine atoms have been detected near the sun's surface, where the temperature is about 6000 °C. At more normal temperatures, chlorine atoms react with each other to form diatomic molecules, as follows:

$$2\text{Cl}(g) \longrightarrow \text{Cl}_2(g)$$

No further reactions occur among chlorine molecules. The same behavior is characteristic of the other halogens.

Apparently, the diatomic molecules of the halogens achieve some of the stability characteristic of the noble gas electron arrangements. How is this possible? How can one chlorine atom (with 17 electrons) approach the argon arrangement when argon has one more electron? We answer this question by suggesting that the two chlorine atoms share two electrons, each atom contributing one. With the two atoms close together, each gains an "interest" in an eighteenth electron by sharing the one electron contributed by the other atom. Thus, each chlorine atom achieves a stable electron arrangement that is like the argon arrangement in many ways. The same argument can be made to explain the diatomic molecules of the other halogens. Because such molecules are bonded by a *shared pair* of electrons, the bond is called a *covalent bond,* as shown in Figure 9-5. We have already indicated that diatomic halogen molecules are far less reactive than halogen atoms. This stability can be explained by using the concept of the covalent bond.

EXERCISE 9-2

Two argon atoms will *not* form a covalent bond to form Ar_2. Why?

The melting and boiling points of the halogens (given in Table 9-7 and Figure 9-6) show the *same* trends as those found for the noble gases. However, the melting and boiling points of the halogens are all somewhat higher. It is logical to relate melting and boiling points of substances such as the halogens and noble gases to the forces acting between molecules. Thus, we would say that as the halogen molecules become heavier, the forces holding

Table 9-7

Some properties of the halogens

Element	Molecular formula	Atomic number	Atomic weight	Boiling point		Melting point	
				(°C)	(K)	(°C)	(K)
fluorine	F_2	9	19.0	−188	85	−218	55
chlorine	Cl_2	17	35.5	−34.1	238.9	−101	172
bromine	Br_2	35	79.9	58.8	331.8	−7.3	265.7
iodine	I_2	53	127	184	457	114	387
astatine*		85					

*Since there is probably less than 30 g of astatine in the entire crust of the earth, it has been studied very little and not much is known about its properties. All of its isotopes are radioactive, with the longest-lived one having a half-life of 8.3 hours.

them together become greater. Similarly, we note that as the noble gas atoms get heavier, the forces between molecules become stronger. We also note that the forces acting between noble gas *atoms,* such as krypton (Kr) atoms, are not as great as the forces acting between halogen *molecules,* such as bromine (Br_2) molecules. The facts that fluorine and chlorine are gases, bromine is a liquid, and iodine is a solid at room temperature do not lessen their basic similarity. If room temperature on this planet were 464 K

Figure 9-6

Trends in the boiling points of the alkali metals, halogens, and noble gases.

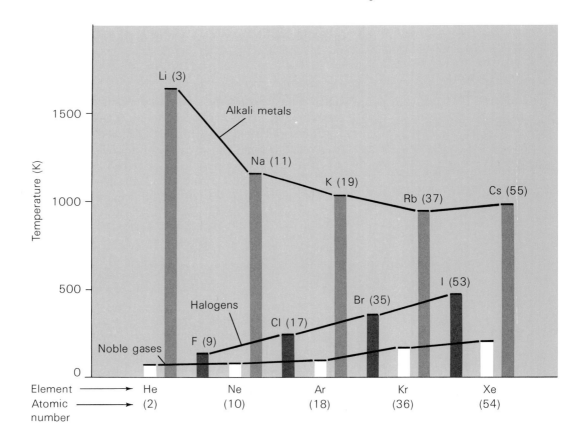

($190\,^\circ$C), all the halogens would be gases. All halogen gases are highly toxic and thus hazardous to handle.

Among the alkali metals, a new and different trend is observed. As the alkali metal atoms get heavier, their melting and boiling points fall and the forces acting between atoms become weaker. These facts suggest that the forces acting between Br_2 molecules are similar to the forces acting between Kr atoms but different from the forces acting between alkali metal atoms. The model we suggested earlier to explain the electrical conductivity of alkali metals indicates clear differences between such metals and covalently bonded materials such as the halogens, which have discrete, or separate, molecular units. Alkali metals do not have separate molecular units.

CHEMICAL PROPERTIES

Earlier, we used the reactions of the alkali metals with chlorine to show the similarities among the alkalies (see page 221). In the same way, we can use the reactions of the halogens with one of the alkali metals, such as sodium, to show the chemical similarities among the halogens. The reactions that occur may be represented by the following equations:

$$Na(s) + \tfrac{1}{2}F_2(g) \longrightarrow NaF(s) + energy$$
$$Na(s) + \tfrac{1}{2}Cl_2(g) \longrightarrow NaCl(s) + energy$$
$$Na(s) + \tfrac{1}{2}Br_2(g) \longrightarrow NaBr(s) + energy$$
$$Na(s) + \tfrac{1}{2}I_2(g) \longrightarrow NaI(s) + energy$$

These reactions proceed readily. They produce ionic solids with the general empirical formula NaX. Indeed, if we let the symbol X represent any halogen, the above equations could all be represented by the single equation

$$Na(s) + \tfrac{1}{2}X_2(g) \longrightarrow NaX(s) + energy$$

Each solid product has a crystalline structure similar to that of $NaCl(s)$, as described on page 221. Each contains positively charged sodium ions and negatively charged halogen ions. The negative ions, F^-, Cl^-, Br^-, and I^-, are called **halide ions.** The structure of these ions can be related to the structure of the corresponding noble gas electron arrangements.

The halogens also react with hydrogen gas to form covalent hydrogen halides, as represented by the following equations:

$$H_2(g) + F_2(g) \longrightarrow 2HF(g)$$
$$H_2(g) + Cl_2(g) \longrightarrow 2HCl(g)$$
$$H_2(g) + Br_2(g) \longrightarrow 2HBr(g)$$
$$H_2(g) + I_2(g) \longrightarrow 2HI(g)$$

The reaction between F_2 and H_2 will usually proceed explosively when the gases are mixed at room temperature. Hydrogen and chlorine can be mixed at room temperature in the dark without reaction, but ultraviolet light will initiate an explosion. The bonds in the Cl_2 molecule can be broken by a beam of ultraviolet light or by high temperatures. The breaking of the bonds starts the reaction. Once started, the combination of hydrogen and chlorine tends to proceed rapidly or explosively without further addition of external energy. The reactions between hydrogen and bromine and hydrogen and iodine are similar but *much* less energetic.

CHEMISTRY OF THE HALIDE IONS

Since all halide ions carry -1 charges, they combine in a one-to-one ratio with metal ions that carry $+1$ charges, such as alkali metal ions. Thus, the single formula MX can represent the empirical formula for any alkali halide. M^+ is the alkali metal ion and X^- is the halide ion. Because the structures of all alkali halides are so similar, we expect their properties to be similar. As expected, all alkali halides resemble sodium chloride in many ways. They are all white solids and are readily soluble in water. They are relatively unreactive toward air and many other reagents. When molten or in aqueous solutions, all compounds of the group MX are good conductors of electricity.

Except for solutions of the alkali metal fluorides, aqueous solutions of all alkali metal halide salts produce precipitates when aqueous silver nitrate ($AgNO_3$) is added to them. We discussed the precipitation reaction for chloride ions, $Cl^-(aq)$, in Chapter 7. Note the similarities in solubility behavior of $Cl^-(aq)$, $Br^-(aq)$, and $I^-(aq)$ ions:

$$Ag^+(aq) + Cl^-(aq) \longrightarrow AgCl(s) \quad \text{WHITE SOLID}$$

$$Ag^+(aq) + Br^-(aq) \longrightarrow AgBr(s) \quad \text{TAN SOLID}$$

$$Ag^+(aq) + I^-(aq) \longrightarrow AgI(s) \quad \text{YELLOW SOLID}$$

As we have indicated, no precipitate forms when $Ag^+(aq)$ ions are added to a solution containing $F^-(aq)$ ions. That is because silver fluoride (AgF) is soluble.

EXERCISE 9-3

Remember that the nitrate ion (NO_3^-) represents a group of four atoms that behave as a single unit similar to the chloride ion. A water solution of silver nitrate conducts electric current. Write an equation for the process that occurs when $AgNO_3(s)$ dissolves in water.

EXERCISE 9-4

Letting X^- represent Cl^-, Br^-, and I^- ions, write a single equation summarizing the process that occurs when aqueous $AgNO_3$ solution is added to a solution of MX in water (M is an alkali metal).

All the hydrogen halides are gases at room temperature, but hydrogen fluoride (HF) liquefies at 19.9 °C, or 292.9 K, and 1 atmosphere pressure.

NEIL BARTLETT

A native of England, Neil Bartlett received his Ph.D. in chemistry from King's College, Newcastle-upon-Tyne, in 1957. After a year as senior chemistry master at the Duke's School, Alnick, he joined the faculty of the University of British Columbia. Since 1969, Bartlett has been Professor of Chemistry at the University of California at Berkeley.

Bartlett is known for showing that xenon, a noble gas formerly considered inert, has a rather extensive chemistry. He first prepared the yellow solid $XePtF_6$ from xenon and the very reactive red compound PtF_6 in 1962. Bartlett succeeded in preparing the first compound of a noble gas. He described this classic experiment as follows: "The predicted interaction of xenon and platinum hexafluoride was confirmed in a simple and visually dramatic experiment. The deep red platinum hexafluoride vapor, of known pressure, was mixed, by breaking a glass diaphragm, with the same volume of xenon, the pressure of which was greater than that of the hexafluoride. Combination to produce a yellow solid was immediate at room temperature and the quantity of xenon which remained was commensurate with a combining ratio of 1 to 1." (See Figure 9-7, on page 227.)

The other hydrogen halides liquefy at much lower temperatures. Hydrogen halides dissolve in water to form aqueous solutions that conduct electric current. Unlike the alkali halides, the compounds do *not* conduct as liquids, which suggests that the pure hydrogen halides are *not* ionic compounds. To account for the conductivity of their water solutions, we assume that they somehow react with water to form ions not previously present. The reactions may be written as follows:

$$HF(g) + \text{water} \longrightarrow H^+(aq) + F^-(aq)$$
$$HCl(g) + \text{water} \longrightarrow H^+(aq) + Cl^-(aq)$$
$$HBr(g) + \text{water} \longrightarrow H^+(aq) + Br^-(aq)$$
$$HI(g) + \text{water} \longrightarrow H^+(aq) + I^-(aq)$$

These solutions have similar properties and are known as *acid solutions*. The common species in *all* the solutions is the aqueous hydrogen ion, $H^+(aq)$. Thus, the similar properties of aqueous acid solutions are attributed to this ion. We will investigate such solutions more fully in Chapter 17.

EXERCISE 9-5

HCl is often prepared by the action of concentrated H_2SO_4 on NaCl:

$$NaCl(s) + H_2SO_4(l) \longrightarrow HCl(g) + NaHSO_4(s)$$

Write an equation for the preparation of HF from NaF.

9.5 Hydrogen: A family in itself

We have already seen that hydrogen is an unusual element. In some respects, it resembles the halogens. Like the halogens, hydrogen has one less electron than its neighboring noble gas, helium. In addition, it forms compounds similar to some halogen compounds. In other ways, however, hydrogen resembles the alkali metals. The removal of one electron from an alkali metal atom leaves an electron arrangement like that of a noble gas. The removal of one electron from a hydrogen atom also leaves an ion with a charge of +1. Closer examination of its chemistry, however, shows that hydrogen is unique. The removal of one electron from a hydrogen atom leaves it with *no* electrons, which is not the case for any other atom. The stable hydrogen ion is simply a proton. We shall see that this element is a family in itself.

Hydrogen, like the halogens, is a diatomic gas at STP. Some of its physical properties are listed in Table 9-8. Like the lighter halogens, hydrogen has a very low melting and boiling point. In fact, hydrogen has the lowest boiling point of any element except helium. It is useful to look at the chemistry of hydrogen as we try to classify this unique element.

One of the most distinctive reactions characterizing *both* the alkalies and the halogens is their reaction with each other. The most familiar example of such a reaction is that between sodium and chlorine to form sodium chloride. Hydrogen also reacts with chlorine to form a compound with a similar formula: hydrogen chloride. The equations for the two reactions are

$$Na(s) + \tfrac{1}{2}Cl_2(g) \longrightarrow NaCl(s)$$
$$\tfrac{1}{2}H_2(g) + \tfrac{1}{2}Cl_2(g) \longrightarrow HCl(g)$$

These equations are similar, but the *products* are not. NaCl is an ionic solid with a high melting point. HCl is a colorless gas. Despite these differences, both products react with water to produce ions, according to similar equations:

$$NaCl(s) + water \longrightarrow Na^+(aq) + Cl^-(aq)$$
$$HCl(g) + water \longrightarrow H^+(aq) + Cl^-(aq)$$

The chemical properties summarized in these four equations suggest similarities between hydrogen and the alkali metals.

Hydrogen can also resemble the halogens in its chemistry. Consider the reactions represented in these equations:

$$Na(s) + \tfrac{1}{2}Cl_2(g) \longrightarrow NaCl(s)$$
$$Na(s) + \tfrac{1}{2}H_2(g) \longrightarrow NaH(s)$$

Here, the products have the same empirical formulas and are both white, crystalline solids. However, they show strikingly different behavior toward water. NaCl quietly dissolves. NaH reacts vigorously, releasing hydrogen:

$$NaCl(s) + H_2O \longrightarrow Na^+(aq) + Cl^-(aq)$$
$$NaH(s) + H_2O \longrightarrow H_2(g) + Na^+(aq) + OH^-(aq)$$

It is clear that while hydrogen resembles both the alkalies and the halogens in some respects, it differs from them in others. Hydrogen is a stable diatomic gas like the halogens, but its chemistry is more like that of the alkalies. It is usually shown above the alkali metals on the left side of the periodic table and sometimes also above the halogens. But it is separated from both families to show its distinctive character.

9.6 Horizontal trends: The second row

The chemistry of the elements discussed so far has been *similar.* Each vertical column that we have considered in our study of the periodic table is made up of elements with strikingly similar chemistry. We acknowledge these similarities by referring to each vertical group as a *chemical family.* Now, let us look *across* a horizontal row of the periodic table. Since the first row consists of only two elements—hydrogen and helium—we will study the second row—from element 3 (lithium) to element 10 (neon). (See Figure 9-9.) Rather than looking for similarities, we now expect *differences.* After all, each element in the second row belongs to a different chemical family.

PHYSICAL PROPERTIES

Lithium is a typical metal. A clean surface of the metal is bright and shiny in appearance and lithium is a very good conductor of heat and electricity—one sixth as good as copper. Lithium can be rolled into sheets or

227

Table 9-8
Some properties of hydrogen

atomic number	1
atomic weight	1.008
boiling point (°C)	−252.8
(K)	20.4
melting point (°C)	−259.2
(K)	14.0
molecular formula	H_2

Figure 9-7
Red platinum hexafluoride gas is shown (above) before reaction with xenon, a noble gas, and (bottom) after reaction. The yellow solid, xenon hexafluoroplatinate ($XePtF_6$), was the first compound formed from a reaction with a noble gas.

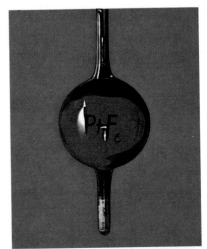

1 H																	2 He
3 Li	4 Be											5 B	6 C	7 N	8 O	9 F	10 Ne
11 Na	12 Mg											13 Al	14 Si	15 P	16 S	17 Cl	18 Ar
19 K	20 Ca														35 Br	36 Kr	
37 Rb	38 Sr														53 I	54 Xe	
55 Cs	56 Ba														85 At	86 Rn	
87 Fr																	

Figure 9-8

The placement of the second row of the periodic table.

Figure 9-9

The physical and chemical properties of second-row elements vary greatly. Lithium has the lowest density of any metal. Here, lithium is floating on oil, which is floating on water in this fish bowl. Carbon is a nonmetal that exists in two forms as diamond and graphite. Diamond does not conduct electricity but graphite is a good conductor of electricity along its layers.

formed into wire: it is malleable and ductile. It has a low melting point and is soft enough to be cut with a knife. All these properties, except softness and low melting point, are characteristic of the metals we see around us. Table 9-9 lists some of the physical properties of the second-row elements. (The last element of this row, neon, is not included, because its properties are listed in Table 9-4.) Lithium is truly metallic, yet because of its extreme chemical reactivity, it is not used as a metal in manufacturing. We never see objects made of metallic lithium.

Beryllium looks like a metal. It is steel gray and very brittle but hard enough to scratch glass. The brittle property is not a metallic characteristic. Beryllium is a fair conductor of electricity—about half as good as lithium— and combines with other metals to form metallic alloys. Beryllium is still quite metallic, although definitely less metallic than lithium.

Elemental, or crystalline, boron has almost no metallic characteristics. About its only metallic property is the luster of its red-to-black crystals. Boron is a semiconductor. For this reason, boron has potential value in the electronics industry. On the other hand, it is a nonconductor when compared to lithium. Its conductivity rises with temperature. Pure boron is very hard, only slightly softer than diamond. It cannot be rolled into sheets or drawn into wires. However, rods of boron can be very strong and flexible and may be useful in manufacturing.

Carbon exists in two forms: diamond and graphite. Although it is shiny, diamond is strictly nonmetallic. It is brittle, does not conduct electricity, and has the highest known melting and boiling points. (See Table 9-9.) Graphite, on the other hand, is less brittle, black but somewhat shiny, slippery, and easily broken into layers. It is used to make pencil "lead." The slippery property of graphite makes it useful as a dry lubricant. Graphite is also added to oils. In contrast to diamond, graphite is a good conductor of electricity along its layers. However, it is a poor conductor at right angles to the layers. You are probably familiar with other forms of carbon, such as soot and charcoal. For the most part, they are composed of very small, randomly

Table 9-9

Some properties of the second-row elements

Description	Lithium	Beryllium	Boron	Carbon		Nitrogen	Oxygen	Fluorine
				Graphite	Diamond			
form and appearance	silvery metal	steel-gray metal	black and red shiny crystals	gray-black plates	shiny diamond	colorless diatomic gas	colorless diatomic gas	yellow diatomic gas
hardness (Mohs scale)	very soft (0.6)	scratches glass (6–7)	very hard (9.3)	soft (1.0)	extremely hard (10, standard)	gas	gas	gas
melting point (°C)	180	1283	2370	3600	3730 (estimate)	−210	−218	−223
boiling point (°C)	1326	2970	—	—	4830 (estimate)	−196	−182	−188
electrical conductance*	excellent	good	poor†	good	no	no	no	no
ductile* and malleable‡	yes	no; brittle	no	no	no	no	no	no
metallic luster*	yes	yes	yes	faint	no	no	no	no
density (g/cm³)	0.53	1.86	2.30	2.1	3.5	0.96 (solid)	1.43 (solid)	1.1 (liquid)
interesting properties	reactive metal	makes copper hard; alloys valuable	semi-conductor; useful in electronics; rods very strong	soft and slippery	hardest material known; nonmetal; low reactivity	nonmetal of low reactivity	moderately reactive nonmetal	most reactive non-metal known

*Characteristic of metal.
†Semiconductor of p-type (see page 356).
‡Refers to property that permits substance to be drawn into wires or rolled into sheets.

arranged graphite crystals. Graphite has a few metallic characteristics, such as electrical conductivity, but it is not classified as a metal.

The elements nitrogen, oxygen, and fluorine are diatomic gases with low boiling points. Besides neon, fluorine is probably the best example of a nonmetal. As you can see, the transition from metallic lithium to nonmetallic fluorine and neon occurs in distinct steps.

CHEMICAL PROPERTIES

To see the trends in the chemistry of the second-row elements, we can examine the compounds that each element forms with hydrogen, chlorine, and oxygen. These compounds are called hydrides, chlorides, and oxides.

Their formulas and the combining ratios of the atoms they contain are summarized in Table 9-10. Neon, the final element in the second row, is not included. Being inert, neon forms no compounds. The ratios below each formula are determined by dividing the number of hydrogen, chlorine, or oxygen atoms by the number of atoms of each second-row element in the formula. For example, B_2H_6 contains six atoms of hydrogen and two of boron. We therefore divide 6 by 2 to get the ratio 3. Note the strikingly regular trends among these ratios for the hydrides and chlorides.

Perhaps the greatest value of Table 9-10 is that it gives formulas for several compounds of one element from each of seven important families. We have already discovered that if we know the equation or formula associated with a compound of one element, we can usually predict correctly the equation or formula for other elements in the same family. For example, recall the equations on page 221 that used MCl as the formula for *any* alkali chloride, M representing any one of the alkali metals. By providing the formulas $BeCl_2$ and BeO, Table 9-7 enables us to predict the formula of *any* alkaline earth chloride or oxide. The name **alkaline earth family** is given to the beryllium family of elements.

The periodic table allows us to make other predictions as well. Given the formulas of second-row chlorides and oxides, we can predict the formulas of the halogen and oxygen family compounds of each second-row element. Finally, because we know these formulas for each second-row element, we know them for each member of each second-row element's vertical family. For example, having deduced the formulas $MgCl_2$ and MgO in the above exercise, we can predict the formulas $MgBr_2$ and MgS. We can predict these formulas because bromine is in the chlorine family and sulfur is in the oxygen family.

EXERCISE 9-6

Use Table 9-10 (page 231) and the periodic table (pages 232–233) to write possible formulas for the following compounds:

(a) a hydride of sodium, element 11
(b) a chloride of aluminum, element 13
(c) an iodide of oxygen, element 8
(d) a fluoride of lead, element 82
(e) a hydride of gallium, element 31
(f) a chloride of silver, element 47
(g) an oxide of cesium, element 55
(h) a fluoride of strontium, element 38

9.7 The importance of the periodic table

You are now in a position to appreciate the importance of the periodic table to the chemist. It helps in writing formulas and equations for over 100 elements. To know where an element fits into the table is to know, for example, whether it is a metal, a nonmetal, or a noble gas.

Table 9-10

Formulas of some compounds of the second-row elements

Group	I Li	II Be	III B	IV C	V N	VI O	VII F
hydrides ratio H/element	LiH 1	BeH_2 2	B_2H_6 3	CH_4 4	NH_3 3	H_2O 2	HF 1
chlorides ratio Cl/element	LiCl 1	$BeCl_2$ 2	BCl_3 3	CCl_4 4	NCl_3 3	Cl_2O 2	ClF 1
oxides ratio O/element	Li_2O 0.5	BeO 1.0	B_2O_3 1.5	CO_2 2	N_2O_5 2.5	O_2 1	F_2O 0.5

OPPORTUNITIES

Computer and information science

Opportunities for chemistry students in computer and information science will be increasing rapidly over the next decade. You read in this chapter how the periodic table is used to classify and interpret chemical information. Similarly, the accumulation of information generated daily by complex industrial societies requires qualified people to collect, classify, and interpret it.

Computer and information scientists design systems that provide employers with access to specific data from computer storage. Such scientists may develop plans for retrieval of information when needed and may be required to determine the feasibility of starting or continuing a specific line of research. Computer and information scientists are often called upon to adapt scientific and engineering formulas to computer systems and may be asked to solve symbolic logic problems.

The complex skills required for this job must be accompanied by an enjoyment of solving problems somewhat in the manner of a detective. When a researcher asks a particular question about a project, the computer scientist must decide where to find the answer and what tool to use. The tools available to these workers include stored data of *Biological* and *Chemical Abstracts* and the *Derwent Central Chemical Patents Index*. These sources of chemical information are often used to uncover previous research in a specific field.

Working with computers can be stimulating and creative for a person who enjoys working in service-oriented technical areas where evaluation and decision making are part of the daily routine. Those who hope to succeed in this field must be able to communicate orally and in writing. They must have above average mathematical skills and a good sense of spatial relationships.

Computer and information scientists may be employed by chemical companies, large pharmaceutical houses, or specialized information exchanges such as the Smithsonian Information Exchange in Washington, D.C. They may also find work with organizations such as the publishers of *The Chemical Abstracts Service* of the American Chemical Society.

A bachelor's degree in chemistry is required for starting positions in this career. However, a master's

degree is often preferred. Prospective computer and information scientists should also take as many computer programming and data processing courses as possible. For additional information contact the American Federation of Information Processing Societies, Inc., 210 Summit Avenue, Montvale, New Jersey 07645.

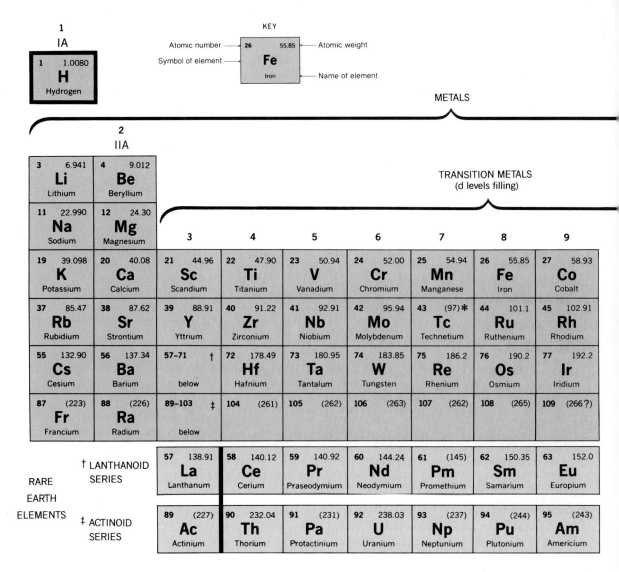

KEY

Atomic number → | 26 | 55.85 | ← Atomic weight
Symbol of element → **Fe** ← Atomic weight
Iron ← Name of element

METALS

TRANSITION METALS
(d levels filling)

1 IA									
1 1.0080 **H** Hydrogen									

2 IIA		**3**	**4**	**5**	**6**	**7**	**8**	**9**
3 6.941 **Li** Lithium	**4** 9.012 **Be** Beryllium							
11 22.990 **Na** Sodium	**12** 24.30 **Mg** Magnesium							
19 39.098 **K** Potassium	**20** 40.08 **Ca** Calcium	**21** 44.96 **Sc** Scandium	**22** 47.90 **Ti** Titanium	**23** 50.94 **V** Vanadium	**24** 52.00 **Cr** Chromium	**25** 54.94 **Mn** Manganese	**26** 55.85 **Fe** Iron	**27** 58.93 **Co** Cobalt
37 85.47 **Rb** Rubidium	**38** 87.62 **Sr** Strontium	**39** 88.91 **Y** Yttrium	**40** 91.22 **Zr** Zirconium	**41** 92.91 **Nb** Niobium	**42** 95.94 **Mo** Molybdenum	**43** (97)* **Tc** Technetium	**44** 101.1 **Ru** Ruthenium	**45** 102.91 **Rh** Rhodium
55 132.90 **Cs** Cesium	**56** 137.34 **Ba** Barium	**57–71** † below	**72** 178.49 **Hf** Hafnium	**73** 180.95 **Ta** Tantalum	**74** 183.85 **W** Tungsten	**75** 186.2 **Re** Rhenium	**76** 190.2 **Os** Osmium	**77** 192.2 **Ir** Iridium
87 (223) **Fr** Francium	**88** (226) **Ra** Radium	**89–103** ‡ below	**104** (261)	**105** (262)	**106** (263)	**107** (262)	**108** (265)	**109** (266?)

RARE EARTH ELEMENTS

† LANTHANOID SERIES

57 138.91 **La** Lanthanum	**58** 140.12 **Ce** Cerium	**59** 140.92 **Pr** Praseodymium	**60** 144.24 **Nd** Neodymium	**61** (145) **Pm** Promethium	**62** 150.35 **Sm** Samarium	**63** 152.0 **Eu** Europium

‡ ACTINOID SERIES

89 (227) **Ac** Actinium	**90** 232.04 **Th** Thorium	**91** (231) **Pa** Protactinium	**92** 238.03 **U** Uranium	**93** (237) **Np** Neptunium	**94** (244) **Pu** Plutonium	**95** (243) **Am** Americium

Figure 9-10

The modern periodic table is organized by atomic number. This arrangement of elements helps simplify our understanding of different chemical properties. Elements represented by the same color in this table have similar chemical properties. In order to further simplify the table, the American Chemical Society has recommended that the vertical columns be numbered 1 through 18. Two other changes also should be noted in the table. The naming of elements 104–109 is not yet settled. Therefore, only the reported atomic number and atomic weight are given here. The -oid ending for the rare earth elements is preferred by IUPAC because the -ide ending can indicate a negative ion.

So far, you have studied elements 1 through 20. Element 21, scandium, is quite similar to aluminum. However, moving on to elements 22, 23, 24, and so on, the vertical relationships become indistinct. For example, metallic manganese, number 25, is markedly different from the nonmetal chlorine, which is the halogen just above bromine in the table. Starting with element 21 and extending to zinc, element 30, is a group of ten metallic elements that exhibits very gradually changing properties. After this group of elements, the vertical relationships appear again. Elements 21 through 29, 39 through 47, 57 through 79, and 89 through 103 are known as the **transition elements.** Their electron arrangements differ from those in the alkali and halogen families, as we shall see in Chapter 21.

The arrangement of elements in the periodic table simplifies the task of understanding the variety of chemical properties found in nature. For example, the elements grouped in a vertical column have definite similarities. General statements can be made about their chemistries and about the compounds they form. Furthermore, the formulas for the compounds and the nature of the chemical bonds that hold them together can be described in

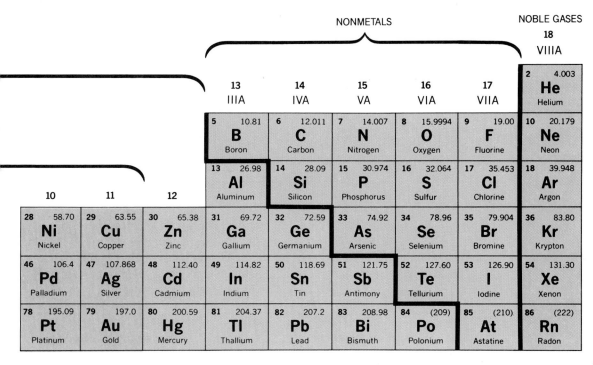

NONMETALS

NOBLE GASES
18
VIIIA

13	14	15	16	17	
IIIA	IVA	VA	VIA	VIIA	
					2 4.003 **He** Helium
5 10.81 **B** Boron	6 12.011 **C** Carbon	7 14.007 **N** Nitrogen	8 15.9994 **O** Oxygen	9 19.00 **F** Fluorine	10 20.179 **Ne** Neon
13 26.98 **Al** Aluminum	14 28.09 **Si** Silicon	15 30.974 **P** Phosphorus	16 32.064 **S** Sulfur	17 35.453 **Cl** Chlorine	18 39.948 **Ar** Argon

10	11	12					
28 58.70 **Ni** Nickel	29 63.55 **Cu** Copper	30 65.38 **Zn** Zinc	31 69.72 **Ga** Gallium	32 72.59 **Ge** Germanium	33 74.92 **As** Arsenic	34 78.96 **Se** Selenium	35 79.904 **Br** Bromine	36 83.80 **Kr** Krypton
46 106.4 **Pd** Palladium	47 107.868 **Ag** Silver	48 112.40 **Cd** Cadmium	49 114.82 **In** Indium	50 118.69 **Sn** Tin	51 121.75 **Sb** Antimony	52 127.60 **Te** Tellurium	53 126.90 **I** Iodine	54 131.30 **Xe** Xenon
78 195.09 **Pt** Platinum	79 197.0 **Au** Gold	80 200.59 **Hg** Mercury	81 204.37 **Tl** Thallium	82 207.2 **Pb** Lead	83 208.98 **Bi** Bismuth	84 (209) **Po** Polonium	85 (210) **At** Astatine	86 (222) **Rn** Radon

* For those elements all of whose isotopes are radioactive, parentheses indicate the isotope with the longest half-life. (For a definition of *half-life*, see page 198.)

| 64   157.25 **Gd** Gadolinium | 65   158.93 **Tb** Terbium | 66   162.50 **Dy** Dysprosium | 67   164.93 **Ho** Holmium | 68   167.26 **Er** Erbium | 69   168.93 **Tm** Thulium | 70   173.04 **Yb** Ytterbium | 71   174.97 **Lu** Lutetium | 4f levels filling |
| 96   (247) **Cm** Curium | 97   (247) **Bk** Berkelium | 98   (251) **Cf** Californium | 99   (254) **Es** Einsteinium | 100   (257) **Fm** Fermium | 101   (258) **Md** Mendelevium | 102   (259) **No** Nobelium | 103   (260) **Lr** Lawrencium | 5f levels filling |

terms of the stable electron arrangements of the noble gases. How electron arrangements and bonding patterns explain the organization of the periodic table will be discussed in the next chapter.

SUMMARY

The periodic table is a classification of the elements in order of increasing atomic number. Properties differ across each horizontal row of the table but are similar in each vertical column. Each vertical column of chemically similar elements is known as a chemical family. Such families include the noble gases, the halogens, and the alkali metals.

Atomic number, used to position each element in the table, is the number of protons in an atomic nucleus. It is also the number of electrons surrounding the nucleus of a neutral atom. Since electrons of atoms interact directly

alkali metal family

alkaline earth family

chemical families

diatomic

halide ion

halogen family

inert

lattice

malleable

noble gas family

periodic table

transition elements

with other atoms, electrons determine the chemistry of an atom. Thus, the number of electrons in an atom determines the position of the element in the periodic table.

The electron arrangements of the noble gases are relatively stable. The low reactivity of noble gases indicates that atoms with electron populations of 2, 10, 18, 36, 54, and 86 are relatively stable.

The alkali metals form $+1$ ions in all their reactions by *losing* one electron per atom. Thus, they achieve an electron arrangement similar to that of the adjacent noble gas atom. All alkali metals react vigorously with oxygen, chlorine, and water. Their reactions with chlorine produce a regular arrangement, or lattice, of ions, such as Na^+ and Cl^- in sodium chloride. The formulas of compounds, such as $NaCl$, showing only the simplest ratio of atoms, are called empirical formulas. The $+$ and $-$ ions in these compounds are held together by ionic bonds: the attractions of oppositely charged ions for each other.

Halogens form -1 ions (halide ions) by gaining one electron per atom. Thus, halide ions display the stable, noble gas arrangement of electrons. The halogen elements form stable, diatomic molecules, such as F_2, Cl_2, Br_2, and I_2. Each atom in such an arrangement achieves a noble gas electron arrangement by *sharing* the one electron contributed by the other atom. The bond by which these molecules share a pair of electrons is called a covalent bond.

Hydrogen resembles both alkali and halogen elements in some respects but differs from them in others. It is therefore not considered part of either family but, rather, a family in itself.

There are regular trends among the dissimilar elements in a horizontal row of the periodic table: in the formulas of their hydrides, chlorides, and oxides. The trends recur in each row and allow us to predict formulas for similar compounds of different family members.

QUESTIONS AND PROBLEMS

A

1. You are asked to sort all of the elements into two groups.
   (a) What two groups would be most obvious?
   (b) Are there elements on the borderline of these groups?
   (c) What properties would help you decide into which group to put a given element.
   (d) Classify each of the following elements in one or the other of your two groups: (1) Na, (2) Au, (3) S, (4) Mg, (5) Cl, (6) F, (7) Al, (8) Pb, (9) Cr, (10) N, (11) O, (12) Cs, (13) Ca.
   (e) Are there limitations to this system of classification? Explain.
2. Mendeleev was successful with his periodic table—much more so than his contemporaries who were trying to organize the elements. What did he do that made his work so dramatically successful?
3. Today we arrange elements in order of increasing atomic number. (a) What did Mendeleev use to arrange elements in his table? (b) Explain why Mendeleev did not use atomic numbers.
4. (a) Write formulas for the *fluorides* formed by Li, C, Na, Cs, Be, K. (b) Write formulas for the *hydrides* formed by O, N, F.

5. (a) What elements make up the noble gas family? (b) Why are these elements considered a chemical family? (c) Which of the elements of the noble gas family are hardest to convert to liquids? In other words, which element has the lowest boiling point? (d) How does the molecular formula of argon differ from the molecular formula of fluorine? Why? (e) What electron populations are characteristic of the noble gas family? Write the number of electrons for each noble gas element. (f) Can you make a correlation between your answers to (c) and (e)?

6. (a) If a sodium atom loses an electron, what atom does its electron population now resemble? (b) If a fluorine atom gains an electron, what atom does its electron population now resemble? (c) Describe the "atomic makeup" of a piece of sodium fluoride. (d) Write a balanced equation to represent the reaction between cesium and chlorine.

7. How does the periodic table tell you which elements belong to the same chemical family?

8. (a) Explain in terms of electronic structure patterns why Rb loses an electron when it reacts with iodine while iodine gains an electron. (b) What holds the resulting ions together?

9. (a) Which one of the following elements has the highest melting point? (1) Li, (2) Na, (3) K, (4) Rb, (5) Cs. (b) How does the trend for the alkali metals compare with the trend for the noble gases?

10. (a) Write the equation representing the reaction of Cs with H_2O. (b) It can be dangerous. Why?

11. (a) Does the *trend* in melting points for the halogens resemble that of the noble gases or that for the alkali metals? Explain. (b) Are forces acting between xenon (atomic number 54) molecules as great as those acting between iodine (atomic number 53) molecules? What kind of evidence supports your answer?

12. A number of formulas are written below, but only some are correct. Which of the formulas written here are *incorrect* based on your application of the periodic table? (a) KCl, (b) $RbCl_2$, (c) $AlCl_3$, (d) CCl_4, (e) CCl_2, (f) NF_3, (g) $CsCl_3$, (h) $MgCl_3$, (i) $SrCl_2$, (j) $PbCl_6$, (k) SrO, (l) BaO_3, (m) $AgCl$, (n) AgF_2.

13. What explanation based on atomic structure can be given for the fact that the alkali metals are good conductors of electricity?

14. Why do we propose that hydrogen is a family by itself?

15. What are the two most common forms of the element carbon? How do they differ in properties?

16. Compare the chemical properties of ordinary table salt (sodium chloride) with the properties of sodium metal and chlorine gas.

17. What does the empirical formula NaCl tell you about the compound it represents?

18. Sodium chloride contains sodium ions and chloride ions. How does (a) a sodium ion differ from a sodium atom and (b) a chloride ion differ from a chlorine atom?

B

19. Given that fluoride combining ratios equal the number of fluorine atoms divided by the number of atoms of the other element in the compound, find combining ratios for the fluorides of elements 3 through 9 and 11 through 17.

20. Use formulas of hydrides of the first twenty elements to show a periodicity in the chemical properties of these elements. Repeat for the oxides.

21. Examine the arrangement of elements in Table 9-2. What do you notice about the properties of elements (a) in a vertical column and (b) in a horizontal row? (c) What do we call the elements in a vertical column and in a horizontal row?

22. Explain how Mendeleev was able to predict the properties of "missing" elements such as "ekasilicon," "ekaboron," and "ekaaluminum."

23. (a) What happens to the electrical conductivity of the elements as we go across a horizontal row from left to right?
    (b) What happens to the "metallic character" of the elements as we go across the table from left to right?
    (c) Which is the more metallic element, boron or aluminum? Beryllium or magnesium? Carbon or lead?
    (d) (1) On the basis of your answer to (c), what happens to the metallic character of the elements as we go down in a column of the table? (2) Which is more metallic, diamond or silicon? (Both have the same crystal structure.) (3) Does silicon's role as a "semiconductor" make sense in view of its position in the periodic table? Explain.

24. Which of the following systems should be good conductors of electricity? Explain in each case. (a) solid sulfur, (b) solid strontium (atomic number 38), (c) solid KI, (d) KI in water solution, (e) molten KI, (f) graphite, (g) liquid bromine, (h) molten barium hydride, (i) solid silver bromide, (j) molten silver bromide, (k) solid silicon, (l) solid tin, (m) liquid HCl, (n) aqueous solution of HCl, (o) liquid xenon.

25. Why does knowing the equation for the reaction between sodium and water allow you to write equations for the reactions between potassium and water, rubidium and water, and cesium and water?

26. Compare the electron populations of the halide ions F^-, Cl^-, Br^-, and I^- with those of the noble gas atoms Ne, Ar, Ke, and Xe. How does this comparison "explain" the stability of the halide ions?

27. Use specific examples to show how hydrogen is similar in its chemistry to (a) an alkali metal and (b) a halogen.

28. How does an alkali halide such as sodium chloride differ from (a) an alkali–hydride and (b) a hydrogen halide?

C

29. Why is the element hydrogen placed above both the halogen family and the alkali metal family of elements? Support your answer with some specific examples.

30. Compare the forces of attraction between alkali metal atoms with those between noble gas atoms, with respect to (a) strength and (b) similarity. Use the information in Tables 9-4 and 9-6 to justify your answers.

31. (a) Explain, in terms of stable electron populations, why the element fluorine exists as diatomic F_2 molecules rather than as single F atoms. (b) Look at the electronic structure of the helium atom. In view of this structure and your answer to part (a), would you expect sodium vapor to be Na, Na_2, or Na_3? Explain.

32. Write empirical formulas, using the information given in Table 9-10 and the periodic table, for the compounds formed from each of the following combinations of elements:
    (a) barium, element 56, with chlorine, element 17
    (b) francium, element 87, with iodine, element 53

    (c) strontium, element 38, with oxygen, element 8

    (d) cesium, element 55, with selenium, element 34

    (e) germanium, element 32, with oxygen, element 8

    (f) arsenic, element 33, with chlorine, element 17

    (g) silicon, element 14, with hydrogen, element 1

    (h) gallium, element 31, with oxygen, element 8

33. After looking at the periodic table, write a formula for (a) an oxide of scandium, (b) an oxide of titanium, (c) a fluoride of lead, (d) a sulfide of hydrogen, (e) a sulfide of magnesium, and (f) a hydride of boron.

34. Write balanced equations for each of the following processes:

    (a) magnesium metal burns in bromine vapor

    (b) hydrogen gas and fluorine gas are allowed to mix

    (c) hydrogen gas and chlorine gas are illuminated with ultraviolet light

    (d) rubidium metal is mixed with chlorine

    (e) beryllium metal is mixed with fluorine

    (f) aluminum metal burns in air

    (g) barium metal reacts with hydrogen gas

    (h) silver nitrate solution is poured into a solution of sodium iodide. Repeat for a solution of sodium fluoride

    (i) sodium hydride is mixed with warm water (reaction goes faster in warm water)

    (j) sodium fluoride is heated with concentrated sulfuric acid (H_2SO_4)

35. On the basis of your knowledge of the periodic table, determine which of the following are *not* reasonable reactions. For any which are *not* reasonable, rewrite the equation correctly.

    (a) $K(g) + \text{energy} \longrightarrow K^+(g) + e^-$

    (b) $F(g) + e^- \longrightarrow F^-(g) + \text{energy}$

    (c) $Cl_2(g) \longrightarrow 2Cl^+(g) + 2e^- + \text{energy}$

    (d) $Ne(g) + \text{energy} \longrightarrow Ne^+(g) + 2e^-$

    (e) $Kr(g) + \text{energy} \longrightarrow Kr^-(g) + e^-$

36. On the basis of the information in Tables 9-6 and 9-7, predict the physical properties of francium and astatine. For each element, write three chemical reactions that are typical of the element.

37. To which chemical family (IA, IIA, VIA, VIIA, or VIII) must each of the following belong: (a) element A combines with oxygen to form compound AO; (b) element B is gaseous and chemically nonreactive; (c) element C combines with element D to form compound CD; element C also reacts vigorously with water, releasing hydrogen gas; (d) element E reacts with element C to form compound C_2E; (e) if these elements are found successively in the periodic table, what is their order?

38. Assume for the moment that you are Marie Curie, faced with the problem of separating ions of radium (atomic number 88) from a water solution. Write a balanced equation for the process which you would try. (Hint: refer to problems 24, 25, and 31 in Chapter 7.)

It is the behavior and distribution of the electrons around the nucleus that gives the fundamental character of an atom: it must be the same for molecules.

CHARLES ALFRED COULSON

10

ATOMIC STRUCTURE, BONDING, AND THE PERIODIC TABLE

Objectives

After completing this chapter, you should be able to:

- Use the noble gas electron populations to predict the makeup of relatively stable electron groups.

- Differentiate between noble gas core electron groups and valence electrons.

- Draw Lewis structure diagrams for the first 20 elements in the periodic table.

- Show that each chemical family is made up of atoms with identical numbers of valence electrons.

- Predict which elements tend to form ionic or covalent bonds with each other.

- Explain how ionization energies furnish experimental support for the concepts of valence electrons and noble gas core electrons.

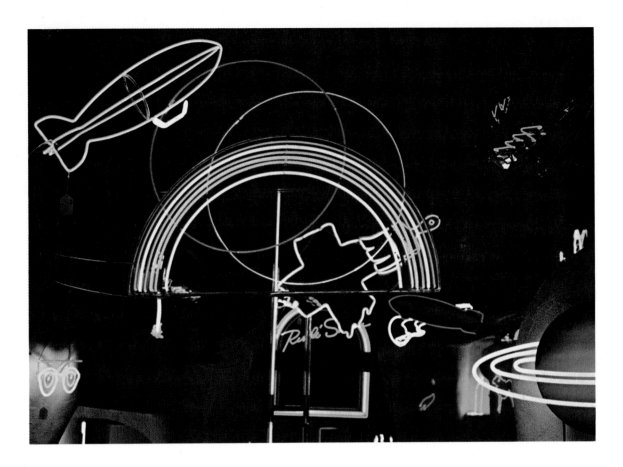

With the aid of electricity and sculptured glass tubes, low pressure gas discharge tubes are used to make glowing works of art. Different colors are due to different gases *in* and different coatings *on* the glass.

We have seen that when many elements react to form stable compounds, ions with a noble gas electron pattern are formed. Some atoms achieve the stable ion arrangements by gaining electrons. Others lose electrons. Still other atoms achieve noble gas arrangements by sharing electrons. The formation of $Cl_2(g)$ molecules from separate chlorine atoms (discussed on page 222) is an example of electron sharing. The formation of table salt, sodium chloride, from its elements in an example of electron gain and loss. Each sodium atom loses an electron to a chlorine atom, forming the Na^+ and Cl^- ions that combine to make up the salt crystals.

The organization of the periodic table seems closely related to the structure of the nuclear atom model discussed in Chapter 8. The modern table arranges the elements by their atomic numbers. Atomic number is the number of protons in the nucleus—and also the number of electrons surrounding the nucleus—of each atom of an element. Therefore, we shall look again at the nuclear atom model to study the arrangement of electrons around a nucleus. We shall find that the formulas of compounds, as well as the nature of the bonds holding them together, can frequently be understood in terms of electron arrangements resembling those of the noble gases.

Use atomic structure diagrams and equations similar to those on pages 188 and 189 to illustrate the change of (a) a neutral F atom to an F⁻ ion, and (b) a neutral Na atom to a Na⁺ ion. (c) Diagram the atomic structure of neon (Ne) and compare it to the structures of F⁻ and Na⁺.

10.1 The noble gases and stable groups of electrons

We begin our study by reexamining several noble gas electron populations. (See Table 9-7, page 223.) The first noble gas is helium, which has two electrons. Since helium is stable and nonreactive, two electrons must form a stable group. Lithium, with three electrons, reacts by losing one electron per atom to generate the Li⁺ ion. The Li⁺ ion, like helium, is nonreactive. It is very difficult to pull an electron from Li⁺. Li⁺ has the *same* electron arrangement as the neutral helium atom.

Figure 10-1

Atomic structure diagrams of elements 1 through 10 show the stable, noble gas subgroups of electrons. Except for hydrogen, each element through fluorine contains the helium core of 2 electrons, which is enclosed by the heavy black line in each structure. Neon has 2 electrons plus 8 electrons.

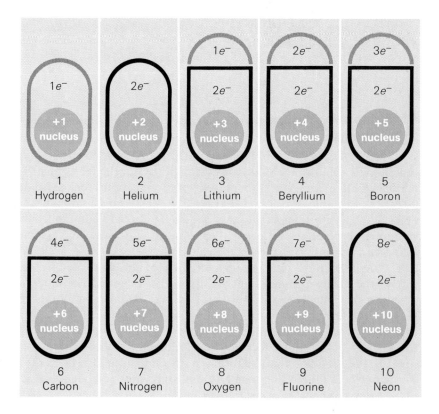

*Remember that the nucleus is actually very, very small in relation to the size of the atom. These drawings are not to scale. If they were to scale, the circled nucleus in each diagram would be an invisibly small dot.

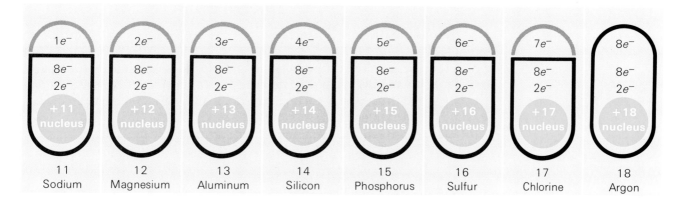

Figure 10-2

Atomic structure diagrams of elements 11 through 17 show the stable, noble gas subgroups of electrons. These elements contain the neon core of 10 electrons, which is enclosed by the heavy black line in each structure. Argon begins a new core grouping of 2, 8, and 8 electrons.

Examining models of atoms and ions helps us understand chemical behavior. Imagine that you are building models of neutral atoms of various elements. You would begin with the tiny, heavy, positive nucleus of each atom. Then you would surround each nucleus with enough electrons to neutralize its charge. In each case, *the first two electrons that you would add would form a stable group* known as the **helium core.** Regardless of how many additional electrons are required to form electrically neutral atoms, all atoms except hydrogen contain the helium core of two electrons.

The second noble gas is element 10, neon. Apparently, ten electrons also form a stable population. Each atom of sodium, element 11, tends to lose one electron during compound formation, to form a stable Na^+ ion. Atoms of element 9, fluorine, tend to pick up one electron, to form stable F^- ions. (These halide ions were discussed in Chapter 9.) Figure 10-1 illustrates the atomic structures of the first ten elements.

Both of the stable ions described above have the same electron populations as the second noble gas, neon. We can picture the ten electrons comprising the neon population as being composed of two subgroups: the helium core of two and an additional eight electrons. We assume that the eight additional electrons, when added to the helium core, form a second stable group. When combined, these two stable subgroups form a total stable population of ten electrons known as the **neon core.** Regardless of how many additional electrons are required to form electrically neutral atoms, the atoms of all elements beyond neon in the periodic table contain the neon core. Figure 10-2 shows the atomic structures of elements 11 through 18.

Argon, element 18, is the third noble gas in the periodic table. Its inert character reveals the special stability associated with a population of 18 electrons. The K^+ and Cl^- ions found in the compound potassium chloride (KCl) both display the argon arrangement of electrons. We can assume that 8 more electrons, added to the neon core of 10, form a third stable electron population of 18, known as the **argon core.** The atoms of all elements beyond argon in the periodic table contain the argon core of 18 electrons in subgroups of 2, 8, and 8. Figure 10-3 illustrates the atomic structures of elements 19 and 20.

Figure 10-3

Atomic structure diagrams of elements 19 and 20 show the stable, noble gas subgroups of electrons. Both elements contain the argon core of 18 electrons, in subgroups of 2, 8, and 8, which is enclosed by the heavy black line in each structure diagram.

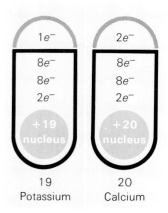

> Use atomic structure diagrams and equations similar to those on pages 188 and 189 to illustrate the change of (a) a neutral K atom to a K^+ ion, and (b) a neutral Cl atom to a Cl^- ion. (c) Diagram the atomic structure of argon (Ar) and compare it to the structures of Cl^- and K^+.

NOBLE GAS CORE STRUCTURES AND THE PERIODIC TABLE

Let us rearrange the 20 atomic structures shown in Figures 10-1, 10-2, and 10-3 by placing them in the positions that the elements they represent occupy in the periodic table. The result is shown in Figure 10-4. Examine the table carefully. Recall that the atoms in each vertical column have similar

Figure 10-4

Below, atomic structure diagrams of the first 20 elements appear in their periodic table positions.* The atom of each element is represented as follows: the nucleus is circled and its charge is given, but neutrons are not shown; the noble gas core (nucleus and noble gas subgroups of electrons) is enclosed by a black line; valence electrons are enclosed by a gray line above the noble gas core.

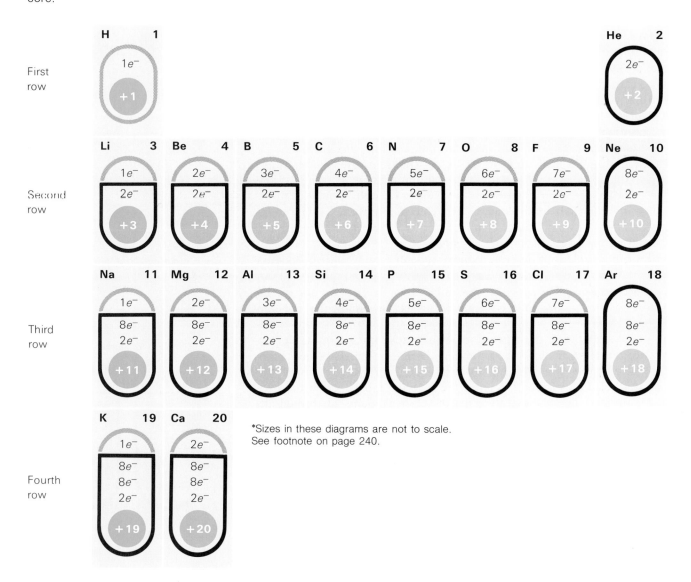

First row

Second row

Third row

Fourth row

*Sizes in these diagrams are not to scale. See footnote on page 240.

chemistry. Does dividing the total electron populations into smaller, stable subgroups help explain why the similarities exist?

We know that elements 3, 11, and 19—the chemically similar alkali metals—each have one more electron per atom than the noble gas preceding each one. This fact is emphasized in Figure 10-4 by the placement of the "extra" electrons outside the boxed noble gas cores. Perhaps the alkali metals are similar because their atoms each have one electron outside their noble gas electron core. The loss of this electron would leave behind a noble gas core.

Those electrons outside the noble gas core are called **valence electrons.** Because they do not belong to a stable group, we assume that valence electrons take part in chemical reactions. They are believed to be responsible for forming chemical bonds between atoms.

The electrons comprising noble gas core groups are frequently too stable and too rigidly held by their atoms to play any role in chemical reactions (although exceptions are known). Thus, ions with such groups tend to be stable and nonreactive. This assumption forms the basis for a much-abbreviated system of structural diagrams. They are known as Lewis structures, in honor of the American chemist Gilbert Newton Lewis (1875–1946), who proposed that the electron arrangements of atoms provide a basis for predicting what chemical reactions can occur. The **Lewis structure** of an atom uses the symbol of the element to represent the nucleus *and* the noble gas core of electrons. Any additional, or valence, electrons outside the noble gas core are represented by dots or small *x*'s placed around the atom's symbol. Lewis structures of the alkali metals in Figure 10-4 would simply be Li·, Na·, and K·, where the single (·) represents the one valence electron of each atom. Remember that the symbol of the element represents the entire remaining structure, as shown in Figure 10-5.

Now look at the structures of elements 4, 12, and 20 in Figure 10-4. These elements, beryllium, magnesium, and calcium, are the first three members

Figure 10-5

Compare atomic and Lewis structure diagrams of the elements potassium and chlorine. The symbols K and Cl represent the noble gas cores. The cores include the nuclei and the noble gas subgroups of electrons, enclosed by the heavy black line in each structure.

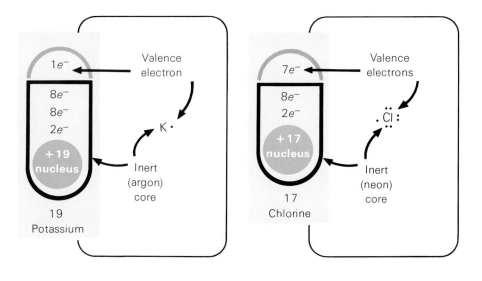

of the **alkaline earth family** of metallic elements. Each alkaline earth atom, with two more electrons than the preceding noble gas, has *two* valence electrons outside its noble gas core. The Lewis structures of the first three alkaline earth elements are Be·, Mg·, and Ca·. We assume that their similar chemistry results from their having an identical number of valence electrons. Figure 10-4 shows that each vertical column, or each chemical family, is made up of atoms with identical numbers of valence electrons. Thus, the number of valence electrons must be responsible for the similarities within chemical families and, therefore, for the chemistry of the elements.

Only valence electrons are involved in chemical reactions, which is a principal reason for using Lewis structure diagrams. Figure 10-6 gives Lewis structure diagrams for the atoms of the first 20 elements, placing them in the positions they occupy in the periodic table. Figure 10-6 is identical to Figure 10-4, except that the full electron-arrangement diagrams have been replaced by the simpler Lewis structure diagrams. As expected, Figure 10-6 reveals the same patterns found in Figure 10-4.

Lewis structure diagrams of noble gas atoms could be drawn without any dots, since the noble gases have *no* valence electrons. It is more conventional, however, to show noble gas atoms with the entire subgroup of electrons around the previous noble gas core. Thus, neon and argon are shown with eight dots, to represent the eight electrons of the last group added. Remember, though, that these dots represent a noble gas subgroup rather than valence electrons. This practice more clearly reveals the trend of adding one more electron (thus, one more dot) to each atom, moving from left to right across each horizontal row of the periodic table.

Figure 10-6

Above are Lewis structure diagrams of the first 20 elements in the periodic table. The symbols of the elements represent the nucleus and noble gas core electrons. Each dot (·) represents 1 valence electron.

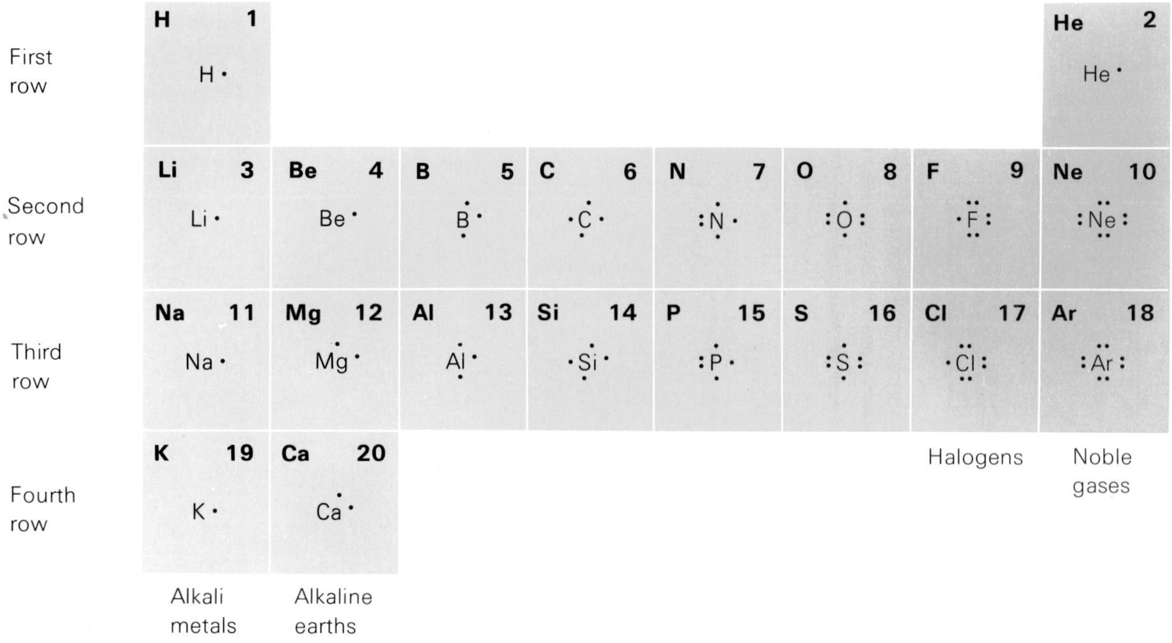

PREDICTING THE EXISTENCE OF STABLE IONS

Table 10-1 lists the more stable ions of the first 20 elements. Such ions are commonly found in the aqueous solutions of the ionic solids they form. Ionic solids were discussed in Chapter 6. A more complete listing of common aqueous ions of many elements is given in Appendix 4. As expected, lithium (Li), sodium (Na), and potassium (K)—elements 3, 11, and 19—each form +1 ions. This occurs as the neutral atoms lose one valence electron. The result is an ion with only relatively stable groups of electrons. Each alkali metal ion achieves the electron population of the preceding noble gas atom in the periodic table. The atoms of fluorine (F) and chlorine (Cl)—elements 9 and 17—form −1 ions. This occurs as each atom gains one valence electron. Each halide ion achieves the electron population of the adjacent noble gas atom in the periodic table.

Look at some of the alkaline earth family elements in Figure 10-4. The elements represented are number 4, beryllium (Be); number 12, magnesium (Mg); and number 20, calcium (Ca). The atoms of each of these elements have two more electrons than the preceding noble gas atom. We can *predict* that their chemistry will involve the *loss* of two electrons per atom. The resulting +2 ions have, in each case, only noble gas core electron populations. Table 10-1 indicates the existence of the stable, aqueous ions Mg^{++} and Ca^{++}.*

Examination of the *complete* periodic table (see pages 230–231 or the inside back cover) reveals additional alkaline earth elements. These are number 38, strontium (Sr); number 56, barium (Ba); and number 88, radium (Ra). Such elements would also be expected to form +2 ions for the reasons mentioned above. Radium is intensely radioactive and relatively rare. It is of interest primarily because of its radioactivity. But chemically stable aqueous Sr^{++}, Ba^{++}, and Ra^{++} ions do exist.

The **oxygen family** of elements, represented in Figure 10-4 by elements number 8, oxygen (O), and number 16, sulfur (S), each need two electrons to

*The Be^{++} ion is also known to exist, but it reacts further with water in neutral solution.

Table 10-1
The more stable aqueous ions of elements 1 through 20

Positive ions (cations)			Negative ions (anions)	
lithium Li$^+$	beryllium* Be^{++}		oxide* O^{--}	fluoride F$^-$
sodium Na$^+$	magnesium Mg^{++}	aluminum Al^{+++}	sulfide S^{--}	chloride Cl$^-$
potassium K$^+$	calcium Ca^{++}			

*These ions are not stable in aqueous solution.

GERHARD HERZBERG

Gerhard Herzberg was born in Hamburg, Germany, a few years after the discovery of the electron. After completing his studies in physics, he began independent research at the Universities of Darmstadt and Göttingen. In 1935, Professor Herzberg joined the staff of the University of Saskatchewan in Canada. In 1945, he became professor at the Yerkes Observatory of the University of Chicago, where he applied spectroscopic techniques to astronomy. He returned to Canada in 1948 and is presently Distinguished Research Scientist of the National Research Council of Canada.

Professor Herzberg has made many major contributions to physics, chemistry, and astronomy. Early in his career, his studies with W. Heitler of the spectrum of N_2 presented the first serious doubts about the presence of electrons in atomic nuclei—the standard model at that time. In more recent years, Herzberg has established accurate values for the dissociation energy of molecular hydrogen and for the ionization potential of helium. For his spectroscopic studies of the electronic structure and geometry of molecules and free radicals—important intermediates in many chemical reactions—he was awarded the 1971 Nobel Prize for Chemistry.

achieve the electron populations of the neighboring noble gases, neon and argon. Adding two electrons to its six valence electrons brings the oxygen family atom up to a valence electron population of eight, which is identical to the population of the neighboring noble gas.

Use atomic structure diagrams and equations to illustrate the change of (a) a neutral Mg atom to a Mg^{++} ion, (b) a neutral O atom to an O^{--} ion, (c) a neutral Ca atom to a Ca^{++} ion, and (d) a neutral S atom to a S^{--} ion. (e) Diagram the atomic structure of neon (Ne) and compare it to the structures of Mg^{++} and O^{--}. (f) Diagram the atomic structure of argon (Ar) and compare it to the structures of Ca^{++} and S^{--}.

Figure 10-7

Atomic structure diagrams can be used to show the formation of Na^+ and Cl^- ions during the bonding of NaCl.

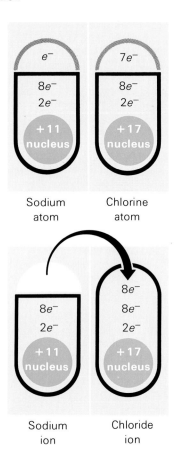

Neutral atoms do not gain or lose large numbers of electrons during chemical reactions. Once neutral atoms have gained or lost several electrons, the resulting electric charges make it difficult for positive ions to lose electrons or for negative ions to gain electrons. This is understandable in terms of the attraction of positive ions for their remaining electrons and the repulsion of an electron for a negatively charged ion. Thus, ions with charges greater than +2 or −2 are relatively rare, although not unknown. Aluminum atoms, for example, contain three more electrons than neon atoms. The loss of three electrons from an aluminum atom or scandium atom will result in a stable, neon-like Al^{+++} ion, or an argon-like Sc^{+++} ion.

10.2   Gaining and losing electrons: The ionic bond

The formation of ionic compounds, as the name suggests, involves the joining of positive and negative ions. The alkali and alkaline earth metals and all halogen family members form stable ions during compound formation. Thus, we expect ion formation by elements occupying the two vertical columns in the periodic table closely preceding and following the noble gases. Many other elements with atomic numbers greater than 20 also form stable ions.

Oppositely charged ions move toward each other. Thus, stable ionic aggregates, or clusters, can form. Such clusters are known as **ionic compounds.** An **ionic bond** is simply the force of attraction that oppositely charged ions exert on one another. As we have indicated (see page 221), each ionic solid consists of a regular array, or lattice, of alternately placed + and − ions. Each ion is surrounded by as many oppositely charged ions as is possible in such an arrangement (see Figure 7-8b, on page 169, and Figure 9-3, on page 221). Many ionic solids have the sodium chloride type of lattice.

The initial process suggested for the formation of sodium chloride is illustrated in Figure 10-8(a). Each sodium atom transfers a single electron to a chlorine atom. A Na^+ ion and a Cl^- ion result. Both ions display a noble-gas

Lewis structure diagrams

(a)
NaCl

(b)
MgCl$_2$

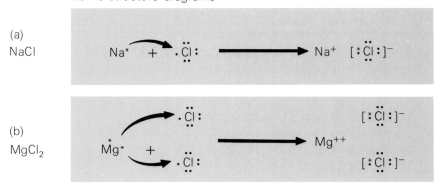

Figure 10-8

Lewis structure diagrams can be used to show the formation of (a) NaCl and (b) MgCl$_2$. Symbols of the elements represent the entire noble gas core; dots (·) represent valence electrons; and asterisks (*) represent transferred valence electrons.

electron configuration. The asterisk in the diagram helps us identify the transferred electron, but it does *not* indicate a difference among electrons. All electrons are the same. The initial process for the formation of MgCl$_2$ is indicated in Figure 10-8(b). Each magnesium ion transfers two electrons to become a Mg^{++} ion. The two electrons are gained by two chlorine atoms that become two chloride ions. One Mg^{++} ion plus two Cl$^-$ ions give an electrically neutral combination. All of the ions formed have a noble-gas electron configuration.

Some comments on energy changes in the overall process are appropriate here. Recall that in Chapter 8 (page 189) you learned that the removal of an electron from a sodium atom to give a sodium ion is an *endothermic* process. Energy is required to *pull* the electron from the neutral atom even though a noble-gas type ion is formed. In forming magnesium salts, more than twice as much energy is required to pull two electrons from a Mg atom to give a Mg^{++} ion. On the other hand, very simple observation indicates that the burning of sodium metal in chlorine gas is a strongly *exothermic* process. Energy is dramatically released. The burning of Mg in Cl$_2$ releases even more energy. Lots of heat and light are given off. Where does all of this energy come from? You know that the gaining of an electron by a fluorine or chlorine atom to give a fluoride or chloride ion is a somewhat exothermic process, but this step cannot even provide enough energy to pull off the electrons from the metal.

The major exothermic process in the formation of NaCl is the formation of the solid NaCl lattice from the gaseous Na$^+$ and Cl$^-$ ions. Many Na$^+$ and *an equal number* of Cl$^-$ ions combine in a 1 to 1 ratio to give a continuous NaCl solid lattice. There are no distinct NaCl molecules. The **empirical formula** NaCl indicates the 1:1 ratio of Na$^+$ and Cl$^-$ ions in the solid. Similarly, the empirical formula MgCl$_2$ indicates a ratio of one Mg^{++} ion to two Cl$^-$ ions in the neutral MgCl$_2$ lattice. The energy released when the lattice forms is a direct result of the attraction between the positive and negative ions and of the movement of these ions to form an ordered cluster.

Thus, the empirical formula of any ionic compound can be obtained by combining the ions in a ratio that will result in zero charge for the resulting solid. Frequently, the charges on the ions themselves can be determined from the number of electrons that must be lost or gained to achieve the noble-gas electron configuration. For example, magnesium oxide (MgO) has Mg^{++} and O^{--} ions arranged in an NaCl 1:1 type of lattice. Sodium sulfide (Na$_2$S) has Na$^+$ and S^{--} ions arranged in a 2:1 ratio in the lattice.

For each of the following pairs of elements, determine (1) the charge on each ion formed, and (2) the empirical formula of the ionic compound you expect the pair of elements to form:

(a) lithium (Li) and sulfur (S)
(b) potassium (K) and chlorine (Cl)
(c) lithium (Li) and oxygen (O)
(d) aluminum (Al) and fluorine (F)
(e) aluminum (Al) and oxygen (O)
(f) calcium (Ca) and sulfur (S)
(g) calcium (Ca) and iodine (I)
(h) barium (Ba) and oxygen (O)

10.3 Sharing electrons: The covalent bond

Our previous discussion has been focused on the electronic composition of ionic solids. While these are important and common compounds, such as salt and lime, they represent only a minor part of the total group of compounds known in the world. In fact, most of the substances that are important to life and to much of the world around us belong to a different group of materials called *covalent compounds*. All of the substances found in natural gas and petroleum-based products; substances in foods, such as sugar, protein, starch and fats; substances in our clothing, such as cotton, silk, wool, linen, and artificial fibers; many kinds of building materials, such as wood; and catalytic enzymes are included in this group. The type of bonding in these compounds is very important and is essential to life as we know it.

Many elements do *not* form stable ions but they *do* form stable compounds. Carbon is a good example. It is known for its ability to form large and important molecules but not for its ability to form ions. It is easier, however, to start our discussion of the type of bond in carbon molecules with simpler molecules such as the diatomic chlorine molecule or the diatomic hydrogen molecule. Two chlorine atoms combine to give the Cl_2 molecule and two hydrogen atoms give H_2. Each chlorine atom in Cl_2 contributes one outer electron to form a pair of electrons that has a high probability of moving between the nuclei of the two atoms to form a chemical bond. We do not have positive chlorine ions in the molecule. Instead, the nucleus of each chlorine atom contains a positive charge, which holds the pair of electrons between the nuclei and creates a bond. In such cases, two or more atoms can share electrons rather than gain or lose them. As discussed on page 222, the *sharing* by two atoms of a pair of electrons is known as a **covalent bond.**

The simplest covalent bond is found in the $H_2(g)$ molecule. We assume that each hydrogen atom contributes its electron to the bond. Thus, the hydrogen nuclei share the two electrons to generate a stable structure. Both nuclei are bonded together in the H_2 molecule by the attraction of the two positive nuclei for the two negative electrons, as shown in Figure 10-9a. The covalent bond results because two valence electrons spend a very significant part of their time *between* the two bonded nuclei.

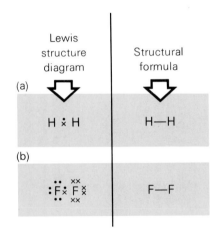

Lewis structure diagram	Structural formula
(a) ⇩	⇩
H ⋅ H	H—H
(b)	
:F⋅ F⋅	F—F

Figure 10-9

(a) In the Lewis structure diagrams of the H_2 and F_2 molecules, the symbols H and F represent the chemically inert cores (the nucleus for H, and the nucleus and two electrons of a helium core for F). Dots and small *x*'s represent valence electrons. Two different symbols are used, to indicate the atom to which each electron belonged before bonding. In reality, however, all electrons are identical. (b) The structural formulas for H_2 and F_2 incorporate a dash (—) to represent the two electrons of each bond. Other electrons, if any, are not shown.

Each of the diatomic halogen molecules—F_2, Cl_2, Br_2, and I_2—is assumed to contain a covalent bond. The formation of $Cl_2(g)$ molecules from chlorine atoms was outlined on page 222 and illustrated in Figure 9-5.* Similarly, Figure 10-9 shows that $F_2(g)$ molecules form because each fluorine atom, with nine electrons of its own, contributes one electron to generate a stable bond. Through sharing an electron pair, each atom gains a portion of a tenth electron. Each is considered to achieve a stable neon-type electron arrangement. Using Lewis structure diagrams, we write

$$\ddot{\underset{\cdot\cdot}{F}}\cdot \ + \ \cdot\ddot{\underset{\cdot\cdot}{F}}: \ \longrightarrow \ :\ddot{\underset{\cdot\cdot}{F}}:\ddot{\underset{\cdot\cdot}{F}}:$$

The formation of other non-ionic compounds can be accounted for in a similar manner. Hydrogen fluoride, for example, forms because both the hydrogen and fluorine atoms achieve a more stable electron arrangement through sharing. Using Lewis structure diagrams to represent the reaction between hydrogen and fluorine atoms, we write

$$H\cdot \ + \ \cdot\ddot{\underset{\cdot\cdot}{F}}: \ \longrightarrow \ H:\ddot{\underset{\cdot\cdot}{F}}:$$

Here, too, the positive H nucleus and the positive $\ddot{\underset{\cdot\cdot}{F}}:$ fragment are both attracted to, and held together by, the pair of electrons in the bond. *The mutual attraction of both nuclei for the shared electron pair constitutes the covalent bond.* The bonding in hydrogen chloride can be similarly explained.

When hydrogen combines with oxygen, each atom can also be considered to approach a noble gas electron arrangement through sharing. But oxygen atoms, with only six valence electrons, must share *two* electrons from other atoms to acquire a stable arrangement of eight valence electrons. Since each hydrogen atom can contribute only one electron, we predict that two hydrogen atoms must bond to each oxygen atom. Thus, the formula for water, H_2O, is explained. Using Lewis structure diagrams, the reaction is illustrated in Figure 10-10a on page 250. During the reaction, one hydrogen atom forms a covalent bond with an oxygen atom. The hydrogen atom thus achieves a stable arrangement with two electrons, but the oxygen atom then has only seven valence electrons. For this reason, we predict that a second hydrogen atom will also bond with the still unstable HO to form H_2O.

EXERCISE 10-5

Draw the Lewis structure diagram and structural formula for H_2O_2. Hint: There is an O—O bond in H_2O_2.

Examination of the other second row elements—nitrogen and carbon—indicates that their atoms have seven and six electrons, respectively. Thus, nitrogen atoms must form *three* covalent bonds to gain a stable arrangement of eight valence electrons. (Recall that the helium core of two electrons is included in each symbol.) In the case of carbon, with four valence electrons,

*Figure 9-5 (page 222) shows that covalent bonds hold atoms together because the pair of electrons between the two positive nuclei *attract* both nuclei and give each atom a stable electron population.

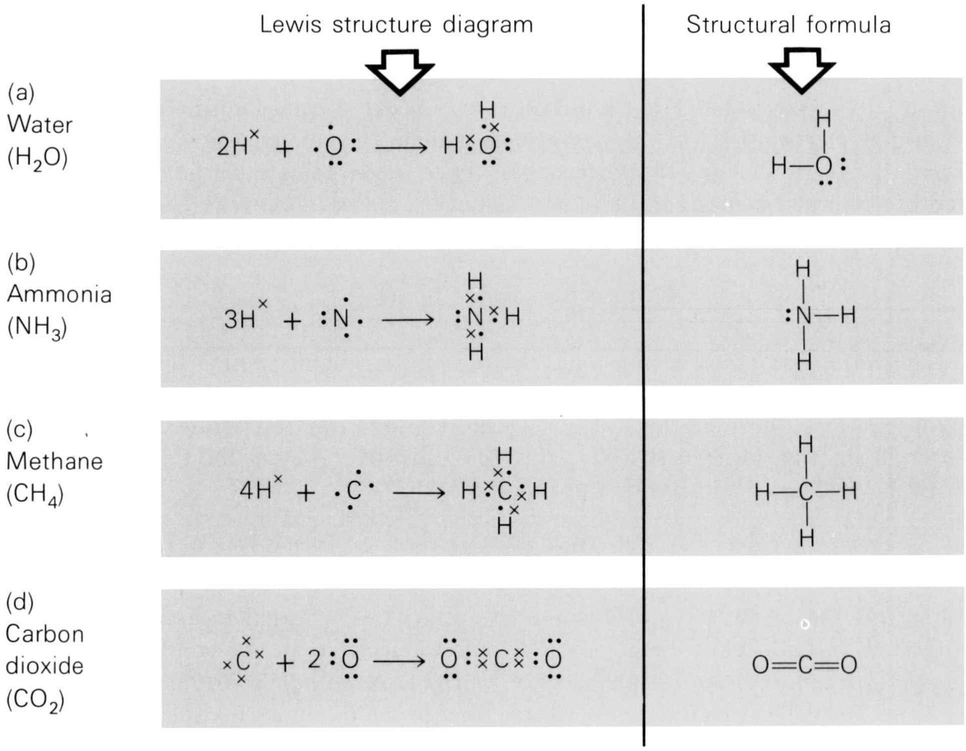

Figure 10-10

Compare the Lewis structure diagrams and structural formulas of (a) water, (b) ammonia, (c) methane, and (d) carbon dioxide. Note that there is a double bond between each oxygen atom and the carbon atom in CO_2. The ($\cdot$) and ($\times$) indicate to which atom each electron belonged before bonding. The dots shown in the structural formulas for water and ammonia represent unshared pairs of electrons.

we predict formation of *four* covalent bonds before the carbon atoms achieve their most stable electron pattern. Since hydrogen and fluorine atoms need to form only *one* covalent bond to acquire stable electron populations, it follows that three hydrogen or three fluorine atoms would bond with each nitrogen atom, and four hydrogen or four fluorine atoms would bond with each carbon atom. Bonding capacities and the formulas of possible compounds can be predicted using the concept of electron sharing. Figures 10-10b and 10-10c give Lewis structure diagrams and structural and molecular formulas for the compounds we would expect to obtain using hydrogen.

Thus, we see that atoms can achieve a more stable electron arrangement by sharing electrons with and from other atoms. An atom that needs one electron to achieve a stable arrangement must share one electron from another atom. An atom lacking one electron will form only *one* covalent bond. An atom that is two electrons short of a stable arrangement must share two electrons from other atoms and will form *two* covalent bonds. An atom lacking three electrons will need to share three electrons and will form *three* covalent bonds.

Sometimes, more than one covalent bond is formed beween two atoms. For example, in the case of carbon dioxide (CO_2), the carbon atom needs four more electrons to achieve stability. Each oxygen atom needs two more electrons. Thus, the carbon atom must form four covalent bonds, and each oxygen atom must form two. The carbon atom forms *two* covalent bonds by sharing two pairs of electrons with *each* oxygen atom, as shown in Figure 10-10d. Because all the bonding electrons are shared by all three bonded atoms, each atom achieves a stable arrangement of eight valence electrons. We say that a *double* bond exists between carbon and oxygen.

EXERCISE 10-6

> Using the concept of covalent bonding, (1) draw Lewis structure diagrams and structural formulas, and (2) predict the molecular formulas of stable compounds for each of the following combinations of elements: (a) oxygen and fluorine, (b) sulfur and hydrogen, (c) phosphorus and hydrogen, (d) chlorine and fluorine, (e) silicon and fluorine, (f) carbon and chlorine.

10.4 Ionization energy: Evidence of electron grouping

We have mentioned the tendency of elements to achieve stable, noble gas electron arrangements through loss or gain of one or more valence electrons. For example, the 19 electrons needed to neutralize the 19 protons in a potassium nucleus are arranged in groups. The first 2 electrons are most tightly held. The next 8 electrons, when added to the first 2, form the neon core and are somewhat less tightly held. The next 8, bringing the total to 18 electrons, form the argon core and are less tightly held than the electrons of the neon core. The nineteenth electron, the valence electron, is not as tightly held as any of the 3 core groups. Thus, the nineteenth electron is the most easily lost. Sodium atoms have the stable, tightly held neon core plus 1 (more easily lost) valence electron. Lithium atoms have the helium core plus 1 (more easily lost) valence electron. An easily lost valence electron is the common feature of lithium, sodium, and potassium atoms, and accounts for similarities in their chemistry.

Like the alkali metals, each family of elements in the periodic table is pictured as having a tightly held, stable group of electrons and one or more less tightly held valence electrons. The atoms of each family have the *same* number of valence electrons. Though they are less tightly held, energy is needed to remove such electrons from the atoms.

Let us now examine the ionization energies of the various elements in order to gain insight into the electronic structure of atoms. **Ionization energy** is defined as *the amount of energy required to remove an electron from a gaseous atom.* We can represent this process by the word equation

gaseous atom + energy ⟶ gaseous ion + gaseous electron

The following equation represents the process for a gaseous atom of sodium:

$$Na(g) + energy \longrightarrow Na^+(g) + e^-$$

Figure 10-11

The brightness of a dim television screen can be increased by raising the kinetic energy and number of bombarding electrons. This is done by increasing the accelerating voltage.

The energy necessary to remove the first electron from an atom, forming an ion, is called the atom's **first ionization energy** (represented as E_1).

The term *ionization energy* is also applied to the removal of an electron from an *ion*. For example, the ionization energy of Na^+ is the energy necessary for the process

$$Na^+(g) + energy \longrightarrow Na^{++}(g) + e^-(g)$$

Since this process removes a second electron from a sodium atom after the first electron is gone, the ionization energy of Na^+ is called the **second ionization energy** of sodium (represented as E_2). This process requires a very large amount of energy.

The ionization energy for certain atoms and molecules may be determined by bombarding the atomic or molecular vapor with electrons whose kinetic energy is accurately known. How do we obtain a beam of electrons of known kinetic energy? On page 204 you were told that an electric discharge tube gives a beam of electrons. The kinetic energy of the electrons can be increased by increasing the voltage used on the tube. The adjustment is not too different from the adjustment for picture brightness on your television tube. In order to get a brighter picture—more energetic electrons and a stronger current—you turn up the voltage on the picture tube. When the kinetic energy of the bombarding electrons used in the measurement is increased to a certain critical value, singly charged positive ions can be determined electrically.*

These ions result from collisions between the atoms being studied and the bombarding electrons. When the bombarding electrons have the same kinetic energy as the energy needed to separate the most loosely bound electron in the atom, a collision will knock the electron from the atom. The energy required to break an electron away from a given target atom is characteristic of that atom—it is a measure of the atom's ionization energy. We can read the voltage on the tube generating the electron beam as soon as positive ions are obtained. The value can be expressed in terms of calories, kilocalories, or joules.

Ionization energies can also be determined even more precisely by using methods of atomic spectroscopy described in Chapter 11. The method to be used is determined by the system studied.

The first ionization energies of the first 20 elements, in kilocalories per mole, are listed in Table 10-2. When ionization energies are plotted against atomic number, they show a regularly repeated, or periodic, pattern. The

*In actual operation such measurements are made in a mass spectrometer (page 206) where somewhat more elaborate velocity selectors may be used.

Table 10-2

First ionization energies of elements 1 through 20

Atomic number	Element	First ionization energy (kcal/mole)
1	hydrogen (H)	313.6
2	helium (He)	566.7
3	lithium (Li)	124.3
4	beryllium (Be)	214.9
5	boron (B)	191.2
6	carbon (C)	259.5
7	nitrogen (N)	335.4
8	oxygen (O)	313.8
9	fluorine (F)	401.5
10	neon (Ne)	497.0
11	sodium (Na)	118.4
12	magnesium (Mg)	175.2
13	aluminum (Al)	137.9
14	silicon (Si)	187.9
15	phosphorus (P)	254.1
16	sulfur (S)	238.8
17	chlorine (Cl)	300.1
18	argon (Ar)	363.2
19	potassium (K)	100.0
20	calcium (Ca)	141.0

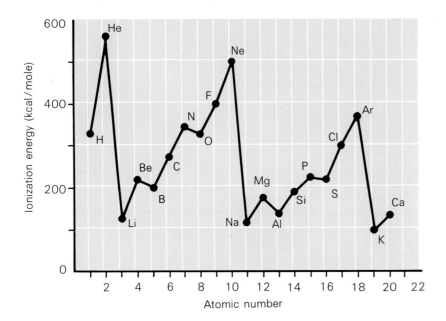

Figure 10-12
Ionization energy as a function of atomic number.

graph in Figure 10-12 reveals a dramatic drop in ionization energy as we go from a noble gas to the element following it. This drop is followed by a slow rise in ionization energy, another repeating regularity, as we proceed across a row of the periodic table. We see that ionization energy increases more or less regularly across the first two rows of the table, in each case reaching a maximum energy level at a noble gas. As soon as we encounter an alkali metal, we notice that the ionization energy falls dramatically. In subsequent elements, the general upward trend repeats itself. Values for additional ionization energies are shown in Figure 10-13. High points in the curve appear for the noble gas elements 2, 10, 18, 36, 54, and 86. As we might expect, the alkali metals rubidium (37) and cesium (55) both have low ionization energies.

There is a startling similarity between the regularities revealed by ionization energies and the periodicity of chemical properties of the elements. Chemical periodicity is emphasized by the placement of elements in the periodic table. The trends we observe in chemical properties can be explained in part by the trends in ionization energies.

For example, contrast the first ionization energies of sodium and chlorine:

$$Na(g) + 118.4 \text{ kcal/mole} \longrightarrow Na^+(g) + e^-(g)$$
sodium atom + ionization energy $\longrightarrow$ sodium ion + electron

$$Cl(g) + 300.1 \text{ kcal/mole} \longrightarrow Cl^+(g) + e^-(g)$$
chlorine atom + ionization energy $\longrightarrow$ chlorine ion + electron

Since sodium has a low ionization energy, 118.4 kilocalories per mole, a relatively small amount of energy is required to remove an electron. This is consistent with the chemical evidence that sodium tends to form compounds involving the ion Na^+. The ease of forming Na^+ ions can be explained in terms of the low ionization energy of the sodium atom. In contrast, a large amount of energy, 300.1 kilocalories per mole, is required to remove an electron from a chlorine atom. It is not surprising, then, that chlorine shows little tendency to lose eletrons in chemical reactions. Instead, chlorine frequently acquires electrons to form negative chloride (Cl^-) ions—a process that gives off energy.

Between sodium and chlorine, there is a slow rise in ionization energy, as shown in Figure 10-12 and in Table 10-2. For magnesium and aluminum, the elements following sodium, the first ionization energy is low. As the ionization energy rises, the chemistries of silicon, phosphorus, and sulfur show a trend toward electron sharing. Those elements usually do not reach a stable arrangement by losing electrons because the ionization energy required is too high. Instead, they "seek" stability by sharing electrons, or alternatively, as for sulfur and chlorine, by acquiring electrons to form negative ions.

The correlations detailed above confirm the idea that the electronic structure of an element is closely related to its chemical behavior. In particular, the most weakly bound electrons, the so-called valence electrons, are of the greatest importance in this respect. Valence electrons are viewed as being outside the stable core of an atom's electrons. They now can be defined as the most loosely bound electrons of an atom.

If the energy of bombarding electrons in ionization energy experiments is increased sufficiently, it becomes possible to remove two or more electrons from a multi-electron atom. Of course, it is always harder to remove the

Figure 10-13

A plot of ionization energy versus atomic number.

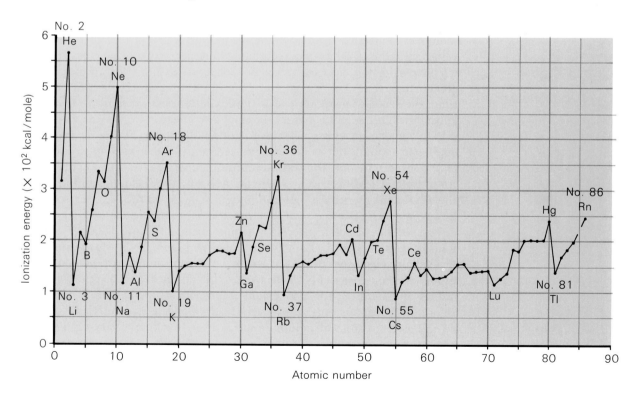

second electron than the first. This is partly due to the additional electrical attraction of a positive ion for an electron, compared to the attraction exerted by a neutral atom for an electron.

The values for the first three ionization energies for sodium are shown below. Note that the first ionization energy for sodium is relatively small, but values for the second and third ionization energies are large. Why is this so?

$$Na(g) + 118 \text{ kcal } (E_1) \longrightarrow Na^+(g) + e^-(g)$$
$$E_1 = \text{first ionization energy}$$

$$Na^+(g) + 1,091 \text{ kcal } (E_2) \longrightarrow Na^{++}(g) + e^-(g)$$
$$E_2 = \text{second ionization energy}$$

$$Na^{++}(g) + 1,653 \text{ kcal } (E_3) \longrightarrow Na^{+++}(g) + e^-(g)$$
$$E_3 = \text{third ionization energy}$$

It is instructive to consider ionization energies for other elements such as magnesium and aluminum in answering this question. The experimental values of the ionization energies for the three elements are shown in Table 10-3. Let us begin by comparing sodium and magnesium. For each, the first ionization energy (E_1) is low, although the value for magnesium is higher than the value for sodium. The difference is caused by the higher nuclear charge of magnesium. Magnesium, element 12, has 12 protons in its nucleus, compared to 11 protons in the sodium nucleus. We would expect a valence electron to be more strongly attracted to the +12 charge of the magnesium nucleus than to the +11 charge of the sodium nucleus.

The second ionization energy (E_2) reverses the situation, however. About three times as much energy is needed to remove the second electron from sodium as is needed to remove the second electron from magnesium. The values in Table 10-3 are understandable in terms of two principles. First, as we have said, the electrical attraction of a positive ion for an electron is greater than that of a neutral atom for an electron. In the case of magnesium, the second ionization energy (345 kilocalories) exceeds the first (175 kilocalories) because it is harder to remove an electron from a Mg^+ ion than it is to remove an electron from a neutral Mg atom. The same principle applies to sodium, but it alone cannot account for the *much greater* difference between the first and second ionization energies (E_1 and E_2) observed for sodium. The difference is dramatic. According to the data in Table 10-3, the E_2 for magnesium is *not quite twice* as large as the E_1. For sodium, however, the E_2 is *more than nine times* as large as the E_1.

The second principle necessary to understand the much larger second ionization energy for sodium is illustrated in Figure 10-2. Magnesium has *two* valence electrons—that is, it has two electrons more than the neon core of ten. We would expect both electrons to be loosely held. The relatively low values of E_1 and E_2 for magnesium confirm this expectation. Sodium, on the other hand, has only *one* loosely held valence electron. Removal of a *second* electron from a sodium atom involves breaking a very tightly held electron from the very stable neon core arrangement. As we would expect, it is difficult to remove that electron.

Table 10-3
Successive ionization energies of sodium, magnesium, and aluminum

Element	Ionization energy (kcal/mole)			
	E_1	E_2	E_3	E_4
sodium	118	1,091	1,653	—
magnesium	175	345	1,839	2,526
aluminum	138	434	656	2,767

The stable core and valence electron diagrams in Figure 10-2 show that, for magnesium, the *third* electron to be removed comes from the stable neon core. Therefore, we expect E_3 for magnesium to be *much* greater than E_2. The data in Table 10-3 show that this is indeed the case. The data also support the Figure 10-2 picture of aluminum. Aluminum, element 13, has *three* more electrons than the noble gas neon. Accordingly, we expect the first *three* electrons of aluminum to be loosely held. The *fourth* electron would come from the very stable neon core pattern. Therefore, a very large increase in ionization energy for aluminum would be predicted between the third and fourth ionization energies (E_3 and E_4). The data from Table 10-3 are consistent with this prediction.

Ionization energy data give striking confirmation of the grouping of electrons in atoms, as shown in Figures 10-1, 10-2, and 10-4. Ionization energy measurements for silicon and phosphorus show large jumps between E_4 and E_5 for silicon and between E_5 and E_6 for phosphorus. This is in accord with the fact that silicon has four valence electrons and phosphorus has five valence electrons, as shown in Figure 10-2. Each of the noble gases, with all their electrons in stable groups, show the expected very high first ionization energies. (See Table 10-2.) Finally, the ionization energies for the second-row elements, lithium through neon, show similar trends, as expected from drawings of their electron arrangements.

Table 10-4
Successive ionization energies of lithium, beryllium, and boron

Element	Ionization energy (kcal/mole)			
	E_1	E_2	E_3	E_4
lithium	124	1,748	2,815	—
beryllium	215	418	3,532	5,004
boron	191	577	872	5,964

Use the periodic chart in Figure 10-4 to predict at which ionization energies large jumps would be expected for (a) element 3, lithium; (b) element 4, beryllium; and (c) element 5, boron. (d) Are your predictions confirmed by the data in Table 10-4?

Key Terms

alkaline earth family

argon core

covalent bond

empirical formula

first ionization energy

helium core

ionic bond

ionic compound

ionization energy

Lewis structure

neon core

oxygen family

second ionization energy

valence electrons

SUMMARY

The noble gases helium, neon, and argon illustrate the stability of electron populations of 2, 10, and 18. Atoms of all elements beyond helium in the periodic table have the stable 2-electron group known as the helium core. The 10-electron neon population, known as the neon core, contains the 2-electron helium core plus 8 more electrons. Thus, 10 electrons form a second stable group. Atoms of all elements beyond neon in the periodic table contain the neon core of 10 electrons in subgroups of 2 and 8. Argon's 18 electrons are known as the argon core and indicate the stability of 8 more electrons added to the 10 electrons of neon. Atoms of all elements beyond argon in the periodic table contain the argon core of 18 electrons in subgroups of 2, 8, and 8.

Atomic structure diagrams can be divided into a noble gas core and electrons outside this core, called valence electrons. Only valence electrons are assumed to take part in chemical reactions. The Lewis structure diagram of an atom uses the element's symbol to represent the nucleus and noble gas core of electrons. The additional valence electrons are usually represented by dots around the symbol of the atom.

Each alkali metal atom has one valence electron in addition to its noble gas core group and forms +1 ions. Each halogen atom has seven valence electrons in addition to its noble gas core group and forms −1 ions. Each alkaline earth atom has two valence electrons in addition to its noble gas core group and forms +2 ions. In a similar fashion, each vertical column of the periodic table consists of elements whose atoms have identical numbers of valence electrons. The number of valence electrons is responsible for the similarities within chemical families and for the chemistry of the elements.

Stable electron arrangements can be achieved by different types of bonding. Ionic bonding results when oppositely charged ions attract each other and form electrically neutral aggregates, or clusters, known as ionic compounds. Covalent bonding is the sharing of a pair of electrons by two atoms.

Ionization energy is the amount of energy required to remove an electron from a gaseous atom or ion. The first ionization energy is the energy required to remove a first electron from the neutral atom. The second ionization energy is the energy required to remove a second electron from the ion with a +1 charge. Third and fourth ionization energies are known for many elements. Ionization energy measurements show that valence electrons are relatively easy to remove from atoms. Removal of noble gas core electrons requires much more energy. Measurements of ionization energy tend to support the idea of stable, noble gas core groups of electrons within atoms.

QUESTIONS AND PROBLEMS

A

1. (a) Indicate the meaning of each of the following terms: (1) helium core, (2) neon core, (3) argon core, (4) valence electrons. (b) Use these designations, as appropriate, to show the *electronic structure* of: (1) Be (2) N (3) F (4) Na (5) F^- (6) Na^+ (7) Ne (8) Mg (9) Ca (10) Cl (11) Al.

2. The Lewis structure for an atom is M : (a) What ion would you expect this atom to form? (b) Write the formula of the chloride of this element. (c) Write the formula for the oxide of the element. (d) Which of the following atomic numbers might be appropriate for the element M: 3, 11, 12, 21, 35, 36, 37, 55, 56? Explain your choice.

3. Draw Lewis structure diagrams for: (a) H_2O (b) HBr (c) NH_3 (d) $TiCl_4$ (e) ClF (f) CO_2 (g) CH_4 (h) $MgCl_2$ (i) CF_2Cl_2 (j) C_2H_6 (k) PCl_3 (l) $SiCl_4$.

4. Why do two atoms that are linked by a covalent bond stay together? Explain in terms of attractive forces.

5. What is meant by the ionization energy of an element? Write equations for the first, second, and third ionization energies of magnesium.

6. Which of the following ionization energy values would be the largest?
   (a) The first ionization energy for Na; the first ionization energy for Mg; the first ionization energy for Si. Explain your answer.
   (b) The second ionization energy for Na; the second ionization energy for Mg; the second ionization energy for Si. Explain your answer.
   (c) The first ionization energy for Mg; the first ionization energy for C; the third ionization energy for Be.

7. Element Y has the Lewis structure Y : and element X has the Lewis structure : X . . Write the Lewis structure formula for the compound that would be formed between Y and X.

8. Which one of the following atoms would have the *largest* first ionization energy: (a) Na (b) Rb (c) Ca (d) C (e) Ne (f) Ba?

9. What experimental facts suggest that the first noble-gas type core for an atom has two electrons in it?

10. Write the equation for the formation of the most stable ion that would be expected for each of the following: (a) Sr (b) F (c) Cs (d) S (e) Sc.

B

11. Draw the Lewis structural formula for: N_2, SO_4^{--}, CsBr, K_2S, Sr^{++}, Mg^+. What is shown in the diagram for N_2?

12. The ionization energies for a given element, M, are listed as: 140 kcal, 274 kcal, and 1,178 kcal. Given this information, write the formulas of an oxide, a phosphide, a sulfide, a fluoride, and a nitride of M.

13. Which of the following species have the same electronic structural arrangement? (a) Na (b) K (c) Na^+ (d) Mg^{++} (e) F^- (f) Ne (g) Rb^+ (h) Al^{+++}.

14. Describe how to predict the empirical formula of any ionic compound.

15. Explain the difference between an ionic bond and a covalent bond.

16. Where are those elements most likely to form ionic compounds located in the periodic table? Explain your answer.

17. Predict empirical formulas for compounds formed by each of the following pairs of elements: (a) sodium and fluorine, (b) sodium and sul-

fur, (c) cesium and bromine, (d) magnesium and iodine, (f) magnesium and sulfur, (e) potassium and nitrogen.

18. Draw Lewis structure diagrams, indicating the electrons involved in forming covalent bonds, for each of the following: (a) H_2 (b) F_2 (c) HCl (d) Cl_2 (e) H_2S (f) CH_4 (g) NH_3 (h) SiF_4.

19. Explain why the second ionization energy for magnesium is relatively low, while the second ionization energy for sodium is very high.

20. More and more energy is required to remove successive electrons from an atom. How many electrons would be relatively easy to remove from a (a) lithium atom, (b) potassium atom, (c) calcium atom, and (d) barium atom? Explain your answers.

21. Write balanced equations for each of the following processes: (a) lithium reacting with chlorine, (b) potassium reacting with water, (c) magnesium burning in fluorine, and (d) hydrogen reacting with liquid bromine.

22. Use Lewis structure diagrams to represent each process in problem 21.

C

23. On the basis of your knowledge of the periodic table and electronic structure, which of the following formulas are not reasonable? (a) Li_2O (b) RbI (c) CaO (d) Mg_2O (e) Al_2O_3 (f) BeCl (g) KN.

24. Several elements are described as follows:
    (1) Element M: Silvery gray in color; can be drawn into wires and rolled into sheets; first ionization energy = 175 kcal/mole; second ionization energy = 345 kcal/mole; third ionization energy = 1,840 kcal/mole; does not react with cold water.
    (2) Element P: Red solid; exists as powder; first ionization energy = 252 kcal/mole; second ionization energy = 454 kcal/mole; third ionization energy = 690 kcal/mole.
    (3) Element X: Yellowish gas; *very* irritating when inhaled; first ionization energy 400 kcal/mole; one mole of X gas will react with one mole of H_2 to give two moles HX. Following data available: $X(gas) + e \longrightarrow X^- +$ energy.
    (4) Element Z: Colorless gas; very light; a balloon filled with molecules of the element Z will rise rapidly; first ionization potential = 313 kcal/mole; no second ionization potential yet reported.
    (5) Element L: A very light, silvery substance that floats on water and reacts violently with water; tarnishes quite rapidly in air; first ionization energy = 99 kcal/mole; second ionization energy = 732 kcal/mole.
    (a) Write the formula for the most ionic substance that can be formed from the five elements shown.
    (b) Write the formula for a covalent compound that can be formed from any two of the above five elements.
    (c) Would you expect compound LZ to be a solid, liquid, or gas? An ionic or covalent substance?
    (d) Write a formula for a compound formed between elements M and X; elements L and P, elements Z and X.

25. Write Lewis structures for each of the elements listed below.
    (a) Element A combines with oxygen to give the compound A_2O_3.
    (b) Element B is gaseous and chemically nonreactive.

(c) Element C combines with element D to form the compound CD. (Element C also reacts vigorously with water to give H_2 gas.)

(d) Element D in its molecular form will react with one mole of H_2 to give two moles of HD.

(e) Element E reacts with element C to give a compound C_2E.

26. Under appropriate conditions, cesium will react with oxygen to give a compound, Cs_2O_2. Under somewhat different conditions the unusual "superoxide" CsO_2 is formed. (a) Write Lewis structures for both Cs_2O_2 and CsO_2. (b) Which of the two compounds does not show agreement with the octet rule (that is, the noble gas electron configuration)? (c) Would the O—O bond in the negative peroxide ion (O_2^{--}) be a double bond? Explain.

27. Simple sulfides such as potassium sulfide (K_2S) will pick up elemental sulfur to form compounds with formulas such as K_2S_4, K_2S_5, and so on. Write a Lewis structure for K_2S_4 and show that it could be expanded to give K_2S_n where n is any whole number. Note that sulfurs form a chain.

28. Write balanced equations for each of the following processes:

(a) Magnesium metal reacts with aqueous hydrochloric acid (HCl) to give H_2 gas and magnesium chloride in solution.

(b) Strontium metal (atomic number 38) reacts with water to give hydrogen gas and another product.

(c) Elemental silicon (atomic number 14) reacts with elemental fluorine (F_2) to give a compound containing both elements.

(d) Elemental chlorine and elemental fluorine combine to give a new molecule containing both elements.

(e) Potassium metal reacts with oxygen gas to give potassium superoxide (KO_2).

29. The art of prediction is not alway easy. The following ionization energy values are known:

Element	Atomic Number	Ionization Energy
Oxygen	8	314 kcal/mole O atoms
Fluorine	9	402 kcal/mole F atoms
Neon	10	497 kcal/mole Ne atoms

(a) Using this information, *predict* the ionization energy for sodium, atomic number 11.

(b) The actual value of the ionization energy for sodium is 118 kcal/mole. What factor enters with sodium that makes the prediction of part (a) erroneous.

(c) Values for the *second ionization* energies are:

Element	Atomic Number	Ionization Energy
Oxygen	8	807 kcal/mole O atoms
Fluorine	9	805 kcal/mole F atoms
Neon	10	943 kcal/mole Ne atoms
Sodium	11	1,088 kcal/mole Na atoms

Which value for which atom is now unpredictable? Clearly, additional factors must be considered. Do you have any clues?

Treat electrons like light waves. Though they are bullets, they are also waves.

LOUIS DE BROGLIE

11

LIGHT, COLOR, AND ATOMIC STRUCTURE

Objectives

After completing this chapter, you should be able to:

- Explain the difference between infrared rays, ultraviolet rays, X-rays, and visible light, in terms of wavelength, frequency, and energy.

- Understand the meaning of the *s, p, d,* and *f* orbitals of quantum mechanics.

- Understand and use energy level diagrams to describe the hydrogen atom and atoms with many electrons.

- Describe the relationship between an element's placement in the periodic table and its ionization energy, as represented in an energy level diagram.

White light contains all the colors of the visible spectrum.

Our studies so far have used a relatively simple picture of the atom. We have accepted a model in which a very dense, positively charged nucleus, composed of protons and neutrons, is surrounded by rapidly moving electrons. But to better understand the nature of chemical bonding, we must refine and improve this model. In the process, we shall examine the areas of *spectroscopy* and *quantum mechanics.*

We begin by asking three important questions about atomic structure that have intrigued chemists since the nuclear atom model was first proposed in 1912:

1. What is the detailed arrangement of electrons outside the nucleus? (The question is important because it relates chemistry to atomic structure.)
2. What is the detailed structure of the nucleus?
3. What holds nuclear particles together? Why do protons that are crammed tightly in a very small nucleus not fly apart? Why is the nucleus so stable?

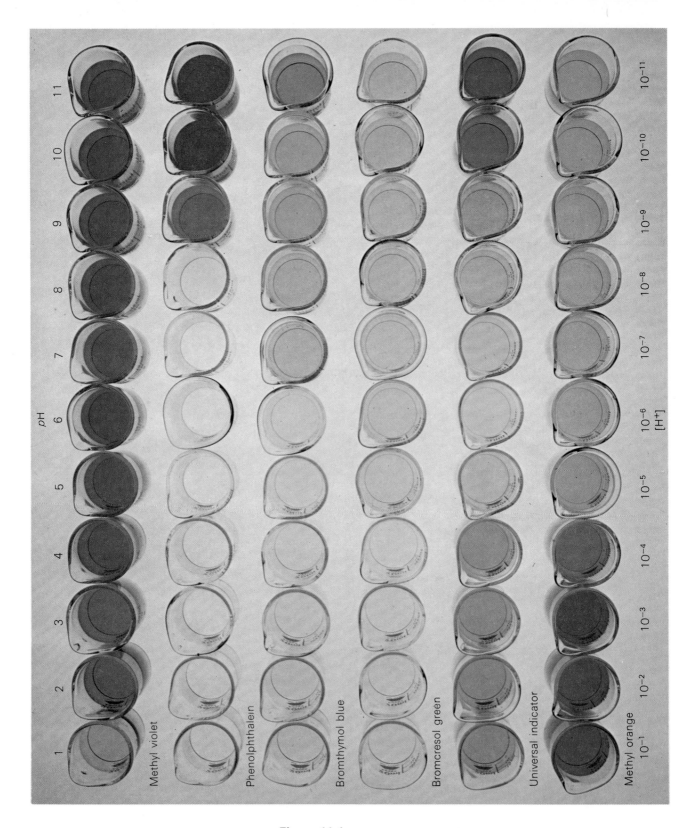

Figure 11-1
Various acid-base indicators in solution produce different colors at different hydrogen ion concentrations. Refer to Chapter 17 for a discussion of ionic equilibrium of acids and bases.

At present, scientists have a good deal of information about question 1. We shall examine much of that information in this chapter. Progress is being made toward answering question 2. Evidence suggests that protons and neutrons are arranged in stable shells, or energy levels, in the nucleus, much like electrons are arranged outside the nucleus.* Although much of the emerging information on nuclear structure reminds us in some ways of data on electronic structure, the nuclear situation is far more complicated. Many new and strange particles are being found as a result of work in high-energy physics.

The answer to question 3 remains unclear. We still are not sure what holds a nucleus together. Very strong, short-range forces clearly exist in the nucleus. To study such forces, scientists need large and expensive particle accelerators. The work is slow, difficult, expensive, and important.

In this chapter, and in the Extension beginning on page 292, we will focus on electronic structure and some of the experiments that led to our current picture of the atom. It is interesting that much of what we now know about atomic structure was obtained from the study of light.

11.1   Light, color, and electromagnetic waves

Why are the solutions in Figure 11-1 different colors? Why does a sodium vapor lamp glow with a yellow light, while a neon sign is bright red? Why does a color television tube glow red, green, and blue? In seeking answers to such questions, scientists developed a more detailed model of the atom.

All of us associate color with light. In some cases, color is best displayed under white light, as with the acid-base indicators in various solutions shown in Figure 11-1. (Acid-base indicators in solutions are discussed in Chapter 17). In other cases, a substance gives off colored light when excited by a source of energy. The source could be the flame of a Bunsen burner, a laser, or the electric current in a neon tube, sodium vapor lamp, or television tube. For example, flame tests for various ions show deep red for lithium chloride, brick red for calcium chloride, green for barium chloride, and blue-green for copper chloride.

What is the nature of color and light?† Is light made up of little particles, or is it a continuous, wavy stream? Is light a solid ray, as a beam of sunlight appears to be? Some simple, everyday observations are useful in answering these questions. For example, an oil film floating on water shows constantly changing colors as we watch it. A soap bubble floating in air exhibits fleeting and fragile displays of color when sunlight strikes the bubble, as shown in Figure 11-2. Perhaps you have seen "diffraction jewelry" such as the pin in Figure 11-2, or "diffraction signs," such as those fastened to cars by enthusiastic owners. The surfaces of these items display a variety of colors that change as you look at them from different angles. The surfaces are marked with many very closely spaced parallel lines. The color arises from the interaction of white light with the closely spaced lines.

*There is now evidence to suggest that even protons may have a nuclear structure. See Maurice Jacob and Peter Landshoff, "The Inner Structure of the Proton," *Scientific American,* vol. 242 (March 1980) pp. 66–75.

†An excellent article is Kurt Nassau, "The Causes of Color," *Scientific American,* vol. 243 (October 1980) pp. 124–154.

The colors displayed by an oil film, a soap bubble, diffraction jewelry, and diffraction signs are called **interference,** or **diffraction, colors.** They can best be explained by visualizing light as a wavelike disturbance. Because light may be described, in part, in terms of waves, we should examine waves in more detail.

Figure 11-2

White light striking a soap bubble is separated into its various colors. Closely spaced parallel lines on diffraction jewelry also produce the spectral colors from white light.

WAVES: TERMS AND SYMBOLS

A beach is one of the best places in the world to study waves. On some days, waves are small. The distance between the **crest,** or top part of the wave, and the **trough,** or bottom part of the wave, is small. We say that small waves have a small **amplitude.** On good surfing days, however, the waves are tall. We say that tall waves have a large amplitude. The distance between two corresponding points on successive waves is the **wavelength** of the wave. On the average, large ocean waves have a wavelength of several metres. Wavelengths may also be small.

Another characteristic of waves, **frequency,** is of special interest. Frequency may be determined by counting the number of crests or troughs that pass a given fixed point in 1 minute (or some other unit of time). By counting the number of times a float or bob rises per minute, or by counting the number of waves breaking on the shore in 1 minute, you can determine the frequency of the ocean waves. (See Figure 11-3, where frequency is expressed in bobs per minute.) Frequency always has units of time in the denominator.

The symbol for the frequency of light waves is the Greek letter nu (ν). The unit used to express the frequency of light waves is 1/sec or sec $^{-1}$. The wavelength of light is given by the Greek letter lambda (λ) and expressed in units of length. The metre is appropriate for ocean waves, while the nanometre (10^{-9} metre) is the preferred unit for the short wavelengths of light.*

What is the relationship between frequency and wavelength? The number of waves breaking on the shore in a given period of time is determined by how fast the waves are coming in (wave velocity) and by the distance between crests or troughs (wavelength). We can express this relationship in an equation as

$$\text{frequency of waves} = \frac{\text{velocity of wave}}{\text{wavelength of wave}}$$

For light, this relationship is described by an almost identical equation:

$$\text{frequency of light} = \frac{\text{velocity of light}}{\text{wavelength of light}}$$

or
$$\nu = \frac{c}{\lambda} = \frac{3.00 \times 10^8 \text{ m/sec}}{\lambda}$$

where ν = frequency of light, c = velocity of light, and λ = wavelength of light.

*Other units of length can also be used. For example, the centimetre (10^{-2} metre), the micron (10^{-6} metre), the millimicron (10^{-9} metre), and the angstrom (10^{-10} metre) still appear in scientific literature.

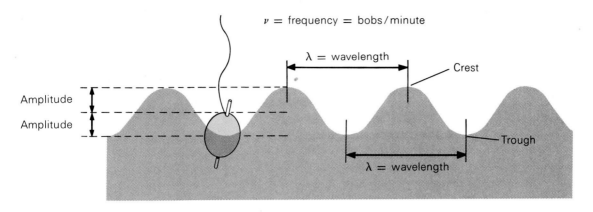

ν = frequency = bobs/minute

λ = wavelength

Crest

Amplitude

Amplitude

Trough

λ = wavelength

Figure 11-3

Light travels in waves, as does the energy in the sea.

EXERCISE 11-1

Determine the wavelength of light having a frequency (ν) of 7.5 $\times$ 10^{14} waves/sec, using the relationship $\nu = c/\lambda$, in which the velocity of light (c) = 3.00 $\times$ 10^8 m/sec.

EXERCISE 11-2

What is the frequency of light with a wavelength of 5,000 $\times$ 10^{-10} m?

WAVES AND ENERGY: ELECTROMAGNETIC RADIATION

Visible light from a desk lamp, infrared rays from a heat lamp, microwaves from a microwave oven, radio waves from a broadcasting tower, and X-rays from a dentist's X-ray machine are all different forms of a wavelike phenomenon called **electromagnetic radiation.** These forms of radiation differ from each other in the *frequency* of electromagnetic waves. In electromagnetic radiation, electric and magnetic fields oscillate (move back and forth) in a regular pattern to define a wavelength and a frequency for the radiation. The oscillating fields exert alternating pushes and pulls on a charged particle or detector placed in the fields.

Sunlight is separated into the colors of the rainbow when it passes through a prism. The prism separates visible light into different colors, each with its own special frequency. Red light has the lowest frequency and is bent the

Table 11-1

Visible spectral colors and the wavelengths of various regions of the electromagnetic spectrum

Color	Wavelength (λ) ($\times 10^{-7}$ metre)	Frequency (ν) (vibrations/sec)
ultraviolet	2.0 to 4.0	15 to 7.5 $\times 10^{14}$
violet	4.0 to 4.2 (4.1)*	7.5 to 7.1 $\times 10^{14}$
blue	4.2 to 4.9 (4.7)	7.1 to 6.1 $\times 10^{14}$
green	4.9 to 5.8 (5.2)	6.1 to 5.2 $\times 10^{14}$
yellow	5.8 to 5.9 (5.8)	5.2 to 5.1 $\times 10^{14}$
orange	5.9 to 6.5 (6.0)	5.1 to 4.6 $\times 10^{14}$
red	6.5 to 7.0 (6.5)	4.6 to 4.3 $\times 10^{14}$
infrared	longer than 7.0	less than 4.3 $\times 10^{14}$
radio	2200	1.4 $\times 10^{12}$

*The values in parentheses represent the wavelengths most characteristically recognized as the pure colors.

least in passing through a prism. Violet light has the highest frequency and is bent the most. The bending of visible light is shown in Figure 11-4. Thus, a beam of white light is separated into all its different frequencies. Table 11-1 gives the frequency and wavelength for each color of what we know as the **visible spectrum.**

EXERCISE 11-3

The light produced by a sodium vapor lamp shows two lines with wavelengths of 5.89 $\times 10^{-7}$ m and 5.90 $\times 10^{-7}$ m. Identify the color of the sodium vapor lamp using the data assembled in Table 11-1. What color is the light from a potassium vapor lamp if the spectral lines for potassium have wavelengths of 4.044 $\times 10^{-7}$ m and 4.047 $\times 10^{-7}$ m? What color is a mercury vapor lamp if the frequencies of the lines are 6.6 $\times 10^{14}$ waves/sec and 5.5 $\times 10^{14}$ waves/sec? Estimate the wavelength of one of the most intense lines in the *neon* spectrum.

White light, or any multi-color light, is a composite of all the separate colors it contains. Whenever light is separated into individual colors, the result is known as the spectrum of that light. In order to separate a beam of light into its spectral colors in a laboratory, an elaborate prism arrangement is usually used. Such an instrument is known as a **spectrometer**—an instrument that measures spectra.

Figure 11-4
The effect of a prism on white light.

Table 11-1 shows that visible light has wavelengths ranging from 4.0×10^{-7} metres to 6.5×10^{-7} metres. This is a very narrow range of wavelengths, and it would be strange indeed if longer and shorter wavelengths did not exist. Of course, they do exist, but our eyes are sensitive to

Table 11-2
Spectrum of electromagnetic radiation

Name of radiation	Approximate range of wavelength
radio waves	a few metres and up
microwaves	a few millimetres to a few metres
infrared waves	$7,500 \times 10^{-10}$ metre to 10^{-4} metre
visible light	$4,000$ to $7,500 \times 10^{-10}$ metre
ultraviolet light	100 to $4,000 \times 10^{-10}$ metre
X-rays	0.1 to 500×10^{-10} metre
gamma rays	less than 0.5×10^{-10} metre

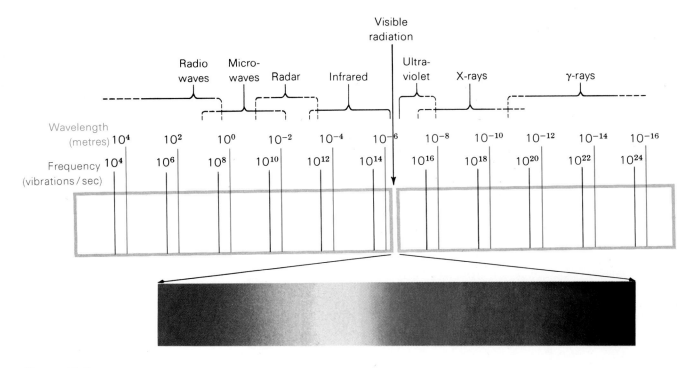

Figure 11-5

The entire electromagnetic spectrum.

only a very narrow band in the complete electromagnetic spectrum. The spectrum of sunlight shows dark regions at positions beyond the location of visible light in both directions. Frequency and wavelength approximations for the dark regions are shown in Figure 11-5. The complete spectrum is represented in Figure 11-5 and is given in tabular form in Table 11-2.

Many of the terms in Table 11-2 are familiar. **Radio waves** have very long wavelengths and low frequencies. **Microwaves** are somewhat shorter in wavelength. **Infrared** radiation has somewhat longer wavelengths than red light in the visible spectrum. **Ultraviolet** light, which causes sunburn, is found just beyond the violet end of the visible spectrum.* **X-rays** have very short wavelengths and very high frequencies. A large body of evidence indicates that excessive exposure of animals to X-rays increases their chances of getting cancer. For this reason, people are cautioned against undue exposure to both X-rays and ultraviolet rays. Higher-frequency radiation can cause extensive damage to most living things.

The German physicist Max Planck (1858–1947) developed a theory to explain the relationship between the frequency and energy of electromagnetic radiation. In 1900, Planck suggested, that light is made up of bundles of energy called *quanta*. These bundles are also called *photons*. The energy in any one bundle, or **quantum,** depends on the color of the light. Violet light has more energy per quantum than does red light. Red light has more energy per quantum than does infrared radiation, and so on. Furthermore, *the amount of energy per quantum is directly proportional to the frequency of the radiation.* We can express this relationship as an equation:

*The earth is protected from ultraviolet light by the "ozone layer" in the stratosphere. Excessive exposure to ultraviolet light has been connected with the development of skin cancer. Thus, the possible destruction of stratospheric ozone by fluorocarbons, such as Freon, caused much concern at one time, but current evidence indicates that the danger is much less than was originally believed.

(energy per quantum of light or radiation) =

$$\text{(a constant)} \times \text{(frequency of light)}$$

Using symbols, we write

$$E = h\nu$$

The letter h stands for **Planck's constant of action,** which is usually referred to simply as Planck's constant. It is equal to 1.58×10^{-37} kilocalorie $\times$ seconds but is sometimes expressed in other units, such as 6.63×10^{-34} joules $\times$ seconds. The equation $E = h\nu$ is one of the most important and powerful statements of modern science. It marks the birth of the **quantum theory,** which underlies all our ideas about atomic structure.

EXERCISE 11-4

Using Planck's relationship, $E = h\nu$, find the energy of red light that has a frequency of 4.6×10^{14} waves/sec. Use a value of h that will give an answer in kilocalories.

11.2 The hydrogen spectrum and hydrogen atoms

You may be wondering what light has to do with atomic structure. To answer such a question, let us look at the light given off by a "hydrogen light." A hydrogen light is related to a neon light. You will recall that a tube containing low-pressure neon gas glows with a beautiful red color when electricity is passed through it. This red light is the heart of the electric sign business. If the light from the neon tube is passed through the spectrometer, bright red and even green lines appear. If a tube containing low-pressure hydrogen is used instead of low-pressure neon, light with a purplish color is emitted. When this light is passed through the spectrometer, the series of lines shown in Figure 11-6c appears. Each line in the spectrum corresponds to a given pure color or frequency given off by hydrogen atoms. Every hydrogen discharge lamp emits the same small group of "lines" or frequencies. The other spectra in Figure 11-6 are discussed in the Extension (pages 292-297).

The Extension of this chapter (pages 292-297) describes in some detail how our ideas about light and observations of the hydrogen spectrum led to what is known as the Bohr model of the hydrogen atom. The Danish scientist Neils Bohr (1885-1962) was a contemporary of Max Planck. The Bohr model pictures each hydrogen atom as having one electron. Initially, the electron in each atom resides at the most stable, or lowest, level of energy. The lowest level is called the **ground state.**

Energy added to a collection of hydrogen atoms is absorbed by the electrons as they jump from the ground state to higher **energy levels.** Many electrons jump to a second level, while others jump to a third level. Other electrons jump to still higher levels. According to Bohr, because electrons are unstable at higher levels of energy, they will eventually fall back to their stable, ground state level.

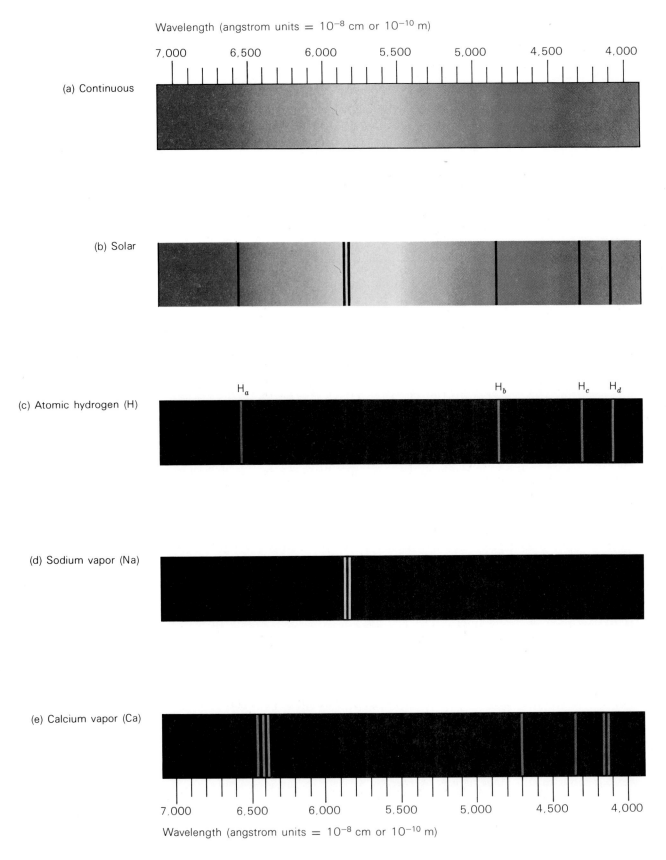

Wavelength (angstrom units = 10^{-8} cm or 10^{-10} m)

7,000   6,500   6,000   5,500   5,000   4,500   4,000

(a) Continuous

(b) Solar

(c) Atomic hydrogen (H)

H_a H_b H_c H_d

(d) Sodium vapor (Na)

(e) Calcium vapor (Ca)

7,000   6,500   6,000   5,500   5,000   4,500   4,000

Wavelength (angstrom units = 10^{-8} cm or 10^{-10} m)

THE QUANTUM MODEL FOR THE HYDROGEN ATOM

The light given off by an excited hydrogen atom corresponds to a change in the energy of the hydrogen atom from a condition of higher to a condition of lower energy. Since only certain lines appear in the spectrum, the hydrogen atom seems to be using only *a limited number of energy levels.* Bohr's model correctly predicted the limited number of lines observed in the hydrogen spectrum. A familiar analogy will help explain this limitation. Suppose a bookcase is standing on the floor, as shown in Figure 11-7, on page 274. A paperweight is shown in what we can call its ground state. Since the paperweight is on the floor, it can fall no further. To place the paperweight on a shelf, work must be done to lift it. This work is stored in the paperweight as *potential* energy. If the weight falls back to the floor, its potential energy will be released as *kinetic* energy. Also, heat energy is released as the weight hits the floor. The amount of energy released will be exactly equal to the amount of energy used to raise the paperweight to the shelf.

Our picture of the hydrogen atom is much like our picture of the bookcase. The electron in a hydrogen atom can exist at *any one* of many different and distinct energy levels, as shown in the energy level diagram in Figure 11-8. The electron can be placed at levels 1, 2, 3, 4, and so on, but not at intermediate points where no levels exist. In our bookcase analogy, the paperweight could be placed only on a shelf or on the floor. It could not hang suspended in mid-air between shelves. When the paperweight falls, energy is released. When the electron "falls" from level 2 to level 1 in Figure 11-8, energy is released as electromagnetic radiation. The amount of energy released is equal to the difference in energy between level 2 and level 1 and is expressed in kilocalories per mole. Light of the same frequency (or color) will be given off every time an electron falls from level 2 to level 1. Light of a different frequency (or color) will be given off every time an electron falls from level 3 to level 1. Another spectral line will be seen.

Figure 11-8 gives an accurate picture of the limited number of energy levels an electron can occupy. But there are other things about the electron that cannot easily be pictured. A fundamental philosophical principle known as the **uncertainty principle** limits the accuracy of any measurement. For example, if we try to measure the position and velocity of an electron, some probe, such as a light beam, must be used. Because the energy of the light beam makes the electron move, we cannot know exactly where the electron would have been if the light beam had not been used. Thus, the position of the electron must remain somewhat uncertain.

The Bohr model of the hydrogen atom described in the Extension on pages 292–297 conflicts with the uncertainty principle. Therefore, Bohr's mechanical model—placing electrons in circular orbits—is no longer accepted. We do, however, accept Bohr's scheme for quantization of energy in the atom. This idea provides the basis for the diagram in Figure 11-8.

Figure 11-6

Opposite are the visible portions of various spectra, showing only the principal lines: (a) continuous spectrum of a hot solid; (b) the dark absorption lines are caused by cooler gases at the sun's surface; (c) emission spectrum of atomic hydrogen; (d) emission spectrum of sodium vapor; and (e) emission spectrum of calcium vapor. Spectral wavelengths are measured in angstrom units. (One angstrom equals 10^{-8} cm, or 10^{-10} m.)

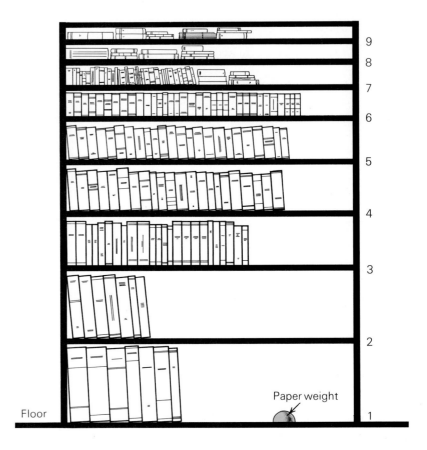

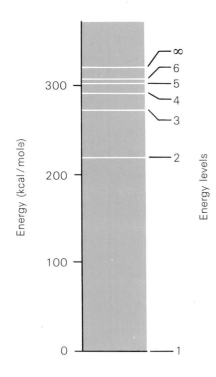

Figure 11-8

The electron in a hydrogen atom can occupy any one of these energy levels.

Figure 11-7

A paperweight can rest on the floor or on the shelves of a bookcase but at no place in between.

A key question not answered by the Bohr model is why electrons are restricted to certain energy levels. Such restrictions made no sense until the French physicist Louis de Broglie suggested in 1923 that electrons had wavelike properties. After all, we picture light as wavelike when we explain its interference effects (see Figure 11-2). And we picture it as particles when we speak of quanta, or photons, of light energy. Why, asked de Broglie, could not electrons, as well as light, have *both* wavelike and particle-like properties?

De Broglie's brilliant suggestion, which seemed contradictory at first, was verified in 1927 by Clinton Davisson and Lester Germer at Bell Telephone Laboratories in the United States. Davisson and Germer found that electrons reflected from a crystalline surface produced interference patterns similar to those formed by electromagnetic waves, such as X-rays. Such patterns, like those shown in Figure 11-9, are the classic sign of wavelike behavior. The equation derived by de Broglie and verified by Davisson and Germer is:

$$\frac{\text{wavelength of the matter-wave}}{\text{associated with a particle}} = \frac{\text{Planck's constant}}{\text{mass of particle} \times \text{velocity of particle}}$$

or

$$\lambda = \frac{h}{mv}$$

where λ = wavelength, h = Planck's constant, m = mass of particle, and v = velocity of particle. This equation expresses the relationship between the wavelike and particle-like properties of electrons.

If we use the de Broglie relationship to calculate the wavelength of an electron in an atom, we find that the electron's wavelength is comparable to the size of the atom. But standing waves of any size can exist as complete waves only within distances that are one, two, three, or some other whole-number multiple of their wavelength. Thus, electrons can occur *only* at certain discrete energy levels: those levels at which the wavelength "fits" an integral number of times. It was de Broglie's idea of the *wave nature of matter* that provided the key to this understanding. For this reason, de Broglie was awarded the Nobel Prize for Physics in 1929.

Why do apples, oranges, and baseballs not show wavelike behavior too? The answer is that their wavelengths are too small to be detected. The mathematics of de Broglie's suggestion show that only objects of very small mass have detectable wavelengths. It is the large mass of apples, oranges, and baseballs that makes their wavelengths too short to be detected. They appear only as particles and not as waves.

11.3 Wave mechanics: The quantum theory

For the reasons we have mentioned, atoms are best described by a new form of mechanics called **wave mechanics,** or **quantum mechanics.** (The mechanics describing the movement of baseballs, cars, and other large objects are not applicable to electrons.) The first term refers to the wavelike nature of the electron. The second term reminds us of the limitations, or *quantization,* of electron energy states. Although highly mathematical in nature, the predictions of quantum mechanics about atoms can be summarized in nonmathematical terms.

One result of using the quantum theory is that the distinct energy levels we have mentioned can be identified by *quantum numbers.* The bookshelf numbers in Figure 11-7 and the energy level numbers in Figure 11-8 are called **principal quantum numbers.** Each principal quantum number represents and identifies a particular energy level. For the hydrogen atom, the formula is

$$E_n = -\frac{313.6}{n^2} \text{ kcal/mole}$$

where n is the principal quantum number. Here, E_n is the energy of the level with the principal quantum number n. Notice that for the hydrogen atom *with only a single electron* all orbitals with the same value of n have the same energy.

The negative sign in front of 313.6 is significant and deserves comment. Simple algebra shows that as n becomes larger, the energy becomes less negative and as n approaches infinity, the energy approaches zero. Why is the energy of a hydrogen atom negative in its lowest state? To answer this question, look at Figure 11-10. You will notice two energy scales on the diagram. In the scale on the left, we have assigned zero to the *lowest* energy level, or ground state. In the scale on the right, we have assigned zero to the *highest* energy level, which corresponds to the ionization energy. Physical

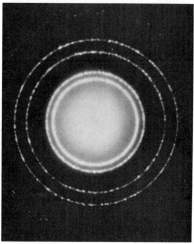

Figure 11-9

The diffraction pattern made by a beam of X-rays passing through thin aluminum foil is shown at top. Compare it with the diffraction pattern made by a beam of electrons passing through the same foil, at bottom.

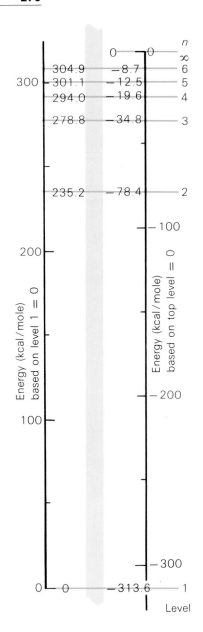

Figure 11-10

The energy level scheme of the hydrogen atom.

scientists have determined that calling the highest level zero offers a number of advantages. If this scheme is used, then by definition *the hydrogen atom has zero energy when the electron and nucleus are infinitely far apart.* They are infinitely far apart when the hydrogen atom is ionized. As the electron approaches the proton, energy is released. The atom now has *less* energy than it did at the defined zero energy. Thus, all energies recorded for the hydrogen atom must carry a negative, or minus, sign.

EXERCISE 11-5

Using the equation

$$E_n = -\frac{313.6}{n^2} \text{ kcal/mole}$$

determine the energy level values when $n = 1, 2, 3, 4$, and infinity. Compare your answers to the values given in the scale on the right side of Figure 11-10.

ELECTRON DISTRIBUTION: ORBITALS

The principal quantum number, n, defines the energy of the electron *but not its path.* According to quantum mechanics, it is possible to identify only a region of space around the atomic nucleus where an electron of given energy is most likely to be found. The distribution of all possible locations of an electron in a given energy level is called an **orbital.**

The boundaries of an orbital are based on probability. Two analogies may help clarify this idea. When you pluck a guitar string and watch it closely, you see a blur. You know that, at any instant, the string is someplace in that blurry space. However, you cannot know exactly where it is. In the guitar analogy, the blur made by the plucked string might be called the orbital of the string.

A better analogy is a bird-feeding station in a large, snow-covered park. A single bird's motion around the feeder can be studied by placing a camera directly above the feeder and snapping a picture once every minute during daylight hours. Each picture shows the bird in one position, sometimes at or near the feeder, sometimes farther away. If we stack all the pictures on top of each other, making sure each is a transparency so that we can look through all of them, we obtain the composite picture shown in Figure 11-11. The picture is dark around the feeding station, where the bird spent much time, and lightens gradually toward the edges of the park, where the bird spent less time. Thus, the composite print gives a probable distribution of the bird's position around the feeder.

The composite photo could be called the bird's orbital. By analogy, the feeder is an atomic nucleus and the bird is an electron. Quantum mechanics can give the same kind of information about the electron as the composite print does about the bird. The composite print does not tell us where the bird is at any moment or how the bird got from one spot to another. In the

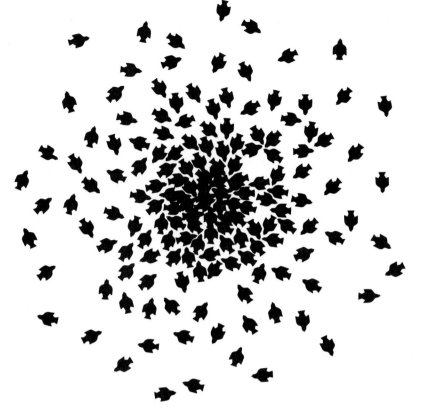

Figure 11-12

A computer generated a plot of the 1s orbital by plotting the position of a 1s electron at 4,000 consecutive instants of time. The computer also generated plots of the 2s orbital and the 3s orbital.

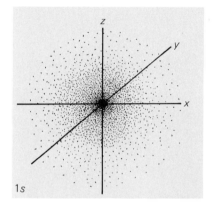

1s

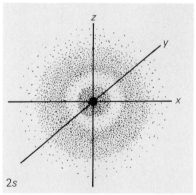

2s

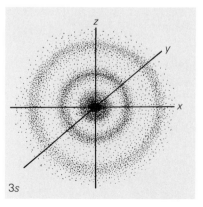

3s

Figure 11-11

A composite picture of a bird at a bird feeder taken over a period of time.

same way, we cannot know where an electron is at any instant or how it moves from one place to another.

The quantum number n gives us the energy of orbitals in the hydrogen atom. For example, if $n = 1$, $E = 313.6$ kcal, if $n = 2$, $E = -78.4$ kcal, if $n = 3$, $E = -34.8$ kcal, and so on. The value of n also gives us the number of orbitals with that value of n. *For each value of* n *there are* n^2 *orbitals with that* n *value.* For example, if $n = 1$ there is only one orbital; if $n = 2$, there are 4 orbitals; if $n = 3$, there are 9 orbitals, and so on. *For the hydrogen atom,* the n^2 orbitals associated with a given n all have the same energy. In general, the electron is farther from the nucleus as n gets larger. But orbitals with the same n can differ in the spatial distribution of electrons. "Orbital shapes" are quite different. Let us examine characteristic orbital shapes for hydrogen.

s ORBITALS

The lowest energy level of a hydrogen atom has an n value of 1 and an energy of -313.6 kcal/mole. Quantum mechanics tell us that there is only one orbital at this level, since $1^2 = 1$. This orbital corresponds to an

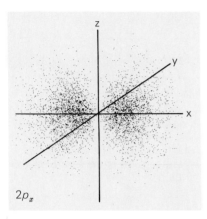

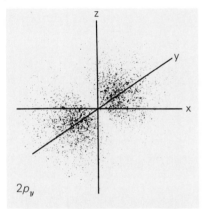

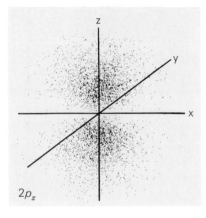

Figure 11-13

Computer-generated plots of the three 2p orbitals: $2p_x$, $2p_y$, and $2p_z$ orbital.

electron distribution that is spherically symmetrical around the nucleus. A "composite picture" of this orbital is shown in Figure 11-12. Notice how closely it resembles the distribution pattern of the bird feeding in the park (Figure 11-11). We call this a *1s orbital,* and the electron moving in this orbital is called a 1s electron. Thus, the probability of finding a 1s electron in a given unit of volume decreases in a spherical pattern as we go away from the nucleus. This suggests that an atom does not have a definite size, but "fades away" as the distance from the nucleus increases and the likelihood of the electron's being there decreases.

The next energy level corresponds to $n = 2$. For hydrogen, quantum mechanics tell us that there are n^2, or four different orbitals with the same energy. The energy value is $-313.6/4$, or -78.4 kilocalories per mole. One of these orbitals is spherically symmetrical and is named the *2s orbital.* (See Figure 11-12.) The composite of its probable distribution shows that a 2s electron spends more time farther away from the nucleus than does a 1s electron. This is consistent with the higher energy of the 2s electron. Thus, the composite picture of the 2s orbital shows a faint ring out from the nucleus where the probability of finding the electron is slight. When $n = 3$, there are nine orbitals. One of those is the *3s orbital,* with the probable distribution shown in Figure 11-12.

p ORBITALS

As we have explained, the 2s orbital is one of four orbitals with the principal quantum number 2. The other three are known as *2p orbitals.* Each 2p orbital shows electron probability concentrated along the direction of one of three coordinate axes: x, y, and z. As Figure 11-13 shows, a $2p_x$ orbital concentrates electron density along the x axis in a dumbbell-shaped pattern. Note that the probability of finding a p electron at the nucleus is zero. The probability of finding an s electron at the nucleus, as shown in Figure 11-12, is significant. The p_y and p_z orbitals have dumbbell-like electron distributions similar to the p_x orbital, but these are distributed along the y and z axes (Figure 11-13). *Every* energy level with n above 1 has three p orbitals. For example, when $n = 3$, there are three *3p orbitals.* Except for being farther from the nucleus as n increases, the general outward dumbbell-shape of p orbitals does not change.

d AND f ORBITALS

The information about orbitals can be summarized by constructing an **energy level diagram** for hydrogen such as the one in Figure 11-14. Each orbital is represented as a circle into which electrons can be placed. As we have indicated, there are n^2 orbitals for each value of n. When $n = 3$, there are 3^2, or nine, orbitals. We have already mentioned the one 3s orbital and the three 3p orbitals. The five remaining orbitals in the third quantum level are called *3d orbitals.* Seven f orbitals appear at the fourth quantum level, when $n = 4$. They are called *4f orbitals.* We will consider the somewhat more complicated spatial distributions of the 3d and 4f orbitals in Chapter 21, on the transition elements.

How many orbitals would be expected for $n = 4$? How many orbitals are accounted for by the $4s$, $4p$, and $4d$ groups? The remaining electrons in the fourth quantum level are f electrons. How many $4f$ electrons are there?

11.4 Electronic structure and the periodic table

In Chapter 10, it was noticed that the noble gases are not at all reactive. We associated their low reactivity with a stable electronic structure. For example, we noted that helium with 2 electrons is inert, thus a 2-electron pattern is very stable. Similarly, neon with 10 electrons is inert; thus a 10-electron pattern is very stable. We noted that sodium, with one more electron than neon, will lose that electron easily in chemical reactions to give the neon-

Figure 11-14

The energy level scheme of the hydrogen atom, a single electron atom.

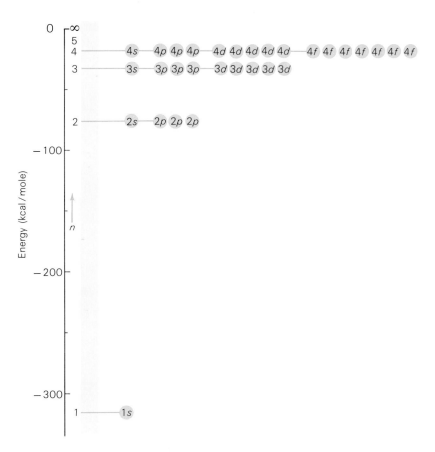

like Na$^+$ ion. The sodium electron structure is *not* stable, but the 10 electron Na$^+$ pattern *is* stable. We will now try to tie these general observations into the more detailed patterns available from our orbital model for electronic structure.

MANY-ELECTRON ATOMS

On page 277 you were told that the energy of a hydrogen orbital is given by the *n* value for that orbital, and the number of orbitals for a given *n* is n^2. For $n = 2$ we have 4 orbitals, all with the same energy. For $n = 3$ we have nine orbitals with identical energy. This is shown in Figure 11-14. This simple energy level pattern for hydrogen made its spectrum simple and permitted Bohr to first develop his model for the hydrogen atom. The simplicity is a direct result of the fact that we have only *a single electron outside of the hydrogen atom nucleus*. Other atoms with more than one electron outside the nucleus have a more complicated energy level pattern because the electrons repel each other. They interact with each other. The arrangement is shown in Figure 11-15. Notice that in general, *s* orbitals are lower than *p* orbitals *of the same* n; *p* orbitals are lower than *d* orbitals *of the same* n. For example, 4*s*, 4*p*, and 4*d* orbitals represent the same principal quantum number, $n = 4$. This change from the hydrogen energy level scheme is a direct result of electron-electron repulsion in many electron atoms.

With this refinement in the energy level scheme and two other rules, we can construct the periodic table from electronic structure arguments. These rules and their explanation are given here without experimental justification, but you can rest assured that ample evidence for these rules exists. The two additional rules are:

1. A single orbital of any element can accommodate, at most, two electrons. Each electron in an orbital may be visualized as spinning around its own axis. The first electron in an orbital spins in one direction, and the second electron must spin in the reverse direction. Such electrons are said to differ in spin, or to have "paired spins." Since there are only two directions of spin, each orbital can accommodate only two electrons.
2. All orbitals of equal energy will accept one electron each before any orbital accepts a second electron. This is a result of the fact that two electrons in the same orbital repel each other. If there is an empty orbital of equal energy, the electron will go in there first, rather than go into an occupied orbital.

THE BUILD-UP OF THE PERIODIC TABLE:
THE FIRST EIGHTEEN ATOMS

Let us now use these rules to suggest the electron arrangement for an atom in its lowest energy state or its "ground state." We do this by filling orbitals with electrons until the positive nuclear charge is just neutralized by the negative charge of all the extranuclear electrons. We shall always place each additional electron in the lowest *unfilled* orbital available. Thus, in

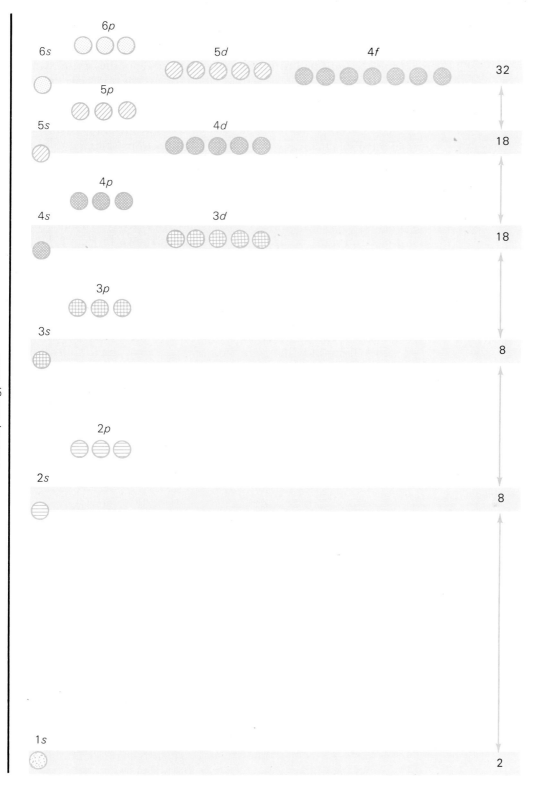

Figure 11-15

The energy level scheme of an atom with many electrons. Note that this scheme differs markedly from that for a one-electron atom shown on page 279. The differences arise from electron-electron repulsion in the atom with many electrons.

hydrogen with only a single electron, the electron is placed in the $1s$ level as shown in Figure 11-16. The number of electrons in a particular orbital, or set of orbitals, is indicated by a superscript to the right of the orbital symbol. So we write $1s^1$ as the **electron configuration** for hydrogen. For helium, with a nuclear charge of $+2$, we need two electrons. Both electrons can be accommodated in the $1s$ orbital. For helium, therefore, we write $1s^2$.

Electron arrangements for the first 11 elements of the periodic table are represented schematically in Figure 11-16. Electrons spinning in one direction are shown by one slash mark. Electrons spinning in the opposite direction are shown by a second slash at right angles to the first. Only two electrons are allowed per orbital.

For lithium, with a nuclear charge of $+3$, there are three electrons outside the nucleus. Two occupy the $1s$ orbital. Since two electrons fill an orbital, the third electron must occupy the $2s$ orbital, as shown in Figure 11-16. The electron configuration of lithium is then $1s^2 2s^1$. Because the $2s$ electron spends most of its time farther away from the nucleus than the $1s$ electrons, it is less tightly bound to the rest of the atom. The $2s$ electron should be relatively easy to lose, thus forming the Li^+ ion. This is consistent with the idea of valence electrons, proposed in Chapter 10, and with the relatively low ionization energy observed for lithium, as shown in Table 10-4.

The beryllium atom has one more electron than the lithium atom. The fourth electron for beryllium occupies the $2s$ level, making its electron configuration $1s^2 2s^2$. In a boron atom, five electrons are needed to balance the nuclear charge. Thus, boron's electron configuration is $1s^2\ 2s^2 2p_x^{\,1}$. Continuing this process, we arrive at the following configurations:

Carbon atom	$1s^2\quad 2s^2 2p_x^{\,1} 2p_y^{\,1}$
Nitrogen atom	$1s^2\quad 2s^2 2p_x^{\,1} 2p_y^{\,1} 2p_z^{\,1}$
Oxygen atom	$1s^2\quad 2s^2 2p_x^{\,2} 2p_y^{\,1} 2p_z^{\,1}$
Fluorine atom	$1s^2\quad 2s^2 2p_x^{\,2} 2p_y^{\,2} 2p_z^{\,1}$
Neon atom	$1s^2\quad 2s^2 2p_x^{\,2} 2p_y^{\,2} 2p_z^{\,2}$

Note that the stable electron configurations of the noble gases helium and neon correspond to the *complete* filling of first and second energy level orbitals, as shown in Figure 11-16. The especially stable electron populations of the first two noble gases seem clearly related to the number of electrons required to fill the $1s$, $2s$, and $2p$ orbitals. These numbers are two, two, and six, respectively. Thus, neon has eight more electrons than helium.

If we proceed beyond neon, we come to sodium. Again, we are forced to use an orbital of higher quantum number, the $3s$:

Sodium atom	$1s^2\quad 2s^2 2p_x^{\,2} 2p_y^{\,2} 2p_z^{\,2}\quad 3s^1$

The $3s$ electron is at a higher energy level than the electrons in any other orbital. Therefore, it should be more easily removed to form Na^+. Perhaps

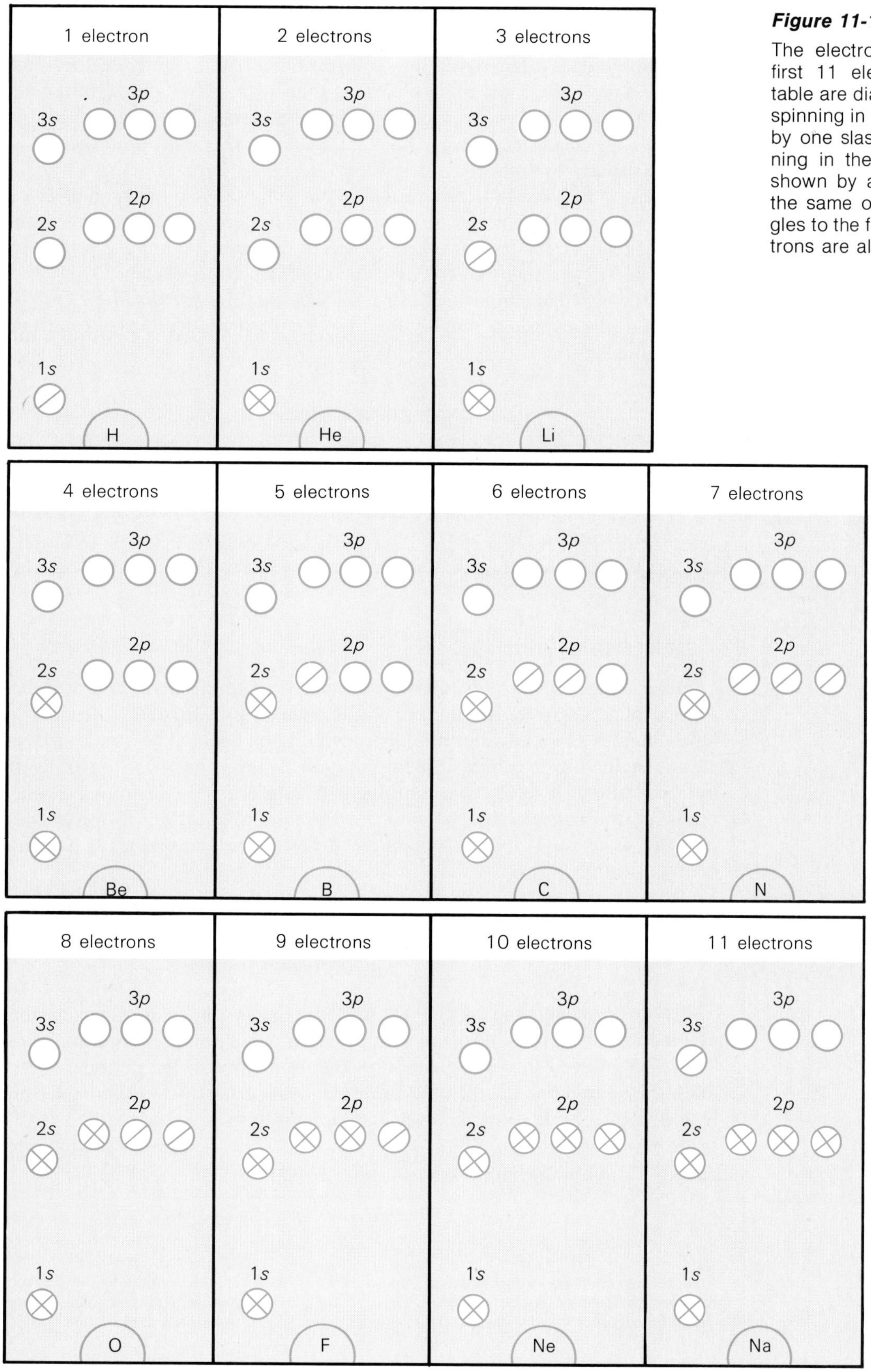

Figure 11-16

The electron arrangements of the first 11 elements of the periodic table are diagramed here. Electrons spinning in one direction are shown by one slash mark. Electrons spinning in the opposite direction are shown by a second slash mark in the same orbital circle at right angles to the first slash. Only two electrons are allowed per orbital.

the chemistry of sodium is like the chemistry of lithium. A review of Chapters 9 and 10 shows that this is so.

The noble gas argon represents completion of the $3s$ and the three $3p$ orbitals. Argon has eight more electrons than neon. The "magic number" eight—the number of electrons separating neon and helium as well as argon and neon—is clearly related to the eight electrons required to fill the one s and the three p orbitals.

In many other cases (already indicated in Chapter 10), the properties of the first several elements are suggested by their electron patterns. We now see that the electron structures introduced in Chapter 10, and given here in greater detail, are themselves based on available energy levels. The close agreement between similarities in chemical properties predicted by energy level diagrams and those actually found by experiment increases our faith in these models.

EXERCISE 11-7

Magnesium has one more electron than sodium. Write the electron configuration for magnesium. What ion is suggested for magnesium?

ELEMENTS BEYOND ARGON

Elements beyond argon follow the pattern of electron distribution described above. The only modification is in the appearance of the energy level diagram as we advance to higher levels. Look again at Figure 11-15. As we go beyond the $3p$ orbitals, we find that the $3d$ levels lie *above* the $4s$ levels but *below* the $4p$ levels.* If we continue to add electrons to the lowest available level, the first electron beyond the closed configuration of argon goes into the $4s$ level. Thus, the potassium atom (atomic number 19) has the configuration

$$\text{Potassium atom} \qquad \underbrace{1s^2 \quad 2s^22p^6 \quad 3s^23p^6}_{\text{ARGON CONFIGURATION}} \quad 4s^1$$

The portion designated as the argon configuration is also known as the argon core. Thus, the potassium configuration is often written as argon core $+4s^1$. Potassium is the first element of the fourth row of the periodic table. With the next element, calcium (atomic number 20), the twentieth electron is added to complete the $4s$ level. Its configuration is written either as

$$\text{Calcium atom} \qquad \underbrace{1s^2 \quad 2s^22p^6 \quad 3s^23p^6}_{\text{ARGON CONFIGURATION}} \quad 4s^2$$

---

*The arrangement of levels shown in Figure 11-15 is an acceptable simplification. In fact, however, the relative position of levels changes as nuclear charge and number of electrons change. The order given in Figure 11-15 accounts for the form of the periodic table.

or abbreviated [Argon core] $4s^2$. When another electron is added to form the next element, scandium, the available orbital of lowest energy is one of the $3d$ levels (remember, the $3d$ levels are a little lower in energy than the $4p$ levels). Scandium is thus represented as

Scandium atom $1s^2$ $2s^22p^6$ $3s^23p^63d^1$ $4s^2$

ARGON CONFIGURATION

As electrons are added to form other elements, they enter the $3d$ orbitals until these orbitals are filled with their maximum number of ten electrons.

Scandium is the first of the group of elements known as transition metals. Chapter 21 deals with the transition elements and presents their electron configurations in detail.

IONIZATION ENERGIES AND ENERGY LEVELS

We have seen that the rows of the periodic table correspond to the filling of orbitals in order of increasing energy. The noble gases at the end of each row correspond to the filling of the outer s and p orbitals for that row. After each noble gas, the next electron must be placed in an s orbital of a higher principal quantum number. This starts a new row of the table. The relationship between the number of elements in each row and the number of available orbitals of increasing energy is summarized in Table 11-3.

Ionization energies were described in detail in Chapter 10. Let us repeat our definition: the *first ionization energy* of an atom is the amount of energy required to remove the most loosely bound electron from the gaseous atom.

Table 11-3

Number of elements in each row of the periodic table

Row of table	Number of elements	Lowest energy available orbitals			
1	2	1s			
2	8	2s	2p		
3	8	3s	3p		
4	18	4s	3d	4p	
5	18	5s	4d	5p	
6	32	6s	4f	5d	6p
7		7s	5f	6d	7p

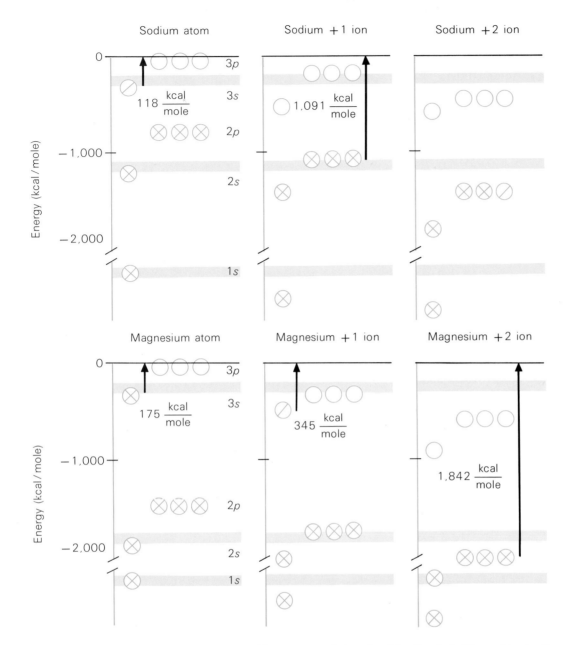

Figure 11-17

These energy level diagrams of the ionization of sodium and magnesium show that much more energy is needed to remove two electrons from sodium than two electrons from magnesium. Removing more than one electron from sodium or more than two electrons from magnesium is difficult because lower, more stable orbitals are involved. Thus, sodium has *one* valence electron, and magnesium has *two* valence electrons.

Since energy is involved in the definition of ionization, we should be able to come full circle and represent ionization energy on an energy level diagram. Let us attempt this by using the energy level diagram for a hydrogen atom, as shown in Figure 11-14. As our eyes travel up the diagram, the levels become closer together. As the principal quantum numbers grow larger, the differences between levels become smaller, until differences are too small to see on the diagram: the levels run together, and the energy of the system approaches zero. Remember, the energy was negative in the original atom.

As you will recall, zero was arbitrarily selected as the energy value for the state in which an electron is completely removed from an atom. Thus, if we add enough energy to raise the electron from its ground state orbital to the zero energy level (as we have defined it), we have ionized the atom. In terms of the energy level diagram, the ionization energy *is the energy needed to lift an electron from the highest occupied orbital to the zero energy level.* The zero level of energy is where the electron is *completely* detached from the nucleus. The zero energy level in an energy level diagram corresponds to $n = $ infinity (∞).

These arguments can be summarized in a quantitative manner. We are seeking the difference between the highest, or ionization, level of energy (represented as zero kilocalories per mole) and the ground state, or $1s$, level (represented as -313.6 kilocalories per mole). The ionization energy for hydrogen, therefore, is $0 - (-313.6) = 313.6$ kilocalories per mole. This is the experimental *value*. The ionization energy required to remove an electron from any energy level is thus the energy of that orbital. The first ionization energies and electron arrangements of the first 20 elements are given in Table 11-4.

As you read in Chapter 10, the term *ionization energy* can also be applied to ions, which are atoms that have already lost an electron. For example, the ionization energy of $Al^+(g)$ is the energy needed in the process

$$Al^+(g) + energy \longrightarrow Al^{++}(g) + e^-(g)$$

Table 11-4

A proposed electron arrangement for the first 20 elements

Atomic number	Element	First ioniza- tion energy (kcal/mole)	Electrons in first level	Electrons in second level	Electrons in third level	Electrons in fourth level
1	hydrogen (H)	313.6	1			
2	helium (He)	566.7	2 (major level full)			
3	lithium (Li)	124.3	2	1 (lost easily)		
4	beryllium (Be)	214.9	2	2		
5	boron (B)	191.2	2	3		
6	carbon (C)	259.5	2	4		
7	nitrogen (N)	335.4	2	5		
8	oxygen (O)	313.8	2	6		
9	fluorine (F)	401.5	2	7		
10	neon (Ne)	497.0	2	8 (major level full)		
11	sodium (Na)	118.4	2	8	1 (lost easily)	
12	magnesium (Mg)	175.2	2	8	2	
13	aluminum (Al)	137.9	2	8	3	
14	silicon (Si)	187.9	2	8	4	
15	phosphorus (P)	254.1	2	8	5	
16	sulfur (S)	238.8	2	8	6	
17	chlorine (Cl)	300.1	2	8	7	
18	argon (Ar)	363.2	2	8	8 (major level full)	
19	potassium (K)	100.0	2	8	8	1 (lost easily)
20	calcium (Ca)	141.0	2	8	8	2

Since this process removes the *second* electron from an aluminum atom, the ionization energy of Al$^+$ (g) is called the *second ionization energy* of aluminum. Because the second electron must be pulled from a *positive ion* (Al$^+$) rather than from a *neutral atom* (Al), the second ionization energy for an atom is *always larger* than the first ionization energy. For example, the first ionization energy for aluminum is 138 kilocalories per mole:

$$Al(g) + 138 \text{ kcal} \longrightarrow Al^+(g) + e^-(g)$$

while its second ionization energy is 434 kilocalories per mole:

$$Al^+(g) + 434 \text{ kcal} \longrightarrow Al^{++}(g) + e^-(g)$$

The effect of the unbalanced positive charge on the Al$^+$ ion is apparent.

The concept of *valence electrons,* presented in Chapter 10, is also illustrated in energy level diagrams. On page 251, valence electrons were defined as the most loosely bound electrons of atoms. They are represented as occupying the highest energy levels used in energy level diagrams. For example, the valence electrons of sodium and magnesium occupy the 3s orbital. In Figure 11-17, these valence electrons are shown to be much closer to the ionization level than are the stable "neon core" electrons in the 1s, 2s, and 2p orbitals. The valence electrons of aluminum occupy the 3s and 3p levels. It is also important to remember that the detailed position of energy levels is determined by the protons in the nucleus as well as by the given level. Levels change for different atoms.

SUMMARY

Light exhibits wavelike properties. All waves—large and small—have a top, or crest, followed by a bottom, or trough. The distance between corresponding points on successive waves is called the wavelength (λ) of that wave. The number of crests or troughs that pass a particular point in a given time is known as the frequency (v) of that wave.

Light passing through a prism is separated into all the colors that compose it. This display is known as the spectrum of that light. Visible light is a small part of a total spectrum, ranging from gamma rays to radio waves. This spectral range is called the electromagnetic spectrum, and the radiation it covers is called electromagnetic radiation. All such radiation travels at the speed of light, differing only in wavelength and frequency. The *shorter* the wavelength, the *higher* the frequency and the higher the energy of the radiation per quantum. The *longer* the wavelength, the *lower* the frequency and the lower the energy per quantum. The energy of such radiation is directly proportional to its frequency, as given by the relationship $E = hv$, where E is the energy per quantum of light, or radiation, v is the frequency of light, and h is the proportionality constant known as Planck's constant of action.

A model of the hydrogen atom pictures its single electron at one of a limited number of energy levels. These levels are represented by horizontal lines in a so-called energy level diagram. Initially, all electrons in a collection of hydrogen atoms are at the lowest energy level, known as the ground state. When electrons absorb energy, they jump to higher levels. As they fall back to the ground state, the absorbed energy is released, producing the observed hydrogen spectral lines.

We can understand the limitations an energy level model places on electrons—allowing them only certain levels of energy—by assuming that electrons have wavelike properties. This assumption forms the basis of wave mechanics, also known as quantum mechanics. Quantum theory identifies energy levels by quantum numbers. The number of each energy level corresponds to a principal quantum number (n).

The *region* in space around the atomic nucleus where an electron of given energy is most likely to be found is an atomic orbital. There are n^2 orbitals available to an electron of given energy. The *s* orbitals have spherical symmetry around the nucleus, and there is *one s* orbital for each value of n. The *three p* orbitals for each value of n above one have a dumbbell-like shape along x, y, and z axes. There are also five *d* orbitals for each value of n above two, and seven *f* orbitals for each value of n above three.

For elements composed of atoms with many electrons, the orbitals associated with a particular principal quantum number (n) are separated into several sublevels of energy. By using an energy level diagram and by assuming that each orbital can accommodate two electrons, we can now explain the entire periodic table. We do this by adding electrons to empty orbitals around a nucleus, filling the lowest energy levels first, until an electrically neutral atom results. Energy level diagrams for atoms with many electrons can also be used to illustrate ionization energies and valence electrons.

QUESTIONS AND PROBLEMS

A

1. (a) We know relatively little about the forces holding particles together in the nucleus, yet this is not a serious problem for chemists. Why? (b) What detailed atomic structural information is important to chemists?
2. (a) What is the accepted symbol for the wavelength of electromagnetic radiation? (b) What is the accepted symbol for the frequency of the radiation? (c) What is the symbol for the velocity of light? (d) Write the expression that shows the relationship between wavelength, frequency, and the velocity of light for electromagnetic radiation. (e) What is the amplitude of a wave?
3. Imagine yourself on a small elevated peninsula above a beach. Out in the water a colored float tied to a lobster trap bobs up and down as the waves go by. The frequency of the ocean waves is determined by the *number of times* the float rises with a crest in a period of 10.00 minutes. The velocity of the incoming waves is determined by how long it takes a given crest to move between two poles that are 30.5 metres apart in the direction of the wave motion. The wavelength is the distance between two crests.
(a) Suppose the velocity of the waves remains constant, but the *distance between crests* is cut in half. Using common sense as a guide, what happens to the number of crests going past the float in a minute? What is this called?
(b) Suppose the distance between crests is kept constant but the velocity with which the waves approach the shore is doubled. What happens to the number of crests going past the float in a minute?

Key Terms

amplitude
crest
electromagnetic radiation
electron configuration
energy level
energy level diagram
frequency
ground state
infrared
interference, or diffraction, colors
microwave
orbital
Planck's constant of action
principal quantum number
quantum
quantum mechanics
quantum theory
radio wave
spectrometer
trough
ultraviolet
uncertainty principle
visible spectrum
wavelength
wave mechanics
X-ray

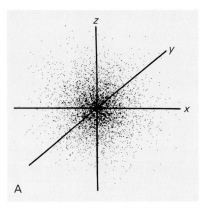

A

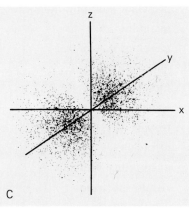

B

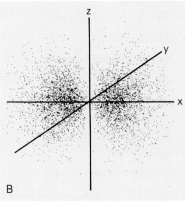

C

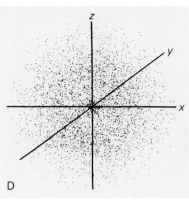

D

(c) Are your observations consistent with the expression given as an answer to 2(d)? Explain.

4. (a) Arrange the following types of radiation in order of *decreasing* wavelength: visible light, infrared, X-rays, ultraviolet, gamma rays, microwaves, radio waves, radar waves. (b) Which of the types of radiation are most penetrating? (c) According to our current information, which of the types of radiation do the most damage to the human body?

5. The spectrum of atomic hydrogen contains a sharp red line. In general terms, describe how this line is produced.

6. (a) What color of light in the visible spectrum has the longest wavelength? (b) Which color has the shortest wavelength? (c) Is the visible spectrum a very large segment of the total electromagnetic spectrum? Explain why it is so important.

7. (a) What is the *ground state* of a hydrogen atom? (b) What is an excited state? (c) How is a hydrogen atom in the ground state transformed into a hydrogen atom in an excited state?

8. Why is it that electrons must be treated as waves in some experiments, but an apple is never treated as a wave in any experiment?

9. What information about an electron in a hydrogen atom is represented by its principal quantum number?

10. (a) How many orbitals are there with an *n* value of 4 in a hydrogen atom? (b) Do these orbitals differ in energy in the single-electron hydrogen atom? (c) Why is the hydrogen spectrum relatively simple?

11. How does the *energy level diagram* for a hydrogen atom serve to represent each of the following?
    (a) The ionization of a *ground state* hydrogen atom.
    (b) The ionization of an excited hydrogen atom. (Use any excited hydrogen atom as your example.)
    (c) A hydrogen atom that is giving off radiation such as the red line in the hydrogen spectrum.
    (d) The absorption of energy by a hydrogen atom.
    (e) How is Planck's expression, $E = h\nu$, important in the above?

12. (a) How does the energy level diagram for a multi-electron atom such as sodium differ from the energy level diagram for the hydrogen atom? (b) Write the energy level diagram for a multi-electron atom such as Na.

13. What orbital designation is appropriate for each of the electron distributions shown in the margin?

14. What information is implied by the quantum number 5 for an electron?

15. Write the detailed ground state electronic configuration, that is, $1s^2$, $2s^2$, $2p^3$, and so on for each of the following: (a) He, (b) C, (c) F, (d) Na$^+$, (e) F$^-$, (f) Ne, (g) Ar, (h) Ca, (i) Sc.

16. (a) What electronic configuration is associated with the nonreactive Ne, Ar, Kr atoms? What about He? (b) What negative ion has the He configuration? What positive ions have the He configuration?

B

17. The most intense line in the visible spectrum of cesium has a wavelength of 4.56×10^{-7} metre. (a) What is the frequency of this line? Give units for your answer. Remember that the velocity of light, *c*, is 3.00×10^8 m/sec. (b) What color is the flame test for Cs? (Remember Na is yellow.) (c) What is the *energy* in *one quantum* of this radiation? Planck's constant, *h*, is 1.58×10^{-34} calorie $\times$ seconds. Give units.

18. Two sets of waves are traveling at the same speed. The first set has a longer wavelength than the second. Which set will have the higher frequency? Explain.

19. The electron in a hydrogen atom falls from level 2 to level 1. (See page 275.) (a) What is the numerical value of the energy given off in this transition? (b) Using the value of Planck's constant given in 17(c), calculate the frequency of the radiation given off. (c) Could you see the radiation given off? If so, what color would it be?

20. Why was the *Heisenberg Uncertainty Principle* important in our decision to abandon the Bohr atom?

21. What is indicated about electrons by the terms (a) *wave mechanics,* and (b) *quantum mechanics?*

22. It is suggested that our picture of an electron in an atom resembles a vibrating guitar string in some ways. Explain.

23. Describe the shape of (a) an *s* orbital, and (b) a *p* orbital.

24. Make a table listing the principal quantum numbers 1 through 4, the types of orbitals, and the number of orbitals of each type.

25. (a) Describe how an energy level diagram of an atom with many electrons *differs* from the energy level diagram of the hydrogen atom. (b) What causes the difference? (c) Would the energy level diagram of He^+ be like that of H or that of Li? Explain.

26. The electron configuration of sodium (element 11) can be written: neon core $+ 3s^1$. Write out in detail what the words *neon core* represent.

27. What is meant by the term *argon core?*

28. What must be done to a $2s$ electron to make it a $3s$ electron? What happens when a $3s$ electron becomes a $2s$ electron?

29. Name the element that corresponds to each of the following electron configurations: (a) $1s^1$, (b) $1s^2\ 2s^22p_x{}^1$, (c) $1s^2\ 2s^22p_x{}^22p_y{}^12p_z{}^1$, (d) $1s^2\ 2s^22p^6\ 3s^23p_x{}^1$, (e) $1s^2\ 2s^22p^6\ 3s^23p^6$.

30. Explain why a minus sign $(-)$ is placed in front of the energy level values of the hydrogen atom.

31. Write the electron configuration of each of the following elements in its ground state: (a) hydrogen, (b) boron, (c) nitrogen, (d) sulfur, (e) magnesium, (f) potassium, (g) iron, (h) copper, (i) bromine.

C

32. Given the successive ionization energies for the elements listed below, determine the number of valence electrons for each.
    Element A 259.5 562.3 1,104 1,487 9,042 11,298
    Element B 100.1 733.8 1,061 1,405 1,905 2,300
    Element C 150.8 295.2 571 1,704 2,122 2,560

33. Elements A, B, C, and D have electronic structures as shown:
    A $1s^2\ 2s^22p^2$, B $1s^2\ 2s^22p^6\ 3s^1$, C $1s^2\ 2s^22p^6\ 3s^23p^6\ 4s^2\ 3d^5$,
    D $1s^2\ 2s^22p^6\ 3s^23p^5$
    Answer the following questions:
    (a) Which of the elements, if any, are metals?
    (b) Which of the elements, if any, is (or are) transition metal(s)?
    (c) What *ionic* compound (or compounds) could be formed from the above elements?
    (d) What covalent compound (or compounds) could be formed from the above elements?
    (e) Which elements above would react most dramatically with water?

Spectra and the Bohr atom

What experimental observations concerning light and the hydrogen spectrum led to the development of the Bohr model of the hydrogen atom? Let us examine these experiments and note the influence that Rutherford's nuclear atom model and the newly developed quantum theory had on Bohr.

Emission and absorption spectra We have mentioned that the red glow of a neon sign results when an electric current passes through a tube filled with neon gas under low pressure.* The first reported studies of the light emitted by various excited gases were made by the Scottish physicist Thomas Melvill (1726–1753). Melvill heated various substances to incandescence in a flame and allowed the light emitted to pass through a small hole and then through a prism. The prism separated the light into its various component colors.

Beginning in 1752, Melvill found that the spectrum of light produced by a hot gas was very different from the familiar continuum of rainbow colors obtained when light from the sun was analyzed in the same way. What Melvill found became known as *emission spectra* in later work. Instead of a continuous band of colors (as shown in Figure 11-6a), the emission spectrum consisted of a small number of individual spots separated by dark gaps. If the light was allowed to pass through a narrow slit rather than a small hole, the spots became lines separated by dark gaps, as shown in Figure 11-18. Each spot or slit had the color of that part of the spectrum in which it was located. The dark gaps indicated that many of the rainbow colors were missing.

Melvill also noted that the colors and locations of the bright spots or lines were different when different substances were put into the flame.† For example, ordinary table salt produced a bright yellow color, now known to be characteristic of the element sodium. The visible spectrum of sodium consists of two yellow lines, very close together, as shown in Figure 11-6d. Note the difference between that

spectrum and the other emission spectra shown in Figures 11-6c and 11-6e.

In 1802, the British scientist William Wollaston noticed seven irregularly spaced dark lines across the otherwise continuous spectrum of the sun (Figure 11-6b). Then, in 1823, British astronomer John Herschel suggested that each gas could be identified by its unique emission spectrum. It has since been shown that this can be done, just as a person can be identified by his or her fingerprints. But the key observations that led to a better understanding of spectra were made by the German physicist Gustav Robert Kirchhoff (1824–1887) in 1859. By that time, it was known that the two prominent yellow lines in the emission spectrum of sodium had the *same* wavelengths as two prominent dark lines crossing the sun's spectrum. (See Figures 11-6b and 11-6d.) It was also known that light emitted by a very hot, glowing solid formed a perfectly continuous spectrum with no dark lines.

Kirchhoff showed that if light from a glowing solid passed through cool sodium vapor *before* it was spread out by a prism, the resulting spectrum showed two prominent dark lines at exactly the places they occurred in the sun's spectrum. When this experiment was repeated with different gases or vapors in place of sodium, each was found to produce its own characteristic pattern of dark lines. An example of a dark-line pattern is shown in Figure 11-18b. Patterns of dark lines against an otherwise continuous spectrum are known as *absorption spectra*. Finally, Kirchhoff showed that the wavelength corresponding to each dark absorption line was the *same* as that of a bright line in the emission spectrum of the same substance. Evidently, a substance could *absorb* only the frequency of light that it emitted when excited by heat or electricity. This finding allowed Kirchhoff to identify the sources of certain dark solar lines as absorption by sodium and calcium (Figures 11-6d and 11-6e), and to conclude that those elements are in the sun's atmosphere.

The spectrum of atomic hydrogen Many researchers tried in vain to find regularities in emission spectra in the period following Kirchhoff's discoveries. Finally, in 1885, the Swiss schoolteacher Johann Jakob Balmer (1825–1898) published a paper that contained the first important breakthrough in this area. By trial and error, Balmer had

---

*This was discussed on page 271 and in the Extension of Chapter 8, pages 203–209.

†Recall that emission spectra also result when gases or vapors under low pressure are excited by a continuous electric current. One example is the emission spectrum of atomic hydrogen, shown in Figure 11-6c.

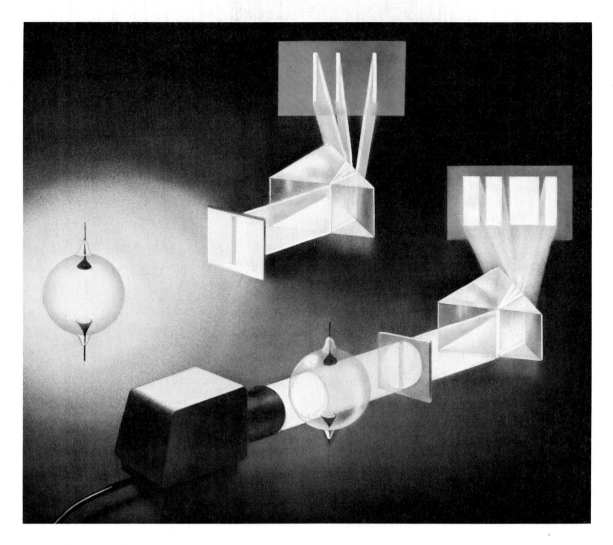

Figure 11-18

At top, a *bright line* spectrum is produced by hot gases, which emit only certain wavelengths of light. The shape of the slit dictates the shape of the lines on the screen. At bottom, a *dark line* spectrum is produced by cool gases, which absorb only certain wavelengths of light. This happens when white light from a hot solid is passed through the cool gas.

hit upon a simple formula that gave the observed wavelengths of the visible portion of the emission spectrum of atomic hydrogen. His formula was

$$\lambda = b \left(\frac{n^2}{n^2 - 2^2} \right)$$

where λ is the wavelength, b is a constant with the value 3.645×10^{-5} centimetres, and n is a whole number, different for each line. (Note that n does *not* stand for the principal quantum number in this case.)

Figure 11-6c shows the visible portion and part of the ultraviolet portion of the emission spectrum of atomic hydrogen. When Balmer gave n a value of 3, his expression correctly predicted the wavelength of the first red line, labeled H_a in Figure 11-6c. With n equal to 4, his expression gave the wavelength of the second line, H_b, which is light blue in color. With n equal to 5, Balmer got the wavelength of the third line H_c, which is dark blue in color. With n equal to 6, the result was the wavelength of the fourth line, H_d, which is violet in color. The series of lines described by Balmer's formula is called the *Balmer series*. The shortest wavelength possible from the formula is 3.645×10^{-5} centimetres, when n equals infinity. This value is known as the *series limit*. All the values were in close agreement (about 0.02 percent) with the actually observed values.

In a paper written in 1885, Balmer speculated that there might be other series of lines in the hydrogen spectrum besides the one shown in Figure 11-6c. He suggested that other wavelengths in the hydrogen spectrum could be predicted by modifying his original formula. Thus, he simply replaced the 2^2 in his equation with 1^2, 3^2, 4^2, and so on. Balmer's suggestions intensified the search for other series of spectral lines.

A modification of Balmer's formula was proposed by the Swedish spectroscopist Johannes R. Rydberg (1854–1919) in 1889, when he rewrote the Balmer expression in terms of $1/\lambda$, a quantity known as the *wave number*. This quantity is directly related to the energy of the light it describes. The larger the value of the wave number, the greater the energy of the light. Rydberg's modification took the form

$$\frac{1}{\lambda} = R\left(\frac{1}{2^2} - \frac{1}{n^2}\right)$$

Here, λ is the wavelength of a particular line in an atomic spectrum and R is a constant, known as the Rydberg constant, with a value of 109,677.6 centimetres^{-1}. As before, n has the values 3, 4, and 5 for the first three lines of the Balmer series.

Improvements in spectroscopy soon made possible the exploration of the ultraviolet and infrared portions of the spectrum. As Balmer had predicted, additional series of atomic hydrogen emission lines were soon found. Between 1906 and 1914, at Har-

vard, Theodore Lyman found such a series in the ultraviolet region, subsequently named for him. In 1908, the German physicist Friedrich Paschen found several lines of another series in the infrared region, later named for him. Other series in the far infrared region were discovered at Johns Hopkins University in 1922 and 1924.

The Rydberg modification was able to predict correctly every observed wave number in every one of the hydrogen spectral series. To predict the wave numbers in the ultraviolet region (Lyman) the term $1/2^2$ was replaced by the term $1/1^2$, and n had whole number values beginning with 2 instead of 3. For the infrared (Paschen) series, the first term was replaced by the term $1/3^2$, and n was given whole number values beginning with 4. The wave numbers of *all* possible emission lines in *all* the known spectral series could therefore be summarized in *one* formula:

$$\frac{1}{\lambda} = R\left(\frac{1}{n_1^2} - \frac{1}{n_2^2}\right)$$

As impressive as these expressions are—and they certainly work—it is important to realize that neither Balmer nor Rydberg knew *why* their empirical formulas worked or what they meant.

Bohr's model of the hydrogen atom The nuclear atom model was developed by Rutherford in 1911. This model was described on page 185 and the experiments suggesting the model were detailed in the Extension of Chapter 8. Rutherford's model for the simplest atom, hydrogen, is shown in Figure 11-19. It is called the planetary model because of its resemblance to planets moving around the sun. The model assumed that the electron was attracted to the oppositely charged nucleus but that its rapid motion around the nucleus kept the electron from falling toward it. Since the electron was forced into a circular path by this attraction, it was continually changing direction. According to classical physics, such a moving electron must continually *lose* energy, slow down, and thus fall into the nucleus. Since this does *not* happen, either Rutherford's model or classical physics must be in error.

Niels Bohr, having just completed a period of study with Rutherford, attacked this problem in 1913. He decided to try to *combine* Rutherford's classical model with the newly developing quantum ideas. Quantum theory, a somewhat radical innovation in physics, had been introduced in 1900 by Max Planck. As you know, this theory assumed that radiant energy—such as ultraviolet radiation, visible light, and infrared radiation—was emitted in tiny but definite packets called *quanta.* Bohr boldly assumed that hydrogen's orbiting electron *did not normally radiate energy.* Even though this assumption contradicted classical physics, it was necessary in order to explain why the electron does not fall into the nucleus. Thus, radiation of light from hydrogen atoms is expected *only* in special circumstances.

Such a special circumstance, suggested Bohr, might be in an electrical discharge tube, where the electrons of hydrogen atoms receive enough energy to move them into larger-than-normal orbits. He called these larger orbits *excited states.* This excited electron, said Bohr, would later "fall" or jump back from the larger orbit to its original orbit. It should have *less* energy in its normal orbit than in the larger, excited orbit. If we represent the electron's normal energy as E_1 and its energy in the excited state as E_2, then the change in energy, ΔE, can be represented by the expression

$$\Delta E = E_2 - E_1$$

Bohr assumed that this change in energy was radiated as a *photon,* the name given to a quantum of visible or ultraviolet light.

If an electron could occupy an infinite number of possible orbits, its jumps from those orbits should give rise to an infinite number of different radiation frequencies. This would mean that the hydrogen spectrum should contain all possible colors—in other words, it should be continuous. But hydrogen gives off only *certain* wavelengths of light, as shown in Figure 11-6c. Bohr assumed that when an electron was excited and thus moved into larger orbits, only certain orbits were available. Let us label the available orbits 1, 2, 3, 4, 5, and 6, as in Figure

Figure 11-19

Diagram of the Rutherford atom.

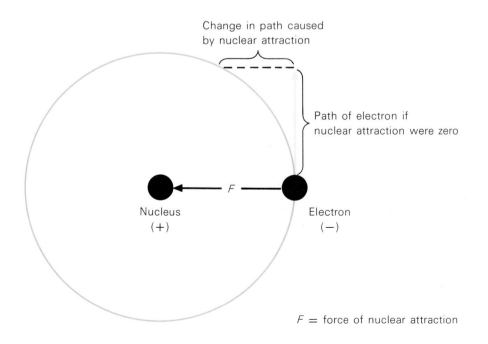

Change in path caused by nuclear attraction

Path of electron if nuclear attraction were zero

Nucleus
(+)

F

Electron
(−)

F = force of nuclear attraction

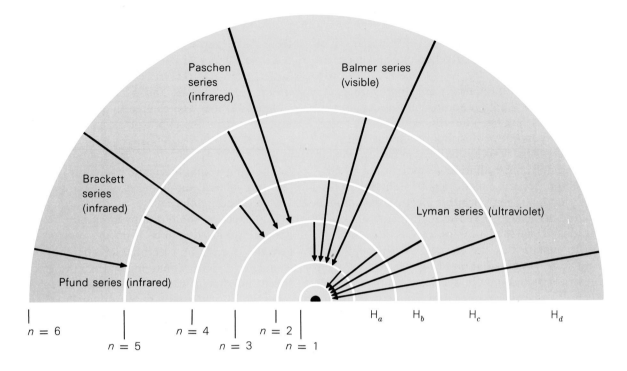

Figure 11-20

Bohr model of the hydrogen atom, showing available electron orbitals.

11-20, and the energies of the electron in those orbits E_1, E_2, E_3, E_4, E_5, and E_6. An electron jumping from orbit 3 to orbit 2 would radiate a photon of energy, represented as $\Delta E = E_3 - E_2$. Thus, Bohr's model had a mechanism to account for the fact that hydrogen emitted light of only certain wavelengths. Further, by combining quantum theory with classical physics, Bohr showed that his available orbits had radii that were multiples of a constant squared. The available radii (r) have values given by the expression $r = k \times n^2$, where k is a constant and n represents a quantum number with the value—1, 2, 3, and so on—associated with each orbit.

Energy-level diagrams Since the electron has a particular energy that is different for each different orbit it occupies, an alternative view of Bohr's model concentrates on the possible energy states, or *levels*, available to the electron. According to this view, as shown in Figure 11-21, the electron normally occupies the lowest, or ground state, level. It can absorb only the amounts of energy needed to raise it to some higher level. When it later returns to

lower levels and, eventually, to the ground state, it must emit *exactly the same* amount of energy it absorbed in jumping to the higher level. Thus, the diagram explains why hydrogen, when excited, should show dark absorption lines at the *same* wavelength it shows bright emission lines. The dark and bright lines are shown in Figures 11-6b and 11-6c.

Using a combination of classical and quantum physics, Bohr derived a formula that would predict the *frequencies* of the different kinds of light the hydrogen electrons could emit. Frequency (ν), like wave number, is a quantity directly related to the energy of the light it describes. Bohr's formula for these frequencies was

$$\nu = \text{a constant} \times \left(\frac{1}{n_2{}^2} - \frac{1}{n_1{}^2} \right)^*$$

where the first term (n_2) represents the energy of the electron in the larger, excited orbit and the second

---

*The n values as used here give a negative frequency. This makes no physical sense, but it gives the correct sign for the energy change when used in Planck's equation, $E = h\nu$.

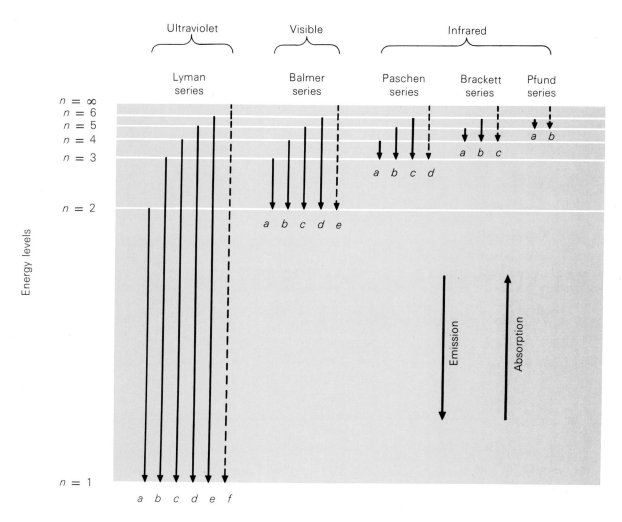

Ultraviolet   Visible   Infrared

Lyman series   Balmer series   Paschen series   Brackett series   Pfund series

$n = \infty$
$n = 6$
$n = 5$
$n = 4$

$n = 3$

$n = 2$

$n = 1$

Energy levels

a b c d

a b c

a b

a b c d e

a b c d e f

Emission   Absorption

Figure 11-21

An alternative view of the Bohr model.

term (n_1) gives the energy in the final, lower orbit. Notice that this formula, derived from Bohr's model, is identical to Rydberg's empirical formula on page 294. (Recall the relationship between frequency and wavelength given on page 266.) Furthermore, the value of the constant in Bohr's formula was identical to the value of Rydberg's constant, R. Bohr could use his hydrogen atom model to derive and explain Balmer's formula. Thus, Bohr was able to explain what Balmer and Rydberg could not. He was able to explain why hydrogen atoms emit and absorb only certain kinds of light, and why the Balmer and Rydberg expres-

sions could so accurately predict the wavelengths or wave numbers of that light.[*]

The Bohr model of planetary, circular orbits for electrons does *not* work for atoms with *more* than one electron. But his idea concerning energy levels, when suitably modified (as shown in Figure 11-15), does work beautifully. It provides us with a most useful scheme for explaining the bonding capabilities of all the elements in the periodic table.

[*]An excellent review of this historical background and present-day research is Theodor W. Hansch, Arthur L. Schawlow, and George W. Series, "The Spectrum of Atomic Hydrogen," *Scientific American*, vol. 240 (March 1979) pp. 94–110.

The nature of the chemical bond is the problem at the heart of all chemistry.

BRYCE CRAWFORD, JR.

12

MOLECULAR ARCHITECTURE: GASEOUS MOLECULES

Objectives

After completing this chapter, you should be able to:

- Explain how atoms combine to form molecules.
- Distinguish between ionic and covalent bonds.
- Identify partly ionic covalent bonds.
- Predict the shape of a molecule from the valence-shell electron-pair repulsion (VSEPR) theory.
- Relate molecular shape to the orbitals used in forming the bonds.
- Predict whether a molecule will be polar, and identify which end will be positive and which negative.

Researchers work with high-vacuum apparatus to study the reactions of volatile compounds that are easily decomposed by contact with air.

By now, you have become familiar with the atom as a positively charged nucleus surrounded by negatively charged electrons. Because electrons repel each other, the prospect of two atoms approaching each other to form a molecule might seem remote. Yet, molecules do exist. Forces of attraction, therefore, must draw atoms to one another. Further, the forces of attraction must be greater than the forces of repulsion exerted by electrons.

In Chapter 10, we began our discussion of the bonding of atoms in molecules. Since then, we have learned more about atomic structure and the fundamental particles that make up atoms. Now it is time to apply this knowledge in the development of a more detailed view of chemical bonds. In this chapter, we explore the forces of attraction and repulsion within molecules, and we look at their relationship to the geometry of gaseous molecules. In the next chapter, we will consider such forces as they apply to liquids and solids. Our discussion will focus on two important questions: why do atoms cluster and stay together, and why do the clusters have characteristic properties?

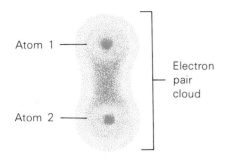

Atom 1 ——

Atom 2 ——

Electron pair cloud

Figure 12-1
There is a high probability that electrons of two hydrogen atoms will be found between two nuclei.

12.1 Covalent bonds

In Chapter 10, we stated that diatomic molecules such as H_2 and F_2 are held together by a pair of shared electrons. This link is called a *covalent bond*. Let us look more carefully at this bond in gaseous H_2 molecules.

HYDROGEN MOLECULES

When a diatomic hydrogen molecule is made from two hydrogen atoms a great deal of heat is released:

$$H(g) + H(g) = H_2(g) + 103 \text{ kcal}$$

Thus, the hydrogen molecule is more stable (has less stored energy) than the two separate hydrogen atoms. Why should the hydrogen molecule be more stable than the original, isolated hydrogen atoms? To answer this question, we must examine the interactions between the electric charges of the atoms themselves.

A sizable amount of evidence tells us that the electron cloud of an isolated hydrogen atom *before reaction* is spherical like a ball (has spherical symmetry). But if two hydrogen atoms are brought together, the electron of atom 1 will be attracted by and pulled toward the nucleus of atom 2. In the same way, the electron of atom 2 will be pulled toward the nucleus of atom 1. Electrons 1 and 2 will spend a considerable amount of time in the space *between* the two nuclei, as illustrated in Figure 12-1. While in this space, each electron is attracted to *both* nuclei. Such attraction is the "glue" that holds the two atoms together. *Thus, we see that the chemical bond in H_2 forms because each of the two electrons is attracted to the protons of both atoms at the same time. This arrangement is more stable than the arrangement in separated atoms, where each electron is attracted to only one proton.*

Nevertheless, repulsions do occur when two hydrogen atoms approach each other. The two electrons repel each other, as do the two protons. The repulsions tend to push the two atoms apart. The length of the stable bond in the hydrogen molecule is determined by a balance between the forces of attraction and the forces of repulsion. The attractive forces are stronger.

We can represent the chemical bond in the hydrogen molecule by using notations from quantum mechanics, such as those shown in Figure 12-2. Figure 12-2a shows the electron distribution of an isolated hydrogen atom in cross section. The electron distribution extends far from the nucleus. It is uniform in all directions but is most concentrated near the nucleus. Therefore, we must find a uniform way to represent the electron concentration in the central region of the $1s$ orbital. We do this by depicting the $1s$ orbital as a circle with a radius large enough to contain most of the electron distribution. An orbital can accommodate either one or two electrons, *but no more than two.* Figure 12-2b shows a $1s$ orbital that is empty, then half filled with one electron, and finally filled with two electrons.

Figure 12-2c shows the interaction of two hydrogen atoms. Each atom has a single electron in the $1s$ orbital. As the two hydrogen atoms approach, the orbitals tend to overlap. *In the region of overlap, the two electrons are shared by the two protons,* as shown by the darkened area. When sharing occurs, both electrons can be near *both* protons much of the time. The sharing of electrons forms a covalent bond.

Figure 12-2
Schematic representation of an atom and the interaction between two atoms.

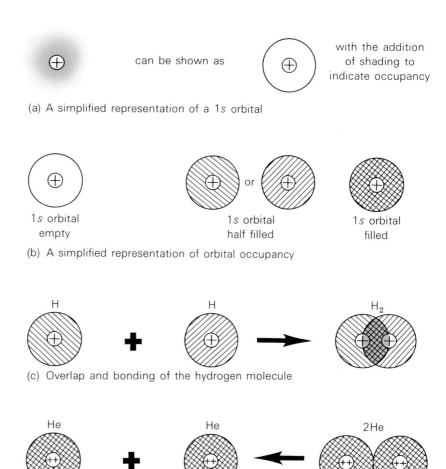

(a) A simplified representation of a 1s orbital

1s orbital
empty

1s orbital
half filled

1s orbital
filled

(b) A simplified representation of orbital occupancy

(c) Overlap and bonding of the hydrogen molecule

(d) Absence of overlap for two helium atoms

HELIUM ATOMS: NO BOND FORMATION

Measuring the density of helium gas shows that it consists of single atoms. Molecules of helium (He$_2$) do *not* form. What difference between hydrogen and helium atoms accounts for the absence of bonding between helium atoms?

Figure 12-2d is a simplified diagram of the interaction between two helium atoms. Unlike the diagram for hydrogen (Figure 12-2c), the representation of each helium atom is darkly shaded *before* the two atoms approach. This indicates that there are already two electrons in each 1s orbital. The rule for orbital occupancy tells us that the 1s orbital can contain only two electrons. Thus, when two helium atoms approach, their orbitals cannot overlap because *each is already filled.* Filled orbitals cannot overlap enough to share electrons. As a result, a chemical bond does *not* form.

REPRESENTING CHEMICAL BONDS

Chemical bonds can form between two atoms if they can share valence electrons using partially filled orbitals. A shorthand notation to illustrate this proposed rule is called a *representation of the bonding.* For hydrogen and helium, our rule about covalent bond formation can be shown quite simply through an **orbital representation:**

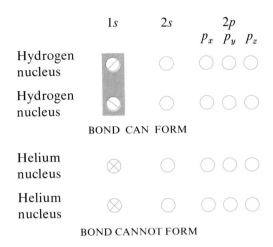

As indicated in Chapter 10, the sharing of electrons can be shown by representing valence electrons as dots placed between the atoms:

$$\text{H} \cdot + \text{H} \cdot \longrightarrow \text{H} : \text{H}^*$$

In this *Lewis structure diagram*, the symbol H represents the bare proton. The symbol of an atom in a dot formula represents the atomic core—that is, the atom minus its valence electrons. Valence electrons are shown by the dots. We shall use both *orbital* and *Lewis structure diagrams* to show chemical bonding. In addition, chemists often represent a shared pair of electrons between atoms with a dash (—): H—H = H : H.

FLUORINE ATOMS AND MOLECULES

A fluorine atom has seven valence electrons, occupying the outermost, partially filled cluster of energy levels. The electron dot (Lewis structure) diagram for fluorine is

$$\cdot \ddot{\underset{\cdot\cdot}{\text{F}}} :$$

*The use of color here is convenient for purposes of electron bookkeeping. But remember, all electrons in a molecule are indistinguishable except for their energy. Instead of the small *x*'s used in Chapter 10, color will be used to differentiate the valence electrons of each atom.

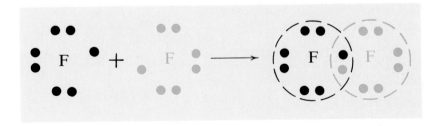

Figure 12-3

Sharing of electrons in the fluorine molecule. The rings shown here represent, not discrete electron orbits, but a region in which electrons are located about 90 percent of the time.

You may recall from Chapter 8 on page 189 that a gaseous fluorine atom *releases* energy when it gains an electron to form a gaseous fluoride ion:

$$\cdot\ddot{\underset{\cdot\cdot}{F}}\colon + e^- \longrightarrow \colon\ddot{\underset{\cdot\cdot}{F}}\colon^- + E$$

The energy change associated with this process is known as the **electron affinity** of fluorine. It is symbolized above by the letter E. The release of energy indicates that a fluoride *ion*, with a *completed* energy level 2, is more stable than a fluorine atom, with its *incomplete* second level. We shall be concerned here with the different ways two fluorine atoms can gain stability when they form a bond.

To better understand why such a bond will form between fluorine atoms, examine the orbital representation of a single fluorine atom:

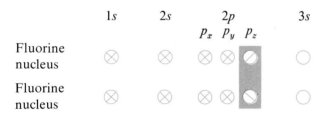

Each atom has a valence electron in a half-filled p_z orbital. Now, suppose two gaseous fluorine atoms interact. We can imagine the two atoms positioning themselves so that the half-filled p_z orbitals overlap in space. The half-filled p_z orbital of atom 1 shares the one valence electron from the half-filled p_z orbital of atom 2. Atom 2 does the same thing with the electron from atom 1. Thus, part of the electron affinity of atom 1 is satisfied without pulling an electron away from atom 2. Atom 2, of course, derives the same energy benefit from the electron of atom 1, which it now shares. Each fluorine atom has acquired part interest in the other's electron. The orbital representation of the two fluorine atoms in the F_2 molecule is

The overlapping, half-filled p_z orbitals are enclosed in the shaded area. A chemical bond forms between two fluorine atoms because the shared electrons are attracted to both positive nuclei simultaneously. In fact, *all chemical bonds form because one or more electrons are placed so that they are attracted to two or more positive nuclei simultaneously.*

Figure 12-4

Sharing of electrons in hydrogen fluoride.

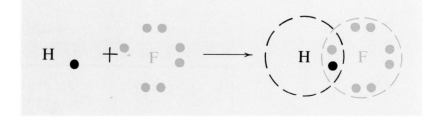

The interaction between fluorine atoms can also be represented by an electron dot diagram. The electron sharing is shown in Figure 12-3. Regardless of how the interaction is represented, only electrons in the $n = 2$ energy level are involved. We have not mentioned the $1s$ orbital in this discussion. The $1s$ electrons are too tightly bound to play a significant role in the chemistry of fluorine. Only electrons in the $n = 2$ energy level are involved.

All methods of representation show that when two atoms of fluorine share electrons, they have effectively filled their valence orbitals. *They have no additional bonding capacity.* F_2 does not gain a third or fourth atom, to form F_3 or F_4. Fluorine is **univalent:** each atom has a bonding capacity of one.

The compound hydrogen fluoride can also be represented by an electron dot diagram, as shown in Figure 12-4. A count of the number of valence electrons represented by the dots surrounding each atom in the hydrogen fluoride (HF) molecule shows that the hydrogen atom has only two electrons near it, while fluorine has eight. The molecule is stable, however, because hydrogen has only one valence orbital: the $1s$ orbital. Two electrons completely fill this orbital, as indicated by the orbital diagram for a hydrogen fluoride molecule:

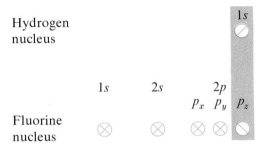

A chemical bond results from the sharing.

EXERCISE 12-1

> With the aid of the orbital diagram for hydrogen fluoride, shown above, explain why compounds with the formulas H_2F and H_3F do not form.

12.2   Bonding capacities of second-row elements

In Chapter 9, we identified several compounds of second-row elements. We wrote formulas for HF, H_2O, NH_3, and CH_4. We can now explain the formation of such compounds in more detail.

EXERCISE 12-2

> Draw Lewis electron dot diagrams to represent the molecules H_2O, H_2O_2, F_2O, NH_3, CH_4, and BF_3. Refer to Chapter 10 if necessary.

Figure 12-5

The placement of the second row of the periodic table.

OXYGEN ATOMS

A neutral oxygen atom has eight electrons. Six of them occupy the $2s$ and $2p$ orbitals. Those six electrons are much more easily removed than the two electrons in the $1s$ orbital. The six *valence electrons* occupy a higher energy level, which is not yet filled. Thus, for oxygen, the $2s$ and $2p$ orbitals are *valence orbitals*. Two ways in which these orbitals can accommodate the six valence electrons are shown below:

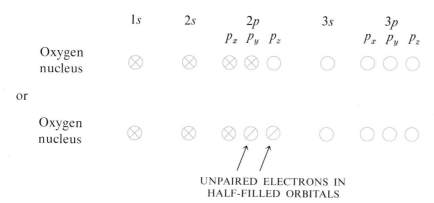

Since electrons repel each other, the configuration in which electrons are *farther apart* has *lower energy*. A configuration with one electron in each of two separate orbitals keeps electrons farther apart than a configuration with two electrons in a single orbital. This is so because each orbital occupies a different region in space. Therefore, we would expect the configuration with two unpaired electrons to have lower energy than the one in which all electrons are paired. Experiments confirm this expectation. Much of the chemistry of oxygen can be most easily interpreted using orbital model 2. In that model, two unpaired electrons occupy separate, half-filled orbitals. The difference in configuration between the two models of oxygen can also be represented by an electron dot diagram. The $\cdot \overset{\cdot\cdot}{\text{O}} \colon$ arrangement has lower energy than the $\overset{\cdot\cdot}{\underset{\cdot\cdot}{\text{O}}} \colon$ arrangement.

Suppose a single hydrogen atom approaches an oxygen atom in its lower energy state. Since each atom has partially filled valence orbitals, electron sharing can take place. Each shared electron becomes close to two nuclei at the same time:

$$H \cdot \; + \; \cdot \ddot{\underset{\cdot\cdot}{O}} : \longrightarrow H : \ddot{\underset{\cdot\cdot}{O}} :$$

HALF-FILLED ORBITAL

Thus, a *stable bond* is formed between hydrogen and oxygen atoms. But the species HO* still has one unpaired electron, which can form a second bond with another hydrogen atom to produce water.

$$H \cdot \; + H : \ddot{\underset{\cdot\cdot}{O}} : \longrightarrow H : \ddot{\underset{\cdot\cdot}{\underset{H}{O}}} :$$

The electron dot model suggests that the residual bonding capacity of HO has been used up. We are not surprised to find that H_2O is very stable. This is in striking contrast to the very reactive HO molecule. Oxygen is said to be **divalent** in the compound H_2O. Each atom in this molecule has filled its valence orbitals by electron sharing.

What happens when two HO units approach each other? An electron dot diagram suggests that a covalent bond will form. It also predicts a formula for the resulting compound. The equation is

$$: \ddot{\underset{\cdot\cdot}{\underset{H}{O}}} \cdot \; + \; \cdot \underset{H}{O} : \longrightarrow \left(: \ddot{\underset{H}{O}} : \;\; O : \right)$$

The compound produced when two HO units combine is the well-known, reactive bleaching agent *hydrogen peroxide* (H_2O_2). Our model accurately predicts the existence of an O—O bond in hydrogen peroxide.

When applied to the chemistry of oxygen and fluorine, the type of analysis used above predicts the existence of a compound with the formula F_2O. In the electron dot representation

$$: \ddot{\underset{\cdot\cdot}{O}} : \underset{:\,F\,:}{F} :$$

oxygen is again divalent.

EXERCISE 12-3

Draw orbital and electron dot representations for each of the following molecules: OF, F_2O_2, HOF, and HFO_2. Which would you expect to be most reactive? Why? Which would be free radicals?

---

*A molecular species with a half-filled valence orbital is frequently called a **free radical**. The HO· unit is called the **hydroxyl radical**. It has been identified and studied as an intermediate species in reactions occurring in high-temperature flames and in the stratosphere.

NITROGEN ATOMS

Like oxygen, the nitrogen atom is most stable when it has the maximum number of partially filled valence orbitals. Its electrons are then as far apart as possible. The most stable state for the five valence electrons of nitrogen is

From this representation we can predict that nitrogen will form stable hydrogen and fluorine compounds with formulas NH_3 and NF_3. The electron dot formulas for these compounds are

Occasionally, chemists use structural formulas instead of electron dot representations. Nonbonding electrons are not shown. Structural formulas for the two nitrogen compounds considered above are

The formulas show that the nitrogen atom is **trivalent** in both compounds: it has a bonding capacity of three.

EXERCISE 12-4

The molecule NH_2 has residual bonding capacity and is very reactive. The hydrazine molecule (N_2H_4) is more stable. Draw electron dot and structural formulas for N_2H_4. (*Hint:* see H_2O_2 on page 306.)

CARBON ATOMS

Like the oxygen and the nitrogen atom, the carbon atom is most stable when it has the maximum number of partially filled p orbitals. Experiments show that when there are two electrons in the $2s$ orbital and one electron in each of the $2p$ orbitals, the carbon atom is in its lowest energy state. In that state, the valence electrons are at a maximum average separation from each other.

The orbital diagram above suggests that carbon is divalent. It should form compounds with formulas such as CH_2 and CF_2. Study the electron dot and orbital diagrams for the molecule CH_2 (shown below). Note that the orbitals used for bonding are enclosed in shaded rectangles.

$$H : \overset{\cdot\cdot}{C} : H$$

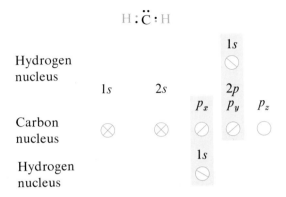

The two diagrams show a situation we have not found before. After the two available partially filled $2p$ orbitals of the carbon atom form covalent bonds with the hydrogen atoms, one carbon p orbital still *remains vacant*. In addition, there is a filled, nonbonding $2s$ orbital.

What are the consequences of this arrangement? Our representations help explain some of the properties of CH_2 and CF_2. For instance, both CH_2 and CF_2 exist as unstable reaction intermediates at high temperatures. Neither molecule has ever been found in a pure state or in high concentrations at room temperature. Both CH_2 and CF_2 are very reactive molecules.

Such experimental data can best be understood if we picture one of the electrons in the carbon $2s$ orbital as moving up to the empty $2p_z$ orbital. This process is called **promoting the electron.** It results in only a slight increase in the energy of the system. As a result of promotion, the carbon atom gains two more half-filled orbitals. The new orbital diagram for carbon is

	$1s$	$2s$	$2p$		
			p_x	p_y	p_z
Carbon nucleus	⊗	⊘	⊘	⊘	⊘
		↑	↑	↑	↑

It thus gains the capacity to form *two* more covalent bonds. The electron dot representation for the additional reaction of CH_2 with hydrogen is

$$\overset{\cdot}{\underset{\cdot\cdot}{C}} : H + H_2 \longrightarrow H : \overset{\overset{\displaystyle H}{\cdot\cdot}}{\underset{\cdot\cdot}{\underset{H}{C}}} : H$$

The bonding orbitals are indicated with arrows in the orbital diagram. Each covalent bond significantly increases the stability of the system. The decrease in the energy of the system due to new bond formation more than compensates for the small amount of energy required to promote the $2s$ electron to the $2p_z$ level. The net result is that carbon is **tetravalent:** its bonding capacity is four.

EXERCISE 12-5

Draw an electron dot representation of the reaction between CF_2 and two F atoms. Draw an electron dot diagram of the CF_2 molecule both *prior to* and *after* electron promotion.

EXERCISE 12-6

Draw electron dot representations of each of the following molecules: CH_3, CF_3, CHF_3, CH_2F_2, CH_3F, CCl_4, and $CFCl_3$. Which are highly reactive? Why?

EXERCISE 12-7

Draw electron dot and structural formulas of the ethane molecule (C_2H_6). This molecule forms when two CH_3 molecules are brought together. Explain why C_2H_6 is much more stable than the highly reactive CH_3 molecules.

EXERCISE 12-8

Phosphorus forms the compound PF_3. Draw the electron dot representation for PF_3. How does it differ from the compounds of carbon in 12-6 and 12-7 with respect to the number of electron pair bonds around the phosphorus atom? What orbitals of P might be used?

BORON ATOMS

The boron atom presents the same option for orbital occupancy as does the carbon atom. This is shown by the following orbital diagrams:

The promoted electron configuration is somewhat higher in energy than its predecessor. Before promotion, the low-energy 2s orbital is fully occupied. Although higher in energy, the promoted boron atom gains bonding capacity when the three electrons are in 2s and two 2p orbitals. The boron atom shown in the first configuration can form only *one covalent bond*. The boron atom shown in the second configuration can form *three covalent bonds*. Since each bond lowers the energy of the atom, the chemistry of boron is determined by the promoted electron configuration. Thus, we expect boron to be trivalent and to form three bonds, as in the BF_3 molecule:

$$\longrightarrow \quad \overset{\displaystyle :\ddot{F}:}{\underset{\displaystyle :\ddot{F}:}{B:\ddot{F}:}}$$

The *empty* orbital remaining in BF_3, indicated by the arrow, suggests the possibility of additional reactivity for this species. But BF_3 is a stable molecule. It will not gain another fluorine *atom* to form BF_4. But what can be done with the *empty* orbital?* The open orbital permits BF_3 to combine readily with a molecule such as $:NH_3$, which has an unused electron pair. The resulting compound can be represented as

$$\overset{\displaystyle F \quad H}{\underset{\displaystyle F \quad H}{F:B:N:H}}$$

Similarly, a fluoride *ion* can combine with a BF_3 molecule to form a BF_4^- *ion:*

$$\overset{\displaystyle :\ddot{F}:}{\underset{\displaystyle :\ddot{F}:}{:\ddot{F}:B}} \;+\; :\ddot{F}:^- \longrightarrow \left[\,\overset{\displaystyle :\ddot{F}:}{\underset{\displaystyle :\ddot{F}:}{:\ddot{F}:B:\ddot{F}:}}\,\right]^-$$

To use the empty orbital, *two* electrons must be donated, not one, as in the formation of the regular bond between a boron atom and a fluorine atom. In this sense, BF_3 and CF_2 differ sharply.

Let us review the differences between CF_2 and BF_3.

$$\cdot\overset{\displaystyle}{C}:\ddot{F}: \qquad\qquad \overset{\displaystyle :\ddot{F}:}{\underset{\displaystyle :\ddot{F}:}{B:\ddot{F}:}}$$

The carbon atom in CF_2 has *one* empty valence orbital and *one* unused valence electron pair. When one electron of the unused pair moves into the empty orbital, two half-filled orbitals are formed. The half-filled orbitals make CF_2 very reactive toward any species with one half-filled valence orbital. For example, CF_2 combines with additional fluorine atoms to form CF_4 or with other CF_2 groups to form the polymer† Teflon. In contrast, the

*Remember that the 2s electrons are *too low* in energy for promotion to the valence shell, so the third p orbital is empty.

†Polymers are long-chained carbon molecules of high molecular weight. See Chapter 19 for a discussion of polymers.

boron atom in BF_3 has *one* empty valence orbital but *no* unused valence electron pairs. Therefore, the boron atom cannot obtain half-filled orbitals through the promotion of an unused electron. Instead, BF_3 combines with molecules or ions that can provide an electron pair.

You may be wondering about the hydrides of boron. The line of reasoning we used in considering compounds of hydrogen with fluorine, oxygen, nitrogen, and carbon suggests that hydrogen and fluorine behave similarly. A compound with the formula BH_3 should exist, just as BF_3 exists. We have discussed the formation of H_2O, F_2O, HF, F_2, CH_4, CF_4, and so on. Why have we not discussed BH_3? Strangely, no stable BH_3 species is known. The simplest boron hydride is B_2H_6, known as diborane. This molecule and other boron hydrides have challenged many chemical theories, but are now well understood. They will be considered later.

EXERCISE 12-9

> What reaction, if any, would occur between an HO molecule and a CH_3 molecule? Illustrate, using electron dot diagrams.

BERYLLIUM AND LITHIUM ATOMS

In its lowest energy state, the beryllium atom has two electrons in the $2s$ orbital. It also has two very tightly held $1s$ electrons. All $2p$ orbitals are vacant. Like the cases of boron and carbon, one $2s$ electron from the beryllium atom can be promoted to an empty $2p$ orbital:

	$1s$	$2s$	$2p$		
			p_x	p_y	p_z
Beryllium nucleus	⊗	⊘	⊘	○	○

After electron promotion, we would expect beryllium to form two bonds, which, in fact, it does. In combination with fluorine, beryllium forms BeF_2. BeF_2 molecules are found at temperatures above 1000 K. At lower temperatures, experiments show that BeF_2 molecules tend to combine with each other to form a three-dimensional ionic solid structure. It will be considered in the next chapter. BeH_2 is also a solid at room temperature.

The bonding capacity of a lithium atom is almost predictable from the foregoing discussion. Since the lithium atom has just one valence electron, its bonding capacity should be one:

	$1s$	$2s$	$2p$		
			p_x	p_y	p_z
Lithium nucleus	⊗	⊘	○	○	○

We expect lithium to form the fluorine compound LiF, which it does at very high temperatures. Like BeF_2, ionic LiF molecules combine at lower temperatures, forming a three-dimensional ionic solid.

Write an electron dot formula for the simplest *gaseous* LiCH₃ compound.

12.3 Bond types among the second-row fluorides

All chemical bonds are a result of electrons occupying space between two nuclei. Yet, it is often true that electron sharing is not exactly *equal* sharing. Sometimes, the electrons tend to be somewhat nearer one of the nuclei than the other. This is clearly illustrated by comparing covalent bonding in gaseous fluorine (F_2) to ionic bonding in gaseous lithium fluoride (Li^+F^-).

GASEOUS LITHIUM FLUORIDE

We have already discussed the bonding in a fluorine (F_2) molecule. Since neither fluorine atom can completely pull an electron away from the other, the atoms share a pair of electrons equally in a covalent bond. How does bonding in lithium fluoride compare to that in a fluorine molecule?

As we have said, a lithium atom has one valence electron. Accordingly, it can share a pair of electrons with only one fluorine atom. We would thus expect the formation of a stable gaseous lithium fluoride molecule.* The electron dot diagram would be

$$Li : \ddot{\underset{..}{F}} :$$

However, lithium and fluorine atoms attract electrons with unequal force. This is shown by the ionization energies of the two atoms and by the electron affinity of fluorine.

$$401.5 \text{ kcal} + F(g) \longrightarrow F^+(g) + e^-(g)$$
$$124.3 \text{ kcal} + Li(g) \longrightarrow Li^+(g) + e^-(g)$$
$$F(g) + e^- \longrightarrow F^-(g) + 79 \text{ kcal}$$

*It is important to note that the term "molecule" has nothing to do with bonding type. We have three-dimensional covalent structures such as diamond and three-dimensional ionic structures such as NaCl. We have "one-dimensional" covalent molecules such as Cl_2 and "one-dimensional" ionic molecules such as LiF. The term "molecule" refers to the actual unit that can be identified experimentally as a discrete species under the conditions used.

Figure 12-6

Electron distribution in various bond types: (a) electrons equally shared between F atoms, (b) electrons distorted toward fluorine, and (c) ionic LiF linkage.

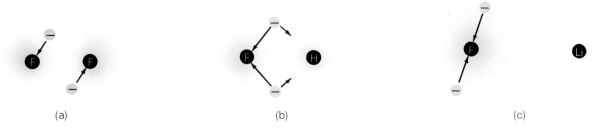

(a)   (b)   (c)

Clearly, the fluorine atom holds electrons much more strongly than the lithium atom. As a result, the electron pair in the gaseous LiF bond is more strongly attracted to the fluorine than to the lithium atom. Energy is lowered when the electrons move toward the fluorine atom. *When the bonding electrons move closer to one of the two atoms, the bond is said to have* **ionic character,** *or to be* **polar.**

In the most extreme cases, bonding electrons move so close to one of the atoms that the atom develops virtually the same electron distribution as its negative ion. This situation occurs in the case of gaseous lithium fluoride. Using an electron dot representation, we might write

$$Li^+ \quad :\ddot{F}:^-$$

or

$$Li^+ \ F^-$$

The bond in this kind of molecule is an *ionic bond.**

Remember, only one principle governs the formation of any chemical bond between two atoms: *all chemical bonds form because electrons occupy space between two positive nuclei* and are attracted to both. The phrase *covalent bond* indicates that, in terms of energy, the most stable distribution of electrons between two atoms is symmetrical. When the bonding electrons are somewhat closer to one of the atoms than to the other, the bond is said to have *ionic character*, or to be *polar*. The term *ionic bond* indicates that the electrons are displaced so much toward one atom that the bonded atoms must be considered a pair of adjacent ions. Figure 12-6 shows schematically the distribution of bonding electrons in covalent F—F, partially ionic H—F, and ionic Li—F bonds. The figure also shows where the bonding electrons might be located in a stop-action snapshot. In another snapshot, the bonding electrons may be in a different location. In each type of bond, electron-nucleus attractions account for the energy and stability of the molecules.

THE ELECTRIC DIPOLE OF IONIC BONDS

The movement of a negative electric charge (an electron) from one atom toward another atom is represented crudely in Figure 12-6c, in the lithium fluoride (LiF) molecule. The lithium end is positive and the fluorine end is negative. It is said to possess an **electric dipole,** and the molecule is called a **polar molecule.**

The forces of attraction between neighboring polar molecules are much stronger than those between nonpolar molecules. An arrow like the one in Figure 12-7 is commonly used to indicate that a polar bond has both a negative and a positive end. The head of the arrrow indicates the negative end. Because the polar bond has two oppositely charged ends, it is called a *dipole.*

Polar molecules are attracted to each other when the positive end of one molecule approaches the negative end of a neighboring molecule. Of course, the *force* one polar molecule exerts on another depends on the *alignment* of the approaching molecules. If ends of like charge approach each other, the force exerted is one of repulsion.

*Remember that the term "molecule" refers to the experimentally identifiable unit under the conditions used. For example, Ar, F_2, H_2, and LiF are molecules regardless of their bond type. In fact, Ar has no bonds!

Figure 12-7

Representations of the electric dipole of gaseous lithium fluoride.

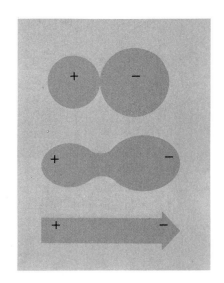

Table 12-1

Bond types in some fluorine compounds

Compound	Bond	Ionization energies (kcal/mole)		Bond type
		Element bonded to F	Fluorine	
FF	F—F	F　401.5	401.5	COVALENT
F_2O	O—F	O　313.8	401.5	increasingly ionic
NF_3	N—F	N　335.4	401.5	
CF_4	C—F	C　259.5	401.5	
BF_3	B—F	B　191.2	401.5	
BeF_2	Be—F	Be　214.9	401.5	increasingly covalent
LiF	Li—F	Li　124.3	401.5	IONIC

IONIC CHARACTER OF FLUORINE AND HYDROGEN BONDS

We can expect the ionic character of lithium fluoride to appear in other fluorides as well. Ionization energies give us a rough clue as to the nature of the electron-nucleus attractions in such bonds. Table 12-1 compares the ionization energies of each second-row element with that of fluorine. The last column describes the type of chemical bond we expect to find between the fluorine atom and atoms of other second-row elements. The trend from covalent to ionic bond type shown in Table 12-1 greatly influences the trend in properties of the fluorine compounds listed. The trend in bond type is the result of increasing differences in ionization energies between the two bonded atoms. The differences in melting and boiling points among the fluorine compounds, shown in Table 12-2, illustrate a corresponding trend in properties.

In Chapter 9 (on pages 226–227), hydrogen was described as a family in itself. Its chemistry often separates it from the rest of the elements in the periodic table. Thus, it is not surprising that an attempt to predict the ionic character of bonds between hydrogen and other atoms often does not produce the expected result. Let us look at some examples.

The measured ionization energy of the hydrogen atom, 313.6 kilocalories per mole, is quite close to that of the fluorine atom. This might lead us to expect a covalent bond between the hydrogen and fluorine atoms in the hydrogen fluoride (HF) molecule. The measured properties of hydrogen fluoride, however, show that the molecule has a significant electric dipole. This indicates a significant ionic character in the bond. The same holds true for the O—H bonds of water and, to a lesser extent, for the N—H bonds of ammonia. Examination of the properties of various compounds that include hydrogen indicates that the ionic character of bonds to hydrogen is similar to that of bonds to an element with an ionization energy in the region of 200 kilocalories per mole. Because of its unique electronic pattern, hydrogen forms bonds that are more ionic in character than we would have predicted based on its measured ionization potential.

Two specific examples of the problem of determining the ionic character of bonds to hydrogen involve the hydrides of carbon and lithium. The C—H bond seems to have only a slightly ionic character. At the other end of the periodic table, gaseous lithium hydride is known to have a considerable electric dipole, but with the polarity reserved. In lithium hydride, the electrons move toward the hydrogen atom, leaving the lithium atom with a partial positive charge. This result would be expected from comparing lithium's ionization energy (124.3 kilocalories per mole) to 200 kilocalories per mole instead of to hydrogen's measured ionization energy (313.6 kilocalories per mole). We thus find it useful *to consider the bonding of hydrogen in terms of an apparent ionization energy near 200 kilocalories per mole.*

12.4 Molecular geometry

Molecular properties result from both the identity of the atoms in a molecule and their geometrical arrangement. Shapes of many molecules have been determined experimentally. On the basis of this structural information, a number of electron models have been proposed to help us understand and even predict molecular geometry.

Chemists disagree over the relative value of these models. The truth is that we do not have a "correct" model for a system as complicated as a many-electron atom. With each model, a general theoretical picture is constructed *after* a large amount of experimental information about structure has been accumulated. The theory is then used to predict an unknown structure. Experimental information about the structure of a particular molecule is obtained and compared with that predicted by the general theory. The value of the theory is judged by its ability to correlate and anticipate new experimental results.

Table 12-2

Melting and boiling points of some fluorine compounds

Compound	Melting point (°C)	Boiling point (°C)
F_2	−220	−188
F_2O	−224	−145
NF_3	−206	−129
CF_4	−184	−128
BF_3	−127	−100
BeF_2	800 (sublimes)	—
LiF	842	1676

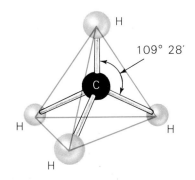

109° 28'

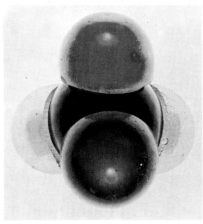

Figure 12-8

The geometry of the methane molecule (CH_4) and a space-filling model of the methane molecule.

In this book, we shall focus on a relatively simple, but very effective, old theoretical model that has been vigorously revived in the past 20 years.* It is known as the **valence-shell electron-pair repulsion (VSEPR) model.** The effectiveness of this model is matched only by its simplicity.

ARRANGEMENT OF ELECTRONS IN ORBITALS

In our discussion of orbital shapes for the gaseous hydrogen *atom,* we recognized one spherical $2s$ and three dumbbell-shaped $2p$ orbitals. (See pages 277–278.) The $2p$ orbitals are at 90° angles to each other. This known orbital geometry is restricted to the hydrogen atom because of its makeup: one proton and one electron. Additional electrons and protons create electrical fields that greatly distort the electron or orbital arrangement. For example, in CH_4, we might expect an H—C—H angle of 90°, since the $2p$ carbon orbitals forming the C—H bonds presumably make a 90° angle. (See pages 307–308.) In fact, the experimental H—C—H bond angle is 109°28', not the 90° predicted by p-orbital geometry. Further, we do not find that the three hydrogen atoms bound to p orbitals differ from the hydrogen bound to an s orbital. *Experiment shows that every hydrogen atom in CH_4 is the same. All H—C—H angles in the molecule CH_4 are 109°28'.* The molecule has four hydrogen atoms arranged at the corners of a regular tetrahedron. The carbon atom is in the center of the tetrahedron, as shown in Figure 12-8. What kind of model can interpret these and related observations?

The theory we shall consider proposes that *the arrangement of atoms around any given central atom is determined primarily by repulsions between electron pairs in the bonding and free valence shell of that central atom.* For example, the geometry of CH_4 would be determined by the repulsion between electron pairs making up each of the four C—H bonds. The electron pairs move as far apart as possible, and the tetrahedron results.†

METHANE AND CARBON TETRAFLUORIDE: CH_4 AND CF_4

The bonding in methane (CH_4) and in carbon tetrafluoride (CF_4) involves the *four* valence orbitals of carbon. All four orbitals are identical. They include the $2s$ orbital and all three $2p$ orbitals in atomic-orbital notation. It is impossible to tell the one $2s$ bond from the three $2p$ bonds. Four bonds are formed by carbon; the four bonding orbitals are called sp^3 hybrids. We are not surprised that in CH_4 the angle between *any two* C—H bonds is 109°28' and that in CF_4 the angle between *any two* C—F bonds is 109°28'. A **tetrahedral molecule** is formed. The four hydrogen atoms occupy positions at the corners of a *regular tetrahedron.* The structure for CH_4 is shown in Figures 12-8 and 12-9.

*One of the most articulate and able spokesmen for this model is Professor Ronald J. Gillespie of McMaster University in Canada.

†The four bonding orbitals are identical. They are called **hybrid orbitals** and are labeled sp^3 hybrids. They are made by mixing, or "hybridizing," the one s and three valence p orbitals to obtain four orbitals pointing toward the corners of a regular tetrahedron. The mathematics of quantum mechanics indicates that the combination of one s and three $2p$ orbitals should yield four equivalent orbitals.

The geometry of CH_4 is fairly obvious. But, can the theory predict the geometry of molecules such as NH_3, H_2O, NF_3, BF_3, BeF_2, and LiF?

AMMONIA: NH_3

The NH_3 molecule is like the CH_4 molecule except that the extra valence electron of nitrogen results in a free electron pair in place of one of the C—H bonds. (See Figure 12-9.)

You might expect the H—N—H bond angle in NH_3 to be 109°28′. The measured value is 107°. To explain the small difference between 107° and 109°, let us consider the formation of an ammonia molecule from three protons and a nitride ion.* The nitride ion has eight valence-shell electrons arranged in four equivalent orbitals. We shall assume that electrons in each of the four orbitals of the nitride ion repel each other and that each free pair occupies an orbital forming an angle of 109°28′ with every other orbital. In short, the valence electron pairs of N^{---} assume the geometry of CH_4. To form an NH_3 molecule from an N^{---} ion, three protons must be added. As the first proton approaches the electron cloud of the bonding orbital of the N^{---} ion, the cloud is pulled out. It becomes *longer* and *thinner*. As the orbital shrinks in diameter, the other electron clouds expand to utilize newly available space. This process is repeated as the second and third protons are added. At the end, we would expect the free electron cloud to occupy more volume than each of the three bonding clouds, as shown in Figure 12-9. Thus, the H—N—H angle should be somewhat less than 109°28′! Our prediction was experimentally verified by the angle of 107°. Therefore, we conclude that free electron pairs occupy more space than bonding pairs.

*It is convenient to imagine the interaction of protons or of *positive* fluorine *ions* with a hypothetical nitride ion (N^{---}). This does not imply that we have F^+ ions or N^{---} ions in the final molecule. Electrons move out to form a covalent bond. Ultimately, the cloud is probably closer to the fluorine than to the nitrogen.

Figure 12-9

Two-dimensional representations of the geometries of CH_4, NH_3, and H_2O. All electron pair clouds repel each other. Four pairs of electrons are shared in the tetrahedral-shaped CH_4 molecule. The pyramidal structure of the NH_3 molecule results when there is *one* pair of free electrons. The angular structure of the H_2O molecule results when there are *two* pairs of free electrons.

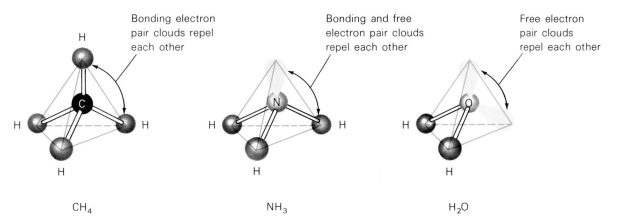

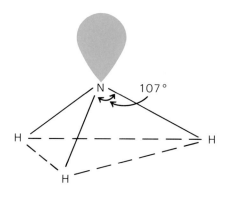

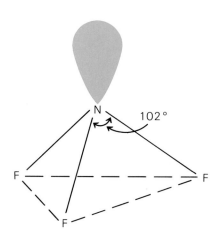

Figure 12-10

The geometries of NH_3 and NF_3 show a pyramidal structure with an equilateral triangle as a base and one free electron pair cloud.

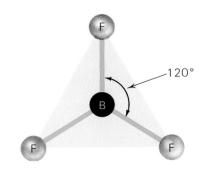

Figure 12-11

The geometric structure of BF_3.

WATER: HOH

In water, we have *two* free electron pairs and two O—H bonds. Because the two free electron pairs occupy more space than the bonding clouds, we would expect the H—O—H angle in water to be less than that in NH_3, which is 107°. The value of 104°30′ is reasonable.

In atomic-orbital notation a *divalent* atom such as oxygen, using two $2p$ orbitals as bonding orbitals, forms an **angular molecule.** Thus, the geometry of a water molecule is described as angular. The molecule can be conveniently represented using the structural model:

$$H \overset{O}{\frown} H \qquad \angle H\text{—}O\text{—}H = 104°30′$$

A line is drawn between the oxygen atom and each hydrogen atom to indicate that a chemical bond holds the two together. No line is drawn between the two hydrogen atoms, since we do not think they are directly bonded.

EXERCISE 12-11

If an F^+ were bonded to an O^{--}, the electron cloud would be pulled strongly toward the F^+ unit, and the cloud would become thinner. (The bond would become a polar covalent bond with the negative end toward fluoride.) Estimate the F—O—F bond angle in F_2O.

NITROGEN FLUORIDE: NF_3

The explanations used for the geometry of ammonia (NH_3) also apply to the geometry of nitrogen fluoride (NF_3). Those explanations indicate that NF_3, having three fluorine-nitrogen bonds and one free electron pair, should have a bond angle smaller than the 107° bond angle of NH_3. The reason for the smaller bond angle is that the fluorine atom pulls the electron cloud into a longer and thinner cloud than does the proton. The cloud becomes thinner because fluorine attracts electrons more strongly than does hydrogen. The measured value is 102°, as shown in Figure 12-10. One prediction based on our theory—a smaller angle in NF_3 than in NH_3—is verified.

According to atomic-orbital notation, a *trivalent* atom such as nitrogen, using three $2p$ orbitals as bonding orbitals, forms a **pyramidal molecule** that is an *irregular tetrahedron.* Thus, the geometry of NF_3 and NH_3 is described as pyramidal. These molecules can be represented as shown in Figure 12-10.

BORON TRIFLUORIDE: BF_3

The boron atom in boron trifluoride (BF_3) uses one $2s$ and two $2p$ orbitals in bonding in atomic-orbital notation. Therefore, the bonding is called sp^2 and involves only *three* pairs of electrons. Experiments confirm that the structure of BF_3 is that of an *equilateral triangle,* shown in Figure 12-11. Each electron pair forms a bond with fluorine. The arrangement permitting maximum distance between three electron pairs is a **planar molecule.** The BF_3 molecule has a fluorine atom at each vertex of an equilateral triangle. All atoms of this molecule lie in a plane, with each of the three fluorine atoms the same distance from the boron atom.

In atomic-orbital notation, each of the three bonds to boron is equivalent. Apparently, the combination of one electron in a $2s$ orbital and two electrons in two $2p$ orbitals produces three identical sp^2 hybrid orbitals. Thus, sp^2 bonding orbitals produce planar, triangular molecules.

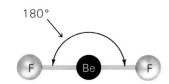

Figure 12-12
The geometric structure of BeF_2.

BERYLLIUM FLUORIDE AND LITHIUM FLUORIDE: BeF_2 AND LiF

In gaseous BeF_2 there are only *two* electron pairs to keep apart and no free electron pairs. Each electron pair forms a bond with fluorine. Experiments indicate that BeF_2 is a **linear molecule** and is symmetrical, as shown in Figure 12-12. Such a geometric arrangement gives maximum distance between electron clouds. By analogy to BeF_2, the structure of gaseous BeH_2 is also considered to be linear and symmetrical. Of course, the lithium fluoride (LiF) molecule can only be linear, because it consists of only two atoms. *Diatomic molecules can only be linear.*

In atomic-orbital notation, bonding between the beryllium atom and fluorine atoms in gaseous beryllium fluoride (BeF_2) involves one electron in a $2s$ orbital and one electron in a $2p$ orbital. This type of bonding is sometimes called sp bonding. A linear molecule is implied by this notation.

A SUMMARY OF MOLECULAR GEOMETRY

In summary, the structure of most molecules can be derived by using the *valence-shell electron-pair repulsion (VSEPR) model.* The following procedure permits application of this theory.

1. Write a Lewis electron dot formula for the molecule, showing electron pair bonds (electron pairs bonded between atoms) and free electron pairs.
2. Count the number of electron-pair bonds and the number of free electron pairs around the central atom.
3. Refer to Figure 12-13 on page 321 for the general shapes of molecules. If the number of atoms in the molecule is two, the molecule must be linear. If a central atom containing only two valence electrons and no free electron pairs is bonded to two other atoms using its two valence electrons, the molecule must be *linear* (gaseous BeF_2, CO_2). If a central atom has three valence electrons and no free electron pairs and it forms regular covalent bonds with three other atoms, the structure will be a *planar equilateral triangle* (BF_3). If the central atom has two valence electrons forming bonds to two other atoms and has, in addition, a free electron pair, the structure will be *angular* with an X—M—X bond angle less than 120° ($SnCl_2$ in gas phase). If the central atom has four valence electrons with which it forms four covalent bonds to four other atoms and if it has no free electron pairs, the structure will be a regular *tetrahedron* (CF_4, CH_4, $SiCl_4$, GeF_4). If the central atom has three valence electrons with which to form covalent bonds to three other atoms and has, in addition, a free electron pair, the molecule will be a low *trigonal pyramid* (NH_3, PCl_3, PH_3). If the central atom has two

Table 12-3

Bonding data and molecular geometry for some fluorine compounds in the gaseous state

Element	Bonding orbitals	Bonding orbitals after promotion			Bonding capacity	Number of free electron pairs in valence shell	Shape of fluoride molecule	Formula
		1s	2s	2p (P_x P_y P_z)				
He	none	⊗	○	○ ○ ○	0	0	no fluoride known	He
Li	s	⊗	⊘	○ ○ ○	1	0	linear diatomic	LiF
Be	sp	⊗	⊘	⊘ ○ ○	2	0	linear triatomic	BeF_2
B	sp^2	⊗	⊘	⊘ ⊘ ○	3	0	planar triangular	BF_3
C	sp^3	⊗	⊘	⊘ ⊘ ⊘	4	0	tetrahedral	CF_4
N	sp^{3*}	⊗	⊗	⊘ ⊘ ⊘	3	1	pyramidal	NF_3
O	$sp^{3\dagger}$	⊗	⊗	⊗ ⊘ ⊘	2	2	angular	F_2O
F	p	⊗	⊗	⊗ ⊗ ⊘	1	3	linear diatomic	F_2
Ne	none	⊗	⊗	⊗ ⊗ ⊗	0	4	no fluoride known	Ne

*Nearly sp^3, with one orbital containing a free pair of electrons (can be labeled p^3).
†Nearly sp^3, with two orbitals containing free pairs of electrons (can be labeled p^2).

valence electrons with which to form covalent bonds to two other atoms and has, in addition, two free electron pairs, the molecule will be *angular* (H_2O, F_2O, H_2S). If the central atom has five valence electrons with which to form five covalent bonds to other atoms and it has no free electron pairs, the geometry will be that of a *trigonal bipyramid* (PCl_5). If the central atom has free electron pairs replacing one or more of the five covalent bonds, structures such as those indicated in Figure 12-13(d) will be formed. Finally, if the central atom has six valence electrons with which to form six covalent bonds and it has no free electron pairs, the geometry will be that of a regular *octahedron* (SF_6). Replacement of one or more of the covalently bonded atoms by a free electron pair on the central atom gives structures such as these indicated in Figure 12-13(e).

In addition to Figure 12-13, Table 12-3 illustrates a summary of our discussion of procedure 3. Even though the table involves shapes of only the second row fluorides, the principles and VSEPR model illustrated in this table can be used for all other molecules as well. This table also indicates the hybrid atomic orbital symbolism associated with each geometrical pattern. It is assumed that electrons are promoted to orbitals that are required for the hybridization indicated. Our discussion has implied that the symbolism, such as sp = linear, sp^2 = triangular, and sp^3 = tetrahedral, arose because

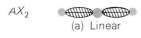

these orbitals were used in known structures. But it should be pointed out that quantum mechanical analysis also indicates that these very geometries are the ones that arise from the mixing of the orbitals indicated. The theoretical analysis in these cases is in agreement with our experimental experience.

12.5 Molecular shape and electric dipoles

Consider the fluorides of the second-row elements. There is a continuous increase in the ionic character of the bonds formed by fluorine with elements F, O, N, C, B, Be, and Li. (Look up the ionization energies of these elements in Table 12-1, on page 314.) This ionic character results in an electric dipole in the bond. In other words, the centers of positive and negative charge are separated. The *molecular dipole* will be determined by the sum of the dipoles of all bonds *if the sum takes into account the geometry of the molecule.* Since the properties of a molecule are strongly affected by the molecular dipole, we shall investigate its relationship to molecular architecture and the ionic character of individual bonds. For this study, we shall begin at the left side of the periodic table. Information on geometry and bonding is summarized in Table 12-3.

THE MOLECULAR DIPOLE OF LiF

The lithium-fluorine bond is highly ionic because of the large difference in the electron attracting ability of lithium and fluorine (see Table 12-1). Consequently, gaseous lithium fluoride has an unusually high electric dipole. It is a highly polar molecule in which the lithium atom carries a positive charge and the fluorine atom carries a negative charge.

THE MOLECULAR DIPOLE OF BeF$_2$

The beryllium-fluorine bond is also highly ionic. Unlike LiF, however, there are two bonds to fluorine in the BeF$_2$ molecule. The electrical properties of the entire molecule depend on how those two bonds are oriented in relation to each other. We need to find the "geometrical sum" of the two bond dipoles in such a molecule.

The geometrical sum of two dipole arrows is illustrated for several different cases in Figure 12-14 (see page 322). Figure 12-14a shows how two arrows that point in the same direction combine to form a longer arrow. Figure 12-14b shows how two arrows of different length that point in opposite directions combine to form a shorter arrow. Figure 12-14c shows how

Figure 12-13

In these two-dimensional representations of the general shapes of molecules of the nontransitional elements, darker circles represent the central atom (*A*), and lighter circles represent the attached atoms (*X*). Each hatched area represents a bonding electron pair cloud, and each colored area represents a free electron pair cloud (*E*).

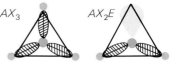

AX_2
(a) Linear

AX_3 AX_2E
(b) Triangular

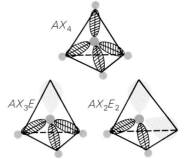

AX_4

AX_3E AX_2E_2
(c) Tetrahedral

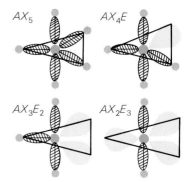

AX_5 AX_4E

AX_3E_2 AX_2E_3
(d) Trigonal bipyramidal*

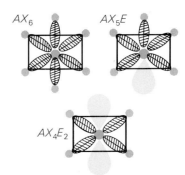

AX_6 AX_5E

AX_4E_2
(e) Octahedral†

*Describes two pyramids with a common triangular base and six plane surfaces.

†Describes two pyramids with a common rectangular base and eight plane surfaces.

321

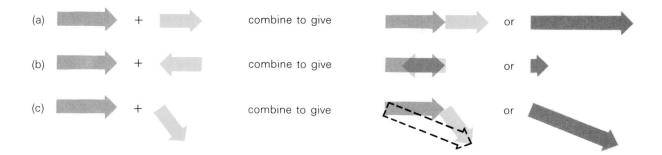

(a)

+

combine to give

or

(b)

+

combine to give

or

(c)

+

combine to give

or

Figure 12-14

The geometric sum of dipoles.

two arrows that are not aligned add up to form an arrow that represents a combination of the two directions.

In the linear, symmetrical BeF_2 molecule, the two bond dipoles point in *opposite* directions. Since the two bonds are equivalent, their sum is zero, as shown in Figure 12-15. Thus, even though the bonds in BeF_2 have dipoles, the molecule has no *net* dipole. The molecular dipole is zero, which is represented by the equivalent arrows.

EXERCISE 12-12

Decide whether each of the following gaseous molecules is polar or nonpolar: H_2O, CH_4, CH_2F_2, NH_3, and BeH_2. If the molecule is polar, indicate which end will be positive and which negative. Refer to Tables 12-1 and 12-3 and recall that the effective ionization energy of hydrogen (H) is 200 kcal/mole.

THE MOLECULAR DIPOLES OF BF₃ AND CF₄

Both the boron trifluoride (BF_3) and carbon tetrafluoride (CF_4) molecules are thought to be moderately ionic in their bonds. Yet, each overall molecular dipole is zero. Careful consideration of the molecular geometry shows that in each case there is complete cancellation of individual bond dipoles. This cancellation is shown in Figure 12-16 for the BF_3 molecule, whose shape is planar triangular.

Figure 12-15

The absence of a molecular dipole in BeF_2.

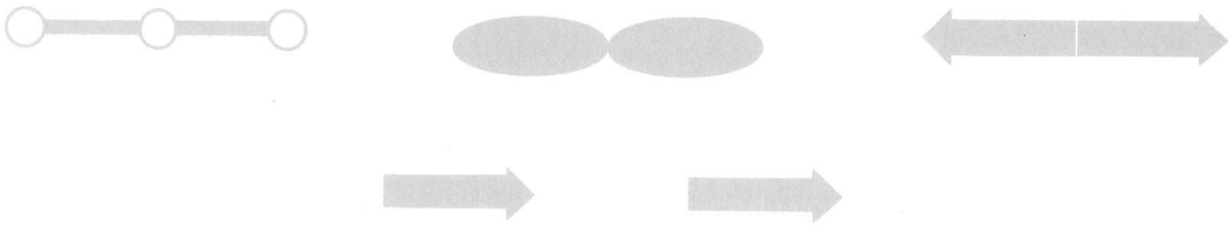

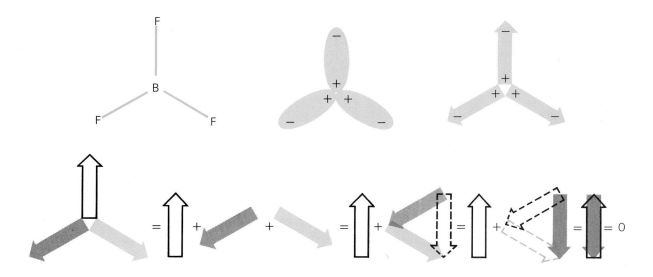

Figure 12-16

The absence of a molecular dipole in BF_3.

THE MOLECULAR DIPOLE OF NH_3

With ammonia, a new factor is introduced. Each N—H bond forms a dipole with H as the positive end. The free electron pair projecting from the nitrogen core forms a dipole with the nitrogen as the positive end. The molecular dipole will be obtained by adding these dipoles appropriately. The addition is shown in Figure 12-17. As the figure shows, the NH_3 molecule has a fairly large dipole. We assume that the free electron pair contributes to both the geometry and the dipole of the molecule. (The geometry of the NH_3 molecule is shown in Figure 12-10 and discussed on pages 316–317.)

EXERCISE 12-13

Would you expect H_2O to have a larger or smaller dipole moment than NH_3? Explain.

12.6 Single and multiple bonds

In determining the bonding capacity of a given atom from the second row of the periodic table, we counted the number of hydrogen or fluorine atoms with which the given atom can combine. Oxygen combines with *two* hydrogen atoms to form water. Thus, oxygen is described as *divalent*. This means that oxygen has a bonding capacity of two. Carbon combines with *four* hydrogen atoms to form methane (CH_4). Carbon is *tetravalent*. It has a bonding capacity of four. In each case, the bond formed between the hydrogen atom and the central atom is a single bond. In other words, the two atoms share one pair of electrons.

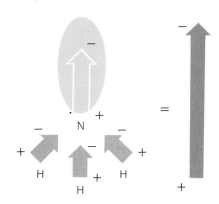

Figure 12-17

The molecular dipole of NH_3.

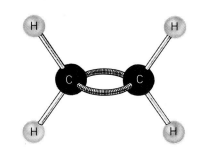

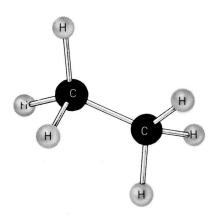

Figure 12-18

Models of an ethylene molecule (C_2H_4) and an ethane molecule (C_2H_6).

Another familiar carbon compound that has single bonds made with one pair of electrons is ethane (C_2H_6). As its electron dot structure indicates, each bond is made with a single pair of electrons.

$$
\begin{array}{cc}
H & H \\
H : \overset{\cdot\cdot}{\underset{\cdot\cdot}{C}} : \overset{\cdot\cdot}{\underset{\cdot\cdot}{C}} : H \\
H & H
\end{array}
$$ *

THE CARBON-CARBON DOUBLE BOND

Another compound that contains two carbon atoms bonded to hydrogen atoms is ethylene (C_2H_4). Note that ethylene has two fewer hydrogen atoms than ethane (C_2H_6). Suppose we write an electron dot representation of ethylene following the pattern established for ethane:

$$
\begin{array}{cc}
H & H \\
H : \overset{\cdot\cdot}{\underset{\cdot\cdot}{C}} : \overset{\cdot\cdot}{\underset{\cdot\cdot}{C}} \cdot H
\end{array}
$$

TWO HYDROGEN ATOMS REMOVED LEAVE
TWO HALF-FILLED ORBITALS

This formula has two unpaired electrons, indicated by the arrows. Thus, the arrangement has unused bonding capacity. Such an unstable arrangement suggests an alternative structure. The two unused electrons could be paired to form an additional shared-electron pair bond between the carbon atoms. The atoms would be joined by *two* shared-electron pair bonds, known as a **double bond** and written as C=C.

$$
\begin{array}{cc}
H & H \\
\overset{\cdot\cdot}{C} : : \overset{\cdot\cdot}{C} \\
H & H
\end{array}
$$

Much experimental evidence supports the idea of double-bond formation. Indeed, the study of carbon chemistry, presented in Chapter 19, includes many compounds whose molecules contain double bonds.

STRUCTURAL ISOMERS

Experiments show that the ethylene (C_2H_4) molecule is planar. That is, the four hydrogen atoms and the two carbon atoms all lie in one plane. If the CH_2 groups could twist around the C=C axis, the molecule would not retain its flat form. But this does not tend to occur. The ball-spring-and-stick model shown in Figure 12-18 (top) shows the rigidity of the double-bonded structure. Compare it with the ball-and-stick model for ethane (C_2H_6), in which rotation around the C—C bond is relatively easy.

---

*Remember again that color is only a bookkeeping convenience. All hydrogen atoms are identical, and all carbon atoms are identical.

It is possible to replace the hydrogen atoms of ethylene with halogen atoms. One such compound has the formula $C_2H_2Cl_2$. Examination of the model for ethylene suggests three possible ways in which the two chlorine atoms could be arranged in the compound $C_2H_2Cl_2$. The three configurations are shown in Figure 12-19. Three different compounds with this formula have indeed been identified experimentally, and structural studies confirm the existence of the three configurations. All three compounds have the same formula, and all are called dichloroethylene. *Different compounds with the same formula are called* **isomers.**

Look at Figure 12-19 again. In one $C_2H_2Cl_2$ molecule, two chlorine atoms are attached to the *same* carbon atom, (Figure 12-19a). In the remaining two molecules (Figure 12-19b and c), the chlorine atoms are attached to *different* carbon atoms. Isomers in which the actual bonding arrangements among atoms differ—for example, two chlorine atoms to a single carbon atom in one case, and one chlorine atom to each of two carbon atoms in the other— are called **structural isomers.** The structural isomer with chlorine atoms attached to two different carbon atoms has two different forms. One is the so-called *cis* form: two chlorine atoms on the *same side* of the double bond, as illustrated in Figure 12-19b. The other is the so-called *trans* form: two chlorine atoms on *opposite sides* of the double bond, as illustrated in Figure 12-19c. Structural isomers are the geometric consequences of the carbon-carbon double bond.

THE TRIPLE BOND IN NITROGEN

Multiple bonds are not restricted to carbon compounds. Let us investigate another type of multiple bond by writing an electron dot formula for nitrogen. Suppose nitrogen had a single bond. The structure would be

$$: \! \overset{\cdot}{N} : \overset{\cdot}{N} :$$

The problem we recognized with C_2H_4 also exists here. It is worse, however, because each nitrogen atom has *two* half-filled orbitals. The solution suggested for ethylene can be applied here to suggest the formation of a **triple bond,** written as N≡N or

$$\cdot N : : : N :$$

The triple bond between nitrogen atoms in elemental nitrogen is very strong and accounts for the very high dissociation energy of N_2.

$$N_2(g) + 226 \text{ kcal} \longrightarrow 2N(g)$$

The energy of dissociation is six times larger than the energy of the single N—N bond in hydrazine (N_2H_4):

$$H_2N—NH_2 + 38 \text{ kcal} \longrightarrow 2H_2N$$

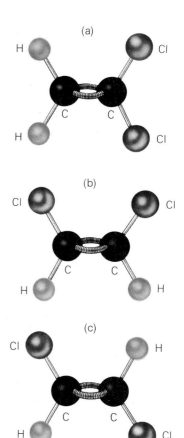

Figure 12-19

Models of the three isomers of dichloroethylene ($C_2H_2Cl_2$). Top (a) configuration is unsymmetrical, middle (b) is the *cis* form, and bottom (c) is the *trans* form.

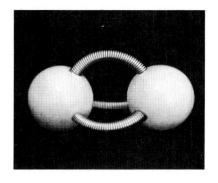

Figure 12-20

A ball-and-spring model of a triple bond.

325

Indeed, we can understand the very inert nature of nitrogen at low temperatures in terms of the very strong triple bond in the elemental nitrogen molecule. Similarly, a good part of the high reactivity of fluorine can be ascribed to the very weak single bond between two fluorine atoms:

$$F_2(g) + 37 \text{ kcal} \longrightarrow 2F(g)$$

Key Terms

angular molecule

divalent

double bond

electric dipole

electron affinity

free radical

hybrid orbital

hydroxyl radical

ionic character

isomer

linear molecule

orbital representation

planar molecule

polar

polar molecule

promoting the electron

pyramidal molecule

structural isomer

tetrahedral molecule

tetravalent

triple bond

trivalent

univalent

valence-shell
electron-pair repulsion
(VSEPR) model

SUMMARY

The chemical bond between hydrogen atoms in H_2 forms because each of the two electrons is attracted simultaneously to two protons. In terms of energy, this arrangement is more stable than are two separate, isolated hydrogen atoms. Any chemical bond arising from the sharing of two electrons is known as a covalent bond. Such bonds can form if two atoms share valence electrons using partially filled orbitals. Helium atoms do not bond together because their orbitals are filled.

All chemical bonds form because one or more electrons are simultaneously attracted to two or more positive nuclei. The bonding capacity of the second-row elements is determined by the maximum number of half-filled valence orbitals that can be created for each element's atoms. This number is calculated by assuming that one electron occupies each valence orbital before a second electron can fill any valence orbital. The placement of valence electrons in separate orbitals, where possible, is known as promoting the electrons. The resulting bonding capacities are one for fluorine and lithium, two for oxygen and beryllium, three for nitrogen and boron, and four for carbon. Using either an orbital representation or an electron dot representation of bonding, we can discuss the bonding capacities of the second-row elements—and any other elements—and thereby understand their chemistry.

The nature of the bond between two atoms is related to the ionization energy of each. If the sharing of electrons is unequal, the bonding electrons move closer to one of the atoms and thus form a bond with ionic character, known as a polar bond. A polar bond gives one bonded atom a negative charge and the other a positive charge. The condition gives rise to an electric dipole.

The shape of a molecule can be obtained by the valence-shell electron-pair repulsion model. Molecular shapes are summarized in Table 12-3. Both bonded electron pairs and free electron pairs must be considered. Symmetrical molecules with polar bonds can be nonpolar when the electric dipoles cancel each other. If the polar bonds do not cancel each other, the molecule is called a polar molecule.

Multiple bonds form between atoms sharing more than one pair of electrons. Double bonds result from the sharing of two electron pairs. Double bonds between carbon atoms can result in two or more molecules with identical formulas but different structures. Such molecules are called isomers. Triple bonds result from the sharing of three electron pairs. Double bonds are stronger than single bonds, and triple bonds are stronger than double bonds.

QUESTIONS AND PROBLEMS

A

1. (a) Explain, using attractive and repulsive forces between atomic particles, why the covalent bond forms and binds two atoms together?
   (b) Is the combination of two ground state hydrogen atoms an endothermic or exothermic process?
   (c) Show, using the orbital notation of Figure 12-2, how two hydrogen atoms form a bond.
   (d) Would you expect gaseous sodium vapor to have molecules containing one atom or two?
   (e) Draw the orbital model for sodium. Refer to Figure 12-2.
   (f) Would you expect the vapor molecules of copper to have one or two atoms present? Explain.

2. (a) Show the bond between two fluorine atoms using the Lewis electron dot model. (b) Show the bond between two fluorine atoms using the orbital designation. (c) Show the bond in HF using the orbital designation. (d) Would you expect a molecule with the formula H_2Cl to form? Explain.

3. What reaction would be expected between a C_2H_5 unit and an F atom? Draw Lewis electron dot structures to represent this reaction.

4. (a) Write the ground state electronic configuration of the oxygen atom. Why do two electrons tend to avoid the same orbital in the oxygen atom?
   (b) Use an orbital model to show how the oxygen atom combines with hydrogen atoms to give water.
   (c) Write the electron dot structure for the oxygen atom.
   (d) Write the electron dot structure for the CH_3 group.
   (e) What compound would you expect to form between O and CH_3 groups? Write the Lewis electron dot formula for this compound.

5. What is meant by the electron affinity of the fluorine atom? Write the defining equation including the energy term. Be sure to indicate the physical state of each species.

6. Two OH groups combine to give hydrogen peroxide. Would you expect two HF groups to combine in the same way to give an F—F bond? Explain.

7. (a) Write the ground state electronic configuration of the carbon atom.
   (b) Why is a carbon atom changed to a promoted state in the reaction process?
   (c) What compound would you expect to form between carbon and fluorine if one of the $2s$ electrons in carbon was not promoted?
   (d) Draw electron dot representations of two isomers of $C_3H_6F_2$. (*Hint:* there are a total of four different isomers.)

8. (a) Why was the structure of the compound formed between boron and hydrogen a problem?
   (b) Does the same problem exist with the compound formed between fluorine and boron? Explain.
   (c) Draw the electron dot structure for the negatively-charged anion of formula $B(CH_3)_3F$.

9. (a) What is meant by a bond dipole? (b) Arrange the following bonds in

decreasing order of bond dipole (or in decreasing order of polarity): F—F, C—F, HF, C—H, NaF(g).

10. Use the valence-shell electron-pair repulsion (VSEPR) model for molecular structure to draw or describe the structures of the following gaseous molecules: (a) BeF_2, (b) BF_3, (c) CF_4, (d) H_2O, (e) NH_3, (f) PF_5, (g) SF_6, (h) C_2H_6, (i) C_2H_5Cl, (j) two isomers of C_3H_7Cl.

11. (a) It is known that the C—F bond is quite strongly polar but CF_4 has no molecular dipole. Explain these two apparently contradictory facts. (b) Would you expect H_3CF to be a polar molecule? In other words, would you expect H_3CF to have a molecular dipole? Explain. (c) Did you think H_2O and NH_3 have molecular dipoles? Explain.

12. (a) What is a double bond? (b) Would the carbon-carbon distance be shortest in a double bond, triple bond, or single bond?

13. Which of the following compounds contain a double bond: CH_4, C_2H_6, C_2H_4, $C_2H_2Cl_2$, C_2H_3Cl, LiF, NH_3?

14. Which of the following molecules would have distinct *cis-* and *trans-* isomers? Draw electron dot structural formulas for the following isomers: (a) CH_4, (b) C_2H_6, (c) $C_2H_2Cl_2$, (d) CO_2, (e) BF_3, (f) C_2H_4.

B

15. Draw an electron dot diagram to show why the neutral species BF_4 is not stable but the compound KBF_4 is stable.

16. On the basis of electron dot formulas write the most reasonable structural formulas for: (a) H_2O_2, (b) OFH, (c) $(CH_3O)_2$, (d) $(NH_2)_2$, (e) F_2O_2, (f) BH_3, (g) Which one of the above "reasonable formulas" turns out to be wrong? Explain.

17. Beryllium forms the molecule $Be(CH_3)_2(g)$. (a) Write the electron dot formula that shows the structure. (b) On the basis of ionization potentials from page 287, would you expect the Be—C bond to be polar or non-polar? Explain. (c) Would the molecule $Be(CH_3)_2(g)$ be a polar *molecule*? Explain.

18. (a) The compound $(CH_3)_2O$ reacts with BF_3 to form a relatively stable product. Draw a complete electron dot structure for the product. (b) Could we expect a product to be $BF_3(O(CH_3)_2)_2$? Explain. (c) Should BF_3 react with hydrazine, N_2H_4? If reaction occurs, draw structures. If reaction does not occur, explain why.

19. The molecule acetylene (C_2H_2) will react vigorously with chlorine gas to give $C_2H_2Cl_4$. (a) Write electron dot formulas to show this reaction. (b) Write formulas of two possible isomers for the products. (c) Which isomer would have a measurable dipole?

20. Chemical bonds are indicated below for some gaseous substances. Indicate which atom will be *more* positive in each bond. If both atoms have equal positive character, indicate that the bond is nonpolar: (a) Na—Cl, (b) C—F, (c) P—O, (d) K—H, (e) C—C, (f) the two bonds in

$$
\begin{array}{ccc}
 & H & H \\
 & | & | \\
H- & C- & C-H \\
 & | & | \\
 & H & H
\end{array}
$$

21. Why is elemental $N_2(g)$ nonreactive at room temperature?

22. Draw orbital electron configurations for each of the following elements

in its lowest energy state: (a) potassium, (b) phosphorus, (c) chlorine, (d) aluminum, (e) calcium, (f) silicon, (g) sulfur.

23. Draw an electron dot diagram for each of the following molecules or ions: (a) AlF_3, (b) SiH_4, (c) GeH_4, (d) $Ca(OH)_2$, (e) OH^+, (f) H_2NOH (Hydroxylamine) (g) ClF.

24. Which of the following molecules has a molecular dipole? (a) NaF(g), (b) MgO(g), (c) $MgI_2(g)$, (d) $GaCl_3(g)$, (e) $CO_2(g)$, (f) $H_2S(g)$, (g) $CCl_4(g)$, (h) $PF_3(g)$?

25. Would dimethyl hydrazine, $H_2N_2(CH_3)_2$, have *cis*- and *trans*-isomers? Draw electron dot formulas and justify your answer.

C

26. (a) What causes two atoms such as two hydrogen atoms to bond together by a covalent bond?
    (b) Suppose the electron density between two nuclei were increased, what effect on bond strength would be anticipated?
    (c) What effect on internuclear distance would be expected if electron density between two nuclei were to be increased?
    (d) The C—C distance in

$$H-\underset{\underset{H}{|}}{\overset{\overset{H}{|}}{C}}-\underset{\underset{H}{|}}{\overset{\overset{H}{|}}{C}}-H$$

is 1.53 Å; the distance in

$$\underset{H}{\overset{H}{\diagdown}}C=C\underset{\diagdown H}{\overset{\diagup H}{}}$$

is 1.34 Å; the distance in H—C≡C—H is 1.20 Å. Are these numbers in agreement with predictions made in (c) above? Explain. One angstrom unit (Å) is equal to one hundred-millionth of a centimetre or $1\ \text{Å} = 10^{-8}\ \text{cm}$.

27. The distance between oxygen atoms in the oxygen molecule is 1.21×10^{-8} cm (1.21 Å). The distance between oxygen atoms in H_2O_2 is 1.48×10^{-8} cm (1.48 Å). (a) Explain these differences in the oxygen-oxygen bond length on the basis of Lewis electron dot formulas. (b) What value would you suggest for the single bond-covalent *radius* of the oxygen atom?

28. The distance between nitrogen atoms in elemental nitrogen is 1.10 Å. The distance between nitrogen atoms in H_4N_2 is 1.47 Å. Suggest a reasonable value for the *double bond* radius of the nitrogen atom?

29. How does the covalent B—F bond in BF_3 differ in form from the covalent B—N bond in F_3BNH_3?

30. Consider the species NO_2^+, NO_2, NO_2^-. Select from the following O—N—O bond angles the approximate value that would be expected for *each* of the NO_2 species: 180°, 130°, 120°.

31. Look at the data of problem 31 for O—O distances in O_2 and H_2O_2. The molecule ozone, O_3, is a triangular molecule with an O—O bond distance of 1.28 Å and an O—O—O bond angle of 117°. What kind of hybridization is indicated for the orbitals of the central oxygen?

We are all agreed that the theory is crazy. The question that divides us is whether it is crazy enough to have a chance of being correct.

NIELS BOHR

13

MOLECULAR ARCHITECTURE: LIQUIDS AND SOLIDS

Objectives

After completing this chapter, you should be able to:

- Locate in the periodic table those elements that form molecular liquids and solids, network solids, and metals.

- Distinguish between network solids and molecular solids by noting differences in their melting and boiling points.

- Explain metallic properties such as electrical conductivity and malleability by using the atomic model of metallic structure.

- Describe the atomic structure of an ionic compound, explain why it is likely to be chemically stable, and account for its solubility in water.

- Predict from a compound's structural formula whether it can form hydrogen bonds, and describe the effect of such bonds on the compound's boiling point.

Each snowflake is a single crystal of ice that is highly symmetrical but different in shape.

In Chapter 12, you read that the diatomic molecule of fluorine (F_2) does not form larger molecules such as F_3 or F_4. We explained this by noting that each fluorine atom has only *one* partially filled valence orbital. When two atoms combine by sharing electrons from two formerly half-filled orbitals, the valence orbitals in both fluorine atoms are completely filled. A very stable electron configuration has been achieved. The stable F_2 molecule exists as a unit.

At sufficiently low temperatures, however, fluorine molecules condense to form a liquid. And at still lower temperatures, a solid may form. The observation is general. *Any pure gas, when cooled sufficiently under appropriate pressure, will condense to form a liquid. At lower temperatures, it will form a solid.* Even helium gas will condense, forming liquid helium.

Two or more atoms remain near each other in a particular arrangement because energy is lowest in that arrangement. This is true whether the arrangement is regular, as in a crystal, or irregular, as in a liquid. The cluster of atoms is stable if, and only if, the energy is lower when the atoms are together than when the atoms are apart. Let us examine bonding between atoms in liquids and solids in more detail.

Figure 13-1

These elements form molecular crystals bound by van der Waals forces.

Figure 13-2

Melting and boiling points of noble gases and halogens.

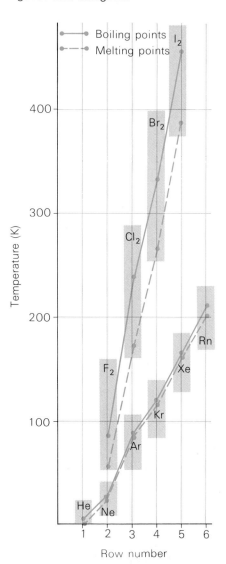

13.1   Molecular solids: Van der Waals forces

The same forces that cause molecular fluorine to condense at 85 K cause noble gases such as helium to condense. These forces are named **van der Waals forces,** after Johannes D. van der Waals (1837–1923), the Dutch physicist who studied them. The forces are weak and poorly understood but very important. Gasoline is a liquid because of van der Waals forces. Molecules with filled valence orbitals, such as Cl_2, F_2, or C_8H_{18} (one component of gasoline), cannot interact strongly with each other because the electrons of one molecule cannot closely approach the nuclei of another molecule. All outer valence orbitals are filled in the separate molecules. On the other hand, weak interactions that lower the energy of the system by a few kilocalories per mole are possible. Liquids or molecular crystals with low melting points will result. We can state a generalization: *van der Waals forces are the basic forces of attraction among molecules in which all valence orbitals of all atoms are filled.*[*] The *elements* forming van der Waals liquids and solids are concentrated in the upper right-hand corner of the periodic table, as shown in Figure 13-1.

NUMBER OF ELECTRONS

In Chapter 9, we said that the boiling points of the noble gases increase as their atomic numbers increase. Boiling points range from 4.2 K for helium (atomic number 2) to 211 K for radon (atomic number 86). A similar trend is observed for the halogens. Their boiling points increase from 85 K for F_2 (atomic number 9) to 457 K for I_2 (atomic number 53). These data, along with corresponding melting points, are plotted in Figure 13-2. The horizontal axis shows the row each element occupies in the periodic table. Helium occupies the first horizontal row, number 1, in the table. Neon and fluorine occupy the second horizontal row, number 2, and so on.

Figure 13-3 plots melting and boiling points against row number for compounds formed by carbon with halogen atoms. These compounds have the general formula CX_4, where X represents any halogen atom. As in Figure 13-2, the horizontal axis shows the row each halogen occupies in the periodic table. In CX_4, it is the halogen atoms that touch the atoms of neighboring molecules. Such outer atoms seem to determine the size of intermolecular forces. This fact suggests that intermolecular forces are only effective over very short distances. Atoms must touch.

[*]Number of electrons and molecular shape are factors affecting van der Waals forces.

As Figures 13-2 and 13-3 indicate, van der Waals forces in substances in closely related groups increase as the number of electrons in the outermost atoms increases. Thus, the boiling point of a substance depends on the number of electrons in the interacting outermost atoms. A related generalization can be made for similar molecules: the larger the molecule, the higher its boiling point. For example, the exterior atoms of both methane (CH_4) and ethane (C_2H_6) are hydrogen atoms. Yet, the boiling point of ethane is 185 K, which is higher than the boiling point of methane, 112 K. The difference is thought to be due to the greater amount of contact surface between two ethane (C_2H_6) molecules than between two methane (CH_4) molecules. The same effect is found for C_2F_6 (boiling point 195 K) and CF_4 (boiling point 145 K).*

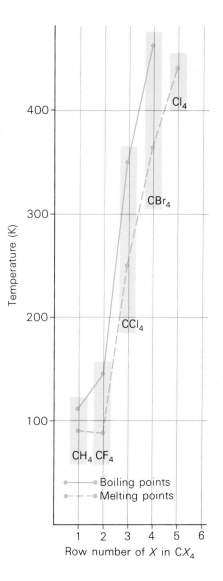

Figure 13-3

Melting and boiling points of CX_4 molecules.

EXERCISE 13-1

Gaseous phosphorus is made up of P_4 molecules with four phosphorus atoms arranged at the corners of a regular tetrahedron. With such a geometry, each phosphorus atom is bound to three other phosphorus atoms. Would you expect this gas to condense to a solid with a low, intermediate, or high melting point? After making a prediction on the basis of the valence orbital occupancy, check the melting point of phosphorus in a handbook. (*Hint:* Compare with I_2.)

EXERCISE 13-2

Natural gas, widely used as a fuel, is mostly methane (CH_4). The gasoline used in automobile engines is a mixture of similar molecules whose "average" formula is C_8H_{18}. Explain why natural gas does not condense to a liquid even in winter, while gasoline remains as a volatile liquid even in summer.

MOLECULAR SHAPE

A substance whose structure has a high degree of symmetry generally has a higher melting point than a closely related compound whose structure lacks symmetry. For example, consider two of the three structural isomers of formula C_5H_{12}: pentane and neopentane. Their molecular shapes differ drastically, as Figure 13-4 shows. The zigzag shape of normal pentane

---

*Notice that these two factors—number of electrons and molecular size—might lead to another generalization: boiling point increases in proportion to molecular weight. Molecular weight, molecular size, number of electrons, and boiling point all tend to increase together. This molecular weight–boiling point correlation has some usefulness with molecules of similar composition and general shape, but there is no direct causative relation between molecular weight and boiling point.

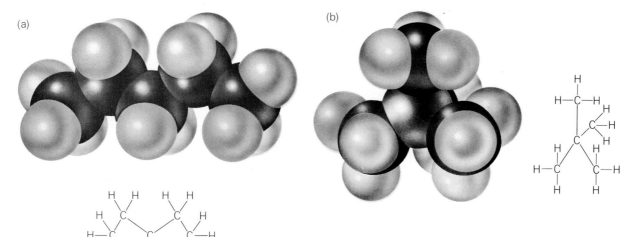

(a)

(b)

Figure 13-4

Molecular shape influences melting and boiling points of two structural isomers of C_5H_{12}. (a) Normal pentane melts at $-130\ °C$ and boils at $36\ °C$. (b) Neopentane melts at $-20\ °C$ and boils at $9\ °C$.

allows van der Waals forces to act between the external envelope of hydrogen atoms of one molecule and those of adjacent molecules. This large surface contact results in a relatively *high boiling point*. On the other hand, this flexible, snakelike molecule does not pack readily in a regular lattice, so its crystal has a *low melting point*.*

Compare normal pentane with highly compact, symmetrical neopentane. The ball-like neopentane readily packs in an orderly crystal lattice that, because of its stability, has a rather *high melting point*. Once melted, however, neopentane forms a liquid that boils at a temperature lower than that at which normal pentane boils. Neopentane has less surface contact with its neighbors and thus is more volatile, with a relatively *low boiling point*.

Most carbon compounds condense to form molecular liquids and solids. Their melting points are generally low (below about 300 °C), and many carbon compounds boil below 100 °C. The similar chemistry of the liquid and solid phases suggests that the basic geometry of the organic molecule is not changed as we go from the liquid to the solid state. Van der Waals forces hold both liquids and solids together.

EXERCISE 13-3

The two isomers of butane (C_4H_{10}) have boiling points of 0 °C and -12 °C. Their molecular structures are

(a)   (b)

$$H-\overset{\displaystyle H}{\underset{\displaystyle H}{C}}-\overset{\displaystyle H}{\underset{\displaystyle H}{C}}-\overset{\displaystyle H}{\underset{\displaystyle H}{C}}-\overset{\displaystyle H}{\underset{\displaystyle H}{C}}-H \quad \text{and}$$

Which boiling point would you assign to isomer (a), and which would you assign to isomer (b)? Explain your answers.

*A discussion of the crystal lattice arrangement is found in Chapter 9, page 221.

A MODEL FOR VAN DER WAALS FORCES

It is not easy to construct a model illustrating van der Waals forces. We know that such forces are effective over very short distances and that they increase as the number of electrons increases. We also know that they become stronger as outer electrons are held less tightly and as the amount of intermolecular contact increases. Ball-shaped molecules with minimal intermolecular contact have weaker van der Waals forces than do wildly intertwining molecular chains. These facts can be explained by a model that pictures van der Waals forces as arising from vibrations of electron clouds.

To simplify the picture, let us consider the van der Waals forces between spherical xenon atoms. If the electron cloud around a spherical xenon atom begins to vibrate, the heavier xenon nucleus cannot follow instantaneously. As a result, an "instantaneous dipole" is created. This "dipolar molecule" could attract another dipolar molecule, as shown in Figure 13-5. Such attraction would occur if all clouds vibrated in unison in a correlated manner. Because the electrons move back and forth, the molecules do not have a permanent or measurable dipole. However, the instantaneous dipole does permit attraction between molecules. For bonding, it is essential that all electron motions occur together. Clusters of molecules form liquids or solids because the energy of such clusters is slightly lower than the energy of separate gaseous molecules. Liquids and solids with low boiling and melting points result.

Figure 13-5

A model for van der Waals forces. (a) Spherical xenon atom. (b) Instantaneous dipole resulting from electron cloud vibrations. (c) Attraction between molecules as all clouds vibrate together in a correlated manner. The positive and negative charges shown are, not permanent, but instantaneous dipoles.

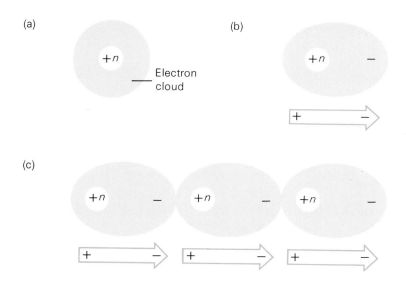

13.2 Covalent bonds and network solids

The substances we have considered so far in this chapter all consist of small molecules in which the bonding capacity of each atom has been completed. The chemical bonds holding these atoms in molecules are strong. But no

WILLIAM N. LIPSCOMB, JR.

Born in Cleveland, Ohio, William Lipscomb graduated from the University of Kentucky. His Ph.D. was granted by the California Institute of Technology in 1946. After receiving his Ph.D., Lipscomb taught at the University of Minnesota, where he rapidly rose to the position of professor. In 1959, he went to Harvard University, where he is currently the Abbott and James Lawrence Professor of Chemistry.

Lipscomb's early research used X-ray crystallography to define the structures of the formerly mysterious boron hydrides. He then developed theoretical models to explain bonding in these network solids and to tie them into the main fabric of chemistry. In more recent years, Lipscomb has turned his attention to the structure and operation of enzymes and other large molecules.

Lipscomb received the Nobel Prize for Chemistry in 1976 for his work on the structure and bonding of boron hydrides and their derivatives. The awarding of this honor recognized Lipscomb's high achievement in advancing our fundamental understanding of the chemical bond and stability in molecules.

strong bonds are left to join the separate molecules. Only weak van der Waals forces are present.

The situation involving the second-row element carbon, however, is very different. As noted in Chapter 9, solid carbon exists in two principal forms: diamond and graphite. Diamond is a classic example of a **three-dimensional network solid,** as shown in Figure 13-6a. Graphite is a typical **two-dimensional network solid,** as shown in Figure 13-6b. Let us examine their structures in some detail.

THREE-DIMENSIONAL NETWORK SOLIDS

In diamond, each carbon atom is bonded to four others in a tetrahedral arrangement. Because hybrid orbitals made from one s and three p orbitals are used in bonding, such bonding is known as sp^3. See Chapter 12. Because each atom is bonded to four others, the portion of the diamond structure represented in Figure 13-6a is not a complete molecule. The outermost carbon atoms represented are each bonded to additional carbon atoms, which in turn are bonded to still other carbon atoms, and so on. In other words, the entire diamond structure is one huge molecule with strong covalent bonds throughout. All the valence electrons of each carbon atom are firmly held between the atoms. The very high melting point (above 3500 °C), extreme hardness, and low electrical conductivity of diamond can be interpreted in terms of this continuous three-dimensional structure.

The compound silica (SiO_2), known as quartz, also has a covalent network structure. In one form, silicon atoms replace carbon atoms in a diamond-like arrangement. An oxygen atom is located between every pair of adjacent silicon atoms, as shown in Figure 13-8. The resulting structure is hard and has a very high melting point. The elements that tend to form three-dimensional network solids are shown in the darkest shaded portion of the periodic table in Figure 13-7.

TWO-DIMENSIONAL NETWORK SOLIDS

In contrast to diamond and silica, the graphite form of carbon has a two-dimensional, layered structure. As shown in Figure 13-6b, each carbon atom is surrounded by three others in a *plane* (instead of four others in a

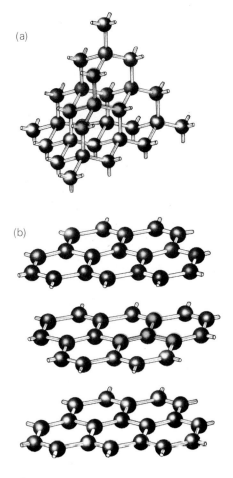

Figure 13-6

Carbon: (a) diamond; (b) graphite.

Figure 13-7

These elements form network solids with covalent bonding.

H₂																	He
												B	C	N₂	O₂	F₂	Ne
													Si	Pₙ	S₈	Cl₂	Ar
													Ge	As	Se	Br₂	Kr
													Sn	Sb	Te	I₂	Xe
														Bi	Po	At₂	Rn

tetrahedron, as in diamond). Such geometry suggests the sp^2 bonding we discussed for boron trifluoride (BF), on pages 318–319. Let us draw the electron dot representation for graphite. We see that each carbon atom is using only three valence electrons and orbitals. We are short one valence electron per carbon atom*:

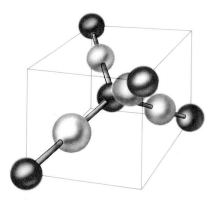

Figure 13-8

Cubic form of silicon dioxide. One small segment of a diamond-like lattice in which silicon atoms occupy the places of the carbon atoms and oxygen atoms lie between the silicon atoms. To get the formula for this solid, consider the central silicon atom as a unit. There are four oxygen atoms bonded to that silicon, but each of these oxygen atoms must be shared equally with another silicon atom. Thus, you can say that the central silicon atom has a one-half interest in the four oxygen atoms to which it is bonded. Based on this information, the formula for silicon dioxide is Si × $4O \times \frac{1}{2}$ = SiO_2.

What should we do with the extra electron and orbital for each atom? One solution would be to use the extra electrons to form multiple bonds, as we did with C_2H_4 and N_2 (see pages 324–325). Then there would be ten extra electrons (indicated below by X's) for the ten carbon atoms in the two rings:

Since there is only one extra orbital and one extra electron per carbon atom, *only one of the three bonds per carbon atom can become a double bond at any one time.* Therefore, one possible graphite structure might be represented as shown here:

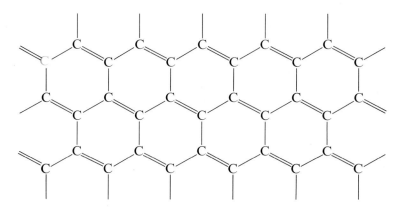

*Remember, each carbon atom has 4 valence electrons. But in the diagram, we have pictured 14 carbon atoms and 42 electrons. (Count them.) On the average, each carbon is short 1 electron.

Figure 13-9
Silicate mineral, talc, breaks into thin sheets, feels "soapy," and is used to make talcum powder.

By using that arrangement, all carbon atoms have filled $2s$ and $2p$ valence levels. But that is not the only arrangement that provides carbon atoms with filled valence orbitals. There is a second arrangement:

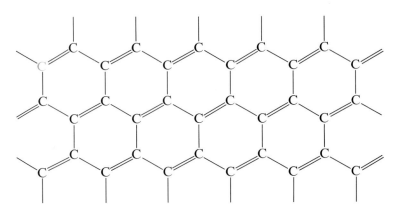

A third representation is equally valid:

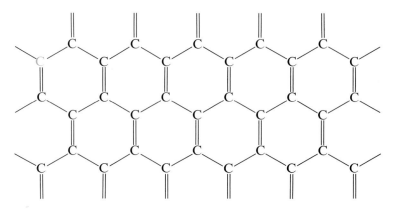

Notice that the double bond can move around. *Any two adjacent atoms in the plane can be joined by a double bond.* This means that the electrons making up the double bonds can move easily between atoms in the plane. Such mobile electrons are described as **delocalized electrons.**[*] They are responsible for the high electrical conductivity of graphite *along* the plane. On the other hand, there is no easy route for electrons to move between layers. Graphite is a poor electrical conductor in a direction *perpendicular* to the layers. Thus, the bonding within any one plane, or layer, is strong, giving rise to the name *two-dimensional network*. The separate layers are held together by weak van der Waals forces and can slide past each other. Thus, graphite breaks easily into layers.

Another two-dimensional network structure is the silicate mineral talc. If you have ever handled talc, you know how it can break into thin sheets and how "soapy" it feels. It is an ingredient in talcum powder, which is used as a skin lubricant. Such solids are like a book with very strong pages. Each

---

[*]Graphite can also be described in another way. Note that the s, p_x, and p_y orbitals are used to make the planar layers, with three triangularly arranged bonds around each carbon atom. This is an sp^2 structure. The p_z orbitals of each carbon atom remain perpendicular to the graphite layer. The p_z orbitals of adjacent carbon atoms can overlap to form giant orbitals extending over the whole layer. Electrons can then move easily over the layers using these giant π *orbitals*.

page is hard to tear or rip apart, but it can easily be separated from the other pages. Carbon is the only element known to form this layered structure, but (as noted) layered silicates are common in nature. No graphite analogues of Si, Ge, Sn, or Pb exist.

13.3 The metallic state

How do we recognize a metal? Let us review some properties shared by all metals. First and foremost, metals have a characteristic metallic luster: they are bright and shiny. With few exceptions—such as gold, copper, bismuth, and manganese—metals have a silvery-white color, because they reflect *all* frequencies of light.

All metals are excellent conductors of electricity and heat. All true metals can be drawn into wire or hammered into sheets without shattering. In other words, metals are ductile and malleable.

Elements that border the division between metals and nonmetallic elements may have *some* characteristic properties of metals but not others. For example, antimony looks like a metal, but it is brittle instead of malleable and is a poor conductor of electricity and heat.

Metals have strikingly similar properties that differ from the properties of the nonmetallic elements. The similarities among metals suggest that they might all have somewhat similar electron configurations and structures. Further, we find metals on the left-hand side of the periodic table, as shown in the darkest-shaded portion in Figure 13-10. Metallic characteristics increase as we look down each group of the table. For example, the fourth group goes from nonmetallic carbon to metallic lead. What orbital descriptions are suggested by all these characteristics of metals?

Figure 13-10

The elements with the lightest shading are molecular liquids and solids bound by van der Waals forces. The elements with medium shading are covalently bonded network solids. The metallic elements are indicated by the darkest shading.

H₂																	He
Li	Be											B	C	N₂	O₂	F₂	Ne
Na	Mg											Al	Si	Pₙ*	S₈	Cl₂	Ar
K	Ca	Sc	Ti	V	Cr	Mn	Fe	Co	Ni	Cu	Zn	Ga	Ge	As	Se	Br₂	Kr
Rb	Sr	Y	Zr	Nb	Mo	Tc	Ru	Rh	Pd	Ag	Cd	In	Sn	Sb	Te	I₂	Xe
Cs	Ba	La–Lu	Hf	Ta	W	Re	Os	Ir	Pt	Au	Hg	Tl	Pb	Bi	Po	At₂	Rn
Fr	Ra	Ac–Lr															

	La	Ce	Pr	Nd	Pm	Sm	Eu	Gd	Tb	Dy	Ho	Er	Tm	Yb	Lu
	Ac	Th	Pa	U	Np	Pu	Am	Cm	Bk	Cf	Es	Fm	Md	No	Lr

*Phosphorus has several different molecular forms. Volatile white phosphorus is P₄. Non-volatile, insoluble red phosphorus is polymeric Pₙ.

BONDING IN METALS

Table 13-1 summarizes information about the elements in the third row of the periodic table. Notice that the characteristics shared by the three metallic elements sodium (Na), magnesium (Mg), and aluminum (Al) are few valence electrons, vacant valence orbitals, and low ionization energies. Vacant valence orbitals and low ionization energies are the two conditions necessary for metallic bonding. We shall see how those characteristics are essential to the existence of metals.

Like other bonds, the metallic bond forms because electrons are strongly attracted by two or more positive nuclei simultaneously. Our problem is to obtain some insight into the special way in which electrons in metals behave. Consider the lithium atom. Since each lithium atom has a single $2s$ electron, we would expect a single covalent bond between two lithium atoms to form Li_2. This bond is represented as

$$Li : Li$$

Indeed, Li_2 is observed at high temperatures in the vapor phase.

Notice, however, that gaseous Li_2 still has *three* unused valence orbitals in each atom. The existence of just one unused or empty orbital per carbon atom confers a special reactivity upon the CH_2 molecule. In Chapter 12, we read that one electron of an unused pair in the carbon atom is promoted to the empty valence orbital and that two new covalent bonds then form. CH_2 is a very reactive molecule. This argument suggests that Li_2 could be a reactive molecule. With Li_2, three empty valence orbitals are available.

However, the molecule Li_2 lacks bond-forming electrons. The valence electron for each atom is already involved in a bond between two lithium atoms. Thus, there are no free pairs. If extra bonds *between Li_2 molecules* are

Table 13-1

Bonding and energy data for some third-row elements

Type of element	Bond type	Element	Atomic number	Bonding orbitals after promotion				First ionization energy (kcal/mole)	Boiling point (°C)
				3s	3p P_x	P_y	P_z		
metal	metallic	sodium	11	⊘	○	○	○	118	889
		magnesium	12	⊘	⊘	○	○	175	1120
		aluminum	13	⊘	⊘	⊘	○	138	2327
	network	silicon	14	⊘	⊘	⊘	⊘	188	2355
nonmetal	molecular	phosphorus	15	⊗	⊘	⊘	⊘	242	280
		sulfur	16	⊗	⊗	⊘	⊘	239	445
		chlorine	17	⊗	⊗	⊗	⊘	300	−34
		argon	18	⊗	⊗	⊗	⊗	363	−186

to form, the valence electrons in Li_2 must do double duty. They must spend some time in the empty valence orbitals of lithium atoms from other Li_2 molecules. Simple logic suggests that *chances for this electron sharing would be best if the electrons between lithium atoms were rather weakly held.* Sharing would be achieved most easily if the atoms had a low ionization energy, which they do. Under these conditions, the electron pair between nuclei could wander toward other nuclei, all of which have empty orbitals.

Experiments suggest that this is indeed what happens. Everywhere an electron moves, it finds itself between two positive nuclei with orbitals available to form a bond. Orbital geometry adjusts to permit relatively high electron density between any two atoms. (Remember that in CH_4 and C_2H_4, the orientation of orbitals was determined by the arrangement of nuclei and electrons surrounding a given atom.) The space around a central atom is a region of almost uniformly low potential energy. Each valence electron is virtually free to make its way throughout the crystal.

This valence orbital argument leads us to picture a metal as an array of positive ions located at crystal lattice sites and immersed in a "sea" of mobile electrons. A metallic structure of this type will be formed by atoms having many vacant valence orbitals and low ionization energy for valence electrons. These are the very features that identify metals.

PROPERTIES OF METALS AND THE METALLIC BOND

The discussion of electron location emphasizes an important difference between metallic and covalent bonding. In covalent bonds, the electrons are concentrated in certain regions of space between two atoms. The bonds are localized. In contrast, the valence electrons in a metal are spread almost uniformly throughout the crystal. The electrons are delocalized. The carbon atoms in diamond are held together by strongly localized covalent bonds. Diamond is very hard and rigid. Metals, on the other hand, are malleable and ductile because of the mobility of the electrons—metals can be worked. Under stress, one plane of atoms may slip over another. But, as it does so, electrons can move easily between planes to maintain bonding. See Figure 13-11.

If the easy movement of electrons is blocked, metals become hard and brittle. Metals can be hardened by alloying them with elements that have a strong attraction for electrons or that lack empty bonding orbitals. Thus, the process of alloying helps reduce the easy mobility of metallic electrons. Often, just a trace of carbon, phosphorus, or sulfur will turn a relatively soft

Figure 13-11

(a) Planes of metal atoms can slip over each other but (b) bonding is still maintained as electrons move easily between planes.

(a)
(b)

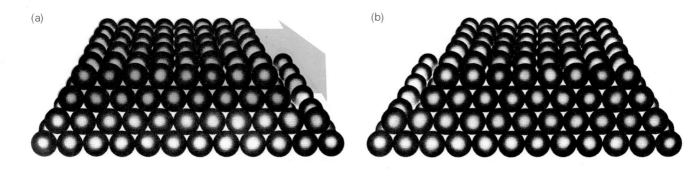

and workable metal into a brittle solid. This effect is similar to that of a dam, which is able to control water movement over a whole stream, even though it occupies a very small fraction of the volume of the stream bed. In a sense, carbon, phosphorus, and sulfur atoms can be regarded as dams in the mobile electron sea.

The excellent heat conductivity of metals is also due to the mobility of their electrons. Metal valence electrons in regions of high temperature can acquire large amounts of kinetic energy. These energized electrons move through the metal very rapidly and transfer their energy to the ions in the cooler regions. In contrast, when electron movement is restricted between atoms, heat energy can be passed stepwise from atom to atom by vibration. This is a much slower process than that occurring in pure metals.

Why do metallic characteristics gradually disappear as we look to the right across the periodic table? One reason is that, as the nuclear charge of an atom increases, so does the ionization energy of its valence electrons. Table 13-1 shows this trend for third-row elements. Moving to the right across the row, it becomes more and more difficult for the increasingly nonmetallic atoms to release electrons that can move freely throughout the crystal to form delocalized bonds.

A second reason for the gradual disappearance of metallic characteristics is that more orbitals are filled as we look toward the right-hand side of the table. The number of empty orbitals, needed to form delocalized bonds, decreases. Specific, localized electron-pair bonds (covalent bonds) between two nuclei become the more likely bonding pattern as we move across the table from left to right. Consequently, the possibility for delocalization and metallic-bond formation decreases.

EXERCISE 13-4

Lithium is far more malleable than aluminum. Explain why this is true.

STRENGTH OF THE METALLIC BOND

Is a metallic bond weak because electrons are mobile? We can get some idea of the effectiveness of this electron sea in binding atoms by comparing the energy necessary to vaporize 1 mole of a metallic element with the energy required to vaporize 1 mole of a covalently bonded solid element such as silicon or germanium. Free atoms are present in the vapor in each case.

The heat of vaporization for covalently bonded silicon is 85 kilocalories per mole. As Table 13-2 shows, alkali metals such as sodium and potassium require only about one fourth to one third as much energy for vaporization as does silicon. The four valence electrons of each silicon atom form strong localized bonds between atoms. With sodium or potassium, only one electron is available per atom, and it must move between many nuclei. Thus, the binding energy between electrons and nuclei in the alkali metal crystals is rather small, and the resulting metallic bonds are rather weak.

Table 13-2

Heats of vaporization of metals

Row of periodic table	Heats of vaporization (kcal/mole)		
second	Li 32.2	Be 53.5	B 129
third	Na 23.1	Mg 31.5	Al 67.9
fourth	K 18.9	Ca 36.6	Sc 73
fifth	Rb 18.1	Sr 33.6	Y 94
sixth	Cs 16.3	Ba 35.7	La 96

For metals with a greater nuclear charge and more valence electrons than the alkalis, the bonds should be stronger. In such metals, there are more electrons in the sea and a stronger attraction to the nucleus because of increased nuclear charge. Our expectation of stronger bonds in such cases is supported by the experimental heats of vaporization shown in Table 13-2. For example, let us compare the heats of vaporization for magnesium and aluminum. Not only are both values higher than the heats of vaporization for alkali metals, but the value for aluminum is higher than that for magnesium. The higher value for aluminum shows that the metallic bond is indeed stronger when both the number of valence electrons and the charge on the nucleus are increased. Thus, the strength of the metallic bond tends to increase from left to right across a row of the periodic table. The transition metals are harder and melt and boil at higher temperatures than the alkali or alkaline earth metals. This is explained by the fact that transition elements have *d* electrons and *d* orbitals that may be used in forming metal-metal bonds.

GEOMETRY OF METALS

Metals, like molecules, have a characteristic geometry. For example, the metals lithium, sodium, potassium, rubidium, and cesium all have a structure that is described as body-centered cubic. A portion of this structure is shown in Figure 13-12. Notice that the structural unit is a simple cube with an atom at each corner and another atom inside the cube at its center. The name **body-centered cubic arrangement** describes what we see: an atom is centered in the body of each cube. Since the actual structure of a metal consists of thousands of cubes, each atom in the structure is the same. There is no difference between central and corner atoms. *Each atom, regardless of position in the alkali metal structure, has eight neighbors.* We can see this by focusing on the atom in the center of the cube in Figure 13-12. The body-centered cubic arrangement is typical of the alkali metals, as well as of one of our most important metals: iron. Pure iron has a body-centered cubic structure.

The structure of the metals beryllium and magnesium can be obtained by packing balls of equal size as close together as possible in a continuous pattern. One of the resulting structures, the **hexagonal close-packed arrangement,** is shown in Figure 13-13 on page 344. In this pattern, each atom is surrounded by 12 identical metal atoms. Six atoms are in a plane around any given atom, with three atoms above the plane and three below. A close-packed arrangement is what we would expect for metals, since atoms of identical size are packed together. In contrast to the localized electron-pair bonds of diamond, the electrons in a metal can adjust to any pattern the atoms take. Compared with atoms in a body-centered arrangement, atoms in a close-packed arrangement have more valence electrons and a larger number of neighbors with which electrons can be shared. The number of atoms per unit volume is slightly greater in the hexagonal close-packed arrangement than in the body-centered cubic structure. With atoms of identical mass, the hexagonal close-packed arrangement would have a higher density than the body-centered cubic form.

Aluminum displays another close-packed geometry very similar to that described for beryllium and magnesium. The structure found in aluminum

Figure 13-12

In a body-centered cubic cell, the atoms actually touch each other. The structure is "opened up" here to make it easier to see.

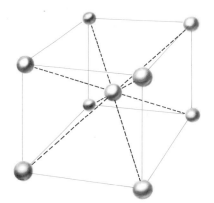

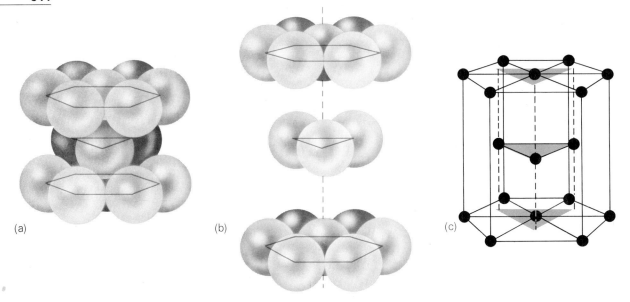

Figure 13-13

Hexagonal close-packed structure: normal and exploded views characteristic of beryllium and magnesium.

is known as the **cubic close-packed arrangement,** shown in Figure 13-14. This structure is also known as a face-centered cubic structure because the repeating cell is a cube with an extra atom in the center of each face. As in the hexagonal close-packed structure, each atom has 12 identical neighbors. Structural differences between hexagonal and cubic close-packed patterns occur only in the arrangement of atoms in the second shell of neighbor atoms. The number of atoms per unit volume is the same for both cubic and hexagonal close-packed structures.

Figure 13-14

Cubic close-packed structure characteristic of aluminum.

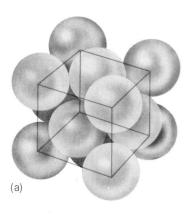

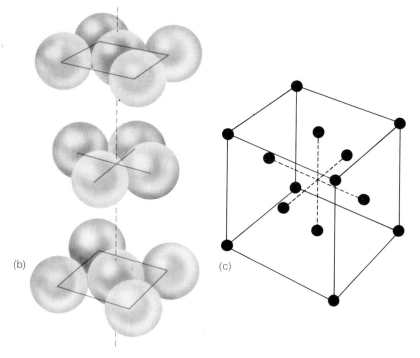

Except barium, all the metals of groups II (Be to Ra) and III (Al to Ac) have one of the close-packed arrangements at room temperature. To determine which structure a given metal takes requires an analysis of a number of factors, but nearly all pure metals adopt one of the close-packed patterns or the body-centered cubic form.

13.4 Ionic structures: The ionic bond

Evidence for the existence of ions was introduced in Chapter 7. Compounds such as common table salt (NaCl) are pictured as clusters of positive and negative ions, as shown in Figures 7-8b (page 169) and 9-3 (page 221). A model of a sodium chloride crystal is shown in Figure 13-15 (page 346). Such ionic solids are usually neatly stacked so that each positive ion has only negative-ion neighbors and each negative ion has only positive-ion neighbors. In Chapter 9, we read that any alkali metal (M) will react with any halogen (X_2) to form a solid product with the general formula MX. All MX-type compounds, of which NaCl is the most familiar, have strikingly similar physical properties. Next, we will examine some of those properties, explore how the ionic model proposed in Chapter 7 can explain them, and try to explain why such compounds form.

PROPERTIES AND STABILITY OF IONIC CRYSTALS

Ionic solids such as sodium chloride (NaCl) and lithium fluoride (LiF) form regularly shaped crystals with well-defined crystal faces. Pure samples of these solids are usually transparent and colorless, but color may be caused by traces of impurities or crystal defects. Most ionic crystals have high melting points, and their liquids have high boiling points. Sodium chloride melts at 801 °C and boils at 1413 °C. Lithium fluoride melts at 842 °C and boils at 1676 °C.

As indicated in Chapters 7 and 9, all alkali halides are soluble in water. They dissolve to form solutions that conduct electricity. Molten sodium chloride and lithium fluoride also conduct electricity, but their electrical conductivities are much lower than those of metals. The conductivity of copper metal at room temperature is 100,000 times greater than that of molten sodium chloride at 801 °C. This suggests that an electric charge moves through molten sodium chloride by a different mechanism than that by which it moves through solid copper. Experiments similar to those described in Chapter 7 indicate that the charge is carried in molten NaCl by slow-moving Na^+ and Cl^- ions. (See Section 7.4.) Some electrical conductivity in the liquid state is a property characteristic of molten ionic substances. In contrast, molecular crystals generally melt to form molecular liquids that do not conduct electricity. Metals have higher conductivity because electrons move more rapidly and easily than Na^+ and Cl^- ions.

To form ions, metal atoms lose electrons and nonmetal atoms gain electrons to give species that frequently have noble gas electron configurations. For example, in LiF, each Li^+ ion has the helium core of 2 electrons, while each F^- ion has the neon core of 10 electrons. In NaCl, the Na^+ ion also has the neon arrangement of 10 electrons. The Cl^- ion has the argon arrangement of 18 electrons. Al^{+++} has the neon core of 10 electrons.

ISABELLA L. KARLE

In 1944, Isabella Karle received her Ph.D. from the University of Michigan at the age of 22. Her research, under Dr. Lawrence Brockway, involved electron diffraction on the structure of molecules.

Dr. Karle developed the symbolic addition method of determining molecular structures directly from X-ray diffraction experiments. This method has made possible the structure determination of complex molecules that previously could not be tested experimentally. Before she developed this method, crystal structure analysis was difficult, time-consuming, and available for only a few kinds of material. Over half the investigations using X-ray structure analysis now employ the analytical techniques developed by Dr. Karle.

Dr. Karle has continued to make significant contributions in crystallography by her analyses of important materials in both organic chemistry and biochemistry. Currently, she is head of the X-ray analysis section of the Naval Research Laboratory in Washington, D.C., where she began working almost 35 years ago. She was elected to the National Academy of Sciences in 1978. She is married to Dr. Jerome Karle, recipient of the 1985 Nobel Prize in Chemistry. See page 349.

Figure 13-15

In the model of a portion of a sodium chloride crystal, the dark spheres are
Na⁺ ions. This so-called rock salt structure is typical of more than 140
known compounds, of which a large number are ionic salts. The photograph
is of natural rock salt structure, called halite.

Why do metals such as lithium and sodium lose electrons, while non-metals such as fluorine and chlorine gain them? Experimentally measured ionization energies, discussed in Chapter 10, reveal relatively low first ionization energies for lithium and sodium. This indicates that the valence electrons of these atoms are loosely held. The relatively high ionization energy values for fluorine and chlorine and their high electron affinity values indicate that their atoms should strongly attract additional electrons. In solid LiF, the Li^+ and F^- ions hold their electrons tightly. Therefore, the characteristic electron mobility of metals is not present in ionic solids.

The stability of an ionic solid is due primarily to the electrostatic attractions between its oppositely charged ions. The energy of several oppositely charged ions is lower when the ions cluster together than when they remain apart. Energy is released when the ions come together and is therefore required to separate them.

In an actual reaction between sodium and chlorine, many positive and negative ions form. In the arrangement with lowest energy, each positive ion is surrounded by as many negative ions as possible, and vice versa. For example, if gaseous LiF molecules were to form and approach each other, the electrostatic attraction between positive Li^+ and negative F^- ions in the molecule would generate pairs of lithium fluoride ions. These can be pictured as Li^+F^- and F^-Li^+. Repetition of the process could easily form larger clusters. In these clusters, each Li^+ ion would be surrounded by six F^- ions and each F^- ion by six Li^+ ions. (See the NaCl structure in Figure 13-15.)

SOLUBILITY OF IONIC SOLIDS IN WATER

We have looked at covalent molecules and ionic crystal lattices. Most molecules fall between these two extremes. They are held together by covalent bonds that still have enough charge separation to influence the proper-

Figure 13-16

Hydration of ions, showing orientation of water dipoles.

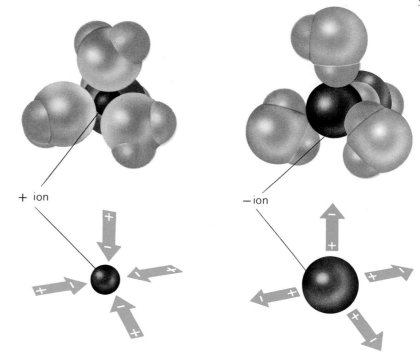

ties of the molecular substance. In Section 12.3, we referred to such molecules as *polar molecules.*

Chloroform ($CHCl_3$) is an example of a polar molecule. It has four tetrahedrally oriented bonds, as shown in Figure 12-8. There are three C—Cl bonds and one C—H bond. The C—Cl bonds do not have the same dipole as does the C—H bond. When the dipoles of the C—Cl and C—H bonds are added, there is a molecular dipole remaining. Molecular dipoles were discussed on pages 321–323. Electric dipoles are important to chemists because dipoles affect chemical properties such as solvent action.

Why does a strongly polar solvent such as water dissolve an ionic solid such as NaCl? Crystalline NaCl is quite stable, as indicated by its high melting point, yet the lattice is broken up when NaCl dissolves in water. To break up the stable crystal arrangement requires strong interaction between water molecules and the ions that make up NaCl. This interaction can be explained in terms of attraction between the ions and the dipolar water molecules. Remember that unlike charges attract so the negative ends of water molecules are pulled in toward the positive sodium ions and the positive ends of water molecules are pulled in toward the negative chloride ions. Energy is released. See Figure 13-16. The process is called **hydration.** This hydration process provides the energy to break up the NaCl lattice. The energy released in hydration is very close to the ionic lattice energy that holds NaCl crystals together. The hydrated ions in solution are separated and insulated from each other by the envelopes of water surrounding each ion. For this reason they can move freely in the solution.

Another factor furthers solubility. There is a marked increase in the randomness of the system when the regular NaCl lattice is broken up. The Na^+ and Cl^- ions are distributed randomly throughout the liquid phase. The entropy of the system increases (see pages 441–443). Because of ion hydration, there is little tendency to re-form the crystal.

These two effects of ion hydration—compensation for the lattice energy of the crystalline solid and stabilization of the ions in a more random

348

arrangement—give water distinctive properties as a solvent for ionic solids and for other electrolytes such as hydrogen chloride (HCl) and sulfuric acid (H_2SO_4). The tendency toward maximum randomness is the dominant factor promoting solubility of all solutes in liquid solvents.

13.5   Hydrogen bonding

Figures 13-2 and 13-3 reveal how, in similarly shaped molecules, boiling points and melting points of related substances increase regularly as the number of electrons increases. We concluded that, for similarly shaped molecules, more electrons mean more van der Waals forces and, thus, higher boiling and melting points.

Figure 13-17 plots the boiling points of the *hydrides* of four neighboring chemical families in the periodic table. Note that the boiling points of the carbon, silicon, germanium, and tin hydrides increase with the number of electrons. Each of these elements represents a different horizontal row of the periodic table. In Figure 13-17 these rows are represented as 2, 3, 4, and 5. Apparently, van der Waals forces hold these molecules together in their liquid states: the pattern is the normal and expected one.

Now, consider the boiling points of hydrogen iodide (HI), hydrogen bromide (HBr), hydrogen chloride (HCl), and hydrogen fluoride (HF). Hydrogen fluoride is far out of line, boiling at 19.9 °C instead of below −95 °C, as we would predict from an extrapolation of the other three boiling points. There is an even larger discrepancy between the boiling point of H_2O and

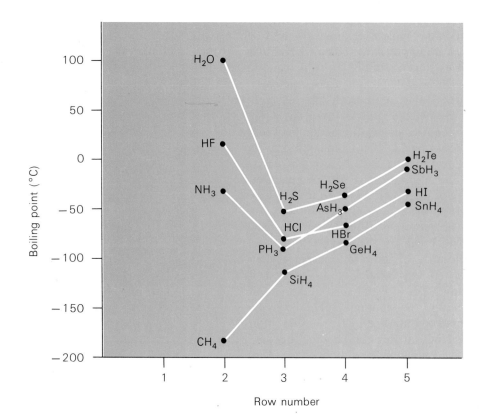

Figure 13-17

Trends in boiling temperatures of some hydrides.

the value we would predict from the trend suggested by the boiling points of hydrogen telluride (H_2Te), hydrogen selenide (H_2Se), and hydrogen sulfide (H_2S). Both HF and H_2O have a relatively low number of electrons per molecule: ten. Yet, each has an unexpectedly high boiling point. How can we account for these abnormally high values? Clearly, additional new forces—more powerful than van der Waals attractions—must be holding HF and H_2O molecules together in their liquid states.

The new forces are also evident with solid compounds. The most familiar example is solid H_2O, or ice. Ice has a structure in which the oxygen and hydrogen atoms are distributed in a regular hexagonal crystal lattice. Each O atom, like each C atom in diamond, is surrounded by four other O atoms in a tetrahedral arrangement. The H atoms are found on the lines extending between the oxygen atoms:

$$\diagdown O-H\cdots O\diagup$$

The force of attraction between O—H and O must be the bond that joins the water molecules together in the crystal lattice of ice. This bond is a **hydrogen bond**.

LOCATION AND ENERGY OF HYDROGEN BONDS

Hydrogen bonds are found between only a few elements of the periodic table. Studies of boiling points show that such bonds are present mainly between molecules in which hydrogen is connected to fluorine (F), oxygen (O), or nitrogen (N) atoms. Fluorine, oxygen, and nitrogen have strong attraction for electrons. The shared electrons in their covalent bonds to hydrogen move over to make fluorine, oxygen, or nitrogen atoms somewhat negative in charge. The hydrogen atom then has a small positive charge.

The hydrogen bond results when the positively charged hydrogen atom interacts with a free electron pair on the negatively charged fluorine, oxygen, or nitrogen atom. The hydrogen bond is usually represented by the notation O—H····O, in which the solid line represents the original O—H bond in the parent compound, as in water, and the dotted line represents the second bond formed by hydrogen. The line is dotted to indicate that the bond is much weaker than a normal covalent bond. Consideration of the boiling points plotted in Figure 13-17, on the other hand, shows that the interaction must be much stronger than that involving van der Waals forces. The energy of the hydrogen bond places it between van der Waals attractions and covalent bonds in terms of strength. Roughly speaking, the energies are in the ratio

van der Waals attractions		hydrogen bonds		covalent bonds
1	:	10	:	100

Since hydrogen bonds are much stronger than van der Waals forces, their presence greatly influences the properties of a substance. Figure 13-18 illustrates how hydrogen bonding can produce clusters of strongly bound molecules of hydrogen fluoride. The hydrogen bond to fluorine is clearly evident

JEROME KARLE

Jerome Karle earned graduate degrees at Harvard and the University of Michigan. In 1943 he received his Ph.D. in physical chemistry from the University of Michigan.

Shortly after graduation, Dr. Karle joined the Naval Research Laboratory in Washington, D.C. He spent a short time working for the Manhattan Project to develop the first atomic bomb. He is now Chief Scientist of the Laboratory for the Structure of Matter at the Naval Research Laboratory.

Dr. Karle was awarded the 1985 Nobel Prize in Chemistry, along with Dr. Herbert A. Hauptman, for developing key mathematical techniques through which X-ray crystallography can be used directly to determine the three-dimensional structure of natural substances vital to the internal chemistry of the human body and of antibiotic drugs. Many thousands of three-dimensional structures have been determined in recent years using the techniques developed by these two scientists.

For many years their work was considered controversial and unaccepted by scientists until Dr. Karle's wife, Isabella, studied the mathematics of the research and pointed out to others its potential applications in the understanding of crystals and the development of drugs. See her biography on page 345.

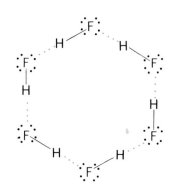

Figure 13-18

A descriptive formula of the H_6F_6 species.

in most of the properties of hydrogen fluoride. The high boiling point of HF, compared with the boiling points of the other hydrogen halides, is one of several pieces of data indicating that HF does not exist in the liquid phase as separate HF molecules but as *aggregates of molecules*, which we describe in general terms as $(HF)_x$. Hydrogen fluoride gas contains the "floppy ring" species H_6F_6 shown in Figure 13-18. Gaseous HF also contains some single HF molecules and perhaps some molecules of an intermediate type, such as H_2F_2.

SIGNIFICANCE OF HYDROGEN BONDS

Hydrogen bonds are extremely important. They play a central role in biological systems and are of major importance in defining many of the properties of water. They are of great importance in defining the structure of a variety of solids ranging from copper sulfate pentahydrate to DNA, the major carrier of genetic information in living systems. See Chapter 20. The point is easily illustrated by considering the properties of water. Much of the chemistry of life takes place in water solutions. When water in living systems freezes, plants and animals die. Hydrogen bonds determine the freezing and boiling points of water, they determine its role as a solvent, and even its chemistry. For example, sugars are organic molecules that dissolve in water only because of hydrogen bond interactions. In addition, many nitrogen compounds that are essential to the building of proteins dissolve only through hydrogen bonding. Interactions of these types of bonds are indicated below:

HYDROGEN BOND

WATER   PART OF SUGAR

HYDROGEN BOND

PART OF AMINO ACID   WATER

We will find the importance of hydrogen bonds growing as we learn more about all aspects of science.

SUMMARY

Gases condense because of forces of attraction between their molecules. These forces, known as van der Waals forces, are the result of the temporary attraction between electrons of one molecule and the nuclei of other, adjacent molecules. Van der Waals forces increase as the number of electrons in an atom increases and as the area of contact between two molecules in the system increases.

Elements such as carbon (diamond) and silicon, located in the middle section of the periodic table, form network solids in which the entire structure is one huge molecule. Three-dimensional network solids have very high melting points and low electrical conductivity.

The graphite form of carbon and the silicate mineral talc form two-dimensional networks with strong covalent bonds within a layer, but with only weak van der Waals forces between layers. Each carbon atom in a graphite layer forms three bonds. Its additional valence electron can move freely within the layer, giving graphite its high electrical conductivity along the planes.

Metallic elements are found on the left-hand side of the periodic table. Metallic properties increase as we look down each group on the table. All metals are lustrous, are excellent conductors of electricity and heat, and are ductile and malleable. The low ionization energy, few valence electrons, and many vacant valence orbitals common to all metals are responsible for their well-known properties. Metals are pictured as clusters of positive ions arranged in one of several possible regular patterns called lattices and held together by a "sea" of valence electrons. The strength of the metallic bond increases as nuclear charge increases and as more valence electrons become involved in bonding.

Ionic compounds form when metals located at the extreme left-hand side of the periodic table react with nonmetals located at the far right-hand side. Most alkali halides form cubic crystals in which each alkali ion is surrounded by six halide ions and each halide ion is surrounded by six alkali ions. The oppositely charged ions are held together by electrostatic attraction. Ions in most alkali halides frequently have stable, noble gas electron configurations.

Hydrogen bonds can form between some molecules in which hydrogen atoms are bonded to fluorine, oxygen, or nitrogen atoms. They are stronger than van der Waals forces, but weaker than covalent bonds. Hydrogen bonds are responsible for the abnormally high boiling point of water, and for many important events in biological chemistry.

Key Terms

body-centered cubic arrangement
cubic close-packed arrangement
delocalized electron
hexagonal close-packed arrangement
hydration
hydrogen bond
three-dimensional network solid
two-dimensional network solid
van der Waals forces

QUESTIONS AND PROBLEMS

A

1. (a) When we want to convert a gaseous substance to a liquid, we cool the gaseous substance and reduce its volume. Why? (*Hint:* consider the forces involved.) (b) Substances that obey the ideal gas law over a very wide temperature range, such as H_2, He, and so on, are usually very hard to liquify. Why?
2. Differentiate between bonds that hold carbon and hydrogen together in methane, CH_4, and the "bonds" that hold methane molecules together in liquid methane (b.p. = -161 °C). Consider names, relative bond strengths, and electronic nature of linkages.
3. (a) Which should have the higher boiling point, krypton or neon? Offer a reasonable explanation based on electronic models. (b) Consider the following facts: (1) Liquid propane (C_3H_8), which is used for torches, stoves, lanterns, and so on, is sold in steel tanks and when the valve is opened the propane vaporizes; (2) Gasoline (C_8H_{10}) can be handled in open cans, but a strong odor is detectable above the can and a spark gives a disastrous explosion and fire; (3) Lubricating oil ($C_{20}H_{42}$) can be handled in open cans, there is no odor above the oil, and a spark above the surface is harmless; (4) All three substances can be ignited by a direct flame. Explain in terms of the forces acting between molecules.

4. A given hydrocarbon has the composition C_6H_{14}. There are several isomers with this composition. The first isomer (L) boils at 49 °C; the second isomer (H) boils at 69 °C. Suggest structural formulas for (L) and (H) that would be consistent with these facts.

5. Diamond, a three-dimensional network solid, is a covalently bound substance that is the hardest material known. Diamond powder is a very effective abrasive. Graphite, another form of carbon, is used as a lubricant in locks, and so on. Explain these differences in the properties of graphite and diamond.

6. Arrange the following compounds in order of increasing melting point: $SnCl_4$, $SnBr_4$, and SnI_4. Explain why each melting point of these compounds is different.

7. Why is graphite a good lubricant for locks and talcum powder a good lubricant for the skin? Could you reverse these roles?

8. (a) How could you determine experimentally whether a substance is a metal or a nonmetal? You should give at least four characteristics of metals as opposed to nonmetals. (b) Explain these key properties of metals in terms of electronic structure. For example, why is sodium a solid and chlorine a gas?

9. Graphite is a good electrical conductor *along the planes*. In what way does the electronic structure of graphite resemble the electronic structure of a metal and in what way does it differ?

10. (a) What characteristics of an atom, such as ionization energy, make it adopt the metallic structure? (b) Which element would have the higher heat of vaporization, K or Sc? Explain in terms of the electronic model.

11. (a) Draw a body-centered cubic arrangement. (b) How many nearest neighbors does each atom have in a body-centered cubic arrangement? (c) What is a face-centered cubic arrangement? (d) How many nearest neighbors does each atom have in a face-centered cubic arrangement? (The face-centered cubic arrangement is also called the cubic close-packed arrangement.) (e) Why are these descriptions important in a study of metals?

12. In order to dissolve an ionic solid in a liquid, the energy released when the lattice was formed must be provided. What process provides this energy when NaCl dissolves in water?

13. Water boils at 100 °C and H_2S boils at −60 °C. Would we expect this trend for boiling points as we go down this vertical family in the periodic table if van der Waals forces were the controlling factor? Explain.

14. Would you expect the boiling point of PH_3 to be above or below the boiling point of NH_3? Explain.

15. Hydrogen bonds are extremely important in biology. Give one illustration of this point.

B

16. Differentiate between and give examples of: (a) a molecular solid; (b) a three-dimensional network solid; (c) a two-dimensional network, or layered, solid; (d) a metal; and (e) an ionic solid.

17. Describe several physical and/or chemical properties for each of your examples in problem 16.

18. The electrical conductivity of solid silver is much better than that of molten AgCl. In turn, the electrical conductivity of molten AgCl is much

better than that of solid AgCl. Explain these facts using appropriate models.

19. Explain why a sharp blow with a hammer will deform a metal such as aluminum but will shatter an ionic crystal such as NaCl.

20. Which type of bonding (van der Waals, covalent network solid, metallic, or ionic) is involved in each of the following substances: (a) He(l), (b) Fe(s), (c) diamond(s), (d) $CH_4(g)$, (e) KCl(s), (f) Ag(s), (g) Si(s), (h) $SiO_2(s)$, (i) $I_2(g)$, (j) $C_2H_2(l)$, (k) $SO_2(l)$, (l) SiC(s), (m) MgO(s)?

21. Describe the process by which ions of a solid such as NaCl go into solution when the crystal is placed in water.

22. H_2SO_4 dissolves in water to liberate a large amount of heat. What are the forces that further this process?

23. Explain why NaCl dissolves readily in water but very poorly in gasoline (C_8H_{18}) or benzene (C_6H_6).

24. The compound CH_4 boils at $-161\ °C$ while H_3CF boils at $-68\ °C$. Offer a reasonable explanation for these two facts.

C

25. Unlike carbon, silicon does not form double bonds except under very special circumstances. Using this information, can you justify the fact that CO_2 is a gas but SiO_2 (quartz) is a hard solid with a high melting point?

26. With reference to question 25, would a research project to prepare a graphite-like form of silicon have a very good possibility of success? Explain.

27. You are given a sample of a white solid. Describe some experiments you can perform to help you decide whether the solid is held together by ionic bonds, covalent bonds, van der Waals forces, or metallic bonds.

28. A silvery, shiny substance melts at $-39\ °C$ and boils at $357\ °C$. It is an excellent conductor of electricity. It is insoluble in water and has a vapor pressure of 1.7×10^{-3} mm at $25\ °C$. Do these data suggest that the substance is (a) a three-dimensional network substance, (b) a two-dimensional network substance, (c) an ionic substance, (d) a hydrogen-bonded liquid, or (e) a metal? Can you identify the material?

29. Consider the successive elements sodium, magnesium, and aluminum. Considering sodium as the standard, copy the following table and fill in the blank spaces with *higher, lower, more,* or *less,* as appropriate.

	Sodium	Magnesium	Aluminum
ionization energy	XXX		
heat of vaporization	XXX		
metallic character	XXX		
hardness	XXX		

30. Each of three bottles on a chemical stockroom shelf contains a colorless liquid. The labels have fallen off the bottles but are still on the shelf. They read as follows:

1. *n*-butanol, mol wt = 74.12
   structure:

$$H-\overset{\overset{\displaystyle H}{|}}{\underset{\underset{\displaystyle H}{|}}{C}}-\overset{\overset{\displaystyle H}{|}}{\underset{\underset{\displaystyle H}{|}}{C}}-\overset{\overset{\displaystyle H}{|}}{\underset{\underset{\displaystyle H}{|}}{C}}-\overset{\overset{\displaystyle H}{|}}{\underset{\underset{\displaystyle H}{|}}{C}}-O$$

2. *n*-pentane, mol wt = 72.15
   structure:

$$H-\overset{\overset{\displaystyle H}{|}}{\underset{\underset{\displaystyle H}{|}}{C}}-\overset{\overset{\displaystyle H}{|}}{\underset{\underset{\displaystyle H}{|}}{C}}-\overset{\overset{\displaystyle H}{|}}{\underset{\underset{\displaystyle H}{|}}{C}}-\overset{\overset{\displaystyle H}{|}}{\underset{\underset{\displaystyle H}{|}}{C}}-\overset{\overset{\displaystyle H}{|}}{\underset{\underset{\displaystyle H}{|}}{C}}-H$$

3. diethyl ether, mol wt = 74.12
   structure:

$$H-\overset{\overset{\displaystyle H}{|}}{\underset{\underset{\displaystyle H}{|}}{C}}-\overset{\overset{\displaystyle H}{|}}{\underset{\underset{\displaystyle H}{|}}{C}}-O-\overset{\overset{\displaystyle H}{|}}{\underset{\underset{\displaystyle H}{|}}{C}}-\overset{\overset{\displaystyle H}{|}}{\underset{\underset{\displaystyle H}{|}}{C}}-H$$

The three bottles are marked A, B, and C. In an attempt to identify the liquid in each bottle, a series of measurements was made on each. The results are recorded below:

Liquid	Melting point (°C)	Boiling point (°C)	Density (gm/cm³)	Solubility (gm/100 ml H₂0)
A	−130	36	0.63	0.036
B	−116	34.6	0.71	7.5
C	−89.2	117.7	0.81	7.9

Based on these data, match each liquid with the proper label. Explain how each type of measurement influenced your choice.

31. Which of the following compounds are likely to exhibit hydrogen bonding in their liquid and solid phases:

(a) CH_4, (b) CH_3OH, (c) NH_3, (d) $H_3CC\overset{\displaystyle \nearrow O}{-}OH$, (e) C_8H_{18}, (f) $H_3C-O-CH_3$?

Silicon chips

The dramatic increase in the number of practical uses for transistors in recent years is due to a series of rapid developments in the production of the semiconductors from which they are made. A semiconductor is a material that falls somewhere between conductors and insulators in its ability to conduct an electric current. Semiconductors tend to conduct electricity better than insulators like glass, but not as well as conductors like copper. The elements silicon and germanium are the two most widely used semiconductor materials.

Silicon and germanium are quite similar in their structure and chemical behavior. They are members of the same chemical family, and their atoms have *four* valence electrons. Silicon and germanium atoms bond together in a crystal lattice similar to that of diamond. Each atom is bonded to four neighboring atoms through shared electrons, as shown in the pure silicon crystal structure at top. In its pure state, silicon is not a conductor, because the four valence electrons of each atom are employed in the formation of the crystal. Semiconductors are made by adding small amounts of an impurity to extremely pure samples of silicon or germanium. The presence of such impurities permits the partial flow of electric current because the impurities act as either electron donors or acceptors.

Arsenic and antimony are typical donor impurities. These elements have *five* valence electrons. When very small quantities of arsenic are added to silicon, for example, each arsenic atom joins the crystal lattice and *donates* one electron to the crystal structure. Since four out of the five electrons are paired, the fifth is relatively free to wander through the crystal lattice as do the free electrons of a metallic conductor. The detached electron leaves behind a donor arsenic atom with a positive charge bound into the crystal lattice. Silicon with this type of crystal structure is called an *n*-type, or negative-charge, conductor. An *n*-type semiconductor, as illustrated at right, consists of silicon and the arsenic impurity, which adds a few extra electrons.

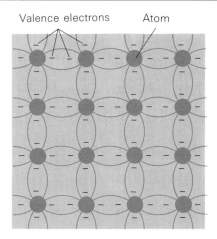

Pure silicon crystal structure with four valence electrons.

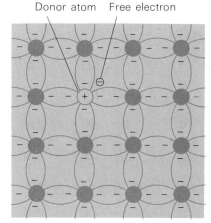

Silicon semiconductor with an *n*-type crystal structure containing donor atoms and free electrons.

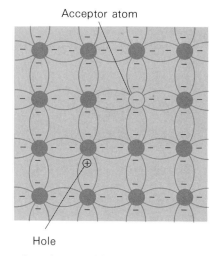

Silicon semiconductor with a *p*-type crystal structure containing acceptor atoms and free (positive) holes.

The size of a computer chip compared to a pin and a needle. A chip may consist of over a million transistors.

Aluminum, boron, and gallium are typical acceptor impurities. These elements have *three* valence electrons. When very small quantities of aluminum are added to silicon, for example, each aluminum atom joins the crystal lattice and *accepts* one electron from a neighboring silicon atom. This leaves an empty space, or *hole,* in the bond where the electron was removed. This hole represents a positive charge because it acts as a trap into which an electron can fall. As an electron fills a hole, it leaves another hole into which yet another electron can fall. Thus, the hole can pass from one atom to another. The hole leaves behind an acceptor aluminum atom with a negative charge bound into the crystal lattice. Silicon with this type of crystal structure is called a *p*-type, or positive-charge, conductor. This type of semiconductor, as illustrated above, consists of silicon plus the aluminum impurity, which adds a few extra holes to the crystal.

Electricity flows as a movement of electrons in *n*-type structures and as a movement of holes in *p*-type structures. The proper combination of *n*-type and *p*-type semiconductors has permitted the production of many very useful electronic devices that make possible most of the electronic equipment we now take for granted. Incredibly small, yet effective, devices can be made for changing an alternating current to a direct current, for amplifying electrical signals, and for storing information.

A *transistor,* or electronic switch, consists of layers of *p*-type and *n*-type semiconductors. We can think of the *p*-type semiconductor as an electrically positive zone with a deficiency of electrons due to the presence of impurities. The *n*-type semiconductor is an adjacent area that receives the surplus of electrons and is the electrically negative zone. Two *n* zones separated by a *p* zone can function as a transistor. A small voltage in the *p* zone can control the flow of current between the *n* zones and thus act as a switch.

Silicon has replaced germanium as the semiconductive material most commonly used in making transistors, calculators, electric watches, and other electronic devices. Silicon is used instead of germanium for two main reasons. First, the melting point of silicon is 1420 °C; that of germanium is 937 °C. The higher melting point of silicon results in more efficient diffusion of acceptor or donor atoms throughout the crystal structure. Second, the oxide of silicon is insoluble in water. This oxide, which forms on the outside layer, prevents the vital impurities from diffusing out of the crystal and the oxide is an electrical insulator. On the other hand, the oxide of germanium is water soluble and does not function well as a diffusion mask.

To make a transistor, manufacturers "grow" ultra-pure silicon crystals and cut them into thin slices, or wafers. The first step in growing silicon crystals is the reaction of silica (SiO_2) with carbon:

$$SiO_2(s) + C(s) \xrightarrow{\Delta} CO_2(s) + Si(s)$$

The SiO_2 is kept in excess to prevent the formation of SiC. The silicon produced by this process is further purified. It is then treated with hydrogen chloride to produce trichlorosilane ($SiHCl_3$), which is subsequently decomposed in a hydrogen atmosphere to form very pure silicon. The pure silicon is used to make the rod-shaped crystals from which ultra-thin, ultra-pure wafers can be sliced.

The crystals are made by heating pure polycrystalline silicon to 1400 °C in a quartz crucible in the presence of a protective argon atomosphere. A seed crystal of pure silicon is immersed in the melt and rotated in one direction while the crucible is rotated in the other direction. The seed crystal provides the foundation upon which the large crystal is built. It is slowly withdrawn from the melt as the crucible rotates. Rotating the seed while slowly withdrawing it permits the crystal to grow. The crystals produced in this manner most commonly have diameters of from 75 to 100 millimetres. About 28 wafers can be sliced from about 1 inch of crystal. The wafers are polished with fine abrasives in a liquid solution. They must be perfectly smooth on all surfaces. Even the edges are rounded to minimize chipping.

The wafers are then heated and exposed to impurities to form *n*-type and *p*-type layers. The next processing step is the formation of the protective oxide coating. This is done by heating a batch of wafers up to 1200 °C in an oxygen atmosphere. The SiO_2 covering acts as the mask that prevents diffusion of impurities through the surface of the wafer. Several masking layers are needed to adequately protect the wafer. Therefore, several oxidations are usually required.

Etching of the necessary integrated circuit patterns, or "chips," onto the surface of the wafer is then done, according to a predetermined diagram. The circular wafer shown in the figure on this page contains about 230 individual chips. Such etching can be done by a wet chemical process that uses hydrofluoric acid. However, it is more commonly done by an ion-implantation technique. This technique depends on the acceleration of ions of the material to be implanted. These ions are then "fired" at the wafers. The etching pattern on the surface of the oxide permits the ions to enter the crystal lattice at selective sites. The higher the energy of the ion beam, the greater the depth of penetration. Selective penetration of impurities can also be accomplished by varying the thickness of the oxide masking layers. An 8,500 Å* layer stops the penetration of boron ions; a 2,000 Å layer halts the penetration of arsenic ions. When completed,

*One angstrom (Å) is equal to 1×10^{-8}cm, which is one hundred-millionth of a centimetre. The angstrom is widely used to describe atomic and molecular sizes.

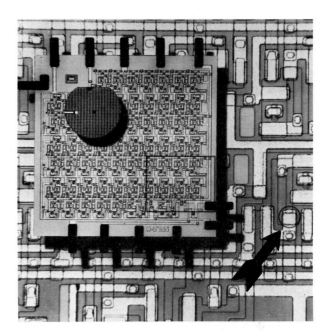

The circular wafer in this composite photograph contains about 230 integrated circuit patterns called chips. The white square on the wafer is a single chip. Magnified 60 times, this chip is also shown as the square behind the wafer. Each chip contains 452 transistor switches that regulate semiconductor charges. The area on the magnified chip enclosed by the drawn rectangle is further magnified 300 times and appears as the background of the entire photo. This enlargement shows the arrangement of the circuit components. The circle drawn on the enlarged background shows a single transistor switch. An arrow points to this circle.

the chips are thoroughly tested to ensure that they are uniform and perfect.

The practical possibilities for use of silicon chips seem limitless. Such chips have already been used in pocket calculators, electric watches, and miniature radios. Miniaturization of such large objects as television cameras is already a reality. We can look forward to dramatic improvements in all forms of electronics because of continued developments in the field of semiconductors and integrated circuits.

Although a typical chemical reaction . . . may appear far removed from the working of an engine, the same fundamental principles of heat and work apply to both.

FREDERICK THEODORE WALL

14

ENERGY CHANGES IN CHEMICAL AND NUCLEAR REACTIONS

Objectives

After completing this chapter, you should be able to:

- Understand the concept of heat content and its changes.
- Understand how reaction heats are measured in a calorimeter.
- Understand and use Hess' Law to calculate reaction heats.
- Construct enthalpy change diagrams and relate them to the heat absorbed or released during a chemical reaction.

Chemical reactions produce energy changes in which metals such as magnesium, and nonmetals such as sulfur combine very rapidly with oxidizing agents such as potassium nitrate or perchlorate to produce this fireworks display. Colors arise from the presence of salts such as chlorides of strontium, copper, and so on.

Why does the fuel you burn to cook your food or warm your house give off heat? Why does a burning candle give off heat? Where does the heat come from? How does heat get into the candle or the fuel in the first place?

Every chemical reaction is associated with energy changes. Often, the energy changes are the most important feature of a chemical process. This is certainly true of burning home-heating fuel. We do not burn fuels to make CO_2 and H_2O. We burn it for heat. This is also the case for the reaction occurring in the cylinders of an automobile engine. A mixture of gasoline and air is ignited in the engine, releasing energy. The heat energy is directly responsible for moving the car. Similarly, a piece of chocolate is able to provide your body with energy as it is "burned up" with oxygen in the cells of your body.

14.1 Heat and chemical reactions

In the laboratory, we studied the heat—or, more precisely, the thermal energy—released when a given mass of candle burned. The amount of heat was measured in calories. One *calorie* is the quantity of heat required to raise the temperature of 1 gram of liquid water 1 °C. In measuring large amounts of heat, we frequently use the kilocalorie. One **kilocalorie** equals 1,000 calories. The total amount of heat released when a candle burns depends on the amount of candle burned up. The amount of heat energy associated with any process depends on the quantities of chemical substances involved. Therefore, we specify both the amount of heat and the amounts of all chemical substances involved. We express these quantities as a certain number of calories (or kilocalories) per gram (or per mole) of chemical substance.

A few years ago the world faced an energy crisis. Today we have enough energy for our current needs but tomorrow is uncertain. We do know, however, that we are using oil and gas much, much more rapidly than they are being formed. Oil and gas supplies can not last forever. A possible solution is a mixture of hydrogen and carbon monoxide gases, known as water gas, made from coal and water. Water gas has been used widely as fuel. Let us look at the manufacture and use of this former, and possibly future, fuel.

ENERGY CHANGES AND WATER GAS

To produce water gas, steam at a temperature of 600 °C is passed over hot coke. Coke is almost pure carbon. It is made by heating coal in the absence of air. Equation *1a* describes the process that occurs:

$$H_2O(g) + C(s) \xrightarrow{600\ °C} CO(g) + H_2(g) \tag{1a}$$

Heat is absorbed during this process. In the commercial preparation of water gas, the equipment operator must allow for such absorption. The operator does this by periodically turning off the steam supply and burning the coke in air in order to again get it red hot. Careful measurement reveals that 31.4 kilocalories of heat are absorbed per mole of carbon converted to CO. Since heat must be supplied to the reactants, we can *rewrite* equation *1a*, placing the heat term on the *left* side of the equation, along with the chemical reactants:

$$31.4\ \text{kcal} + H_2O(g) + C(s) \longrightarrow CO(g) + H_2(g) \tag{1b}$$

The water gas can now be burned instead of natural gas to give heat energy. It is important to know how much heat can be obtained from burning water gas. Two chemical reactions are involved, one describing the burning of CO and the other describing the burning of H_2:

$$CO(g) + \tfrac{1}{2}O_2(g) \longrightarrow CO_2(g) \tag{2a}$$

$$H_2(g) + \tfrac{1}{2}O_2(g) \longrightarrow H_2O(g) \tag{3a}$$

Both reactions release heat. Careful measurements show that 67.6 kilocalories of heat are released per mole of $CO(g)$ burned (as shown in *2a*) and 57.8 kilocalories are released per mole of $H_2(g)$ burned (according to *3a*). Since heat is produced in the above reactions, it can be placed on the *right-hand* side of the equations, along with the chemical products. Rewriting equations *2a* and *3a*, we get

$$CO(g) + \tfrac{1}{2}O_2(g) \longrightarrow CO_2(g) + 67.6 \text{ kcal} \qquad (2b)$$

$$H_2(g) + \tfrac{1}{2}O_2(g) \longrightarrow H_2O(g) + 57.8 \text{ kcal} \qquad (3b)$$

Suppose the chemical plant that produces the water gas also uses some of it to provide heat to the plant. The plant's business manager, interested in economy, thinks in terms of gains and losses. In looking at equations *1a, 2a,* and *3a,* the manager observes first that carbon and steam are used up to produce water gas. Then, the water gas is burned, producing carbon dioxide and steam. Without knowing much chemistry, the manager can see that what is finally accomplished is the burning of carbon to form carbon dioxide. This observation is confirmed by adding equations *1a, 2a,* and *3a:*

$$H_2O(g) + C(s) \longrightarrow CO(g) + H_2(g)$$

$$CO(g) + \tfrac{1}{2}O_2(g) \longrightarrow CO_2(g)$$

$$\underline{H_2(g) + \tfrac{1}{2}O_2(g) \longrightarrow H_2O(g)}$$

$$C(s) + O_2(g) \longrightarrow CO_2(g) \qquad (4a)$$

Hoping to save money, the business manager asks: "Why not burn the carbon directly and save the cost of producing the water gas?" To answer this question, the chemical engineer goes to the laboratory and measures the heat released per mole of carbon when the carbon is burned directly (as in equation *4a*). The engineer's measurements show that reaction *4a* releases 94.0 kilocalories of heat per mole of carbon burned. Rewriting *4a,* we get

$$C(s) + O_2(g) \longrightarrow CO_2(g) + 94.0 \text{ kcal} \qquad (4b)$$

The engineer can now point out that burning 1 mole of carbon directly releases 94.0 kilocalories of heat. The same mole of carbon, if first converted to water gas, eventually releases the *sum* of the heats evolved in reactions *2b* and *3b* when the water gas is finally burned. The sum of 67.6 kilocalories and 57.8 kilocalories is 125.4 kilocalories. The chemical engineer thus concludes that water gas is a better energy source for many reasons than is carbon in the form of coal.

With new respect for the process, the business manager asks: "Where did this extra heat come from? Did we get something for nothing?" The answer is no. The water gas releases more heat per mole of carbon because the operator added that amount of heat during reaction *1b*. Somehow, the 31.4 kilocalories of heat absorbed during that reaction were stored in the water gas. The final heat energy "ledger" is shown in Table 14-1 (page 362).

The quantitative value 31.4 kilocalories in equation *1b* indicates that a definite amount of energy was indeed "stored" in the water gas. The amount of energy stored depends on the reactants and products involved. A fixed amount of heat energy must be added to coal and steam to make a specified amount of carbon monoxide and hydrogen. We can picture this heat as being part of the CO and H_2 molecules. Those molecules "contain" heat energy, just as they contain carbon, oxygen, and hydrogen atoms. We might say that reaction *1b* increases the "heat content" of the molecules by rearranging the atoms to form the products. Thus, a mole of each molecular substance can be thought of as having a characteristic heat content, just as it has a characteristic mass. Heat content is, by definition, the measure of the energy stored in a substance during its formation.

Table 14-1

Some energy changes in the manufacture and use of water gas

Reaction	Debit	Credit
Energy absorbed: $H_2O(g) + C(s) + 31.4 \text{ kcal} \longrightarrow CO(g) + H_2(g)$	31.4 kcal	
Energy released: $CO(g) + \frac{1}{2}O_2(g) \longrightarrow CO_2(g) + 67.6 \text{ kcal}$		67.6 kcal
Energy released: $H_2(g) + \frac{1}{2}O_2(g) \longrightarrow H_2O(g) + 57.8 \text{ kcal}$		57.8 kcal
Overall reaction (total of above): $C(s) + O_2(g) \longrightarrow CO_2(g)$	31.4 kcal absorbed	125.4 kcal released
Net:		125.4 −31.4 94.0
Experimental value for $C(s) + O_2(g) \longrightarrow CO_2(g)$		94.0 kcal released

The heat we can measure in a chemical reaction is the heat absorbed or released in that reaction. When heat is *absorbed*, as in reaction *1b*, the temperature of the surroundings decreases because the surroundings provide the heat absorbed by the reactants. That is why the chemical engineer had to *reheat* the coke continually in order to keep the water gas reaction going. When heat is *released*, as in reactions *2b* and *3b*, the temperature of the surroundings increases. It is the temperature changes of the surroundings that enable us to calculate the heat effect during a chemical reaction.

HEAT CONTENT OF A SUBSTANCE

The energy changes observed in the production of water gas are easy to understand if we assume that every substance has a characteristic heat content. When one substance is converted to another, heat must be either lost or gained, depending on the process. Our study of water gas reactions provides several convenient examples. Equation *1b* shows that 1 mole of steam reacts with 1 mole of carbon by absorbing 31.4 kilocalories of heat energy. The products (which make up water gas) are 1 mole of carbon monoxide and 1 mole of hydrogen. Those products contain 31.4 kilocalories of heat more than the reactants: coke and steam.

Chemists symbolize heat content, also known as **enthalpy**, by the letter H. The *difference* in heat content—the difference between the H's of the products and the H's of the reactants—is called the **heat of reaction.** It is often written ΔH, the Greek letter delta signifying "difference" or "change." Using symbols,

$$\Delta H = H \text{ products } - H \text{ reactants}$$

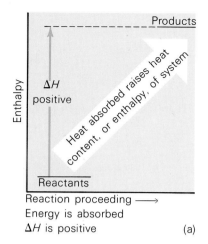
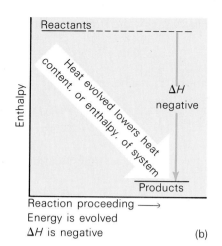

Energy is absorbed
ΔH is positive   (a)

Energy is evolved
ΔH is negative   (b)

Figure 14-1

Enthalpy changes during reactions.

For reaction *1b*, ΔH is +31.4 kilocalories. It is positive (+) in value because the heat content has been raised. More heat is stored in the products than was stored in the reactants, as shown in the enthalpy change diagram in Figure 14-1a. Such reactions are said to be **endothermic.** Endothermic reactions always have *positive* ΔH values.

We can express reaction *1a,* along with its ΔH value, by writing

$$H_2O(g) + C(s) \longrightarrow CO(g) + H_2(g)$$
$$\Delta H = +31.4 \text{ kcal as written} \qquad (1c)$$

This reaction is exactly equivalent to reaction *1b*:

$$31.4 \text{ kcal} + H_2O(g) + C(s) \longrightarrow CO(g) + H_2(g)$$

If the energy stored in the reactants of a given reaction is greater than that stored in the products, heat will be released during that reaction. The heat of reaction, ΔH, will be negative because the heat content has been lowered. Such reactions, which evolve, or release, heat energy, are known as **exothermic** chemical reactions. According to the conventions described above, exothermic reactions must have *negative* ΔH values.

Consider reaction *2,* in which the 1 mole of carbon monoxide gas produced in reaction *1* is burned to form 1 mole of carbon dioxide. In this reaction, 67.6 kilocalories of heat are released, or evolved. Thus, less heat is stored in the product, $CO_2(g)$, than was stored in the reactants, C(s) and $O_2(g)$. Because 67.6 kilocalories were released, that amount of heat is no longer stored as heat content. Thus, 67.6 kilocalories less heat are stored in the product. Therefore, ΔH is −67.6 kilocalories. The negative sign tells us that the change in heat content, or enthalpy, is negative, since ΔH has gone from a higher to a lower value, as shown in the enthalpy change diagram in Figure 14-1b. Equation *2b* can thus be rewritten as

$$CO(g) + \tfrac{1}{2}O_2(g) \longrightarrow CO_2(g)$$
$$\Delta H = -67.6 \text{ kcal as written} \qquad (2c)$$

This has exactly the same meaning as

$$CO(g) + \tfrac{1}{2}O_2(g) \longrightarrow CO_2(g) + 67.6 \text{ kcal}$$

Equation *3b* can be rewritten as

$$H_2(g) + \tfrac{1}{2}O_2(g) \longrightarrow H_2O(g)$$
$$\Delta H = -57.8 \text{ kcal}$$

(3c)

This has exactly the same meaning as

$$H_2(g) + \tfrac{1}{2}O_2(g) \longrightarrow H_2O(g) + 57.8 \text{ kcal}$$

To summarize, the heat effect in a chemical reaction is a measure of the *difference* between the heat content of the products and the heat content of the reactants as written. The value of ΔH is positive when the heat content, or enthalpy, of the system is rising (heat absorption). It is negative when the enthalpy is dropping (heat release).

When expressing heats of reaction, it is important to specify the physical state—solid, liquid, or gas—of each reactant and each product. For example, ΔH for the process

$$H_2(g) + \tfrac{1}{2}O_2(g) \longrightarrow H_2O(g)$$

is -57.8 kilocalories. The ΔH for the process

$$H_2(g) + \tfrac{1}{2}O_2(g) \longrightarrow H_2O(l)$$

is -68.3 kilocalories. The values differ because heat is required to change liquid water to gaseous water.* Thus, less heat is given off if steam is the product.

MEASURING HEATS OF REACTION

Measuring reaction heats is called **calorimetry,** a name related to the unit of heat, the calorie. You already have some experience in calorimetry. In Experiments 9 and 12 in the Laboratory Manual, you measured the heat of combustion of a candle and the heat of solidification of wax. In Experiment 19, you measured the heat evolved when NaOH reacted with HCl. The device you used was a simple *calorimeter.*

Calorimeters vary in the details of their construction and are adapted to the particular reaction being studied. Figure 14-2 shows the general plan of a calorimeter that might be used in measuring the heat evolved during a combustion reaction. It would give more reliable answers than the cruder tin-can and polyethylene calorimeters you were asked to use in your experiments.

The calorimeter shown in Figure 14-2 consists of an insulated vessel containing a known mass of water. Known masses of the substances to be burned and an excess of oxygen are introduced under pressure into a reaction chamber placed in the vessel. The reaction mixture is ignited by means of an electrical resistance wire sealed into the reaction chamber. The heat produced by the reaction changes the temperature of the water, which is stirred to keep its temperature the same throughout. From this temperature

---

*The heat of vaporization of $H_2O(l)$ at 25 °C (298 K) is 10.5 kcal/mole. At 100 °C (373 K) the heat of vaporization of $H_2O(g)$ is 9.7 kcal/mole.

Figure 14-2

Schematic diagram of a calorimeter.

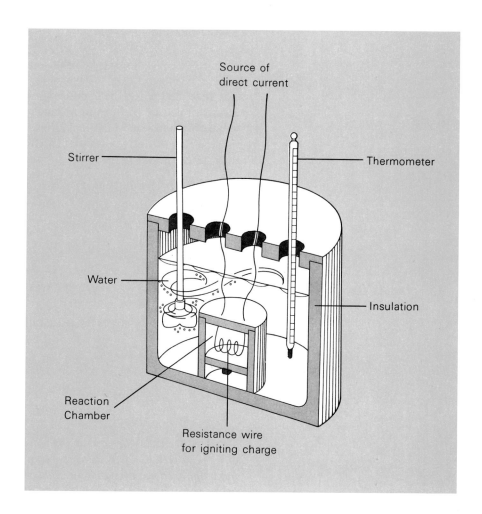

change, and from the amount of heat required to raise the temperature of the calorimeter and its contents by 1°C, the heat of combustion per mole of substance burned can be calculated.

ADDITIVITY OF HEATS OF REACTION

Let us return to the study of the economics of our water gas fuel problem, as summarized in Table 14-1. The business manager's original suggestion—to burn carbon directly to obtain carbon dioxide—is understandable when reactions *1a, 2a,* and *3a* are added algebraically (as on page 361). The substances $H_2O(g)$, $CO(g)$, and $H_2(g)$ appear on both the left and right sides of those equations and thus are canceled out. They are then subtracted from both sides of the final, overall equation, *4a.*

If we rewrite and add the equations, adding heat terms to each, we obtain the following:

$$31.4 \text{ kcal} + H_2O(g) + C(s) \longrightarrow CO(g) + H_2(g) \qquad (1b)$$

$$CO(g) + \tfrac{1}{2}O_2(g) \longrightarrow CO_2(g) + 67.6 \text{ kcal} \qquad (2b)$$

$$H_2(g) + \tfrac{1}{2}O_2(g) \longrightarrow H_2O(g) + 57.8 \text{ kcal} \qquad (3b)$$

$$\overline{31.4 \text{ kcal} + C(s) + \tfrac{2}{2}O_2(g) \longrightarrow CO_2 + 125.4 \text{ kcal}}$$

This reduces to

$$C(s) + O_2(g) \longrightarrow CO_2(g) + 94.0 \text{ kcal}$$

You read that when the chemical engineer went back to the lab to measure the heat released by burning 1 mole of carbon directly to carbon dioxide, the value obtained was 94.0 kilocalories. That is precisely the value we have just obtained by adding equations *1b*, *2b*, and *3b*: 94.0 kilocalories. Earlier, we rewrote equation *4a* as

$$C(s) + O_2(g) \longrightarrow CO_2(g) + 94.0 \text{ kcal} \qquad (4b)$$

If we use ΔH notation, equation *4b* can be rewritten, with exactly the same meaning, as

$$C(s) + O_2(g) \longrightarrow CO_2(g)$$
$$\Delta H = -94.0 \text{ kcal} \qquad (4c)$$

By adding the ΔH values for reactions *1c, 2c,* and *3c*, we obtain the ΔH value for reaction *4c*:

$$\begin{aligned} \Delta H_4 &= \Delta H_1 + \Delta H_2 + \Delta H_3 \\ &= +31.4 + (-67.6) + (-57.8) \text{ kcal} \\ &= -94.0 \text{ kcal} \end{aligned}$$

Thus, we see that *when a reaction can be expressed as the algebraic sum of two or more reactions, its heat of reaction is the algebraic sum of the heats of the individual reactions that were added together.* This generalization holds for every set of reactions ever tested. Because the generalization has been so widely tested, it is called a law—the **Law of Additivity of Heats of Reaction.** It was first proposed in 1840 by G. H. Hess on the basis of his experimental measurements of reaction heats. It is also known as **Hess' Law of Constant Heat Summation.**

PREDICTING HEATS OF REACTION

The heats of many reactions have been measured. About a dozen such reactions are listed in Table 14-2. Notice that each of the 14 reactions listed in the table represents the formation of 1 mole of product. Furthermore, each product is a compound, formed from the elements it contains. Such heats of reaction are called **heats of formation** and are symbolized as ΔH_f. *By convention, the ΔH_f of all elements in their standard states at 25°C and 1 atmosphere is zero.* See Appendix 2 on pages 641 to 643.

Using measured values, many unmeasured heats of reaction can be predicted by applying the Law of Additivity of Heats of Reaction. The chemical engineer in the previous example could have used this law (as we did on pages 363–366) to predict the heat of reaction *4c* without ever measuring it. Of course, this could be done only if the engineer knew the heats of reactions *1c, 2c,* and *3c.*

Let us try a few other examples. Suppose we are interested in the heat of combustion of nitric oxide, $NO(g)$. The appropriate equation is

$$NO(g) + \tfrac{1}{2}O_2(g) \longrightarrow NO_2(g)$$
$$\Delta H_5 = ? \tag{5}$$

Note that, while equation 5 is not listed in Table 14-2, both compounds—$NO(g)$ and $NO_2(g)$—are represented there. If we can combine the equations in Table 14-2 that contain $NO(g)$ and $NO_2(g)$ to obtain equation 5, we can use Hess' law to find the heat of reaction of that equation.

The relevant equations from Table 14-2 are

$$\tfrac{1}{2}N_2(g) + \tfrac{1}{2}O_2(g) \longrightarrow NO(g)$$
$$\Delta H_6 = +21.6 \text{ kcal/mole NO} \tag{6}$$

and

$$\tfrac{1}{2}N_2(g) + O_2(g) \longrightarrow NO_2(g)$$
$$\Delta H_7 = +8.1 \text{ kcal/mole NO}_2 \tag{7}$$

We note that $NO(g)$ is a reactant in reaction 5, but a product in reaction 6. Obviously, the addition of equations 6 and 7 will not give

$$NO(g) + \tfrac{1}{2}O_2(g) \longrightarrow NO_2(g)$$

Since NO is a reactant in this process, we need to use the reverse of the reaction producing NO. We obtain the heat of the reverse reaction by changing the algebraic sign of ΔH_6. Thus, if 21.6 kilocalories of heat are *absorbed* when 1 mole of $NO(g)$ is formed, then 21.6 kilocalories of heat will be *released* when 1 mole of $NO(g)$ is decomposed in the reverse reaction:

$$NO(g) \longrightarrow \tfrac{1}{2}N_2(g) + \tfrac{1}{2}O_2(g)$$
$$\Delta H_8 = -21.6 \text{ kcal/mole NO} \tag{8}$$

Now we can add reactions 7 and 8 to obtain reaction 5, which is the equation for oxidation of NO to NO_2:

$$\tfrac{1}{2}N_2(g) + O_2(g) \longrightarrow NO_2(g)$$
$$NO(g) \longrightarrow \tfrac{1}{2}N_2(g) + \tfrac{1}{2}O_2(g)$$
$$\overline{NO(g) + \tfrac{1}{2}O_2(g) \longrightarrow NO_2(g)}$$

$$\Delta H_7 = +8.1 \text{ kcal/mole NO}_2$$
$$\underline{\Delta H_8 = -21.6 \text{ kcal/mole NO}}$$

$$\Delta H_5 = -13.5 \text{ kcal/mole NO}$$

The usefulness of Table 14-2 is shown by the above examples. Many reaction heats can be predicted from the table, provided that every compound in a reaction is included in the table in the phase indicated. Consider

Table 14-2

Heats of reaction between elements ($T = 25$ °C, $P = 1$ atm)

Elements	Compound (product)		Heat of reaction (kcal/mole of product) ΔH
	Formula	Name	
$H_2(g) + \frac{1}{2}O_2(g) \longrightarrow$	$H_2O(g)$	water vapor	-57.8
$H_2(g) + \frac{1}{2}O_2(g) \longrightarrow$	$H_2O(l)$	water	-68.3
$S(s) + O_2(g) \longrightarrow$	$SO_2(g)$	sulfur dioxide	-71.0
$H_2(g) + S(s) + 2O_2(g) \longrightarrow$	$H_2SO_4(l)$	sulfuric acid	-194.0
$\frac{1}{2}N_2(g) + \frac{1}{2}O_2(g) \longrightarrow$	$NO(g)$	nitric oxide	$+21.6$
$\frac{1}{2}N_2(g) + O_2(g) \longrightarrow$	$NO_2(g)$	nitrogen dioxide	$+8.1$
$\frac{1}{2}N_2(g) + \frac{3}{2}H_2(g) \longrightarrow$	$NH_3(g)$	ammonia	-11.0
$C(s) + \frac{1}{2}O_2(g) \longrightarrow$	$CO(g)$	carbon monoxide	-26.4
$C(s) + O_2(g) \longrightarrow$	$CO_2(g)$	carbon dioxide	-94.0
$C(s) + 2H_2(g) \longrightarrow$	$CH_4(g)$	methane	-17.8
$2C(s) + 3H_2(g) \longrightarrow$	$C_2H_6(g)$	ethane	-20.2
$3C(s) + 4H_2(g) \longrightarrow$	$C_3H_8(g)$	propane	-24.8
$\frac{1}{2}H_2(g) + \frac{1}{2}I_2(g) \longrightarrow$	$HI(g)$	hydrogen iodide	$+6.2$
$S(s) + \frac{3}{2}O_2(g) \longrightarrow$	$SO_3(g)$	sulfur trioxide	-94.4

a more complicated example—the oxidation of ammonia (NH_3)—as represented by the equation

$$NH_3(g) + \tfrac{7}{4}O_2(g) \longrightarrow NO_2(g) + \tfrac{3}{2}H_2O(g)$$
$$\Delta H_9 = ? \tag{9}$$

In reaction 9, we find three compounds: $NH_3(g)$, $NO_2(g)$, and $H_2O(g)$. They are all found in Table 14-2. Thus, we can calculate ΔH_9 for the oxidation of ammonia. Note that the value of ΔH shown in the table for $H_2O(g)$, 57.8 kilocalories, corresponds to the formation of 1 mole of $H_2O(g)$. Since equation 9 shows the formation of $\frac{3}{2}$ moles of $H_2O(g)$, the coefficient of all the terms in the equation shown in the table, as well as the value of ΔH, must be multiplied by $\frac{3}{2}$.

Thus, when we wish to predict the heat of some reaction, we refer to Table 14-2. Of course, the list in the table includes only a small fraction of the known values. Many more heats of reaction are tabulated in Appendix 2 as "Heats of Formation." The reported heats are ΔH values for the formation of 1 mole of the *compound* from its elements, taken at 25 °C and 1 atmosphere. (As already indicated, the ΔH value for the formation of *all elements* at these conditions is taken to be zero.) This convention allows us to use a shortcut when predicting heats of reaction. The result of adding equations to obtain a net equation is identical to the result of adding the heats of formation of all product compounds and subtracting the sum of the heats of formation of all reactant compounds. In the case of reaction 9, we can write

$$\Delta H_9 = \{\Delta H_{NO_{2(g)}} + \tfrac{3}{2}\Delta H_{H_2O_{(g)}}\} - \{\Delta H_{NH_{3(g)}} + \tfrac{7}{4}\Delta H_{O_{2(g)}}\}$$
$$\Delta H_9 = [8.1 + \tfrac{3}{2}(-57.8)] - [-11.0 + 0] = -67.6 \text{ kcal}$$

As before, the -57.8 in the second term is for 1 mole of $H_2O(g)$. Since equation 9 shows the formation of $\tfrac{3}{2}$ mole of $H_2O(g)$, the -57.8 must be multiplied by $\tfrac{3}{2}$. The general statement is

$$\Delta H_{process} = \text{sum of } \Delta H_{\substack{\text{formation values} \\ \text{for products}}} - \text{sum of } \Delta H_{\substack{\text{formation values} \\ \text{for reactants}}}$$

EXERCISE 14-1

Predict the heat of reaction for

$$CO(g) + \tfrac{1}{2}O_2(g) \longrightarrow CO_2(g)$$

from two appropriate reactions listed in Table 14-2. Compare your results with ΔH_2 on page 366.

EXERCISE 14-2

Predict the heat of reaction for

$$CH_4(g) + 2O_2(g) \longrightarrow CO_2(g) + 2H_2O(g)$$

Which is the more energetic fuel: 1 mole of CO or 1 mole of CH_4?

EXERCISE 14-3

Show that the value of ΔH for the oxidation of 1 mole of ammonia (NH_3) is -67.6 kcal.

THE SEARCH FOR NEW FUELS

Ultimately, the sun is the source of all the energy stored in fuels other than those used for nuclear processes. Fossil fuels provide us with energy obtained from the sun thousands to millions of years ago. Because fossil fuels will ultimately run out, chemists are searching for ways to use the sunshine of *today* to generate hydrogen and carbon monoxide. These gases can then be stored and burned, using standard combustion processes, to provide energy on demand.

A process developed at Oak Ridge National Laboratory, in Tennessee, is an example of research efforts to produce H_2 from heat. The heat comes from focusing sunlight through lenses. The process involves heating a mixture of cerium oxide (CeO_2) and sodium hydrogen phosphate (Na_2HPO_4). The equation is

$$2CeO_2 + 6Na_2HPO_4 \xrightarrow[\text{sunlight}]{750\text{–}950\ °C}$$
$$2Na_3Ce(PO_4)_2 + 2Na_3PO_4 + \tfrac{1}{2}O_2 + 3H_2O$$

DOUBLE SALT

If the solid products of this reaction are heated with Na_2CO_3 and H_2O, hydrogen gas is obtained. The equation is

$$2Na_3Ce(PO_4)_2 + 2Na_3PO_4 + 3Na_2CO_3 + H_2O \xrightarrow[\text{sunlight}]{650\text{–}950\ °C}$$
$$2CeO_2 + 6Na_3PO_4 + 3CO_2 + H_2$$

Note that the net result of these two heating steps is the generation of 1 mole of H_2 and $\tfrac{1}{2}$ mole of O_2 from H_2O. To make the process cyclic, the original starting materials for the heating steps must be regenerated using solar en-

Figure 14-3

This apparatus is used at Oak Ridge National Laboratory to demonstrate the generation of hydrogen from water by using energy available from sunshine. Reactants such as CeO_2 and Na_2HPO_4 are placed inside a platinum (Pt) boat in a furnace that can be heated to 1000 °C. Steam is introduced into the furnace. The result of the complex series of reactions that follows is the production of hydrogen gas, a fuel, at the gas exit. In practice, this system would be heated by the sun, not by a furnace.

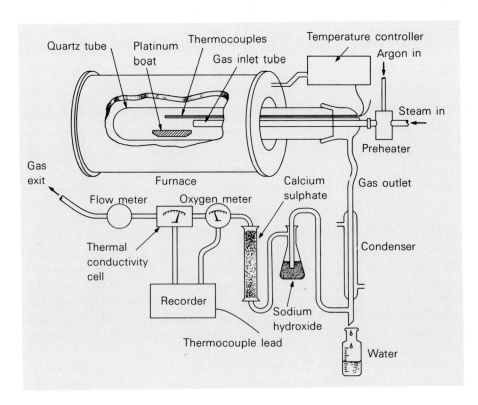

ergy. We see that, other than H_2 and O_2, the materials produced are CeO_2, Na_3PO_4, CO_2, and H_2O. The materials consumed are CeO_2, Na_2HPO_4, Na_2CO_3, and H_2O. It is essential to generate Na_2HPO_4 and Na_2CO_3 from the products Na_3PO_4 and CO_2. The process can be carried out by heating with CO_2 the solid $Na_3PO_4 \cdot CeO_2$ mixture obtained from the hydrogen generation step. The equation is

$$6Na_3PO_4 + 2CeO_2 + 6H_2O + 6CO_2 \xrightarrow{25\,°C}$$
$$\text{SOLID MIXTURE}$$

$$6Na_2HPO_4 + 2CeO_2 + 6NaHCO_3$$

To convert the $NaCHO_3$ obtained above to the needed Na_2CO_3, the $NaHCO_3$ is heated:

$$6NaHCO_3 \xrightarrow{200\,°C} 3Na_2CO_3 + 3H_2O + 3CO_2$$

To see the cyclic nature of the process, we must add all the equations together. We obtain

$$2CeO_2 + 6Na_2HPO_4 \xrightarrow{\Delta^*}$$
$$2Na_3Ce(PO_4)_2 + 2Na_3PO_4 + \tfrac{1}{2}O_2 + 3H_2O$$

$$2Na_3Ce(PO_4)_2 + 2Na_3PO_4 + 3Na_2CO_3 + H_2O \xrightarrow{\Delta}$$
$$2CeO_2 + 6Na_3PO_4 + 3CO_2 + H_2$$

$$6Na_3PO_4 + 2CeO_2 + 6H_2O + 6CO_2 \xrightarrow{\Delta}$$
$$6Na_2HPO_4 + 2CeO_2 + 6NaHCO_3$$

$$6NaHCO_3 \xrightarrow{\Delta} 3Na_2CO_3 + 3H_2O + 3CO_2$$

The net process is

$$H_2O \xrightarrow{\Delta} H_2 + \tfrac{1}{2}O_2$$

Thus, heat from sunlight could be used to generate a gaseous fuel. Although the equations add up, such processes are still difficult and inefficient. We have a long way to go in the search for new fuels.

14.2 Conservation of energy

The Law of Additivity of Reaction Heats is a very useful and reliable generalization. Why does it work? The explanation can be found by connecting the behavior of a chemical system to the behavior of other, more familiar systems. We did this when we used the example of a bookcase to develop the concept of potential energy. See Figure 11-7 on page 274. We also did it when we used the ping-pong ball model to develop the concept of rapidly moving gas molecules exerting pressure. We now find it useful to look at the behavior of rubber bands in seeking an explanation for the Law of Additivity of Reaction Heats.

*The Δ symbol in all equations means that heat is applied.

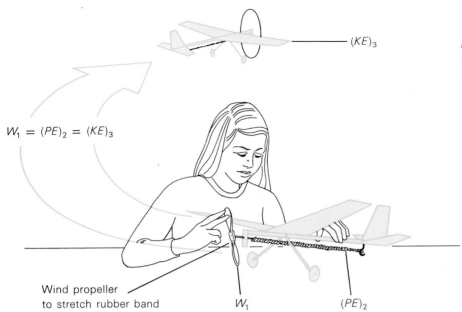

$(KE)_3$

$W_1 = (PE)_2 = (KE)_3$

Wind propeller
to stretch rubber band

W_1

$(PE)_2$

Figure 14-4

Conservation of energy in a stretched
rubber band.

CONSERVATION OF ENERGY IN A STRETCHED RUBBER BAND

Figure 14-4 shows what occurs when a rubber band is stretched. The rubber band is installed in the model airplane. It is stretched by turning the propeller, as shown in the figure. Work must be done to stretch the band. Let us call the amount of work W_1. When the propeller is released, it turns, the airplane moves, and the rubber band returns to its original, unstretched length. But now the plane has energy of motion, *kinetic energy*, represented as $(KE)_3$. The size of $(KE)_3$ depends on how much work, W_1, was done in stretching the rubber band. Indeed, we may write

$$W_1 = (KE)_3$$

The energy of the work done, W_1, equals the kinetic energy, $(KE)_3$, imparted to the model plane. But was energy conserved? If so, where was the energy once the work, W_1, had been done but before the propeller was released? When the rubber band was stretched, there was no outward sign of energy. The rubber band was motionless and had no kinetic energy. We can say that energy was conserved if we assume that energy was present in a different form in the stretched rubber band. That form of energy is known as *energy of position*, or **potential energy,** represented as $(PE)_2$. We say that as the work, W_1, is performed, potential energy is stored in the rubber band.

We assume that the energy of the work of stretching the rubber band is stored in the rubber band as potential energy. This assumption can be summarized as

$$W_1 = (PE)_2 = (KE)_3$$

While we did not measure $(PE)_2$, all the original energy, W_1, was recovered eventually as $(KE)_3$. It is reasonable to conclude that this is possible only because all the original energy was stored in the stretched rubber band.

This idealized analysis of a stretched rubber band neglects friction, heat losses through propeller inefficiencies, and so on. A more detailed analysis would have to include all such terms. The general conclusions remain the same.

CONSERVATION OF ENERGY IN CHEMICAL REACTIONS

Figure 14-5 shows an apparatus in which an electric current can be passed through water. The electric current causes decomposition of the water. As work (electrical work) is done, hydrogen gas and oxygen gas are produced. Measurements of the electric current and voltage show that 68.3 kilocalories of electrical work (W_1) must be done to decompose 1 mole of water. The equation for this reaction is

$$\underset{\substack{\text{ELECTRICAL} \\ \text{WORK}}}{68.3 \text{ kcal}} + H_2O(l) \longrightarrow H_2(g) + \tfrac{1}{2}O_2(g)$$

Now suppose we measure the heat of reaction of hydrogen and oxygen in a calorimeter such as that shown in Figure 14-2. This experiment has been performed many times. The result is that 68.3 kilocalories of heat energy, Q_4, are produced for every mole of water formed. The equation for this reaction is

$$H_2(g) + \tfrac{1}{2}O_2(g) \longrightarrow H_2O(l) + 68.3 \text{ kcal}$$
$$\Delta H = -68.3 \text{ kcal}$$

In certain respects, this situation reminds us of the rubber band. As before, we put a readily measured amount of energy, which we will once again label W_1, into the system. We can recover this energy at any time we choose. In this case, it is recovered as heat energy and labeled Q_4.

$$W_1 = Q_4$$

The electrical energy, W_1, is stored in the chemical substances $H_2(g)$ and $O_2(g)$. Hydrogen and oxygen have more potential energy than water, just as the stretched rubber band has more potential energy than the unstretched rubber band. As we shall see later, the energy stored in H_2 and O_2 is truly energy of position—energy due to the position of atoms in each substance. Energy is released when the atoms take up new positions to form water. The relationship between atom position and energy content of the system is illustrated in Figure 14-6. The overall result is that energy is *conserved*. The result is summarized in the figure. Note that hydrogen and oxygen atoms have different relative positions after each process.

The relationships depicted in Figure 14-6 can be represented schematically by the enthalpy change diagram, shown in Figure 14-7, which is similar

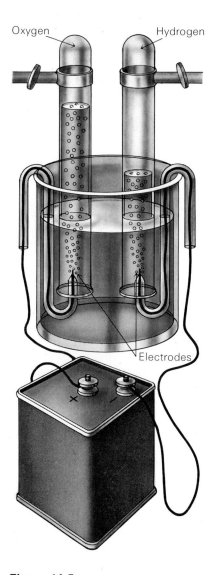

Figure 14-5

Apparatus for the electrolytic decomposition of water.

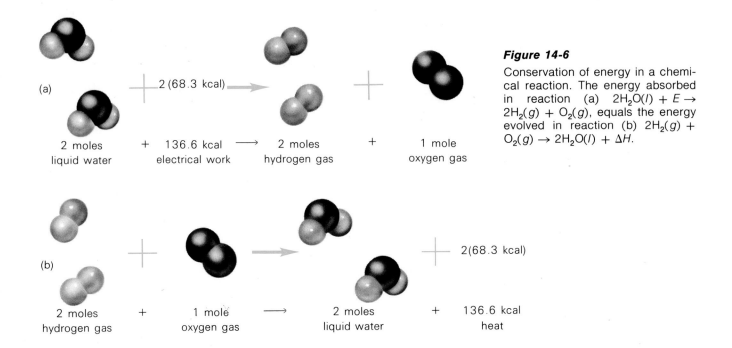

(a)

2 moles
liquid water
+ 136.6 kcal
electrical work
⟶ 2 moles
hydrogen gas
+ 1 mole
oxygen gas

(b)

2 moles
hydrogen gas
+ 1 mole
oxygen gas
⟶ 2 moles
liquid water
+ 136.6 kcal
heat

Figure 14-6

Conservation of energy in a chemical reaction. The energy absorbed in reaction (a) $2H_2O(l) + E \rightarrow 2H_2(g) + O_2(g)$, equals the energy evolved in reaction (b) $2H_2(g) + O_2(g) \rightarrow 2H_2O(l) + \Delta H$.

to that in Figure 14-1. If the enthalpy of 2 moles of water is represented by a line on the diagram, then the energy of 2 moles of hydrogen plus 1 mole of oxygen should be represented by a line 136.6 (or 2 × 68.3) kilocalories higher. The diagram shows that, when water is decomposed, energy must be supplied to raise the enthalpy of the system to that of 2 H_2 plus O_2. When hydrogen burns in oxygen, the enthalpy of the system drops to that of water. *Energy is released.*

Energy is conserved in chemical processes as well as in a system such as the rubber band. An important idea arises from this discussion. Each molecule has kinetic energy, due to its movement through space, as indicated by its temperature. In addition, each molecule has potential energy, due to the arrangement of atoms in the molecule. In short, each molecule has a certain capacity to store energy—it has a definite heat content, or enthalpy, at a given temperature. As the temperature is raised, more energy is stored.

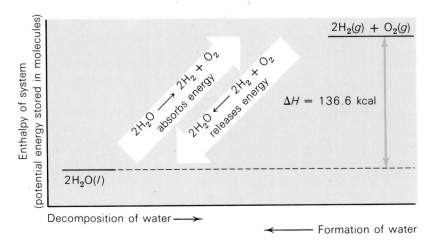

Figure 14-7

Enthalpy changes in the decomposition and formation of water.

THE LAW OF CONSERVATION OF ENERGY

Kinetic and potential energy were conserved in each of the systems we studied. Our specific observations can be generalized. For all systems, energy is always conserved. (As you will read in Section 14.4, this statement is also true of nuclear systems when the equivalence of mass and energy is recognized.) This generalization is known as the *Law of Conservation of Energy*. It is confirmed by thousands and thousands of experiments and can be used to make very accurate predictions. It requires only that we recognize kinetic and potential energy and keep a careful account of both.

The Law of Additivity of Heats of Reaction for chemical systems, discussed in Section 14.1, is simply a special case of the more general Law of Conservation of Energy. Predictions based on the Law of Additivity of Heats of Reaction have always agreed with experiment. Conservation of energy has *always* been in agreement with experiment whenever a careful energy balance has been obtained.

14.3 Molecular energy

We have found it useful to talk about various energy forms and changes associated with quantities of matter that can easily be handled in the laboratory. We find that all such macroscopic forms of energy can be discussed in

OPPORTUNITIES

Industrial hygiene

An industrial hygienist is responsible for establishing and maintaining safe working conditions for those involved in all phases of chemical production. The Federal Occupational Safety and Health Act of 1970 required the Department of Labor to set standards ensuring healthful conditions for employees working where toxic materials are used. So the number of employment opportunities in industrial hygiene have increased. Industries especially in need of hygienists are those involved with petrochemicals, mining, and heavy manufacturing. The chemical industry, however, is the chief employer of such workers.

Industrial hygienists may conduct health programs designed to eliminate and control occupational health hazards. They collect and analyze samples of dust, gases, water, and any other materials that may be potential sources of pollution in the working environment. When hazards are identified, industrial hygienists often must help design equipment that will minimize the hazards. They must train fellow employees to follow safety procedures and must also be able to answer questions about potential hazards in the work place.

People employed in the field of industrial hygiene must be able to communicate effectively. They work closely with toxicologists and physicians in order to keep well informed of possible detrimental effects of chemicals used in specific manufacturing processes. Attention to detail is essential for success in this field.

Industrial hygienists must be certified by the American Board of Industrial Hygiene. Certification is obtained after working in the field for a specified period of time and after passing two examinations.

A bachelor's degree in chemistry is needed for beginning positions in

this field and to prepare for the examinations. Graduate work is essential for advancement to more responsible positions. In addition to a good background in chemistry, a prospective industrial hygienist should take courses in biology and human physiology. For further information about this career, you may contact the American Industrial Hygiene Association, 66 South Miller Road, Akron, OH 44313.

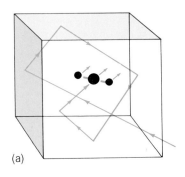

(a)

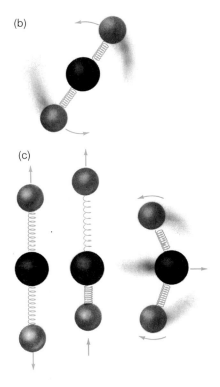

(b)

(c)

Figure 14-8

The carbon dioxide molecule exhibits three types of motion: (a) translational motion is the movement of the molecule from place to place, (b) rotational motion is the rotation of the molecule around its center axis, and (c) vibrational motion is the alternating movement of the atoms toward and away from the center of mass. The arrows show motion at a given instant.

terms of the two kinds of energy in our rubber band: kinetic and potential energy. We are able to explain all macroscopic forms of energy with a microscopic model involving only the energy of motion and the energy of position of atomic and molecular particles. Our knowledge about the properties of familiar objects such as stretched rubber bands can be clarified by considering the energy of an individual molecule.

MOLECULAR ENTHALPY

Let us picture a molecule in terms of a ball-and-spring model. The springs represent the bonds between atoms. We start the springs vibrating and then toss the entire assembly through space in an end-over-end motion. There are now three kinds of kinetic energy associated with our model, as pictured in Figure 14-8. They are called translational, rotational, and vibrational energies and can be described as follows. The *energy of translation*, illustrated in Figure 14-8a, is the motion of the entire molecular assembly through space. This is the translational kinetic energy of the gas laws. The *energy of rotation*, illustrated in Figure 14-8b, is the end-over-end motion of the molecule as it twirls through space. This is rotational kinetic energy. The *energy of vibration*, illustrated in Figure 14-8c, is the vibration of the balls on the springs. This is vibrational kinetic energy.

The model describes reasonably well a molecule in the gaseous state, but in the liquid and solid states all molecular motions are severely restricted. In the condensed phases, translational molecular motion is restricted almost completely to a back-and-forth motion of the balls around a given point. Rotation and vibration are also frequently restricted.

Besides the various kinds of kinetic energy shown in Figure 14-8, we also find potential energy, which is related to the forces of attraction acting *between* molecules. The potential energy of molecules in the gaseous state is larger than the potential energy of molecules in the liquid state because energy (heat of vaporization) was required to separate the molecules and overcome the forces holding the molecules together. Similarly, the potential energy in the liquid state is higher than that in the solid state (heat of melting or fusion).

Chemical-bond energy is also present *within* the molecule. Such energy is related to the forces that link the atoms together in the molecule. Stable systems have lower potential energy. In addition, each atom has energy associated with the electrons and with the nucleus. The store of energy in the nucleus is related to the forces holding the nuclear particles together. Because the nucleus remains intact and is apparently unaffected by chemical reactions, nuclear energy does not change in chemical processes.

Molecular heat content, or molecular enthalpy, is the sum of all forms of molecular energy. Adding the *molecular* enthalpy, or "heat content," of 6.02×10^{23} molecules of a given kind, gives the *molar* enthalpy of that substance. We do not have absolute values for this quantity, but changes are easily measured and are very useful.

ENERGY CHANGES ON WARMING

Now that we have seen what makes up the enthalpy of a substance, we can visualize the effects produced by warming the substance. At a low temperature, the substance is a solid. Warming the solid increases the kinetic energy of the back-and-forth motions of the molecules around their regular positions in the solid lattice. As the temperature rises, the motions disturb the regularity of the solid to a greater and greater extent. Too much of this random movement destroys the lattice completely. A phase change occurs at the temperature above which the kinetic energy of the particles causes so much random movement that the lattice is no longer stable: the solid melts.

In the liquid phase, each molecule has more freedom of movement, particularly for translation and rotation. Such movements are enhanced as the liquid is warmed. As the kinetic energy rises, more of the molecules move away from the liquid region, with its relatively low potential energy. Another phase change occurs: the liquid vaporizes. The system is going from a state of lower potential energy to a state of higher potential energy.

If we continue warming the vapor, we reach a point at which the kinetic energies of vibration, rotation, and translation (Figure 14-8) become equal to chemical-bond energies. Molecules begin to break down into the individual atoms that make them up. That is why only a very few diatomic molecules are found on the sun. Its surface temperature is so high (6000 K) that more complex molecules cannot survive.

To conclude this study, let us consider the magnitudes of the energy effects discussed above. Phase changes usually involve energies of several kilocalories per mole. Chemical reactions usually involve energies of from 50 to several hundred kilocalories per mole. Thus, energies involved in chemical reactions are usually from 10 to 100 times larger than those involved in phase changes.

14.4 Nuclear energy

Now let us consider nuclear changes. The fact that nuclei do remain intact during chemical reactions suggests that much larger energies are required for nuclear processes. Experiments prove this to be true. Nuclear reactions usually involve energy changes over 1 million times greater than those found in chemical reactions. This enormous difference in energy level accounts for our interest in nuclear reactions as a source of energy.

NUCLEAR FISSION

One of the first nuclear processes used on earth for the production of energy was the **nuclear fission** of uranium-235.* Nuclear fission is *the splitting of a heavy nucleus into two lighter nuclei of comparable mass.* In Chapter 8, we looked at the natural radioactivity of uranium-238. We found that fission occurs in the lighter isotope, U-235, when a stray neutron enters a mass of pure U-235. The nuclear equation for a mole of $^{235}_{92}U$ is

$$^{235}_{92}U + ^1_0n \longrightarrow ^{141}_{56}Ba + ^{92}_{36}Kr + 3^1_0n + \text{about } 4.5 \times 10^9 \text{ kcal}$$

Notice that this equation conserves both charge ($92 = 56 + 36$) and the number of nucleons (protons plus neutrons: $235 + 1 = 141 + 92 + 3$).

However, the fission of U-235 does not have to give barium and krypton as fission products. Any two fragments of roughly equal mass are possible products, as long as their atomic numbers add up to 92. Each fragment is an

*Lise Meitner, an Austrian physicist whose 1939 calculations theoretically supported the idea of fission, is generally credited for establishing that U-235 does undergo fission when it absorbs neutrons.

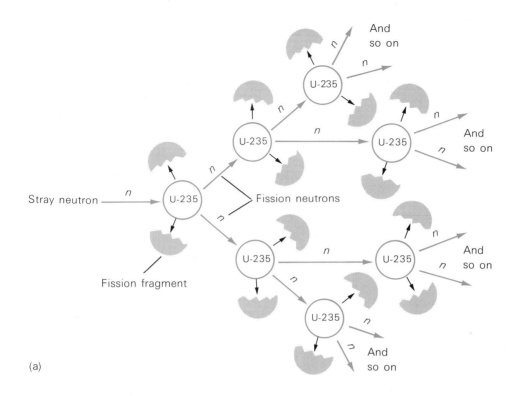

(a)

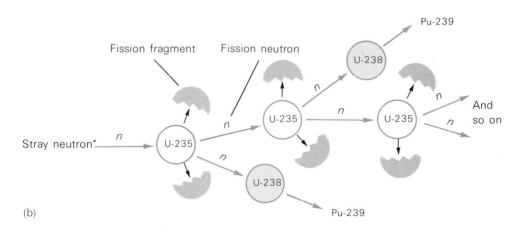

(b)

Figure 14-9

Comparison of (a) an uncontrolled nuclear reaction and (b) a controlled nuclear reaction.

*Students frequently ask "Where does the first neutron come from?" The answer is that in a radioactive mass where alpha, beta, and gamma rays are being emitted, reactions to generate a few neutrons occur constantly. For example, Sir James Chadwick, an English physicist, discovered neutrons in 1932 when he bombarded beryllium atoms with alpha particles. In addition, there are infrequent, spontaneous decay processes for U-235 that give off small numbers of neutrons that are enough to maintain a very low neutron concentration.

isotope with an excess of neutrons. Stable barium, for example, is barium-137, with 81 neutrons. Barium-141 has 85 neutrons and is radioactive, as are fission fragments in general.

If the original sample of U-235 contains enough atoms, the released neutrons go on to cause other fissions, resulting in the familiar **chain reaction.** The energy released is on the order of 4.5×10^9 kilocalories per mole of uranium. Compare that figure to the molar heat of combustion of carbon, which is 94 kilocalories per mole of carbon burned: the energy obtained from a mole of uranium is roughly 50 million times the energy released by burning a mole of carbon to produce carbon dioxide!

If a chain reaction is *uncontrolled,* a tremendous explosion can occur. An uncontrolled chain reaction will occur when highly purified radioactive substances such as U-235 and plutonium (Pu)-239 exceed a certain **critical mass.** Put another way, if there is enough U-235 or Pu-239 packed into a given space, the neutrons given off will continue to shatter more and more nuclei until there is an explosion. Of course, this all happens in microseconds. The process is self-sustaining. Figure 14-9a illustrates this process. U-235 and Pu-239 were components of the first two atomic bombs.

If the amount of material present is less than the critical mass, neutrons released in early fissions tend to "wander" out of the material before striking other nuclei. In such a case, there will not be an explosion. The fission reaction is no longer self-sustaining. Clearly, uncontrolled nuclear chain reactions cannot be used as a source of energy.

A *controlled* chain reaction occurs when a mixture of U-238 isotopes is *enriched* by the addition of U-235. The neutrons released in the fission of the U-235 in the mixture are absorbed by U-238 nuclei. The U-238 nuclei are thereby transmuted into Pu-239. This process is illustrated in Figure 14-9b. However, the neutrons emitted by U-235 fission are very fast-moving. Under ordinary conditions, they move too fast to cause other nuclei of U-235 to split. Instead of being absorbed by the nuclei, they bounce off. In such a case, a **nuclear reactor** could not operate efficiently. To overcome this problem, the neutrons from U-235 fissions are slowed down by the use of a *moderator.* One common moderator is ordinary water, which is placed around the enriched uranium "fuel" elements to slow down emerging neutrons. *Slow* neutrons are then captured by other U-235 atoms that undergo more fission.

With so many fissions under way, you might think that a nuclear reactor would explode like an atomic bomb. It cannot, however, for the following reason. One kind of nuclear bomb contains only U-235. All the neutrons that strike U-235 nuclei cause fissions, which release more neutrons. Since each U-235 fission produces two neutrons, the chain reaction proceeds geometrically. One neutron causes two fissions, which cause four fissions, which trigger eight, then 16 fissions, and so on through succeeding generations. The tenth generation involves more than a thousand fissions. The twentieth generation includes more than a million fissions. All this occurs virtually instantaneously.

In a nuclear reactor, however, only *one* neutron from each U-235 fission is allowed to cause other fissions. The "extra" neutrons are absorbed by U-238 nuclei, which retain the neutrons and turn into Pu-239, as shown in Figure 14-10. To further control this process, rods of boron or cadmium are inserted into the reactor. These *control rods* absorb any stray neutrons that might cause unwanted fissions, which could send the reaction out of control.

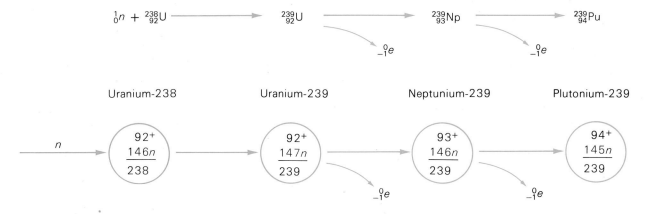

Figure 14-10

Neutron bombardment of uranium-238 forms uranium-239, which is unstable because it has too many neutrons. Beta-decay reduces the number of neutrons and increases the number of protons, creating new heavy elements.

It is often difficult to visualize the magnitude of the energy of such fission processes. If we could properly harness the energy released in the fission process, less than 250 grams of uranium-235 would keep an automobile running for about 100 years! Where does all this energy come from? The answer can be derived from the Special Theory of Relativity, formulated by the great German-American physicist Albert Einstein in 1905. Mass and energy, according to the theory, are different forms of the same thing. They are related by the equation $E = mc^2$, where E is energy, m is mass, and c is the velocity of light (3×10^{10} centimetres per second). In a nuclear process, mass can be converted to energy. However, energy and mass together must be conserved.

Because c^2 is so large, a very small change in mass corresponds to a staggering energy change. Mass changes in chemical reactions are too small to be measured on even the most sensitive balances. But nuclear reactions *do* involve measurable mass changes, as we shall see.

NUCLEAR FUSION

Energy is released when a heavy nucleus such as U-235 undergoes fission to give two lighter nuclei. Energy is also released when neutrons and protons *combine* to form nuclei of somewhat heavier atoms such as helium or lithium. Such a combination is called **nuclear fusion.** Let us examine mass relationships in the hypothetical process

$$2{}_1^1\text{H} + 2{}_0^1n \longrightarrow {}_2^4\text{He}$$

Very precise measurements indicate that the mass of 2 moles of hydrogen atoms is 2.01565 grams, and that the mass of 2 moles of neutrons is 2.01614 grams. The expected mass of the helium atom—two protons, two electrons, and two neutrons—would be the sum of the masses of the individual particles. Thus, the expected mass of 1 mole of helium atoms should be the sum of 2 moles of hydrogen atoms and 2 moles of neutrons.

$$\begin{array}{r} 2.01565 \text{ g} \\ +2.01614 \text{ g} \\ \hline 4.03179 \text{ g} \end{array}$$

But, in fact, the measured mass of 1 mole of helium atoms is 4.00260 grams! Simple subtraction (4,03179 g − 4.00260 g) shows that 0.02919 g of matter is converted to energy for each mole (4.00 g) of helium formed. This mass change is equivalent to 628,000,000 kcal per *mole* of helium formed* or 157,000,000 kcal per *gram* of helium formed. This is a staggering amount of energy. This loss of mass is the source of energy in the sun and other stars. For comparison, the burning of coal to get carbon dioxide yields about 7.8 kcal per gram. Breaking the helium atom up into two hydrogen atoms and two neutrons would require the same amount of energy as is released when helium forms. This quantity of energy is called the **nuclear binding energy** of helium.

We can make a significant comparison between nuclear binding energies by dividing the total binding energy of each nucleus by the number of nucleons (protons plus neutrons) in the nucleus. The resulting quantity—the binding energy per particle in the nucleus—varies systematically as the mass number of the nucleus increases. The variation is plotted in Figure 14-11.

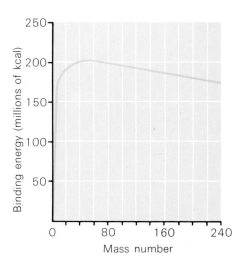

Figure 14-11
The binding energy per nuclear particle.

EXERCISE 14-4

If a nitrogen atom has a mass of 14.003 074 38 on the atomic weight scale, how much mass is converted to energy in the formation of 1 mole of nitrogen from hydrogen atoms and neutrons? If 1 g of mass is equivalent to 2.16×10^{10} kcal, what is the binding energy of the nitrogen nuclei per mole?

The nuclei with mass numbers around 60 have the highest binding energy per heavy particle. Therefore, they are the most stable nuclei. The graph in Figure 14-11 can help us understand the processes of nuclear fission and fusion. Nuclear fusion is a process by which nuclei of small mass fuse to form a nucleus of greater mass. When light nuclei such as $_1^1H$ or $_1^2H$ are fused, the binding energy per nucleon increases and energy is released. We see from the graph that the energy released *per nucleon* is much greater in the fusion process than it is in the fission process. That is, the curve is much steeper on the left side than it is on the right. If the nuclei of the heavier elements, such as uranium and plutonium, are split into two smaller fragments, energy is again released. The binding energy is a maximum near atomic number 60. In considering the amount of energy released in fission and fusion processes, we must consider the energy released per gram (or per ton) of material transformed. Clearly, in Figure 14-11, fission releases more energy *per gram* than does fusion (6.3×10^8 kcal/g compared to 1.9×10^7 kcal/g).

*The energy conversion is done through the use of Einstein's equation, $E = mc^2$:

$$E = 0.2919 \text{ g} \times \left(3.000 \times 10^{10} \frac{\text{cm}}{\text{sec}}\right)^2 = 2.627 \times 10^{19} \frac{\text{ergs}}{\text{mole}} \text{ of He}$$

Since 1.00 kcal is equal to 4.18×10^{10} ergs, we get

$$2.627 \times 10^{19} \text{ ergs} \times \frac{1 \text{ kcal}}{4.18 \times 10^{10} \text{ ergs}}$$

$$= 6.28 \times 10^8 \text{ kcals or } 628,000,000 \text{ kcal/mole of He formed}$$

Unfortunately, scientists have not yet been able to bring about the direct combination of two protons and two neutrons on earth to produce a helium nucleus. They have, however, been able to bring about the combination of heavy hydrogen-2 ($_1^2H$) to produce heavier nuclei. A typical reaction is

$$_1^2H + _1^2H \longrightarrow _1^3H + _1^1H + 7.5 \times 10^7 \text{ kcal per 2 moles of } _1^2H$$

The hydrogen-3 nucleus can then combine with another hydrogen-2 nucleus to form helium:

$$_1^3H + _1^2H \longrightarrow _2^4He + _0^1n + 4.0 \times 10^8 \text{ kcal per mole } _2^4He \text{ formed}$$

The energy evolved is tremendous. But it is very difficult to start these reactions because temperatures of 10 million to 100 million °C are required. If a fission-bomb reaction is used as a trigger, the fusion process can be initiated to give an uncontrolled hydrogen-bomb explosion. Such weapons may produce unlimited explosive power, but they are of no value in the production of controlled power.

To achieve the much more difficult objective of a controlled release of energy from fusion demands alternative ways to heat, contain, and compress the hydrogen reactants. Very strong magnetic fields might be used toward this end. However, the problem of obtaining useful energy from fusion is still very difficult.

EXERCISE 14-5

In the radioactive decomposition of radium-226, the equation for the nuclear process is

$$_{88}^{226}Ra \longrightarrow _{86}^{222}Rn + _2^4He$$

The mass of the $_{88}^{226}Ra$ atom is 226.025 360 mass units, the mass of $_{86}^{222}Rn$ is 222.017 530, and the mass of $_2^4He$ is 4.002 603 61. How much mass is converted to energy in the radioactive decay process? If 1 g of mass is equivalent to 2.16×10^{10} kcal, how many kilocalories of energy are released for every mole of $_2^4He$ liberated?

Let us review the difference between nuclear and chemical reactions. In representing the fission of uranium-235, *the reaction takes place inside the nucleus*. The equation for the fission process is written in terms of the nuclei and particles associated with them. A nuclear equation tells us nothing about what compounds were used or were formed in the process. It summarizes only the nuclear changes. During nuclear change, there is much disruption of other atoms because of the tremendous amounts of energy liberated. We do not know in detail what happens, but eventually the process results in electrically neutral substances (chemical compounds or elements), and the neutrons are absorbed by other nuclei. Thus, in nuclear reactions, changes take place in the nuclei. In chemical reactions, the nuclei remain intact and the changes are explainable in terms of the electrons outside the nucleus.

SUMMARY

Both the amount of heat and the amount of chemical substances are specified when the heat of a reaction is measured. The heat term (frequently in kilocalories) can be inserted in a chemical equation. If heat is used up, the heat term is placed to the left of the equation's arrow, along with the reactants. If heat is released, the heat term is placed to the right of the equation's arrow, along with the products. *Endothermic* reactions absorb energy. *Exothermic* reactions release energy. Enthalpy, or heat content (H), is a measure of the energy stored in a substance. The change in heat content, or enthalpy, when a chemical reaction takes place is symbolized by ΔH. The change in enthalpy of the system is given by

$$\Delta H = \text{(enthalpy of products)} - \text{(enthalpy of reactants)}$$

A positive value of ΔH means that during the process, energy from the surroundings is stored in the products. The reaction is endothermic. A negative value of ΔH means that during the process, energy is liberated and joins the surroundings so that the products of the reaction have less energy than the original reactants. The process is exothermic. By proper addition of ΔH values for known reactions, ΔH values for new reactions can be predicted with a high degree of accuracy.

When a reaction can be expressed as the algebraic sum of two or more equations, the heat of the reaction is the algebraic sum of the heats of the reactions added. This generalization is known as Hess' Law or the Law of Additivity of Heats of Reaction.

The energy of a system can be classified as kinetic and potential. The total energy of mechanical and chemical systems is always conserved. Energy is stored in a molecule as *kinetic energy* (energy of motion) and as *potential energy* (energy of position). Potential energy is associated with attractions and repulsions among the molecules, atoms, or charged particles making up a substance. In addition, each nucleus contains a tremendous amount of stored energy.

Phase changes involve energies of several kilocalories per mole. Chemical changes involve ten to one hundred times more energy than phase changes. Nuclear changes involve energies that are millions of times greater than chemical energies. The very large energy changes associated with nuclear processes have verified the Einstein relationship $E = mc^2$.

Nuclear fission is the splitting of a heavy nucleus into two lighter nuclei of comparable mass. The splitting of 1 mole of U-235 atoms releases some 50 million times more energy than the burning of 1 mole of carbon to form carbon dioxide. Nuclear fusion is the combining of protons and neutrons to form nuclei of somewhat heavier atoms, such as helium nuclei.

Key Terms

calorimetry
chain reaction
critical mass
endothermic
enthalpy
exothermic
heats of formation (ΔH_f)
heat of reaction (ΔH)
Hess' Law of Constant Heat Summation
kilocalorie
Law of Additivity of Heats of Reaction
nuclear binding energy
nuclear fission
nuclear fusion
nuclear reactor
potential energy

QUESTIONS AND PROBLEMS

A

1. (a) How many calories would be required to raise the temperature of 1.00 gram of water through 10.0 °C? (b) How many calories would be required to raise the temperature of 10.0 g of water through 10.0 °C? (c) How many calories would be required to raise the temperature of 5.0 g of H_2O through 5.0 °C? (d) How much energy would be released in cooling 5.0 g of H_2O through 5.0 °C? (e) What is a kilocalorie?

2. Identify each of the following reactions as exothermic or endothermic. Which processes store more energy in the products than in the reactants?

   (a) $2Hg(l) + I_2(s) \longrightarrow Hg_2I_2(s)$
   $\Delta H = -28.9$ kcal

   (b) $\frac{1}{2}N_2(g) + \frac{3}{2}F_2(g) \longrightarrow NF_3(g)$
   $\Delta H = -27.2$ kcal

   (c) $Si(s) + 2Br_2(l) \longrightarrow SiBr_4(l)$
   $\Delta H = -95.1$ kcal

   (d) $Na(s) \longrightarrow Na(g)$
   $\Delta H = +25.98$ kcal

   (e) $NH_4NO_3(s) \xrightarrow{H_2O} NH_4^+(aq) + NO_3^-(aq)$
   $\Delta H = +6.1$ kcal

3. (a) Write the equation with appropriate energy terms to show how "water gas" is made. (b) Write equations with appropriate energy terms to show the burning of 2.00 "moles of water gas" (or 44.2 litres of "water gas" at STP, which is 1.00 mole of CO plus 1.00 mole of H_2). (c) What is (a) + (b) equivalent to in terms of energy?

4. Briefly describe the operation of a calorimeter, indicating the measurements you would make when using one.

5. The heat of formation of each compound is associated with a specific equation. For example, the heat of formation for water is associated with the equation $H_2(g) + \frac{1}{2}O_2(g) \longrightarrow H_2O(l)$ $\Delta H = -68.3$ kcal. Write the equation and the ΔH value for the heat of formation of each of the following substances: (a) $SO_2(g)$, (b) $C_3H_8(g)$, (c) $N_2O(g)$, (d) $Fe_2O_3(s)$ $\Delta H = -196$ kcal, (e) $CO_2(g)$, (f) $N_2(g)$, (g) $O_2(g)$, (h) any element in its standard state (25 °C, 1 atm).

6. Add equations and ΔH_{form} values to get ΔH for the burning of 1.00 mole of $C_3H_8(g)$ to give $H_2O(l)$ and $CO_2(g)$. Be sure to write all equations out so that you can show that addition and cancellation gives you the desired equation.

7. What is ΔH for the process:

$$2Al(s) + Fe_2O_3(s) \longrightarrow Al_2O_3(s) + 2Fe(s)?$$

ΔH_{form} $Al_2O_3 = -399$ kcal/mole; ΔH_{form} $Fe_2O_3 = -196$ kcal/mole

8. What is ΔH for the process:

$$SiO_2(s) + C(s) \longrightarrow CO_2(g) + Si(s)?$$
$$\Delta H_{form}\ SiO_2(s) = -205\ kcal/mole$$

   (This is the process by which elemental Si for transistors is made.)

9. What is ΔH for the formation of $H_2SO_4(l)$ from $H_2O(l)$ and $SO_3(g)$. See Table 14-2 on page 368.

10. State and illustrate with a simple example the law of conservation of energy.

11. What is ΔH for the process:

$$P_4O_{10}(s) + 6H_2O(l) \longrightarrow 4H_3PO_4(s)?$$
$$\Delta H_{form}\ P_4O_{10} = -720\ kcal/mole;\ \Delta H_{form}\ H_3PO_4(s) = -303\ kcal/mole$$

   See Table 14-2. Note: this value is *not* the ΔH_f for H_2SO_4.

12. A baseball is thrown up into the air then falls to earth. (a) Where in the path of the baseball is its *potential* energy a maximum? (b) Where in its path is its *kinetic* energy a maximum? (c) What happens to its kinetic energy when it hits the ground?

13. Why do we say that chemical energy stored in a molecule is *potential energy*? Give your answer from an atomic standpoint.

14. Name and briefly describe three kinds of kinetic energy associated with an individual gaseous molecule. Refer to Figure 14-8.

15. Which has the highest "heat content" or enthalpy: 10.0 g of ice at 0 °C, 10.0 g of liquid H_2O at 100.0 °C, or 10.0 g of steam at 100.0 °C? How is the energy stored in the sample with the highest heat content?

16. When tritium ($_1^3H$) and deuterium ($_1^2H$) combine in a nuclear process, a neutron ($_0^1n$) is one of the products. What is the other product?

17. The process of problem 16 is strongly exothermic. Will the products have exactly as much *total mass* as the reactants? Explain your answer.

18. If conversion of one gram of mass to energy releases 2.16×10^{10} kcal, how much energy would be released in the following process:

$$3_1^1H + 4_0^1n \longrightarrow {}_3^7Li?$$

   Use the following information: mass hydrogen atom = 1.00782; mass $_0^1n$ = 1.00807; mass Li atom = 7.01601; and refer to pages 380–381.

B

19. Careful measurement reveals that 31.4 kcal of heat are absorbed per mole of carbon reacting to form water gas (see equation *1b*, on page 360). How much heat would be absorbed during the reaction if the carbon used had a mass of (a) 6.0 g (b) 12.0 g and (c) 24.0 g?

20. (a) Write an equation for each of the following, and include the heat term in the equation: (1) 26.4 kcal quantity of heat is released when 1.00 mole of carbon, $C(s)$, burns to form carbon monoxide, $CO(g)$; (2) 16.2 kcal quantity of heat is absorbed (used up) when 1.00 mole of nitrogen, $N_2(g)$, burns to form nitrogen dioxide, $NO_2(g)$. (b) Determine ΔH for each of the reactions described in (a). For each, include both numerical value *and* the sign + or −. (c) Which ΔH value represents a "heat of formation" value?

21. Convert each of the following equations to the ΔH notation. Indicate whether each reaction is exothermic or endothermic.

(a) $Cd(s) + \frac{1}{2}O_2(g) = CdO(s) + 63.3$ kcal

(b) $Cl_2(g) + \frac{7}{2}O_2(g) + 65$ kcal $= Cl_2O_7(g)$

(c) $2Au(s) + \frac{3}{2}O_2(g) = Au_2O_3(s) + 2$ kcal

(d) $2B(s) + 3H_2(g) + 7.5$ kcal $= B_2H_6(g)$

(e) $\frac{1}{2}Br_2(l) + \frac{1}{2}Cl_2(g) + 3.5$ kcal $= BrCl(g)$

22. For each type of reaction carried out inside a calorimeter, indicate whether the water temperature will rise or fall: (a) an exothermic reaction, (b) an endothermic reaction.

23. Given the following equation $C(s) + O_2(g) \longrightarrow CO_2(g)$ $\Delta H = -94.0$ kcal: (a) How many *moles* of carbon must be burned to produce 611 kcal of energy? (b) How many *grams* of carbon must be burned to produce 611 kcal of energy?

24. Using the data in Table 14-2, determine ΔH for the following processes (Indicate the sign of ΔH in each case.):

(a) $SO_2(g) + \frac{1}{2}O_2(g) \longrightarrow SO_3(g)$

(b) $4NH_3(g) + 5O_2(g) \longrightarrow 4NO(g) + 6H_2O(l)$

(c) The decomposition to the elements 25 °C of 2 moles of gaseous hydrogen iodide.

(d) For each of the reactions above indicate whether the process is exothermic or endothermic.

25. (a) Water gas is 50 percent H_2 and 50 percent CO. How much energy is given off if 22.4 litres at STP of water gas is burned to form $CO_2(g)$ and $H_2O(l)$ using O_2 in the air? (b) How much energy is given off if 22.4 litres at STP of methane, or natural gas, (CH_4) is burned to form $CO_2(g)$ and $H_2O(l)$ using O_2 in the air? (c) What is the ratio between these two numbers? Refer to Table 14-2. (d) Which is more efficient for heating your home: water gas or methane?

26. For a sample of water, indicate whether each of the following increases the kinetic energy or the potential energy or both: (a) warming ice from -10 °C to 0 °C; (b) melting ice at 0 °C; (c) warming water from 0 °C to 100 °C; (d) vaporizing water at 100 °C; (e) warming water vapor from 100 °C to 110 °C.

27. Match the following ΔH values with the appropriate reaction: $\Delta H_1 = -57.8$ kcal; $\Delta H_2 = +10.5$ kcal; $\Delta H_3 = -7.5 \times 10^7$ kcal. Explain your answers.

(a) $H_2O(l) \longrightarrow H_2O(g)$

(b) $H_2(g) + \frac{1}{2}O_2(g) \longrightarrow H_2O(g)$

(c) $^2_1H + ^2_1H \longrightarrow ^3_1H + ^1_1H$

28. Identify each of the missing products (name the element and give its atomic number and mass number) by balancing each of the following nuclear equations: (a) $^{235}_{92}U + ^1_0n \longrightarrow ^{143}_{54}Xe + ? + 3^1_0n$ (b) $^{235}_{92}U + ^1_0n \longrightarrow ? + ^{102}_{39}Y + 3^1_0n$ (c) $^{14}_7N + ^1_0n \longrightarrow ? + ^1_1H$.

29. (a) Assuming it could be properly harnessed, how long would the energy involved in the fission of 1 pound (465 g) of U-235 keep an automobile running? (b) How would you word the advertising blurb that would give the "range" of this car on a tank of fuel? You need the following information. The energy released in the fission of 235 g of $^{235}_{92}U$ is about 4.5×10^9 kcal that is equivalent to the energy released in burning 140,000 gal of gasoline. Assume the car uses 15.0 gallons of gasoline per week.

C

30. When 5.6-g sample of solid KOH was added to 100 g of water, the solid dissolved completely. This was accompanied by a rise of 13 °C in the water temperature. (a) Calculate the heat of solution of KOH, in kilocalories per mole of KOH. (b) Determine ΔH for the reaction, in kilocalories per mole of KOH.

31. A 6.4-g sample of sulfur is introduced into a calorimeter containing 200 g of water. When the sulfur sample is burned completely to form sulfur dioxide, $SO_2(g)$, the water temperature is observed to rise by 71 °C. (a) Calculate the amount of heat released inside the calorimeter. (b) Determine the heat of combustion of sulfur, in kilocalories per mole of sulfur. (c) Determine ΔH for the reaction, in kilocalories per mole of sulfur.

32. When a 20.0 g sample of NH_4NO_3 was added to 100.0 g of water in a calorimeter, the solid dissolved completely and the temperature of the water was *lowered* by 15 °C. (a) Calculate the heat of solution of NH_4NO_3 in kcal/mole. (b) Write the balanced equation for the dissolving of NH_4NO_3 in water and give the ΔH value for the process.

33. Butyric acid ($C_3H_7C\overset{O}{\overset{\|}{-}}OH$) is the terrible, smelly substance in rancid butter. It has a ΔH_{form} of -125 kcal/mole. Calculate ΔH for the complete combustion of 2 moles of butyric acid to $CO_2(g)$ and $H_2O(l)$. Refer to page 367.

34. Suppose methane (CH_4) burns in a shortage of oxygen so that the products of combustion are $H_2O(l)$ and $CO(g)$ rather than $CO_2(g)$. What fraction of the total energy released in combustion to CO_2 would be obtained by the above combustion process?

35. When $^{235}_{92}U$ is struck by a neutron, the unstable isotope $^{236}_{92}U$ results: $^{235}_{92}U + ^{1}_{0}n \longrightarrow ^{236}_{92}U$. When $^{236}_{92}U$ undergoes fission, there are numerous possible products. If strontium-90 ($^{90}_{38}Sr$) and 3 neutrons are the results of one such fission, what is the other product?

$$^{236}_{92}U \longrightarrow ^{90}_{38}Sr + ? + 3^{1}_{0}n$$

36. The following masses are useful: neutron, $^{1}_{0}n = 1.0087$; proton, $^{1}_{1}H = 1.0073$; electron $= 0.00055$; and 1.00 g of mass $= 2.16 \times 10^{10}$ kcal. (a) If the copper atom $^{65}_{29}Cu$ has an actual mass of 64.955, what is the binding energy of $^{65}_{29}Cu$ in kcal/mole of Cu atoms? (b) What is the binding energy in kcal/nucleon?

37. Most chemical properties of atoms show a periodic variation with atomic number. Is this also true with nuclear binding energy? Explain.

A molecular system [passes] . . . from one state of equilibrium to another . . . by means of all possible intermediate paths, but the path most economical of energy will be more often traveled.

HENRY EYRING

15

THE RATES OF CHEMICAL REACTIONS

Objectives

After completing this chapter, you should be able to:

● Understand how the rate of a reaction depends on the nature of the reactants, their concentration, and their temperature.

● Describe the reaction mechanism and the sequence of steps in a complex chemical reaction.

● Analyze a potential energy diagram in terms of reactants, products, activation energy, and the activated complex.

● Understand the suggested mechanism by which catalysts speed up certain reactions.

Differences in rates of running are very important to both these animals.

A candle does not burn until it is lighted with a match. A match does not burn until you strike it on the proper surface. The chemicals in the candle and the match, and the oxygen in the air, are present at all times. But a candle may be surrounded by air for an unlimited period of time, and not burn. So may a match. Why? Once started, the burning of a candle or match continues until the candle or match is consumed. No further lighting or striking is needed. Why?

Like an unlighted candle or match, natural gas from a stove may never react with the air in a closed room. But the mixture may explode suddenly if a glowing match is brought into the room. Aluminum powder will also burn rapidly when ignited in air. On the other hand, iron reacts quite slowly with air, forming rust. But very finely divided "pyrophoric iron" bursts into flame as soon as it is exposed to air. At times, large chunks of white phosphorus can ignite spontaneously. For that reason, the substance should be handled with great care.

Figure 15-1
Very finely divided pyrophoric iron bursts into flame as soon as it is exposed to air.

The reactions just mentioned all require oxygen from the air. Yet, they require very different amounts of time to take place. Such examples illustrate the fact that *reactions proceed at different rates.* The study of reaction rates is called **chemical kinetics.**

15.1 Measuring reaction rates

Reaction rates are important for many reasons. Let us look at an example. The oxides of nitrogen produced at ground level by automobile engines contaminate our air. Those oxides can decompose spontaneously to form harmless oxygen and nitrogen gas. However, the *rate* at which they decompose at normal air temperatures is so slow that their concentrations can build up in the atmosphere and become a pollution problem. One way to solve the problem would be to speed up the decomposition reaction at normal air temperatures:

$$2NO(g) \longrightarrow N_2(g) + O_2(g)$$

We can say the rate of the decomposition reaction should be increased.

What is the exact, quantitative meaning of the expression *rate of reaction?* How can we measure an increase or decrease in reaction rate? Let us look at the reaction between carbon monoxide and nitrogen dioxide:

$$CO(g) + NO_2(g) \longrightarrow CO_2(g) + NO(g)$$

If we heat a mixture of $CO(g)$ and $NO_2(g)$ to 200 °C, we observe a gradual disappearance of the reddish-brown color of $NO_2(g)$. This change tells us that a reaction is taking place. Since both products are colorless, the rate at which $NO_2(g)$ is used up is revealed by the rate at which its reddish-brown color disappears. Utilizing this information, we can develop a quantitative method for determining the rate of the reaction. We find the rate of reaction by measuring the *change* in color during successive one-minute periods. The color change is proportional to the change in the number of moles of $NO_2(g)$ present in a 1-litre volume. *The number of moles of a substance present in 1 litre is known as the* **concentration** *of that substance.* The rate of reaction can thus be found by dividing the change in $NO_2(g)$ concentration by the time interval in which the change occurs. To summarize,

$$\text{rate of reaction} = \frac{\text{change in } NO_2(g) \text{ concentration (moles/litre)}}{\text{time interval (minutes)}}$$

We can express the rate of a reaction in several ways. It can be expressed as the rate of consumption of a *reactant*—in our example, $CO(g)$ or $NO_2(g)$. It can also be expressed as the rate of formation of a *product*—in our example, $CO_2(g)$ or $NO(g)$. The substance whose rate of change is easiest to measure is chosen to express the reaction rate. In our example, that substance is $NO_2(g)$, because of its characteristic color. If we prefer to measure the rate at which $CO_2(g)$ is produced, the rate is expressed in the form

$$\text{rate of reaction} = \frac{\text{change in } CO_2(g) \text{ concentration}}{\text{time interval}}$$

The change in concentration can be expressed in any set of convenient units. If the substance is a gas, changes in partial pressure can express changes in gas concentration. If the substance is in solution, changes in moles per litre (or changes in molarity) can be used. The time intervals can also be expressed in whatever units are appropriate to the reaction under study. We might use microseconds for the explosion of natural gas from a stove, seconds or minutes for the burning of a candle, and days or years for the rusting of iron.

15.2 Factors affecting reaction rates

In the laboratory, you may have observed the reaction between ferrous ions, $Fe^{++}(aq)$, and permanganate ions, $MnO_4^-(aq)$. You may also have seen the reaction between oxalate ions, $C_2O_4^{--}(aq)$, and permanganate ions. Both reactions illustrate the concept that *the rate of a reaction depends on the nature of the reacting substances.* The reaction between $IO_3^-(aq)$ ions and $HSO_3^-(aq)$ ions, the so-called iodine-clock reaction, highlights another concept: that *the rate of a reaction depends on the concentrations of reactants and on the temperature.* We will examine these factors one at a time.

NATURE OF REACTANTS

Let us take a look at two reactions. Both occur in water solutions:

$$5Fe^{++}(aq) + MnO_4^-(aq) + 8H^+(aq) \longrightarrow$$
$$5Fe^{+++}(aq) + Mn^{++}(aq) + 4H_2O(l) \quad \text{VERY FAST}$$
$$5C_2O_4^{--}(aq) + 2MnO_4^-(aq) + 16H^+(aq) \longrightarrow$$
$$10CO_2(g) + 2Mn^{++}(aq) + 8H_2O(l) \quad \text{SLOW}$$

At room temperature, both ferrous ions, $Fe^{++}(aq)$, and oxalate ions, $C_2O_4^{--}(aq)$, decolorize a solution of purple permanganate ions, $MnO_4^-(aq)$. But the time required for the decoloration is vastly different for each ion. Since the reactants are identical except for the $Fe^{++}(aq)$ and $C_2O_4^{--}(aq)$ ions, the time difference must be due to the different characteristics of those two ions.

The following reactions also involve gases at 20 °C:

$$2NO(g) + O_2(g) \longrightarrow 2NO_2(g) \quad \text{MODERATELY FAST}$$
$$CH_4(g) + 2O_2(g) \longrightarrow CO_2(g) + 2H_2O(g) \quad \text{VERY SLOW}$$

The *oxidation* of nitric oxide, $NO(g)$, is moderately fast at temperatures of 10 °C to 100 °C at one atmosphere pressure. The oxidation of methane, $CH_4(g)$, the main component of natural gas, however, is very slow at room temperature and one atmosphere pressure if no spark is present. It is so slow at this temperature that, for all practical purposes, no reaction occurs. At higher temperatures, of course, methane burns readily. Again, the difference in reaction rates must be due to the different characteristics of $NO(g)$ and $CH_4(g)$. The other reactant, $O_2(g)$, is the same in each case.

One interesting frontier of chemistry is the identification of characteristics that will cause different molecules to react at different rates. Chemical reactions that involve breaking several chemical bonds and forming new ones tend to proceed slowly at room temperature. Thus, we might expect that examining the equation of a reaction would be all we need to determine its general rate. In the reaction involving $C_2O_4^{--}(aq)$ and $MnO_4^-(aq)$ ions, many bonds are broken in the $5C_2O_4^{--}(aq)$ and $2MnO_4^-(aq)$ ions. Many new bonds are formed as the $10CO_2(g)$ and $2Mn^{++}(aq)$ products form. As expected, the reaction proceeds slowly.

The reaction between $CH_4(g)$ and $O_2(g)$ also involves breaking bonds in the reactant molecules and forming new bonds in the $CO_2(g)$ and $H_2O(g)$ product molecules. This reaction, illustrated in Figure 15-2, is also slow at room temperature. But notice that, in the reaction between $NO(g)$ and $O_2(g)$, one bond is broken (the O—O bond in O_2) and two new bonds are formed (the N—O bonds in NO_2). In spite of that, the reaction proceeds at a moderate rate at room temperature. It is slow at low temperatures but becomes rapid at high temperatures.

The reaction between $Fe^{++}(aq)$ ions and $MnO_4^-(aq)$ ions does not follow the same "rule." Since it also involves breaking chemical bonds, we would expect it to be slow. But instead, it is very rapid. Evidently, we cannot predict the rate of a reaction by looking only at its equation.

A fair question to ask at this point is, do all chemical reactions involve bond breaking? Not always. For example, the reaction

$$Fe^{++}(aq) + Ce^{++++}(aq) \longrightarrow Ce^{+++}(aq) + Fe^{+++}(aq) \quad \text{VERY RAPID}$$

appears to involve the transfer of one electron from one reactant ion to the other. Note that losing one electron would cause $Fe^{++}(aq)$ ions to become $Fe^{+++}(aq)$ ions. If the other reactant, $Ce^{++++}(aq)$, gained this electron, it would become $Ce^{+++}(aq)$. Because the reaction involves no other changes—no bond breaking—we expect it to be very rapid and it is.

Figure 15-2

The balanced chemical equation and structural equation for the combustion of CH_4.

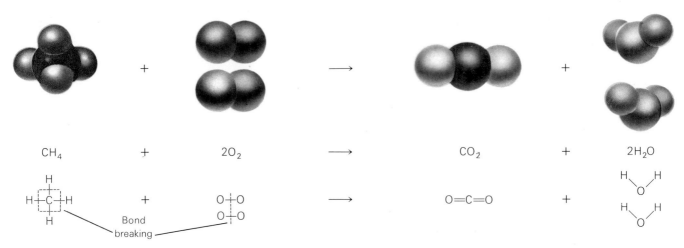

These and other examples suggest the following rules:

1. Reactions are usually rapid at room temperature if they do not involve bond breaking.
2. Reactions are usually slow at room temperature if they do involve bond breaking.
3. It is usually impossible to predict the rate of a reaction merely by examining its chemical equation.

We can say little more about how the nature of reactants determines reaction rate until we consider in detail how some reactions take place. For the time being, we shall simply say that this is an active field of study and that much remains to be learned.

EXERCISE 15-1

Are any of the following three reactions likely to be extremely rapid at room temperature? Are any likely to be extremely slow at room temperature? Explain.

(a) $Cr^{++}(aq) + Fe^{+++}(aq) \longrightarrow Cr^{+++}(aq) + Fe^{++}(aq)$

(b) $3Fe^{++}(aq) + NO_3^-(aq) + 4H^+(aq) \longrightarrow$
$$3Fe^{+++}(aq) + NO(g) + 2H_2O$$

(c) $C_8H_{18}(l) + 12\frac{1}{2}O_2(g) \longrightarrow 8CO_2(g) + 9H_2O(g)$

EFFECT OF CONCENTRATION: COLLISION THEORY

In our search for a model to explain reaction rates, it will help if we look at only one reaction process. We can then focus on other factors affecting reaction rates. The first such factor is *concentration*.

A great number of experiments have shown that for many reactions, increasing the concentration of a reactant increases the reaction rate. However, sometimes there will be no such effect. In this section, we shall explore how increasing the concentration of a reactant increases the reaction rate. In the next section, we shall try to explain why some reactions proceed at a rate independent of the concentrations of the reactants. Both explanations are based on a model of the way in which chemical reactions take place.

Our picture of chemical reactions involves collisions between reacting particles. The particles may be atoms, molecules, or ions. Just as an increase in the number of cars on a highway leads to a higher rate of car collisions, an increase in the number of particles in a given volume leads to more frequent particle collisions and more rapid reaction. (See Figure 15-3. The higher frequency of collisions in the lower bulb tends to cause a higher rate of reaction.) This model is the **collision theory** of reaction rates.

Collision theory states that increasing the concentration of one or more reactants in a **homogeneous system** will increase the rate of reaction. Lowering the concentration will have the opposite effect. That is exactly what can

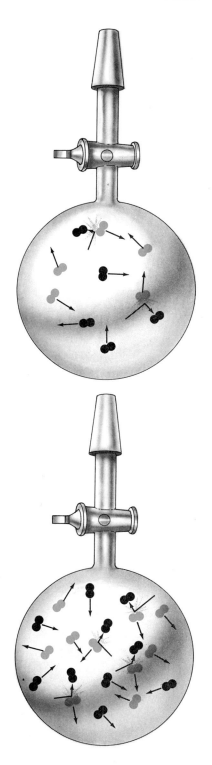

Figure 15-3

The number of collisions per second depends upon the concentration of each reactant.

be observed in the reaction between dissolved $HSO_3^-(aq)$ and $IO_3^-(aq)$ ions in Experiment 20 in the Laboratory Manual. The rate of the reaction varies with the concentration of either ion. In gaseous mixtures, the concentration of an *individual* reactant can be raised by admitting more of that gas into the mixture. The concentrations of *all* gaseous components can be raised at the same time by simply decreasing the volume of the container. Decreasing the volume by compressing the gas will raise the concentration of all reactants. (Recall that concentration is the number of moles of a substance per litre. A concentration of 1.0 mole/1.0 litre, for example, would obviously be greater than a concentration of 1.0 mole/2.0 litres.) Decreasing the volume, and thus increasing the concentration, does increase the reaction rate. Increasing the volume by allowing the gas to expand has the opposite effect on concentration. Predictably, it also has the opposite effect on reaction rates.

In a **heterogeneous system,** the components are in two or more different phases. For example:

$$wood(s) + oxygen(g) \xrightarrow{\text{burning}} carbon\ dioxide(g) + water(g)$$

In a system of this kind, the rate of reaction depends on the extent of the area of contact between or among the reactants. For example, a large log burns slowly, but if it is ground up into a fine powder and the powder is suspended in air, the log may burn explosively. A number of disastrous explosions have occurred in grain storage elevators when a spark would ignite grain dust suspended in the air inside the elevator. Such explosions can be very powerful. In a sense, the surface area of a solid, or a pure liquid, determines its "effective concentration." This effective concentration, or area, is important in determining the number of molecular collisions that involve the solid as one reactant.

As the size of particles is reduced the surface area of the solid goes up rapidly. You have applied this idea many times in your life. When you chopped wood into kindling to start a fire, you increased the surface area of the wood so it would burn faster. When the fire was going well, you added a large log of reduced surface area so the fire could be controlled.*

Some numbers will help us to see how surface area changes as particle size decreases. Consider a single cube that is 10 cm on a side. Each face of the cube has an area of 10 cm $\times$ 10 cm or 100 cm². Since every cube has six faces, the total area of that cube will be

$$100\ \frac{cm^2}{face} \times 6\ \frac{faces}{cube} = 600\ \frac{cm^2}{cube}$$

The volume of the cube will be 10 cm $\times$ 10 cm $\times$ 10 cm = 1,000 cm³. Now let us break this cube into cubes that are only 1 cm on a side. Each of these smaller cubes now has a volume of 1 cm³, and there are 1,000 of them. Each small cube has a total area of

$$1\ \frac{cm^2}{face} \times 6\ \frac{faces}{cube} = 6\ \frac{cm^2}{cube}$$

Since we have 1,000 cubes, the total surface area of the smaller cubes is 6,000 cm². If we now break each of the small 1 cm cubes into 1,000 cubes of 0.1 cm on a side, we get a million cubes with a total area of 60,000 cm². The area is *100 times* that of the larger cube. A process that is moderately slow

*It is also obvious that the concentration of oxygen in the air is an important factor in the rate of a combustion process. A splinter of wood that is glowing in air of 20% oxygen will burst into flames in pure oxygen.

with the large cube would be 100 times faster using the finer material.

Collision theory provides a good explanation of reaction rate behavior. It is reasonable that reaction rate should depend on the number of collisions among the reactant molecules. It is so reasonable, in fact, that we are left wondering why the concentrations of some reactants in certain reactions do *not* affect the rate. We find the explanation by considering how a reaction takes place.

REACTION MECHANISM

Let us take another look at the reaction between $Fe^{++}(aq)$ and $MnO_4^-(aq)$ ions in acid solution. The equation is

$$5Fe^{++}(aq) + MnO_4^-(aq) + 8H^+(aq) \longrightarrow$$
$$5Fe^{+++}(aq) + Mn^{++}(aq) + 4H_2O(l)$$

This equation indicates that one MnO_4^- ion, 5 Fe^{++} ions, and 8 H^+ ions (a total of 14 ions) must react with one another. For this reaction to take place in a single step, the 14 ions would have to collide with each other *simultaneously*. The probability of such an event is extremely small. It is so small that a reaction depending on such a collision would proceed at an immeasurably slow rate. Since the reaction occurs at an easily measured, rapid rate, it must proceed by a *sequence of steps*, rather than by a single step involving the simultaneous collision of 14 ions.

Chemists regard the simultaneous collision of even four molecules or ions as an extremely improbable event if the molecules or ions are at low concentration or if they are in the gas phase. We thus conclude that a complex chemical reaction proceeding at a measurable rate probably takes place in a series of simpler steps that usually involve two-body collisions. Such a series of steps is called the **reaction mechanism.**

EXERCISE 15-2

> Write a balanced equation for the burning of a candle in air, assuming that the compound that makes up the candle is $C_{25}H_{52}$. Assume that the products of burning are $CO_2(g)$ and $H_2O(g)$. Since a candle burns at a measurable rate of speed, could your equation represent the actual mechanism of the combustion reaction? Explain.

Consider the oxidation of gaseous hydrogen bromide, $HBr(g)$. It is a relatively rapid reaction at temperatures between 400 °C and 600 °C. The equation for the reaction is

$$4HBr(g) + O_2(g) \longrightarrow 2H_2O(g) + 2Br_2(g)$$

According to the collision theory, we would expect increasing the partial pressure (and thus the concentration) of either reactant to speed up the reaction. Experiments show that this is indeed the case. Quantitative studies have been made of the reaction rate at various pressures and with various mixtures of the two reactants. The studies show that changes in either the oxygen or hydrogen-bromide concentration *can equally affect* the rate of reaction. This observation raises a question. Since the reaction requires four molecules of HBr for every one molecule of O_2, why does changing the HBr concentration have the same effect as making an equal change in the O_2 concentration?

The explanation is found in the process by which the oxidation reaction occurs. The overall reaction brings together five molecules, four of HBr and one of O_2. But the chance that five gaseous molecules will collide simultaneously is practically nil. As noted earlier, the reaction must occur in a sequence of simpler steps, preferably two-body collisions. All experimental observations of this process are explained by the following series of reactions:

$$HBr(g) + O_2(g) \longrightarrow \text{HOOBr}(g) \qquad \text{SLOW} \qquad (1)$$

$$\text{HOOBr}(g) + HBr(g) \longrightarrow 2\text{HOBr}(g) \qquad \text{FAST} \qquad (2)$$

$$\text{HOBr}(g) + HBr(g) \longrightarrow H_2O(g) + Br_2(g) \quad \text{FAST} \qquad (3)$$

Since 2 HOBr(g) molecules are made in reaction (2), reaction (3) must occur again to get rid of the other HOBr(g) molecule.

$$\frac{\text{HOBr}(g) + HBr(g) \longrightarrow H_2O(g) + Br_2(g) \quad \text{FAST}}{4HBr(g) + O_2(g) \longrightarrow 2H_2O(g) + 2Br_2(g)} \qquad (3)$$

Observe that adding reactions *1*, *2*, and twice *3* gives the overall oxidation reaction. Notice too that each step in the sequence involves a collision of only *two* molecules. (Reaction *3* is taking place twice.) Finally, note that the proposed reaction *1* is slow, whereas reactions *2* and *3* are fast. As we shall see, this explains why HBr and O_2 have the same effect on the reaction rate.

Because it is slow, reaction *1*, forming HOOBr, is a "bottleneck" in the oxidation of hydrogen bromide. The HOOBr formed by this slow reaction is immediately consumed in reaction *2*, which is rapid. But no matter how fast reactions *2* and *3* are, they can produce H_2O and Br_2 only as fast as the slowest reaction in the sequence. Thus, the factors that determine the rate of reaction *1* determine the rate of the overall process. Because reaction *1* requires one O_2 molecule for every one HBr molecule, both substances influence the rate equally. Reaction *1* is the **rate-determining step** in the reaction mechanism.

Several features of the mechanism of the oxidation of hydrogen bromide are typical of other reaction mechanisms. First, none of the steps in the reaction mechanism requires the collision of more than two particles. *Most chemical reactions proceed by sequences of steps, each involving only two-particle collisions.* Second, the overall, or net, reaction does not show the stepwise mechanism. In general, *the net equation for a reaction is not the mechanism, nor can the mechanism be deduced from the net equation.* Therefore, the net equation cannot be used to predict the rate of the reaction it represents. This generalization holds true except in the rare case in which the reaction is so simple that the net equation also represents the mechanism. An example of such a simple reaction is the one discussed on page 392, in which the iron (Fe^{++}) ion reacts with the cesium (Ce^{++++}) ion. In most cases, the various steps by which atoms are rearranged and recombined—the reaction mechanism—can be determined only by experiment. Thus, the rates at which reactions occur must also be determined by experiment.

EXERCISE 15-3

Imagine that a telegram is to be sent on a winter night from New York City to a rancher who lives near the top of a mountain in Nevada. The steps in sending the telegram are the following: (1) The New Yorker calls the telegraph office and dictates a ten-word message. (2) The New York telegraph operator transmits the message to the Reno office. (3) The Reno office types out the message and hands it to a deliverer. (4) The deliverer gets into a car and struggles 20 miles up the snow-covered road to the rancher's house. How great an increase in the rate of transmitting a message would be observed if (a) the secretary receiving the message in New York could type twice as fast, (b) electricity (by some undefined and miraculous process) could move twice as fast in the wire, and (c) the Reno office could telephone the message to the rancher rather than send the deliverer? What is the rate-determining step in the original process?

SOME SPECIFIC REACTION MECHANISMS

Let us now take a look at some specific reaction mechanisms. Such an examination will illustrate how seemingly similar reactions are, in reality, dramatically different. For example, the reactions of hydrogen gas (H_2) with the halogens chlorine (Cl_2) and iodine (I_2) seem very much the same. The equations for the reactions are

$$H_2(g) + Cl_2(g) \longrightarrow 2HCl(g)$$
$$H_2(g) + I_2(g) \longrightarrow 2HI(g)$$

Yet, in the presence of light, even at room temperature, the reaction of hydrogen with chlorine is very fast; the gases explode. By contrast, the rate of the reaction of hydrogen with iodine at room temperature in the presence of light is extremely slow. Although the equations for the reactions are similar, experimental studies of the rates suggest that the reactions must occur by very different mechanisms.

The reaction between hydrogen and chlorine appears to proceed through a series of steps. First, light energy causes a Cl_2 molecule to split into two very reactive Cl atoms. Each is reactive; each is one electron short of the stable argon configuration. Next, a reactive Cl atom collides with an H_2 molecule, forming an HCl molecule and a reactive H atom. The reactive H atom reacts with a Cl_2 molecule, forming another HCl and a reactive Cl atom. The reactive Cl atom again reacts with another H_2 molecule forming another HCl molecule and a reactive H atom. To summarize,

$$Cl_2(g) + light \longrightarrow 2Cl(g)$$
$$Cl(g) + H_2(g) \longrightarrow HCl(g) + H(g)$$
$$H(g) + Cl_2(g) \longrightarrow HCl(g) + Cl(g)$$

The last two reactions keep repeating, producing a reactive hydrogen atom and chlorine atom, respectively. A stepwise reaction in which each step generates the reactive substance necessary for the next step is called a *chain reaction*. Uncontrolled chain reactions usually occur almost instantaneously. Since the above reactions are exothermic, a large amount of energy is liberated very quickly, making the reactions potentially explosive.

By contrast, the reaction of H_2 with I_2 is *not* a chain reaction. The best current idea of the reaction mechanism suggests that H_2 and I_2 molecules collide, but that the I_2 molecules tend to dissociate into two iodine atoms just before the reaction with H_2 occurs. The mechanism is written as

$$I_2 \longrightarrow 2I$$
$$H_2 + 2I \longrightarrow 2HI$$

The difference between the reactions of Cl_2 and I_2 with H_2 is that in the Cl_2 reaction, the chlorine molecule dissociates by itself to give chlorine atoms. One of the latter then collides with an H_2 molecule to give HCl and a hydrogen atom. The chain then proceeds. In contrast, the I_2 molecule does not dissociate until it is so close to the H_2 molecule that both iodine atoms react with a single H_2 molecule. Thus, no additional atoms are left to maintain a chain.

It is the very different mechanisms described above that account for the very different rates of the two reactions under consideration. We see once more that the overall equations tell us nothing that would suggest such different rates. Information about rates of reaction is obtainable only from laboratory studies, not from overall equations.

EFFECT OF TEMPERATURE

In the laboratory, you have discovered that temperature has a marked effect on the rate of chemical reactions. For example, $C_2O_4^{--}(aq)$ ions decolorize $MnO_4^-(aq)$ ions faster at 50 °C than at 20 °C. Increased temperature also speeds up the reaction between $IO_3^-(aq)$ and $HSO_3^-(aq)$ ions. The situation is analogous to lighting a candle with a match to start a combustion reaction. The match raises the temperature of the candle at the wick, thus starting the reaction. Once started, the reaction continues to release enough heat to keep the temperature high. This keeps the reaction going at a relatively rapid rate. Thus, raising the temperature speeds up the burning of a candle. At room temperature, the reaction of candle wax with oxygen is so slow that, for all practical purposes, it does not occur.

Figure 15-4

Temperature has a marked effect on the rate of chemical reactions.

EXERCISE 15-4

Explain why a mixture of one part CH_4 gas (natural gas) and two parts oxygen can stand quietly for years if kept from sparks or flame but will explode violently if a spark is passed through the container.

In nearly all known reactions, an increase in temperature causes a pronounced increase in reaction rate. Two questions come to mind: why does a rise in temperature speed up a reaction, and why does a rise in temperature have such a large effect? To answer these questions, we look again at the collision theory.

Suppose we have a mixture of household gas (mostly methane, CH_4) and air at 25 °C and 1 atmosphere pressure. From what we know about molecular sizes, we can calculate that a particular CH_4 molecule collides with an O_2 molecule about once every one-thousandth of a microsecond (10^{-9} second). In other words, the methane molecule encounters 10^9, or 1,000,000,000, oxygen molecules every second. Yet, the reaction does not proceed noticeably. That is, in the absence of increased temperature, there is no flame or explosion. We can conclude either that most of the collisions are ineffective or that the collision theory is not as good an explanation of reaction rates as we thought. We shall see that the former conclusion is valid. *Most molecular collisions are ineffective.*

Chemists have learned that chemical reactions do not occur with every collision but only *when a collision involves more than a certain amount of energy.* We can explain this phenomenon by using the analogy of cars bumping each other on a highway. In a line of heavy traffic, you may sometimes gently bump the car in front of you or receive such a bump from the car behind you. No damage is done to the cars. On the other hand, a higher-speed collision occasionally occurs. It is the high-energy collisions that cause damage. It is also the high-energy molecular collisions that cause a chemical reaction. A certain amount of energy is required to bring about molecular rearrangements involved in a chemical reaction. If the amount of energy present exceeds the "threshold," the reaction can occur. If the available energy is less than the threshold amount, the reaction cannot occur. The geometry of the collision is also important. A low-velocity collision with a car's bumper is much less effective than such a collision with a fender or door. A head-on collision is much more likely to be destructive than a glancing one. For a very effective collision, a combination of high velocity and proper geometry is necessary.

THRESHOLD ENERGY AND REACTION RATE

The above observations on threshold energy make us wonder what energy molecules possess at a given temperature. We have already compared gas molecules to ping-pong balls. (See pages 19–21 and pages 34–42.) We have also discussed the idea that at any given temperature, gas molecules in a given volume have a distribution of various kinetic energies. Some move rapidly and have high kinetic energy. Some move slowly and have low kinetic energy. Most molecules fall somewhere in between. At a given temperature, the average kinetic energy of a gas sample depends on that temperature (pages 118–121). Thus, we already picture any sample of gas as having molecules with different kinetic energies at a given temperature. Experiments support this model.

Figure 15-5 (on page 400) shows a device for measuring the distribution of atomic or molecular velocities. It consists of two disks, D_1 and D_2, rotating rapidly on the same axle. The disks rotate in a vacuum chamber in front of

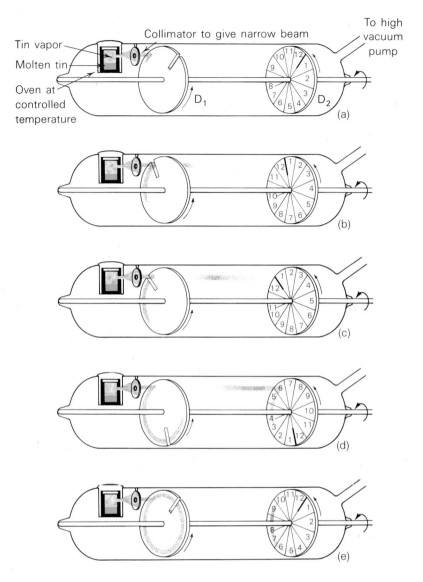

Figure 15-5
A rotating disk can be used to measure atomic or molecular velocities.

a small oven containing molten tin. The entire system is held at a controlled temperature. The temperature is high enough to vaporize some of the tin. Tin vapor streams out of the small opening in the oven and strikes the rotating disk, D_1. When the disk has rotated to the position shown in Figure 15-5b, a small amount of tin vapor passes through the slot in disk D_1. A short time later, as shown in Figure 15-5c, the atoms of tin have traveled part of the way toward the second rotating disk. The fastest-moving atoms have traveled farther than the others and are "leading the way." The slowest-moving atoms are beginning to lag behind. Still later, the atoms have spread out even more, as shown in Figure 15-5d. There, the fastest atoms have already reached the second rotating disk. As the atoms reach disk D_2, they condense on its cool surface. The place on disk D_2 where a given atom condenses depends on how long that atom took to travel from D_1 to D_2. It also depends on how fast disk D_2 is rotating.

As the slotted disk, D_1, lets through burst after burst of tin atoms, a layer of tin builds up on the surface of disk D_2. The pattern of this layer is determined by the distribution of velocities of the atoms escaping from the oven at temperature T. Figure 15-6 shows a close-up of disk D_2 divided into sections, like slices of a pie. The fastest-moving atoms are condensed on pie slices 3, 4, and 5. The slowest-moving atoms are condensed on pie slices 10 and 11. If the disk is cut up and each slice placed on a sensitive laboratory balance, the mass of each slice can be accurately measured. The mass of tin atoms deposited on each slice is then calculated by subtracting the mass of an "empty" slice. A plot of the mass of tin against the pie-slice number indicates the distribution of atomic velocities in the tin-vapor sample.

EXERCISE 15-5

Why is there no tin deposited on pie slices 1 and 2? on slice 12? What would be the effect of speeding up the rotation of the disks on the amount of tin deposited on each slice? Which slices would have more tin deposited on them? Which would have less? Which would have the same amount?

You read in Chapter 5 that the relationship between the velocity of a gas molecule, its mass, and its kinetic energy is expressed by the equation

$$KE = \tfrac{1}{2}mv^2$$

Clearly, the curve shown in Figure 15-6 contains information about the distribution of kinetic energies. From the rate of rotation of the disks and the distance between them, we can calculate the velocity a tin atom must have

Figure 15-6

The relative distribution of atomic or molecular velocities obtained from a rotating disk.

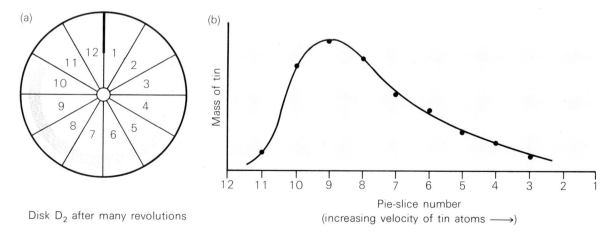

Disk D_2 after many revolutions

Pie-slice number
(increasing velocity of tin atoms ⟶)

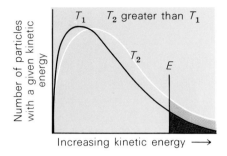

Number of particles with a given kinetic energy

T_1 T_2 greater than T_1

T_2

E

Increasing kinetic energy $\longrightarrow$

Figure 15-7

The effect of temperature on the relative distribution of atomic or molecular kinetic energies.

to condense on a particular pie slice.* From the atomic mass and the velocity, we learn the kinetic energy of the atom. Figure 15-7 shows the result of such calculations, after running the rotating disk experiment at *two* different temperatures, T_1 and T_2. At temperature T_1, a few atoms have very low kinetic energies. Some have very high kinetic energies. Most have intermediate kinetic energies, as shown by the black line. At a higher temperature, T_2, the energy distribution is altered, as indicated by the white line. As you can see, *increasing the temperature causes a general shift of the molecular distribution toward higher kinetic energies.* Notice especially that in going from T_1 to T_2, there is a large increase in the number of molecules having kinetic energies above a certain value, E.

We can use the curves shown in Figure 15-7 to help answer our questions concerning the effect of temperature on reaction rates. Suppose a reaction can proceed only if two molecules collide with a total kinetic energy exceeding a certain threshold energy, E. Figure 15-7 shows a typical situation. At T_1, the lower temperature, the dark-shaded area in the "tail" of the curve is proportional to the number of molecules possessing at least the threshold energy. These molecules have enough energy for effective collisions even with molecules of very low kinetic energy. Since only a small number of molecules have this much energy, few collisions are effective. The reaction is slow. But if we raise the temperature to T_2, the number of molecules with energy E or greater is raised. The area that is proportional to the number of molecules having the increased amount of energy includes the dark-shaded area *and* the light-shaded area. *Only a small temperature change is needed to make a large change in the number of molecules with more than the threshold energy.* Reaction rate is very sensitive to temperature change. For example, an increase of 10 °C will usually double the number of molecules exceeding the threshold energy, E.

It should be noted also that raising the temperature also increases the frequency of collisions. This is, however, a very small effect compared with that caused by increasing the number of molecules with sufficient energy to react. (An increase of 10° from 273 K to 283 K will increase the number of collisions by less than 2 percent.)

15.3   The role of energy in reaction rates

We have discussed the effect of temperature in increasing the fraction of molecular collisions having a total energy greater than the threshold energy. Now our understanding of the role of energy in fixing reaction rates can be discussed in more detail.

ACTIVATION ENERGY

Suppose you wished to make a bicycle trip from city R to city P. As shown in Figure 15-8, the trip requires that you take a route over a mountain range between the two cities. If it were not for the mountains, you could almost

*Here is an example. Suppose the disks are 1.0 cm apart and are rotating at a rate of 1,200 revolutions per minute, or $\frac{1,200}{60} = 20$ revolutions per second. One-twentieth of a second is required for a complete revolution. There are 12 sections on the disk; thus, each section will be in the target area for $\frac{1}{12} \times \frac{1}{20} = \frac{1}{240}$ second. The atoms collecting on slice 1 require $\frac{1}{240}$ second to travel the distance of 1.0 cm between disks. Their average velocity will be 1.0 cm/$\frac{1}{240}$ sec = 240 cm/sec. Those on slice 2 require $\frac{2}{240}$ of a second; they are going 120 cm/sec; and so on.

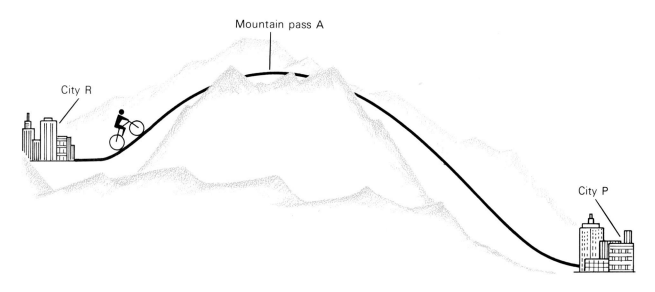

Figure 15-8

A bicyclist crossing mountains en route from City R to City P.

coast all the way to your destination, for P is nearer to sea level than is R. The slowest and most difficult part of the trip is crossing the mountains. You need to exert a great deal of energy in order to get to the top of the mountain pass. However, once your bike reaches the top of the pass, A, in Figure 15-8, you can relax and coast the rest of the way to city P. Given this route, only bicyclists who have enough energy to reach point A will be able to complete the journey. Those who do not will have to turn back and return to the starting point, R.

Chemical reactions are similar to the bicycle-trip analogy. As particles collide and reactions take place, atoms temporarily form bonding arrangements that are less stable—but of higher potential energy—than those of reactants or products. The high-energy molecular arrangements are like the mountain pass, A. They place an energy "barrier" between reactants and products. In our road map, city R represents reactants and city P represents products. The road itself represents the path the reactants must travel if they are to become products. Only if the colliding molecules have enough energy to cross the barrier imposed by the unstable arrangements can the reaction take place. The height of the barrier determines the threshold energy, or minimum energy, necessary to permit a reaction to occur. It is called the **activation energy.**

The energy barrier for a chemical reaction is shown graphically in Figure 15-9, on page 404. The figure represents the reaction between carbon monoxide (CO) and nitrogen dioxide (NO_2) to produce carbon dioxide (CO_2) and nitric oxide (NO), all of which are gases. The diagram is the equivalent of a side view of the road map in our bicycle-trip analogy. It is known as a **potential energy diagram.**

The horizontal axis of Figure 15-9 is called the *reaction coordinate*. It shows the progress of the reaction. This is analogous to the road trip of our bicyclist in Figure 15-8. As we move from left to right along this reaction

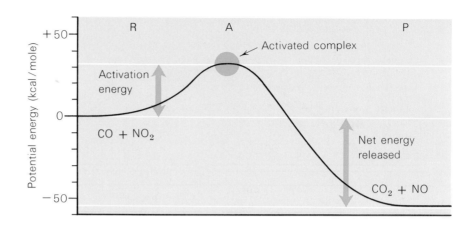

Figure 15-9

Potential energy diagram for the reaction $CO + NO_2 \longrightarrow CO_2 + NO$.

coordinate, we travel through the different "parts" of the reaction. Between R and A, the CO and NO_2 molecules are approaching each other. They collide at point A. Here they form an unstable intermediate arrangement known as the **activated complex.** The space between A and P is where the activated complex separates into the product molecules, CO_2 and NO.

The vertical axis of Figure 15-9 represents the total potential energy of the system. Thus, the curve provides a history of the potential energy change during a collision that results in a reaction. The energy required to overcome the potential energy "barrier" to the reaction is usually provided by the kinetic energies of the colliding particles. As we have seen, kinetic energies depend on the temperature.

Let us move again from left to right along the curve of Figure 15-9. This time, we will describe in detail the events that occur. Along the flat region at the left, near R, CO and NO_2 are approaching each other. In this region, they possess kinetic energy as they speed toward each other. Their total potential energy shows no change. The beginning of the rise in the curve signifies that the two molecules have come sufficiently close to have an effect on each other. As they approach one another, the molecules slow down. This is due to the mutual repulsion of their surface electrons and to the deformation of both molecules. As they slow down, their kinetic energies convert to potential energy to "climb the curve." If they have sufficient kinetic energy, they can climb the left side of the "barrier" all the way to the summit at A. Reaching the summit is interpreted as follows: CO and NO_2 molecules had enough kinetic energy to overcome the forces of repulsion that develop as the molecules come very close to each other. Here at the summit, the molecular cluster is unstable with respect to either the forward reaction, which forms CO_2 and NO, or the reverse reaction, which returns the molecules to their original state as CO and NO_2. The intermediate arrangement, the activated complex, is of key importance. Its potential energy determines the activation energy.

Once the activated complex forms, there are two possibilities:

1. The activated complex can separate into the original CO and NO_2 molecules, which would then return to their starting point, R, on the curve.
2. The activated complex can separate into the product molecules, CO_2 and NO.

The latter possibility is represented by moving down the right side of the "barrier" toward P. In the flat region beyond P, CO_2 and NO have separated so much that they do not have any effect on each other. The potential energy of the activated complex has once again been converted to kinetic energy.

If the CO and NO_2 molecules from R do not have sufficient energy to reach the summit, A, they reach a point that is only part way up the left side of the "barrier." Repelling one another, they separate again, going back "downhill" to the left.

We have called the difference between the high potential energy of the activated complex and the lower energy of the reactants the activation

OPPORTUNITIES

Laboratory technology

There are many employment opportunities in the chemical industry for high school graduates who do not intend to continue their studies beyond high school, as well as for those who choose to delay college for a time. People who enjoy working in a laboratory might find satisfying jobs as chemical operators or laboratory technicians.

Industrial chemical firms generally hire potential operators directly out of high school and train them on the job. Training periods can range from several days to several years, depending on the degree of skill needed. Full-fledged chemical operators are almost always trained by an apprentice system within a company and are rarely recruited from other companies. Laboratory technicians may receive training at technical schools or junior colleges. However, high school graduates with courses in chemistry can qualify for such work with companies that provide on-the-job training.

Chemical operators have the responsibility for controlling and monitoring manufacturing machinery in the production area of a chemical plant. They may also monitor the rate of flow of raw materials into reaction vessels, regulate safety devices such as pressure and temperature gauges, and sample materials during the manufacturing process. Their job is to see that proper conditions for producing specific chemical products are maintained. For example, operators in pharmaceutical companies may measure coating materials for tablets, or they may control the mixing operations for the production of creams, ointments, liniments, and other products. They may operate autoclaves, water stills, and many other types of equipment.

Laboratory technicians may perform simple tests on materials being used or produced by a chemical company. They assist in such areas as quality control, and often set up equipment and instruments needed by the chemist in charge of the laboratory. The degree of responsibility assumed by laboratory technicians is directly related to their education and experience.

Many firms offer financial aid to employees attending college on a

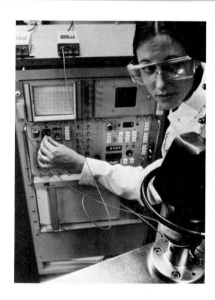

part-time basis. The chemical industry is interested in finding bright, ambitious people who aspire to challenging positions. It is to the advantage of a company to encourage this type of employee. The position of chemical operator or laboratory technician may be the first step in a productive career. For further information, you may contact the Chemical Manufacturers Association, 2501 M Street NW, Washington, D.C. 20037.

energy. *Activation energy is the minimum energy necessary to transform reactants into an activated complex.* In the process, bonds may be weakened or broken as reactants are forced close together in opposition to forces of repulsion. Also, energy may be stored in a vibrating molecule. Later, this energy can cause a reaction when molecules collide. *Increasing the temperature affects reaction rate by increasing the number of molecular collisions that have enough energy to form an activated complex.* The activation energy required to start a reaction can be determined experimentally by measuring the change in reaction rate with temperature.

HEAT OF REACTION

We can deduce the heat of reaction (discussed in Chapter 14) from Figure 15-9. In our example, the reactants have a higher total energy than do the products. This means that in the course of the reaction, there will be a net *release* of energy. Therefore, the reaction is *exothermic*. Figure 15-9 shows that the reaction releases 54 kilocalories of heat per mole of $CO(g)$ consumed. Notice that the height of the energy barrier between reactants and products has no effect on the net amount of heat released. To get to the top of the barrier, we must put in an amount of energy equal to the activation energy. We get all the energy back on the way down the other side.

Now let us consider the *reverse* reaction. See Figure 15-10. Here, we are interested in the reaction between CO_2 and NO to produce CO and NO_2:

$$CO(g) + NO_2(g) \longleftarrow CO_2(g) + NO(g)$$

Several points are of interest in Figure 15-10. First, the process is *endothermic*. The products have more energy than the reactants. Secondly, the activation energy is higher than the activation energy of the $CO + NO_2$ reaction. The difference is the ΔH of the reaction of $CO + NO_2$. The reaction between CO_2 and NO will be slower than the reaction between CO and NO_2 if concentrations are comparable.

Figure 15-10

Potential energy diagram for the reaction CO_2 + NO $\longrightarrow$ CO + NO_2.
Note that this is just the reverse of Figure 15-9.

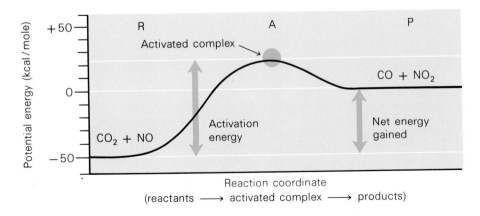

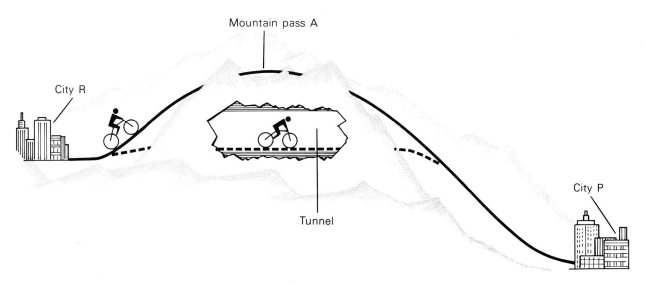

Figure 15-11

An easier, alternative route by tunnel through the mountains from City R to City P.

ACTION OF CATALYSTS

Some substances are able to make a slow reaction go more rapidly without themselves undergoing permanent change. Such substances are called **catalysts.** You may recall the effect of $Mn^{++}(aq)$ ions in speeding up the reaction between $C_2O_4^{--}(aq)$ and $MnO_4^-(aq)$ ions. $Mn^{++}(aq)$ was a catalyst.

The action of a catalyst can be explained in terms of our mountain-pass analogy. In that analogy, the traveler wishing to go from city R to city P had to cross a mountain range. Energy was required to get to the top. Let us now imagine that highway engineers build a tunnel through the mountains, as shown in Figure 15-11. Now, travelers have their choice of either the rugged, but more scenic route, or the easier, alternative route by tunnel. The alternative route allows more bicyclists to make the trip from R to P. The trip is easier; more people make it in a day.

Figure 15-12 shows the analogous situation for a chemical reaction. The solid curve includes the activation energy barrier that must be overcome in order for the reaction to take place. When a catalyst is added, a new reaction path is provided. It is analogous to our alternative bicycle route through the tunnel. The catalyzed path has a different activation energy barrier, as shown by the dashed curve in Figure 15-12. The new reaction path corresponds to a new reaction mechanism in which a different activated complex is formed. Because the "barrier" is lower, more particles can get over it during any given time period. Thus, the rate of the reaction is increased.

Note that the activation energy of the reverse reaction in catalysis is lowered by exactly the same amount as the activation energy for the forward reaction. This accounts for the fact that a catalyst has an equal effect on

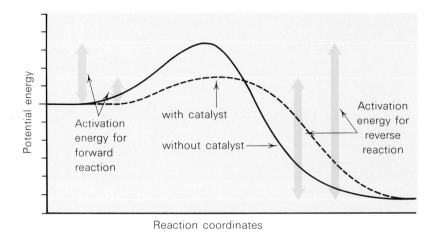

Figure 15-12

The effect of a catalyst on a reaction and its reverse.

reverse and forward reactions. Both forward and reverse reactions are speeded up by the same factor. If a catalyst doubles the rate of reaction in one direction, it doubles the rate of reaction in the reverse direction.

EXAMPLES OF CATALYSIS

In all cases of catalysis, the catalyst acts by inserting intermediate steps in a reaction. The steps would not occur without the catalyst. In the process, the catalyst itself is regenerated, or remade. An example of this phenomenon is the catalytic action of acid on the decomposition of formic acid (HCOOH). A model of formic acid is shown in Figure 15-13. The carbon atom has attached to it a hydrogen atom, an oxygen atom, and an OH group.

Figure 15-14 shows how the molecule might decompose. If the hydrogen atom attached to the carbon atom moves over to the OH group, the carbon-oxygen bond can break to give a molecule of water and a molecule of carbon monoxide. Such movement, shown in the center drawing, requires a large amount of energy. This indicates that there is a high activation energy, which only a few HCOOH molecules can overcome. Thus, the reaction occurs very slowly.

$$HCOOH \longrightarrow H_2O + CO$$

However, if sulfuric acid (H_2SO_4) is added to an aqueous solution of formic acid, carbon monoxide bubbles out rapidly. This also occurs if phosphoric acid (H_3PO_4) is added. The common factor is that both acids release hydrogen ions, $H^+(aq)$, often abbreviated H^+. Although H^+ makes a difference, careful analysis shows that the concentration of hydrogen ions is constant during the rapid decomposition of formic acid. H^+ does *not* get used

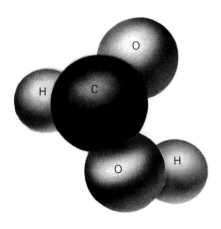

Figure 15-13

A space-filling model of formic acid (HCOOH).

up. Evidently, the hydrogen ion acts as a catalyst. Any substance that is permanently used up is a reactant, not a catalyst.

Chemists have a rather clear picture of how H^+ catalyzes the decomposition of formic acid. The availability of H^+ in the solution makes a new reaction path available. Figure 15-15 shows this reaction path. The new reaction mechanism begins with the addition of a hydrogen ion to formic acid. Thus, the catalyst is consumed at first, forming a new species, $(HCOOH_2)^+$. In this species, one of the carbon-oxygen bonds is weakened. With only a small expenditure of energy, the next reaction can occur. The latter produces $(HCO)^+$ and H_2O. Finally, $(HCO)^+$ decomposes to produce carbon monoxide (CO) and H^+. The last reaction of the sequence regenerates the catalyst, H^+. Summarizing the steps,

$$HCOOH + H^+ \longrightarrow (HCOOH_2)^+$$
$$(HCOOH_2)^+ \longrightarrow (HCO)^+ + H_2O$$
$$\underline{(HCO)^+ \longrightarrow CO + H^+}$$
$$HCOOH \longrightarrow CO + H_2O$$

Each step in the new reaction mechanism is governed by the same principles that govern a simple reaction. Each subreaction has its own activation energy. The overall reaction has a potential energy diagram that is merely a composite of the simple energy curves of the succeeding steps. This is also shown in Figure 15-15.

The highest energy required in the new reaction path is much less than the value for the uncatalyzed reaction in Figure 15-14. Thus, the rate of decomposition is much faster when acid is present. Notice that catalytic action does not *cause* the reaction. A catalyst *speeds up* a reaction that might

Figure 15-14

Potential energy diagram for the uncatalyzed decomposition of formic acid.

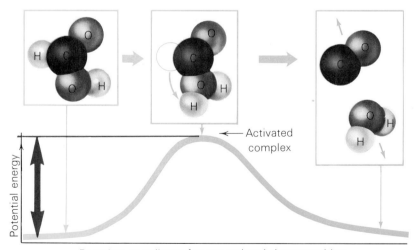

Reaction coordinate for uncatalyzed decomposition

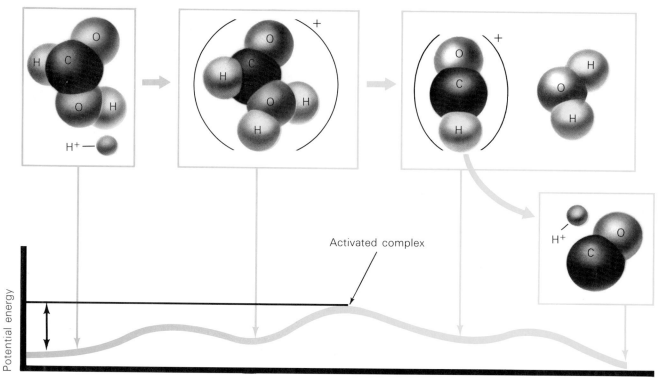

Figure 15-15

Potential energy diagram for the catalyzed decomposition of formic acid.

take place anyway, but at a much slower rate.

In some cases, a catalyst is a solid substance on whose surface a reactant particle can be adsorbed, or held. The adsorbed particle stays in place until a particle of another reactant reaches the same point on the solid. This brings the two particles into a position that favors reaction. Metals such as iron, nickel, platinum, and palladium seem to work in this way to catalyze reactions between gases. Bonds may be weakened or actually broken. A reactant particle will then react more easily with another reactant particle.

A very large number of catalysts, called *enzymes,* are found in living tissues. Among the best known examples of these are digestive enzymes, such as ptyalin in saliva and pepsin in gastric juice. A common function of these two enzymes is to speed up the breakdown of large molecules, such as starch and protein, into smaller molecules that can then be used by body cells. Enzymes of many kinds speed up a multitude of chemical reactions within living things. Life as we know it would be impossible without enzymes.

In most cases, the specific methods by which catalysts work are not clearly understood. Thus, a long period of laboratory experimentation is often needed to find a substance that catalyzes a specific reaction. Someday, perhaps, catalysts will be tailored to particular needs. This exciting prospect has stimulated great activity on the frontier of chemical catalysis.

SUMMARY

Reactions proceed at different rates. The study of reaction rates is called chemical kinetics. The rate of a reaction is found by dividing the change in concentration of a reactant or product by the time interval in which the change occurs:

$$\text{reaction rate} = \frac{\text{concentration change}}{\text{time interval}}$$

The factors that affect reaction rates are (1) the nature of the reactants, (2) their concentration, (3) their temperature, and (4) catalysts. The idea that chemical reactions result from collisions between reacting particles is called collision theory.

The series of steps by which a reaction is thought to occur is called its reaction mechanism. The slowest step in this sequence is called the rate-determining step. Most reactions proceed by sequences of steps involving two-particle collisions. The overall, net equation almost never represents the mechanism by which a reaction occurs. Thus, net equations cannot be used to predict rates of reaction. Although the equations for reactions may be similar, experimental studies of their rates may suggest that the reactions actually occur by very different mechanisms.

Relatively few collisions between particles are effective. Effective collisions are those that produce chemical changes. Only molecules above a certain threshold kinetic energy and with proper collision geometry cause effective collisions. Raising the temperature increases the kinetic energy of molecules and, thus, the rate of reaction between molecules.

High-energy, intermediate molecular arrangements cause a potential energy barrier between reactants and products. The height of the barrier, known as the activation energy, is a measure of the minimum energy needed for the reaction to occur.

A potential energy diagram shows the path of a reaction along a horizontal axis known as a reaction coordinate. The highest point on the path, the point of maximum potential energy, corresponds to formation of the unstable intermediate known as the activated complex. Its potential energy determines the height of the barrier, which is the activation energy.

A catalyst speeds up a chemical reaction without being consumed. A catalyst provides an alternative, easier reaction path, with a lower activation energy than that required by the uncatalyzed path.

Key Terms
activated complex
activation energy
catalysis
catalyst
chemical kinetics
collision theory
concentration
heterogeneous system
homogeneous system
potential energy diagram
rate-determining step
reaction mechanism

QUESTIONS AND PROBLEMS

A

1. (a) What kind of a unit is always in the denominator of a rate expression? (b) A man runs 16 kilometres in 4 hours. What is his average rate of travel? (c) A process generates 0.5 mole of CO_2 in 10 minutes. What is the rate of the CO_2 formation process? Show units.

2. Express the rate of the reaction $2HI(g) \longrightarrow H_2(g) + I_2(g)$ in terms of H_2 concentration buildup.

3. Gaseous NH_3 reacts very slowly with O_2. A catalyst is needed to get a reasonable rate at workable temperatures. The somewhat related compound PH_3 (see periodic table) reacts quite rapidly with oxygen at room temperature. What causes the difference?

4. Several processes are listed below. Some are fast and some are slow. Indicate *why* you would expect each process to be fast or slow at room temperature.

   (a) $5C_2O_4^{--} + 2MnO_4^- + 16H^+ \longrightarrow$
   $$10CO_2 + 2Mn^{++} + 8H_2O \quad \text{SLOW}$$

   (b) $2C_2H_6 + 7O_2 \longrightarrow 4CO_2 + 6H_2O$ SLOW

   (c) $Ag^+ + Cl^- \longrightarrow AgCl$ FAST

   (d) $Ce^{++++} + Fe^{++} \longrightarrow Fe^{+++} + Ce^{+++}$ FAST

   (e) $H^+ + OH^- \longrightarrow H_2O$ FAST

   (f) $2NH_3 + 2\frac{1}{2}O_2 \longrightarrow 2NO + 3H_2O$ SLOW

5. Explain, in terms of the kinetic theory, why an increase in concentration of reactants should increase the rate of a reaction.

6. In recovering copper from a solution obtained by passing water through mine dumps, iron scrap is used. The equation is

$$Fe(s) + Cu^{++}(aq) \longrightarrow Fe^{++}(s) + Cu(s)$$

How might you *increase* the rate of the process?

7. A wooden splint that glows in air will burst into flames if it is plunged into a tube full of O_2 gas. Why?

8. (a) A reaction mechanism that involves the simultaneous collision of just 4 molecules is not normally proposed. Why?

   (b) Explain why the following equation does not represent the *mechanism* for the oxidation of HBr by O_2:

$$4HBr + O_2 \longrightarrow 2H_2O + 2Br_2$$

   (c) Write the suggested reaction mechanism for the reaction given in (b) above. What piece of experimental information suggests this mechanism? Refer to page 396.

9. (a) Why does milk keep better in a refrigerator than it does if allowed to stand on the cupboard shelf? Explain. Consider the energetics of the process. (b) Why does a match ignite a mixture of propane and oxygen?

10. Why is lump sugar satisfactory for use in hot chocolate but granulated sugar preferable for use in iced tea?

11. Examine Figures 15-5 and 15-6. Which "pie slice" contains the *slowest-moving* tin atoms?

12. Refer to Figure 15-7. Describe the shape of the plot obtained at a third temperature T_3, where T_3 is *lower* than the original temperature, T_1.

13. Figure 15-8 is a "road map" for the trip from city R to city P via mountain pass A. Assuming such a trip is similar to a chemical reaction, what is represented by city R, city P, and mountain pass A?

14. Consider the reaction $A + B \rightarrow C + D + 70$ kcal. The activation energy for the process between A and B is 22 kcal. (a) Sketch the potential

energy diagram for the process between C and D. Label both axes. (b) Sketch the potential energy diagram for the reverse process. (c) Sketch the potential energy diagram for the forward process in the presence of a good catalyst.

15. Write the equation for the decomposition of formic acid (HCOOH). What is the catalyst for this process?

16. Refer to Figure 15-9. Where does the activated complex form? What two possible "futures" exist for the activated complex?

B

17. An aircraft flies between Toronto and Windsor, Canada, (a distance of 200 miles) in an hour and 10 minutes (gate to gate). (a) What is the effective overall *rate of travel* of the airplane? Give units. (b) Was the airplane going faster than the overall rate at some point on its trip? Explain. (c) What is frequently the rate-determining step when the airport is very busy or is closed by a storm?

18. A mixture of isooctane (C_8H_{18}) and oxygen in an automobile cylinder explodes in accordance with the equation

$$C_8H_{18}(g) + 12\tfrac{1}{2}O_2(g) \longrightarrow 8CO_2(g) + 9H_2O(g)$$

What *units* might be used to describe the reaction? Refer to page 391.

19. (a) The reaction of a gaseous chlorine atom with a hydrogen molecule (H_2) is very fast. Knowing this fact, explain why ultraviolet light will initiate the reaction of a mixture of H_2 and Cl_2. (b) Refer to the reaction between H_2 and I_2 on page 398. Would you expect ultraviolet light to be as effective in initiating this reaction as it is in the $H_2 + Cl_2$ reaction? Explain.

20. What is the function of the pilot light in a gas stove? Explain by drawing the approximate potential energy diagram for the reaction

$$CH_4(g) + 2O_2(g) \longrightarrow CO_2(g) + 2H_2O(g)$$

and indicating with an arrow the function of the pilot light.

21. A manufacturer of flashcubes containing magnesium powder and oxygen gas wants to achieve a somewhat longer duration for the flash period. Can you suggest a possible way to do this?

22. An old rule of thumb states that, for many common reactions, an increase in temperature of 10 °C doubles the rate of the reaction. Does this mean that an increase of 10 °C doubles the average kinetic energy of the molecules? Explain, using a curve like that shown in Figure 15-7.

23. Explain why a *mechanism* for ammonia formation such as

$$3H_2 + N_2 \longrightarrow 2NH_3$$

is highly unlikely.

24. Given the kinetic energy distribution curves and the threshold energies for reactions A and B, at right: (a) Which reaction will be faster at room temperature? (b) Which reaction will show the greater rate increase if the temperature is raised 10 °C?

25. Given the potential energy diagram at right: (a) Is this reaction exothermic or endothermic? (b) Would you expect this reaction to occur rapidly

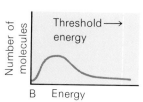

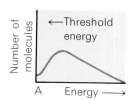

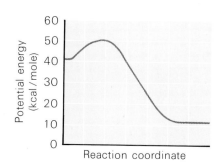

at room temperature? Why or why not? (c) What is ΔH for this reaction? (d) What is the activation energy for the forward reaction? (e) At what point on the graph might an activated complex exist? (f) Which of the following reactions could this diagram most likely represent?

(1) $H_2SO_4(l) \xrightarrow{H_2O} 2H^+(aq) + SO_4^{--}(aq)$

(2) $NH_4NO_3(s) \xrightarrow{H_2O} NH_4^+(aq) + NO_3^-(aq)$

(3) $(NH_4)_2Cr_2O_7(s) \xrightarrow{heat} N_2(g) + Cr_2O_3(s) + 4H_2O(g)$

Explain your reasoning in each case.

26. Consider the potential energy diagram at left. (a) Is this reaction exothermic, endothermic, or neither? (b) Would you expect this reaction to occur rapidly at room temperature? Explain your answer. (c) Of the following reactions, which one might this diagram represent?

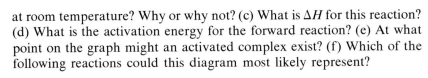

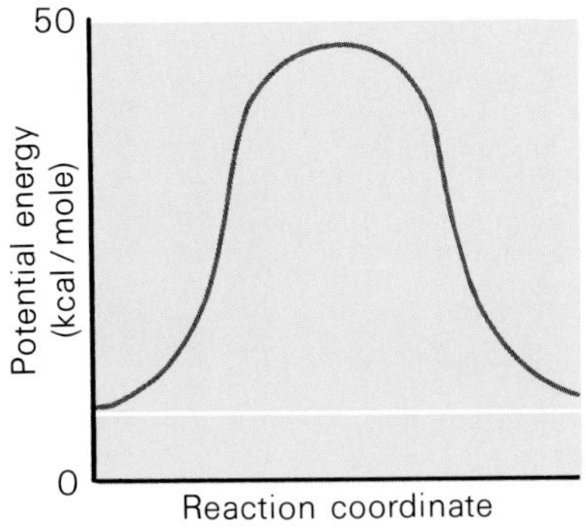

(1) $H_2SO_4(l) \longrightarrow 2H^+(aq) + SO_4^{--}(aq)$
(2) $NH_4NO_3(s) \longrightarrow NH_4^+(aq) + NO_3^-(aq)$
(3) $CH_4(g) + 2O_2(g) \longrightarrow CO_2(g) + 2H_2O(g)$
(4) $C/(graphite) \longrightarrow C/(diamond)$, where $\Delta H = +0.45$ kcal

Reaction (4) takes place only at high temperatures and pressures.

27. For a particular reaction, ΔH is -60 kcal, and the activation energy for the forward reaction is $+30$ kcal. (a) Draw the potential energy curve for the reaction. Be sure to label the axes and include a scale on the potential energy axis. (b) Label the parts of the curve representing (1) the reactants, (2) the products, (3) the activation energy of the forward reaction, (4) the activated complex, and (5) the net energy absorbed or released (give the amount of energy, and note whether it is absorbed or released).

C

28. In Experiment 10, hydrogen gas is produced by the reaction of magnesium ribbon with 15 ml of 3 M hydrochloric acid added to about 40 ml of water:

$$Mg(s) + 2H^+(aq) + 2Cl^-(aq) \longrightarrow Mg^{++}(aq) + 2Cl^-(aq) + H_2(g)$$

(a) How can you express the rate of the reaction? (b) What units, if any, should be used? (c) What effect on rate would you expect to result from using 15 ml of 6 M HCl in 40 ml of water? (d) Is your answer qualitative or quantitative? Why?

29. Black powder is a mixture of charcoal, sulfur, and the compound saltpeter, or potassium nitrate. A laboratory mixture of the powdered ingredients may not ignite when touched with the flame of a match, but, after processing through a powder mill, it explodes violently when ignited by a match. Can you suggest why?

30. One of the important processes in the formation of photochemical smog is

$$2NO(g) + O_2 \xrightarrow{\text{light}} 2NO_2(g)$$

NO comes from automobile exhaust products. The rate of NO_2 formation is studied at constant O_2 concentration and variable NO concentration. Data are shown below:

Relative initial concentration of NO	Relative initial rate of process
1	1
2	4
3	9

What would the relative initial rate of the process be at an NO concentration of 5? Write a generalized expression for the rate of NO_2 formation when NO concentration is a variable. Would these data be consistent with the following mechanism? Explain your answer.

$$NO + NO \longrightarrow N_2O_2 \quad \text{FAST}$$
$$N_2O_2 + O_2 \longrightarrow 2NO_2 \quad \text{SLOW}$$

Explain your answer.

31. The rate law for the reaction

$$NO + O_3 \longrightarrow NO_2 + O_2$$

is rate $= k[NO][O_3]$. Which of the following mechanisms is consistent with this rate law? Explain your answer.

(a) $NO \longrightarrow N + O$ SLOW
$N + O_3 \longrightarrow NO_2 + O$ FAST
$O + O \longrightarrow O_2$ FAST
$NO + O_3 \longrightarrow NO_2 + O_2$ OVERALL

(b) $NO + O_3 \longrightarrow NO_2 + O_2$ SLOW

(c) $O_3 \longrightarrow O_2 + O$ SLOW
$O + NO \longrightarrow NO_2$ FAST
$O_3 + NO \longrightarrow NO_2 + O_2$ OVERALL

32. Suppose the disks on the apparatus in Figure 15-5 are 1.0 cm apart and rotating at a rate of 1,200 revolutions per minute to yield the "pie-slice" distribution shown in Figure 15-6. Suppose the rate of rotation of the disks is increased to 2,400 revolutions per minute. What are the velocities of the tin atoms that arrive on pie slices 1 and 2? Refer to the footnote on page 402.

33. For the reaction

$$H_2 + Cl_2 \longrightarrow 2HCl + 44\,kcal$$

the energy of activation for the process as written is 37 kcal. What is the energy of activation for the decomposition of HCl to give H_2 and Cl_2?

By . . . equilibrium, we mean a state in which the properties of a system, as experimentally measured, would suffer no further observable change even after the lapse of an indefinite period of time. It is not intimated that the individual particles are unchanging.

GILBERT NEWTON LEWIS AND MERLE RANDALL

16

EQUILIBRIUM IN CHEMICAL AND PHASE CHANGES

Objectives

After completing this chapter, you should be able to:

- Explain phase, solubility, and chemical equilibrium in terms of equal rates of forward and reverse processes.

- Apply Le Chatelier's principle to predict how an equilibrium system will respond to various changes.

- Write equilibrium expressions for various reactions.

- Recognize the significance of large and small values of the equilibrium constant, K.

- Understand the connection between equilibrium, minimum energy, and maximum randomness.

- Apply equilibrium constants in solving various types of equilibrium problems.

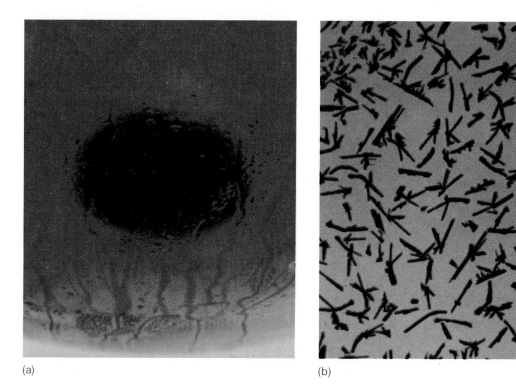

(a)   (b)

Evaporation-condensation equilibrium(a): bromine vapor in equilibrium with liquid bromine. Note the very thin streamlets of reddish-brown liquid bromine on the glass. (A small pool of liquid bromine appears black in this photograph.) Sublimation-condensation equilibrium(b): purple iodine vapor in equilibrium with crystals of solid iodine. Note the grayish-black crystals of solid iodine on the glass.

In the previous chapter, we looked at the rate of the reaction between carbon monoxide, $CO(g)$, and nitrogen dioxide, $NO_2(g)$:

$$CO(g) + NO_2(g) \longrightarrow CO_2(g) + NO(g)$$

If we mix $CO(g)$ and $NO_2(g)$ in a sealed container at room temperature, the red-brown color of $NO_2(g)$ fades, indicating that some of it is being converted to colorless $NO(g)$. You might expect that the $NO_2(g)$ would be consumed and that all the color would eventually disappear. But this does *not* happen. Instead, the color fades to a light brown and then shows no further changes. Apparently, the relative amounts of $NO_2(g)$ and $NO(g)$ become constant. But what is happening on the molecular level? To better understand this chemical system, it will be helpful to review some physical systems you studied earlier.

16.1 Equilibrium in phase changes

The *equilibrium* state was introduced in Chapter 6, where we discussed vapor pressure above a liquid. Let us review that discussion as we consider equilibrium in phase changes.

LIQUID-VAPOR EQUILIBRIUM

In Chapter 6, we described apparatus used for determining the vapor pressure above liquid water at various temperatures. See Figure 6-6, on page 141. If the sealed, evacuated flask containing liquid water is removed from a constant temperature bath at 0 °C and placed in a constant temperature bath at 25 °C, the observed vapor pressure increases to 23.8 mm Hg. It then remains constant. The same, constant vapor pressure—23.8 mm Hg—is also reached if the flask is removed from a 50 °C constant temperature bath and cooled to 25 °C. *If the same vapor pressure is reached whether a system approaches equilibrium from a pressure above or below the equilibrium vapor pressure, the pressure obtained is the equilibrium value.*

In describing molecular activity at equilibrium, we visualized a system in which some molecules leave the surface of the liquid and enter the vapor phase while other molecules return from the vapor phase to the surface of the liquid. We defined equilibrium as the condition existing in a system when the rate at which molecules leave the surface equals the rate at which they return. By this definition, vigorous molecular activity goes on *even after all external signs of change have disappeared.* Double arrows ($\rightleftharpoons$) are frequently used to represent equilibrium. For example, water in equilibrium with its vapor is represented as $H_2O(l) \rightleftharpoons H_2O(g)$.

EXERCISE 16-1

> Would water boiling in an open pan on a stove represent an equilibrium system? Explain.

Our molecular model for equilibrium has interesting and important consequences. It permits us to describe equilibrium as that state in which the *rate* of the forward reaction (evaporation) equals the *rate* of the reverse reaction (condensation).

SOLUTION-SOLUTE EQUILIBRIUM

In Chapter 6, we considered a number of different solutions, including a solution of iodine in alcohol. Figure 16-1 shows the result of adding solid iodine in small increments to an alcohol-water mixture at room temperature. (See page 419.) When iodine no longer dissolves and the color of the solution becomes constant, we assume that the solution is *saturated:* it has reached equilibrium. To test whether our solution really has reached equi-

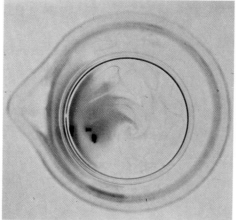 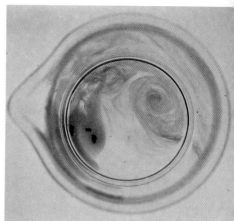

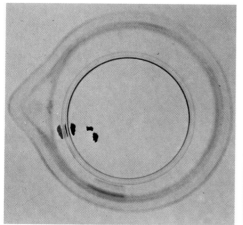

Figure 16-1
When iodine dissolves in an alcohol-water mixture, equilibrium can be recognized by the constant color of the solution.

librium, a hot alcohol-water saturated iodine solution can be cooled to room temperature. Crystals will form until at equilibrium the final solution is identical in color to that prepared by dissolving iodine. The equilibrium condition (solubility) is the same whether we dissolve solid I_2 in an unsaturated solution or crystallize I_2 from a supersaturated solution. Our basic test for the equilibrium condition has been passed.

What does the iodine-alcohol system look like on the molecular level? Iodine molecules leave the surface of the crystal at a given rate as the crystals are placed in the water-alcohol solution. The rate at which molecules return to the crystal surface is low at first, since there are very few molecules of I_2 in the alcohol (as shown by lack of color) and there is very low probability that those leaving the crystal will return immediately. As time passes, however, more and more I_2 molecules are in the solution phase. The probability that a molecule in solution will encounter the surface of an iodine crystal and become bonded to the crystal increases. As a result, *the rate of crystallization** increases until it becomes *equal to* the *rate of solution*. The system is then at equilibrium. No net change is apparent even though action on the molecular level is vigorous. The equation for the equilibrium processes is $I_2(s) \rightleftharpoons I_2(l)$. Note the use of double arrows to indicate that I_2 is crystallizing and dissolving at the same rate.

What determines the equilibrium point, or the solubility, of iodine in the alcohol? In Chapter 6, we found that solubilities ranged from very low (sodium chloride in ethyl alcohol) to high (sodium chloride in water). Clearly, the *type of solid and liquid* is important in determining the concentration at which solubility equilibrium is established. Further, we found that solubilities vary with temperature. Thus, *temperature* is another important factor in determining the equilibrium concentration.

Our analysis of two physical systems at equilibrium also provides useful generalizations for a study of chemical systems at equilibrium. Several questions have been answered:

    1. *How do we recognize equilibrium?* We recognize equilibrium by the constancy of observable properties in a closed system at a uniform

---

*Rate of crystallization is the rate at which molecules or ions bond to crystals.

temperature. It is important that we specify a *closed* system. The evaporation of water from an open dish does *not* represent equilibrium even though the *rate* of evaporation is *constant*. Further, we found that the equilibrium condition is the same regardless of the side from which it is approached. This knowledge helps greatly in recognizing equilibrium.

2. *What are the molecules doing at equilibrium?* The molecules participate in both forward and reverse reactions. *At equilibrium, the rate of the forward reaction is equal to the rate of the reverse reaction.*

3. *How can we change the equilibrium condition?* So far, we have identified temperature and the nature of the reactants as two of the important variables determining equilibrium.

16.2 Equilibrium in chemical systems

Can our observations on physical equilibrium be applied to chemical systems? Let us look at two chemical equilibrium systems in order to determine whether they are really at equilibrium, or whether they are simply changing very slowly.

THE N_2O_4-NO_2 EQUILIBRIUM

In many ways, the N_2O_4-NO_2 equilibrium system can be compared to the liquid-vapor equilibrium system. As before, it is useful to start with experiments. First, we place identical amounts of red-brown nitrogen dioxide (NO_2) in two bulbs. Next, we immerse bulb A in an ice bath and bulb B in boiling water, as shown in Figure 16-2. Bulb A becomes lighter in color, while bulb B turns deep red-brown. The deep red-brown color suggests a high concentration of NO_2 molecules. But why is bulb A light in color? If we determine the molecular weight of the colorless vapor or nearly colorless liquid present at even lower temperatures, we find a value equal to *twice* the molecular weight of NO_2. The formula of this substance is $(NO_2)_2$ or N_2O_4. As we noted, N_2O_4 is colorless and NO_2 is red-brown.

Now let us transfer these two bulbs to a bath at room temperature, as shown in Figure 16-3. Immediately, bulb A becomes a darker brown while bulb B becomes lighter in color. Our observations can be summarized by two equations:

BULB A (COLD BULB WARMING UP)

$$N_2O_4(g) \longrightarrow 2NO_2(g)$$

BULB B (WARM BULB COOLING OFF)

$$2NO_2(g) \longrightarrow N_2O_4(g)$$

Finally, as the two bulbs approach the same temperature, the colors stop changing. The bulbs are now *identical* in color: both are the same shade of brown. The system is at chemical equilibrium. If we heat the bulb that was cold and cool the bulb that was not, we can repeat the cycle.

We have watched the two bulbs approach equilibrium from opposite directions. As bulb A was warmed, equilibrium was approached by the

Figure 16-2

Nitrogen dioxide gas at different temperatures. Bulb A: N_2O_4 at 0 °C (almost colorless). Bulb B: NO_2 at 100 °C (red-brown).

Figure 16-3

Nitrogen dioxide gas at room temperature: bulb A and bulb B after transfer to water bath at 25 °C.

422

dissociation of N_2O_4 to form NO_2. As bulb B was cooled, equilibrium was approached by NO_2 molecules combining to form N_2O_4. If we join the two equations, we can write

$$N_2O_4(g) \longrightarrow 2NO_2(g)$$
$$N_2O_4(g) \longleftarrow 2NO_2(g)$$
$$N_2O_4(g) \rightleftharpoons 2NO_2(g)$$

COLORLESS   RED-BROWN

The double arrows indicate that both *dissociation* and *formation* of N_2O_4 are taking place and, thus, that chemical equilibrium prevails.* Just as with phase changes at equilibrium, the *microscopic* processes of chemical reactions at equilibrium continue at the same rates, but no observable, or *macroscopic,* changes are apparent.

As we noted earlier, equilibrium is approached very rapidly in some systems, while in others it may be approached so slowly that years are required to reach the equilibrium condition. In the latter case, it is difficult to be sure that a system is really at equilibrium. Small changes are hard to recognize. How can we be sure, then, that a chemical system is really at equilibrium and is not just changing very slowly? *One of the best ways to be sure that we have a true equilibrium condition is to show that we can get the same equilibrium point from either the reactants or the products—that is, we can arrive at the same equilibrium value by starting from either side.* This is what we have done for the N_2O_4-NO_2 system.

THE CaCO₃-CaO EQUILIBRIUM

Equilibrium in chemical systems can be illustrated further by considering the formation of quicklime, or calcium oxide (CaO). Millions of tons of CaO are made each year in the United States for plaster, mortar, and for agricultural purposes. Limestone, known chemically as calcium carbonate ($CaCO_3$), is heated in a blast of hot air (800 °C), and CaO and CO_2 are formed. The equation for this reaction is $CaCO_3(s) \xrightarrow{heat} CaO(s) + CO_2(g)$. The CO_2 is constantly carried away in the hot air stream and can be recovered. If, however, $CaCO_3$ is heated in a closed, evacuated system, the pressure of CO_2 builds up above the solid CaO-$CaCO_3$ mixture. A given equilibrium pressure is reached at a given temperature. If the temperature is raised, a higher equilibrium pressure is observed. If the temperature is lowered, the equilibrium pressure falls.

How does the equilibrium state differ experimentally from the *steady-state* process by which CaO is made commercially? In the commercial steady-state process, solid $CaCO_3$ is fed in at one end of the furnace and solid CaO and gaseous CO_2 are removed at the other end as quickly as they are formed. If we look solely at the reactor, the system does not seem to be undergoing any change. However, if we look at the disappearing piles of $CaCO_3$ and the growing piles of CaO, it is evident that a continuing change is taking place.

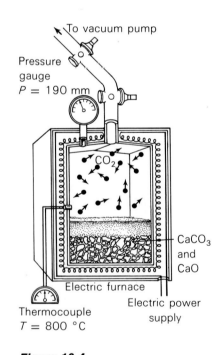

Figure 16-4

Closed system at *equilibrium.* Calcium carbonate heated in evacuated, closed vessel. Pressure reaches a constant value at a given temperature.

*In some cases, an equals sign (=) replaces the double arrow to indicate equilibrium.

If the system described above is to reach equilibrium, the reactor must be *closed* so that CO_2 can no longer escape. A suitable vessel is shown in Figure 16-4. $CaCO_3$ is placed in the vessel, which has an attached pressure gauge. The air in the vessel is pumped out, and the gas exit valve is closed. If the $CaCO_3$ and the container are heated to 800 °C, the pressure of CO_2 inside the vessel builds up until the pressure gauge reads 190 millimetres. As long as the temperature is held constant at 800 °C, the pressure will remain steady at 190 millimetres. The system appears to be at equilibrium since no changes in macroscopic properties can be detected and the system is closed.

If we want to be sure that the system is at equilibrium when the pressure is at 190 millimetres, we can carry out the reverse reaction in the closed vessel. CaO is placed in the reactor, and air is removed. Then CO_2 is pumped into the system until the pressure reads well above 190 millimetres. For example, suppose CO_2 is added until the pressure is 400 millimetres. (The exact value is not important.) The valve is now closed to seal the system and the temperature is raised. Pressure begins to fall. Finally, at 800 °C, the pressure gauge again becomes steady at 190 millimetres. *The same equilibrium pressure of CO_2 has been reached starting with either $CaCO_3$ or with CaO and CO_2,* as shown in Figure 16-5.

Both the decomposition and formation of $CaCO_3$ are taking place in the container:

$$CaCO_3(s) \longrightarrow CaO(s) + CO_2(g)$$
$$CaCO_3(s) \longleftarrow CaO(s) + CO_2(g)$$
$$\overline{CaCO_3(s) \rightleftharpoons CaO(s) + CO_2(g)}$$

At equilibrium, the rates of formation and decomposition of $CaCO_3$ are equal.

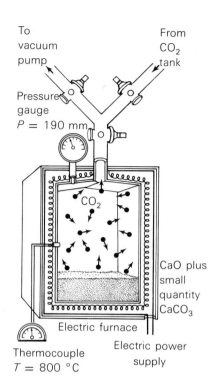

Figure 16-5

CO_2 added to CaO in closed system. Equilibrium established between CaO, CO_2, and $CaCO_3$.

EXERCISE 16-2

Which of the following systems constitute steady-state situations, and which are at equilibrium? For each, a constant property is indicated. (a) An open pan of water is boiling on a stove. The temperature of the water is constant. (b) A balloon contains air and a few drops of water. The pressure in the balloon is constant. (c) A colony of ants is going about its daily routine in an anthill. The population of the anthill is constant. (d) A Bunsen burner burns in the laboratory to give a well-defined flame. Supplies of gas and air are constant.

EQUILIBRIUM AND THE MEANING OF AN EQUATION

Much of our work in chemical equilibrium utilizes chemical equations. Let us review what is meant by an equation such as the one we discussed in this section: $N_2O_4(g) \rightleftharpoons 2NO_2(g)$. The equation tells us that *one* molecule of N_2O_4 can decompose to produce two molecules of NO_2. It also tells us that two molecules of NO_2 can combine to form one molecule of N_2O_4. The equation does *not* tell us anything about the *relative amounts* of NO_2 and N_2O_4 present in a vessel at equilibrium. It definitely does *not* say that at equilibrium there will be 2 moles of NO_2 for each mole of N_2O_4. *It does say that for every mole of N_2O_4 that decomposes, 2 moles of NO_2 will be formed.* Relative concentrations are determined by factors controlling equilibrium.

Chemists often find it convenient to systematically record information about such chemical processes in a table. For example, let us consider the following process:

$$PCl_3(g) + Cl_2(g) \rightleftharpoons PCl_5(g)$$

All reagents and products are gases at 250 °C and 1 atmosphere pressure. If we start with 1 mole of PCl_3 and 1 mole of Cl_2 in a vessel under a pressure of 1 atmosphere and a temperature of 250 °C, we may bring the system to equilibrium; we can then measure the moles of free PCl_3, the moles of free Cl_2, and the moles of free PCl_5 present *at equilibrium*. At 250 °C and 1 atmosphere pressure, such measurements will show 0.71 mole of PCl_3, 0.71 mole of Cl_2, and 0.29 mole of PCl_5. If this is true equilibrium, we shall obtain exactly the same mixture by starting with 1 mole of PCl_5 instead of with 1 mole of PCl_3 and 1 mole of Cl_2. We do, in fact, obtain the same mixture. The above observations are summarized in Table 16-1.

Table 16-1

Number of moles present in a chemical system before reacting and at equilibrium ($T = 250$ °C, $P = 1$ atm)

Moles	$PCl_3(g)$ +	$Cl_2(g)$ $\rightleftharpoons$	$PCl_5(g)$
initial, forward reaction	1.0	1.0	0.0
at equilibrium	0.71	0.71	0.29
initial, reverse reaction	0.0	0.0	1.0

The composition of the equilibrium mixture does *not* depend on the direction from which equilibrium is approached. Further, the equation does *not* indicate relative concentrations at equilibrium. The state of equilibrium for a system is characterized by the relative amounts of products and reactants present. In the decomposition of PCl_5, any change in conditions causing a larger percentage of the PCl_5 to dissociate would shift the state of equilibrium in favor of the formation of more PCl_3 and Cl_2. The equation is

$$PCl_5(g) \rightleftharpoons PCl_3(g) + Cl_2(g)$$

16.3 Altering equilibrium

Concentration and *temperature,* factors that affect the rate of reaction, also affect equilibrium. Equilibrium is attained when the rates of opposing reactions become equal. Any condition that changes the rate of one of the reactions involved in the equilibrium may affect the chemical composition of the system at equilibrium.

CONCENTRATION

Consider the reaction between ferric ion (Fe^{+++}) and thiocyanate ion (SCN^-), which was demonstrated in the laboratory:

$$Fe^{+++}(aq) + SCN^-(aq) \rightleftharpoons FeSCN^{++}(aq)$$

We have visual evidence of concentration at equilibrium, since the intensity of the color is fixed by the concentration of $FeSCN^{++}$ ions. Increasing either the concentration of ferric ions—by adding a soluble salt such as ferric nitrate, $Fe(NO_3)_3$—or the concentration of thiocyanate ions—by adding, say, sodium thiocyanate, NaSCN—changes the concentration of one of the reactants, as shown in Figure 16-6. Immediately, the color of the solution darkens, indicating an increase in the concentration of the colored ions, $FeSCN^{++}$. *The equilibrium concentrations are affected if the concentrations of reactants (or products) are altered.*

TEMPERATURE

We have already considered an example of a change in an equilibrium system as temperature is altered. The relative amounts of NO_2 and N_2O_4 are readily and obviously changed by a change in temperature. *The equilibrium concentrations are changed if the temperature is altered.*

CATALYSTS: WHY THEY DO *NOT* ALTER EQUILIBRIUM

Although catalysts increase the rate of reactions, it has been found through experiment that *adding a catalyst to a system at equilibrium does not alter the equilibrium state.* Thus, it must be true that any catalyst has the

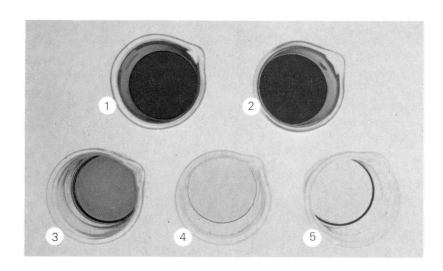

Figure 16-6

Equilibrium concentrations are affected by reactant concentrations. As the concentration of $FeSCN^{++}$ decreases, the color of the solution lightens: (1) 0.0010 *M* KSCN, 0.10 *M* $Fe(NO_3)_3$; (2) 0.0010 *M* KSCN, 0.040 *M* $Fe(NO_3)_3$; (3) 0.0010 *M* KSCN, 0.016 *M* $Fe(NO_3)_3$; (4) 0.0010 *M* KSCN, 0.0064 *M* $Fe(NO_3)_3$; (5) 0.0010 *M* KSCN, 0.0026 *M* $Fe(NO_3)_3$.

same relative effect on the rate of forward and reverse reactions. You will recall that the effect of a catalyst on reaction rates can be discussed in terms of lowering the activation energy. *Such lowering increases the rate of both forward and reverse reactions.* Thus, a catalyst produces no *net* change in the equilibrium concentrations, even though the system may reach equilibrium much more rapidly.

16.4 Predicting new equilibrium concentrations: Le Chatelier's principle

We are not satisfied with the conclusion that certain changes affect equilibrium concentrations. We would also like to predict the *direction* of the effect (does the change favor products or reactants?) and the *magnitude* of the effect (how much does it favor products or reactants?). The first desire—to know the qualitative effects—is satisfied by a generalization first proposed by French chemist Henry Louis Le Chatelier (1850–1936). That generalization is now called **Le Chatelier's principle.**

Le Chatelier looked for regularities in a large amount of experimental data on equilibrium systems. Summarizing the regularities, he made the following generalization: *if an equilibrium system is subjected to a change, processes occur that tend to counteract partially the imposed change.* To clarify the meaning of this generalization, or principle, we shall restate it as it applies to changes in temperature, concentration, and pressure.

THE EFFECT OF TEMPERATURE CHANGES

According to Le Chatelier's principle, *if the temperature of a system at equilibrium is raised by adding heat (thermal energy), a reaction will take place that will absorb energy. If the temperature of a system at equilibrium is lowered by removing heat, a reaction will take place that will give off energy.*

How does this generalization apply to a specific system? Consider the reaction of N_2O_4 to form NO_2:

$$14.1 \text{ kcal} + N_2O_4(g) \rightleftharpoons 2NO_2(g)$$

Our experimental observations indicated that warming a bulb containing NO_2 and N_2O_4 causes a shift of the equilibrium state in favor of the formation of NO_2 (the reddish-brown color deepened). We note that this is in accord with Le Chatelier's principle. An increase in temperature yields more NO_2. The formation of NO_2 from N_2O_4 *absorbs* heat. An increase in the temperature of the equilibrium system results in the formation of more NO_2 by the reaction that absorbs heat. The reverse process, the formation of more N_2O_4 as the temperature is lowered, liberates heat. This is also in accord with Le Chatelier's principle.

In the water vapor equilibrium system, raising the temperature of liquid water raises its vapor pressure. This is also in accord with Le Chatelier's principle, since heat is absorbed as the liquid vaporizes. The absorption of heat, which accompanies the change to the new equilibrium condition, *partially* counteracts the temperature rise causing the change. This may be seen in the equation $H_2O(l) + 9.7 \text{ kcal} \rightarrow H_2O(g)$. As heat is added to a closed system, more $H_2O(g)$ is formed *and* the temperature goes up. However, it rises less than it would if water did not vaporize.

THE EFFECT OF CONCENTRATION CHANGES

Le Chatelier's principle can be stated specifically for changes in concentration. *If a reactant or product is added to a system at equilibrium, that reaction will occur that partially uses up the added reactant or product. If a reactant or product is removed, that reaction will occur that produces more of the removed reactant or product.*

We have considered this principle in laboratory experiments. If a solution containing $SCN^-(aq)$ ions is added to an equilibrium solution containing both $Fe^{+++}(aq)$ and $SCN^-(aq)$ ions, the color of the solution deepens:

$$Fe^{+++}(aq) + SCN^-(aq) \rightleftharpoons FeSCN^{++}(aq)$$

COLORLESS   RED

A new state of equilibrium is then attained in which the $FeSCN^{++}$ concentration is higher than it was before the addition of SCN^-. Increasing the concentration of SCN^- has increased the concentration of $FeSCN^{++}$. This agrees with Le Chatelier's principle. The change imposed on the system was an increase in the concentration of SCN^-. This change can be counteracted in part by the reaction of some Fe^{+++} and SCN^- to form more $FeSCN^{++}$, thus using up some of the added SCN^-. The same argument applies to an addition of a solution containing $Fe^{+++}(aq)$ ions. The formation of $FeSCN^{++}$ *uses up* a *portion* of the added reactant, *partially* counteracting the change.

Le Chatelier's principle can also be applied to gaseous systems. Consider the equilibrium system $PCl_3(g) + Cl_2(g) \rightleftharpoons PCl_5(g)$. If we add more Cl_2 while the volume and temperature of the system are held constant, the pressure goes up, the concentration of Cl_2 goes up, and more $PCl_5(g)$ is produced. This observation is in agreement with Le Chatelier's principle: when a reactant is added, that reaction occurs that uses up some of the reactant. Similarly, if more PCl_5 is added at fixed volume and temperature, additional PCl_3 and Cl_2 are produced. Although pressure goes up in both these systems as more reagents are added, the direction of the equilibrium shift is determined by the changes in *concentration* resulting from added reagents.

THE EFFECT OF VOLUME CHANGES

Is adding or subtracting a reagent the only way to change concentration? If we remember that concentration is number of moles *per unit volume,* we see that changes in the volume also change the concentrations in a gaseous system. Consider again the PCl_3-Cl_2-PCl_5 equilibrium. Suppose the volume of the equilibrium system is reduced suddenly to *half* its original volume. The pressure on the system goes up. The reduction in volume doubles the initial concentration of the $PCl_3(g)$ and the $PCl_5(g)$. The concentration changes compensate for each other because they appear on opposite sides of the equation. We have, however, an additional concentration change for which there is no compensation. The change in volume also doubles the concentration of Cl_2. The additional increase in concentration has the effect of shifting the equilibrium toward the production of more PCl_5. The expected formation of PCl_5 does indeed occur. Thus, changes in *volume* of gaseous systems bring about changes in concentrations that shift the equilibrium if the number of moles of gaseous products is different from the number of moles of gaseous reactants. See $PCl_3 + Cl_2$ above.

THE EFFECT OF PRESSURE ON CONCENTRATION

Changes in volume always bring about changes in the *pressure* of closed gaseous systems; for this reason, people sometimes prefer to consider that pressure causes a shift in equilibrium. *Pressure effects are always those that result from changes in effective concentration.*

It is possible, however, to analyze the set of changes in the gaseous PCl_3-Cl_2-PCl_5 system in a somewhat different (and perhaps easier) manner if we apply Le Chatelier's principle while considering for convenience that pressure is the changing factor. We know from earlier in this section that a reduction in the volume of the system causes the formation of more $PCl_5(g)$: $PCl_3(g) + Cl_2(g) \rightarrow PCl_5(g)$. The basic cause of the shift is a change in concentration of all components as a result of a volume reduction. The same fact can also be stated in terms of the pressure: if pressure is *increased* on a system at equilibrium, that reaction takes place that tends to *decrease* total pressure. In the case just discussed, an increase in pressure resulted in the reaction that converts 1 mole of PCl_3 (22.4 litres at STP) and 1 mole of Cl_2 (22.4 litres at STP) to 1 mole of PCl_5 (22.4 litres at STP). A reduction in the total number of moles per unit volume reduces the pressure. The converse is also true. If pressure is *reduced* on a system at equilibrium, that reaction takes place that tends to *increase* the total pressure of the system. A decrease in pressure should favor the conversion of $PCl_5(g)$ to $PCl_3(g)$ and $Cl_2(g)$.

What happens if the reaction does not involve an increase or decrease in the total volume of the system? Our statement would seem to indicate that no change is to be expected. Experiment confirms such a prediction. Consider the first reactions mentioned in this chapter.

$$CO(g) + NO_2(g) \rightleftharpoons CO_2(g) + NO(g)$$

If we increase the pressure on this system by reducing the volume, the concentrations of all four gases goes up, but the composition of the gaseous mixture does not change. When the number of moles of gaseous reactants is the same as the number of moles of gaseous products, no equilibrium shift is observed as pressure is increased.

EXERCISE 16-3

Consider the equilibrium for making methyl alcohol:

$$2H_2(g) + CO(g) \rightleftharpoons H_3COH(g) + energy$$

(a) What would happen to the system if more H_2 was added at constant volume? (b) What would happen to the equilibrium if the volume of the system was reduced and the pressure on the system increased? (c) What would happen if the temperature of the system was raised and the volume kept constant?

16.5 Application of equilibrium principles: The Haber process

The control of equilibrium by the application of Le Chatelier's principle is literally feeding and clothing a large segment of the world's population. The large-scale production of ammonia (NH_3) is possible today because we can apply the principles proposed by Le Chatelier. Liquid ammonia for fertilizer, as well as ammonia for the synthesis of nylon and hundreds of other chemicals can be obtained from the air. Before people learned to control equilibrium, they depended on the erratic whims of nature and international politics for their essential nitrogen compounds.

We observed in Chapter 9 that the nitrogen of the earth's atmosphere is of low reactivity at low temperatures. The nitrogen atoms combine so strongly with each other that the N_2 molecule does not react easily at 25 °C. Under proper conditions, however, we can make nitrogen molecules combine with hydrogen molecules to form ammonia:

$$N_2(g) + 3H_2(g) \rightleftharpoons 2NH_3(g) + 22 \text{ kcal}$$

Can our knowledge of equilibrium and kinetics (Chapter 15) be used to decide what conditions would produce the most ammonia? First, we must consider pressure. In the formation of ammonia, *one* volume of nitrogen and *three* volumes of hydrogen yield *two* volumes of ammonia (four volumes reduce to two). Le Chatelier's principle tells us that high pressure (small volume, thus high concentration) should favor the production of ammonia. Next, we consider temperature. We note that heat is *given off* when ammonia is formed from N_2 and H_2. Le Chatelier's principle tells us that low temperatures will favor a reaction liberating heat; therefore, we want a low temperature. It begins to look as though our preferred reactor should be a *refrigerated* tank containing N_2 and H_2 under thousands of pounds of *pressure*. If such a tank is prepared and the N_2-H_2 mixture is allowed to stand for several days, *no NH_3 is found*. If the mixture is allowed to stand for several weeks, *no NH_3 is found*. Even after several years, the yield of NH_3 cannot be detected. Apparently, N_2 reacts *very* slowly with H_2 at the low temperature that favors production of NH_3. The situation reminds us of the reaction of H_2 and O_2 mentioned earlier. In the H_2-O_2 case, a spark made the reaction proceed explosively. We have no such result when we pass a spark through the N_2-H_2 mixture. Other ways must be found to accelerate the process.

We remember from Chapter 15 that reactions can be made to go faster by raising the temperature. We must now compromise: low temperature is required for a desirable equilibrium state, and high temperature is necessary for a satisfactory rate. The compromise used industrially involves an intermediate temperature, about 400 °C. Even then, the success of the process depends on the presence of a suitable catalyst to achieve a reasonable reaction rate.* Very high pressures are desirable, but equipment that will stand up under both high pressure and high temperature is expensive to build. A pressure of about 300 atmospheres is actually used. Under these conditions, 300 atmospheres and 400 °C, only slightly more than 30 percent of the reactants are converted to NH_3. The NH_3 is removed from the mixture by liquefying it under conditions at which N_2 and H_2 remain as gases. The unreacted N_2 and H_2 are then recycled until the total percentage conversion of expensive hydrogen to ammonia is very high.

Before World War I, the principal sources of nitrogen compounds were nitrate deposits in Chile. The German chemist Fritz Haber (1868–1934)

---

* As catalysts for the process get better, the conditions needed for the conversion get milder; that is, lower temperatures and pressures can be used. Since high temperature and pressure are expensive, better catalysts yield less expensive ammonia.

successfully developed the process we have just described, thus allowing the world to use its almost unlimited supply of atmospheric nitrogen.

EXERCISE 16-4

> Nitrogen compounds can be made commercially from the reaction of nitrogen with oxygen. The essential equation is
>
> $$N_2(g) + O_2(g) + 21.5 \text{ kcal} \longrightarrow 2NO(g)$$
>
> Using Le Chatelier's principle, select conditions that would maximize the yield of NO(g) in the process.

16.6 Quantitative aspects of equilibrium

Le Chatelier's principle permits us to make qualitative predictions about equilibrium states. For example, we know that raising the pressure will favor the production of NH_3 from N_2 and H_2. We must now ask, by *how much* will an increase in pressure favor NH_3 formation? Will the yield change by a factor of 10 or by only 0.01 percent? To control a reaction, we need *quantitative* information about equilibrium. Experiments show that quantitative predictions are possible. Furthermore, we are delighted to learn that they can be explained in terms of our models for the reaction of materials on the molecular level.

THE EQUILIBRIUM CONSTANT

As you remember, we used laboratory colorimetric observations (based on estimation of color) to measure the concentration of $FeSCN^{++}$ in solutions containing ferric and thiocyanate ions (Fe^{+++} and SCN^-). The reaction is $Fe^{+++}(aq) + SCN^-(aq) \rightleftharpoons FeSCN^{++}(aq)$. From $[FeSCN^{++}]$* and the initial values of $[Fe^{+++}]$ and $[SCN^-]$ we calculated the values of $[Fe^{+++}]$ and $[SCN^-]$ at equilibrium. We then made calculations for various combinations of those values. Many such experiments show that the ratio

$$\frac{[FeSCN^{++}]}{[Fe^{+++}][SCN^-]}$$

comes closest to being a fixed value. Note this ratio carefully. The reaction *product* appears on top.

Colorimetric analysis is not very exact. More accurate data on another system at equilibrium are shown in Table 16-2. The reaction is

$$2HI(g) \rightleftharpoons H_2(g) + I_2(g)$$

---

*Hereafter, we shall regularly use the square bracket notation to indicate concentration of the substance inside the bracket. We read $[FeSCN^{++}]$ as *concentration of FeSCN^{++}(aq) ions in solution.*

Table 16-2

Equilibrium concentration at 698.6 K of hydrogen, iodine, and hydrogen iodide*

Experiment number	$[H_2]$ (moles/litre)	$[I_2]$ (moles/litre)	$[HI]$ (moles/litre)
1	1.8313×10^{-3}	3.1292×10^{-3}	17.671×10^{-3}
2	2.9070×10^{-3}	1.7069×10^{-3}	16.482×10^{-3}
3	4.5647×10^{-3}	0.7378×10^{-3}	13.544×10^{-3}
4	0.4789×10^{-3}	0.4789×10^{-3}	3.531×10^{-3}
5	1.1409×10^{-3}	1.1409×10^{-3}	8.410×10^{-3}

*Values above the line were obtained by heating hydrogen and iodine together; values below the line, by heating pure hydrogen iodide.

The data have been expressed in concentrations, although pressure units would be equally good for a reaction involving gases.

EXERCISE 16-5

For experiments 4 and 5 in Table 16-2, why is $[H_2]$ equal to $[I_2]$? For experiment 1 in Table 16-2, what were the concentrations of H_2 and I_2 before the reaction occurred to form HI?

Let us work with these data to compute the value of the ratio

$$\frac{[H_2][I_2]}{[HI]}$$

Note again that the products are on top. From our computations, we obtain the numbers shown in Table 16-3. In view of the precision of the data from which these ratios are derived, the ratios are far from constant. Now let us try the ratio

$$\frac{[H_2][I_2]}{[HI]^2}$$

These calculations are summarized in Table 16-4. The results are most encouraging. They imply that with a fair degree of accuracy, we can write

$$\frac{[H_2][I_2]}{[HI]^2} = \text{a constant} = 1.835 \times 10^{-2} \text{ at 698.6 K}$$

Look at this ratio in terms of the reaction $2HI(g) \rightleftharpoons H_2(g) + I_2(g)$. The ratio is the product of the equilibrium concentrations of the substances produced, $[H_2] \times [I_2]$, divided by the *square* of the concentration of the reacting

Table 16-3

Values of $[H_2][I_2]/[HI]$ for data in Table 16-2

Experiment number	$\dfrac{[H_2][I_2]}{[HI]}$ (moles/litre)
1	32.429×10^{-5}
2	30.105×10^{-5}
3	24.866×10^{-5}
4	6.495×10^{-5}
5	15.477×10^{-5}

Table 16-4

Values of $[H_2][I_2]/[HI]^2$ for data in Table 16-2

Experiment number	$\dfrac{[H_2][I_2]}{[HI]^2}$ (moles/litre)
1	1.8351×10^{-2}
2	1.8265×10^{-2}
3	1.8359×10^{-2}
4	1.8390×10^{-2}
5	1.8403×10^{-2}
average	1.835×10^{-2}

substance, $[HI]^2$. The squaring of the $[HI]$ term may seem more reasonable if we rewrite the above equation as

$$HI(g) + HI(g) \rightleftharpoons H_2(g) + I_2(g)$$

The power to which we raise the concentration of each substance is equal to its coefficient in the reaction.

LAW OF MASS ACTION

We have considered precise equilibrium data for the reaction

$$2HI(g) \rightleftharpoons H_2(g) + I_2(g)$$

The concentrations of the molecules in the reaction of 2 HI to produce H_2 and I_2 were found to have a simple relationship:

$$\frac{[H_2][I_2]}{[HI]^2} = \text{a constant}$$

The ratio is constant at constant temperature, providing a specific illustration of a more general relationship known as the **Law of Mass Action.** For the general process $aA + bB \rightleftharpoons eE + fF$, when equilibrium exists, there will be a simple relation between the concentrations of products, $[E]$ and $[F]$, and the concentrations of reactants, $[A]$ and $[B]$:

$$\frac{[E]^e[F]^f}{[A]^a[B]^b} = K = \text{a constant} \quad \text{(at constant temperature)}$$

In this generalized equation, we see again that the numerator is the product of the equilibrium concentrations of the species formed, each raised to the power equal to the number of moles of that species in the chemical equation. The denominator is again the product of the equilibrium concentrations of the reacting species, each raised to a power equal to the number of moles of the species in the chemical equation. The quotient of these two quantities remains constant. The constant K is called the **equilibrium constant.** The preceding generalization is one of the most useful in all chemistry. From the equation for any chemical reaction, we can immediately write an expression, in terms of the concentrations of reactants and products, that will be constant at any given temperature. If this constant is known (from measuring concentrations in a particular equilibrium system), it can be used in calculations for any other equilibrium state of that system at the same temperature.

Table 16-5 lists some reactions and the numerical values of the equilibrium constants. First, let us verify the forms of the equilibrium constant relationships for the concentrations. Some entries in the table have an unexpected form. For the reaction.

$$Cu(s) + 2Ag^+(aq) \rightleftharpoons Cu^{++}(aq) + 2Ag(s)$$

you do not find $\dfrac{[Cu^{++}][Ag]^2}{[Ag^+]^2[Cu]} = K$

but, rather, you find $\dfrac{[Cu^{++}]}{[Ag^+]^2} = K$

Table 16-5

Some equilibrium constants

Reaction	Equilibrium law relationship	K at stated temperature	Comments
$Cu(s) + 2Ag^+(aq) \rightleftharpoons Cu^{++}(aq) + 2Ag(s)$	$K = \dfrac{[Cu^{++}]}{[Ag^+]^2}$	2×10^{15} at 25 °C	Solids not included
$Ag^+(aq) + 2NH_3(aq) \rightleftharpoons Ag(NH_3)_2^+(aq)$	$K = \dfrac{[Ag(NH_3)_2^+]}{[Ag^+][NH_3]^2}$	1.7×10^7 at 25 °C	NH_3 really ties up Ag^+
$N_2O_4(g) \rightleftharpoons 2NO_2(g)$	$K = \dfrac{[NO_2]^2}{[N_2O_4]}$	0.87 at 55 °C	Reactions do not go to completion
$HSO_4^-(aq) \rightleftharpoons H^+(aq) + SO_4^{--}(aq)$	$K = \dfrac{[H^+][SO_4^{--}]}{[HSO_4^-]}$	0.013 at 25 °C	
$PbCl_2(s) \rightleftharpoons Pb^{+++}(aq) + 2Cl^-(aq)$	$K = [Pb^{++}][Cl^-]^2$	2×10^{-5} at 25 °C	$PbCl_2$ has low solubility
$CH_3COOH(aq) \rightleftharpoons H^+(aq) + CH_3COO^-(aq)$	$K = \dfrac{[H^+][CH_3COO^-]}{[CH_3COOH]}$	1.8×10^{-5} at 25 °C	Acetic acid does not ionize much
$AgCl(s) \rightleftharpoons Ag^+(aq) + Cl^-(aq)$	$K = [Ag^+][Cl^-]$	1.7×10^{-10} at 25 °C	AgCl is quite insoluble
$H_2O \rightleftharpoons H^+(aq) + OH^-(aq)$	$K = [H^+][OH^-]$	10^{-14} at 25 °C	H^+ and OH^- combine to give water
$AgI(s) \rightleftharpoons Ag^+(aq) + I^-(aq)$	$K = [Ag^+][I^-]$	1.5×10^{-16} at 25 °C	AgI is even less soluble than AgCl

This occurs because the concentrations of solid copper and solid silver are incorporated into the equilibrium constant. The concentration of solid copper (moles of copper per litre) is fixed by the density of the metal—it cannot be altered either by the chemist or by the progress of the reaction. The same is true of the concentration of solid silver. Since neither of these concentrations varies (no matter how much solid is added), there is no need to write them each time an equilibrium calculation is made. The last equation will suffice.

EXERCISE 16-6

If we assign the equilibrium constants K' and K to the following expressions, $K' = \dfrac{[Cu^{++}][Ag]^2}{[Ag^+]^2[Cu]}$ $K = \dfrac{[Cu^{++}]}{[Ag^+]^2}$

show that $K = K' \times \dfrac{[Cu]}{[Ag]^2} = K' \times$ constant

Another equilibrium constant of unexpected form applies to the reaction $H_2O \rightleftharpoons H^+(aq) + OH^-(aq)$. For this reaction, we might have written

$$\frac{[H^+][OH^-]}{[H_2O]} = K$$

Instead, Table 16-5 lists the equation as $[H^+][OH^-] = K$. The concentration of water does not appear in the denominator of this expression. Water is frequently omitted in treating aqueous reactions that consume or produce water. This omission is justified because the variation in the concentration of water during reaction is very very small in dilute aqueous solutions. We can treat $[H_2O]$ as a concentration that does not change. Thus, it can be incorporated into the equilibrium constant.

EXERCISE 16-7

> Water has a density of 1 g/ml. Calculate the concentration of water (expressed in moles per litre) in pure water. Now calculate the concentration of water in 0.10 M aqueous solution of acetic acid (CH_3COOH), assuming each molecule of CH_3COOH occupies the same volume as one molecule of H_2O. Are we justified in assuming that the water concentration in 0.10 M acetic acid is the same as that in pure water at the same temperature? Explain.

The equilibrium constants for the process of dissolving a solid also have an unexpected form. For the equation $AgCl(s) \rightleftharpoons Ag^+(aq) + Cl^-(aq)$, the equilibrium constant is $K = [Ag^+(aq)][Cl^-(aq)]$. For the equation $PbCl_2(s) \rightleftharpoons Pb^{++}(aq) + 2Cl^-(aq)$ the equilibrium constant is $K = [Pb^{++}(aq)][Cl^-(aq)]^2$. In both these cases, the simplified form of the equilibrium constant is used because the *concentration* of a pure solid, which is the number of moles per litre, is determined by the density of the solid, not by the mass of solid in the system.

Consider the case of $AgCl(s) \rightleftharpoons Ag^+(aq) + Cl^-(aq)$. The K expression for this reaction is expected to have the form

$$\frac{[Ag^+(aq)][Cl^-(aq)]}{[AgCl(s)]} = K$$

But AgCl, like copper and silver, is a solid with a constant concentration. Thus, we can multiply both sides of the equation by the constant $[AgCl(s)]$: $[Ag^+(aq)][Cl^-(aq)] = K[AgCl(s)] = K_{sp}$. The combined solubility equilibrium constant, which includes in its value the constant concentration of solid AgCl, is given the special name **solubility product constant,** or $\mathbf{K_{sp}}$, for short.

As you can see, the K_{sp} (or solubility product constant) for $PbCl_2$ would be $K_{sp} = [Pb^{++}(aq)][Cl^-(aq)]^2$.

In summary, the concentrations of pure solids and pure liquids are constant and incorporated in the equilibrium constant. Thus, they do not appear in the equilibrium constant relationship. Also, the concentrations of solvents are assumed to be equal to that of the pure solvent and, thus, constant in dilute solution. They do not appear in the equilibrium constant either.

EXERCISE 16-8

(a) The density of solid I_2 is 4.93 g/cm³. How many *grams* of iodine are present in 1,000 cm³ (1 litre) of solid iodine? (b) How many *moles* of iodine are present in a litre of I_2 crystals (mol wt I_2 = 253.81 g/mole)? (c) How many moles of I_2 are present in 2 litres of I_2 crystals? (d) Does the *concentration* of solid I_2, which is the number of moles per litre, change if we take 2 litres of solid I_2 rather than 1? (e) The density of $AgCl(s)$ is 5.56 g/cm³. How many grams of AgCl are present in *1 litre* of solid? (f) How many "moles"* per litre of AgCl are present in 1 litre of solid AgCl?

SIZE OF THE EQUILIBRIUM CONSTANT

Let us now look at the numerical values of the equilibrium constants in Table 16-5. The K's listed range from 10^{+15} to 10^{-16}, so we see that there is a wide variation. We want to acquire a general sense of the relationship between the size of the equilibrium constant and the state of equilibrium. *A large value of* K *means that in the system at equilibrium, there are larger concentrations of products than of reactants.*[†] Remember that the numerator of our equilibrium expression contains the concentrations of the products of the reaction. The value of 2×10^{15} for the K for the reaction of $Cu(s) + Ag^+(aq)$ certainly indicates that if a reaction is initiated by placing metallic copper in a solution containing Ag^+—for example, in silver nitrate solution—silver metal will plate out. When equilibrium is finally reached, the concentration of copper ions $[Cu^{++}]$, will be very much greater than the square of the concentration of silver ions, $[Ag^+]^2$.

A *small value of* K *for a given reaction implies that very little of the products have to be formed from the reactants before equilibrium is attained.*[†] The value of $K = 10^{-16}$ for the reaction $AgI(s) \rightleftharpoons Ag^+(aq) + I^-(aq)$ indicates that very little solid AgI can dissolve before equilibrium concentrations are

---

*For ionic solids, the term *mole* should be replaced by the term *formula weight*. The answer is the same in both cases.

†Note, however, that because equilibrium constants differ in the powers to which numbers are raised, the size of the constant gives us only a rough indication of the equilibrium condition. If we want to make quantitative comparisons, we must calculate the concentrations or consider only strictly comparable reactions.

attained. Silver iodide has extremely low solubility. Conversely, if 0.1 M solutions of KI and $AgNO_3$ are mixed, the values of $[Ag^+]$ and $[I^-]$ are large. Equilibrium cannot be reached until the $[Ag^+]$ and $[I^-]$ have been greatly reduced by the precipitation of AgI.

SOME SAMPLE EQUILIBRIUM CONSTANT CALCULATIONS

The quantitative aspects we have been discussing apply to real systems. Many equilibrium constant calculations are made every day in research on fuels and combustion. In such research, high temperatures ensure that the process considered will go rapidly to equilibrium. Equilibrium constant calculations are also widely used in all branches of chemistry and in chemical engineering. The following problems are examples of this type of calculation.

Example 1: For the Haber process of NH_3 formation,

$$N_2(g) + 3H_2(g) \rightleftharpoons 2NH_3(g)$$

it is found that at a given temperature, the concentrations at equilibrium are: $[N_2] = 0.30$ mole/litre; $[H_2] = 0.10$ mole/litre; and $[NH_3] = 0.20$ mole/litre.

Calculate the value of K. (Some additional information about the Haber process is shown in the margin.)

Answer:

$$K = \frac{[NH_3]^2}{[N_2][H_2]^3} = \frac{[0.20]^2}{[0.30][0.10]^3} = \frac{[4.0 \times 10^{-2}]}{[3.0 \times 10^{-1}] \times [1.0 \times 10^{-1}]^3}$$

$$K = \frac{4 \times 10^{-2}}{3 \times 10^{-4}} = 1.3 \times 10^2$$

Example 2: In the reaction

$$Fe^{++}(aq) + Ag^+(aq) \rightleftharpoons Fe^{+++}(aq) + Ag(s)$$

the value of K is 3.0. If, at equilibrium, the concentration of $Fe^{++}(aq)$ is 0.20 mole per litre and the concentration of $Ag^+(aq)$ is 0.30 mole per litre, what is the concentration of $Fe^{+++}(aq)$ in the system?

Answer:

$$K = \frac{[Fe^{+++}(aq)]}{[Fe^{++}(aq)][Ag^+(aq)]}$$

$$\frac{[Fe^{+++}(aq)]}{[0.20][0.30]} = 3.0$$

$$[Fe^{+++}(aq)] = 3.0 \times 0.20 \times 0.30 = 18 \times 10^{-2}$$

$$[Fe^{+++}(aq)] = 0.18 \text{ mole/litre}$$

Equilibrium Constant and Temperature

Values of K change with temperature. Actual values of K for the Haber process

$$N_2(g) + 3H_2(g) \rightleftharpoons 2NH_3(g)$$

$$K = \frac{[NH_3]^2}{[N_2][H_2]^3}$$

are shown as follows:

°C	K
350	1.94
400	0.506
450	0.160

Note that K (and the yield) fall as temperature rises.

Example 3: It is known that $K = 6 \times 10^{-2}$ at 600 °C for the process

$$CO + Cl_2 \rightleftharpoons COCl_2$$

PHOSGENE

Calculate the equilibrium concentration of Cl_2 that would be needed to convert 50 percent of the added CO to $COCl_2$.
Answer:

$$K = \frac{[COCl_2]}{[CO][Cl_2]} = 6 \times 10^{-2}$$

Since $[COCl_2] = [CO]$, we can write

$$K = \frac{\cancel{[COCl_2]}}{\cancel{[CO]}[Cl_2]} = 6 \times 10^{-2}$$

$$[Cl_2] = \frac{10 \times 10^{-1}}{6 \times 10^{-2}} = 1.7 \times 10^1 = 17 \text{ moles/litre}$$

Clearly, this is approaching an unreasonable concentration for a gas.

Example 4: The effect of volume changes on an equilibrium system was discussed on page 427. The system $PCl_5(g) \rightleftharpoons PCl_3(g) + Cl_2(g)$ has an equilibrium constant of 0.050 at 250 °C. Using Le Chatelier's principle, decide qualitatively what will happen if the volume of the container is reduced from 1 litre to 0.50 litre.

Answer: Qualitatively, a decrease in volume means that the pressure on the gaseous system has been increased. Le Chatelier's principle states, *if pressure is increased on a system at equilibrium, that reaction takes place which tends to decrease total pressure.* Since 1 mole of $PCl_5(g)$ has less volume than 1 mole of $PCl_3(g)$ *plus* 1 mole of $Cl_2(g)$, and will generate less pressure, Le Chatelier's principle indicates that $PCl_3(g)$ and $Cl_2(g)$ will combine to produce more $PCl_5(g)$. The basis for this change in quantitative terms is seen in Table 16-6.

Table 16-6

Effect of pressure increase* on the equilibrium system
$PCl_5(g) \rightleftharpoons PCl_3(g) + Cl_2(g)$ ($T = 250$ °C, $K = 0.050$)

Component	Original equilibrium concentration (moles/litre)	Temporary non-equilibrium concentration (moles/litre)	Final equilibrium concentration (moles/litre)	Change (moles/litre)
PCl_5	0.80	1.60	1.71	+0.11
PCl_3	0.20	0.40	0.29	−0.11
Cl_2	0.20	0.40	0.29	−0.11

*Pressure increased by reducing volume from 1.00 litre to 0.50 litre.

> Using example 4 and Table 16-6, (a) show that the original and final states yield the same equilibrium constant, and (b) show that the values of the changes are in agreement with the chemical equation.

Example 5: Use the K_{sp} value for $PbCl_2$ at 25 °C to calculate the solubility of $PbCl_2$ in pure water at 25 °C. The equation for the reaction is

$$PbCl_2(s) \rightleftharpoons Pb^{++}(aq) + 2Cl^-(aq)$$

The K_{sp} is $[Pb^{++}(aq)][Cl^-(aq)]^2 = 2 \times 10^{-5}$.

Answer: Let X equal the number of moles of $PbCl_2$ that dissolve in one litre of water at equilibrium. The equation then tells us that every mole of $PbCl_2$ that dissolves produces 2 moles of $Cl^-(aq)$. Thus, at equilibrium, the concentration of Cl^- will be $2X$. We then write

$$X[2X]^2 = 2 \times 10^{-5}$$
$$4X^3 = 2 \times 10^{-5}$$
$$X^3 = 0.5 \times 10^{-5} = 5.0 \times 10^{-6}$$
$$X = \sqrt[3]{5} \times 10^{-2}$$
$$X = 1.7 \times 10^{-2} \text{ moles } PbCl_2/\text{litre}$$

> The K_{sp} value for $BaSO_4$ is 1.5×10^{-9}. Calculate the number of moles of $BaSO_4$ that will dissolve in a litre of water at 25 °C.

16.7 Qualitative aspects of equilibrium

The values of K_{sp} provide us with a quantitative means of determining solubilities. In many cases, qualitative information is all that is needed. Fortunately, a set of solubility rules, correlated by the anion (negative ion) of the salt, can provide us with a rapid, qualitative estimate of the solubility of a compound. The rules are given in Table 16-7.*

> Using the rules in Table 16-7, indicate which of the following compounds are soluble: (a) KNO_3 (b) $Mg(OH)_2$ (c) PbS (d) $AgCl$ (e) $MgSO_4$ (f) $BaSO_4$ (g) $CsNO_3$ (h) $NaOH$ (i) Ag_2SO_4 (j) $BaCO_3$ (k) K_2CO_3.

*A different summary of these facts appears in Table 16-8, on page 449.

Table 16-7

General rules for predicting the solubility of common compounds in water

Rule	Negative ions (anions)	Statement of rule
1	NO_3^-	all nitrates are soluble
2	Cl^-	all chlorides are soluble *except* those of Ag^+, Pb^{++}, and Hg^+ ($AgCl$, $PbCl_2$, and Hg_2Cl_2 are insoluble)*
3	SO_4^{--}	all sulfates are soluble *except* Ag_2SO_4, $PbSO_4$, $HgSO_4$, $CaSO_4$, $SrSO_4$, and $BaSO_4$
4	CO_3^{--}	all carbonates are insoluble *except* Li_2CO_3, $NaCO_3$, K_2CO_3, Rb_2CO_3, Cs_2CO_3, and $(NH_4)_2CO_3$
5	OH^-	all hydroxides are insoluble *except* $LiOH$, $NaOH$, KOH, $RbOH$, $CsOH$ (Group 1) and $Sr(OH)_2$ and $Ba(OH)_2$ (heavier Group 2)
6	S^{--}	all sulfides are insoluble *except* Li_2S, Na_2S, K_2S, Rb_2S, Cs_2S, CaS, SrS, and BaS

*The word *insoluble* means it dissolves less than 10^{-4} moles/litre. (See page 154.)

16.8 Factors determining equilibrium

We now have a picture of equilibrium activity on the molecular level: at equilibrium, the rate of the forward reaction is equal to the rate of the reverse reaction. We know how to shift an equilibrium through control of such variables as temperature and concentration. Finally, we know how to express equilibrium relationships in a quantitative fashion through the use of the equilibrium constant.

It is now appropriate for us to raise a more basic question. What factors determine where an equilibrium lies? What determines whether a substance is soluble? To approach this question, some analogies are helpful.

ENERGY

We are not surprised that a skier goes downhill. A skier moves spontaneously from a condition of *higher* potential energy to a condition of *lower* potential energy. The energy of position is less at the bottom of the hill. Similarly, we know in advance that a skateboard will spontaneously carry a rider down a hill. Our everyday experience tells us that systems tend to move spontaneously from a condition of higher potential energy to a condition of lower potential energy. The potential energy is changed to kinetic energy and ultimately to heat in the process. "Burned" elbows and knees offer grim proof of the conversion of kinetic energy to heat in skateboarding.

Chemical systems are not too different from the skateboard or the skis. You read in Chapter 14 that each chemical system has a given energy, or

Figure 16-7

The skier moves spontaneously from a condition of high potential energy to a condition of lower potential energy.

heat, content. A portion of this chemical energy is potential energy resulting from the *position* of atoms relative to each other. Chemical systems, like skateboards, tend to move toward a condition of lower energy. Further, the chemical energy of chemical systems can be converted to molecular kinetic energy, appearing as heat. Chemical systems seem to proceed spontaneously from a state of higher energy to a state of lower energy. The energy involved in the change may appear as kinetic energy. We may see it ultimately as heat. In short, a reaction will proceed spontaneously if the products will have lower energy than the reactants. This generalization is in accord with our experience, particularly for reactions that release a large amount of energy.

Unfortunately, this logical generalization based on analogy soon runs into trouble. It conflicts with experiment. Our generalization predicts that only exothermic reactions (reactions liberating heat) will be spontaneous. *Yet, many endothermic reactions are spontaneous.* Consider, for example, the evaporation of a liquid such as water. This is clearly an endothermic process. The system moves spontaneously from a condition of lower potential energy to one of higher potential energy.

Our generalization encounters still another difficulty: *spontaneous chemical reactions do not always go to completion, but proceed only until equilibrium is reached.* The system does not always go completely to the lower state but seems to stop at an intermediate equilibrium position. Why?

RANDOMNESS

Perhaps a factor *in addition to* the energy of the system is important in determining the equilibrium position. What is this additional factor? It is more subtle than energy content as a driving force for a process and it is harder to recognize; but it is still very familiar to you from your everyday experience. The new factor is simply the general tendency of objects or particles to get mixed up, to become more *randomly* arranged. One of the most fundamental laws of the universe is that a system tends to become disorganized, randomized, or mixed up, unless work is done on it. Look at the young child's room in Figure 16-8. Unless work is done to keep it neat, it will spontaneously approach a state of maximum disorder.

The *tendency toward randomness* is a characteristic of chemical systems as well as of real-life situations. If a tank of nitrogen is connected to a tank of oxygen of equal pressure and the valve between the two is opened, the two gases will mix randomly until a uniform mixture of nitrogen and oxygen is obtained. The mixing process goes on spontaneously because of the random motion of the molecules. Similarly, if we place a few drops of methyl violet solution in a test tube of water, the color will spread gradually through the test tube. (See Figure 16-9.) Because molecular motion is increased by an increase in temperature, the system approaches the uniform color more rapidly as the temperature is raised. However, there is no tendency for the reverse process to occur spontaneously unless work is done on the system. We never see the nitrogen molecules going back into one tank and the oxygen molecules going back into the other tank spontaneously. Machines and work must be used to separate the components of the nitrogen-oxygen mixture; the process is *not* spontaneous. Similarly, methyl violet molecules, once distributed in the water, will have no tendency to re-form crystals. Crystals will form only if external conditions are changed and work is done.

Figure 16-8

A young child's room and chemical systems have in common a tendency toward randomness.

It appears that the natural tendency of the universe to approach a state of maximum disorder can be important in determining the equilibrium condition in a system. The degree of disorder is so important in science that it has been given a special name. A property known as **entropy** measures the randomness of the system on the atomic-molecular level. *As the degree of atomic-molecular disorder increases, the entropy of the system increases.*

How does entropy, or disorder, apply to an equilibrium system? Let us take our simplest example, the vaporization of a liquid. We notice that the vaporization of a liquid is both *spontaneous* and *endothermic*. Our ping-pong ball model for liquids indicated that the balls, or molecules, attract one

Figure 16-9

A few drops of methyl violet in a test tube of water tend to become randomly distributed.

another when close together in the liquid phase. Separating these balls and moving them into the vapor phase required work. Each molecule had to be pushed out of the liquid phase and into the vapor phase, where molecules had higher potential energy. The energy needed to vaporize the liquid came from the kinetic energy of the molecules. The faster-moving molecules either flew off into the vapor phase or knocked other molecules into the vapor phase by collision. We noticed that if heat was not supplied from the outside, the liquid became colder. The faster-moving molecules left, whereas the slower molecules remained in the liquid. The kinetic energy of the faster-moving liquid molecules was converted to the higher potential energy of gaseous molecules. Heat of vaporization had to be supplied to keep the temperature constant.

What is the driving force in this vaporization process? It is clearly the tendency of the molecules to seek a more random distribution throughout all possible space. In our vaporization example, the tendency to seek randomness is *opposed* by the energy effect. The overall process is endothermic.

In other cases, however, both energy effects and entropy effects *combine* to drive a process. An example of such a reaction is the familiar burning of a candle. Energy is given off as the candle burns. Randomness is increased as the solid candle is changed to hot, gaseous products. Since there is *both* a lowering of energy *and* an increase in randomness as this reaction occurs, the formation of products is highly favored. In addition, because entropy effects are dependent upon the motion of molecules, we would expect entropy to become a more important factor in equilibrium as the temperature is raised. More molecules will have the energy to move into the randomized vapor state as the temperature is raised. In fact, *entropy effects become more important as temperature is raised.*

Let us summarize the generalizations our model suggests in relation to chemical equilibrium:

1. All systems tend to approach equilibrium.
2. The first factor important in determining the equilibrium state of a system is its *energy content*. As far as the energy criterion alone is concerned, systems tend to move toward a state of minimum potential energy. The enthalpy or heat content of the system falls. Strongly exothermic reactions tend to be spontaneous.
3. The second factor important in determining the equilibrium state is *the degree of randomness, or the entropy,* of the system. In general, systems tend to move spontaneously toward a state of maximum randomness or disorder: toward maximum entropy.
4. *The equilibrium state is a compromise between these two factors: minimum energy and maximum randomness, or, in the more sophisticated language of the laboratory, minimum enthalpy and maximum entropy.* At very low temperatures, energy tends to be the more important factor. Under such conditions, equilibrium favors the substance with lowest heat content, or lowest energy. At very high temperatures, randomness becomes more important. Under such conditions, equilibrium favors a random distribution of reactants and products without regard to energy differences.

In terms of randomness, we see that dissolving a solid in a liquid results in an increase in the entropy (disorder) of the system. A well-ordered crystal is converted to a thoroughly disordered solution. Clearly, entropy is the major driving force in dissolving a salt in a liquid. If the solution process is endothermic (NH_4NO_3 in water), energy or enthalpy opposes solution. If the endothermic energy term is large enough, the solid will not dissolve. However, if the process is exothermic, both entropy and enthalpy drive the solution process.

SUMMARY

Physical and chemical processes reach a state of equilibrium when the rate of the forward reaction is equal to the rate of the reverse reaction. At equilibrium, no macroscopic changes are visible.

Equilibrium states can be altered as predicted by Le Chatelier's principle: if an equilibrium system is subjected to a change, processes occur that tend to counteract partially the imposed change. Temperature and concentration are the two most important variables involved in determining equilibrium. Catalysts have no effect on a system already at equilibrium. However, they will allow a system not at equilibrium to reach equilibrium more quickly than it otherwise would.

The equilibrium constant, determined from experimental observations, may be summarized as follows. For the reaction $aA + bB \rightleftharpoons eE + fF$, the expression

$$\frac{[E]^e \times [F]^f}{[A]^a \times [B]^b} = K$$

is a constant at constant temperature. The symbol for the equilibrium constant is the letter K.

As the expression for the equilibrium constant suggests, a very large K value indicates that the reaction goes toward completion. Products are favored. Conversely, a very small K value indicates that the reaction does not go very far toward completion. Reactants are favored. (See page 435.)

Only substances with variable concentrations are included in equilibrium constant expressions. Such substances include gases and aqueous ions. The constant concentrations of pure solids and pure liquids are incorporated into the value of K. The concentrations of these substances do not appear in equilibrium constant expressions. The equilibrium constant for the dissolving of a solid is known as its solubility product constant, K_{sp}.

The equilibrium state is often a compromise between two opposing factors: the tendency toward minimum energy and the tendency toward maximum randomness. The energy factor is favored at low temperatures. The randomness factor is favored at high temperatures.

Key Terms

entropy

equilibrium constant (K)

Law of Mass Action

Le Chatelier's principle

solubility product constant (K_{sp})

QUESTIONS AND PROBLEMS

A

1. (a) Describe the processes involved in the equilibrium that exists between a liquid and its vapor. (b) What changes occur in the two processes

indicated above when the temperature is raised? (c) How would changes in the liquid-vapor equilibrium be observed? (d) How could you recognize equilibrium in the liquid-vapor system?

2. (a) Describe the equilibrium processes that occur in a saturated solution of a salt in water. (b) How could you be sure that you have a true equilibrium between a solid salt and its solution and not just a case of slow dissolving of the solid?

3. One drop of water may or may not establish a state of vapor pressure equilibrium when placed in a closed bottle. Explain.

4. What do the following experiments (done at 25 °C) show about the state of equilibrium: (a) one litre of water is added, a few millilitres at a time, to a kilogram of salt, which only partly dissolves; (b) a large saltshaker containing 1 kg of salt is gradually emptied into 1 litre of water; the same amount of salt dissolves as in (a)?

5. A glass bulb is filled with a brownish gas from a cylinder labeled NO_2. When the glass bulb containing the brownish gas is cooled in ice, much of the brown color fades. When the bulb is placed in hot water at about 90 °C, the color becomes very dark brown. (a) Explain these observations in terms of equilibrium. (b) How is this equilibrium like the examples described in 1 and 2?

6. Which of the following are equilibrium situations?
   (a) Grand Coulee dam and the lake behind it. The water level is constant and a river flows into the lake.
   (b) The liquid mercury in a thermometer and the mercury vapor above it in a closed thermometer at constant temperature.
   (c) A well-fed lion in a cage. The lion's weight is constant.
   (d) A solution of yellow sodium chromate (Na_2CrO_4) has been cooled so that a significant number of yellow crystals separate out and the yellow color of the solution remains constant.
   (e) A plant manufacturing CaO from $CaCO_3$. The plant produces 200 tons of CaO each day.

7. When some of the NO_2-N_2O_4 mixture shown in Figure 16-3 is placed in a syringe at room temperature, the following is observed: as the piston is gradually pushed in, the red-brown color of NO_2 first darkens and then becomes progressively lighter; however, it always remains darker than it was originally. The equation for the equilibrium is

$$N_2O_4(g) \rightleftharpoons 2NO_2(g) \quad \Delta H = +14.1 \text{ kcal}$$

COLORLESS   RED-BROWN

Explain the results of this experiment, using Le Chatelier's principle.

8. Does K for the reaction

$$2NO_2 \rightleftharpoons N_2O_4$$

change with temperature? What qualitative evidence can you suggest in support of your answer?

9. Methanol (methyl alcohol) is made according to the following net equation: $CO(g) + 2H_2(g) \rightleftharpoons CH_3OH(g) + \text{heat}$. Predict the effect on equilibrium concentrations of an increase in (a) temperature and (b) pressure.

10. When NH_4NO_3 dissolves in water, the solution becomes cold enough to

freeze ice to the outside of the beaker. What is the driving force for this process? Explain.

B

11. Some solid radioactive iodine-131 is added to a saturated solution of iodine in a beaker. The beaker also contains some nonradioactive solid iodine crystals. Eventually, the initially nonradioactive solution becomes radioactive. How do you account for this observation?

12. If the phase change represented by heat $+ H_2O(l) \rightleftharpoons H_2O(g)$ has reached equilibrium in a closed system: (a) What will be the effect of a reduction of volume, thus increasing the pressure? (b) What will be the effect of an increase in temperature? (c) What will be the effect of injecting some steam into the closed system, thus raising the pressure?

13. How does a catalyst affect the equilibrium conditions of a chemical system?

14. Nitric oxide, NO, releases 13.5 kcal/mole when it reacts with oxygen to give nitrogen dioxide. Write the equation for this reaction and predict the effect of (1) raising the temperature, and of (2) increasing the concentration of NO (at a fixed temperature) on (a) the equilibrium concentrations, (b) the numerical value of the equilibrium constant, and (c) the speed of formation of NO_2.

15. Given: $SO_2(g) + \frac{1}{2}O_2(g) \rightleftharpoons SO_3(g) + 23$ kcal. (a) For this reaction discuss the conditions that favor a high equilibrium concentration of SO_3. (b) How many grams of oxygen gas are needed to form 1.00 gram of SO_3?

16. Water expands when it freezes to become ice. Will putting ice under pressure cause it to melt? Explain, using Le Chatelier's principle.

17. Consider the reaction between N_2 and O_2 that is important in the formation of photochemical smog: $N_2 + O_2 + \text{heat} \rightleftharpoons 2NO$. Why does an automobile engine, which operates at high temperature, generate more nitrogen oxides, which pollute the air than does an engine that operates at lower temperature?

18. Write the equilibrium constant expression for each of the following reactions:

(a) $3O_2(g) \rightleftharpoons 2O_3(g)$

(b) $MgCO_3(s) \rightleftharpoons MgO(s) + CO_2(g)$

(c) $SO_2(g) + NO_2(g) \rightleftharpoons SO_3(g) + NO(g)$

(d) $2Bi^{+++}(aq) + 3H_2S(g) \rightleftharpoons Bi_2S_3(s) + 6H^+(aq)$

19. Using Le Chatelier's principle, explain why increasing the pressure on boiling water causes it to stop boiling until a higher temperature is reached. The equation for the reaction is $H_2O(l) \rightleftharpoons H_2O(g)$ $\Delta H = +9.7$ kcal/mole.

20. Select from each of the following pairs the more random system: (a) your room just after you cleaned it and your room a week later; (b) a beaker just before and just after you have dropped it; (c) logs stacked in your fireplace and their combustion products after burning; (d) a cube of sugar before and after being dropped into hot coffee.

21. Consider the reaction $Y(s) + 2W(g) \rightleftharpoons 2Z(s)$, where $\Delta H = -200$ kcal. (a) Which reaction (forward or reverse) is driven by the tendency

toward minimum energy? (b) Which reaction is driven by the tendency toward maximum randomness?

22. For each of the following reactions, state (1) whether the tendency toward minimum energy favors reactants or products, and (2) whether the tendency toward maximum randomness favors reactants or products. Explain each of your answers.

(a) $CO(g) + \frac{1}{2}O_2(g) \rightleftharpoons CO_2(g)$, where $\Delta H = -67.6$ kcal.

(b) $Mg(s) + \frac{1}{2}O_2(g) \rightleftharpoons MgO(s)$, where $\Delta H = -146$ kcal.

(c) $NaCl(s) \rightleftharpoons NaCl(aq)$, where $\Delta H = +0.4$ kcal.

(d) $H_2SO_4(aq) \rightleftharpoons H_2SO_4(l)$, where $\Delta H = +19.0$ kcal.

(e) $Pb^{2+}(aq) + 2I^-(aq) \rightleftharpoons PbI_2(s)$, where $\Delta H = -14.0$ kcal.

(f) $Br_2(s) \rightleftharpoons Br_2(l)$, where $\Delta H = +16.2$ kcal.

23. The water gas reaction $CO_2(g) + H_2(g) \rightleftharpoons CO(g) + H_2O(g)$ was carried out at 900 °C with the following results.

TRIAL NO.	PARTIAL PRESSURE, ATM AT EQUILIBRIUM			
	CO	H_2O	CO_2	H_2
1	0.352	0.352	0.648	0.148
2	0.266	0.266	0.234	0.234
3	0.186	0.686	0.314	0.314

(a) Write the equilibrium constant expression.

(b) Verify that the expression in (a) is a constant, using the data given.

24. In the reaction $2HI(g) \rightleftharpoons H_2(g) + I_2(g)$ at 448 °C the partial pressures of the gas at equilibrium are as follows: $[HI] = 4 \times 10^{-3}$ atm; $[H_2] = 7.5 \times 10^{-3}$ atm; $[I_2] = 4.3 \times 10^{-5}$ atm. What is the equilibrium constant for this reaction?

25. Reactants A and B are mixed, each at a concentration of 0.80 mole/litre. They react slowly, producing C and D: $A + B \rightleftharpoons C + D$. When equilibrium is reached, the concentration of C is measured and found to be 0.60 mole/litre. Calculate the value of the equilibrium constant.

26. Write the expression indicating the equilibrium law relations for the following reactions.

(a) $N_2(g) + 3H_2(g) \rightleftharpoons 2NH_3(g)$

(b) $CO(g) + NO_2(g) \rightleftharpoons CO_2(g) + NO(g)$

(c) $Zn(s) + 2Ag^+(aq) \rightleftharpoons Zn^{++}(aq) + 2Ag(s)$

(d) $PbI_2(s) \rightleftharpoons Pb^{++}(aq) + 2I^-(aq)$

(e) $CN^-(aq) + H_2O(l) \rightleftharpoons HCN(aq) + OH^-(aq)$

27. Use solubility rules to determine which of the following salts are *insoluble*: (a) $PbCl_2$ (b) CsCl (c) $Mg(NO_3)_2$ (d) AgCl (e) $BaSO_4$ (f) $MgSO_4$ (g) NH_4NO_3 (h) ZnS (i) $Pb(OH)_2$ (j) $Al(OH)_3$ (k) NaOH (l) K_2SO_4.

28. Given the reaction at equilibrium $2A + B \rightleftharpoons C + 3D$ write the equilibrium constant expression and calculate the value of K if the equilib-

rium concentrations are found to be $[A] = 10.0\ M$, $[B] = 15.0\ M$, $[C] = 5.0\ M$, and $[D] = 25.0\ M$.

C

29. The solubility product constant of CuS is 4×10^{-38} at 25 °C. (a) Calculate the number of *moles* of CuS that will dissolve in 1 litre of solution at 25 °C. (b) Calculate the solubility of CuS in water in grams per litre at 25 °C.

30. The solubility of silver sulfide (Ag_2S) is 1.3×10^{-6} mole/litre at 20 °C. (a) Write the equation for dissolving Ag_2S in water. (b) Write the solubility product expression for the reaction. (c) Calculate the value of K_{sp} for the reaction.

31. Reactants A and B are mixed, each initially at a concentration of 1.0 M. They react slowly to produce C according to the equation

$$2A + B \rightleftharpoons C$$

When equilibrium is established, the concentration of C is found to be 0.30 M. Calculate the value of K.

32. Consider the reaction described by the following equation:

$$H_2(g) + CO_2(g) \rightleftharpoons H_2O(g) + CO(g)$$

One mole of $H_2(g)$ and 1 mole of $CO_2(g)$ are allowed to react in a 1-litre container until equilibrium is attained. If the equilibrium concentration of H_2 is 0.44 M, what is the equilibrium concentration of each of the other substances?

33. At 25 °C, K is 2.2×10^{-3} for the reaction

$$2ICl(g) \rightleftharpoons I_2(g) + Cl_2(g)$$

Calculate K for $I_2(g) + Cl_2(g) \rightleftharpoons 2ICl(g)$.

34. Consider the reaction:

$$4HCl(g) + O_2(g) \rightleftharpoons 2H_2O(g) + 2Cl_2(g) + 27\ \text{kcal}$$

What effect would the following changes have on the equilibrium concentration of Cl_2? Give your reasons for each answer.

(a) Increasing the temperature of the reaction vessel.
(b) Decreasing the total pressure.
(c) Increasing the concentration of O_2.
(d) Increasing the volume of the reaction chamber.
(e) Adding a catalyst.

35. What volume of water is necessary to dissolve 0.01 mole of AgCl completely? The K_{sp} for AgCl is 1.7×10^{-10}.

36. The K_{sp} of MnS is 4.9×10^{-9} at 18 °C. How much MnS in grams will be present in 0.100 litre of saturated MnS solution at 18 °C?

Solubility and solubility equilibrium

The oceans are salty because salt dissolves in water. Other minerals, such as limestone ($CaCO_3$), dissolve only slightly: they have low solubility. Much of our fresh water contains small amounts of dissolved $CaCO_3$ and other minerals and is known as hard water. The behavior of $CaCO_3$ and of other slightly soluble substances is controlled by equilibrium processes involving solubility.

Before taking a closer look at solubility behavior, let us review some properties of substances such as NaCl and $CaCO_3$. Both consist of positive and negative ions held together in the solid state by electro-

Figure 16-10
The growth of formations in limestone caves is a direct product of equilibrium processes.

static attraction. Ionic substances conduct electricity when melted or, if sufficiently soluble, when dissolved in water. This conductivity is due to the ions' ability to move past each other and to act independently. The dissolving in water of soluble salts such as NaCl, Na_2CO_3, and $NaNO_3$ can be represented by the equations

$$NaCl(s) \rightleftharpoons Na^+(aq) + Cl^-(aq)$$
$$Na_2CO_3(s) \rightleftharpoons 2Na^+(aq) + CO_3^{--}(aq)$$
$$NaNO_3(s) \rightleftharpoons Na^+(aq) + NO_3^-(aq)$$

Once an ion is freed in this way, it becomes independent of the solid from which it came. For example, a solution made by dissolving NaCl, $NaNO_3$, and Na_2CO_3 in the same container of water contains $Na^+(aq)$, $Cl^-(aq)$, $NO_3^-(aq)$, and $CO_3^{--}(aq)$ ions. All the $Na^+(aq)$ ions are identical, regardless of their source. Similarly, a solution of both NaCl and NH_4Cl contains $Cl^-(aq)$ ions. All the $Cl^-(aq)$ ions show identical behavior, even though they came from two different sources. Because we shall deal only with water solutions, all dissolved ions will be aquated. Thus the term aq will be understood to follow the formula of all dissolved species, and so will not be written.

Ionic solutions are electrically neutral: they do not give an electric shock when touched. This neutrality indicates that the number of positive charges carried by the ions balances the number of negative charges. The three equations above demonstrate this fact. However, while the total charge is conserved, the number of positive ions and negative ions need not be the same. In the second equation, for example, there are twice as many Na^+ as CO_3^{--}.

Predicting solubility behavior Soluble substances generally dissolve enough to give ion concentrations above 0.1 mole per litre (0.1 M) at room temperature. Several positive ions form compounds above 0.1 M with almost all negative ions. The cations H^+, NH_4^+, and the alkali ions (Na^+, Li^+, K^+, Rb^+, Cs^+, Fr^+) form such soluble compounds as indicated by rules 1, 2, and 3 of Table 16-8. Three anions show similar behavior (see rules 4 and 5). Most compounds involving NO_3^-, CH_3COO^-, and ClO_4^- ions are water soluble.

Other anions form compounds of high solubility with some cations, and compounds of low solubility with others. This is shown in Table 16-8 in rules 6 through 10, each of which is divided into two parts. The halide ions Cl⁻, Br⁻, and I⁻, for example, form soluble compounds with all cations except those above the line in rule 6. Table 16-8 lists the behavior of the other commonly encountered anions in a sim-ilar way (rules 7 through 10). Thus, Table 16-8 rather neatly summarizes the solubility behavior of a large number of ionic compounds. Table 16-8 also enables us to predict that certain combinations of aqueous ions form compounds of low solubility. It enables us to predict, for example, that solutions of $NaCl$ and $AgNO_3$, when mixed, form the solid pre-cipitate $AgCl(s)$. This occurs because the combina-

Table 16-8
Solubility of common compounds in water

Rule	Negative ions (anions)	+	Positive ions (cations) $\longrightarrow$	Compounds with the solubility
1	essentially all		alkali ions (Li^+, Na^+, K^+, Rb^+, Cs^+, Fr^+)	soluble
2	essentially all		hydrogen ion [$H^+(aq)$]	soluble
3	essentially all		ammonium ion (NH_4^+)	soluble
4	nitrate, NO_3^-		essentially all	soluble
5	acetate, CH_3COO^-		essentially all	soluble
6	chloride, Cl^- bromide, Br^- iodide, I^-		Ag^+, Pb^{++}, Hg_2^{++}, Cu^+	low solubility
			all others	soluble
7	sulfate, SO_4^{--}		Ca^{++}, Sr^{++}, Ba^{++}, Pb^{++}, Ra^{++}, Ag^+	low solubility
			all others	soluble
8	sulfide, S^{--}		alkali ions, $H^+(aq)$, NH_4^+, Be^{++}, Mg^{++}, Ca^{++}, Sr^{++}, Ba^{++}, Ra^{++}	soluble
			all others	low solubility
9	hydroxide, OH^-		alkali ions, $H^+(aq)$, NH_4^+, Sr^{++}, Ba^{++}, Ra^{++}	soluble
			all others	low solubility
10	phosphate, PO_4^{---} carbonate, CO_3^{--} sulfite, SO_3^{--}		alkali ions, $H^+(aq)$, NH_4^+	soluble
			all others	low solubility

tion of Ag^+ from $AgNO_3$ and Cl^- from NaCl has very low solubility (rule 6). The reaction

$$Ag^+ + Cl^- \longrightarrow AgCl(s)$$

removes the bulk of the ions from solution. Since the remaining Na^+ and NO_3^- ions form only soluble compounds (rules 1 and 4), they remain dissolved as free ions.

The predictability with which precipitates are formed gives the chemist the power to determine what ions may be present in any sample of water. Suppose, for example, you were given a beaker of clear liquid and told only that it was a water solution of one or more ionic compounds. You pour a portion of it into a test tube, add some NaCl solution to the tube, and observe the formation of a white precipitate. By rule 6 of Table 16-8, your "unknown" solution must contain one or more of the ions Ag^+, Pb^{++}, Hg_2^{++}, or Cu^+. If no precipitate had formed, you would know only that none of these ions was present to an appreciable extent.

Now, return to the example of the solution that did form a white precipitate on addition of NaCl. A further look at Table 16-8 suggests other steps that might help you identify the ions present. Rule 7 in Table 16-8 suggests adding Na_2SO_4 solution to a second sample of the original solution. Formation of a second precipitate would indicate the presence of Ag^+ ions or Pb^{++} ions or both, since only such ions show low solubility in both Cl^- and SO_4^{--} ion solutions.

The solubility rules given in Table 16-8 also suggest ways of separating different substances that may be dissolved in the same solution. Consider, for example, a solution known to contain both lead nitrate, $Pb(NO_3)_2$, and magnesium nitrate, $Mg(NO_3)_2$. The lead and magnesium can be separated by removing from solution almost all the Pb^{++} as a solid precipitate. Thus, we must add a test solution that will precipitate only Pb^{++}, and not Mg^{++} as well. Rule 7 in Table 16-8 shows that Pb^{++} is not soluble in SO_4^{--}, but that Mg^{++} is. We can therefore add Na_2SO_4 solution to our solution, and know that it will precipitate as $Pb(SO_4)(s)$ virtually all the Pb^{++} present. The Mg^{++} will remain dissolved. The solid can be removed by simple filtration, producing the desired separation. Mg^{++} can then be precipitated as $MgCO_3(s)$ by rule 10.

Quantitative water pollution analysis Solubility behavior allows us to detect the presence of ions dissolved in water. Water pollution studies require that both the identity and the amount of any species present be specified. Many chemicals, when present in small amounts, are beneficial—even essential—to our health. But the same chemicals in larger quantities may cause sickness or death. For example, in the case of fluoridation of drinking water, it was discovered that water supplies containing close to 1 part of fluoride for each million parts of water (abbreviated *1 ppm,* or 1 part per million, by water-quality chemists) greatly reduced the incidence of dental caries in growing children. More than 1.5 parts per million of fluoride caused severely mottled teeth. In much higher concentrations, fluorides are deadly poisons.

Some impurities are present in water in such minute amounts that solubility behavior cannot detect them. Others are present in detectable amounts. The water supplies of most cities contain from 20 to 50 parts per million of chloride (Cl^-). Water with less than 200 parts per million of chloride is generally acceptable as drinking water.

Determination of the Cl^- ion concentration in drinking water is an excellent example of the application of equilibrium principles to solubility behavior. In essence, the analysis involves precipitation of the Cl^- as $AgCl(s)$ by addition of an $AgNO_3(aq)$ test solution of known concentration. This can be represented by three types of equations:

MOLECULAR
$$AgNO_3(aq) + NaCl(aq) = AgCl(s) + NaNO_3(aq)$$
IONIC
$$Ag^+, NO_3^- + Na^+, Cl^- = AgCl(s) + Na^+, NO_3^-$$
NET IONIC
$$Ag^+ + Cl^- = AgCl(s)$$

Because $AgNO_3$, NaCl, and $NaNO_3$ exist as separate ions in solution, the ionic equation is the most accurate representation. But note that the Na^+ and NO_3^- ions are present in identical form on both sides of the ionic equation. They are called *spectator ions* because they have not changed during the reaction. These ions are not shown in the *net* ionic equation, which represents the chemical changes that have actually occurred.

Since there is no such thing as complete insolubility, however, the above equations are misleading. They seem to imply that all the Ag^+ and Cl^- ions have precipitated. But, as you have read, even AgCl(s) dissolves to some extent. This fact can be shown by placing some of the solid in a large container of distilled water, stirring, and leaving the resulting mixture undisturbed for the next several days. One litre of the liquid phase of this water-AgCl mixture is then poured into a new beaker leaving behind all the remaining undissolved solid. The water is allowed to evaporate. A sensitive balance will show that a residue of almost 0.0020 gram of AgCl(s) remains. This is equivalent to

$$0.0020 \text{ g/litre} \times \frac{1 \text{ mole AgCl}}{143.5 \text{ g}} =$$

$$0.000014 \text{ mole AgCl/litre}$$

or about 1.4×10^{-5} mole of AgCl that must have dissolved in the litre of water at room temperature. Even more careful measurements indicate that this figure is closer to 1.3×10^{-5} mole AgCl per litre.

The dissolving of AgCl(s) in water can be represented by the equation

$$AgCl(s) \rightleftharpoons Ag^+ + Cl^-$$

Chemists have found that only definite amounts of the ions of such slightly soluble solids can remain dissolved. When their concentrations exceed these amounts, the excess ions will precipitate out of the solution. Thus, for AgCl at room temperature, the product

$$[Ag^+] \times [Cl^-] = 1.7 \times 10^{-10}$$

is constant at saturation. This expression is just the equilibrium constant for AgCl(s) in equilibrium with its dissolved ions. You have learned that the value of 1.7×10^{-10} is known as the solubility product constant, or K_{sp}. As long as the product $[Ag^+] \times [Cl^-]$ for a given solution is equal to or less than this number, all ions will remain dissolved. However, if this product exceeds the K_{sp} value, precipitation of AgCl occurs until the remaining Ag^+ and Cl^- ion concentrations just equal K_{sp} when multiplied together.

Note that the above does not mean that the $[Ag^+]$ is necessarily equal to the $[Cl^-]$, but simply that their

product must not exceed the constant 1.7×10^{-10}. Thus, if the $[Ag^+]$ increases, the $[Cl^-]$ must decrease, to keep the product of $[Ag^+] \times [Cl^-]$ from exceeding the K_{sp} value. In the above example, the concentrations of Ag^+ and Cl^- ions would be equal since both ions came from dissolving AgCl(s), and since 1 mole of AgCl(s) yields 1 mole of Ag^+ ion and 1 mole of Cl^- ion. Specifically, they would each equal the amount of AgCl(s) that dissolved per litre at room temperature, or about 1.3×10^{-5} mole per litre. This figure can be used to calculate the K_{sp} value, since

$$K_{sp} = [Ag^+] \times [Cl^-]$$
$$= (1.3 \times 10^{-5}) \times (1.3 \times 10^{-5}) = 1.7 \times 10^{-10}$$

If K_{sp} is known, the solubility can be determined by solving for $[Ag^+]$ or $[Cl^-]$, as indicated in Section 16.6. The K_{sp} values for various slightly soluble compounds are listed in Table 16-9.

Table 16-9

Solubility products of various substances at 25 °C

Compound	K_{sp}
AgCl	1.7×10^{-10}
AgBr	5.0×10^{-13}
AgI	8.5×10^{-17}
$AgBrO_3$	5.4×10^{-5}
$AgIO_3$	2.1×10^{-8}
$SrCrO_4$	3.6×10^{-5}
$BaCrO_4$	8.5×10^{-11}
$PbCrO_4$	2.0×10^{-16}
$CaSO_4$	2.6×10^{-4}
$SrSO_4$	7.6×10^{-7}
$PbSO_4$	1.3×10^{-8}
$BaSO_4$	1.5×10^{-9}
$RaSO_4$	4.0×10^{-11}

Suppose we wish to test 1 litre of drinking water for the presence of Cl^- ions by adding 1 millilitre of $0.1\ M$ $AgNO_3$ solution. Since 1 millilitre is equal to 10^{-3} litre, and $0.1\ M$ means 0.1 mole per litre, we have added

$$10^{-3} \text{ litre} \times 0.1 \text{ mole/litre} = 10^{-4} \text{ mole } AgNO_3$$

Actually, since all the $AgNO_3$ is present as ions, we have added 10^{-4} mole each of Ag^+ and NO_3^- ions. Because the volume of solution being tested is still approximately 1 litre (1.001 litre, to be exact), the $[Ag^+]$ is $10^{-4}\ M$, or 10^{-4} mole per litre. The equilibrium condition allows us to write

$$K_{sp} \text{ AgCl} = 1.7 \times 10^{-10}$$
$$= [Ag^+] \times [Cl^-] = 10^{-4} \times [Cl^-]$$

Solving for the $[Cl^-]$, we find the value could be as high as 1.7×10^{-6} mole per litre without formation of a precipitate of $AgCl(s)$. All chloride in excess of that amount would be precipitated as $AgCl(s)$.

By adding a known amount of potassium chromate (K_2CrO_4) to the water being tested, it becomes possible to determine quantitatively the amount of chloride present in the sample, including the almost negligible amount of Cl^- ions remaining dissolved after addition of the silver nitrate. This is because a brick-red precipitate of $Ag_2CrO_4(s)$ will form, but only after most of the Cl^- ions have already precipitated as white $AgCl(s)$. Thus, if an $AgNO_3$ solution of known concentration is added by drops to a solution of both Cl^- and CrO_4^- ions, first a white precipitate of $AgCl(s)$ will form. The moles of Cl^- that react to form this precipitate will be equal to the moles of Ag^+ added. At the first sign of formation of red $Ag_2CrO_4(s)$, the addition of $AgNO_3$ is stopped. If we note how much $AgNO_3$ has been added, we can then calculate the total amount of Cl^- ions originally present in the sample.

Predicting whether a precipitate will form Since a precipitate of a slightly soluble substance must form if the K_{sp} value for that substance is exceeded, a table of K_{sp} values (such as Table 16-9) can be used to predict whether or not precipitates will form. Suppose, for example, we wish to know whether or not a precipitate will form if 1 millilitre of $0.1\ M$ $AgNO_3$ solution is added to a 1.0-litre sample of Cl^- ions. To answer such a question, we must first know what precipitate may possibly form. In this case, of course, it is $AgCl(s)$, by rule 6 in Table 16-8. Next, we find the concentrations of each ion involved in the K_{sp} expression and compare their product to the known K_{sp} value. As already indicated, if that product is less than or equal to the K_{sp} value, no precipitate will form. However, if the K_{sp} value is exceeded, the excess will precipitate.

Suppose in our example that the value of $[Cl^-]$ is 5×10^{-4} mole per litre. The $[Ag^+]$ is found by determining the number of moles of Ag^+ added and then dividing this by the total volume. If 1 millilitre is equal to 10^{-3} litre, then

$$10^{-3} \text{ litre} \times 0.1 \text{ mole/litre} = 10^{-4} \text{ mole } Ag^+$$

This is also the concentration, since the total volume is still approximately 1 litre. Therefore, $K_{sp} = 1.7 \times 10^{-10}$ is compared to

$$[Ag^+] \times [Cl^-] = (10^{-4}) \times (5 \times 10^{-4}) = 5 \times 10^{-8}$$

Since this is greater than 1.7×10^{-10}, a precipitate of AgCl will form.

As a final example, suppose 0.166 gram of $Pb(NO_3)_2(s)$ is dissolved in 1.0 litre of $0.01\ M$ Na_2SO_4. Will a precipitate form? If so, what is it? A useful approach to such a problem would be:

1. Identify the ions in solution. In this case, they are Pb^{++}, NO_3^-, Na^+, and SO_4^{--} ions.
2. Consult the solubility rules to find which combinations of these ions have low solubility and might precipitate. By rule 7, the possible precipitate is clearly $PbSO_4(s)$.
3. Write the equation for the appropriate solubility equilibrium. Our equation is

$$PbSO_4(s) = Pb^{++} + SO_4^{--}$$

4. Write the K_{sp} expression, giving its numerical value. Here, K_{sp} is equal to

$$1.3 \times 10^{-8}* = [Pb^{++}] \times [SO_4^{--}]$$

5. Find the concentrations, in moles per litre, of each ion involved in the K_{sp} expression. For Pb^{++}, since 1 $Pb(NO_3)_2$ gives 1 Pb^{++}, the moles of Pb^{++} will equal those of $Pb(NO_3)_2$. We therefore convert the given mass of the latter to moles:

$$0.166 \text{ g} \times \frac{1 \text{ mole}}{331 \text{ g}} =$$
$$.0005 = 5 \times 10^{-4} \text{ mole}$$

For SO_4^{--}, 1 Na_2SO_4 yields 1 SO_4^{--}. Thus, the number of moles of Na_2SO_4 equals

*Taken from Table 16-9 on page 451.

1.0 litre $\times$ .01 mole/litre =
$$10^{-2} \text{ mole SO}_4^{--}$$

Since both of these ions are dissolved in 1.0 litre of solution, the concentrations are 5×10^{-4} mole per litre for Pb^{++}, and 1×10^{-2} mole per litre for the SO_4^{--}.

6. Finally, compare the *product* of these ion concentrations to the known K_{sp} value, 1.3×10^{-8}

$$[Pb^{++}] \times [SO_4^{--}] = (4 \times 10^{-4})$$
$$\times (1 \times 10^{-2}) = (5 \times 10^{-4})$$
$$\times (1 \times 10^{-2}) = 5 \times 10^{-6}$$

Since our ion product, 5×10^{-6}, is greater than K_{sp}, a precipitate of $PbSO_4(s)$ will form.

The acids . . . are compounded of two substances . . . the one constitutes acidity, and is common to all acids . . . the other is peculiar to each acid, and distinguishes it from the rest.

ANTOINE LAURENT LAVOISIER

17

IONIC EQUILIBRIUM: ACIDS AND BASES

Objectives

After completing this chapter you should be able to:

- Understand the significance of the constant K_w and its use in calculating the concentrations of $H^+(aq)$ and $OH^-(aq)$ ions in any system.

- Differentiate between operational and conceptual definitions of acids and bases.

- Use the Brønsted-Lowry definitions to determine, for any given reaction, which reactant is the acid, what products will form, and whether reactants or products are favored.

- Use the constant K_A to determine the strengths of acids and the equilibrium concentrations of the species present.

Oranges and other edible citrus fruits are acidic.

In Chapter 16, we studied equilibrium in a variety of systems. In every case, we found that no macroscopic changes were observable at equilibrium. We explained equilibrium as a balance of opposing processes. In equilibrium, the forward reaction occurs at the same rate as the reverse reaction.

In this chapter, we shall explore ionic equilibria involving $H^+(aq)$ ions in solution. You will find that this type of equilibrium provides the key for understanding many familiar chemical reactions that occur in water. Reactions in water involving $H^+(aq)$ and $OH^-(aq)$ ions are classed as acid and base reactions. Acid-base reactions are of great importance, not only to chemists, but to biologists, physicians, and engineers as well.

17.1 Electrolytes: Strong and weak

In Chapter 7, we explained electrical conductivity of aqueous solutions by assuming the existence of ions. We pictured positive and negative ions moving through a solution and carrying electric current. That model suggests

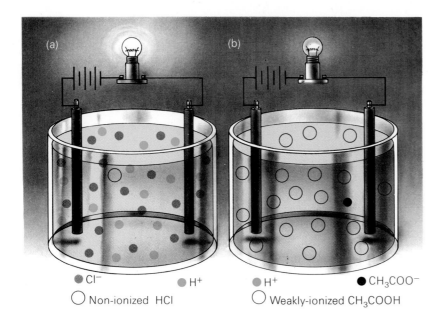

● Cl^- ● H^+
○ Non-ionized HCl

● H^+ ● CH_3COO^-
○ Weakly-ionized CH_3COOH

that solutions with high electrical conductivity contain a great many ions. Solutions with little or no electrical conductivity contain few, if any, ions. Figure 17-1 illustrates a method of measuring the electrical conductivity of a given solution. A brightly lighted bulb indicates high conductivity. A dim glow indicates low conductivity. A completely dark bulb indicates a nonconductor. Quantitative measurements of conductivity can be made by replacing the bulb with a sensitive meter that measures the flow of current.

When solutions of sodium chloride (NaCl) and hydrogen chloride (HCl) are tested, they show high conductivity. The same is true of sodium hydroxide solution, NaOH(aq). Such tests suggest that all three substances dissolve in water to produce high concentrations of ions. The appropriate equations are

$$NaCl(s) \longrightarrow Na^+(aq) + Cl^-(aq)$$
$$HCl(g) \longrightarrow H^+(aq) + Cl^-(aq)$$
$$NaOH(s) \longrightarrow Na^+(aq) + OH^-(aq)$$

Substances that go into water solution almost totally as ions are called **strong electrolytes.**

Some substances dissolve completely in water, producing solutions that show little or no conductivity. For example, acetic acid in water* is a poor conductor of electricity. We therefore assume that acetic acid ionizes only slightly in water:

$$CH_3COOH(aq) \rightleftharpoons H^+(aq) + CH_3COO^-(aq)$$

Note that only a small fraction of the molecules ionize. Most of the dissolved species are neutral molecules. Substances, such as acetic acid, that dissolve in water but produce few ions are called **weak electrolytes.**

*Common vinegar contains 5 percent acetic acid in water.

17.2 Water: A very weak electrolyte

Pure water does not appear to conduct electric current at all. However, when we measure the conductivity of pure water with an extremely sensitive meter, we can observe a tiny electric current. This suggests that water itself forms a few ions. The equation for this process is

$$HOH(l)^* \rightleftharpoons H^+(aq) + OH^-(aq) \qquad (1)$$

The concentration of ions is *very* small. Equilibrium heavily favors non-ionized water. The equilibrium constant for HOH ionization can be written

$$K = \frac{[H^+(aq)] \times [OH^-(aq)]}{[HOH(l)]} \qquad (2)$$

Equilibrium constants for ionization processes are known as **ionization constants**. The expression in (2) was one of the special cases of equilibrium constants considered in Section 16.6.[†] One litre of HOH—1,000 milli-litres—has a mass of 1,000 grams. One mole of water has a mass of 18.0 grams (16 plus 1 plus 1). Therefore, the *concentration* of pure water is

$$1,000 \; \frac{grams}{litre} \times \frac{1 \; mole \; HOH}{18.0 \; grams} = 55.6 \; moles/litre$$

We can incorporate this essentially constant concentration into the equilibrium constant by multiplying both sides of equation 2 by 55.6:

$$K \times [HOH(l)] = \frac{[H^+(aq)] \times [OH^-(aq)]}{[HOH(l)]} \times [HOH(l)]$$

The product $K \times [HOH(l)]$ is a new constant called K_w, in which the subscript w stands for water. Equation 2 thus becomes

$$K_w = [H^+(aq)][OH^-(aq)]$$

The special constant for water, K_w, is known as the *ion-product constant* for water. As you will soon see, it is extremely important to chemistry in water solutions.

K_w AND ITS VARIATION WITH TEMPERATURE

Since water ionizes to a very small extent, both $[H^+(aq)]$ and $[OH^-(aq)]$ are very small. Thus, their product (K_w) should be very small. Experimental measurements confirm this assumption. The constant K_w has a value of 1.0×10^{-14} at 25°C. In pure water, the concentration of $H^+(aq)$ must equal the concentration of $OH^-(aq)$ (see equation *1*). We can therefore write

$$[H^+(aq)] = [OH^-(aq)]$$

---

*Writing water as HOH rather than H_2O makes it easier to visualize the suggested mechanism.

[†]Recall that the concentration of water in the solution is constant and may be included in the equilibrium constant. (See page 434.)

Table 17-1

Values of K_w at various temperatures

Temperature (°C)	K_w
0	0.114×10^{-14}
10	0.295×10^{-14}
20	0.676×10^{-14}
25	1.00×10^{-14}
60	9.55×10^{-14}

The value of K_w is thus $[H^+(aq)][OH^-(aq)] = 10^{-14}$. Substituting $[H^+(aq)]$ for $[OH^-(aq)]$, we write

$$[H^+(aq)][H^+(aq)] = 10^{-14}$$
$$[H^+(aq)] = \sqrt{10^{-14}} = 10^{-7}$$
$$[H^+(aq)] = [OH^-(aq)] = 10^{-7}$$

At equilibrium, both $H^+(aq)$ and $OH^-(aq)$ have a concentration of only 1.0×10^{-7} mole per litre.

The value of K_w, like the value of any equilibrium constant, depends on the temperature. The value of K_w given above (1.0×10^{-14}) holds only at a temperature of 25 °C. What happens to the value if the temperature is changed? Experiments show that the ionization of water absorbs energy:

$$13.7 \text{ kcal} + \text{HOH}(l) \rightleftharpoons H^+(aq) + OH^-(aq)$$

According to Le Chatelier's principle, an *increase* in temperature (adding heat) will shift the equilibrium in the direction that absorbs heat. This shift will *increase* the concentration of ions, and thus *increase* the numerical value of K_w. Conversely, a *decrease* in temperature (removing heat) will shift the equilibrium in the direction that releases heat. This will *decrease* the ion concentrations, and so *decrease* the value of K_w. Experimental values of K_w at various temperatures are given in Table 17-1. The values agree with Le Chatelier's principle.

CONCENTRATIONS OF H⁺ AND OH⁻ IONS IN AQUEOUS SOLUTIONS

As noted above, both $H^+(aq)$ and $OH^-(aq)$ ion concentrations equal 1.0×10^{-7} mole per litre at equilibrium at 25 °C. *Solutions in which* $[H^+(\text{aq})]$ *equals* $[OH(\text{aq})]$ *are called **neutral solutions.*** All *pure water* solutions are neutral, regardless of the temperature.

Does the concentration of $H^+(aq)$ ions *always* equal the concentration of $OH^-(aq)$ ions in water solution? Absolutely not! Suppose we add a solution of $HCl(aq)$ to neutral water. We would be adding $H^+(aq)$ and $Cl^-(aq)$ ions. The equilibrium expression for water alone is

$$\text{HOH}(l) \rightleftharpoons H^+(aq) + OH^-(aq) \qquad (1)$$

Le Chatelier's principle predicts that this equilibrium will shift if $H^+(aq)$ is added to the solution as HCl solution. The reaction

$$H^+(aq) + OH^-(aq) \longrightarrow \text{HOH}(l) \qquad (3)$$

will occur until enough excess $H^+(aq)$ ions are removed to reestablish the equilibrium relationship. Let us see how this works. Suppose 0.01 mole of HCl is dissolved in enough water to produce 1 litre of solution. Since HCl is 100-percent ionized, the concentration of $[H^+(aq)]$ in the solution will be .01, or 10^{-2}, moles per litre. From the value of K_w, $[OH^-(aq)]$ can be easily calculated:

$$[H^+(aq)][OH^-(aq)] = 10^{-14}$$
$$10^{-2}[OH^-(aq)] = 10^{-14}$$
$$[OH^-(aq)] = 10^{-12}$$

The concentration of H⁺(*aq*) is much larger than the concentration of OH⁻(*aq*). *Solutions in which the H⁺(aq) ion concentration is greater than the OH⁻(aq) ion concentration are called acid solutions.* A substance that produces an acidic solution when it is dissolved in water is called an **acid.**

Similarly, when we add NaOH to pure water, we are really adding the ions Na⁺(*aq*) and OH⁻(*aq*). At equilibrium, the concentration of OH⁻(*aq*) will be much greater than the concentration of H⁺(*aq*). Suppose 0.01 mole of NaOH is added to enough water to make 1 litre of solution. Since we assume that NaOH ionizes completely, this procedure will yield an [OH⁻(*aq*)] of 10^{-2} mole per litre. The [H⁺(*aq*)] can now be calculated from K_w:

$$[H^+(aq)][10^{-2}] = 10^{-14}$$
$$[H^+(aq)] = 10^{-12}$$

The [OH⁻(*aq*)] is 10^{10} times larger than the [H⁺(*aq*)] value. *Solutions in which the OH⁻(aq) ion concentration is greater than the H⁺(aq) ion concentration are called basic solutions.* A substance that produces a basic solution when it is dissolved in water is called a **base.**

In the two ordinary laboratory solutions discussed above, the concentration of H⁺(*aq*) ions varied from 10^{-1} to 10^{-12} mole per litre. The range is almost too large to comprehend and would be a mere chemical curiosity but for one fact. The ions H⁺(*aq*) and OH⁻(*aq*) take part in many important reactions that occur in aqueous solutions. For instance, many biological processes are extremely sensitive to the concentration of H⁺(*aq*) ions in

Figure 17-2

The *p*H of human blood must remain nearly constant or serious problems will result. Data from this blood analyzer helps doctors control the blood *p*H of hospital patients.

solution. The concentration of $H^+(aq)$ ions in human blood must stay close to 6.0×10^{-8} mole per litre. Otherwise, severe damage, and even death, may result.

Many reactions in solutions use $H^+(aq)$ or $OH^-(aq)$ ions as reactants. Changing reactant concentrations by factors of from 1 million to 1 trillion would seriously affect both the equilibrium and the course of the reaction. In addition, many reactions involve $H^+(aq)$ or $OH^-(aq)$ as catalysts. The catalytic decomposition of formic acid (discussed in Chapter 15, page 408), is but one example. Indeed, the concentration of $H^+(aq)$ is a most important variable in the study of aqueous solutions.

EXERCISE 17-1

(a) Calculate the concentration of H^+ and OH^- in pure water at 10 °C.
(b) What is the concentration of $OH^-(aq)$ in 0.1 M HCl solution at 10 °C and 25 °C?

EXERCISE 17-2

Suppose 3.65 g of HCl are dissolved in enough water to produce 10.0 litres of solution at 25 °C. Find the value of $[H^+(aq)]$ in the solution. Use the expression $K_w = [H^+][OH^-] = 10^{-14}$ to show that $[OH^-]$ is $1.00 \times 10^{-12}\ M$.

17.3   Acid-base titrations

In Section 17.2, we defined acidic and basic solutions. What happens when an acid is mixed with a base? In this section, we shall look at several examples of such solutions.

HCl AND NaOH IN THE SAME SOLUTION: EXCESS HCl

Suppose we add 0.090 mole of NaOH(s) to 0.100 litre of 1.00 M HCl solution. By adding solid NaOH, we keep the volume of the solution essentially constant. We now have both $H^+(aq)$ and $OH^-(aq)$ in relatively high concentrations in the *same* solution. What will happen? Immediately after the sodium hydroxide dissolves and before reaction, the concentrations of $H^+(aq)$ and $OH^-(aq)$ far exceed the equilibrium values. The **trial product** $[H^+][OH^-]$ far exceeds 1.00×10^{-14} mole^2 per litre^2:

$$\text{initial } [H^+] = 1.00\ M$$

$$\text{initial } [OH^-] = \frac{0.090 \text{ mole}}{0.100 \text{ litre}} = 0.90\ M$$

$$\text{trial product} = [H^+] \times [OH^-] = 9.00 \times 10^{-1} \text{ mole}^2/\text{litre}^2$$

The equilibrium relationship $[H^+][OH^-] = 1.00 \times 10^{-14}$ can be achieved most easily by removing both $H^+(aq)$ and $OH^-(aq)$. The reaction of the two ions to form water is clearly indicated:

$$OH^-(aq) + H^+(aq) \longrightarrow H_2O(l)$$

Since K_w is so small, essentially all the $OH^-(aq)$ is consumed when the HCl is present in excess. In our example, $[H^+]$ initially exceeds $[OH^-]$ by 0.10 mole per litre:

$$\text{initial } [H^+] - \text{initial } [OH^-] = \text{excess } [H^+]$$
$$1.00 \ M \quad - \quad 0.90 \ M \quad = \quad 0.10 \ M$$

If the remaining $[H^+]$ is 0.10 mole per litre at equilibrium, then

$$[H^+][OH^-] = 1.00 \times 10^{-14} \text{ mole}^2/\text{litre}^2$$

$$[OH^-] = \frac{1.00 \times 10^{-14} \text{ mole}^2/\text{litre}^2}{[H^+]}$$

$$= \frac{1.00 \times 10^{-14} \text{ mole}^2/\text{litre}^2}{1.0 \times 10^{-1} \text{ mole}/\text{litre}}$$

$$[OH^-] = 1.0 \times 10^{-13} \text{ mole}/\text{litre}$$

The concentration of $OH^-(aq)$, 10^{-13} mole per litre, is a million times smaller than the $OH^-(aq)$ concentration of pure water. Nearly all the $OH^-(aq)$ has been *neutralized,* or removed, by the HCl present, but a *very* small concentration, 10^{-13} mole/litre, remains.

A helpful way to analyze the reaction
$$OH^-(aq) + H^+(aq) \longrightarrow H_2O(l)$$

Concentrations of reactants and product:

before:	0.90	1.00	0
reacting:	0.90 +	0.90 $\longrightarrow$	0.90
after:	0*	0.10	0.90

*A very small concentration remains. See calculation on this page.

EXERCISE 17-3

Suppose that 0.099 mole of NaOH(s) is added to 0.100 litre of 1.00 M HCl. (a) How many moles of ionized HCl are present in the final solution? (b) From the moles of HCl and the volume, calculate the concentration of excess $H^+(aq)$. (c) Calculate the concentration of $OH^-(aq)$ at equilibrium. (See your calculations for Exercise 17-2.)

HCl AND NaOH IN THE SAME SOLUTION: EXCESS NaOH

Returning to our original 0.100 litre of 1.00 M HCl, let us now consider the addition of 0.101 mole of NaOH(s). Again, we have added both $H^+(aq)$ and $OH^-(aq)$ to the *same* solution, and the concentrations immediately after mixing and before reaction do not satisfy the equilibrium expression,

$$\text{initial } [H^+] = 1.00 \ M$$

$$\text{initial } [OH^-] = \frac{0.101 \text{ mole}}{0.100 \text{ litre}} = 1.01 \ M$$

$$\text{initial product} = [H^+] \times [OH^-] = 1.01 \text{ moles}^2/\text{litre}^2$$

far exceeding 1.00×10^{-14} mole^2 per litre^2.

The resulting solution contains excess hydroxide ions, OH$^-$(aq). Therefore, essentially all the H$^+$(aq) will be consumed, forming water:

$$\text{initial [OH}^-] - \text{initial [H}^+] = \text{excess [OH}^-]$$
$$1.01\ M \quad - \quad 1.00\ M \quad = 0.01\ M$$

The equilibrium concentration of H$^+$ in a solution where [OH$^-$] is equal to 0.010 M is

$$[\text{H}^+] = \frac{1.00 \times 10^{-14}\ \text{mole}^2/\text{litre}^2}{1 \times 10^{-2}\ \text{mole/litre}} = 1 \times 10^{-12}\ \text{mole/litre}$$

HCl AND NaOH IN THE SAME SOLUTION: NO EXCESS

In each example used in this section, a known quantity of NaOH was added to 0.100 litre of 1.00 M HCl. Either HCl or NaOH was in excess. The reaction between H$^+$(aq) and OH$^-$(aq) ions consumes essentially all the constituent ions not in excess. Let us now consider the case in which there is an excess of *neither* HCl *nor* NaOH.

Suppose we add 0.100 mole of NaOH to 0.100 litre of 1.00 M HCl. The initial values of [H$^+$] and [OH$^-$] are equal and their product before reaction far exceeds 1.00×10^{-14}:

$$\text{initial [H}^+] = 1.00\ M$$

$$\text{initial [OH}^-] = \frac{0.100\ \text{mole}}{0.100\ \text{litre}} = 1.00\ M$$

$$\text{initial trial product} = [\text{H}^+] \times [\text{OH}^-] = 1.00\ \text{moles}^2/\text{litre}^2$$

The reaction between H$^+$(aq) and OH$^-$(aq) ions must again occur, forming water: OH$^-$(aq) + H$^+$(aq) = H$_2$O(l). Since 1 mole of OH$^-$(aq) consumes 1 mole of H$^+$(aq), the concentrations [H$^+$] and [OH$^-$] remain equal as the neutralization reaction between H$^+$(aq) and OH$^-$(aq) proceeds. When equilibrium is reached, they will still be equal. This is exactly the situation in pure water. As we saw on page 458,

$$[\text{H}^+] = [\text{OH}^-] = \sqrt{K_w} = 1.00 \times 10^{-7}\ M$$

A solution containing exactly equivalent amounts of HCl and NaOH is neither acidic nor basic; it is neutral. *If equal numbers of moles of HCl and NaOH are added to water, the final solution is neutral.* The point at which the moles of NaOH added become equal to the moles of HCl intially present is called the **equivalence point,** or end point. Adding even a tiny amount of NaOH after the equivalence point has been reached makes the solution strongly basic.

PROGRESSIVE ADDITION OF NaOH TO HCl: A TITRATION

We will now consider the progressive addition of NaOH to a fixed amount of HCl solution. The results appear in Table 17-2 and are plotted graphically in Figure 17-3. We see that [H$^+$] changes by a factor of 10^{10} as the initial

Table 17-2

Changes in $[H^+(aq)]$ and $[OH^-(aq)]$ during the titration of 100 ml of 1.00 M HCl

Initial Number of moles of HCl	Number of moles of NaOH(s) added	Excess moles of		Equilibrium		Acidic or basic
		HCl	NaOH	$[H^+(aq)]$	$[OH^-(aq)]$	
0.100	0.000	0.100	none	1.00	1.00×10^{-14}	acidic
0.100	0.090	0.010	none	1.0×10^{-1}	1.0×10^{-13}	↑
0.100	0.099	0.001	none	1×10^{-2}	1×10^{-12}	
0.100	0.100	none	none	1.0×10^{-7}	1.00×10^{-7}	neutral
0.100	0.101	none	0.001	1×10^{-12}	1×10^{-2}	
0.100	0.110	none	0.010	1.0×10^{-13}	1.0×10^{-1}	↓
0.100	0.200	none	0.100	1.00×10^{-14}	1.00	basic

$[OH^-]$ is changed from 0.99 M to 1.01 M. Since $[H^+]$ can be easily followed experimentally, it provides a sensitive means of observing the mixing of an acid and a base. *The progressive addition of a base to an acid or of an acid to a base is an example of a* **titration process.**

In an acid-base titration, carefully measured amounts of a basic solution of known concentration are added to a known volume of an acidic solution.

Figure 17-3

A titration curve showing pH as a function of volume added when 50 ml of 0.100 molar HCl is titrated with 0.100 molar NaOH.

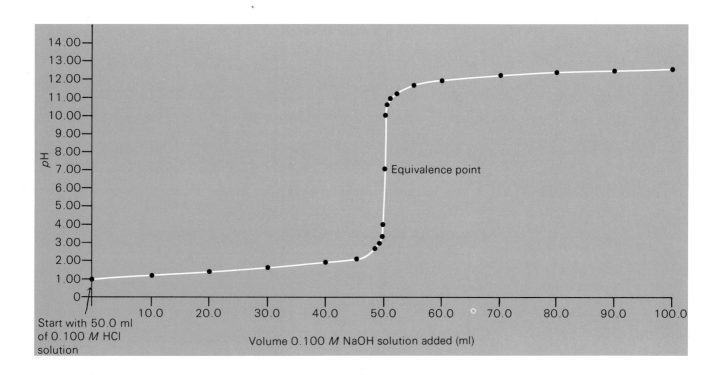

Start with 50.0 ml of 0.100 M HCl solution

Volume 0.100 M NaOH solution added (ml)

Figure 17-4

The concentrations of acidic and basic solutions can be determined by titration. By careful use of a buret, students can accurately measure the volume of solution being added.

The acidic solution contains a substance that changes color at or near the equivalence point to provide visual evidence of the magnitude of [H⁺]. The dye called **litmus** is one such substance. It is red in solutions with excess $H^+(aq)$ and blue in solutions with excess $OH^-(aq)$. Fortunately, the dye color is quite sensitive to very small amounts of $H^+(aq)$. In solutions containing [$H^+(aq)$] of 10^{-6} M, the dye becomes red. In solutions containing [$H^+(aq)$] of 10^{-8} M, the dye becomes blue. Note from Table 17-2 that a slight addition of excess $OH^-(aq)$ causes a great change in the concentration of $H^+(aq)$. That is what causes the marked change in dye color.

There are many other dyes that are very sensitive to the concentration of $H^+(aq)$. Such dyes are called **acid-base indicators.** Not all of them change color where [$H^+(aq)$] equals [$OH^-(aq)$]. Some change on the acidic side, some on the basic side. Litmus and the dye bromthymol blue change at the neutral point at which [$H^+(aq)$] equals [$OH^-(aq)$]. Figure 11-1 (page 264) shows the colors of several acid-base indicators in solutions of varying [$H^+(aq)$].

Acid-base indicators can help us find the concentration of solutions. For example, suppose we add a few drops of bromthymol blue to a 0.100-litre sample of HCl. Carefully, we add an NaOH solution. The color change indicating the equivalence point occurs after 0.050 litre of 0.200 M NaOH has been added. Apparently, this quantity of NaOH was required to neutralize all the added HCl. What was the concentration of the original HCl solution?

For HCl and NaOH at the equivalence point, the number of moles of acid equals the number of moles of base: $H^+, Cl^- + Na^+, OH^- \rightarrow HOH + Na^+, Cl^-$. Therefore,

$$\text{moles of HCl} = \text{moles of NaOH}$$

Using the definition of molarity, we can write

$$\text{number of moles} = \text{molarity} \times \text{volume in litres}$$

At the equivalence point,

$$(\text{molarity} \times \text{volume in litres})_{acid} = (\text{molarity} \times \text{volume in litres})_{base}$$

Then,

$$(\text{molarity} \times 0.100 \text{ litres})_{acid} = (0.200 \ M \times 0.050 \text{ litres})_{base}$$

If three of the four quantities at the equivalence point are known, the fourth can be calculated. Thus,

$$\text{molarity of HCl} = \frac{0.0100 \text{ mole}}{0.100 \text{ litre}} = 0.100 \ M$$

EXERCISE 17-4

A titration similar to the one described in the sample problem above shows that 0.250 litre of 0.200 M NaOH is needed to completely neutralize 1.00 litre of an HCl solution. What was the molarity of the original HCl solution?

EXERCISE 17-5

In another titration, 0.300 litre of 0.150 M KOH is needed to completely neutralize 0.500 litre of a fresh HCl solution. Find (a) the number of moles of NaOH needed to neutralize the HCl; (b) the number of moles of HCl initially present in the HCl solution; and (c) the initial concentration of the HCl solution, in moles per litre, before addition of the NaOH.

17.4   The pH scale

Chemists and biologists have found it convenient to use a quantity known as the **pH** to indicate in abbreviated numerical form the hydrogen-ion concentration in any solution. The pH is officially defined by the expression

$$p\mathrm{H} = -\log_{10}[\mathrm{H}^+(aq)]$$

The pH values for the solutions mentioned above can be obtained from their equilibrium $[\mathrm{H}^+(aq)]$ values. For example, if a solution contains 0.01 M HCl(aq), $[\mathrm{H}^+(aq)] = 0.01\ M = 10^{-2}\ M$. Recall that the HCl$(aq)$ has completely ionized to give $\mathrm{H}^+(aq)$ and $\mathrm{Cl}^-(aq)$. Once the $[\mathrm{H}^+(aq)]$ has been expressed in its exponential form (10^{-2}),* pH is found by taking the exponent of 10 (-2 in this case) and changing its sign. In this example, therefore, pH is equal to $+2$, or simply to 2. The pH values of some common materials with varying $[\mathrm{H}^+(aq)]$ are given in Figure 17-5.

It is important to note the following:

1. A pH value of less than 7 indicates an *acidic* solution. In acidic solutions, $[\mathrm{OH}^-(aq)]$ is less than $[\mathrm{H}^+(aq)]$.
2. A pH value of 7 indicates a *neutral* solution. In neutral solutions, $[\mathrm{OH}^-(aq)]$ is equal to $[\mathrm{H}^+(aq)]$.
3. A pH value of more than 7 indicates a *basic*, or alkaline, solution. In basic solutions, $[\mathrm{OH}^-(aq)]$ is greater than $[\mathrm{H}^+(aq)]$.

17.5   Strengths of acids

As we discussed earlier in this chapter, HCl ionizes completely in dilute water solution. The equation for the process

$$\mathrm{HCl}(aq) \rightleftharpoons \mathrm{H}^+(aq) + \mathrm{Cl}^-(aq)$$

has a *very large* equilibrium constant for HCl ionization because the concentration of HCl(aq) remaining in the solution is *extremely small*. An acid that has a large equilibrium constant is known as a strong acid. We can write K_A for the **acid equilibrium constant** as

$$K_A = \frac{[\mathrm{H}^+(aq)][\mathrm{Cl}^-(aq)]}{[\mathrm{HCl}(aq)]} = \frac{[\text{large}][\text{large}]}{[\text{very small}]} \quad \text{A VERY LARGE NUMBER}$$

*A treatment of logarithms as they relate to pH notation is given in Appendix 2.

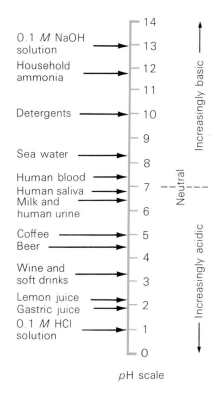

Figure 17-5

The average pH values of some common solutions are shown on a scale of 0 to 14. While nearly all commonly encountered solutions have pH values in that range, values for strongly acidic solutions can be less than 0, and values for strongly basic solutions can be greater than 14. The range of pH values of these solutions can be very narrow, such as in human blood, or very wide, such as in human urine, which has a range from pH 5 to almost pH 8.

For acetic acid (CH₃COOH), very little ionization occurs. This is indicated by the relatively poor conductivity of the solution. The equation for this reaction is

$$CH_3COOH(aq) \rightleftharpoons H^+(aq) + CH_3COO^-(aq)$$

The equilibrium constant for the ionization of this acid is

$$K_A = \frac{[H^+][CH_3COO^-(aq)]}{[CH_3COOH(aq)]} = 1.8 \times 10^{-5} \quad \text{A SMALL NUMBER}$$

The *small value* of the equilibrium constant indicates that very little acetic acid is ionized. As we noted earlier, this value (sometimes called K_A, or the acid ionization constant) is a precise way of indicating acid strength. It is clear from the examples above that a large value of K_A indicates a strong acid, and a small value of K_A indicates a weak acid. Values of K_A for various acids are listed in Table 17-3.

Table 17-3

Relative strengths of acids in aqueous solution at room temperature, $K_A = [H^+][B^-]/[HB]$

Acid	Strength	Reaction	K_A
HCl	very strong	$HCl(l) \longrightarrow H^+(aq) + CL^-(aq)$	very large
HNO₃		$HNO_3(l) \longrightarrow H^+(aq) + NO_3^-(aq)$	very large
H₂SO₄	very strong	$H_2SO_4 \longrightarrow H^+(aq) + HSO_4^-(aq)$	very large
H₃O⁺		$H_3O^+(aq) \longrightarrow H^+(aq) + HOH(l)$	1.0
HSO₄⁻	strong	$HSO_4^-(aq) \longrightarrow H^+(aq) + SO_4^{--}(aq)$	1.3×10^{-2}
HF	weak	$HF(aq) \longrightarrow H^+(aq) + F^-(aq)$	6.7×10^{-4}
CH₃COOH		$CH_3COOH(aq) \longrightarrow H^+(aq) + CH_3COO^-(aq)$	1.8×10^{-5}
H₂CO₃(CO₂ + H₂O)		$H_2CO_3(aq) \longrightarrow H^+(aq) + HCO_3^-(aq)$	4.4×10^{-7}
H₂S	weak	$H_2S(aq) \longrightarrow H^+(aq) + HS^-(aq)$	1.0×10^{-7}
NH₄⁺		$NH_4^+(aq) \longrightarrow H^+(aq) + NH_3(aq)$	5.7×10^{-10}
HCO₃⁻		$HCO_3^-(aq) \longrightarrow H^+(aq) + CO_3^{--}(aq)$	4.7×10^{-11}
H₂O	very weak	$H_2O(l) \longrightarrow H^+(aq) + OH^-(aq)$	$1.8 \times 10^{-16*}$

*The acid equilibrium constant, K_A, for water equals

$$\frac{K_w}{[H_2O]} = \frac{1.00 \times 10^{-14}}{55.6} = \frac{[H^+][OH^-]}{[H_2O]} = 1.8 \times 10^{-16}$$

EXERCISE 17-6

Conductivity measurements show that HF is about ten times more ionized than CH₃COOH. Will K_A for HF be larger or smaller than K_A for CH₃COOH? Try to estimate the value of K_A for HF(aq).

EXERCISE 17-7

When pure acetic acid ionizes, the concentration of H^+ must equal the concentration of CH_3COO^-. If 0.100 mole of CH_3COOH is dissolved in 1.00 litre of water, the $[H^+]$ is found to be about 10^{-3} moles per litre. Show that the K_A for CH_3COOH is very close to 10^{-5}.

EXERCISE 17-8

Which of the following acids is the strongest acid and which is the weakest?

nitrous acid (HNO_2)	K_{HNO_2} $= 5.1 \times 10^{-4}$
benzoic acid (C_6H_5COOH)	$K_{C_6H_5COOH} = 6.6 \times 10^{-5}$
acetic acid (CH_3COOH)	K_{CH_3COOH} $= 1.8 \times 10^{-5}$

DETERMINATION OF K_A

How do we determine values of K_A from laboratory data? The experiments must provide a measurement of hydrogen-ion concentration. A number of procedures are available. One of the most precise methods involves a careful measurement of cell voltage. The method will be discussed in Chapter 18. Acid-sensitive dyes (see Figure 11-1) offer the easiest estimate of $H^+(aq)$ ion concentration when electronic meters are not available. Let us consider a specific example.

Benzoic acid (C_6H_5COOH) is a white solid with moderate solubility in water. The ionization equation and acid equilibrium constant ($K_{C_6H_5COOH}$) are

$$C_6H_5COOH(aq) \rightleftharpoons H^+(aq) + C_6H_5COO^-(aq)$$

$$K_{C_6H_5COOH} = \frac{[H^+][C_6H_5COO^-]}{[C_6H_5COOH]}$$

To determine $K_{C_6H_5COOH}$, we need to obtain experimental values for the concentrations of $H^+(aq)$, $C_6H_5COO^-(aq)$, and $C_6H_5COOH(aq)$ in a given water solution. Once these values are obtained, it is easy to calculate the equilibrium constant. The following experiment gives us the information we need.

A 1.22-gram sample of solid benzoic acid is dissolved in enough water to produce 1.00 litre of solution at 25 °C. Several drops of methyl orange acid-base indicator are added to this solution. The color of the resulting liquid is compared to those of solutions containing methyl orange and known concentrations of $H^+(aq)$ ions (see Figure 11-1). The best color match is obtained with a known solution containing 8×10^{-4} mole $H^+(aq)$ ions per litre. We can therefore conclude that the $H^+(aq)$ ion concentration of the benzoic acid is about 8×10^{-4} M. We write $[H^+] = 8 \times 10^{-4}$ M. The equation tells us that when pure benzoic acid is dissolved in water, the concentration of $H^+(aq)$ ions is the same as that of the benzoate ions, $C_6H_5COO^-(aq)$, since each mole of benzoic acid produces 1 mole of $H^+(aq)$ ions and 1 mole of $C_6H_5COO^-(aq)$ ions when it ionizes. We now have $[H^+] = 8 \times 10^{-4}\ M^*$ and $[C_6H_5COO^-] = 8 \times 10^{-4}\ M$.

The only information still needed is the concentration of non-ionized $C_6H_5COOH(aq)$. We dissolved 1.22 grams of benzoic acid to make 1 litre of solution. Since the molecular weight of benzoic acid is 122, the number of moles of benzoic acid is

$$\frac{1.22\ \text{g benzoic acid}}{122\ \text{g benzoic acid/mole}} = 0.0100\ \text{mole}$$

We have, then, 0.0100 mole of benzoic acid in 1.00 litre of solution. The original benzoic acid concentration is 0.0100 M. To generate an $H^+(aq)$ ion concentration of 8×10^{-4} mole per litre, 8×10^{-4} mole per litre of the original benzoic acid had to ionize. The final concentration of non-ionized benzoic acid in the solution is

$$\begin{pmatrix} \text{original} \\ \text{concentration} \\ \text{benzoic acid} \\ \text{in solution} \end{pmatrix} - \begin{pmatrix} \text{concentration} \\ \text{benzoic acid} \\ \text{lost by} \\ \text{ionization} \end{pmatrix} = \begin{pmatrix} \text{final} \\ \text{equilibrium} \\ \text{concentration} \\ \text{benzoic acid} \end{pmatrix}$$

or $\quad\quad\quad 0.0100\ M \quad - \quad 0.0008\ M \quad = \quad 0.0092\ M$

We can now write

$$[H^+] = 8 \times 10^{-4}\ M$$
$$[C_6H_5COO^-] = 8 \times 10^{-4}\ M$$
$$[C_6H_5COOH] = 0.0092 = 9 \times 10^{-3}\ M$$

Thus,

$$K_{C_6H_5COOH} = \frac{(8 \times 10^{-4})(8 \times 10^{-4})}{9 \times 10^{-3}} = 7 \times 10^{-5}$$

Since the concentration of $H^+(aq)$ ions was measured to only one significant figure, the answer can have no more than one significant figure.

USING K_A VALUES: CALCULATING [H^+] IN AN ACID SOLUTION

Let us use the value of K_A just measured to determine the [H^+] in a solution that contains 0.41 gram per litre of benzoic acid. In this solution (pure benzoic acid in water), the concentration of $H^+(aq)$ ions is again equal to the concentration of benzoate ions, $C_6H_5COO^-(aq)$. The concentration of benzoic acid before ionization took place was 0.0033 mole per litre (0.41 gram per 122 gram per litre). If we remember the relationship

$$C_6H_5COOH \longrightarrow H^+ + C_6H_5COO^-$$

we can write

---

*If you did this experiment, you would find that to determine [H^+] precisely by colorimetric methods is almost impossible. The value given is the best estimate we can make.

$$\left\{\begin{array}{l}\text{original}\\\text{concentration}\\\text{benzoic acid}\\\text{in solution}\end{array}\right\} - \left\{\begin{array}{l}\text{concentration}\\\text{benzoic acid}\\\text{lost by}\\\text{ionization}\end{array}\right\} = \left\{\begin{array}{l}\text{final}\\\text{equilibrium}\\\text{concentration}\\\text{benzoic acid}\end{array}\right\}$$

and the relationship from the above equation

$$\left\{\begin{array}{l}\text{concentration}\\\text{benzoic acid}\\\text{lost by}\\\text{ionization}\end{array}\right\} = \left\{\begin{array}{l}\text{concentration } H^+(aq)\\\text{formed in}\\\text{solution}\end{array}\right\}$$

we can write
$$\underset{0.0033\ M}{\left\{\begin{array}{l}\text{original}\\\text{concentration}\\\text{benzoic acid}\\\text{in solution}\end{array}\right\}} - \underset{[H^+(aq)]}{\left\{\begin{array}{l}\text{concentration}\\\text{benzoic acid}\\\text{lost by}\\\text{ionization}\end{array}\right\}} = \underset{[C_6H_5COOH(aq)]}{\left\{\begin{array}{l}\text{final}\\\text{equilibrium}\\\text{concentration}\\\text{benzoic acid}\end{array}\right\}}$$

We can then write
$$\frac{[H^+][C_6H_5COO^-]}{[C_6H_5COOH]} = 7 \times 10^{-5} = \frac{[H^+][H^+]^*}{0.0033 - [H^+]}$$

The equation
$$\frac{[H^+]^2}{0.0033 - [H^+]} = 7 \times 10^{-5}$$

can be solved, but the algebra can be greatly simplified by making an assumption that permits us to obtain an approximate answer. The quantity 0.0033 minus $[H^+]$ is probably not too different from 0.0033, since a relatively small percentage of benzoic acid is lost by ionization. As you know, in our calculation of K_A we found that $0.0100 - 0.0008 = 0.0092\ M$ benzoic acid at equilibrium. The final value of 0.0092 differs from the original value 0.0100 by less than 10 percent. Let us then assume as a reasonable approximation that 0.0033 minus $[H^+]$ is very close to 0.0033.
We can then write

$$\frac{[H^+]^2}{0.0033} = 7 \times 10^{-5}$$

$$[H^+]^2 = 23 \times 10^{-8}$$

$$[H^+] \approx 5 \times 10^{-4}\ M$$

The symbol $\approx$ means *is about*. The concentration of H^+ is about 5×10^{-4} mole per litre. To complete the calculation, we must check our assumption. We see that 0.0005 is quite small as compared to 0.0033 (the difference is about 15 percent). Thus, the assumption is a reasonably good one.

CALCULATING [H⁺]: ANOTHER EXAMPLE

Other calculations using K_A are frequently made. For example, suppose we need to know the hydrogen-ion concentration in a solution containing both 0.010 M benzoic acid (C_6H_5COOH) and 0.030 M sodium benzoate (C_6H_5COONa). Sodium benzoate is a strong electrolyte. Its aqueous solutions contain sodium ions, $Na^+(aq)$, and benzoate ions, $C_6H_5COO^-(aq)$.

---

*Remember that $[H^+] = [C_6H_5COO^-]$ in pure benzoic acid solution.

Hence, the equilibrium involved is the same as before:

$$C_6H_5COOH(aq) \rightleftharpoons H^+(aq) + C_6H_5COO^-(aq)$$

At equilibrium, concentrations must agree with the equilibrium expression.

$$K_A = \frac{[H^+][C_6H_5COO^-]}{[C_6H_5COOH]} = 6.6 \times 10^{-5}$$

In this case, K_A is more accurately known to be 6.6×10^{-5}, rather than 7×10^{-5}. Again, we neglect the concentration of benzoic acid lost through ionization:

$$[C_6H_5COOH] \approx \left\{\begin{array}{l}\text{original} \\ \text{concentration} \\ \text{benzoic acid} \\ \text{in solution}\end{array}\right\} = 0.010 \ M$$

Further, the concentration of $C_6H_5COO^-(aq)$ ions produced by the ionization of benzoic acid is very small because benzoic acid is a weak acid. However, the concentration of $C_6H_5COO^-(aq)$ ions from the addition of sodium benzoate is relatively large because sodium benzoate is a strong electrolyte and is essentially 100 percent ionized. Thus, we can write

$$[C_6H_5COO^-] \approx \left\{\begin{array}{l}\text{initial concentration} \\ C_6H_5COONa\end{array}\right\} = 0.030 \ M$$

We can then use the expression for the acid ionization constant:

$$K_A = 6.6 \times 10^{-5} = \frac{[H^+][C_6H_5COO^-]}{[C_6H_5COOII]} = \frac{[H^+][0.030]}{[0.010]}$$

Multiplying both sides of the equation by 0.010 and dividing both sides by 0.030, we obtain

$$[H^+] = \frac{[6.6 \times 10^{-5}] \times [0.010]}{[0.030]} = 2.2 \times 10^{-5} \ M$$

The calculation is not complete until we validate our assumptions. Was it reasonable to assume that the concentrations of benzoate ions and benzoic acid were not changed by the ionization of benzoic acid? Let us put numbers into the expression for the concentration of undissociated benzoic acid:

$$\left\{\begin{array}{l}\text{final} \\ \text{equilibrium} \\ \text{concentration} \\ C_6H_5COOH\end{array}\right\} = \left\{\begin{array}{l}\text{original} \\ \text{concentration} \\ C_6H_5COOH\end{array}\right\} - \left\{\begin{array}{l}\text{concentration} \\ C_6H_5COOH \\ \text{lost by} \\ \text{ionization}\end{array}\right\} *$$

$$\left\{\begin{array}{l}\text{final} \\ \text{equilibrium} \\ \text{concentration} \\ C_6H_5COOH\end{array}\right\} = 0.010 \ M - 0.000022 \ M \approx 0.010 \ M$$

*The concentration loss due to ionization is equal to the concentration of $H^+(aq)$ ions, since every $H^+(aq)$ ion formed by ionization takes away one molecule of C_6H_5COOH.

The final concentration of C_6H_5COOH is then 0.010 M within the limits of measurement. The correlation is less than the uncertainty in the measured value of the original concentration of C_6H_5COOH.

EXERCISE 17-9

A certain acid (HB) has an ionization constant of 5×10^{-6}. Show that a 0.2 M solution of this acid has a pH value that is very close to 3.

17.6 Experimental background for acid-base concepts

So far, we have defined an acid aqueous solution as one in which [H$^+$(aq)] is greater than [OH$^-$(aq)]. Similarly, we defined an aqueous basic solution as one in which [OH$^-$(aq)] is greater than [H$^+$(aq)]. Finally, a neutral solution was defined as an aqueous solution in which [H$^+$(aq)] equals [OH$^-$(aq)]. These are useful but arbitrary definitions. We must now look at the background from which ideas about acids and bases developed. Why was it necessary to invent the concept of an acid?

EXPERIMENTAL IDENTIFICATION OF ACIDS

In attempting to classify materials, early experimenters noticed that certain substances had very similar properties. For example, when dissolved in water, one group of substances produces solutions that

1. are electrical conductors;
2. react with an active metal, such as zinc, to liberate hydrogen gas;
3. turn blue litmus solution red and alter the color of other organic dyes, including such common liquids as red cabbage juice;
4. have a sour taste (*tasting is now recognized as a dangerous procedure, since some acids are poisonous and corrosive to the skin; taste should not be used to identify a substance*).

Substances having these properties are called *acids*.

EXPERIMENTAL IDENTIFICATION OF BASES

Another group of substances identified by early experimenters resembles acids in several ways. For example, the other substances are frequently corrosive to the skin. Also, they dissolve in water to give solutions that conduct electricity. On the other hand, solutions of this second group share properties that are very *different* from those of acids. They change red litmus to blue and cause many dyes to assume a color quite different from that found in acidic solutions. The latter substances have a bitter rather than a sour taste and they feel slippery. Finally, when we carefully add one of these compounds to an acid, (in a titration, for example), the identifying properties of both the acid and the compound disappear. Only electrical conductivity remains as a characteristic property of the mixture.

Materials in the second group are called *bases*. Typical bases include sodium hydroxide (NaOH), potassium hydroxide (KOH), calcium hydroxide $Ca(OH)_2$, magnesium hydroxide $Mg(OH)_2$, sodium carbonate (Na_2CO_3), and ammonia (NH_3).

CONCEPTUAL AND OPERATIONAL DEFINITIONS

We now have a set of properties by which we can recognize acids and bases. Such properties are summarized below:

Acid	*Base*
1. electrical conductor	1. electrical conductor
2. reacts with Zn to produce $H_2(g)$	2. destroys the properties of acids
3. makes blue litmus red	3. makes red litmus blue
4. tastes sour	4. tastes bitter
5. frequently corrosive to skin	5. frequently corrosive to skin
	6. feels slippery

Such characteristic properties are often called **operational definitions.** To understand this term, consider the meaning of the word *definition*. According to one dictionary, *definition* means "a statement of what a thing is." By using a definition, we can sort the universe into piles, one containing those objects that fit the definition and another containing those that do not. Our *operational* definition gives the criteria by which this sorting process can be carried out. An operational definition is, then, one that lists the measurements or observations (the *operations*) by which we decide if an object belongs in a given group.

The second type of definition we have used is a **conceptual definition.** It is more concerned with the question, why? It seeks to define the group in terms of *why the class has its properties*. When we state that an acid is a substance that releases hydrogen ions to an aqueous solution, we are using a *conceptual* definition. We are on less secure experimental ground than with the operational definition, but the conceptual definition is far more useful in the construction of chemical models and in *expansion of the definitions to include solvents other than water*. Conceptual definitions lead to new research and to the development of hidden likenesses, or regularities.

Both types of definition have their merits. Neither is *the* correct definition. As we consider more and more complicated systems, the definitions we use must be expanded. In this book, we shall confine our attention to water solutions. Hence, the definitions we propose will make sense when we are discussing water solutions. But there is no *absolute* definition of an acid or a base. Definitions are selected by the chemist on the basis of their ability to explain or simplify the system under study.

PROPERTIES OF ACIDS: A MODEL

What makes an acid behave as it does? What common characteristic or structural feature accounts for the properties of an acid? You are familiar with hydrochloric acid (HCl), nitric acid (HNO_3), acetic acid (CH_3COOH), sulfuric acid (H_2SO_4), and phosphoric acid (H_3PO_4). These acids all contain hydrogen atoms in combined form. Is hydrogen important to the behavior of acids? It might be, since zinc produces hydrogen gas in reacting with acids. Since acids conduct electric current, it is reasonable to propose that acids produce ions when dissolved in water. Given that all acids have similar properties, we might suggest that all contain a common ion. Based on the evidence, the ion $H^+(aq)$ is a likely choice. We therefore postulate: *a substance has the properties of an acid if it can release hydrogen ions in water solution.*

PROPERTIES OF BASES: A MODEL

Using the line of reasoning we applied to acids, we now seek a common factor to account for the similarities of bases. Because of the electrical conductivity of bases, we might look for an ion. Because bases can counteract the properties of acids, we might look for an ion that can remove the acidic hydrogen ion, $H^+(aq)$.

Sodium hydroxide (NaOH), when dissolved in water, produces a solution having basic properties. The hydroxides of many elements from the left-hand side of the periodic table behave in the same way. Perhaps they dissolve to form ions of the following type:

$$NaOH(s) \rightleftharpoons Na^+(aq) + OH^-(aq)$$
$$KOH(s) \rightleftharpoons K^+(aq) + OH^-(aq)$$
$$Mg(OH)_2(s) \rightleftharpoons Mg^{2+}(aq) + 2OH^-(aq)$$
$$Ca(OH)_2(s) \rightleftharpoons Ca^{2+}(aq) + 2OH^-(aq)$$

The hydroxide ion, $OH^-(aq)$, could react with the hydrogen ion to account for the second property of bases, the removal of acidic properties:

$$OH^-(aq) + H^+(aq) \rightleftharpoons H_2O(l)$$

The similarities among the hydroxides are obvious. Let us compare sodium carbonate (Na_2CO_3) and ammonia (NH_3). Na_2CO_3 dissolves in water to produce a solution having the properties of a base. Quantitative studies of the solubilities of carbonates show that the carbonate ion, CO_3^{--}, reacts with water. The reactions are

$$Na_2CO_3(s) \rightleftharpoons 2Na^+(aq) + CO_3^{--}(aq)$$

and

$$CO_3^{--}(aq) + H_2O(l) \rightleftharpoons HCO_3^-(aq) + OH^-(aq)$$

The latter equation indicates that the presence of the carbonate ion, $CO_3^{--}(aq)$, in water increases the hydroxide ion concentration, $[OH^-]$. The hydroxide ion is present in the solutions of NaOH, KOH, $Mg(OH)_2$, and $Ca(OH)_2$. We see that the carbonate ion reacts with water to produce the chemical species $OH^-(aq)$, the same ion found in solutions of the hydroxides. We can postulate that $OH^-(aq)$ ions account for the slippery feel and bitter taste of the basic solutions. The direct reaction between the carbonate ion and a hydrated proton to produce the bicarbonate ion also removes H$^+$ and makes the solution appear to be basic because it removes H$^+$.

$$CO_3^{--}(aq) + H^+(aq) \rightleftharpoons HCO_3^-(aq)$$

We have listed ammonia (NH$_3$) as a base. We also know that ammonia readily forms the ammonium ion, NH$_4^+$. Ammonia reacts with water:

$$NH_3(g) + H_2O(l) \rightleftharpoons NH_4^+(aq) + OH^-(aq)$$

and with hydrogen ions:

$$NH_3(g) + H^+(aq) \rightleftharpoons NH_4^+(aq)$$

The formation of NH$_4^+$(aq) and OH$^-$ in water explains the fact that ammonia gas has basic properties when dissolved in water. The reaction of ammonia with water produces hydroxide ions, which, by our postulate, account for the taste and feel of basic solutions. The direct reaction of an ammonia molecule and a proton to produce the ammonium ion shows how ammonia can act to destroy the acidic properties of any solution that contains hydrogen ions.

When we investigate the reactions of other compounds having basic properties, we find that each compound can produce hydroxide ions in water. The OH$^-$(aq) ions may be produced directly (as when solid NaOH dissolves in water) or by adding a substance that reacts chemically with water (as when solid Na$_2$CO$_3$ and gaseous NH$_3$ dissolve in water):

$$NaOH(s) \rightleftharpoons Na^+(aq) + OH^-(aq)$$
$$CO_3^{--}(aq) + H_2O(l) \rightleftharpoons HCO_3^-(aq) + OH^-(aq)$$
$$NH_3(g) + H_2O(l) \rightleftharpoons NH_4^+(aq) + OH^-(aq)$$

Furthermore, *any substance that can produce hydroxide ions in water also combines with hydrogen ions:*

$$OH^-(aq) + H^+(aq) \rightleftharpoons H_2O(l)$$
$$CO_3^{--}(aq) + H^+(aq) \rightleftharpoons HCO_3^-(aq)$$
$$NH_3(g) + H^+(aq) \rightleftharpoons NH_4^+(aq)$$

Since production of OH$^-$(aq) and reaction with H$^+$(aq) go hand in hand in aqueous solutions, we can describe a base *either* as a substance that produces OH$^-$(aq) *or* as a substance that can react with H$^+$(aq). In solvents other than water, the latter description is generally more useful. Therefore, the more general postulate is: *a substance has the properties of a base if it can combine with hydrogen ions.*

17.7 Expansion of acid-base concepts

We have explained the properties of acids in terms of their ability to release hydrogen ions, H$^+$(aq). Thus, acetic acid is a weak acid because the ionization reaction releases H$^+$(aq) only slightly:

$$CH_3COOH(aq) \rightleftharpoons H^+(aq) + CH_3COO^-(aq) \tag{4}$$

We have explained the properties of bases in terms of their ability to react with hydrogen ions. Ammonia is therefore a base:

$$NH_3(aq) + H^+(aq) \rightleftharpoons NH_4^+(aq) \tag{5}$$

But what happens to these properties when you mix aqueous solutions of acetic acid and ammonia? The reaction that occurs can be broken down into the above sequence of reactions. The net reaction is

$$CH_3COOH(aq) + NH_3(aq) \rightleftharpoons CH_3COO^-(aq) + NH_4^+(aq) \tag{6}$$

This reaction can be explained in terms of the proton-transfer concept of acids and bases.

PROTON-TRANSFER CONCEPT

Practically, the sum of reactions 4 and 5 is reaction 6. *Acetic acid acts as an acid* in giving a proton to ammonia to form the ammonium ion, $NH_4^+(aq)$, just as it gives a proton to water to form the hydronium ion, $H_3O^+(aq)$. In either case, acetic acid releases hydrogen ions. As it ionizes in water, acetic acid releases hydrogen ions and forms $H^+(aq)$. In its reaction with ammonia, acetic acid releases hydrogen ions to NH_3, forming NH_4^+. Similarly, *ammonia acts as a base* by reacting with the hydrogen ion released by acetic acid. The reaction between acetic acid and ammonia is, then, an acid-base reaction, although the net reaction, 6, does not explicitly show $H^+(aq)$.

We can view acid-base reactions even more broadly. Suppose we mix aqueous solutions of ammonium chloride (NH_4Cl) and sodium acetate (CH_3COONa). We can tell by the odor that ammonia has been formed. The following reaction has occurred:

$$NH_4^+(aq),Cl^- + Na^+,CH_3COO^-(aq) \rightleftharpoons$$
$$CH_3COOH(aq) + NH_3(aq) + Na^+,Cl^-(aq)$$

The Na^+ and Cl^- appear on both sides of the equation and may be omitted. This is exactly the reverse of reaction 6. We see that it, too, is an acid-base reaction. Once again, there is an acid that releases $H^+(NH_4^+)$ and a base that accepts H^+ (CH_3COO^-). Once again, the net effect of the reaction is transfer of a hydrogen ion from one species to another.

We see that the acid-base reaction between acetic acid and ammonia yields two products—one an acid (NH_4^+) and the other a base (CH_3COO^-). On reflection, you will realize that every acid-base reaction proceeds in this way. The transfer of a hydrogen ion from an acid to a base necessarily implies that the ion might be given back. Returning the proton, the reverse reaction, is just as much a *hydrogen-ion transfer,* or an *acid-base reaction,* as is the original transfer.

Notice that we are now referring to reactions in which a hydrogen ion is transferred from an acid to a base without specifically involving the aqueous species $H^+(aq)$. A hydrogen ion, H^+, is nothing more than a proton. Consequently, we can frame a more general view of acid-base reactions in terms of **proton transfer.** The main value of this view is that it is applicable to a wider range of chemical systems, *including nonaqueous systems.* This more general view of acids and bases is named the Brønsted-Lowry theory in honor of the two scientists who proposed it independently in 1923.

Let us generalize our view of the acid-base reaction. In our example, the following equation applies:

$$CH_3COOH + NH_3 \rightleftharpoons NH_4^+ + CH_3COO^- \qquad (7)$$
$$\text{ACID} \qquad \text{BASE} \qquad \text{ACID} \qquad \text{BASE}$$

The acetic acid acts as an acid, giving up its proton to form acetate (CH_3COO^-), a substance that can act as a base. We can write the acetic acid-ammonia reaction in a general form:

$$HB_1 + B_2 \rightleftharpoons HB_2 + B_1$$
$$\text{acid}_1 + \text{base}_2 \rightleftharpoons \text{acid}_2 + \text{base}_1$$

The above reversible equilibrium reaction can be viewed as a competition between $base_1$ and $base_2$ for the proton. If $base_1$ is a better competitor for the proton, HB_1 forms and B_2 is left free. (The equilibrium lies to the left in the above equation.) If B_2 is a better competitor, HB_2 forms and B_1 is left free. (The equilibrium lies to the right in the above equation.) These two cases can be illustrated by considering reactions of acetic acid with two bases:

$$H_3CCOOH \quad + \quad \begin{cases} Cl^- \longleftarrow \text{———————} HCl + H_3CCOO^- \\ \text{VERY WEAK BASE} \\ \\ OH^- \text{———————} \longrightarrow HOH + H_3CCOO^- \\ \text{VERY STRONG BASE} \end{cases}$$
$$\text{ACETIC ACID}$$

When acetate ion and chloride ion compete for the proton, acetate as the stronger base gets the H^+ and acetic acid (rather than HCl) forms. When acetate ion and hydroxide ion compete, the very strongly basic OH^- gets the proton, giving water and free acetate ion. In most cases, all four components are present in an equilibrium system.

THE HYDRONIUM ION IN THE PROTON-TRANSFER THEORY

In the proton-transfer view of acid-base reactions, an acid and a base react to form another acid and another base. Let us see how this theory encompasses the elementary reaction between $H^+(aq)$ and $OH^-(aq)$ ions and the reaction in which acetic acid ionizes:

$$H^+(aq) + OH^-(aq) \rightleftharpoons H_2O(l)$$
$$CH_3COOH(aq) \rightleftharpoons H^+(aq) + CH_3COO^-(aq)$$

The theory helps explain these reactions by making a specific assumption about the nature of the species $H^+(aq)$. It assumes that $H^+(aq)$ is more properly written with the molecular formula $H_3O^+(aq)$. Thus, when HCl dissolves in water, the reaction is written

$$HCl(g) + H_2O(l) \rightleftharpoons H_3O^+(aq) + Cl^-(aq)$$

instead of

$$HCl(g) \rightleftharpoons H^+(aq) + Cl^-(aq)$$

Whenever $H^+(aq)$ might appear in an equation for a reaction, it is replaced by the **hydronium ion,** H_3O^+, and a molecule of water is added to the other side of the equation. We write the ionization equation in the form

$$CH_3COOH(aq) + H_2O \rightleftharpoons H_3O^+(aq) + CH_3COO^-(aq)$$

The ionization of acetic acid can now be regarded as an acid-base reaction. The acid CH_3COOH transfers a proton to the base H_2O, forming the acid H_3O^+ and the base CH_3COO^-. The neutralization of $H^+(aq)$ by $OH^-(aq)$ now takes the form

$$H_3O^+(aq) + OH^-(aq) \rightleftharpoons H_2O + H_2O$$

The acid H_3O^+ transfers a proton to the base OH^-, forming an acid, H_2O,* and a base, H_2O, as shown in Figure 17-6. We see that within the proton-transfer theory, the molecule H_2O must be assigned the properties of an acid as well as of a base.

ELECTRON-PAIR DONORS AND ACCEPTORS

We noted earlier that by a slight redefinition of terms, we could expand our concept of acids and bases to include solutions using liquids other than water as a solvent. This idea has been extensively applied by chemists so that we now have at least four additional definitions of acids and bases.

One of the more useful acid-base definitions, suggested by G. N. Lewis, is that a *base* is defined as *an electron-pair donor* and an *acid* as *an electron-*

---

*While abundant evidence indicates a strong interaction between a proton and a water molecule and while some solids such as $[H_3O^+][ClO_4^-]$ clearly show H_3O^+ ions, it is quite certain that when the H^+ ion is placed in liquid water it interacts with many water molecules and not just one.

Figure 17-6

According to the proton-transfer theory of acids and bases, the hydronium ion (H_3O^+) is an acid in this case because it donates a proton to the OH^- ion, which serves as a base.

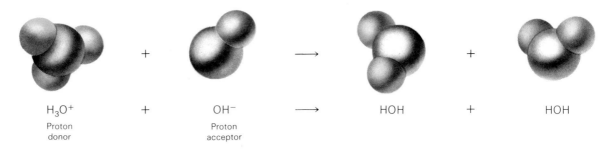

| H_3O^+ | + | OH^- | $\longrightarrow$ | HOH | + | HOH |
| Proton donor | | Proton acceptor | | | | |

pair acceptor. Using these definitions, the concept of acids and bases can be extended to systems not involving protons. As an example, it is convenient to consider the reaction between the fluoride ion and the BF_3 molecule. Electronically, the fluoride ion can be written as

$$: \overset{..}{\underset{..}{F}} : ^-$$

where F represents the fluorine nucleus plus the two electrons in the first electron level. Each dot represents an electron. Remember that this gives an *ion* with the same electron configuration as neon. The compound boron trifluoride (BF_3) can be represented as

$$
\begin{array}{l}
: \overset{..}{\underset{..}{F}} : \\
B : \overset{..}{\underset{..}{F}} : \\
: \overset{..}{\underset{..}{F}} :
\end{array}
$$

where B represents the boron nucleus plus the two electrons of the first electron level and F again represents the fluorine nucleus plus two electrons. We see immediately that boron in BF_3 has only six electrons around it rather than the eight electrons characteristic of a stable configuration for elements in the second row of the periodic table. Can the fluoride *ion* share two of its electrons with the boron of BF_3 to give a more stable structure? The resulting structure would be

$$
\left[
\begin{array}{c}
: \overset{..}{\underset{..}{F}} : \\
: \overset{..}{\underset{..}{F}} : B : \overset{..}{\underset{..}{F}} : \\
: \overset{..}{\underset{..}{F}} :
\end{array}
\right]^-
$$

OPPORTUNITIES

Chemical engineering

Chemical engineers help transform the discoveries of chemists into commercial products and are involved in many phases of the production of chemicals. Not only may such engineers design equipment, but they may also help determine the location and operation of chemical plants. Further, they may be concerned with special problems of manufacturing procedures, such as heat transfer and the mixing, separating, and purifying of chemicals in large volumes.

Employment of chemical engineers is expected to grow through the 1980s as a consequence of industrial expansion. Chemical engineers are needed to design and build plants and equipment, solve environmental problems, develop synthetic fuels, and test chemicals used in the manufacture of plastics, detergents, drugs, and synthetic fibers.

The duties of chemical engineers involve many fields. Consequently, a student interested in the field should have a good background in chemistry, physics, and mechanical and electrical engineering. At present, a graduate with good grades and a bachelor of science degree in chemical engineering can expect a good starting salary. Opportunities for women in this field have greatly expanded in recent years. Currently, one out of every four chemical engineering graduates is a woman. You can obtain additional

career information about chemical engineering by writing to the American Institute for Chemical Engineers, 345 East Forty-seventh Street, New York, New York 10017.

This is just what happens. The complex negative ion (known as the fluoroborate ion) forms. Of course, once the bond between the fluoride ion and boron is formed, all fluorines become equivalent. There is no way to tell which fluorine originated in the fluoride ion.

The fluoroborate ion can be pictured as a tetrahedral arrangement of four F^- ions that have interacted strongly with a central B^{+++} ion. In this process, *the fluoride ion is an electron-pair donor. By the Lewis definitions, F⁻ is a base.* The boron of BF_3 *accepts* the electron pair. We say that BF_3 is an acid. It is an electron-pair acceptor. Using this definition, the concept of acids and bases can be extended to systems not involving protons.

We shall not pursue these ideas further here. However, it is worthwhile to note that the best definition of an acid is dependent upon the system being studied and upon the ideas being considered.

SUMMARY

It is reasonable to write equilibrium constants for ionization processes. Such constants are known as ionization constants. Strong electrolytes have *very large* equilibrium constants. Weak electrolytes such as water have *small* equilibrium constants. The equilibrium constant of an acid is known as its ionization constant, abbreviated K_A. Its numerical value is an indication of the strength of the acid. A large value of K_A indicates a strong acid. A small value of K_A indicates a weak acid. About two water molecules in every billion ionize to give $H^+(aq)$ and $OH^-(aq)$. In every water solution, the special equilibrium constant is

$$K_w = [H^+(aq)][OH^-(aq)]$$

Le Chatelier's principle can be used to show that K_w should increase in value as the temperature rises. The value of K_w can be used to calculate $H^+(aq)$ and $OH^-(aq)$ concentrations in neutral, acidic, and basic solutions. In neutral solutions, $[H^+(aq)]$ equals $[OH^-(aq)]$. The $[H^+(aq)]$ is greater than $[OH^-(aq)]$ in acid solutions and less than $[OH^-(aq)]$ in basic solutions.

The hydrogen-ion concentration of aqueous solutions is an extremely important property. It can be measured by organic dyes, known as acid-base indicators, whose color is sensitive to the $H^+(aq)$ ion. The process of titration involves the mixing of acidic and basic solutions. The acid and the base neutralize each other. Indicators such as litmus change color at the neutral point. The *p*H value is a convenient, shorthand way to represent the concentration of $H^+(aq)$ ions. By definition, *p*H is equal to $-\log_{10}[H^+]$. If $[H^+]$ equals 10^{-7} *M*, the *p*H is 7. The *p*H of acidic solutions is less than 7 and the *p*H of basic solutions is more than 7.

Acids and bases can be defined by using both operational and conceptual definitions. Operational definitions list the observations that define the group. Conceptual definitions explain why the group has its properties. *There is no single, best definition of an acid or a base.*

Other types of acid-base definitions can also be used. A base can be defined as a species that *receives* a proton. An acid can be defined as a species that *donates* a proton. Further definitions may be employed, depending on the system under study.

Key Terms

acid
acid-base indicator
acid equilibrium constant
base
conceptual definition
equivalence point
hydronium ion
ionization constant
litmus
neutral solution
operational definition
proton transfer
*p*H
strong electrolyte
titration process
trial product
weak electrolyte

QUESTIONS AND PROBLEMS

A

1. (a) Is sugar in water a weak electrolyte or a nonelectrolyte? How could you tell? (b) Ethylene glycol ($HOCH_2CH_2OH$) is a common antifreeze. What tests would help to determine whether it is an electrolyte?

2. (a) Write equations, when applicable, to indicate what happens when each of the following substances is dissolved in water. Data indicating the electrical conductivity of 0.001 *M* solutions of each substance are given. (If nothing happens, write *N.R.*, for no reaction, in the equation.)

   (b) Indicate whether each of the substances in part (a) is a strong electrolyte, a weak electrolyte, or a nonelectrolyte.

Substance	Electrical Conductivity $\left(\dfrac{1}{\text{ohms} \times \text{cm}^2}\right)$
(1) HCl	421
(2) Na_2SO_4	256
(3) Acetic acid (H_3CCOOH)	58
(4) HNO_3	over 420
(5) Formic acid (HCOOH)	58
(6) Sugar ($C_{12}H_{22}O_{11}$)	10^{-3}
(7) Water (H_2O)	10^{-4} to 10^{-5}
(8) KCl	147

   (c) It is known that the proton (H^+) is a much, much better electrical conductor than Na^+ or K^+. How is this fact reflected in the above numbers?

   (d) Why does Na_2SO_4 have a higher conductivity than KCl? Explain.

   (e) Why are acetic acid and formic acid low in conductivity even though they give highly-conducting H^+ when they ionize? Explain.

3. (a) What kind of evidence indicates that water ionizes to some very slight degree?

   (b) How many moles of water are contained in 1,000 grams (or 1,000 cc) of water? What is the molar concentration of water in pure liquid water?

   (c) Write the equation for the ionization of pure liquid water.

   (d) Write the equation for the regular *equilibrium constant* for water ionization.

   (e) If you know that the $[H^+]$ in pure water is 10^{-7} moles/liter and the molar concentration of water in pure liquid water is the value that you calculated in part (b), determine the $[OH^-]$ in pure liquid water, and then calculate the regular ionization constant for pure liquid water. See page 457.

   (f) Show from the value in part (e) why we can write $[H^+][OH^-] = 10^{-14}$.

4. Given the equation 13.7 kcal + HOH(l) $\rightleftarrows$ $H^+(aq)$ + $OH^-(aq)$ what happens to K_w as temperature rises? (*Hint:* use Le Chatelier's principle.)

5. Define the term *p*H.

6. Complete the following table.

Solution	[H⁺]	[OH⁻]	pH	Acidic, neutral, or basic
pure water			7	
0.01 M HCl				
0.0001 M NaOH				
M HCl			3	
M NaOH		10^{-5}		
0.01 M NaOH				
M HCl	10^{-6}			
M NaOH			10	
			7	

7. A solution has a pH of 2; a second solution has a pH of 4. The concentration of H^+ in the solution of $pH = 2$ is: (a) 2 times, (b) $\frac{1}{2}$ times, (c) 10 times, (d) 100 times, (e) 10,000 times the concentration of H^+ in the solution of $pH = 4$. Explain your answer.

8. Suppose a 100.00 ml sample of 0.100 molar HCl is mixed with a 100.00 ml sample of 0.100 molar NaOH. (a) What is the $[H^+]$ of the resulting solution? (b) What is the pH of the solution? (c) Suppose HCl is added to this solution until the pH is 2. What is the $[H^+]$ and $[OH^-]$ in the resulting solution? See page 462.

9. What is titration? What is an equivalence point?

10. A sample of apple cider goes sour. When the sour cider is mixed with the purplish juice from red cabbage, the juice goes red in color. When a little lye (NaOH) is mixed with that same purplish cabbage juice the juice goes *blue* in color. The sour cider is a much better electrical conductor than fresh cider. The sour cider reacts with lithium metal to liberate H_2. (a) Is the sour cider acidic, neutral, or basic? (b) What kind of definition have you used? (c) What name can be given to the red cabbage juice? (d) What *conceptual definition* can be applied to the sour cider?

11. When HCl(g) dissolves in water, the water solution becomes a very good electrical conductor. (a) Write an equation to indicate that the dissolving of HCl gas in water is an acid-base reaction. (b) What is the hydronium ion? The ammonium ion? (c) HCl gas will also dissolve in benzene (C_6H_6) but the solution will *not* conduct the electric current. Is this an acid-base reaction? Explain.

12. Write an equation to indicate that the reaction between gaseous NH_3 and gaseous HCl to form crystalline NH_4Cl is an acid-base reaction. What definition is best here?

13. Acid ionization constants for a series of acids of the general form, HX, are given here. Arrange the acids in order of strongest acid to weakest acid. (a) HCN 5×10^{-10} (b) HF 3.5×10^{-4} (c) HClO 3×10^{-8} (d) HOOCCH₃ (acetic) 1.8×10^{-5} (e) C_6H_5COOH (benzoic) 6.5×10^{-5} (f) C_3H_7COOH (*n*-butyric) 1.5×10^{-5}.

B

14. Apply Le Chatelier's principle to the expression for the ionization of water: $13.7 \text{ kcal} + HOH(l) \rightleftharpoons H^+(aq) + OH^-(aq)$. If the tempera-

ture is raised, (a) will the concentration of $H^+(aq)$ at equilibrium increase, decrease, or remain the same? (b) What about the value of $[OH^-(aq)]$ as temperature is raised? (c) What about the value of K_w? (d) At 60 °C, K_w is very close to 10^{-13}. What would be the concentration of H^+ in pure water at 60 °C? Exponential form is acceptable.

15. Can we have a strong electrolyte in dilute solution? Explain and contrast the terms *strong and weak electrolyte* and *dilute and concentrated solution*.

16. Which of the following changes in a system will change the value of K_w (the ion-product constant) for water? Explain your answer. (a) Adding HCl to give a final concentration of H^+ of 10^{-3} M (b) Raising the temperature from 25 °C to 40 °C (c) Adding NaOH to give a final OH^- concentration of 10^{-2} M (d) Adding salt (NaCl) to the water (e) Cooling the water to 10 °C from 25 °C.

17. One mole of nitric acid (HNO_3) and 0.10 mole of sodium hydroxide (NaOH) are added to 1.0 litre of water. (a) Explain why the following statement is false: "The $[H^+]$ of the solution is 1 M and the $[OH^-]$ is 0.1 M." (b) Calculate the actual $[H^+]$ and $[OH^-]$ of the solution.

18. (a) Write the chemical equation for the ionization of one proton from H_2SeO_3. (b) Write the K_A expression for this reaction.

19. List the following acids in order of increasing acid strength: HF, HNO_3, HCO_3^-, H_2S. See page 466.

20. An acid HX has a K_A value of 10^{-7}. (a) What will be the $[H^+]$ of a 0.001 M solution of HX? (b) What will be the pH of the solution?

21. (a) Write the equation that shows the acid-base reaction between hydrogen telluride (H_2Te) and the sulfide ion (S^{--}). (b) What are the two acids competing for H^+? (c) From the values of K_A for these two acids, predict whether the equilibrium favors reactants or products.

22. Write the equations for the reactions between each of the following acid-base pairs (see Appendix 5). For each reaction, predict whether reactants or products are favored: (a) $HBr(aq) + SO_4^{--}(aq)$ (b) $H_3PO_4(aq) + NO_3^-(aq)$ (c) $NH_4^+(aq) + S^{--}(aq)$.

23. A 50.00 ml sample of 0.1250 M HCl solution is titrated with a solution of NaOH. At the equivalence point, 38.72 ml of the NaOH solution have been used. (a) What is the molarity of the NaOH solution? (b) What is the pH of the solution at the equivalence point?

24. An organic acid (HX) has a molar weight of 188.1 g/mole. A 9.405 g sample of this organic acid is dissolved in 500 ml of water. (a) What is the molarity of this acid solution? (b) If a 100.00 ml sample of this acid is neutralized by 76.30 ml of an NaOH solution, what is the molarity of the NaOH solution?

25. The solution of the acid in 24(a) has a pH of 4. (a) What is the concentration of H^+? (b) What is the concentration of X^-? (c) What is the approximate concentration of unionized HX? (d) What is K_A for the acid HX?

26. Consider the acid-base reaction between benzoic acid (C_6H_5COOH) and dihydrogen phosphate ion ($H_2PO_4^-$). (a) Write the balanced equation for this reaction. (b) Label *each* reactant and product as either an acid or a base. (c) Predict whether the products or reactants would be favored at equilibrium. (d) Explain your answer to (c) quantitatively.

27. A student is given three solutions: an acid, a base, and one that is neither acidic nor basic. The student performs tests on these solutions and records their properties. For each property listed below, tell whether it is the property of an acid, the property of a base, the property of neither an

acid nor a base, or whether there is insufficient evidence to decide: (a) The solution has $[H^+] = 10^{-7}\,M.$ (b) The solution has $[OH^-] = 10^{-2}\,M.$ (c) The solution turns litmus paper red. (d) The solution is a good conductor of electricity.

28. When solid NaOH is dissolved in water the solution is a strong electrolyte. When HCl is dissolved in water, the solution is a strong electrolyte. Compare and contrast the role of water in these two solution processes.

29. When excess ammonia is added to a solution of $AgNO_3$ in water, the complex ion $Ag(NH_3)_2^+(aq)$ is formed. Write the equation for this process. Which reagent is the Lewis base? Which is the Lewis acid? Explain.

30. Is the SCN^- ion a Lewis acid or a Lewis base when it reacts with $Fe^{+++}(aq)$ to produce the familiar red solution, $FeSCN^{++}(aq)$?

C

31. One-tenth mole samples of each of the following substances are dissolved in water to make 0.10 molar solutions of each material. Indicate whether each solution would contain a strong electrolyte, a weak electrolyte, or a nonelectrolyte. (a) HNO_3 (b) CsOH (c) acetic acid, H_3CCOOH (d) sugar, $C_{12}H_{22}O_{11}$ (e) HF.

32. What is the pH of a: (a) 1.0 M solution of $HClO_4$? (b) 1.00 M solution of CsOH? (c) What is the $[OH^-]$ of each solution?

33. To 20.0 ml of 0.200 M nitric acid (HNO_3) are added successive amounts of a base of unknown concentration. Neutrality is reached after the addition of 40.0 ml of the base. (a) What was the $[OH^-]$ of the original base solution? Assume that it was 100 percent ionized. (b) Calculate the $[H^+]$ and $[OH^-]$ of the solution when exactly 10.0 ml of the base had been added during the titration. (c) Calculate the $[H^+]$ and $[OH^-]$ of the solution when 20.0 ml of the base had been added during the titration.

34. Suppose 1.00×10^{-2} mole of KOH and 1.0×10^{-4} mole of HCl are both added to a litre of water. Calculate the final $[H^+]$ and $[OH^-]$ of the resulting solution.

35. Calculate the $[H^+]$ of the solution formed by mixing 50 ml of 0.10 molar NaOH and 55 ml of 0.10 molar HCl solution.

36. Exactly 12.0 ml of 0.0240 M NaOH are required to neutralize (completely react with) 20.0 ml of HCl solution. What is the concentration, in moles per litre, of the HCl solution?

37. A 0.1 molar solution of a weak acid (HA) has a pH of 4. Estimate K_A for this acid.

38. The K_A for hypoiodous acid is 2.5×10^{-11}. What is the pH of a solution containing 0.1 M HOI and 0.2 M NaOI? (NaOI is completely ionized to Na^+ and OI^-.)

39. Calculate the $[H^+]$ in 1.00 litre of a solution in which there have been dissolved 0.20 mole of formic acid (HCOOH) and 0.40 mole of sodium formate (HCOONa). The value of K_A for formic acid is 1.8×10^{-4}.

40. When a 100 ml sample of a weak acid (HA) is titrated with a 0.100 M solution of a strong base (NaOH), (a) what is the concentration of the weak acid in the solution when exactly 50.00 ml of the NaOH have been added? (b) What is $[A^-(aq)]$ when exactly 50.00 ml of NaOH have been added? (c) If the acid HA has a K_A of 1.0×10^{-6}, what is $[H^+(aq)]$ when exactly 50.00 ml of NaOH have been added?

Acid-base and reduction-oxidation equilibria are analogous:

$$\text{acid} \rightleftharpoons \text{base} + \oplus$$
$$\text{reducing agent} \rightleftharpoons \text{oxidizing agent} + \ominus$$

In the former case the positive [proton], in the latter the negative electron is involved.

JOHANNES NICOLAUS BRØNSTED

18

OXIDATION-REDUCTION REACTIONS AND ELECTROCHEMISTRY

Objectives

After completing this chapter, you should be able to:

- Determine oxidation numbers of individual elements in both ionically and covalently bonded compounds.

- Construct a series of half-reactions according to decreasing ease of reduction.

- Balance oxidation-reduction reactions by using the oxidation number method and the half-reaction method.

- Diagram and label all parts of an electrochemical cell and explain how the cell operates.

- Explain electrolysis as a special kind of oxidation-reduction process.

An experimental electric car with a nickel-zinc battery pack.

Oxidation-reduction reactions are extremely important both economi-cally and theoretically. When properly controlled, they can provide power for radios, flashlights, calculators, and many other devices that depend on a self-contained, independent power supply. Oxidation-reduction reactions can generate enough power to run an automobile. They can be used to reclaim metals from ores, and, in the hands of an artisan, they can be used to produce silver and gold plating on less costly base metals. Uncontrolled, such reactions can result in significant economic and personal loss due to the corrosion of metal in everything from cars to paper clips and to fires that rage out of control.

In this chapter, you will study oxidation-reduction reactions from a prac-tical as well as theoretical viewpoint. Chemists, of course, are concerned with both aspects. In Chapter 17, we discussed *proton* loss and gain in solu-tions. In this chapter, we will consider oxidation in terms of *electron* loss, and reduction in terms of *electron* gain. We also will discuss the reaction in an electrochemical cell as an oxidation-reduction reaction carried out in a very special way. Since such processes can involve solids as well as solutions, a more general view of chemical equilibrium is needed. We will begin by examining the chemical meaning of the terms *oxidation* and *reduction*.

18.1 Oxidation-reduction processes

Oxidation-reduction processes are all around us. Some are spectacular like Fourth of July fireworks or a raging fire burning out of control. Some are reassuring like the fire burning in a fireplace on a cold night or the candle burning in a room without other light. Other oxidation processes like the corrosion of metal are unspectacular, but effective, in converting metal back to oxides. About one person in every four people employed by the steel industry is working to replace steel lost by corrosion.

SOME IMPORTANT TERMS

When you think about it, it is obvious that the term oxidation is tied in a very direct way to the element oxygen. When magnesium metal burns in air, it combines with oxygen:

$$2Mg(s) + O_2(g) \longrightarrow 2MgO(s)$$

It is logical to say that magnesium is *oxidized* in the process. For reasons given in Chapters 7, 8, and 9, we can picture this reaction as a transfer of two electrons from each magnesium atom to each oxygen atom. Ions of charge $+2$ and -2 are formed. X-ray studies reveal an ionic arrangement similar to that of NaCl. We can then say that a *magnesium atom loses two electrons to an oxygen atom when it is oxidized.* From this fact it is logical to describe oxidation of magnesium as a loss of electrons. Magnesium also loses electrons, or is oxidized, when it combines with chlorine:

$$Mg(s) + Cl_2(g) \longrightarrow MgCl_2(s)$$

Positively charged magnesium ions are again formed. The same type of electron-transfer reaction goes on in both cases even though no oxygen is involved. It was thus logical for earlier chemists to define **oxidation** *as a loss of electrons* and **reduction** *as a gain of electrons,* if they focused on atomic details of the reaction.

EXERCISE 18-1

When a zinc metal strip is placed in a copper sulfate solution, zinc goes into solution as $Zn^{++}(aq)$ and copper deposits out as copper metal. What is oxidized in this process? What is reduced?

The history of the word *reduction* is less obvious. The term reduction was originally applied to reactions in which large amounts of metallic ore were *reduced* to smaller amounts of free metal. As you are now well aware, the conversion of a metal ion to an atom of a free metal requires a gain of one or more electrons. In Chapter 7 this process involving copper ions was described by the equation:

$$Cu^{++}(aq) + 2e^- \longrightarrow Cu(s)$$

The charge on the copper ion was reduced. It is thus logical and consistent to define reduction as a process in which an atom, ion or molecule gains electrons. The electron transfer accounts for the change in properties and the reduction in volume.

As you can see, the old terms oxidation and reduction arose first as strictly operational definitions. As theory developed, new conceptual definitions were adopted. Today oxidation is a loss of electrons; reduction is a gain of electrons.

Further names were also needed. Chemists called oxygen the *oxidizing agent* because it oxidized the magnesium. Since chlorine also oxidized the magnesium, it was also called an oxidizing agent. We can now state that an **oxidizing agent** *brings about the oxidation of another species and is itself reduced.* The same line of reasoning identifies the reducing agent. Magnesium atoms reduce oxygen; two magnesium atoms give four electrons to an O_2 molecule to generate two O^{--} ions. Similarly, magnesium *reduces* chlorine (Cl_2) to give $2Cl^-$. In each case magnesium serves as the reducing agent. A **reducing agent** *brings about the reduction of another species and is itself oxidized.*

EXERCISE 18-2

An oxidation-reduction reaction is represented by the equation

$$Fe^{+++} + Cu^+ \longrightarrow Fe^{++} + Cu^{++}$$

Which species is oxidized? Which species is reduced? Identify both the oxidizing and reducing agents in this process.

OXIDATION NUMBERS

From Chapter 9, we know that potassium has an apparent charge of $+1$ in KCl and that chlorine has a corresponding charge of -1. Since the charges result from an oxidation-reduction process, it is logical to call these numbers **oxidation numbers.** Thus, in KCl the oxidation number of potassium is $+1$, and the oxidation number of chlorine is -1. The same reasoning can be used to explain the oxidation numbers assigned to magnesium and oxygen in MgO: $+2$ and -2 respectively. In MgO, magnesium has an apparent charge of $+2$, and oxygen has an apparent charge of -2. Notice that *the oxidation numbers add up to zero in both compounds. Electrons lost equal electrons gained.* Charge must be conserved.

By convention, we say that the oxidation number of any element is zero. Thus, the oxidation numbers of metallic magnesium and potassium are both zero. Oxygen in gaseous O_2 and chlorine in gaseous Cl_2 have oxidation numbers of zero. Note that we are referring to pure *elements,* not compounds.

Can the idea of oxidation numbers be extended beyond ionic solids? Let us consider that question. If elemental sulfur is burned in air, the equation is

$$S(s) + O_2(g) \longrightarrow SO_2(g)$$

Remembering what happened with magnesium, it is tempting to suggest that each sulfur atom donates or shares a pair of electrons with each of the two oxygen atoms. Extrapolating from what occurred during formation of magnesium oxide, we can *assign* each oxygen atom an oxidation number of -2 and give the sulfur atom an oxidation number of $+4$. As in MgO, the oxidation numbers of the compound add up to zero:

$$\left\{\begin{array}{l}\text{oxidation}\\ \text{number of}\\ \text{sulfur}\end{array}\right\} + 2\left\{\begin{array}{l}\text{oxidation}\\ \text{number of}\\ \text{oxygen}\end{array}\right\} = 0$$

$$+4 \quad + \quad 2(-2) \quad = 0$$

Does this notation mean that each sulfur atom has actually lost four electrons to become an S^{++++} ion? Has each oxygen atom actually gained two electrons to become an O^{--} ion? Not at all. The oxidation numbers in the case of covalently bonded molecules, such as SO_2, are used as convenient bookkeeping devices to keep track of the electron redistribution reactions that take place when the compounds form. Assigning a $+4$ oxidation number to sulfur in sulfur dioxide suggests that four of its six valence electrons are involved in the formation of the molecule.

If four out of six of sulfur's electrons are involved, can we involve the other two electrons and use six out of six electrons? As usual, experiment provides the answer. Under the influence of a catalyst such as NO_2, gaseous SO_2 combines with oxygen to produce SO_3, a liquid whose boiling point is 45 °C: $SO_2(g) + \frac{1}{2}O_2(g) \xrightarrow{NO_2} SO_3(l)$. The SO_2 is oxidized in the classical sense. Again, extrapolation from simpler cases suggests that we can assign an oxidation number of $+6$ to the sulfur of SO_3 if each oxygen atom is assigned its regular oxidation number, -2:

$$\left\{\begin{array}{l}\text{oxidation}\\ \text{number of}\\ \text{sulfur}\end{array}\right\} + 3\left\{\begin{array}{l}\text{oxidation}\\ \text{number of}\\ \text{oxygen}\end{array}\right\} = 0$$

$$+6 \quad + \quad 3(-2) \quad = 0$$

As we have said, the oxidation numbers add up to zero in a neutral molecule. The oxidation number of sulfur goes from $+4$ in SO_2 to $+6$ in SO_3 when SO_2 is oxidized to SO_3 by elemental oxygen. *Apparently, oxidation brings about an increase in oxidation number.*

If SO_3 is allowed to react with calcium oxide, the following process takes place: $CaO(s) + SO_3(g) \rightarrow CaSO_4(s)$ or

$$Ca^{++} \quad :\overset{..}{\underset{..}{O}}:^{--} + \quad \underset{:\overset{..}{O}:}{\overset{:\overset{..}{O}:}{S:\overset{..}{O}:}} \quad \longrightarrow \quad Ca^{++} \quad :\overset{:\overset{..}{O}:^{--}}{\underset{:\overset{..}{O}:}{O:S:\overset{..}{O}:}}$$

Has the oxidation number of sulfur changed in this case? We can check. Calcium, like magnesium, should have an oxidation number of $+2$. According to the rule carried over from the examples of MgO and SO_2 formation, each oxygen atom should have an oxidation number of -2. What, then, is left for sulfur? Remember, the sum of the oxidation numbers in electrically neutral $CaSO_4$ must be zero:

$$
\begin{Bmatrix}\text{oxidation} \\ \text{number of} \\ \text{calcium}\end{Bmatrix} + \begin{Bmatrix}\text{oxidation} \\ \text{number of} \\ \text{sulfur}\end{Bmatrix} + 4\begin{Bmatrix}\text{oxidation} \\ \text{number of} \\ \text{oxygen}\end{Bmatrix} = 0
$$

$$
\begin{aligned}
+2 \quad + \quad x \quad + \quad 4(-2) &= 0 \\
+2 \quad + \quad x \quad + \quad -8 &= 0 \\
x &= +6
\end{aligned}
$$

Sulfur did not change oxidation number when SO_3 combined with CaO. That seems reasonable, since sulfur had already used all six of its outer electrons in SO_3. The oxide ion that combined with sulfur had two electrons from calcium to complete its octet. The reaction between sulfur trioxide and calcium oxide is an example of Lewis acid-base behavior, *not* oxidation-reduction. Assigning oxidation numbers to molecules or ions in such cases enables us to keep track of electrons and to identify an oxidation process. *Oxidation causes an increase in oxidation number. Reduction causes a decrease in oxidation number.* Oxidation-reduction reactions are always identified by a change in oxidation numbers.

Based on arguments of the type outlined above, a set of rules has been developed for assigning oxidation numbers. *It is important to note that oxidation numbers do not represent ionic charges. It is quite probable that ions of the type required by oxidation numbers such as +6 do* not *exist in compounds.* There is no S^{+6} in SO_3, for example. Nevertheless, oxidation numbers provide a convenient way to keep track of electrons in oxidation-reduction reactions. In reading the following rules, you will realize that many were derived from ion-formation studies such as that discussed for MgO:

1. *The oxidation number of a monatomic ion is equal to the charge of the ion.* Chloride (Cl^-), for example, has an oxidation number of -1; oxide (O^{--}), an oxidation number of -2; phosphide in Na_3P, an oxidation number of -3; Fe^{++}, an oxidation number of $+2$; and Fe^{+++}, an oxidation number of $+3$.

2. *The oxidation number of any element is zero.* The oxidation number of elemental chlorine (Cl_2) is zero; of oxygen gas (O_2), zero; of magnesium metal (Mg), zero.

3. *The oxidation number of members of the alkali metal family (Li, Na, K, Rb, Cs, and Fr) in compounds is +1.* The oxidation number of sodium in NaCl is $+1$; of lithium in Li_2O, $+1$. Both have an oxidation number of zero in the metallic form.

4. *The oxidation number of Be, Mg, Ca, Sr, Ba, and Ra in compounds is +2.* The oxidation number of calcium in CaO is $+2$; of Mg in $MgCl_2$, $+2$; of Ba in $BaCl_2$, $+2$.

5. *The oxidation number of oxygen in compounds is taken to be -2* (except in peroxides and superoxides, which contain an oxygen-oxygen bond).

6. *The oxidation number of hydrogen in compounds is taken to be $+1$* (except in hydrides, such as NaH). Hydrogen has an oxidation number of $+1$ in HCl, H_2O, H_2S, and so on.

7. *The oxidation numbers of any other element or elements in a molecule or ion are selected to make the sum of the oxidation numbers equal to the charge on the molecule or ion.* You will recall that for SO_3

$$\left\{\begin{array}{l}\text{oxidation}\\\text{number of}\\\text{sulfur}\end{array}\right\} + 3\left\{\begin{array}{l}\text{oxidation}\\\text{number of}\\\text{oxygen}\end{array}\right\} = \left\{\begin{array}{l}\text{charge on}\\\text{molecule}\\\text{or ion}\end{array}\right\}$$

$$+6 \quad + \quad 3(-2) \quad = \quad 0$$

For the SO_4^{--} ion, with a charge of -2, we can write

$$\left\{\begin{array}{l}\text{oxidation}\\\text{number of}\\\text{sulfur}\end{array}\right\} + 4\left\{\begin{array}{l}\text{oxidation}\\\text{number of}\\\text{oxygen}\end{array}\right\} = \left\{\begin{array}{l}\text{charge on}\\SO_4^{--}\text{ ion}\end{array}\right\}$$

$$+6 \quad + \quad 4(-2) \quad = \quad -2$$

For the ammonium ion (NH_4^+), we can use the rule to write

$$\left\{\begin{array}{l}\text{oxidation}\\\text{number of}\\\text{nitrogen}\end{array}\right\} + 4\left\{\begin{array}{l}\text{oxidation}\\\text{number of}\\\text{hydrogen}\end{array}\right\} = \left\{\begin{array}{l}\text{charge on}\\NH_4^+\text{ ion}\end{array}\right\}$$

$$-3 \quad + \quad 4(+1) \quad = \quad +1$$

8. *In any overall reaction, the net change in oxidation numbers must be zero.* This rule simply states the principle that charge is conserved:

$$2Mg(s) + O_2(g) \longrightarrow 2MgO(s)$$

Mg changes from 0 to $+2 = (+2) \times 2$

O changes from 0 to $-2 = \underline{(-2) \times 2}$

net change $= 0$

$$SO_2(g) + \tfrac{1}{2}O_2(g) \longrightarrow SO_3(l)$$

sulfur changes from $+4$ to $+6 = +2$

one oxygen changes from 0 to $-2 = \underline{-2}$

net change $= \quad 0$

Do not worry about the exceptions noted in rules 5 and 6. We will address those exceptions later, when we consider substances involving them.

EXERCISE 18-3

Use the above rules to obtain the oxidation numbers of each element in (a) Na_3PO_4 (b) K_2CrO_4 (c) HNO_3 (d) $NaBr$ (e) S_8 (f) $Cr_2O_7^{--}$.

USING OXIDATION NUMBERS TO BALANCE
OXIDATION-REDUCTION REACTIONS

Rule 7 for assigning oxidation numbers summarizes a powerful procedure for balancing any oxidation-reduction equation. The rule is based on the principle that electric charge, as well as the number of atoms, must be conserved in any reaction. Remember that the net change in oxidation number in any overall reaction must be zero. Let us look at several examples of how this rule can help us balance equations.

Before an equation can be balanced, we must know the formulas of all reactants and products. This information comes from direct laboratory observation. We find, for example, that when purple potassium permanganate ($KMnO_4$) solution is mixed with aqueous hydrogen sulfide (H_2S), there is a rapid reaction. The purple color fades, and a milky suspension of the free element sulfur forms rapidly in the solution. Tests indicate the presence of $Mn^{++}(aq)$ ions in the final solution. Aqueous potassium permanganate, a strong electrolyte, contains $K^+(aq)$ and $MnO_4^-(aq)$ ions. The $K^+(aq)$ ion is present at the beginning and at the end of the reaction, and therefore undergoes no change in the process. As a "spectator" ion, it is not essential to the equation. The unbalanced equation for the process illustrates these points:

$$MnO_4^- + H_2S \longrightarrow S + Mn^{++} + H_2O$$

To balance an equation, we must first assign oxidation numbers to each element, using rules 1 through 6:

Substance	Element(s)	Oxidation number (per atom)
MnO_4^-	Mn, O	Mn + 7, O − 2
H_2S	H, S	H + 1, S − 2
S	S	0
Mn^{++}	Mn	+2
H_2O	H, O	H + 1, O − 2

Next, we must identify the elements whose oxidation numbers change in the reaction. In this case, they are manganese (Mn) and sulfur (S). The Mn changes from +7 to +2, a change of −5. The S changes from −2 to 0, a change of +2. For the net change in oxidation number for the overall reaction to be zero, the total gain by S must equal the total loss by Mn. The lowest common multiple of 5 and 2 is 10. Therefore, $2MnO_4^-$ must react with $5H_2S$ to make the net change zero. Two Mn atoms change by −10, or 2 times −5, the net change in the oxidation number of manganese. Five S atoms change by +10—or 5 times +2—the net change in the oxidation number of sulfur. The net result—+10 plus −10—is zero. Thus,

$$\overset{\text{Gain } 10e^- \ (5e^-/Mn)}{2MnO_4^- + 5H_2S \longrightarrow 5S + 2Mn^{++}}$$
$$\underset{\text{Lose } 10e^- (2e^-/S)}{}$$

Finally, atoms must also be conserved. The 8 oxygen atoms in the $2MnO_4^-$ on the left-hand side of the equation require the addition of $8H_2O$ molecules to the right-hand side. Since the reaction occurs in aqueous solution, H_2O molecules are present. But $8H_2O$ molecules contain 16H atoms. The $5H_2S$ molecules reacting provide only 10H atoms. Thus, $6H^+$ ions, already present in the acidic solution, must be added to the reactant side:

$$2MnO_4^- + 5H_2S + 6H^+ \longrightarrow 5S + 2Mn^{++} + 8H_2O$$

We can check our work by seeing if electric charge, as well as atoms, have been conserved. To the left of the arrow, on the reactants side, we find $2(-1) + 6(+1) = +4$. To the right of the arrow, on the products side, we find $2(+2) = +4$. The charges, as well as the atoms, have been conserved. The equation is completely balanced.

EXERCISE 18-4

Use oxidation numbers to balance each of the following equations:

(a) $Fe_2O_3 + C \longrightarrow Fe + CO_2$

(b) $Fe_2O_3 + Al \longrightarrow Fe + Al_2O_3$

(c) $Al + Cl_2 \longrightarrow AlCl_3$

(d) $KClO_3 \longrightarrow KCl + O_2$

18.2 Reactions and reduction potentials

In the preceding section our interest was focused on balancing oxidation-reduction reactions through the use of oxidation numbers. In that process, the products had to be identified through laboratory experiments. It would now be helpful if we could find a way to predict the products of the reaction from information about the reactants themselves.

ELECTRON LOSS AND GAIN: HALF-REACTIONS

When copper wire, $Cu(s)$, is placed in silver nitrate solution, $AgNO_3(aq)$, several striking changes are observed. A deposit of needle-like silver crystals appears on the copper wire. The colorless silver nitrate solution gradually turns blue, and the copper wire loses some of its mass. If the masses of copper gained and silver lost are converted to moles, a simple ratio is revealed. Two moles of silver are formed for every 1 mole of copper lost. This was the ratio obtained in Experiment S-3 in the Laboratory Manual. The following Equations represent the reaction:

MOLECULAR $Cu(s) + 2AgNO_3(aq) \longrightarrow Cu(NO_3)_2(aq) + 2Ag(s)$

IONIC $Cu(s) + 2Ag^+(aq) + 2NO_3^-(aq) \longrightarrow$
$Cu^{++}(aq) + 2NO_3^-(aq) + 2Ag(s)$

NET IONIC $Cu(s) + 2Ag^+(aq) \longrightarrow Cu^{++}(aq) + 2Ag(s)$

COLORLESS   BLUE
SOLUTION   SOLUTION

We can see from these equations that copper metal lost two electrons to form $Cu^{++}(aq)$. By the definition of oxidation, *copper was oxidized.* Two $Ag^+(aq)$ ions each gained an electron to form two silver atoms. *Silver ions were reduced.* The oxidation number of copper changed from zero to $+2$, and the oxidation number of the silver ions changed from $+1$ to zero. In accordance with rule 8, the net change in oxidation number is zero. Two silver ions were reduced for every one copper atom oxidized.

The reaction between copper and silver nitrate can be analyzed in terms of electron loss and gain as well as by change in oxidation number. The reaction really consists of two separate parts, represented by the following equations:

OXIDATION "HALF"	$Cu(s) \longrightarrow 2e^- + Cu^{++}(aq)$	(1)
REDUCTION "HALF"	$2e^- + 2Ag^+(aq) \longrightarrow 2Ag(s)$	(2)
NET REACTION	$Cu(s) + 2Ag^+(aq) \longrightarrow Cu^{++}(aq) + 2Ag(s)$	(3)

Since steps 1 and 2 each represent *half* of the total reaction, they are referred to as **half-reactions.** Because the beaker contents remain electrically neutral before, during, and after reaction, we know that the number of electrons *lost* must equal the number of electrons *gained*. Such knowledge provides us with another method (besides change in oxidation number) for balancing oxidation-reduction reactions. Since half-reaction 1 involves a loss of two electrons, half-reaction 2 *must* show a gain of two electrons. Thus, two $Ag^+(aq)$ ions are needed. Adding half-reactions 1 and 2 algebraically, we see that electrons on both sides of the arrow balance, yielding net reaction 3*. More complex examples of this process will be given later in the chapter.

In Chapter 17, we saw that some acids react with certain metals to liberate $H_2(g)$. The reaction between zinc and HCl is typical of this process:

$$Zn(s) + 2H^+ + 2Cl^- \longrightarrow Zn^{++} + 2Cl^- + H_2(g)$$

EXERCISE 18-5

Write oxidation-reduction half-reactions and net reactions for the reaction between zinc, $Zn(s)$, and H^+. Identify the oxidation and reduction half-reactions.

However, not all metals react with strong aqueous acids. Of the common metals, magnesium, aluminum, iron, and nickel liberate H_2 from HCl and H_2SO_4, as does zinc. Other metals, including copper, mercury, silver, and gold, do not produce measurable amounts of hydrogen from HCl or H_2SO_4. This is true, even if we are sure that equilibrium has been attained. Apparently, some metals release electrons to H^+ in water solution (as does zinc) and others do not, as illustrated in Figure 18-1. We can thus say that not all possible oxidation-reduction combinations actually react.

Using a beaker, we can demonstrate another example of oxidation-reduction by placing a strip of metallic zinc in a solution of copper nitrate, $Cu(NO_3)_2$. The strip becomes coated with red-orange metallic copper. The blue color of the original copper nitrate solution disappears. The presence of Zn^{++} ions among the products can be shown by adding a small amount of sodium sulfide solution to the final, colorless solution. White zinc sulfide, $ZnS(s)$, precipitates, indicating that the reaction

$$Zn^{++} + S^{--} \longrightarrow ZnS(s)$$

*The remainder of this chapter considers only aqueous solutions. We shall therefore *not* continue to specify (aq) for each ion, since all ions are in water.

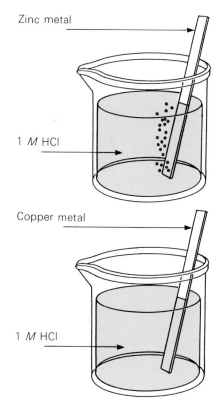

Figure 18-1
Some metals release electrons to H^+ and others do not.

has occurred. The reaction between metallic zinc and aqueous copper nitrate must be

$$Zn(s) + Cu^{++} \longrightarrow Zn^{++} + Cu(s)$$

EXERCISE 18-6

(a) Identify the substance *oxidized* in the above reaction. (b) Identify the substance *reduced*. (c) Indicate both the oxidizing agent and the reducing agent. (d) Write the oxidation half-reaction and the reduction half-reaction. (e) Add the two half-reactions to get the net reaction. (f) Indicate the change in oxidation numbers for both Zn and Cu^{++}, and show that rule 7 (page 490) applies.

What can be said about the state of equilibrium for the reaction between $Zn(s)$ and Cu^{++}? The absence of the blue color of Cu^{++} in the final solution suggests that the reaction has proceeded essentially to completion. Virtually all Cu^{++} ions have become $Cu(s)$. Let us try the reverse process by placing a strip of metallic copper in a zinc sulfate solution. No visible reaction occurs. No trace of Cu^{++} ions or $Zn(s)$ can be detected. Apparently, the state of equilibrium for the reaction greatly favors the products, $Cu(s)$ and Zn^{++}, over the reactants, $Zn(s)$ and Cu^{++}. Zinc metal readily gives electrons to Cu^{++}, but copper metal will *not* give electrons to Zn^{++}.

COMPETITION FOR ELECTRONS

The reaction between $Zn(s)$ and Cu^{++} can be viewed as a competition between Cu^{++} and Zn^{++} ions for a single pair of electrons. Rewriting this equation to indicate the net change:

$$\overset{\text{Lose } 2e^-}{Zn(s) + Cu^{++} \longrightarrow Zn^{++} + Cu(s)}$$

ZINC METAL   COPPER ION   Gain $2e^-$

we see that the Cu^{++} ion gained two electrons from Zn. This suggests that Cu^{++} ions attract electrons more strongly than do Zn^{++} ions. We can show this by writing the half-reaction

$$Cu^{++} + 2e^- \longrightarrow Cu(s)$$

above the half-reaction

$$Zn^{++} + 2e^- \longrightarrow Zn(s)$$

This placement of half-reactions indicates that Cu^{++} is a better competitor than Zn^{++} for electrons.

Perhaps you have noticed the similarity between electron transfer in this chapter and proton transfer in Chapter 17. Recall that our definition of an acid as a *proton donor* allowed us to construct a table of relative strengths of acids (Table 17-3). The best proton donor, or the strongest acid, was placed at the top of the table. The least effective proton donor, or the weakest acid, was placed at the bottom of the table.

We could just as well have constructed a table of relative strengths of bases. There, the strongest *proton acceptor* would be at the top. The weakest proton acceptor would be at the bottom. Each base in such a table would be a stronger base—that is, a better proton acceptor—than the one below it. Let us use the reactions we have been discussing in this chapter to construct a similar table. We wish to rank atoms or ions on the basis of their strength as **electron acceptors.** Our new table will indicate electron-accepting ability rather than proton-accepting ability. The best, or strongest, electron acceptor will appear at the top. The least effective, or weakest, electron acceptor will appear at the bottom. Each substance in the new table will be a better electron acceptor than the substance directly below it.

To find a suitable name for our new table, consider how a species such as Cu^{++} accepts electrons. It accomplishes this by *taking* electrons from another species. In the process, the other species loses electrons. It is oxidized. The electron-acceptor species gains electrons and is thus reduced. The table could therefore show tendency toward reduction, or **reduction potential.** The word *potential* indicates that each species in our table has a certain tendency to be reduced (gain electrons). The term *potential* is a measure of this tendency, but even the species with the highest potential can actually gain electrons only if another species is present to supply them. Let us develop our table from the experimental data we have examined so far.

The reaction between zinc metal and copper ions tells us that Cu^{++} is above Zn^{++} in reduction potential. The known process

$$Cu(s) + 2Ag^+ \longrightarrow Cu^{++} + 2Ag(s)$$

indicates that Ag^+ is able to take electrons from $Cu(s)$. Thus, Ag^+ is stronger than Cu^{++} in electron-gaining ability. Of the three ions, Ag^+ has the highest reduction potential and thus appears at the top of our table:

BEST COMPETITOR FOR ELECTRONS	$Ag^+ + e^- \longrightarrow Ag(s)$
	$Cu^{++} + 2e^- \longrightarrow Cu(s)$
POOREST COMPETITOR FOR ELECTRONS	$Zn^{++} + 2e^- \longrightarrow Zn(s)$

In the net ionic equation

$$Fe(s) + Cu^{++} \longrightarrow Fe^{++} + Cu(s)$$

copper ions (Cu^{++}) take electrons from $Fe(s)$. Thus, Cu^{++} is a better electron acceptor than Fe^{++}. Clearly, the Fe^{++} plus $2e^-$ half-reaction goes below Cu^{++} in our table. But do we place it above or below Zn^{++}? We can answer this question by performing an experiment. When an iron nail is placed in a solution of $ZnSO_4$, there is no observable reaction. The Zn^{++} ions in the solution cannot take electrons from the $Fe(s)$. But when a strip of zinc metal is placed in a solution of $FeSO_4$, a reaction does occur. Some of the zinc dissolves, becoming Zn^{++} ions. A coating of $Fe(s)$ forms on the zinc strip. Hence, Fe^{++} is placed *above* Zn^{++} in our table.

HENRY TAUBE

Henry Taube was born in 1915 in Saskatchewan, Canada, and received his education through the degree of Master of Science there. His work toward the Ph.D., completed in 1940 was done at the University of California, Berkeley, under the guidance of Professor William C. Bray. He continued at Berkeley for one year as instructor, and then took a position at Cornell University. In 1946 he joined the faculty of the Department of Chemistry at the University of Chicago where, in the stimulating atmosphere of that department, his research that eventually led to the Nobel Prize in Chemistry in 1983, properly began.

His work is closely related to the contents of Chapter 18, but deals less with the equilibrium aspect of chemical reactions, as developed in this chapter, than with reaction rates and mechanisms. His major contributions, made with the help of many collaborators, have been to advance our understanding of the reactions of inorganic substances in solutions, especially of oxidation-reduction reactions. In an early effort he developed a correlation between electronic structure and the rates of substitution in transition metal complexes. This study led him to a strategy for exploring the mechanisms of the simplest class of oxidation-reduction reactions, so-called electron transfer reactions, which provided impetus for work in this field by many others.

(a) Use the experimental data given on page 495 to write the half-reactions and the net reaction that summarize the interaction between Zn and Fe^{++}. (b) Identify the oxidation half-reaction and the reduction half-reaction. (c) Decide whether Fe^{++} is a stronger or weaker electron acceptor than Zn^{++}. (d) Does Fe^{++} belong above or below Zn^{++} in our table?

Table 18-1

Qualitative arrangement of half-reactions for reduction processes in aqueous solution. The best competitor for electrons is on top. The poorest competitor for electrons is on the bottom.

Oxidized state			Reduced state
$Cl_2(g) + 2e^-$	$\longrightarrow$		$2Cl^-$
$Br_2(l) + 2e^-$	$\longrightarrow$		$2Br^-$
$Ag^+ + e^-$	$\longrightarrow$		$Ag(s)$
$I_2(s) + 2e^-$	$\longrightarrow$		$2I^-$
$Cu^{++} + 2e^-$	$\longrightarrow$		$Cu(s)$
$2H^+ + 2e^-$	$\longrightarrow$		$H_2(g)$
$Fe^{++} + 2e^-$	$\longrightarrow$		$Fe(s)$
$Zn^{++} + 2e^-$	$\longrightarrow$		$Zn(s)$
$Mg^{++} + 2e^-$	$\longrightarrow$		$Mg(s)$

To decide where to place the aqueous hydrogen ion (H^+) in our table, remember that metals such as zinc and iron react with aqueous acids. (See page 493.) Metals such as copper and silver do not. This information indicates that H^+ can take electrons from $Zn(s)$ and $Fe(s)$, but not from $Cu(s)$ or $Ag(s)$. Thus, H^+ is *below* Ag^+ and Cu^{++} in electron-accepting tendency, but *above* Fe^{++} and Zn^{++}. These and other data are summarized in Table 18-1. Each species on the left-hand side of the table can remove electrons from any species on the right-hand side below it.

18.3 Oxidation-reduction reactions and electrochemical cells

The reaction in an electrochemical cell is an oxidation-reduction reaction carried out in a special way. Electrochemical cells are all around us. One example is the commercial automobile battery, another is the dry cell.

ELECTRICITY FROM OXIDATION-REDUCTION REACTIONS

Our view of oxidation-reduction reactions as a transfer of electrons raises some intriguing questions. We envision electric current in a wire as a flow of electrons. The electricity that lights light bulbs, drives electric motors, and powers pocket radios and calculators is a flow of electrons. Is it possible that chemical substances might be used to provide such electron flow?

In Section 18.2, we mentioned that copper reacts spontaneously with Ag^+ ions. Electrons are transferred from $Cu(s)$ to Ag^+ ions as soon as a copper wire is placed in aqueous $AgNO_3$. Could this reaction be used to force the transferred electrons through an external wire in order to light a light bulb? That depends on how the reaction is carried out. Placing a copper wire in aqueous $AgNO_3$ will *not* cause a bulb to light, because the electrons do not travel through the wire. They take a much shorter and easier route. They transfer from the Cu atoms to the Ag^+ ions in direct contact with them. We know that electricity always takes the "short" circuit when one is available—and one is certainly available here. The Ag^+ ions move to the $Cu(s)$ in solution and pick up electrons directly. There is no need for the electrons to move through the external wire to reach Ag^+ ions.

The only way to make electrons travel through the external wire is to *remove* the "short" circuit path. Clearly, the $Cu(s)$ must be separated from the Ag^+ ions. [The two half-reactions listed in Table 18-1 indicate that the Ag^+-$Ag(s)$ half can accept electrons from the Cu^{++}- $Cu(s)$ half]. We can

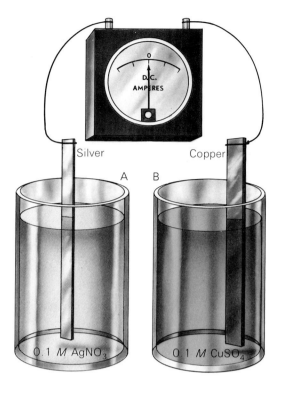

Figure 18-2

An incomplete electrochemical cell. Note that the electrical circuit is not complete.

construct an electrochemical cell using two beakers, as shown in Figure 18-2. In beaker A, we place a strip of metallic silver in a dilute solution (0.1 M) of silver nitrate. In beaker B, we place a strip of copper metal in a dilute solution of aqueous copper sulfate (also 0.1 M). An external wire joins the Cu(s) and Ag(s) strips to an *ammeter*. (Ammeters measure electric current.) The ammeter shows that there is no electron flow. Our set-up of two separate beakers does *not* produce electricity. Why?

The answer is that the circuit is not complete. The solutions are not connected. The electrons needed to make Ag atoms from Ag^+ ions at the silver strip come from the metallic copper. We can visualize the transfer process. As soon as several Cu(s) atoms release electrons and enter the solution as Cu^{++} ions, several things happen. The new Cu^{++} ions make the solution positive, since there are now more Cu^{++} than SO_4^{--} ions. The positive charge tends to prevent more Cu^{++} ions from entering the solution. The released electrons that travel to the Ag(s) attract Ag^+ ions from the $AgNO_3$ solution. However, as soon as several Ag^+ ions combine with these electrons, they form neutral Ag(s) atoms. Since they are no longer in solution, fewer Ag^+ than NO_3^- ions remain. The $AgNO_3$ solution becomes negative. The negative charge attracts the remaining Ag^+ ions, preventing them from leaving the solution. Thus, the expected half-reactions cannot occur.

It is clear that what is needed is a way for Cu^{++} ions to move from around the copper electrode, where we have a surplus of positive charges, over toward the silver electrode, where we have a surplus of negative charges. Negative ions would move in the opposite direction from silver to copper. Such ion movement would eliminate the possibility of the copper sulfate solution becoming positively charged and the silver nitrate solution becoming negatively charged. There are two ways to accomplish such ion transfer:

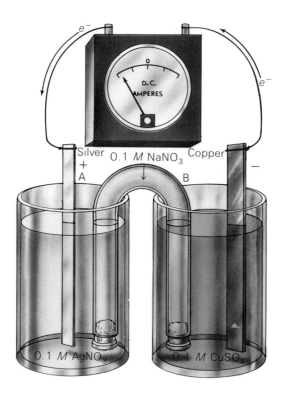

Figure 18-3

An electrochemical cell with a salt bridge in place.

use a salt bridge to connect beakers that each contain half of the reaction, or use a porous cup to separate the solutions. These methods are illustrated in Figures 18-3 and 18-4.

The salt bridge shown in Figure 18-3 consists of a U-shaped glass tube plugged at each end with cotton and containing a 0.1 M sodium nitrate ($NaNO_3$) solution. The solution consists of Na^+ and NO_3^- ions dissolved in water. The tube is inverted into both beakers. The "bridge" completes the electric circuit by furnishing a path through which the positive and negative ions can move between the two beakers.

As soon as the connection is made, the ammeter needle is deflected—electric current will move through the wires. In beaker B, the copper strip will dissolve, and the copper sulfate solution will become a deeper blue. In beaker A, the silver strip will be covered with a loosely adhering deposit of metallic silver crystals. (If the crystals are very small, the deposit of metallic silver may appear black in color.)

The reaction generating the current is familiar to us:

$$Cu(s) + 2Ag^+ \longrightarrow Cu^{++} + 2Ag(s)$$

But some things are different here. The *oxidation* process is taking place in *beaker B* and can be represented by the equation

$$Cu(s) \longrightarrow Cu^{++} + 2e^-$$

The *reduction* process is taking place in *beaker A* and can be represented by the equation

$$2Ag^+ + 2e^- \longrightarrow 2Ag(s)$$

The electrons must then flow from copper to silver through the wire connecting the strips in the beakers. To prevent an accumulation of positive ions around the copper strip or negative ions around the silver strip (see equations), ions must be able to pass between beakers. Our salt bridge provides a path for such movement.

The overall process is a combination of the reactions in the individual beakers:

BEAKER B (OXIDATION)	$Cu(s) \longrightarrow Cu^{++} + 2e^-$
BEAKER A (REDUCTION)	$2e^- + 2Ag^+ \longrightarrow 2Ag(s)$
NET REACTION	$Cu(s) + 2Ag^+ \longrightarrow 2Ag(s) + Cu^{++}$

We see that half the reaction is taking place in beaker A and half in beaker B. The overall reaction is the *sum* of the two *half-reactions*. In studying these half-reactions, we note several interesting features:

1. *The two half-reactions are written separately.* In this electrochemical cell, the half-reactions occur in separate beakers. As the term implies, there must be two such reactions.
2. *Electrons are shown as part of the reaction.* The ammeter shows that electrons are involved. Electrons flow when the reaction starts and stop flowing when the reaction stops. We have already explained why electrons flow from copper to silver through the external wire with this set-up.
3. *New chemical species are produced in each half of the cell.* The copper strip is converted to copper ions (the copper strip loses mass), and the silver ions are changed to metal (the silver strip gains mass). The new species can be explained in terms of *loss of electrons by copper atoms* and *gain of electrons by silver ions.*
4. *The half-reactions, when combined, express the overall reaction.* As you can see from the equations for the two beakers and from the net reaction on page 492, electrons lost by copper must equal electrons gained by silver. Thus, electrons do not appear in our final, overall equation.

OPERATION OF AN ELECTROCHEMICAL CELL: SOME TERMS

A slightly modified electrochemical cell is desirable for the measurements we plan to make. Figure 18-4 shows such a cell in cross-section. The only difference between this new cell and that shown in Figure 18-3 is that the salt bridge has been replaced by a porous porcelain cup. The cup permits ions to pass freely from the copper sulfate solution to the silver nitrate solution. The silver strip and the silver nitrate solution are placed within the cup. The cup is then lowered into a solution of copper sulfate containing a copper strip. Copper ions move from the beaker into the porous cup. Negative nitrate ions move in the opposite direction, from the cup into the beaker. Build-up of a positive charge in the beaker is prevented by the departure of copper ions and the entrance of negative nitrate ions. Negative-charge accumulation is prevented by the simultaneous transfer of NO_3^- and Cu^{++} through the walls of the porous cup in opposite directions. Thus,

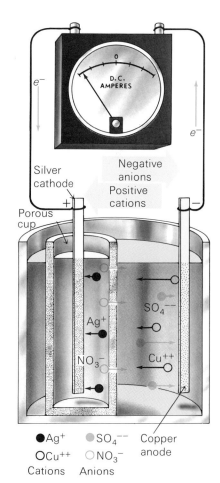

● Ag^+ ● SO_4^{--} Copper anode
○ Cu^{++} ○ NO_3^-
Cations   Anions

Reduction reaction at silver cathode:
$$2Ag^+ + 2e^- \longrightarrow 2Ag(s)$$
Oxidation reaction at copper anode:
$$Cu(s) \longrightarrow Cu^{++} + 2e^-$$

Figure 18-4

The operation of an electrochemical cell using a porous cup.

the cup, while serving as a barrier between the two halves of the cell, is porous enough to permit the passage of ions between the two solutions. The reaction proceeds and electrons flow from the copper to the silver.

Certain terms will help us identify parts of the new cell and discuss its operation more easily. The metal strips of copper and silver placed in the two solutions are known as **electrodes.** Electrons are left on the copper electrodes as copper atoms enter the solution to become ions: $Cu(s) \rightarrow Cu^{++} + 2e^-$. This is an *oxidation* process because the copper metal *releases* electrons as it forms copper ions. *Chemists call the electrode at which oxidation occurs the* **anode.** Because the copper anode provides electrons to the wire, it is known as the negative $(-)$ electrode. Experiments show that, in an electrochemical cell, the copper electrode is negatively charged compared to the silver electrode. At the silver electrode, silver ions make contact with the silver metal surface. Thus, they pick up the electrons necessary to become silver metal: $2Ag^+ + 2e^- \rightarrow 2Ag(s)$. This is a *reduction* process because the silver ions *gain* electrons to become silver metal. *Chemists call the electrode at which reduction occurs the* **cathode.** Because electrons are generated at the copper electrode and used at the silver electrode, the silver is less negative (or more positive) than the copper. The silver strip is known as the positive $(+)$ electrode. Electrons flow through the wire from the negative anode (copper) to the positive cathode (silver).

What happens in the solutions? As new Cu^{++} ions are produced at the anode, they repel other Cu^{++} ions which move toward the cathode. Negative ions move away from the region in which silver ions have been removed and move toward the region in which Cu^{++} ions were generated. The movement of negative ions and positive ions keeps the solution electrically neutral throughout. Positive ions move toward the cathode, negative ions toward the anode. *Because they move toward the cathode, positive ions are called* **cations.** *Because they move toward the anode, negative ions are called* **anions.**

18.4   Electron-attracting power of ions: Half-cell potentials

Voltages from electrochemical cells provide us with quantitative information about the ability of an ion to attract electrons. A standard half-cell must be used as a reference to determine half-cell voltage readings. Standard half-cell potentials, or voltages, are a measure of the electron-attracting ability of an ion. The effects of concentration changes on cell voltages must also be considered.

MEASURING VOLTAGE OF AN ELECTROCHEMICAL CELL

In some ways, the cells we have constructed resemble cells that power flashlights and start cars. If we replace the ammeter of Figures 18-3 or 18-4 with a small light bulb, the bulb will glow, giving off heat and light. The cell is doing electrical work in forcing the electric current through the bulb. As the electrons leave the bulb, they must have lower potential energy than they had when they entered. The drop in potential energy appears as heat and light. The potential-energy change associated with the chemical reaction of the cell is directly related to the voltage of the cell. *The* **voltage** *of a* *chemical cell measures its ability to force electrons through the circuit. As such, it serves as a measure of the tendency of the cell reaction to proceed in one direction.*

Different cells have different voltages. Consider first the cell based on the reactions

$$\begin{array}{r} Zn(s) \longrightarrow Zn^{++} + 2e^- \\ \underline{2e^- + Cu^{++} \longrightarrow Cu(s)} \\ Zn(s) + Cu^{++} \longrightarrow Zn^{++} + Cu(s) \end{array}$$

The experimental arrangement is shown in Figure 18-5a. It involves one half-cell consisting of a zinc rod in 1 M ZnSO$_4$ solution, and another half-cell consisting of a copper rod in 1 M CuSO$_4$ solution. The electron flow will be *from zinc* through the meter *to copper*. Current flowing in that direction causes the **voltmeter** to be deflected to the *right*, as shown. The voltmeter reads close to 1.1 volts when the zinc half-cell is connected. The value is recorded in Table 18-2.

Table 18-2

Half-cell voltages against copper reference cell

Half-reaction in outer beaker	Voltmeter reading
Ag$^+$ + e^- $\longrightarrow$ Ag(s)*	+0.4 volt (left)
Cu^{++} + 2e^- $\longrightarrow$ Cu(s)	0.0 volt (no deflection)
2H$^+$ + 2e^- $\longrightarrow$ H$_2$(g)	−0.3 volt (right)
Ni^{++} + 2e^- $\longrightarrow$ Ni(s)	−0.6 volt (right)
Zn^{++} + 2e^- $\longrightarrow$ Zn(s)	−1.1 volts (right)

*The full equation is Cu + 2Ag$^+$ → Cu^{++} + 2Ag

One interesting feature of the cell we are using is that the Cu^{++}-Cu half-cell is completely contained in the center porous cup, while the Zn^{++}-Zn half-cell is completely contained in the beaker outside the porous cup. Such an arrangement makes it possible to unhook the copper electrode from the voltmeter and *to transfer the entire half-cell to a Ag$^+$-Ag half-cell contained in a similar outer beaker*. The result is shown in Figure 18-5b. Our earlier analysis showed the reaction between copper metal and silver nitrate to be

$$\begin{array}{r} Cu(s) \longrightarrow Cu^{++} + 2e^- \\ \underline{2e^- + 2Ag^+ \longrightarrow 2Ag(s)} \\ Cu(s) + 2Ag^+ \longrightarrow Cu^{++} + 2Ag(s) \end{array}$$

Figure 18-5

Two electrochemical cells involving copper are shown here. In each case, the porous cup containing the copper electrode can be removed and placed in another beaker.

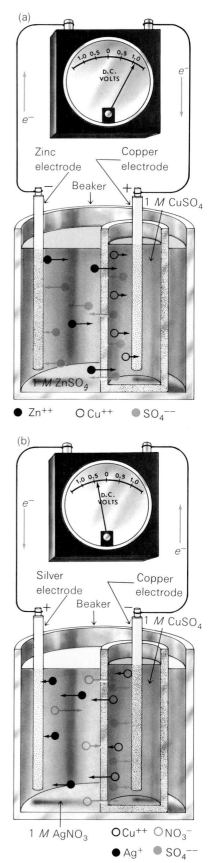

(a)

Zinc electrode   Copper electrode

Beaker

1 M CuSO$_4$

1 M ZnSO$_4$

● Zn^{++} ○ Cu^{++} ● SO$_4$$^{--}$

(b)

Silver electrode   Copper electrode

Beaker

1 M CuSO$_4$

1 M AgNO$_3$ ○ Cu^{++} ○ NO$_3$$^-$

● Ag$^+$ ● SO$_4$$^{--}$

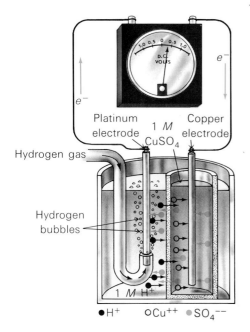

Figure 18-6

An electrochemical cell made of a hydrogen half-cell and a copper half-cell.

We see now that electrons are given up by copper and the electron flow is *from copper* through the meter *to silver*. We observe that the meter is deflected in a direction *opposite* to that of the copper-zinc case described on page 501. The reading on the voltmeter is recorded as 0.4 volt toward the *left*. This value is shown in Table 18-2. If our Cu^{++}-Cu half-cell is placed in a beaker containing an identical Cu^{++}-Cu half-cell, the voltmeter reads 0.0 and no deflection is observed. That observation is also recorded in the table. Table 18-2 can be expanded by systematically placing the Cu^{++}-Cu half-cell in other half-cells contained in external beakers. With a Ni^{++}-Ni half-cell, the reading is 0.60 volt toward the right. Other cells can be constructed and the results added to Table 18-2.

A special cell is required if we want to enter the value for the half-reaction

$$2e^- + 2H^+ \longrightarrow H_2(g)$$

because, in the case of gaseous hydrogen, there is no place to attach the wire, as there is for metallic copper or silver. For this reason, an auxiliary platinum electrode* is placed in the 1 M solution of H^+. Then, hydrogen gas at 1 atmosphere pressure is bubbled around the platinum metal electrode. The arrangement is shown in Figure 18-6. When this electrode is hooked up to the Cu^{++}-Cu half-cell, the reading is 0.3 volt toward the right.

SELECTION OF A STANDARD HALF-CELL

Let us study the results summarized in Table 18-2. In effect, the numbers tell us how each ion competes with Cu^{++} ions for electrons. Silver is clearly a better competitor and is placed first. Instead of indicating "right" and "left" readings, we could designate "left" as positive and "right" as negative. This is a reasonable but *arbitrary* choice. A positive value means that the metal ion is a better competitor than a Cu^{++} ion for electrons. A negative value indicates that a Cu^{++} ion is a better competitor for electrons than the metal ion being considered. The results appear in Table 18-2.

If we review the last experiment, we realize that copper was chosen as a comparison electrode because copper and copper sulfate solution were on

*Platinum is useful as an electrode here because it is chemically inert and because it catalyses the reaction $H + H \rightarrow H_2$. This makes H_2 evolution faster and nonlimiting.

Table 18-3

Half-cell voltages against nickel reference cell

Half-reaction in outer beaker	Voltmeter reading
$Ag^+ + e^- \longrightarrow Ag(s)$	+1.0 volt (left)
$Cu^{++} + 2e^- \longrightarrow Cu(s)$	+0.6 volt (left)
$2H^+ + 2e^- \longrightarrow H_2(g)$	+0.3 volt (left)
$Ni^{++} + 2e^- \longrightarrow Ni(s)$	0.0 volt (no deflection)
$Zn^{++} + 2e^- \longrightarrow Zn(s)$	−0.5 volt (right)

hand. Suppose we had chosen a Ni^{++}-Ni half-cell for comparison. All half-cells in Table 18-3 would have had the same relative positions, but the numbers would have been different. The Ni^{++}-Ni half-cell would obviously be 0.0 volt, and the Cu^{++}-Cu half-cell would read 0.6 volt toward the left because the electrons would be flowing to copper from the new reference half-cell. Electrons would flow out of the half-cell contained in the porous cup. When copper was in the porous cup and nickel was outside it, we had the reverse situation. Values shown in Table 18-3 indicate that all readings would be just 0.6 volt more positive if the Ni^{++}-Ni half-cell was our comparison standard. Any half-cell in the list could have been used as the standard reference for determining the half-cell voltage readings, but *we must use a standard half-cell. Without such use of a second cell, which serves as a standard, there is no place to fasten the second voltmeter wire so that the circuit can be completed and we can get a reading.*

Faced with this situation, chemists have arbitrarily selected the H^+-H_2 half-cell as a standard and have compared all other half-cells to that standard. (See Figure 18-6 for one comparison.) A H^+-H_2 half-cell, measured against an identical H^+-H_2 half-cell used as a standard, generates 0.0 volt. Other half-cells will give the values shown in Table 18-4, *if the measurements are properly made.*

Table 18-4

Selected standard reduction potentials for half-reactions*

Half-reaction		E^0 (volts)
Oxidized state	Reduced state	
$MnO_4^- + 8H^+ + 5e^-$	$\longrightarrow Mn^{++} + 4H_2O$	+1.52
$Cl_2(g) + 2e^-$	$\longrightarrow 2Cl^-$	+1.36
$MnO_2(s) + 4H^+ + 2e^-$	$\longrightarrow Mn^{++} + 2H_2O$	+1.28
$Br_2(l) + 2e^-$	$\longrightarrow 2Br^-$	+1.06
$Ag^+ + e^-$	$\longrightarrow Ag(s)$	+0.80
$I_2(s) + 2e^-$	$\longrightarrow 2I^-$	+0.53
$Cu^{++} + 2e^-$	$\longrightarrow Cu(s)$	+0.34
$2H^+ + 2e^-$	$\longrightarrow H_2(g)$	0.00
$Ni^{++} + 2e^-$	$\longrightarrow Ni(s)$	−0.25
$Co^{++} + 2e^-$	$\longrightarrow Co(s)$	−0.28
$Zn^{++} + 2e^-$	$\longrightarrow Zn(s)$	−0.76

*A more complete list is given in Appendix 6.

EXERCISE 18-8

Suppose a Ag^+-Ag half-cell was chosen as the standard. What would all the values in Table 18-4 be?

THE EFFECT OF CONCENTRATION ON VOLTAGE

The qualification "if the measurement is properly made" suggests that, in defining the reference half-cell and the measured half-cells, we must tell precisely how each half-cell is made. You will remember from your laboratory work (Experiment 28) that the concentration of the solution surrounding the metal electrode is important in determining cell voltage. Studies of a cell based on the reaction

$$Zn(s) + Cu^{++} \longrightarrow Zn^{++} + Cu(s)$$

indicate that if *more* $CuSO_4$ is dissolved in the solution surrounding the copper electrode, the cell voltage *increases*. If a solution containing *less* $ZnSO_4$ is used around the zinc electrode, the voltage *also increases*. A decrease in the concentration of Cu^{++} or an increase in the concentration of Zn^{++} would make the voltage drop. Are not these observations in complete agreement with predictions we would make on the basis of Le Chatelier's principle? The tendency of the reaction to go from left to right is measured by the *positive* voltage value for the cell. Le Chatelier's principle tells us that if the Cu^{++} concentration is reduced—for example, by precipitation of Cu^{++} ions as CuS—there will be less tendency for the reaction to proceed. The equilibrium will shift toward Zn plus Cu^{++}. The voltage should decrease—a prediction in complete agreement with experimental findings.

EXERCISE 18-9

A cell is based on the reaction

$$Ni(s) + 2Ag^+ \longrightarrow Ni^{++} + 2Ag(s)$$

What would be the effect on cell voltage if: (a) more $AgNO_3$ was dissolved in the solution around the silver electrode; (b) more $NiSO_4$ was dissolved in the solution around the nickel electrode; (c) Ag_2S was precipitated by adding H_2S to the Ag^+-$Ag(s)$ half of the cell?

The effects of concentration changes on cell voltage are very familiar to you, although you probably were not aware of the meaning of your observations when you made them. You have all watched a flashlight go dim after continued use or heard the battery of a car gradually run down as someone tried to start it on a cold morning. What happens? Why does the cell or battery stop generating electrical energy?

To answer this question, let us first consider another question. What happens to any cell or battery as it operates? Reactants are used up and products accumulate around the electrodes. Observations reveal that the *voltage also decreases* as the reactants convert to products. Finally, the voltage of the cell reaches zero. We say that the cell is "dead." Equilibrium has been attained, and the reaction that has been producing the energy has the same tendency to proceed as does its reverse. Again, the observed voltage measures the *net* tendency for the reaction to occur. At equilibrium, there is a balance between forward and reverse reactions. Thus, there is *no* tendency for further reaction either way.

STANDARD HALF-CELL POTENTIALS: E^0 VALUES

Since concentrations in a cell do influence its voltage, concentrations of all reagents in a cell should be specified if the measured voltages are going to be fully useful for predicting chemical reactions. Because of this, chemists have introduced the idea of **standard state.** The standard state for gases is taken as a pressure of 1 atmosphere at 25 °C. The standard state for ions is taken as a 1 M solution.* The standard state for other pure substances is taken as the state in which the pure substance exists at 25 °C and 1 atmosphere pressure.

We can now specify more precisely how half-cell potentials should be measured. Values in Table 18-4 were obtained by measuring the cell voltages with all reagents in their standard state. Suppose we set up a cell based on the following reaction, placing the standard hydrogen half-cell to the right in our set-up: $Zn(s) + 2H^+ (1 M) \rightarrow Zn^{++}(1 M) + H_2$ (1 atm). A reading of 0.76 (right) will be made on the voltmeter if the measurement is made at 25 °C. Results for a number of half-cells are listed in Table 18-4. The value of -0.76 volt for the zinc half-cell is known as the E^0 value for the reduction of Zn^{++} ions. The E^0 **value** *is the potential associated with a half-reaction taking place between substances in their standard states.* (The superscript zero means standard state.) By international agreement,[†] the standard H^+-H_2 half-cell is always used as a reference and assigned an E^0 of zero.

A large number of standard half-cell potentials,[‡] E^0 values, have been determined. A more complete list is shown in Appendix 6. What do they mean and what can they be used for? *The standard half-cell potential is a quantitative measure of the electron-attracting ability of an ion in its standard state in the half-cell. In all cases the electron-attracting ability of the hydrogen ion in the standard hydrogen reference half-cell is used as a basis for comparison.* If the sign of E^0 is *positive,* it means that *the ion of the half-cell equation is better at attracting an electron than is the hydrogen ion.* Thus, an E^0 value of $+0.80$ volt for the half-cell defined by the equation $Ag^+ + e^- \rightarrow Ag(s)$ means that silver ions (1 M) in solution have a stronger attraction for electrons than do hydrogen ions (1 M) in solution. We find that silver ions can pull electrons away from hydrogen gas to produce hydrogen ions and silver metal. The overall process then proceeds spontaneously as written with a voltage of 0.80 volt:

$$2Ag^+ + H_2(g) \longrightarrow 2Ag(s) + 2H^+ \qquad E^0 = +0.80 \text{ volt}$$

We have not added 1 M for the concentration of each ionic species, nor have we indicated 25 °C and 1 atmosphere pressure for H_2. All this is implied by the symbol E^0.

---

*Some refinement is needed to define the standard state for ions precisely, but we shall not be concerned with that refinement here.

†The electrode signs used in this text are those adopted by the International Union of Pure and Applied Chemistry.

‡The term *potential* refers to the voltages we have been measuring. It emphasizes the relationship between voltage and potential energy. Voltage is the energy supplied per elementary charge in driving the charge around the circuit of the cell. Voltage multiplied by charge is electrical energy. One volt driving one electron is 1.6×10^{-19} joule. Remember that the joule is a work term.

ALAN G. MACDIARMID

Alan G. MacDiarmid received his undergraduate training at the University of New Zealand. He was awarded Ph.D. degrees in inorganic chemistry from the University of Wisconsin in 1953 and from Cambridge University in 1955. After joining the chemistry faculty at the University of Pennsylvania in 1955, his research focused on the synthesis and characterization of volatile molecular compounds of silicon, sulfur, nitrogen, and fluorine.

He is best recognized for his pioneering role in the discovery that it was possible to "dope" an organic polymer chemically or electrochemically so that it was converted to a metal insofar as its electronic and magnetic properties were concerned while retaining the mechanical properties of an organic polymer. Polyacetylene, $(CH)_x$, the first such polymer studied, was synthesized as silvery, flexible films by the polymerization of acetylene, $HC\equiv CH$. This polymer, as well as several other more recently discovered conducting polymers are now commonly known as "synthetic metals." These metals have opened up a completely new field of nonclassical, interdisciplinary, solid-state science, not only of fundamental scientific interest but also of potential technological importance. Dr. MacDiarmid's highly innovative researches have been widely recognized both nationally and internationally by a variety of prizes and awards.

If the sign of a half-cell is *negative, the hydrogen ion has a stronger attraction for electrons than does the ion of the half-cell.* Thus, for the half-cell with the half-reaction

$$Zn^{++} + 2e^- \longrightarrow Zn(s) \qquad E^0 = -0.76 \text{ volt}$$

the negative sign means that Zn^{++} has less attraction for electrons than does H^+. Accordingly, we would expect that H^+ would be able to take electrons away from metallic zinc to form hydrogen gas. The process has been observed in the laboratory:

$$Zn(s) + 2H^+ \longrightarrow H_2(g) + Zn^{++} \qquad E^0 = +0.76 \text{ volt}$$

For the reverse process, we write

$$Zn^{++} + H_2(g) \longrightarrow 2H^+ + Zn(s) \qquad E^0 = -0.76 \text{ volt}$$

Thus, if only the half-reaction is written, for the two cases we have

$$Zn(s) \longrightarrow Zn^{++} + 2e^- \qquad E^0 = +0.76 \text{ volt}$$
$$Zn^{++} + 2e^- \longrightarrow Zn(s) \qquad E^0 = -0.76 \text{ volt}$$

Note that when the direction of the equation is reversed, the sign of E^0 is reversed.

18.5 Uses of half-cell potentials

Half-cell potential values can be used effectively to predict which half-cell reaction will be favored. They can also be used to balance oxidation-reduction equations. Concentration or temperature changes affect such predictions when E^0 values are applied to conditions other than those of the standard state. E^0 values under standard conditions measure quantitatively the tendency toward minimum energy and the tendency toward maximum randomness.

CALCULATING VOLTAGES OF ELECTROCHEMICAL CELLS

Since experimental cell voltage measurements were the source of our E^0 values, it is easy to calculate cell voltages from tables of E^0 values. Consider the reaction between zinc metal and silver ions:

$$Zn(s) + 2Ag^+ \longrightarrow 2Ag(s) + Zn^{++}$$

The appropriate E^0 values for the two half-reactions are

$$Zn^{++} + 2e^- \longrightarrow Zn(s) \qquad E^0 = -0.76 \text{ volt}$$
$$Ag^+ + e^- \longrightarrow Ag(s) \qquad E^0 = +0.80 \text{ volt}$$

Since Ag^+ has a larger positive E^0 value than Zn^{++}, the silver ion has a greater tendency to pick up electrons than does Zn^{++}. The silver half-reaction will proceed as written above. The half-reaction for Zn^{++} must be reversed in order to write a balanced overall reaction:

$$2Ag^+ + 2e^- \longrightarrow 2Ag(s) \qquad\qquad E^0 = +0.80 \text{ volt}$$
$$\underline{Zn(s) \longrightarrow Zn^{++} + 2e^- \qquad\qquad E^0 = +0.76 \text{ volt}}$$
$$2Ag^+ + Zn(s) \longrightarrow Zn^{++} + 2Ag(s) \qquad E^0 = +1.56 \text{ volts}$$

The positive E^0 value for the overall reaction indicates that the process is spontaneous in the forward direction as written.

Notice that we did *not* double E^0 for the reaction $Ag^+ + e^- \rightarrow Ag$ in obtaining the voltage. The voltage of a half-reaction does *not* depend on how many moles we consider, but it does depend on concentration. Thus,

$$Ag^+ + e^- \longrightarrow Ag(s) \qquad E^0 = +0.80 \text{ volt}$$
$$2Ag^+ + 2e^- \longrightarrow 2Ag(s) \qquad E^0 = +0.80 \text{ volt}$$
$$3Ag^+ + 3e^- \longrightarrow 3Ag(s) \qquad E^0 = +0.80 \text{ volt}$$

The above simply states that the voltage of a cell is not determined by how big the cell is. A tiny cell (battery) for an electric watch can produce a voltage as high as the voltage of a massive storage battery that uses the same reaction.

You might wonder what we would have learned if we had assumed that the cell operates in the reverse direction:

$$Zn^{++} + 2Ag(s) \longrightarrow Zn(s) + 2Ag^+$$

Half-cell values show

$$Zn^{++} + 2e^- \longrightarrow Zn(s) \qquad E^0 = -0.76 \text{ volt}$$
$$\underline{2Ag(s) \longrightarrow 2Ag^+ + 2e^- \qquad E^0 = -0.80 \text{ volt}}$$
$$2Ag(s) + Zn^{++} \longrightarrow 2Ag^+ + Zn(s) \qquad E^0 = -1.56 \text{ volts}$$

The process is spontaneous in the reverse direction, since E^0 is negative.

PREDICTING REACTIONS

Half-cell potentials can be used in general to predict what chemical reactions can occur spontaneously. Suppose someone wishes to know whether zinc can be oxidized when it is placed in contact with a solution of nickel sulfate. The values of E^0 can help that person find out. The standard half-cell potential, or E^0 value, for Zn^{++}-Zn is -0.76 volt. The E^0 value for Ni^{++}-Ni is -0.25 volt. Apparently, Ni^{++} ions have a stronger attraction for electrons than do Zn^{++} ions. We predict from this information that zinc metal will react with nickel ions to produce nickel metal and zinc ions. Zinc is oxidized and nickel is reduced. In fact, that is exactly what happens. Successful predictions save much hard work.

Suppose we want to know whether or not nickel metal will be attacked by chlorine water (Cl_2 gas dissolved in water). Table 18-4 provides the answer. The appropriate half-reactions are

$$Cl_2(g) + 2e^- \longrightarrow 2Cl^- \qquad E^0 = +1.36 \text{ volts}$$
$$Ni^{++} + 2e^- \longrightarrow Ni(s) \qquad E^0 = -0.25 \text{ volt}$$

Since we are interested in the reaction of Cl_2 with metallic nickel, the second half-reaction should be turned around and the sign of E^0 reversed. Then,

$$Cl_2(g) + 2e^- \longrightarrow 2Cl^- \qquad E^0 = +1.36 \text{ volts}$$
$$\underline{Ni(s) \longrightarrow Ni^{++} + 2e^- \qquad E^0 = +0.25 \text{ volt}}$$
$$Ni(s) + Cl_2(g) \longrightarrow Ni^{++} + 2Cl^- \qquad E^0 = +1.61 \text{ volts}$$

The answer to our question is a resounding yes!

We can now make some generalizations on the use of Table 18-4. A substance on the left-hand side of the table reacts by gaining electrons. A substance on the right-hand side reacts by losing electrons. We may draw the following conclusions:

1. *An oxidation-reduction reaction must involve a substance from the left-hand side of the equations in Table 18-4 (something that can be reduced) and a substance from the right-hand side (something that can be oxidized).*
2. *A substance on the left-hand side of the equations in Table 18-4 tends to react spontaneously with any substance on the right-hand side that is lower in the table.*

Applying these rules, we predict that copper metal is oxidized to Cu^{++} ions by $Br_2(l)$ or $MnO_2(s)$, but not by Ni^{++} or Zn^{++} ions. Of course, copper metal cannot be oxidized by either zinc metal or nickel metal because neither zinc nor nickel can accept electrons and thereby be reduced.

EXERCISE 18-10

Use Table 18-4 to decide which of the following substances tend to oxidize bromide ions (Br^-) to produce the element bromine: $Cl_2(g)$, H^+, Ni^{++}, MnO_4^-. Justify your answer in each case.

EXERCISE 18-11

Use Table 18-4 to decide which of the following substances tend to reduce $Br_2(l)$ to $2Br^-$: Cl^-, $H_2(g)$, $Ni(s)$, Mn^{++}. Justify your answers.

CONCENTRATION EFFECTS AND SPONTANEOUS REACTIONS

The foregoing predictions have been based on the values of E^0 and apply only to standard conditions. Yet, we often wish to carry out a reaction at conditions other than standard ones. The prediction must then be adjusted in accordance with Le Chatelier's principle.

For example, we might ask if silver metal will dissolve in $1\ M\ H^+$. According to Table 18-4, the data are

$$2Ag(s) \longrightarrow 2Ag^+(1\ M) + 2e^- \qquad E^0 = -0.80 \text{ volt}$$
$$2e^- + 2H^+(1\ M) \longrightarrow H_2(g) \qquad E^0 = \quad 0.00 \text{ volt}$$
$$\overline{2Ag(s) + 2H^+(1\ M) \longrightarrow 2Ag^+(1\ M) + H_2(g) \quad E^0 = -0.80 \text{ volt}}$$

The negative E^0 shows that the state of equilibrium favors the reactants rather than the products for the overall reaction above. At standard conditions, the reaction will not tend to occur spontaneously. However, if we place Ag(s) in fresh 1 M H$^+$, the initial Ag$^+$ ion concentration will be zero, not 1 M. By Le Chatelier's principle, we know that this will increase the tendency to form products. Will this tendency be enough to counteract the E^0 prediction? Experiments show that some silver will dissolve, though only a *very small amount*. The tendency for silver metal to release electrons is much less than the tendency for H$_2$ to release electrons. *The equilibrium concentration of Ag$^+$ ions is so small that no silver chloride precipitate is formed when the solution is made 0.1 M in Cl$^-$ ions.* This means that the equilibrium concentration of Ag$^+$ ions must be less than 10^{-9} moles per litre. See page 451.

That some silver does dissolve to form Ag$^+$ ions can be shown experimentally by making the solution 0.1 M in I$^-$ ions. Silver iodide has an even lower solubility than does silver chloride. The experiment shows that the amount of dissolved silver is sufficient to precipitate some AgI, but not AgCl. This places the equilibrium Ag$^+$ ion concentration below 10^{-9} M but above 10^{-16} M. Either concentration is so small that we can consider our prediction for the standard state to be generally applicable. Silver metal does not dissolve in 1 M HCl. In general, whether we can apply a prediction based on the standard state to other conditions depends on the size of E^0. If E^0 for the overall reaction is only 0.1 or 0.2 volt (either positive or negative), deviations from standard conditions may invalidate predictions if we do not also consider concentration or temperature changes.

RELIABILITY OF PREDICTIONS

There is one more limitation on the reliability of predictions based on E^0's. Consider the following three reactions:

$$Cu(s) + 2H^+ \longrightarrow Cu^{++} + H_2(g) \quad E^0 = -0.34 \text{ volt}$$

$$Fe(s) + 2H^+ \longrightarrow Fe^{++} + H_2(g) \quad E^0 = +0.44 \text{ volt}$$

$$3Fe(s) + 2NO_3^- + 8H^+ \longrightarrow 3Fe^{++} + 2NO(g) + 4H_2O$$
$$E^0 = +1.40 \text{ volts}$$

The three values of E^0 are easily calculated from half-cell potentials. We can predict with confidence that the reaction of copper metal with dilute H$^+$ will not occur to any appreciable extent. The negative value of E^0 indicates that equilibrium strongly favors the reactants. Moreover, we can predict that the reaction of iron metal with dilute hydrochloric acid *might* occur, and that the reaction of iron metal with dilute nitric acid to produce NO *might* occur. The fairly large positive values of E^0 for the latter two reactions indicate that the products are strongly favored in equilibrium. Experiments may be useful here. First, we immerse a piece of iron in dilute HCl. Bubbles of hydrogen appear; the reaction does occur. Then we immerse a piece of iron in a 1 M nitric acid solution. Although bubbles of hydrogen may appear initially, no nitric oxide (NO) gas appears. The reaction between iron and nitrate ions in acid solution to produce NO does not occur immediately; the reaction rate is very slow. The slow rate could *not* be predicted from the E^0's.

Therefore, the equilibrium predictions based on E^0's do not eliminate the need for experiments. They provide no basis, for example, for anticipating reaction rate; only experiments provide this information. The E^0's do, however, provide definite and reliable guidance concerning the equilibrium state, thus eliminating the need for many experiments. A multitude of reactions do not need to be experimentally tested. We can predict their success or failure by considering the values of E^0.

E^0 AND FACTORS DETERMINING EQUILIBRIUM

We can analyze oxidation-reduction reactions in terms of tendencies toward minimum energy and maximum randomness, just as we did with reactions studied in Chapter 16. We noted that randomness tends toward the maximum and energy tends toward the minimum state. What is the connection between E^0 and these two factors?

Consider two oxidation-reduction reactions for which E^0 shows that products are favored. One reaction is exothermic and the other is endothermic. For the exothermic reaction, reactants, having been mixed, are driven toward equilibrium by their tendency toward minimum energy. Contrast this with the endothermic reaction, for which E^0 shows that equilibrium favors products. When the reactants are mixed, they approach equilibrium *against* the tendency toward minimum energy, since heat is absorbed. The reaction is driven by the tendency toward maximum randomness. Neither the tendency toward minimum energy nor the tendency toward maximum randomness *considered alone* will give us a reliable guide to the direction in which a reaction will proceed. But E^0 does give us the experimental result obtained by combining the two tendencies.

USING HALF-REACTIONS TO BALANCE EQUATIONS

Earlier in this chapter, we used half-reactions to balance simple equations. The process works for more complicated equations too. As a more complex case, we shall consider the example that was used in Section 18-1. Hydrogen sulfide is bubbled into an acidified potassium permanganate solution. From Appendix 6, we see that permanganate is reduced in acid solution as indicated by the equation

$$MnO_4^- + 8H^+ + 5e^- \longrightarrow Mn^{++} + 4H_2O$$

The reduction of hydrogen sulfide is represented as

$$2H^+ + S(s) + 2e^- \longrightarrow H_2S(g)$$

The two equations cannot be added directly as written in Appendix 6. First, the H_2S equation in the appendix must be reversed because $H_2S(g)$ is a reactant in our example. Thus we must write

$$H_2S(g) \longrightarrow 2H^+ + 2S(s) + 2e^-.$$

Next, we must balance the electrons, since the permanganate equation, as written, contains more electrons than the H_2S equation. Using 10 as the lowest common multiple for the number of electrons in each equation, we multiply the H_2S equation by 5 and the MnO_4^- equation by 2:

$$5H_2S(g) \longrightarrow 5S(s) + 10H^+ + 10e^-$$
$$\underline{10e^- + 16H^+ + 2MnO_4^- \longrightarrow 2Mn^{++} + 8H_2O}$$
$$5H_2S(g) + 16H^+ + 2MnO_4^- \longrightarrow 5S(s) + 10H^+ + 2Mn^{++} + 8H_2O$$

Figure 18-7
Electrochemical, or voltaic, cells can be made in various shapes and sizes.

When we add the two equations algebraically, the ten electrons on each side add up to zero. Both atoms and charges are properly balanced in the total equation, but there are H^+ ions included among the reactants as well as among the products. Subtracting 10 H^+ from both sides of the equation, we obtain the final, balanced reaction:

$$5H_2S(g) + 2MnO_4^- + 6H^+ \longrightarrow 5S(s) + 2Mn^{++} + 8H_2O$$

Before leaving the equation, let us check the electric-charge balance:

$$5H_2S(g) + 2MnO_4^- + 6H^+ \longrightarrow 5S(s) + 2Mn^{++} + 8H_2O$$
$$5(0) + 2(-1) + 6(+1) = 5(0) + 2(+2) + 8(0)$$
$$(-2) + (+6) = (+4)$$
$$(+4) = (+4)$$

18.6 Electrolysis

Earlier in this chapter, we dealt with spontaneous processes that released energy as electrical energy. In Chapter 7, we considered the reverse process, where an *externally generated electric current* (electrical energy) was passed through a solution to form products of higher potential energy. Let us compare and contrast the spontaneous process (**electrochemical, or voltaic, cell**) and the nonspontaneous one (**electrolytic cell**).

ELECTROLYTIC AND VOLTAIC CELLS

Many features of voltaic and electrolytic cells are the same. Each cell involves *both* oxidation and reduction processes. In both cells, the *anode* is the electrode at which *oxidation* occurs and the *cathode* is the electrode at which *reduction* occurs. Electric current moves in the external wires of both cells as a result of net *electron* flow in one direction. In both cells, electric current moves through the *solution* of the cell as a result of migrating ions. Because the mechanism for carrying the current through the solution differs from that for carrying the current in a wire, a chemical reaction must take place at *each* junction between metal electrode and solution. Oxidation or reduction processes occur at these junctions. In both cells, *cations* move through the solution toward the *cathode*, and *anions* move through the solution toward the *anode*.

Despite the many common features shared by electrochemical and electrolytic cells, some real and significant differences exist between them. For example, in a voltaic cell, a *spontaneous chemical process* forces electrons around a system. In an electrolysis cell, a dynamo or external voltaic cell *forces* electrons through the system. Consider a cell based on the equation

$$Ni(s) + Cu^{++} \longrightarrow Ni^{++} + Cu(s)$$

The half-reactions and the E^0 values are

$Ni(s) \longrightarrow Ni^{++} + 2e^-$	$E^0 = +0.25$ volt
$2e^- + Cu^{++} \longrightarrow Cu(s)$	$E^0 = +0.34$ volt
$Ni(s) + Cu^{++} \longrightarrow Ni^{++} + Cu(s)$	$E^0 = +0.59$ volt

Nickel is the negative electrode and copper is the positive electrode. The observed voltage is 0.59 volt. The chemical process as written is spontaneous and will force electrons from the nickel electrode to copper with a voltage of 0.59 volt, as shown in Figure 18-8.

Now, what would happen if we hooked up the electrodes to a dynamo that pumps electrons toward the nickel with a potential significantly above 0.59 volt? Experiments show that under these conditions, the nickel cell runs backward, functioning as an electrolysis cell:

$2e^- + Ni^{++} \longrightarrow Ni(s)$	
$Cu(s) \longrightarrow Cu^{++} + 2e^-$	
$Cu(s) + Ni^{++} \xrightarrow[\text{energy}]{\text{electrical}} Cu^{++} + Ni(s)$	$E^0 = -0.59$

If another voltaic cell is used to drive the electrons, rather than a dynamo, the results are similar. Suppose we use a voltaic cell based on the reaction

$$Ni(s) + 2Ag^+ \longrightarrow Ni^{++} + 2Ag(s) \qquad E^0 = +1.05 \text{ volts}$$

to power the electrolysis, as shown in Figure 18-9. The voltage observed in the circuit is the difference between the voltages of the two cells:

$$E^0 = E^0_{\text{Ni-Ag}} - E^0_{\text{Ni-Cu}} = 1.05 - 0.59 = +0.46 \text{ volt}$$

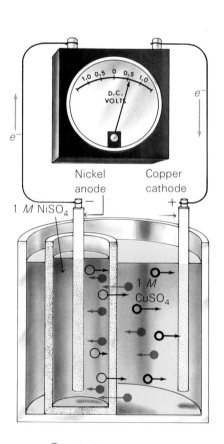

$\bigcirc Ni^{++}$ $\mathbf{O} Cu^{++}$ $\bullet SO_4^{--}$

Figure 18-8

A voltaic cell based on the reaction $Ni(s) + Cu^{++} \rightarrow Cu(s) + Ni^{++}$, where E^0 equals 0.59 volt.

This experimental observation can be understood by considering the net reaction taking place in the entire circuit shown in Figure 18-10. If the net reaction in the electrochemical cell and the net reaction in the electrolytic cell are added together, the net reaction for the combined process is obtained, as shown here:

$$
\begin{array}{ll}
\text{VOLTAIC} & \text{Ni}(s) + 2\text{Ag}^+ \longrightarrow \text{Ni}^{++} + 2\text{Ag}(s) \\
& \qquad\qquad\qquad\qquad E^0 = +1.05 \text{ volts} \\
\text{ELECTROLYTIC} & \text{Cu}(s) + \text{Ni}^{++} \longrightarrow \text{Ni}(s) + \text{Cu}^{++} \\
& \qquad\qquad\qquad\qquad E^0 = -0.59 \text{ volt} \\
\hline
\text{NET REACTION} & 2\text{Ag}^+ + \text{Cu}(s) \longrightarrow \text{Cu}^{++} + 2\text{Ag}(s) \\
& \qquad\qquad\qquad\qquad E^0 = +0.46 \text{ volt}
\end{array}
$$

The E^0 value for the net reaction can be calculated separately from E^0 values for the silver and copper half-cells:

$$
\begin{array}{ll}
2e^- + 2\text{Ag}^+ \longrightarrow 2\text{Ag}(s) & E^0 = +0.80 \text{ volt} \\
\text{Cu}(s) \longrightarrow \text{Cu}^{++} + 2e^- & E^0 = -0.34 \text{ volt} \\
\hline
2\text{Ag}^+ + \text{Cu}(s) \longrightarrow \text{Cu}^{++} + 2\text{Ag}(s) & E^0 = +0.46 \text{ volt}
\end{array}
$$

We see that

1. The E^0 value for the Ni^{++}-Ni half-cell cancels out in considering the net reaction for the overall process.
2. The measured voltage for the overall process involving two cells can be calculated from the net equation for the overall process.

Let us now consider the electrolytic cell, the cell containing nickel and copper half-cells. Electrons forced through the wire from the dynamo or

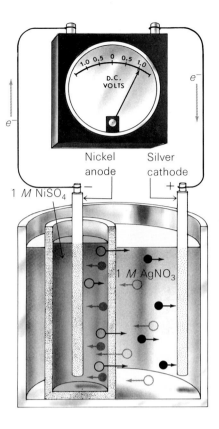

$\bigcirc \text{Ni}^{++}$ $\bullet \text{Ag}^+$ $\bigcirc \text{NO}_3^-$ $\bullet \text{SO}_4^{--}$

Figure 18-9

A voltaic cell based on the reaction $\text{Ni}(s) + 2\text{Ag}^+ \rightarrow 2\text{Ag}(s) + \text{Ni}^{++}$, where E^0 equals 1.05 volts.

Figure 18-10

Two electrochemical cells connected in opposition.

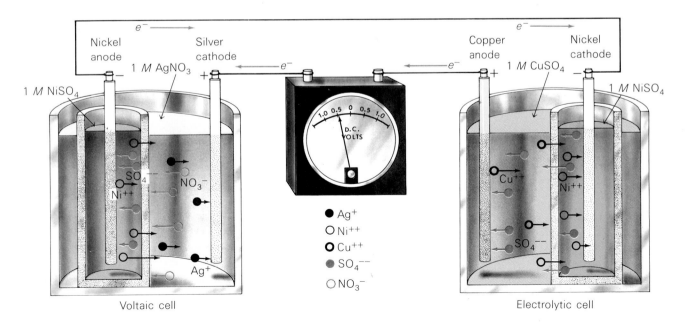

● Ag^+
○ Ni^{++}
◉ Cu^{++}
● SO_4^{--}
○ NO_3^-

Voltaic cell   Electrolytic cell

voltaic cell make the nickel electrode negatively charged. At the nickel electrode, Ni^{++} ions make contact and electrons are transferred to produce nickel metal:

$$\text{REDUCTION} \quad 2e^- + Ni^{++} \longrightarrow Ni(s)$$

At the copper electrode, electrons flow out to the positive electrode of the voltaic cell:

$$\text{OXIDATION} \quad Cu(s) \longrightarrow Cu^{++} + 2e^-$$

Positive ions that formed around the copper electrode now drift toward the negatively charged nickel electrode, while negative ions left in excess around the nickel electrode by removal of Ni^{++} drift toward the positively charged copper electrode. *We see that an electrolytic cell is simply a voltaic cell forced to operate in the reverse direction by a voltage larger than the original cell could produce.*

If more than one reaction is possible in a cell, that reaction will take place that requires the lowest voltage. For example, in the electrolysis of a table salt solution, two reduction reactions are possible at the negative electrode:

$$Na^+ + e^- \longrightarrow Na(s) \qquad E^0 = -2.71 \text{ volts}$$
$$2H^+ + 2e^- \longrightarrow H_2(g) \qquad E^0 = 0.00 \text{ volt}$$

Hydrogen gas is liberated much more easily than sodium if both ions are present at a 1 M concentration. Decreasing the concentration of hydrogen ions should make hydrogen formation more difficult. However, because of the very high voltage needed for sodium deposition (-2.71 volts), hydrogen gas is still more easily produced than sodium metal in any water solution. For this reason, the electrolysis of NaCl in water solution, using platinum electrodes, produces H_2 and Cl_2, *not* Na and Cl_2.

EXERCISE 18-12

Which ion will be deposited first if a solution containing 1 M concentrations of Cu^{++} and Ni^{++} is electrolyzed?

SUMMARY

A chemical reaction involving electron transfer from one species to another can be broken down into two half-reactions. If *oxidation* (loss of electrons) is carried out in one part of a cell and *reduction* (gain of electrons) is carried out in another, electrons can be made to flow through a wire from the part of the cell where electrons are released to the part where they are used. The resulting electric current can be made to do useful work. To prevent charge build-up at any point in the cell, ions must be able to migrate in the liquid phase from one part of the cell to the other. The device used for such an operation is called an *electrochemical,* or *voltaic, cell.*

During its operation, the electrochemical, or voltaic, cell passes from a state of higher energy to a state of lower potential energy, and electrical work is done. The voltage generated by an electrochemical cell is a measure of the force that drives the cell reaction. If the voltage for the $H^+(aq)$-H_2 half-cell is considered to be zero, the voltage associated with other half-cell reactions can be measured under carefully defined experimental conditions.

E^0 values, or standard half-cell potentials, are useful in estimating cell voltages and in deciding whether equilibrium favors a given reaction or its reverse. A number of E^0 values are listed in Appendix 6. A positive E^0 value, favoring a reaction as written, may not guarantee that the process will actually proceed as written. However, a negative E^0 value indicates with certainty that the process will not proceed spontaneously as written.

In balancing oxidation-reduction reactions, both atoms and charge must be conserved. Half-cell reactions can be used to balance oxidation-reduction reactions.

An *electrolytic cell* is a voltaic cell that is forced to operate in the reverse direction by an external voltage greater than that produced by the voltaic cell. If more than one reaction or half-reaction at an electrode is possible in a cell, the process that requires the lowest voltage will take place.

Key Terms
anion
anode
cathode
cation
electrochemical (voltaic) cell
electrode
electrolytic cell
electron acceptor
E^0 value
half-reaction
oxidation
oxidation number
oxidizing agent
reducing agent
reduction
reduction potential
standard state
voltage
voltmeter

QUESTIONS AND PROBLEMS

A

1. Describe three oxidation-reduction processes that you have observed recently.
2. (a) Define oxidation and reduction using the conceptual definition involving electrons.
   (b) Where did the term *oxidation* originate?
   (c) What is the *oxidizing agent* when magnesium burns in air?
   (d) What is the *reducing agent* when Al_2O_3 is converted to metallic Al by using metallic iron?
3. (a) Give the oxidation number of every element in each of the following compounds: (a) MgO (b) SO_3 (c) V_2O_5 (d) H_2SO_4 (e) K_3PO_4 (f) P_2O_3. (Be sure you understand the rules for getting oxidation numbers.) (b) Is an actual ion charge indicated by an oxidation number? Explain.
4. (a) Define oxidation in terms of a change in oxidation number.
   (b) Define reduction in terms of a change in oxidation number.
   (c) Which of the processes indicated below represent an oxidation process; which represent a reduction process, and which are *not* either oxidation or reduction? Use oxidation numbers. Add electrons or atoms (molecules or ions) in the proper place in each half-reaction to balance the charges and atoms.

   (1) $F \longrightarrow F^-$
   (2) $Mg \longrightarrow Mg^{++}$
   (3) $SO_2 \longrightarrow SO_3$
   (4) $MnO_4^- \longrightarrow MnO_2$
   (5) $CaO + SO_3 \longrightarrow CaSO_4$
   (6) $Ti^{+++} \longrightarrow Ti^{++++}$
   (7) $Fe^{+++} \longrightarrow Fe^{++}$
   (8) $2O^{--} \longrightarrow O_2$
   (9) $O_2^{--} \longrightarrow O_2$
   (10) $H_2O + SO_3 \longrightarrow H_2SO_4$

5. A piece of silver (Ag) is placed in a solution of gold chloride ($AuCl_3$). Metallic gold is deposited on the surface of the silver.

   (a) Write the overall *ionic* equation for this process.
   (b) Write the half-reaction representing the *oxidation* process.
   (c) Write the half-reaction representing the *reduction* process.
   (d) From the information given, which ion, Ag^+ and Au^{+++}, is the better competitor for electrons? Explain.

6. The equation used in the commercial "gravity cell" is:

$$Zn + Cu^{++}, SO_4^{--} \longrightarrow Zn^{++}, SO_4^{--} + Cu$$

But when a Zn rod is placed in a solution of copper sulfate, we do not get a shock if we touch the zinc rod, and there is no evidence of electric current.

   (a) Identify the half-reactions in the above process.
   (b) What physical arrangement must be used in carrying out the Zn-Cu^{++} reaction if it is to give electricity? Describe in some detail.
   (c) How do ions move from one half-cell to another? Why is this important?

7. (a) How is the anode defined in an electrochemical cell?
   (b) How is the cathode defined in an electrochemical cell?
   (c) Which metal in an operating Zn-Cu^{++} cell is most likely to have a surplus of electrons and thus have a *negative* charge?
   (d) Which way will electrons flow in the wire connecting the Zn and Cu electrodes? Explain.
   (e) Which way do the negative ions move in the electrochemical cell? The positive ions?

8. Refer to the "gravity cell" mentioned in question 6. The E^0 value for the *oxidation* of Zn is $+0.76$. The E^0 for the *reduction* of Cu^{++} is $+0.34$.

   (a) What is the E^0 for the complete reaction involving the reduction of Cu^{++} by metallic zinc? Write equations and show E^0 combinations.
   (b) Which is the better competitor for electrons, Zn^{++} or Cu^{++}?
   (c) From the equation written in (a) and Le Chatelier's Principle, what would happen to the voltage if the $[Cu^{++}]$ were increased to that of a saturated $CuSO_4$ solution?
   (d) What would happen to the voltage if the solution of $ZnSO_4$ were saturated?
   (e) If the gravity cell is used to generate electricity to run a small motor, we find that on constant running the motor gradually slows down and finally stops. Why? Consider first what happens to the cell voltage then tell why it happens.

9. Potassium dichromate can be used to oxidize ferrous iron. The half-equations are:

$$Cr_2O_7^{--} + 14H^+ + 6e^- \longrightarrow 2Cr^{+++} + 7H_2O$$
$$Fe^{++} \longrightarrow Fe^{+++} + e^-$$

Write the balanced overall ionic equation for the process. Check both sides for charge balance.

10. What happens at each electrode if an electric current is forced through a solution of $CuCl_2$ using inert platinum electrodes?

11. (a) What is meant by E^0 for a half-reaction? (b) What is the reference half-cell?

12. Why do car lights go dim and then go out if they are left on while the motor is not running? Explain in terms of an approach to the equilibrium condition.

B

13. Which *one* of the following equations does *not* describe an oxidation-reduction process? Use oxidation numbers if you are in doubt.

   (a) $2Al + 3Cl_2 \longrightarrow 2AlCl_3$

   (b) $CaO + Mn_2O_7 \longrightarrow Ca(MnO_4)_2$

   (c) $PCl_3 + Cl_2 \longrightarrow PCl_3$

   (d) $3Cu + 2AuCl_3 \longrightarrow Au + 3CuCl_2$

   (e) $2Mn_2O_7 \longrightarrow 4MnO_2 + 3O_2$

14. *Balance* the following oxidation-reduction equations, using the oxidation number method. Be sure charge and atoms are conserved.

   (a) $Cu + Ag^+ \longrightarrow Ag + Cu^{++}$

   (b) $MnO_4^- + H_2S + H^+ \longrightarrow S + Mn^{++} + H_2O$

   (c) $Cu_2S + O_2 \longrightarrow Cu_2O + SO_2$

   (d) $FeCl_3 + SnCl_2 \longrightarrow SnCl_4 + FeCl_2$

15. (a) Determine the oxidation number of manganese in each of the following compounds: (1) $Mn(OH)_2$ (2) MnF_3 (3) MnO_2 (4) K_2MnO_4 (5) $KMnO_4$. (b) Tin normally has oxidation numbers of 0, +2, and +4. Would Sn^{++} be an oxidizing agent, reducing agent, or both? What might the product containing Sn be in each case?

16. In each of the following equations identify: (1) the oxidation numbers of all atoms or ions in each compound, (2) the atom or ion oxidized, (3) the atom or ion reduced, (4) the change in oxidation number associated with the oxidation and the reduction process, (5) the oxidizing *agent*, (6) the reducing *agent*. Use the above information to balance each equation:

   (a) $CH_4 + O_2 \longrightarrow CO_2 + H_2O$

   (b) $Mn + Cl_2 \longrightarrow MnCl_2$

   (c) $PbO_2 + Pb + H_2SO_4 \longrightarrow PbSO_4 + H_2O$

   (d) $NH_3 + O_2 \longrightarrow NO + H_2O$

(e) $HNO_3 + H_2S \longrightarrow NO(g) + S + H_2O$

(f) $CuO + NH_3 \longrightarrow N_2 + H_2O + Cu(s)$

17. In an experiment similar to Experiment 27 in the Laboratory Manual, strips of gold, silver, and tin were placed in beakers containing solutions of Au^{+++}, Ag^+, and Sn^{++} ions. The following results were observed:

> $Au^{+++} + Sn$: metallic gold is deposited on tin strip
> $Au + Ag^+$: no reaction
> $Sn + Ag^+$: metallic silver is deposited on tin strip

Arrange the ions above in order of *increasing* ability to attract electrons (that is, place the ion with the greatest attraction for electrons on the bottom).

18. The following observations involving metals *A, B, C,* and *D* are made: (1) When a strip of *A* is placed in a solution of B^{++} ions, no reaction is observed. (2) Similarly, *A* plus C^+ ions produces no reaction. (3) When a strip of *D* is placed in a solution of C^+ ions, black metallic *C* deposits on the surface of *D* and the solution tests positively for D^{++} ions. (4) When a bar of *B* is placed in a solution of D^{++} ions, metallic *D* appears on the surface of *B* and B^{++} ions are found in the solution. Arrange the ions A^+, B^{++}, C^+, and D^{++} in order of their ability to attract electrons. On top, place the ion that best attracts electrons.

19. Which ion or atom in each set is the better competitor for electrons: (a) Cu^{++} or Zn^{++} (b) Au^{+++} or Cu^{++} (c) O_2 or Mg? See problems 17 and 14.

20. Consider the reaction based on the equation

$$Zn(s) + Cu^{++}(aq) \longrightarrow Zn^{++}(aq) + Cu(s)$$

What experimental evidence can be cited to indicate that the process goes to completion? Is the reaction equilibrium very far to the right?

21. Use the E^0 values in Appendix 6 and the information that many fruits and all carbonated drinks are quite acidic [containing $H^+(aq)$ ions] to explain the following: (a) Cavities in teeth are filled with gold or a silver-mercury amalgam rather than with aluminum or zinc. (b) An unpleasant sensation is felt when aluminum foil accidentally gets in the mouth. (c) Considering only its position in the table of E^0 values, would copper be a satisfactory material for dental fillings? Zinc? Lithium? Explain. Is electrical activity the only important property of filling materials?

22. Use E^0 values to tell whether each of the following processes would go. If they go, indicate the process with a plus sign (+); if they do not go, indicate with a minus sign (−). If it is not oxidation-reduction, write NOR.

(a) $2K + 2(H^+, Cl^-) \longrightarrow H_2 + 2(K^+, Cl^-)$

(b) $Ag^+ + Cl^- \longrightarrow$

(c) $Ni + Br_2 \longrightarrow NiBr_2$

(d) $Zn^{++} + Br_2 \longrightarrow$

(e) $Ca(s) + I_2(s) \longrightarrow$

(f) $Ag(s) + H^+(aq) \longrightarrow$

(g) $Ag(s) + Br_2(l) \longrightarrow$

23. Two voltaic cells defined by equations 1 and 2 are set up in opposition to each other.

$$(1) \quad Ni + 2Ag^+ \longrightarrow Ni^{++} + 2Ag$$
$$(2) \quad Cu + 2Ag^+ \longrightarrow Cu^{++} + 2Ag$$

(a) What E^0 does each cell generate by itself? (b) Which cell will go in its normal way and which cell will be forced backward? (c) What will be the net voltage in the circuit?

C

24. Assuming that all alkali metals have an oxidation number of $+1$, (a) what is the oxidation number of each oxygen atom in Na_2O_2? (b) What is the oxidation number of each oxygen atom in K_2O? (c) What is the oxidation number of each nitrogen in N_2H_4? (d) What is the oxidation number of carbon in C_2H_6? (e) Write Lewis formulas or structural formulas for all these compounds.

25. The E^0 value for the half-reaction $Ag(s) \rightarrow Ag^+ + e^-$ as written is -0.80 volt. What would happen to the measured half-cell value (H^+-H_2 reference) if a salt solution (NaCl) was added to the solution around the silver electrode? Which one of the following values would be most reasonable: -0.40, -0.80, or -1.20? Explain.

26. The overall equation for a relative of the lead storage battery is

$$Pb(s) + PbO_2(s) + 4H^+ \longrightarrow 2Pb^{++} + 2H_2O$$

(a) $Pb(ClO_4)_2$ is water-soluble and $PbSO_4$ is insoluble. Would $HClO_4$ be as useful as H_2SO_4 in making a lead storage battery? Explain, considering only the effect on cell voltage. (b) The E^0 half-cell potential for the reaction $Pb(s) \rightarrow Pb^{++} + 2e^-$ is $+0.13$ volt as written, while for the process $PbO_2(s) + 4H^+ + 2e^- \rightarrow Pb^{++} + 2H_2O$ the value is $+1.46$ volts. What voltage would be expected for the overall PbO_2-Pb cell?

27. A set of reactants and a set of products are listed below for five different systems. Write balanced equations for each system, using half-cell reactions to balance the equations.

(a) $Cu(s) + NO_3^- + H^+$ produces $Cu^{++} + NO_2(g) + H_2O$

(b) $H_2S(g) + NO_3^- + H^+$ produces $S(s) + NO_2(g) + H_2O$

(c) $Cl_2(g) + SO_2(g) + H_2O$ produces $SO_4^{--} + Cl^- + H^+$

(d) $Cr_2O_7^{--} + I^- + H^+$ produces $Cr^{+++} + I_2(s) + H_2O$

(e) $MnO_2(s) + Fe^{++} + H^+$ produces $Mn^{++} + Fe^{+++} + H_2O$

28. A Ce^{++++}-Ce^{+++} half-cell, when connected to a Fe^{+++}-Fe^{++} half-cell, has an overall voltage of $+0.84$ volt. The iron electrode is the anode in the cell. Determine the E^0 value of the Ce^{++++}-Ce^{+++} half-reaction, using the H^+-H_2 standard.

Batteries and electric cars

By the year 2000, electric cars may enjoy as much popularity as they did in 1900. The silent, old electric cars moved slowly and had limited power and range, but they had a unique character and charm. Such cars contained crystal vases and plush cushions and they gave their owners a new and exciting sense of freedom and mobility. By 1912, several thousand electric cars and trucks were in use around the world. But by 1918, the popularity of electric cars was fading. The cars could not compete with the newly emerging gasoline-powered automobiles. The gasoline engine revolutionized life the world over.

Today, shrinking oil supplies and concern about atmospheric pollution threaten the dominance of the gasoline engine. Once again, electric cars are beginning to look attractive. Today, much research focuses on the most serious problem of electric cars: the function of their "battery" as the power storage unit.

Gasoline-powered automobiles are usually started by an electric motor that receives electricity from a "lead storage battery." The lead storage battery in its simplest form involves the reaction between lead metal (Pb) and lead dioxide (PbO_2) to produce two Pb^{++} ions. The appropriate half-reactions are

1. reaction at positive electrode (reduction):

$$PbO_2(s) + 4H^+ + SO_4^{--} + 2e^- \longrightarrow PbSO_4(s) + 2H_2O$$

2. reaction at negative electrode (oxidation):

$$Pb(s) + SO_4^{--} \longrightarrow PbSO_4(s) + 2e^-$$

The overall cell reaction is

$$PbO_2(s) + Pb(s) + 2H_2SO_4 \longrightarrow 2PbSO_4(s) + 2H_2O$$

Because PbO_2, Pb, and $PbSO_4$ are all relatively insoluble in sulfuric acid, the reactants and products remain as solids on each electrode. The equation tells us that a fully charged battery has sulfuric acid as the liquid around the electrodes. A discharged battery has water around the electrodes. Sulfuric acid has a density of about 1.8 grams per cubic centimeter while water has a density of 1.0 gram per cubic centimeter. It is thus easy to see why the service station operator measures the density of the battery fluid in order to determine the degree of charge on the battery. It is also easy to see why a discharged battery can freeze and break in cold weather while a charged battery (containing concentrated H_2SO_4) will not freeze. Concentrated H_2SO_4 solutions do not freeze.

Figure 18-11

Electric cars in the early 1900s could not compete with gasoline-powered cars. These electric cars are now found only in museums.

Applying E^0 values permits us to calculate the voltage of a battery. We write

$$PbO_2(s) + 4H^+ + SO_4^{--} + 2e^- \longrightarrow$$
$$PbSO_4(s) + 2H_2O \quad +1.68 \text{ volts}$$

$$\underline{Pb(s) + SO_4^{--} \longrightarrow PbSO_4(s) + 2e^- \quad +0.35 \text{ volt}}$$

$$Pb(s) + PbO_2(s) + 2H_2SO_4 \longrightarrow$$
$$2PbSO_4(s) + 2H_2O \quad +2.03 \text{ volts}$$

As we can see, only about 2 volts are available from each electrochemical cell. If we need a 12-volt power source, six 2-volt cells must be combined in series to produce the required 12 volts. For a 6-volt battery, only three cells must be combined.

While modern-day lead-acid batteries are well developed and reliable, they still are not well-adapted to use in electric cars. Such batteries are heavy in relation to the energy they store. For example, the very best lead-acid batteries available today store about 40 to 50 watt-hours of electrical energy per kilogram of battery weight. The battery can be charged and discharged about 500 to 800 times. Even the "best batteries" are not well-suited for use in electric cars. The heavy weight of the battery power pack demands a good deal of energy to move it over the ground. For this reason, much of the energy stored is used in moving the battery itself. The car has a short range, poor acceleration, and poor performance. Furthermore, the large battery pack can be quite expensive. Prices of up to $13,000 have been suggested for a reasonable power pack. Even so, the lead-acid battery can be produced using existing production facilities and it is still the cheapest battery available. At present, it is the only battery system available for any kind of electric vehicle. Currently the working range of current electric cars is not much above 100 miles.

The fundamental problem with the lead-acid battery is its weight. We need lighter storage units that hold more watt-hours per kilogram. Several systems are under development. One such system is the nickel-zinc battery. Both nickel and zinc are much lighter than lead. The pertinent electrode reactions are

OXIDATION	$Zn(s) + 2OH^- \longrightarrow$	
	$ZnO(s) + H_2O + 2e^-$	ANODE $(-)$
REDUCTION	$NiOOH(s) + H_2O + e^- \longrightarrow$	
	$Ni(OH)_2(s) + OH^-$	CATHODE $(+)$
OVERALL	$Zn(s) + 2NiOOH(s) + H_2O \longrightarrow$	
	$ZnO(s) + 2Ni(OH)_2(s)$	

A major problem with this system is that the ZnO product is partly soluble in the electrolyte (a concentrated KOH solution). This leads to changes in the anode that tend to shorten cell life. Successful resolution of this cell-lifetime problem may well provide the key to future use of this system.

Another possible battery is based on the reaction between zinc and chlorine:

$$Zn \longrightarrow Zn^{++} + 2e^- \quad +0.76 \text{ volt}$$
$$\underline{Cl_2 + 2e^- \longrightarrow 2Cl^- \quad +1.36 \text{ volts}}$$
$$Zn + Cl_2 \longrightarrow ZnCl_2 \quad +2.12 \text{ volts}$$

However, the difficulty of managing highly poisonous chlorine gas has blocked commercial development. Nevertheless, a still to be verified claim for a battery of this type was made in 1980. According to its developers, the new battery would permit manufacture of a practical electric car able to travel 150 miles or more at a speed of 55 miles per hour without recharge. Evidence for success of this battery is not yet available.

In this battery, a mixture of chlorine and water is cooled slightly to a slushy state, forming a hydrate of chlorine that is stored in an insulated tank. When water is pumped into the tank, it melts the slush and carries it to another tank containing zinc-coated graphite electrodes. There, the chlorine reacts with the zinc, forming zinc chloride and producing an electric current. To recharge, the process is reversed. Batteries using this system are claimed to have a power density of 80 watt-hours per kilogram of battery weight, or two times what a good lead-acid battery now provides.

Another possible system is the nickel-iron battery first developed by Thomas Edison in 1901. It was the most popular kind of rechargeable battery until about 1925. Recent work on this type of battery has generated new interest in its further development.

The battery uses finely divided hydrated nickel peroxide for the positive plate and finely divided iron for the negative plate. The reaction at the iron electrode takes place in two steps. First, iron is oxidized to its $+2$ state and then to its $+3$ state as follows:

FIRST OXIDATION $\quad$ $Fe + 2OH^- \longrightarrow$
$\qquad\qquad\qquad\quad Fe(OH)_2 + 2e^-$ $\quad$ ANODE $(-)$

SECOND OXIDATION $\quad$ $Fe(OH)_2 + OH^- \longrightarrow$
$\qquad\qquad\qquad\quad Fe(OH)_3 + e^-$ $\quad$ ANODE $(-)$

The reaction at the nickel electrode is similar to that which takes place in the nickel-zinc cell:

REDUCTION $\quad$ $NiOOH + H_2O + e^- \longrightarrow$
$\qquad\qquad\qquad Ni(OH)_2 + OH^-$ $\quad$ CATHODE $(+)$

One of the major problems with this battery is that it generates hydrogen gas during charging. The voltages for H_2 generation and iron reduction are very similar. Chemists have proposed that adding a trace quantity of a sulfur-bearing anion will reduce the generation of this highly combustible gas. However, this proposal has yet to be tested fully on large-scale batteries. Another serious problem with the battery is its high purchase price. However, this may be offset by relatively low operating costs during its lifetime. Another drawback is that nickel-iron batteries are not as effective at temperatures below 10 °C as they are at higher temperatures. At 0 °C, they do not operate at all. Vehicles equipped with this kind of battery would have to be supplied with a heater to keep the battery warm in cold climates.

Lithium-iron sulfide batteries have also been developed. An alloy of lithium and aluminum serves as the negative electrode and iron sulfide acts as the positive electrode. The electrolyte is a molten salt consisting of a KCl and LiCl eutectic mixture,* which has a melting point of 352 °C. To operate properly, the cell must be kept above 400 °C at all times. A good insulating jacket will have to be developed to maintain the desired interior operating temperatures. The reactions taking place are

OXIDATION $\quad$ $Li \longrightarrow Li^+ + e^-$ $\qquad$ ANODE $(-)$
REDUCTION $\quad$ $4Li^+ + 4e^- + FeS_2 \longrightarrow$
$\qquad\qquad\qquad\qquad Li_2S + Fe$ $\quad$ CATHODE $(+)$

OVERALL $\qquad$ $4Li + FeS_2 \longrightarrow 2Li_2S + Fe$

There are several serious problems with this battery besides the high operating temperature. Lithium-iron sulfide batteries have relatively short lifetimes and the boron nitride material that is currently used to separate the electrodes is very expensive. Nevertheless, research on this type of battery is intense and improvements are to be expected.

Since about 1970, chemists and engineers in the United States, Great Britain, France, and Japan have been deeply involved in the development of a sodium-sulfur battery. The key component of this power source is the ceramic electrolyte, "beta alumina" ($Na_2O \cdot 11\,Al_2O_3$). This substance is ionically conductive at temperatures between 300 and 350 °C. The reactions are

OXIDATION $\quad$ $Na \longrightarrow Na^+ + e^-$ $\qquad$ ANODE $(-)$
REDUCTION $\quad$ $Na_xS + Na^+ + e^- \longrightarrow$
$\qquad\qquad\qquad\qquad Na_{x+1}S$ $\quad$ CATHODE $(+)$

OVERALL $\qquad$ $Na_x + Na \longrightarrow Na_{x+1}S$

The problems with this battery include the corrosion of the positive electrode, the high operating temperatures, and the low durability of the beta alumina. Table 18-5 summarizes the types of batteries presently available and the number and complexity of the difficulties preventing their full utilization in electric cars.

A new "battery" that "burns" a metal, rather than gasoline, has also been proposed. It differs sharply from commercial rechargeable batteries. The fuel is a metal, such as zinc or aluminum. The metal is oxidized by oxygen in pressurized air that is pumped to the metal electrode. The electric current is generated by the reaction between the metal and oxygen to form the metallic oxide. The reactions are

---

*A mixture whose melting point is lower than that of any other combination of the same substances.

Table 18-5

Number of barriers to successful battery development for electric cars

Battery system	Number of technical barriers	Average difficulty of barriers
nickel-iron	1	medium
lead-acid	2	medium
nickel-zinc	2	difficult
zinc-chloride	5	difficult
lithium and aluminum–iron sulfide	7	difficult
sodium-sulfur (ceramic)	8	difficult

OXIDATION $\quad 2Al(s) \longrightarrow$
$\qquad\qquad\qquad 6e^- + 2Al^{+++}$ ANODE $(-)$

REDUCTION $\quad \frac{3}{2}O_2(g) + 6e^- \longrightarrow$
$\qquad\qquad\qquad 3O^{--}$ CATHODE $(+)$

OVERALL $\quad 2Al(s) + \frac{3}{2}O_2(g) \longrightarrow$
$\qquad\qquad 2Al^{+++} + 3O^{--} \longrightarrow Al_2O_3(s)$

An aluminum-air type battery was tested successfully in mid-1980 at the Lawrence-Livermore Laboratory in California, the center for the United States Department of Energy's metal-air battery research effort. It is claimed that 60 such cells connected in a 970-pound battery could power a full-performance five-passenger electric vehicle for up to 300 miles nonstop at 55 miles per hour. However, new aluminum electrodes must be installed each time the battery is recharged.

Although there are still a considerable number of technical difficulties to overcome, electric cars may eventually take their place in the world's transportation system. They will be especially useful as second cars or as commuting cars. It has been estimated that 90 percent of all car trips made in the United States are less than 20 miles and that 99 percent are less than 100 miles. Electric cars may prove suitable for this kind of travel. Automobile manufacturers are well aware of the potential market for such vehicles and are spending many research dollars to develop the kind of batteries necessary to make electric cars commercially successful.

The property of carbon which distinguishes it most conspicuously is its tendency to form non-ionizing links both with other carbon atoms and with atoms of different elements.

REYNOLDS C. FUSON AND HAROLD R. SNYDER

19

THE CHEMISTRY OF CARBON COMPOUNDS

Objectives

After completing this chapter, you should be able to:

- Assign appropriate names to a number of organic compounds.

- Distinguish between saturated and unsaturated organic compounds and recognize features of the chemistry of each.

- Predict the products of oxidation reactions of primary and secondary alcohols.

- Explain some of the chemistry of selected functional groups.

- Describe the chemical synthesis of a typical polymer.

These swimmers are wearing lightweight suits made of synthetic organic polymers.

The chemistry of carbon is central to all life as we know it. Plants, animals, and people synthesize thousands of organic compounds every day. Organic compounds include sugars, starches, plant oils and waxes, fats, gelatin, dyes, drugs and fibers. Because carbon compounds have their origin in living plants or animals, the chemistry of carbon is called **organic chemistry.** Except for CO, CO_2, Na_2CO_3, $NaCN$, NH_4CNO, and related substances, all compounds containing carbon are called **organic compounds.**

It was once thought that carbon compounds such as those found in living creatures could not be made outside the body of a living creature. This idea was disproved in 1828 when the prominent German chemist Friedrich Wöhler heated the inorganic, ionic salt ammonium cyanate, NH_4CNO, and converted it to the organic compound urea, $(H_2N)_2CO$. Since urea is produced in the bodies of many animals as a product of protein metabolism, its synthesis in the laboratory was very significant. Clearly, the body of a living creature was not the only source of urea.

Wöhler's simple laboratory conversion laid the groundwork for a gigantic chemical industry that today manufactures perfumes, medicines, plastics, fibers, paints, soaps, flavorings, lubricating oils, fuels, and many other products. Where do manufacturers find the enormous quantities of carbon and carbon compounds needed to feed their giant industry? Let us begin our study of carbon chemistry by looking at the chief sources of carbon and carbon compounds.

19.1 Sources of carbon compounds

The raw materials that feed the chemical industry and supply our energy needs come from the remains of plants and animals that lived on the earth millions of years ago. Most of us know these valuable natural materials as coal and petroleum.

COAL

Coal is formed by decayed plant material that accumulated in prehistoric times. During prehistoric days, layers of plant material formed during each growing season and piled on top of existing layers. Soon, air was excluded from the lower layers. Pressure on the lower layers built up. In time, the layers of plant remains were compressed into hard beds of coal. Coal is not pure carbon. It contains many of the elements present in the original plant material, such as hydrogen, nitrogen, oxygen, and sulfur. The hardest coal, anthracite, is 85 to 95 percent carbon by weight. The softest coal, peat, contains much of the original plant material. Peat is about 55 percent carbon.

Although we have known about coal for hundreds of years, its use as a fuel is of relatively recent origin. As late as 1760, wood was the only fuel used in western Europe and North America. However, increasing energy and fiber demands arising from the industrial revolution began to put serious strains on the wood supply. Whole areas of Europe were denuded of trees, and North American forests were cut for fuel and lumber. As wood became harder to obain, coal was seen as a reasonable substitute. By 1900, coal provided about 96 percent of the world's energy needs. Today, petroleum (including natural gas) is the chief fuel in the world, but dwindling petroleum supplies will again direct our attention toward coal.

Although coal is still being formed today, we are consuming it 50,000 times faster than it is being formed. Conservative estimates indicate that our present coal supplies can last for about 200 to 500 years. Since the reserves are finite, long-term solutions to our energy problems are needed.

When coal is heated to a high temperature in the absence of air, coal gas and a complex mixture called coal tar distill away. As discussed in Chapter 14, coke is the solid residue that remains after such heating (see page 360). Coke contains mostly carbon and the mineral materials (ash) present in the original coal. Coke is widely used in the reduction of iron ore and in the preparation of steel.

PETROLEUM AND NATURAL GAS

Petroleum is a complex mixture that may range from a light, volatile liquid to a heavy, tarry substance too viscous to flow. Crudes vary greatly in their value as sources of gasoline. The light, nonviscous crudes, typical of many from the Middle East, are superior to heavier crudes for the prepara-

Table 19-1

Hydrocarbon fractions from petroleum

Fraction	Carbon atoms in molecule	Boiling point range (°C)	Uses
gas	1 to 5	−160 to 30	gaseous fuel, heating
petroleum ether	5 to 6	20 to 60	industrial solvents
light naphtha	6 to 7	60 to 100	industrial solvents
straight-run gasoline	5 to 12	30 to 200	motor fuel
kerosene, fuel oil	12 to 18	180 to 400	diesel fuel, furnace fuel, cracking
lubricants	16 and up	above 350	lubricants
paraffins	20 and up	low-melting solids	candles, matches
asphalt	36	gummy residues	roofing, road construction

tion of gasoline. While the heavy crudes can be used, they cost more to convert to gasoline and similar products.

Petroleum, like coal, has its origin in plant remains that decomposed under special conditions. Petroleum is found in oil pools in porous rock formations. Natural gas is a mixture of low-molecular-weight compounds (CH_4, C_2H_6, C_3H_8, C_2H_4, and so on) found in underground "fields" of porous rock.

Petroleum and natural gas provide the raw materials for thousands of products used in our society. The petrochemical industry is one of the largest industries in the world. Petrochemicals include such items as synthetic rubber, synthetic fibers, plastics, insecticides, and fertilizers.

The tremendous growth of the petrochemical industry has occurred only since 1913. In that year, chemists developed the process for breaking down crude oil into lighter components. Such an operation is called "cracking," since it literally involves cracking larger molecules into smaller ones. *Thermal catalytic cracking* is accomplished by heating crude oil in the presence of a catalyst. At temperatures ranging from 400 to 500 °C, large molecules break up into smaller ones and also tend to *rearrange* into random, branched structures. The branched structures are favored at high temperatures because they have a higher entropy than do straight-chain forms. Lighter, branched hydrocarbons make good fuels because they have a higher octane rating. (Octane rating is discussed on page 538.)

The hydrocarbon components of crude oil are separated in a large fractionating column. Table 19-1 compares the resulting fractions with respect to temperatures at which they are released, length of carbon chain, and uses in industry.

19.2 Naming organic compounds: Basic nomenclature

The simplest kinds of organic molecules consist of a skeleton of carbon atoms linked to one another to form straight or branched chains. The *prefix* (first part) of the name of such a compound is determined by the number of carbon atoms in the *longest straight chain*. The prefixes for compounds containing from one to ten carbon atoms in the longest chain are as follows:

Number of carbons	Prefix	Number of carbons	Prefix
1	*meth-*	6	*hex-*
2	*eth-*	7	*hept-*
3	*prop-*	8	*oct-*
4	*but-*	9	*non-*
5	*pent-*	10	*dec-*

The *suffix* (last part) of the name of the compound varies with the formula of the substance, the type of the bonding between carbon atoms in the chain, and with the nature of atoms and groups of atoms connected to various parts of the chain. The ending *-ane* indicates a family of compounds containing only carbon and hydrogen, with the empirical formula C_nH_{2n+2}. You are probably familiar with *meth*ane (CH_4), *eth*ane (C_2H_6), *prop*ane (C_3H_8), and *but*ane (C_4H_{10}), all of which are gases that are used as fuels. *Oct*ane (C_8H_{18}) is the liquid component of automobile fuel mentioned earlier.

Suppose we wish to construct a molecule whose longest chain contains eight carbons attached to a shorter branch of one carbon:

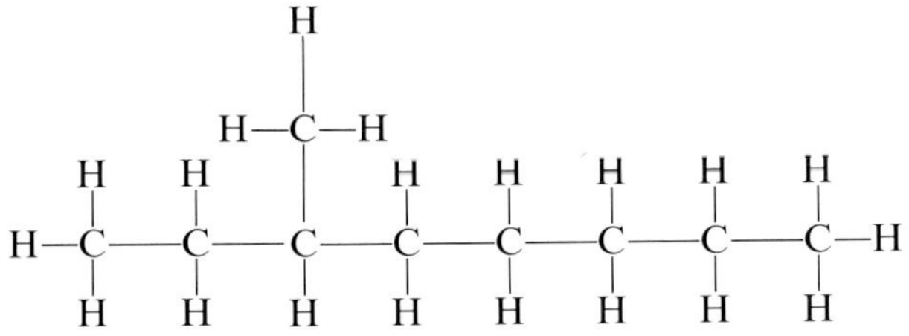

What would we name this compound? We might call it methane octane—methane for its one carbon branch and octane for its eight-carbon chain. Since —CH_3 can be classed as a derivative of methane, we use the stem *meth-* and add *-yl*. Thus, the branch is called *methyl*. The other prefixes we have listed are treated in a similar manner: ethyl, propyl, butyl, and so on. In our example, the methyl group is located on a particular carbon atom of the octane chain. To locate the methyl group, we number the carbons in the long chain starting from the end closest to it:

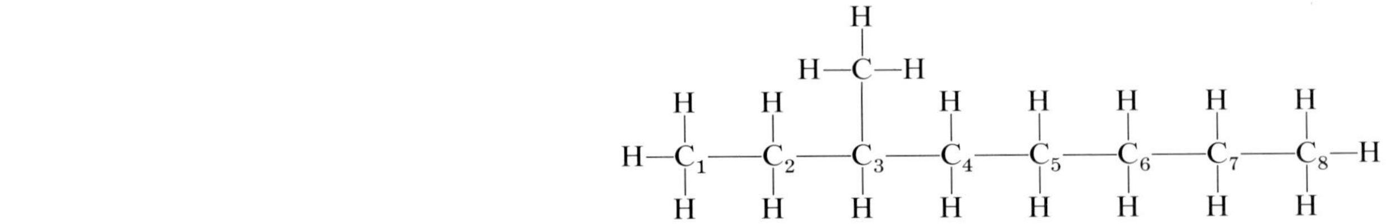

Using the numerical system, we can name the compound 3-methyloctane. The 3 designates the carbon atom to which the methyl branch is bonded.

19.3 Molecular structure of carbon compounds

Carbon compounds vary widely in their properties. To understand the differences among compounds, we must establish the structure of each molecule. We can learn the molecular structure of ethanol, for example, by determining its empirical, molecular, and structural formulas. The **empirical formula** tells only the relative number of atoms of each element in a molecule. For example, the empirical formula of ethanol is C_2H_6O. That of ethane is CH_3.

The first step in determining the empirical formula of a pure substance such as ethanol is to carry out an analysis. A carefully weighed sample of the compound is burned in pure oxygen in a special apparatus. If the compound contains just carbon and hydrogen (or carbon, hydrogen, and oxygen), only carbon dioxide and water are produced. If the compound contains nitrogen as well, nitrogen gas or an oxide of nitrogen is produced. If the compound contains sulfur, oxides of sulfur are generated.* The gases and liquids generated by combustion are carefully trapped, separated, and weighed to determine the composition of the pure substance.

For example, suppose we burn a 46.08-gram sample of pure ethanol. We recover 88.02 grams of pure CO_2 and 54.06 grams of pure H_2O. Nothing else is produced. We can now calculate the empirical formula of ethanol. To do this, we must determine the number of moles of CO_2 and H_2O produced. The number of moles of CO_2 is

$$88.02 \text{ g } CO_2 \times \frac{1.00 \text{ mole } CO_2}{44.01 \text{ g } CO_2} = 2.00 \text{ moles } CO_2$$

The number of moles of H_2O produced is

$$54.06 \text{ g } H_2O \times \frac{1.00 \text{ mole } H_2O}{18.02 \text{ g } H_2O} = 3.00 \text{ moles } H_2O$$

Let us now summarize what we know by writing a tentative equation:

$$C_xH_yO_z + O_2 \longrightarrow 2.00 \, CO_2 + 3.00 \, H_2O$$

Since each mole of CO_2 in the products required 1 mole of carbon atoms from the 46.08-gram sample of ethanol, we can write that x (of C_x) is 2. Similarly, we note that the 6 moles of hydrogen atoms in the water had to originate in the 46.08 grams of ethanol. Therefore, y is equal to 6. We can now write

$$C_2H_6O_z + O_2 \longrightarrow 2.00 \, CO_2 + 3.00 \, H_2O$$

The carbon and hydrogen present in the 46.08-gram sample weigh

$$2 \times 12.01 \text{ g} + 6 \times 1.01 \text{ g}$$

or

$$24.02 \text{ g} + 6.06 \text{ g} = \textit{30.08 g}$$

*Somewhat more complex procedures involving reaction with sodium are frequently used in analyzing organic compounds containing nitrogen or sulfur.

Because only CO_2 and H_2O are formed, the remaining 16.00 grams in the 46.08-gram sample must be oxygen. Converting this to moles of oxygen atoms by the unit factor method* yields

$$16.00 \text{ g oxygen atoms} \times \frac{1.00 \text{ mole O atoms}}{16.00 \text{ g O atoms}} = 1.00 \text{ mole O atoms}$$

We can now write C_2H_6O as the empirical, or simplest, formula of ethanol.

In actual practice, less than a 46.08-gram sample would be used and whole numbers would not be obtained. To get whole numbers, we divide by the smallest subscript. For example, if our numbers were $C_{0.22}H_{0.66}O_{0.11}$, we would divide all subscripts by 0.11 to get the formula C_2H_6O. A typical set of experimental data is given in Exercise 19-1. Data are usually reported as percentage by weight of carbon, hydrogen, oxygen, and so on.

EXERCISE 19-1

Automobile antifreeze often contains a compound called ethylene glycol. Analysis of pure ethylene glycol shows that it contains only carbon, hydrogen, and oxygen. A sample of ethylene glycol with a mass of 15.5 mg is burned and the masses of CO_2 and H_2O resulting are 22.0 mg and 13.5 mg respectively. (a) What is the empirical formula of ethylene glycol? (b) Calculate the percentage by weight of carbon, hydrogen, and oxygen in the sample.

MOLECULAR FORMULAS

The **molecular formula** of a substance gives the *total number of atoms* of each kind in a *molecule* of that substance. To obtain the true molecular formula from the empirical formula, we need an experimental value for the *molecular weight* of the substance under study. We can get the formula weight by adding up the atomic weights in the empirical formula. For ethanol, with an empirical formula of C_2H_6O, the formula weight is 46.02. The experimental molecular weight is found to be 46. Thus, the empirical and molecular formulas of ethanol are identical. However, these formulas are not usually the same. For example, the empirical formula of the fuel gas ethane is CH_3. The formula unit has a mass of 15. The experimental molecular weight of ethane, as found by vapor density, is 30. (The density of the gas at STP is 1.34 g/litre. Multiplying this by 22.4 litres/mole gives a molecular weight of 30.) Thus, there must be *two* empirical formulas in one molecular formula of ethane. Accordingly, the molecular formula of ethane is C_2H_6.

STRUCTURAL FORMULAS: CHEMICAL EVIDENCE AND BONDING RULES

As you will remember from earlier chapters, the **structural formula** of a substance tells us *how atoms are connected* in a molecule of that substance. Chemical facts and bonding rules are very helpful here. For example, we

---

* Remember that the equation for finding the unit factor is 1 mole of oxygen atoms = 16.00 g of oxygen atoms.

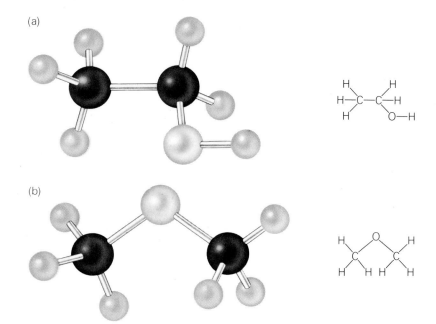

Figure 19-1

The structural formulas and models of the C_2H_6O isomers.

can make reasonable guesses about the structure of ethanol from the bonding rules presented on pages 305–309. The two most obvious choices are shown in Figure 19-1.* Note that every carbon atom has four single bonds, every hydrogen atom has one single bond, and every oxygen atom has two single bonds. The formulas are

$$\underset{(a)}{\overset{\displaystyle H \quad H}{H-\underset{\displaystyle H \ \ H}{C}-C-O-H}} \quad \text{and} \quad \underset{(b)}{\overset{\displaystyle H \qquad H}{H-\underset{\displaystyle H \qquad H}{C}-O-C-H}}$$

We must now appeal to experiment to decide which one of these structures, if either, is correct. It is clear that in structure (a), one carbon atom and one hydrogen atom are bonded to an oxygen atom. Two hydrogen atoms are bonded to one carbon atom and three to another. In structure (b), two carbon atoms are bonded to an oxygen atom and all six hydrogen atoms are bonded to carbons. What information does the chemistry of ethanol provide? Clean sodium metal reacts vigorously with ethanol, producing hydrogen gas and a white, water-sensitive solid with empirical formula C_2H_5ONa. Quantitative experiments show that 2 moles of ethanol generate 1 mole of hydrogen gas. We are reminded of the equation describing the reaction between sodium and water:

$$2Na + 2HOH \longrightarrow 2NaOH + H_2$$
$$2Na + C_2H_5OH \longrightarrow 2C_2H_5ONa + H_2$$

The data are consistent with the structure (a) but not with structure (b) because only $\frac{1}{2}$ mole of H_2 is generated per mole of organic compound. We could also conclude from the data that hydrogen atoms bonded to oxygen

---

*While we cannot positively rule out structures that violate our normal bonding rules, the evidence for such structures would have to be very strong to make us accept them.

532

react with sodium, but that those bonded to carbon do not. Such a conclusion is further strengthened when we learn that another isomer of the compound C_2H_6O does not react with clean sodium. Structure (b) seems to be appropriate for this isomer.

DETERMINING STRUCTURAL FORMULAS: EXPERIMENTAL EVIDENCE

Chemical evidence supports the conclusion that ethanol has the structure C_2H_5OH. Is that structure consistent with the physical and spectral properties of ethanol? Ethanol has a boiling point of 78.5 °C. The isomer to which we have assigned the formula $(CH_3)_2O$ has a boiling point of -25 °C. For reference, we note that ethane (C_2H_6) boils at -88.6 °C and water boils at 100 °C. You will remember that the high boiling point of water is explained by the bonding of hydrogen between molecules. Ethanol with an —OH unit resembles water in some ways and should be able to form hydrogen bonds. The formation of such bonds would account for the relatively high boiling point of ethanol. The absence of the —OH unit in $(CH_3)_2O$ would account for the low boiling point of that structure. Solubility data also are consistent with this information. Ethanol with an —OH unit should be able to form hydrogen bonds to water, such as

$$\begin{array}{ccccc} H & & C_2H_5 & & H \\ | & & | & & | \\ O—H & \cdots & O—H & \cdots & O—H \end{array}$$

Thus, we should anticipate solubility of ethanol in water. Experiments show that water and alcohol dissolve in all proportions. They are completely *miscible,* while dimethyl ether, $(CH_3)_2O$, has limited solubility in water.

Various spectroscopic techniques provide more direct evidence of molecular structure. The *infrared spectrum* of ethanol vapor shows frequencies that can be assigned to the vibration of hydrogen atoms bonded to carbon as well as frequencies that can easily be assigned to the vibration of hydrogen atoms bonded to oxygen. Dimethyl ether, $(CH_3)_2O$, shows only frequencies characteristic of hydrogen atoms bonded to carbon atoms.

EXERCISE 19-2

Ethylene glycol has the empirical formula CH_3O and the molecular formula $C_2H_6O_2$. Using the usual bonding rules (for example, carbon is tetravalent, oxygen is divalent, hydrogen is monovalent), draw some of the structural formulas possible for the compound.

STRUCTURAL COMPONENTS: THE ETHYL GROUP

The chemistry of ethanol described in the preceding section suggests quite clearly that there are two parts to the ethanol molecule. One is the hydroxyl group, or —OH group, which makes ethanol resemble water. The other is the ethyl group, or C_2H_5— group, which gives ethanol a faint resemblance to ethane. In the reactions described above, the —OH group underwent change or substitution, while the C_2H_5— group went through the reactions without change. We note that the C_2H_5— group was a stable unit in a vigorously changing system. We can say that it had *structural integrity* in the processes studied. On the other hand, the —OH group changed markedly. It is thus called a **functional group.**

If we understand the chemistry of a particular functional group, we can usually predict how other compounds containing that group will react. We can place compounds containing a given functional group in a single family. Thus, compounds containing the —OH group are called **alcohols.** The hydrocarbon fragment bonded to the —OH group determines the specific alcohol present. Due to its structural similarity to ethane (H_5C_2)—H, the unit H_5C_2— is called the ethyl group. Thus, H_5C_2OH is ethyl alcohol (or ethanol).

The reaction of ethyl alcohol with HBr shows that it is possible to replace the —OH group and to form new compounds such as ethyl bromide (C_2H_5Br):

$$C_2H_5OH + HBr \longrightarrow C_2H_5Br + HOH$$

By replacing any one hydrogen atom of ethane with an —OH group, we can form ethyl alcohol. By replacing one hydrogen atom of ethane with a bromine atom, we can form ethyl bromide. The structure of that molecule is shown in Figure 19-2. Thus, as noted earlier, ethyl alcohol and ethyl bromide are called *derivatives* of ethane.

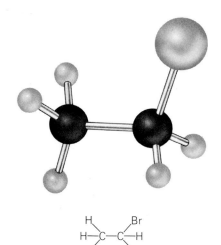

Figure 19-2

The structural formula and model of ethyl bromide (bromoethane).

19.4 Hydrocarbons

As you can see, much organic nomenclature and classification is tied to hydrocarbons. As the name **hydrocarbon** suggests, such substances contain only hydrogen and carbon. Several groups of hydrocarbons can be identified. The **saturated hydrocarbons** have the empirical formula C_nH_{2n+2} and contain only single bonds between carbon atoms. Some members of this series are listed in Table 19-2 on page 537, and some structural formulas are shown in Figure 19-3. The suffix *-ane* identifies members of the group.

A second group of saturated hydrocarbons has the empirical formula C_nH_{2n}. Members of this group contain carbons linked in rings. Such structures, also shown in Figure 19-3, are called **cyclic hydrocarbons.** Thus, C_5H_{10} would be *cyclo*pentane and would be represented as

Note that members of this family also contain only single C—C bonds.

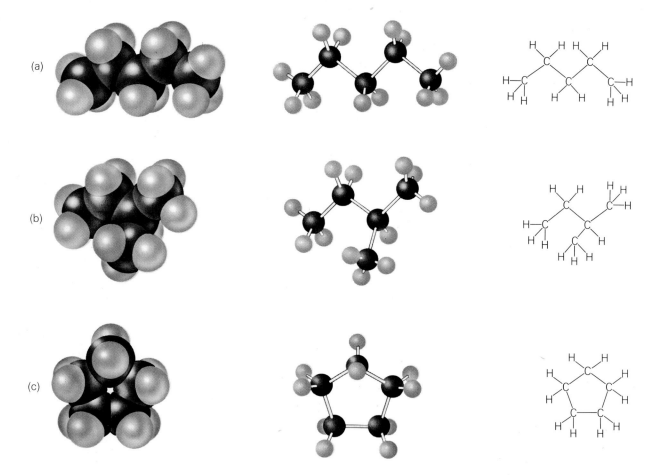

Figure 19-3

Structural formulas and models of some five-carbon saturated hydrocarbons: (a) *n*-pentane, (b) isopentane, (c) cyclopentane.

EXERCISE 19-3

(a) Write the structural formula of cyclopropane. Write the structural formula of cyclohexane. (b) Figure 19-3 shows three five-carbon saturated hydrocarbons. One other is known. It is neopentane. Write its structural formula.

A third hydrocarbon group contains one carbon-carbon double bond in the molecule. The empirical formula for the family is also C_nH_{2n}. It is one of a broader group of molecules called **unsaturated hydrocarbons.** Unsaturated hydrocarbons contain one or more double or triple bonds between carbons. The simplest member of the family, containing one double bond, is ethylene (C_2H_4). It has the formula

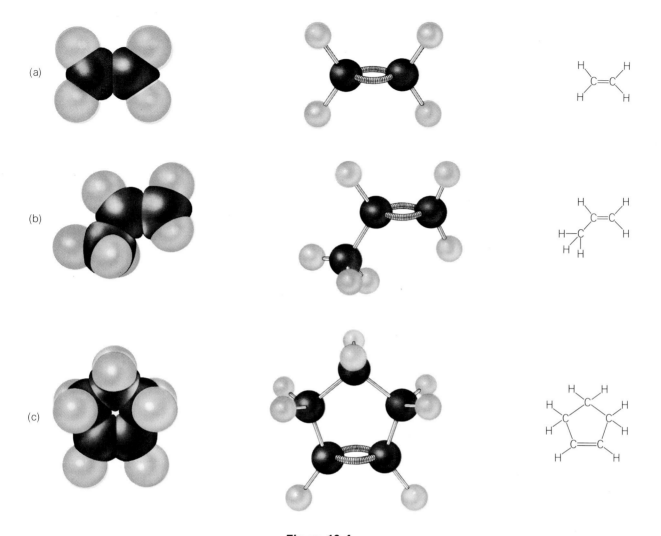

Figure 19-4
Structural formulas and models of some unsaturated hydrocarbons:
(a) ethylene, (b) propylene, (c) cyclopentene.

The next member of the series is propylene (C_3H_6), with the following structural formula:

Note that the propylene molecule is the same as an ethylene molecule in which one hydrogen atom has been replaced by a methyl group. Figure 19-4 shows the structural formulas of some double-bonded molecules. The name of the molecule in each case is obtained by adding *-ene* to the prefix of the name of the parent *hydrocarbon radical*.* For example, ethyl is the parent hydrocarbon radical derived from ethane. Thus, the unsaturated two-carbon

*Note that we use the name of the hydrocarbon radical.

hydrocarbon is ethylene. The three-carbon unsaturated molecule is propylene. Hydrocarbons containing more than one double bond in their structure also exist. Thus, we can have buta*diene*.

$$\begin{matrix} H\diagdown & & H & H & & H \\ & C=C & - & C=C & \\ H\diagup & & & & \diagup H \\ & & & & & H \end{matrix}$$

Butadiene is very important in the manufacture of artificial rubber.

A third, very important type of unsaturated hydrocarbons contains a triple bond. The simplest member of this series has the formula

$$H-C\equiv C-H$$

The systematic name for this molecule is eth*yne*. Its common industrial name is acetylene. The members of this family contain one triple bond per molecule and have the empirical formula C_nH_{2n-2}.

A very important class of cyclic unsaturated molecules contains more than one double bond. The parent compound is named benzene. It has the molecular formula C_6H_6. Benzene will be discussed in much more detail later in the chapter.

SATURATED HYDROCARBONS: ALKANES

Many different structures are found in the saturated hydrocarbons, or **alkanes.** Thus, for the alkane with five carbon atoms, we can draw four different formulas:

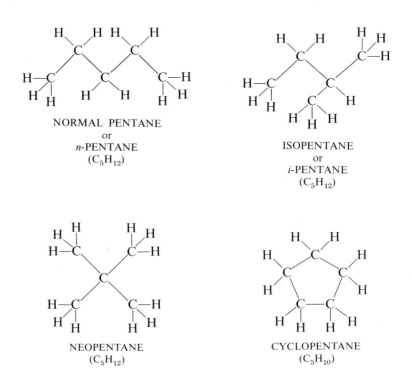

NORMAL PENTANE
or
n-PENTANE
(C_5H_{12})

ISOPENTANE
or
i-PENTANE
(C_5H_{12})

NEOPENTANE
(C_5H_{12})

CYCLOPENTANE
(C_5H_{10})

Table 19-2

Properties of some saturated hydrocarbons

Saturated hydrocarbon	Molecular formula	Melting point (°C)	Boiling point (°C)	Heat of combustion of gas (kcal/mole)
methane	CH_4	−182.5	−161.5	−212.8
ethane	CH_3CH_3	−183.3	−88.6	−372.8
propane	$CH_3CH_2CH_3$	−187.7	−42.1	−530.6
n-butane	$CH_3CH_2CH_2CH_3$	−138.4	−0.5	−687.7
isobutane	$CH_3{-}CH{-}CH_3$ CH_3	−159.6	−11.7	−685.7
n-hexane	C_6H_{14}	−95.3	68.7	−1,002.6
cyclohexane	C_6H_{12}	6.6	80.7	−944.8
n-octane	C_8H_{18}	−56.8	125.7	−1,317.5
n-octadecane	$C_{18}H_{38}$	+28.2	316.1	−2,891.9

Note that three of the compounds have the same molecular formula, C_5H_{12}. Because *n* (the number of carbon atoms) may be very large, the number of saturated alkanes is very, very large. For example, there are 69,491,178,805,831 different possible structures with the *single molecular formula* $C_{40}H_{82}$.

The low-molecular-weight alkanes are gases under normal conditions. (See the boiling points in Table 19-2.) Methane (CH_4) is the principal component of natural gas and the simplest alkane. Gasoline is a mixture of complex, branched and cyclic alkanes with from 6 to 10 carbon atoms. Fuel oil is a mixture containing somewhat longer chains. Paraffin waxes are alkanes with 20 to 35 carbon atoms. Crude oils differ very much in their composition.

Hydrocarbons are used as fuel. Thousands of tons of petroleum-based products are burned in the world every day for the energy they contain. CO_2 and water are the products of combustion. Except for this very important high-temperature process, the alkanes are relatively inert. For example, sodium, a very reactive metal, can be stored under kerosene to protect the metal from water and oxygen.

The relatively inert character of the saturated hydrocarbons shows us that the chemistry of organic compounds is similar in most cases to either the chemistry of the functional groups attached to the saturated hydrocarbon unit, or to the chemistry of double bonds in unsaturated organic molecules. We will develop this idea later. Rather extreme conditions of temperature and pressure are usually needed to bring about controllable reactions of saturated hydrocarbons. Only combustion proceeds easily.

Combustion in automobiles is a very important chemical reaction of alkanes. However, gasoline that contains mainly straight-chain hydrocarbons

Herbert C. Brown was born in London, England, in 1912. He received his Ph.D. degree in 1938 from the University of Chicago. Since 1947, Dr. Brown has been a member of the Department of Chemistry at Purdue University. He presently holds the title of Wetherill Research Professor Emeritus.

Professor Brown is perhaps best known for his research on the importance of boron in organic chemistry. He discovered that the simplest compound of boron and hydrogen, diborane, can be added with remarkable ease to unsaturated organic molecules to form organoboranes. His studies on applications of the borohydrides and diborane to organic synthesis have had a revolutionary impact on synthetic organic chemistry. In 1979, Dr. Brown shared the Nobel Prize for Chemistry with Georg Wittig for developing organoborane compounds that are very useful in the synthesis of organic compounds. Drugs, synthetic sex attractants called pheromones, and various other important compounds can be made more easily using Dr. Brown's methods.

In addition to the Nobel Prize, Dr. Brown has been the recipient of numerous other awards and honors. He won the American Chemical Society Award for Creative Research in Synthetic Organic Chemistry for 1960, the National Medal of Science for 1969, and the Priestley Medal for 1981. Dr. Brown was elected to the National Academy of Sciences in 1957.

is not a useful fuel for automobile engines. The reason is that the gasoline vapor and air mixture burns violently and explosively in the cylinder instead of burning at a uniform rate. We say that the mixture detonates. The result is a "knocking" sound when the piston within the cylinder slams down violently. In proper operation, the piston is pushed down smoothly by the burning mixture.

A gasoline mixture composed mainly of branched and cyclic alkanes and certain unsaturated hydrocarbons burns smoothly with high fuel efficiency. Such a mixture has a higher *octane rating* than the straight-chain alkanes. The octane rating compares the ability of a gasoline to reduce knocking with that of a test fuel given an arbitrary rating of 100.* A higher octane number means that the gasoline has less tendency to knock. The octane rating of a gasoline can be increased by adding substances that improve the rate of smooth burning. The most widely used substance for that purpose was *tetraethyl lead*, $(CH_3CH_2)_4Pb$. The use of this substance has now been reduced because it damages catalytic converters designed to reduce smog formation by cars, and because lead pollutes the environment.

UNSATURATED HYDROCARBONS

Unsaturated hydrocarbons contain double or triple bonds. As mentioned earlier in the chapter, ethylene (C_2H_4) is the simplest unsaturated hydrocarbon. It is much more reactive than its saturated relative ethane. The "extra" reactivity is due to the double bond in the molecule. If ethylene or one of its relatives is shaken with a carbon tetrachloride solution containing elemental bromine, the brown color of the bromine in the solution fades. The ethylene is transformed into a new compound 1,2-dibromoethane. The numbers refer to the carbons to which the bromine atoms are attached. The equation is

$$\underset{\text{ETHYLENE}}{\overset{H}{\underset{H}{>}}C=C\overset{H}{\underset{H}{<}}} + Br_2 \longrightarrow \underset{\text{1,2-DIBROMOETHANE}}{H-\overset{H}{\underset{Br}{C}}-\overset{Br}{\underset{H}{C}}-H}$$

Since the equation shows that bromine clearly *adds* to the ethylene, the reaction is called an **addition reaction.** Many other molecules will add to ethylene, yielding predictable products. Because saturated hydrocarbons cannot add a molecule, such as Br_2, their reactions with halogens can only be carried out under more extreme conditions in the presence of a catalyst. A hydrogen atom is *replaced* by a halogen atom:

$$H-\overset{H}{\underset{H}{C}}-\overset{H}{\underset{H}{C}}-H + Br_2 \xrightarrow{\text{catalyst}} H-\overset{H}{\underset{H}{C}}-\overset{H}{\underset{H}{C}}-Br + HBr$$

Such a process is a **substitution reaction** and is much more difficult to carry out than an addition reaction.

---

*The standard test fuel with an octane rating of 100 is called 2,2,4-trimethylpentane. Since other branched compounds are even more efficient, octane ratings above 100 are possible.

BENZENE: STRUCTURE AND REACTIONS

We now turn to another important class of cyclic hydrocarbon compounds. The simplest example of such a compound is benzene, which has six carbon atoms in a ring and the formula C_6H_6. Experiments show that the **benzene ring** is a planar hexagon with 120° angles between each pair of carbon-carbon bonds. If we count electrons and atoms, we see that there must be three double bonds in a benzene ring. However, we encounter some difficulty when we attempt to represent the bonding in the usual way. If we number the carbons, we can identify two possible structural arrangements:

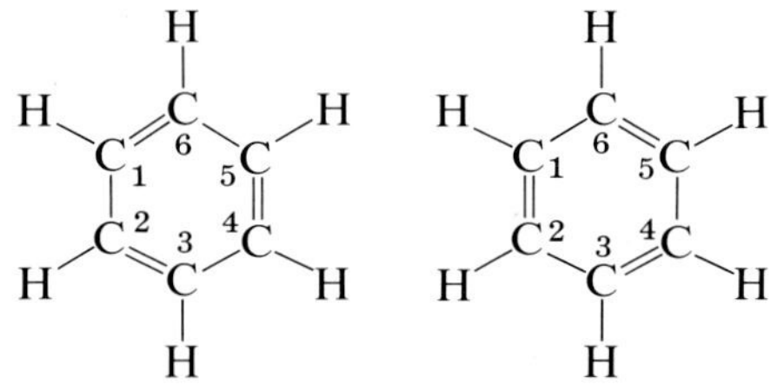

In either arrangement, we should see three long C—C single bonds and three short C=C double bonds. It is important to note that experimental evidence shows *conclusively* that all C—C bonds in benzene are identical in length. On the basis of a large amount of related information, it seems best to describe benzene as the "superposition" or "average" of the two possible structures. We say benzene is a *resonance hybrid* of the following two structures:

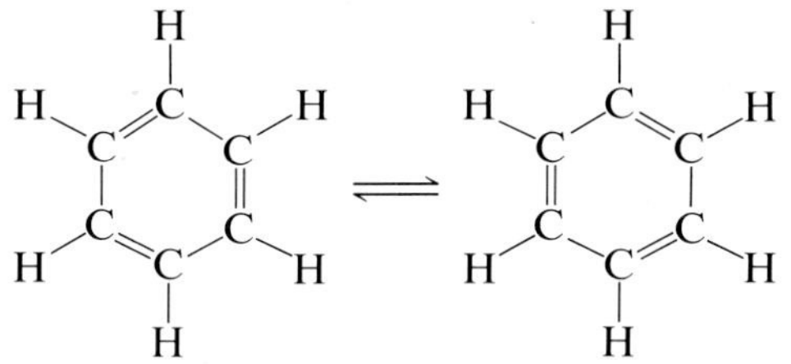

Benzene is not represented accurately by either simple structure, but is a dynamic combination of the two. A convenient way to represent this fact is by using the symbol

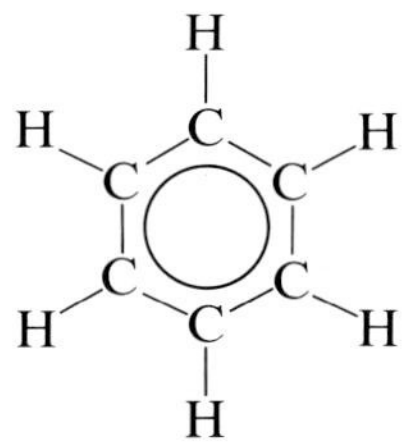

The three "resonating" double bonds are indicated by the circle. Another symbol for benzene omits both the carbons and the hydrogens. We write only

The older symbols

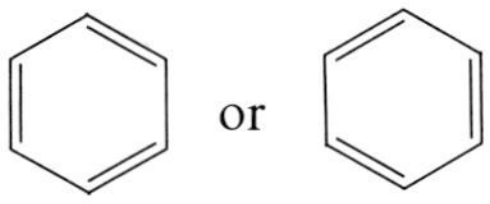

are still used frequently but are not quite right, since they do not indicate that all C—C bonds are equivalent.

Whichever symbol is used, chemists always remember that the carbon-carbon bonds are identical. They have properties unlike those of either simple double or simple single bonds. The situation is very much like the one we faced in Chapter 13 with graphite, where more than one arrangement of the double bonds provided equally good representations of the structure. In the case of graphite, we said that the electrons are "delocalized," since they wander across the surface of the sheet of carbon atoms.

Benzene is like a small fragment of the graphite lattice, and electrical effects are transmitted easily through the delocalized electronic cloud (the π *cloud*, which is above and below the ring surface). Indeed, much of the chemistry of ring compounds reflects this mobility of electrons. All compounds (such as benzene) containing rings with delocalized electrons are called **aromatic compounds.**

Benzene shows neither the typical reactivity nor the usual addition reactions of ethylene. Benzene does *not* react with bromine (Br_2) through addition. Rather, it reacts by substitution:

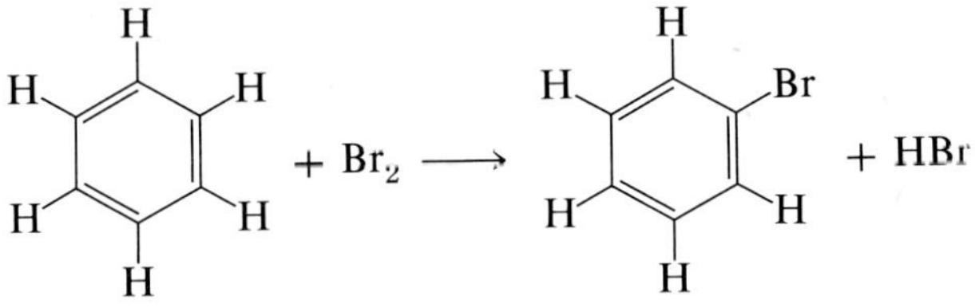

In this reaction, called *bromination*, one of the hydrogen atoms has been replaced by a bromine atom. Notice that the double-bond structure is not affected. The substitution reaction is characteristic of the chemistry of benzene and its derivatives. By substitution, one can introduce functional groups, which can then be modified in various ways.

One of the most important derivatives of benzene is nitrobenzene. The nitro— group is —NO_2. Nitrobenzene is important chiefly because it is readily converted by reduction into an aromatic amine called aniline. One procedure uses zinc as the reducing agent:

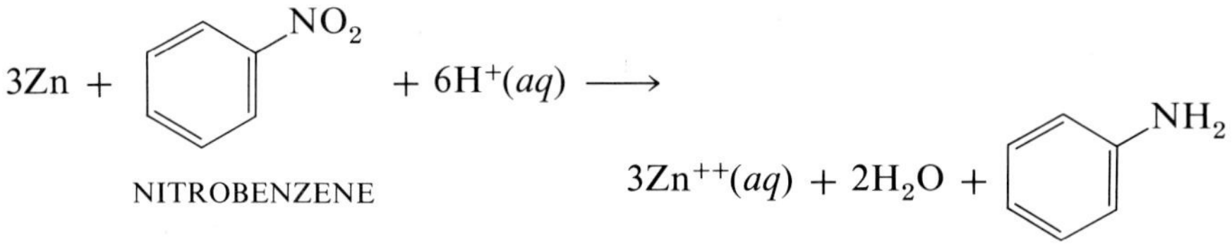

Aniline and other aromatic amines are valuable industrial raw materials. They form an important starting point from which many of our dyestuffs, medicines, and other valuable products are prepared. For example, the indicator methyl orange is an example of an aniline-derived dye. The structure of methyl orange is

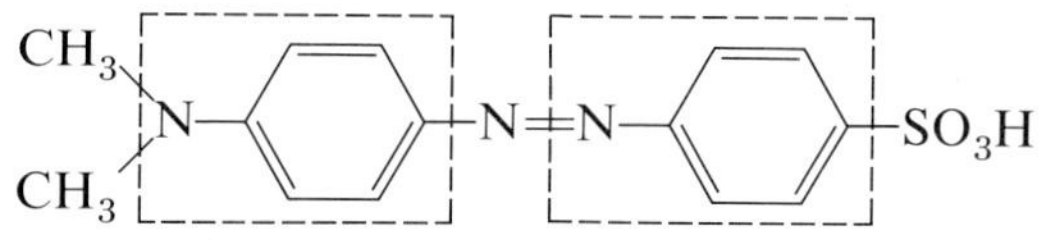

METHYL ORANGE

The portions of the methyl orange molecule set off by the broken lines come from aromatic amines such as aniline. Aniline is the starting material from which the so-called azo dyes, such as methyl orange, are made.

Hydroxybenzene, or phenol, is another important derivative:

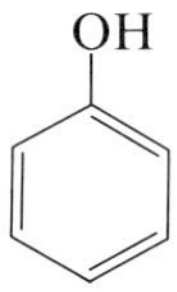

Most phenol is now made industrially from benzene. Phenol, a germicide and disinfectant, was first used by the English surgeon Joseph Lister (1827–1912) in 1867 as an antiseptic in medicine. More effective and less toxic antiseptics have since been discovered (see Table 19-3).

Salicylic acid, which can be prepared from phenol, is quite useful.

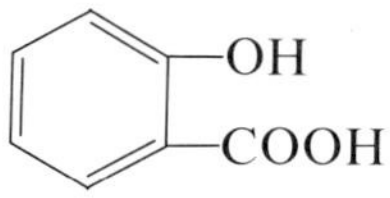

SALICYLIC ACID

A derivative known as its methyl ester, named methyl salicylate, has a pleasant characteristic odor and is called "oil of wintergreen." (See Table 6-6, page 143, for the vapor pressure of methyl salicylate.) The acid itself (or the sodium salt) is a valuable drug in the treatment of arthritis. However, the most widely known derivative of salicylic acid is aspirin, which has the following structure:

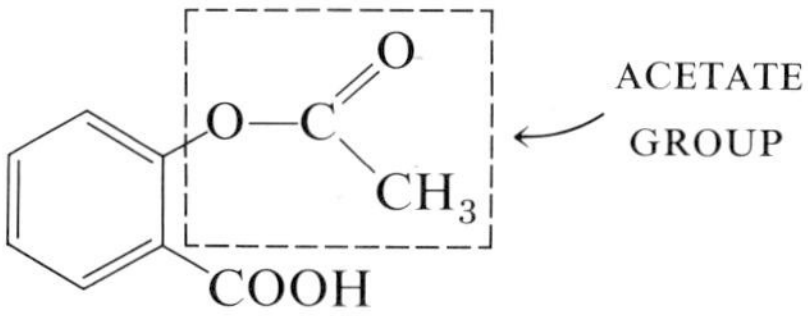

Its chemical name is acetylsalicylic acid. Aspirin is the most widely used pharmaceutical drug in the world. Over 20 million pounds of aspirin, or

Table 19-3

Structures and uses of some benzene derivatives

Structure	Name and use
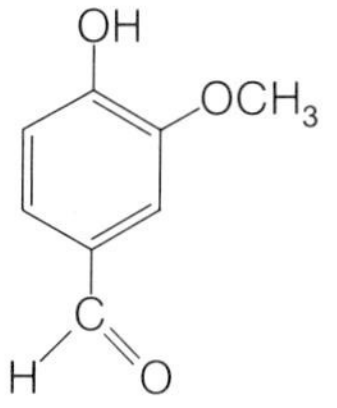	vanillin (vanilla) flavoring material
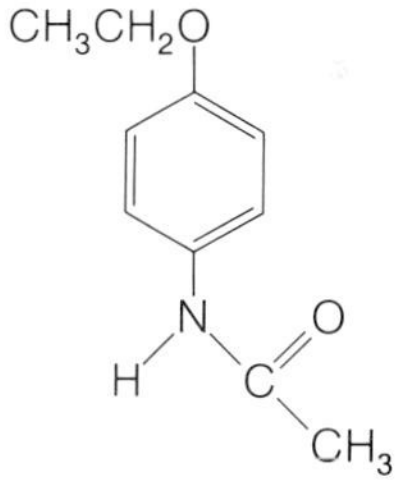	phenacetin pain reliever (in headache remedies)
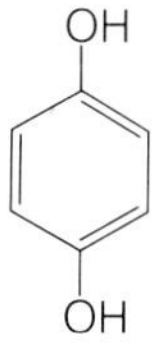	hydroquinone photographic developer
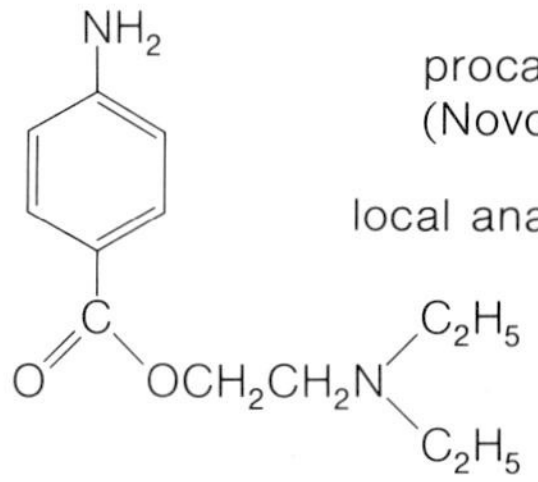	procaine (Novocain) local anaesthetic
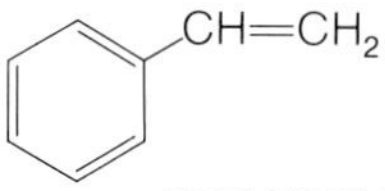	styrene monomer for preparation of polystyrene polymers

about 150 5-grain tablets for every person in the United States, are manufactured each year in the United States alone. Table 19-3 shows the structure of a few simple benzene derivatives that are important commercially. All are prepared from simpler compounds.

19.5 Chemistry of functional groups

We will now summarize some of the chemistry of functional groups. In most cases, the organic hydrocarbon unit is not changed during reaction. The functional groups give molecules their characteristic behavior.

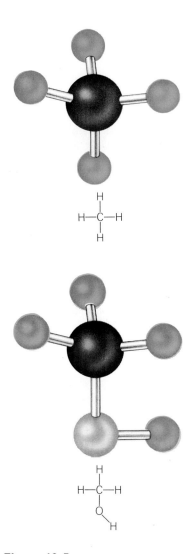

OXIDATION OF METHANOL TO FORMALDEHYDE AND FORMIC ACID: FORMATION OF ALDEHYDES AND ORGANIC ACIDS

We noted earlier that an H atom of methane can be replaced by an —OH radical to produce methanol. (See Figure 19-5.) By making such a substitution, we incorporate oxygen in the molecule. The production of methanol can be regarded as the first step in the complete oxidation of methane to carbon dioxide and water. Oxidation of methanol represents the second step in that oxidation process.

Methanol and other alcohols react with common inorganic oxidizing agents, such as potassium dichromate ($K_2Cr_2O_7$). When an acidic, aqueous solution of potassium dichromate is mixed with methanol, the solution changes from bright orange to muddy green. The green color is that of Cr^{+++} ions. Furthermore, the solution has a strong odor, which is easily identified as that of formaldehyde. The process can be represented by the equation

$$3CH_3OH \quad + \quad Cr_2O_7^{--}(aq) \quad + \quad 8H^+(aq) \longrightarrow$$

METHYL   DICHROMATE
ALCOHOL   ION (ORANGE)

$$2Cr^{+++}(aq) \quad + \quad 7H_2O \quad + \quad 3CH_2O$$

CHROMIC ION   FORMALDEHYDE
(GREEN)

The conversion of $Cr_2O_7^{--}$ to Cr^{+++} strongly suggests that the alcohol was oxidized during the process to form an **aldehyde.** An aldehyde is a compound prepared from an alcohol by oxidation. It has one or more

$$-C\overset{\displaystyle O}{\underset{\displaystyle H}{\big\|}}$$

groups attached to the hydrocarbon group. The structural formula of formaldehyde (CH_2O), the simplest aldehyde, is shown in Figure 19-6.

If methanol is mixed with an aqueous solution of potassium permanganate, the purple color of the latter fades and a product known as formic acid is obtained. The structural formula of this organic acid is shown in Figure 19-7. The balanced equation for the permanganate-methanol reaction is

$$5CH_3OH + 4MnO_4^-(aq) + 12H^+(aq) \longrightarrow$$
$$5HCOOH + 4Mn^{++} + 11H_2O$$

Figure 19-5

At top, the structural formula and model of methane. At bottom, the structural formula and model of methanol.

It is difficult to oxidize formic acid or acetic acid to CO_2 and water in solution. However, burning either formic acid or acetic acid in (see Figure 19-8 on page 544) in oxygen will result in complete oxidation to CO_2 and H_2O.

EXERCISE 19-4

> Propanol has the structural formula $CH_3CH_2CH_2OH$. If it is oxidized carefully, an aldehyde called propionaldehyde is obtained. Vigorous oxidation produces an acid called propionic acid. Draw structural formulas for propionaldehyde and propionic acid.

OXIDATION OF A SECONDARY ALCOHOL: FORMATION OF KETONES

The bonding rules permit us to draw two acceptable formulas for an alcohol that contains three carbon atoms. The formulas are

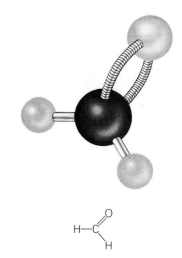

Figure 19-6
The structural formula and model of formaldehyde.

$$
\begin{array}{ccc}
\text{H} & \text{H} & \text{H} \\
| & | & | \\
\text{H}-\text{C}-\text{C}-\text{C}-\text{OH} \\
| & | & | \\
\text{H} & \text{H} & \text{H}
\end{array}
\quad \text{and} \quad
\begin{array}{ccc}
& \text{H} & \\
& | & \\
\text{H} & \text{O} & \text{H} \\
| & | & | \\
\text{H}-\text{C}-\text{C}-\text{C}-\text{H} \\
| & | & | \\
\text{H} & \text{H} & \text{H}
\end{array}
$$

NORMAL-PROPYL ALCOHOL   ISOPROPYL ALCOHOL
(PRIMARY ALCOHOL)   (SECONDARY ALCOHOL)

Note that in the normal alcohol, the —OH group is fastened to an end carbon. Such an alcohol is called a *primary* alcohol. In the isopropyl alcohol, the —OH group is fastened to the middle carbon. If the —OH group is fastened to a

$$
\begin{array}{c}
R \\
| \\
-\text{C}-\text{H} \\
| \\
R
\end{array}
$$

group, the substance is a *secondary* alcohol. Compare Figures 19-9a and 19-9b on page 544. In the alternative naming scheme for the molecules, both would be called propanol, but we would number the carbons in the chain from the end nearest the —OH functional group, and indicate by number the identity of the carbon atom to which the —OH was attached. Thus, *n*-propyl alcohol would also be called 1-propanol. Isopropyl alcohol would be called 2-propanol. An extrapolation of our arguments for the oxidation of the primary alcohols CH_3OH and C_2H_5OH shows that the oxidation of 1-propanol should yield an aldehyde called propionaldehyde.

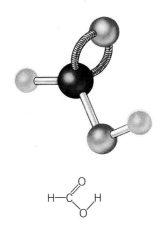

Figure 19-7
The structural formula and model of formic acid.

$$
\begin{array}{ccc}
\text{H} & \text{H} & \text{H} \\
| & | & | \\
\text{H}-\text{C}-\text{C}-\text{C}=\text{O} \\
| & | & \\
\text{H} & \text{H} &
\end{array}
$$

PROPIONALDEHYDE

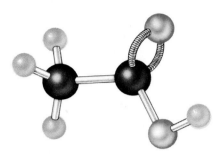

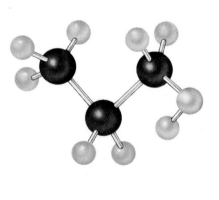

Figure 19-8

The structural formula and model of acetic acid.

The oxidation of 2-propanol, a secondary alcohol, yields a different product:

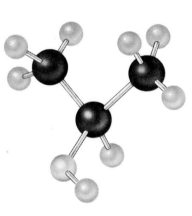

ACETONE, or
DIMETHYL KETONE

The compound obtained is called acetone. Acetone is used frequently as a solvent in the laboratory and in industry. It is a member of a family called **ketones.** Ketones differ from aldehydes in that the former have alkyl groups such as methyl, ethyl, and so on, attached to the

$$-\text{C}\overset{O}{\diagup}-$$

carbon, while an aldehyde has a hydrogen atom in place of one of the alkyls. Compare Figures 19-6 and 19-10. The steps in the oxidation of alcohols are summarized in Figure 19-11. Huge amounts of aldehydes and ketones are used industrially in a variety of chemical processes.

GENERAL FORMULAS OF FUNCTIONAL GROUPS

As our discussion has shown, the chemistry of organic molecules is determined by the functional groups present. In fact, organic molecules are frequently represented by a functional group bonded to the symbol $R-$, where $R-$ represents any hydrocarbon group.

Figure 19-9

(a) The structural formula and model of 1-propanol. (b) The structural formula and model of 2-propanol.

(a)

(b)

The general representation of an alcohol is R—OH. R— can be a methyl group, an ethyl group, a propyl group, or any other hydrocarbon group. Similarly, the general formula for representing an aldehyde is as follows:

$$R-C\overset{\displaystyle O}{\underset{\displaystyle H}{\Big\langle}}$$

The >C=O group is called a **carbonyl group.** The general formula for a ketone is

$$R-C\overset{\displaystyle O}{\underset{\displaystyle R'}{\Big\langle}}$$

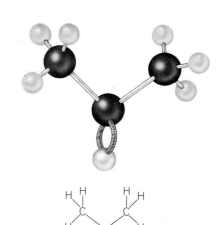

Figure 19-10

The structural formula and model of acetone, the simplest ketone.

Figure 19-11

The successive steps in the oxidation of alcohol to carbon dioxide and water.

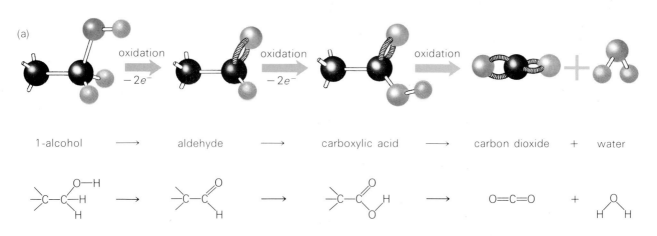

(a)

oxidation $-2e^-$ oxidation $-2e^-$ oxidation

1-alcohol $\longrightarrow$ aldehyde $\longrightarrow$ carboxylic acid $\longrightarrow$ carbon dioxide + water

$O=C=O$

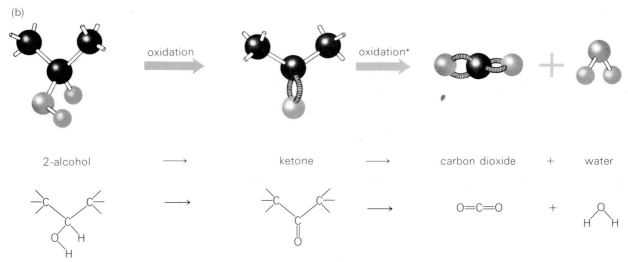

(b)

oxidation   oxidation*

2-alcohol $\longrightarrow$ ketone $\longrightarrow$ carbon dioxide + water

$O=C=O$

*Oxidation of an aldehyde is much easier than the oxidation of a ketone since in the latter case a C—C bond must be broken. Once this process is initiated, it is easy for the reaction to convert all C and H in the molecule to CO_2 and H_2O.

The R— and R'— groups may be the same or they may be different. Note that a ketone also contains a carbonyl group ($\diagup$C=O). The general formula for an organic, or carboxylic, acid is

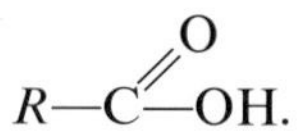

$$R-\underset{}{\overset{\displaystyle O}{C}}-OH.$$

The —COOH group is called a **carboxyl group.** A number of other functional groups and organic compounds are worthy of our attention.

ETHERS

An **ether** is a substance with the general formula R—O—R'. In Section 19.3, we discussed dimethyl ether as an isomer of ethyl alcohol. Because ethers do not have an O—H linkage, they cannot form hydrogen bonds. As a result, they have a higher vapor pressure than the comparable alcohol. They also burn very readily because of the R— groups in the molecule. Diethyl ether was one of the first anaesthetics used in medicine.

Ethers can be made by splitting a water molecule from two alcohol molecules, as shown here:

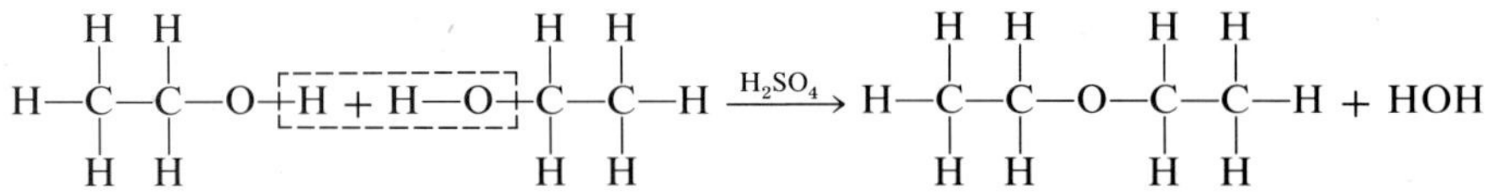

EXERCISE 19-5

Write the structural formula of methylpropyl ether. How does its composition compare with that of diethyl ether?

AMINES

Amines can be prepared by the direct reaction between ammonia and an alkyl halide, such as CH_3CH_2I. A *primary* amine results if *one* proton of ammonia (NH_3) is replaced by an R— group:

$$\underset{\text{ETHYL IODIDE}}{CH_3CH_2I} \quad + \quad 2NH_3 \quad \longrightarrow \quad \underset{\substack{\text{ETHYLAMINE}\\\text{(PRIMARY AMINE)}}}{CH_3CH_2NH_2} \quad + \quad NH_4I$$

$$R-I + 2NH_3 \longrightarrow R-NH_2 + NH_4I$$

The above equations represent the net change that occurs when an excess of ammonia reacts with an alkyl iodide.

Table 19-4

Regularities in the naming of alkanes, alcohols, and amines

Number of carbon atoms	Alkanes	Alcohols	Amines
1	CH_4 methane	CH_3OH methyl alcohol methanol	CH_3NH_2 methylamine
2	CH_3CH_3 ethane	CH_3CH_2OH ethyl alcohol ethanol	$CH_3CH_2NH_2$ ethylamine
3	$CH_3CH_2CH_3$ propane	$CH_3CH_2CH_2OH$ propyl alcohol 1-propanol	$CH_3CH_2CH_2NH_2$ 1-propylamine
4	$CH_3CH_2CH_2CH_3$ butane	$CH_3CH_2CH_2CH_2OH$ butyl alcohol 1-butanol	$CH_3CH_2CH_2CH_2NH_2$ 1-butylamine
8	$CH_3(CH_2)_6CH_3$ octane	$CH_3(CH_2)_6CH_2OH$ octyl alcohol 1-octanol	$CH_3(CH_2)_6CH_2NH_2$ 1-octylamine

A *secondary* amine results if *two* protons of ammonia are replaced by two
R— groups:

$$2C_2H_5 - NH_2 + C_2H_5I \longrightarrow \underset{H}{\overset{C_2H_5}{H_5C_2-N}} + C_2H_5NH_3^+,I^-$$

ETHYL AMINE   ETHYL IODIDE   DIETHYL AMINE

Regularities in the naming of amines, alcohols, and alkanes are shown in
Table 19-4.

ACID DERIVATIVES: ESTERS

In the discussion summarizing functional groups, we noted that a carboxyl (—COOH) group characterizes an organic acid. The hydrogen ion of
the —OH group undergoes mild ionization to make most organic acids
relatively weak in water solution. For example, when dissolved in water,
acetic acid produces a solution that conducts electricity and turns blue litmus paper red. The reaction

$$H_3C-\overset{O}{\underset{OH}{C}} + H_2O \longrightarrow H_3C-\overset{O}{\underset{O^-(aq)}{C}} + H_3O^+(aq).$$

has an equilibrium constant of 1.8×10^{-5}. Thus, acetic acid has a K_A of 1.8×10^{-5}. (See pages 465–471 for a discussion of K_A values.)

In addition to such acidic behavior, another important characteristic of carboxylic acids is that the entire —OH group can be replaced by other groups. The resulting compounds are called *acid derivatives*. We shall consider only two types of acid derivatives: *esters* and *amides*.

Compounds in which the —OH of an acid is transformed into —OR, such as —OCH$_3$, are called **esters**. They can be prepared by the direct reaction between an alcohol and the acid:

$$H_3C-C\underset{OH}{\overset{O}{\Big\langle}} + CH_3OH \longrightarrow H_3C-C\underset{O-CH_3}{\overset{O}{\Big\langle}} + H_2O$$

| ACETIC ACID | + | METHYL ALCOHOL | $\longrightarrow$ | METHYL ACETATE | + WATER |

$$R-C\underset{OH}{\overset{O}{\Big\langle}} + R-OH \longrightarrow R-C\underset{O-R}{\overset{O}{\Big\langle}} + H_2O$$

The reaction between methanol and acetic acid is slow, but it can be greatly accelerated if a catalyst is added. A strong acid, such as HCl or H_2SO_4, is an effective catalyst. Ester names are shown in Table 19-5.

Esters are important substances. The esters of the low-molecular-weight acids and alcohols have fragrant, fruitlike odors and are used in perfumes

Table 19-5

Regularities in the naming of acids, amides, and esters

Number of carbon atoms	Acids	Amides	Esters (acids with methanol)
1	HCOOH formic acid	HCONH$_2$ formamide	HCOOCH$_3$ methyl formate
2	CH$_3$COOH acetic acid	CH$_3$CONH$_2$ acetamide	CH$_3$COOCH$_3$ methyl acetate
3	CH$_3$CH$_2$COOH propionic acid	CH$_3$CH$_2$CONH$_2$ propionamide	CH$_3$CH$_2$COOCH$_3$ methyl propionate
4	CH$_3$CH$_2$CH$_2$COOH butyric acid	CH$_3$CH$_2$CH$_2$CONH$_2$ butyramide	CH$_3$CH$_2$CH$_2$COOCH$_3$ methyl butyrate
8	CH$_3$(CH$_2$)$_6$COOH octanoic acid caprylic acid	CH$_3$(CH$_2$)$_6$CONH$_2$ octanamide caprylamide	CH$_3$(CH$_2$)$_6$COOCH$_3$ methyl octanoate methyl caprylate

and artificial flavorings. Esters are useful solvents. That is why they are commonly found in the paint and adhesive industries and in fingernail polish remover.

EXERCISE 19-6

Write equations for the reactions between (a) ethanol and formic acid, (b) propanol and propionic acid, and (c) methanol and formic acid. Name the esters produced in each case.

ACID DERIVATIVES: AMIDES

A compound in which the —OH group of an acid is replaced by —NH_2 is called an **amide.** When the —OH is replaced by —NHR, the product is called a nitrogen-substituted amide. Regularities in the naming of amides, esters, and acids are shown in Table 19-5.

Amides can be produced from the reaction between ammonia or an amine and an ester:

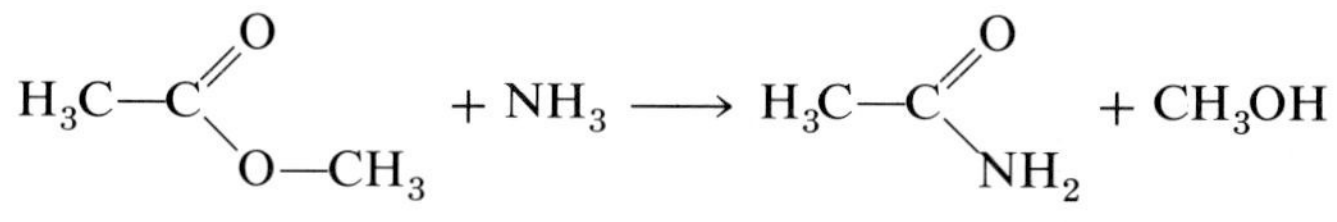

$$H_3C-C\overset{O}{\underset{O-CH_3}{\big|}} + NH_3 \longrightarrow H_3C-C\overset{O}{\underset{NH_2}{\big|}} + CH_3OH$$

METHYL ACETATE   ACETAMIDE

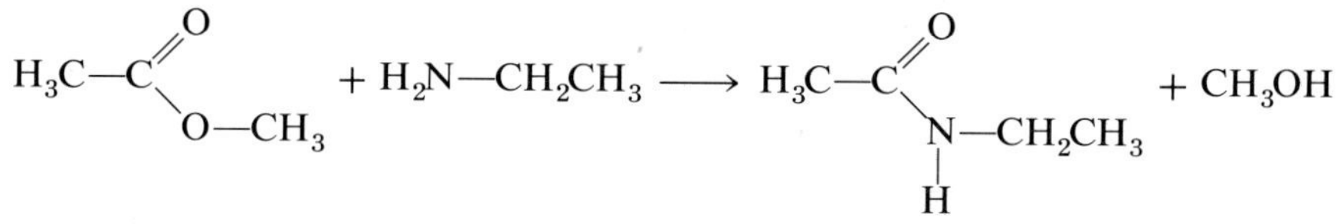

$$H_3C-C\overset{O}{\underset{O-CH_3}{\big|}} + H_2N-CH_2CH_3 \longrightarrow H_3C-C\overset{O}{\underset{\underset{H}{N}-CH_2CH_3}{\big|}} + CH_3OH$$

METHYL ACETATE   ETHYLAMINE   *N*-ETHYL ACETAMIDE

The name *N*-ethyl acetamide indicates that the ethyl group is bonded to the nitrogen. Note the similarity of the two reactions. Amides are important because the amide grouping is the fundamental linkage found in proteins such as enzymes. It is sometimes called the peptide linkage. It links R— groups together.

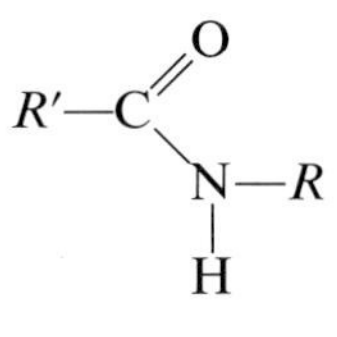

$$R'-C\overset{O}{\underset{\underset{H}{N}-R}{\big|}}$$

Table 19-6 on page 550 summarizes the classes of organic compounds discussed in this chapter. The table lists the structural representations for each functional group as well as the general formula and first member of each group.

Table 19-6

Some classes of organic compounds

Class	Functional group	General formula	First member
alcohol	—OH	R—OH	H_3C—OH
aldehyde	$-\overset{\displaystyle}{\underset{\displaystyle \parallel O}{C}}-H$	$R-\overset{\displaystyle}{\underset{\displaystyle \parallel O}{C}}-H$	$H-\overset{\displaystyle}{\underset{\displaystyle \parallel O}{C}}-H$
carboxylic acid	$-\overset{\displaystyle}{\underset{\displaystyle \parallel O}{C}}-OH$	$R-\overset{\displaystyle}{\underset{\displaystyle \parallel O}{C}}-OH$	$H-\overset{\displaystyle}{\underset{\displaystyle \parallel O}{C}}-OH$
ketone	$-\overset{\displaystyle}{\underset{\displaystyle \parallel O}{C}}-$	$R-\overset{\displaystyle}{\underset{\displaystyle \parallel O}{C}}-R'$	$H_3C-\overset{\displaystyle}{\underset{\displaystyle \parallel O}{C}}-CH_3$
ether	—O—	R—O—R'	H_3C—O—CH_3
amine (primary)	$-NH_2$	$R-NH_2$	H_3C-NH_2
amine (secondary)	—NH—	R—NH—R'	$(CH_3)_2NH$
ester	$-\overset{\displaystyle}{\underset{\displaystyle \parallel O}{C}}-O-$	$R-\overset{\displaystyle}{\underset{\displaystyle \parallel O}{C}}-O-R'$	$H-\overset{\displaystyle}{\underset{\displaystyle \parallel O}{C}}-O-CH_3$
amide	$-\overset{\displaystyle O}{\overset{\displaystyle \parallel}{C}}-NH_2$	$R-\overset{\displaystyle O}{\overset{\displaystyle \parallel}{C}}-NH_2$	$H_3C-\overset{\displaystyle O}{\overset{\displaystyle \parallel}{C}}\diagdown_{NH_2}$

19.6 Polymers

Desired physical properties in an organic material can be obtained by controlling the length of the chain and the types of functional groups. In fact, by adjusting the chain length and molecular geometry, chemists have produced many organic substances called *polymers*.

In order to produce extended chains of atoms, it is necessary to begin with relatively small molecules containing carbon chains of only a few atoms. These small **monomers** must be bonded together, link after link, until the desired range of molecular weight is reached. Often, the desired properties are obtained only with giant molecules, each containing hundreds or even thousands of monomers. The giant molecules are called **polymers,** and the process by which polymers are formed is called **polymerization.**

TYPES OF POLYMERIZATION

The two kinds of polymerization are **addition polymerization** and **condensation polymerization.** *Addition* polymers are formed by the reactions between monomeric units without the elimination of atoms. The monomer is usually an unsaturated organic compound such as ethylene (H_2C=CH_2), which in the presence of a suitable catalyst will undergo an addition reaction to form a long-chain molecule called polyethylene. A general equation for the first stage of such a process is

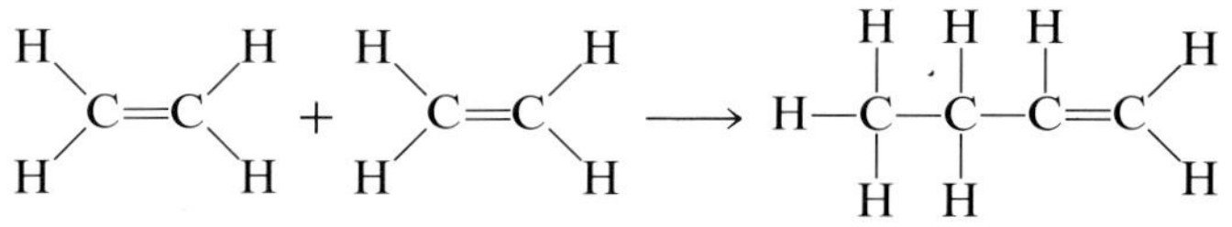

The same process continues, and the final product is the very familiar polymer polyethylene:

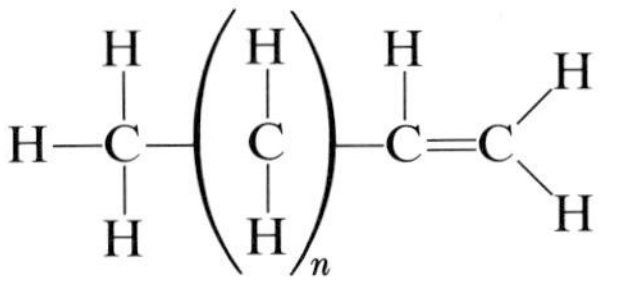

In the above formula, n is a very large number.

When one or more of the hydrogens are replaced by groups such as fluorine (—F), chlorine (—Cl), methyl (—CH_3), or methyl ester (—$COOCH_3$), polymers such as Teflon, Saran, Lucite, and Plexiglas result. It is thus possible to create molecules with custom-built properties for various uses. Table 19-7 lists some common addition polymers and their uses.

Table 19-7

Some common addition polymers

Monomer formula	Monomer name	Polymer name	Uses
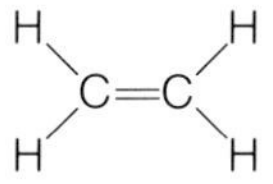 $CH_2{=}CH_2$	ethylene	polyethylene	films, coating for milk cartons, wire insulation, plastic bags, bottles, toys
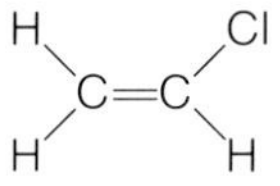 $CH_2{=}CHCl$	vinyl chloride	polyvinyl chloride (PVC)	raincoats, pipes, credit cards, phonograph records, bags, floor tiles, shower curtains, garden hoses, wire insulation, gutters and downspouts
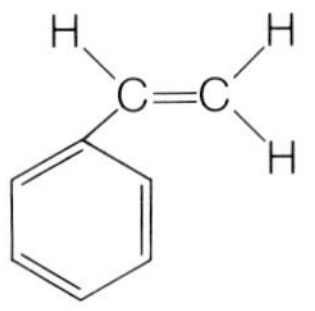 styrene ($CH_2{=}CH$—C_6H_5)	styrene	polystyrene	electrical insulation, packing material, combs
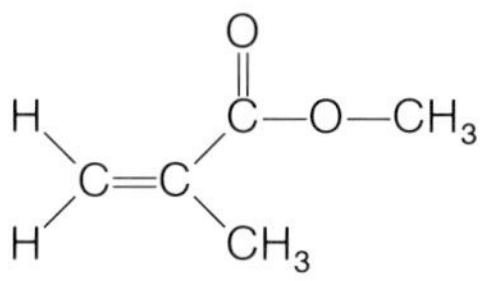 methyl methacrylate ($CH_2{=}C(CH_3)$—$C(O)$—O—CH_3)	methyl methacrylate	Plexiglas, Lucite	glass substitutes, paints
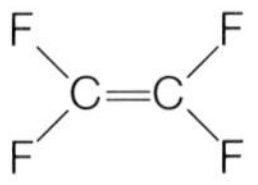 $CF_2{=}CF_2$	tetrafluoroethylene	Teflon	gaskets, bearings, insulation, nonstick pan coatings, chemical resistant films

Condensation polymers are produced by reactions in which some simple molecule, such as water, is eliminated between functional groups. For example, H_2O can be eliminated by combining a carboxylic acid with water:

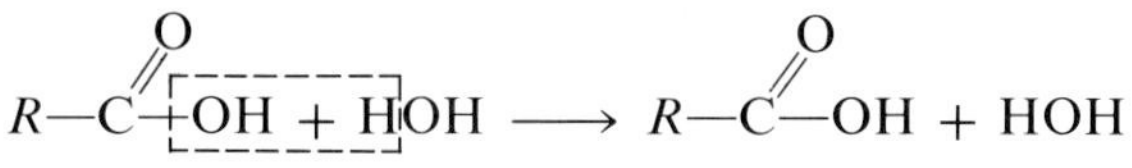

$$R-C\overset{O}{\|}-OH + HOH \longrightarrow R-C\overset{O}{\|}-OH + HOH$$

In order to form long-chain molecules, two or more functional groups must be present in each of the reacting units. For example, when ethylene glycol ($HOCH_2CH_2OH$) reacts with *para*phthalic acid,

$$HOOC-\langle\rangle-COOH$$

a **polyester** of high molecular weight is produced. This synthetic polyester fiber is called Dacron. The following diagram shows the first stages of this process:

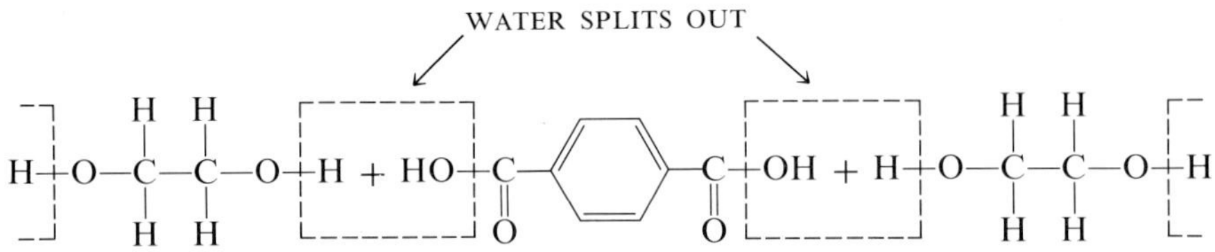

WATER SPLITS OUT

$$H-O-\overset{\overset{H}{|}}{\underset{\underset{H}{|}}{C}}-\overset{\overset{H}{|}}{\underset{\underset{H}{|}}{C}}-O-H + HO-C\underset{O}{\|}-\langle\rangle-C\underset{O}{\|}-OH + H-O-\overset{\overset{H}{|}}{\underset{\underset{H}{|}}{C}}-\overset{\overset{H}{|}}{\underset{\underset{H}{|}}{C}}-O-H$$

NYLON

Nylon, a polymeric amide, is a material widely used in plastics and fabrics. It consists of molecules of extremely high molecular weight made from adipic acid

$$HOOC-CH_2-CH_2-CH_2-CH_2-COOH$$

ADIPIC ACID

and 1,6-diaminohexane

$$H_2N-CH_2-CH_2-CH_2-CH_2-CH_2-CH_2-NH_2$$

1,6-DIAMINOHEXANE

These monomeric molecules can react repeatedly by removal of water. Amide linkages are formed at both ends:

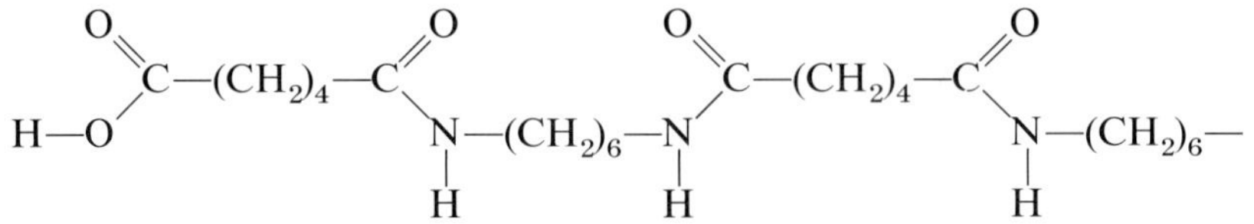

$$H-O\overset{O}{\underset{}{C}}-(CH_2)_4-C\underset{O}{\overset{\|}{}}\overset{}{N}-(CH_2)_6-\overset{}{N}\overset{O}{\underset{}{C}}-(CH_2)_4-C\underset{O}{\overset{\|}{}}\overset{}{N}-(CH_2)_6-$$

Nylon is somewhat similar to natural silk in its structure. Other polyamides can be made from different acids and amines. Polyamides exhibit a variety of properties suited to a variety of uses.

PROTEIN

A very important class of polymeric amides is *protein,* the essential structure of all living matter. Proteins are necessary to the human diet as a source of the "monomeric" units—the amino acids—from which living protein materials are made.

Proteins are polyamides formed by the polymerization, through amide linkages, of α-amino acids. Three of the 25 to 30 important natural α-amino

OPPORTUNITIES

Forensics

Forensic chemists are "police detectives" who work primarily in crime detection laboratories. These chemists analyze many kinds of substances in an effort to solve crimes. They may analyze body fluids (such as blood, perspiration, and saliva), paint, hair, soil, ink, explosives, burned material, paper, and anything else that might provide clues. Frequently, forensic chemists visit the scene of a crime to collect evidence.

A major challenge to forensic chemists is identifying drugs. For example, forensic chemists in the Drug Enforcement Administration recently analyzed an unknown packet of white powder found on a drug overdose victim. Using techniques such as gas chromatography and mass and nuclear magnetic resonance spectroscopy, the chemists were able to identify the molecular structure of the unknown substance. The powder, sold as "China White," was actually a painkilling drug called fentanyl with an additional methyl (CH_3) group added on.

Another important aspect of a forensic chemist's job is testifying in court. Forensic chemists called upon to testify must be able to describe their findings clearly so that nonscientists (such as judges, lawyers, and jury members) can understand the significance of the chemical evidence being presented.

Besides criminalistics (the field where most forensic chemists work), there are other branches of forensic science. These include jurisprudence, pathology-biology, psychiatry, odontology, toxicology, physical anthropology, and questioned documents.

Undergraduates interested in forensic chemistry should study as much analytical chemistry, especially instrumentation, as they can. They also should take courses in biochemistry, bacteriology, botany, biology, microscopy, constitutional law, statistics, and public speaking.

There are entry-level positions available in this field for chemists with a bachelor's degree, but graduate work is essential for advancement. Undergraduate degree holders with the appropriate courses receive on-the-job training for about 1 year. The rapid growth of knowledge in forensic chemistry and the importance of the work done in this area makes it vital that prospective forensic chemists stay thoroughly up-to-date in their knowledge of techniques in the field. Further information about this career can be obtained from the American Academy of Forensic Sciences, 11400 Rockville Pike, Suite 501, Rockville, Maryland 20852.

CHINA WHITE

acids are shown in Figure 19-12. Each acid has an amine group, $-NH_2$, attached to the α-carbon, the carbon atom immediately adjacent to the carboxylic acid group. The protein molecule may involve hundreds of such amino acid molecules, called *peptides,* connected through the amide linkages. A portion of this chain might be represented as in Figure 19-13.

An amide is decomposed by aqueous acids to form an acid and the ammonium salt of an amine:

$$H_3C-\overset{\displaystyle O}{\underset{\displaystyle NHCH_3}{C}} + H_3O^+(aq) \longrightarrow H_3C-\overset{\displaystyle O}{\underset{\displaystyle OH}{C}} + CH_3NH_3^+(aq)$$

N-METHYL ACETAMIDE		ACETIC ACID	METHYL AMMONIUM ION

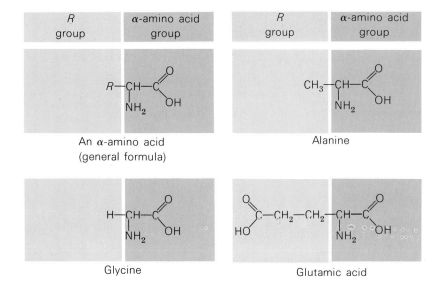

Figure 19-12

Molecular structures of α-amino acids.

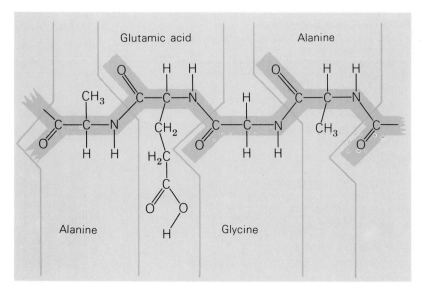

Figure 19-13

The structure of protein, showing the amide chain.

In the same type of reaction, a protein can be broken down into its constituent amino acids. In this way, most of what we call the natural amino acids have been discovered. Studies have been made of proteins from many sources—egg yolk, milk, animal tissues, plant seeds, gelatin, and so on. Such studies have resulted in the identification of about 30 natural amino acids. Consider the many different ways 30 amino acids can be combined in chains of 100 or more units. It is easy to see why there are so many known proteins and why the tissues of different species of plants and animals contain so many different proteins. The determination of amino acid sequences in proteins is a difficult and very important area of science.

Enzymes, the biological catalysts, are also proteins. Each enzyme has its own particular structure, determined by the order and spatial arrangement of the amino acids from which it is formed. Perhaps the most intriguing aspect of the chemistry of living organisms is their ability to synthesize just the right protein structures from the myriad structures possible and to reproduce organisms like themselves. This topic is considered in Chapter 20.

SUMMARY

In order to study the carbon compounds, we must establish their molecular structures. We learn their molecular structures by establishing their empirical, molecular, and structural formulas. The *structural formula* of a compound is a summary of a great deal of explicit information about the compound. From the structural formula, we can deduce many chemical and physical properties. The structural formula is obtained from many laboratory observations.

Hydrocarbons are compounds containing carbon and hydrogen. Straight-chain, *saturated hydrocarbons* have only single bonds in the molecule, and each carbon atom is bonded to four atoms such as carbon and hydrogen. *Unsaturated hydrocarbons* contain one or more multiple (double or triple) bonds. Cyclic hydrocarbons have the carbon atoms arranged in a ring. They may be saturated or unsaturated. Benzene is a cyclic unsaturated compound with the formula C_6H_6. Derivatives of benzene include phenol and acetylsalicylic acid (aspirin).

One or more hydrogen atoms on a hydrocarbon framework can be replaced with a *functional group.* A functional group is a reactive unit such as

$$-\text{OH}, \quad -\text{NH}_2, \quad -\overset{\displaystyle O}{\underset{}{C}}-, \quad \text{and} \quad -\overset{\displaystyle O}{\underset{}{C}}-\text{OH}$$

Much of the chemistry of organic compounds can be discussed in terms of functional groups. Common and important hydrocarbon radicals, or units, are methyl ($-CH_3$), ethyl ($-C_2H_5$), propyl ($-C_3H_7$), butyl ($-C_4H_9$), and phenyl ($-C_6H_5$). New compounds can be made by appropriate reactions carried out on functional groups. For example, oxidation of an alcohol produces an *aldehyde* or a *ketone.* Further oxidation of an aldehyde produces a *carboxylic acid.* The reaction between an alcohol and an organic acid produces an *ester.* Esters partially decompose in water to form the alcohols and

Key Terms

addition polymerization

addition reaction

alcohol

aldehyde

alkane

amide

aromatic compound

benzene ring

carbonyl group

carboxyl group

condensation
  polymerization

cyclic hydrocarbon

empirical formula

ester

ether

functional group

hydrocarbon

ketone

molecular formula

monomer

organic chemistry

organic compound

polyester

polymer

polymerization

saturated hydrocarbon

structural formula

substitution reaction

unsaturated hydrocarbon

acids from which they were synthesized. The reaction between an ester and an amine (R—NH$_2$) produces an *amide*. The amide group is the key structural link in protein molecules.

Very long chains of carbon atoms can be made from small units called *monomers*. When monomers are joined together, they form *polymers*. Various commercial plastics and proteins are examples of polymers.

QUESTIONS AND PROBLEMS

A

1. (a) Describe the process by which coal or petroleum forms. (b) It is frequently said that except for nuclear energy all of our energy comes from the sun. Explain why this is or is not so. (c) How does anthracite coal differ from peat? (d) How does a Middle East crude differ from a heavy domestic petroleum?
2. A compound was found to contain 24.0 g of carbon, 4.0 g of hydrogen, and 32.0 g of oxygen.
   (a) What is the empirical formula of the compound?
   (b) What is the weight of the empirical formula unit?
   (c) The molecular weight is measured to be 60. What is the actual formula of the compound?
   (d) When the compound is dissolved in water then treated with base, one molecule of the compound reacts with *only* one mole of base. Water is a product. Write a structural formula for the compound.
   (e) Could this substance form hydrogen bonds? Explain.
3. Write structural formulas for: (a) ethyl bromide, (b) methyl iodide, (c) cyclopentane, (d) ethy*lene*, (e) acetylene, (f) methyl acetate, (g) dimethyl ether.
4. Draw all of the structural formulas for the compounds with the formula C$_2$H$_3$Cl$_3$. Repeat for the formula C$_3$H$_4$Cl$_4$. What do you call two compounds with the same molecular formula, but with different structural formulas?
5. There are three different forms of the ring compound C$_6$H$_4$Cl$_2$. Draw the structural formula of each.
6. Write structural formulas for (a) *n*-propyl chloride, (b) iso-propyl alcohol, (c) acetone, (d) general formula for a ketone, (e) formaldehyde, (f) acetic acid, (g) propionic acid, (h) diethyl ether, (i) *n*-propyl acetate, (j) acetamide, (k) ethylamine.
7. (a) What is the product when an alcohol such as *n*-propyl alcohol is oxidized with potassium dichromate? (b) What is *n*-propionic acid? (c) What are the products when *n*-propionic acid is oxidized?
8. (a) Write the Lewis structure for ethylene, C$_2$H$_4$. (b) Write a Lewis formula for ethylene that would not include a double bond. (c) Show how polymerization to polyethylene would be expected from the formula in (b).

B

9. A 100 mg sample of a compound containing only C, H, and O was found by analysis to produce 149 mg of CO$_2$ and 45.5 mg of H$_2$O when burned completely in oxygen. Calculate its empirical formula.

10. A compound with the molecular formula $C_4H_{10}O$ reacts with sodium to liberate $\frac{1}{2}$ mole of H_2. (a) Write a structural formula that would be consistent with this information. (b) Write a structural formula for an isomer that would also be consistent with the data given. (c) Write a structural formula for an isomer of this compound that would not react with sodium.

11. A liquid is known to be either an aldehyde or a ketone. What simple test based on oxidation processes could be used to tell which it is? Explain.

12. A saturated fatty acid salt has the formula

$$NaO\overset{O}{\underset{}{\diagdown}}C-CH_2CH_2(CH_2)_nCH_3.$$

A "polyunsaturated" fatty acid salt (used in advertising about cooking oils) has a general formula

$$NaO\overset{O}{\underset{}{\diagdown}}C-CH_2\overset{H}{\underset{}{C}}{=}\overset{H}{\underset{}{C}}(CH_2)_n-\overset{H}{\underset{}{C}}{=}\overset{H}{\underset{}{C}}-CH_3.$$

Describe a simple test that would permit you to differentiate between the saturated and unsaturated acid.

13. How much methyl amine can be made from a 75.0 g sample of methyl iodide and an excess of NH_3? Assume 100 percent yield.

14. You have encountered each of the following names in your daily life. Indicate what the repeating unit characteristic of each polymer is: (a) polyethylene, (b) polyester, (c) nylon, (d) Teflon.

C

15. When 0.601 g of a sample having empirical formula CH_2O was vaporized at 200 °C and 1 atm pressure, the volume occupied was 388 ml. The same volume was occupied by 0.301 g of ethane under the same conditions. (a) What is the molecular formula of CH_2O? One mole of the sample, when allowed to react with zinc metal, liberated (rather slowly) 0.5 mole of hydrogen gas. (b) Write the structural formula.

16. When a worker bee stings, it emits a substance with molecular formula $C_7H_{14}O_2$. The substance attracts other bees to the site and causes them to sting too. The compound is a sweet-smelling ester. When it is treated with sodium hydroxide, sodium acetate and 2-methyl-butanol-1 are formed. What is the structure of the attractant?

17. An ester is formed by the reaction between an acid, $RCOOH$, and an alcohol, $R'OH$. Water is the other product. The reaction is carried out in an inert solvent. (a) Write the equilibrium relation among the concentrations, including the concentration of the product water. (b) Calculate the equilibrium concentration of the ester if $K = 10$ and the concentrations *at equilibrium* of the other constituents are: $[RCOOH] = 0.1\ M;$ $[R'OH] = 0.1\ M;\ [H_2O] = 1.0\ M.$
(c) Repeat the calculation of part (b) if the equilibrium concentrations are: $[RCOOH] = 0.3\ M;\ [R'OH] = 0.3\ M;\ [H_2O] = 1.0\ M.$

Thus, if the structure is helical, we find that the phosphate groups lie on a helix of a diameter about 20 Å, and the sugar and base groups must accordingly be turned inwards towards the helical axis.

ROSALIND FRANKLIN

SOME ASPECTS OF BIOCHEMISTRY

Objectives

After completing this chapter, you should be able to:

- Describe how the major classes of biochemicals—sugars, fats, and proteins—differ.
- Explain the mechanisms and properties of enzymes.
- Discuss the processes and substances through which genetic information passes from one generation to the next.
- Discuss some of the implications and techniques of gene splicing.

558

Biochemical mechanisms determine the replication of living things.

Since we ourselves are living organisms, the chemistry of living things is of special interest to us. Is the chemistry of life processes unique, or can we apply our knowledge of the chemistry of nonliving things to the chemistry of life?

In Chapter 19, we noted the great number and complexity of organic compounds. We also learned some of the techniques that enable chemists to study such compounds. The compounds that make up living things are among the most complex substances known. Nevertheless, the discovery of the structure of many proteins and the synthesis of a virus and complex biochemical substances from nonliving materials confirm our belief that a more complete understanding and control of biochemical reactions are within our ability. These accomplishments, together with the development of gene-splicing techniques, are yielding new weapons against diseases that afflict human beings, especially those diseases involving abnormal body chemistry.

One hundred and fifty years ago, the chemistry of living organisms was regarded as something distinct from the chemistry of rocks, minerals, and other nonliving things. But as our understanding of chemical principles increased, so did our knowledge about the chemistry of life. Compounds that had once been obtained only from plants and animals were produced in the laboratory from ordinary inorganic substances. As you read in Chapter 19, Friedrich Wöhler was the first to do this when he synthesized the organic compound urea, $(H_2N)_2CO$, by heating the inorganic salt ammonium cyanate, NH_4CNO. Today, most chemists believe it is possible to understand the chemistry of living organisms.

This belief received dramatic support in the fall of 1967 when L. A. Kornberg and M. Goulian of Stanford University reported the laboratory synthesis of a biologically active virus from nonliving compounds (see Section 20.3). Like the natural living virus, the synthetic virus could attack and destroy bacterial cells.* The work of Kornberg and Goulian suggests that biological chemistry is not fundamentally different from other chemistry.

Nevertheless, we do refer to a large area of chemical study as **biochemistry.** This is because a chemist working effectively in a *particular* area of science must devote special, but not exclusive, attention to what is known about that field of knowledge. Biochemists are primarily concerned with the chemical processes that occur in living organisms. Still, such scientists must use information from all branches of chemistry to answer questions such as the following:

1. What kind of molecules make up living systems?
2. What are the structures of biologically significant materials?
3. How is biologically significant information passed from parents to offspring?

20.1   Molecular composition of living systems

The chemical makeup of even the smallest plant or animal is extremely complex. We shall focus on those aspects of biochemistry that shed light on the applicability of general chemical principles to the study of biological processes. *All our knowledge of biochemistry has been obtained through use of the same basic ideas and experimental methods you have learned in this course.* Specifically, we shall now consider four classes of compounds that have great importance in biochemistry: sugars, fats, proteins, and cellulose.

SUGARS

The word *sugar* brings to mind the sweet, white, crystalline grains we find on any dinner table. The chemist calls this substance **sucrose** and knows it as just one of many "sugars" that are classed together because they have a related composition and undergo similar reactions. Sugars are part of the

*These experiments raise a fascinating philosophical question. Did Kornberg and Goulian synthesize life in a test tube? Certainly, the synthesized virus, produced from nonliving compounds by the action of a biological catalyst (an enzyme) and a template-pattern molecule, has all the characteristics of a natural virus, which is usually defined as a living body. The identity of the natural and synthetic virus was established by Robert Sinsheimer of the California Institute of Technology in 1967 (see Section 20.3).

larger family of **carbohydrates.** The name is derived from the fact that many such compounds have the empirical formula $C(H_2O)_n$, or hydrate of carbon $(C \cdot_n H_2O)$.

EXERCISE 20-1

Glucose, a sugar that is simpler than sucrose, has a molecular weight of 180 and the empirical formula CH_2O. What is the molecular formula of glucose?

The structure of the **glucose** molecule was deduced by a series of steps similar to those described in Chapter 19 for ethanol. Glucose was found to contain one aldehyde group,

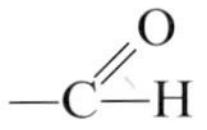

$$-\overset{\displaystyle O}{\underset{\displaystyle}{C}}-H$$

and five hydroxyl groups (—OH). These functional groups exhibit their typical chemistry. The aldehyde part can be oxidized to an acid group. If a mild oxidizing agent (such as the hypobromite ion in bromine water) is used, the aldehyde group can be oxidized without oxidizing the hydroxyl groups. If all the oxygen-containing groups are reduced, *n*-hexane,

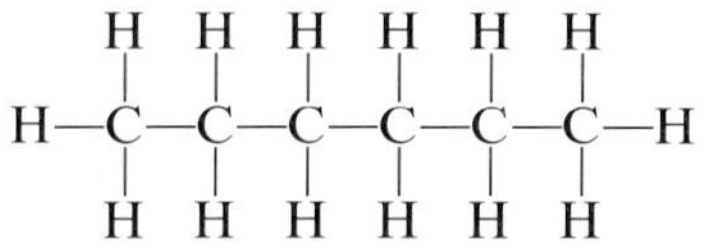

results. This test helps establish that the glucose molecule has a chain structure. But glucose, like other simple sugars, can exist as a straight chain in equilibrium with a cyclic structure. In solutions, the cyclic structure prevails.
  The following equation shows both forms of glucose:

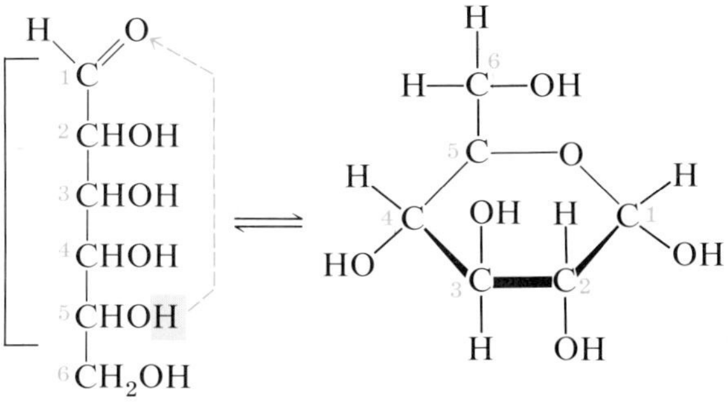

In forming a ring, the proton of the OH group on carbon 5 moves over to the aldehyde oxygen of carbon 1. Then a ring structure is formed through carbon 5, the O from the OH of carbon 5 and carbon 1.

The sugar **fructose** also occurs naturally and also has the molecular formula $C_6H_{12}O_6$. It is an isomer of glucose with the carbon of the $C{=}O$ group at the second position in the carbon chain instead of at the end. Such an arrangement makes fructose a ketone. Ketones were discussed in Section 19.5. The structure of sugar will be important in the discussion in Section 20.3.

EXERCISE 20-2

Draw a structural formula for the fructose molecule. (Remember that fructose is an isomer of glucose.)

The two sugars we have discussed are **monosaccharides**—they have a single, simple sugar unit in each molecule. The sugar on your table is a **disaccharide**—it has two units. Each molecule of sucrose contains one molecule of glucose hooked together with one molecule of fructose. A molecule of water is lost when the two simple sugars link. Fructose has a slightly different ring structure because the $>C{=}O$ group is not on the end carbon. The formation of sucrose is

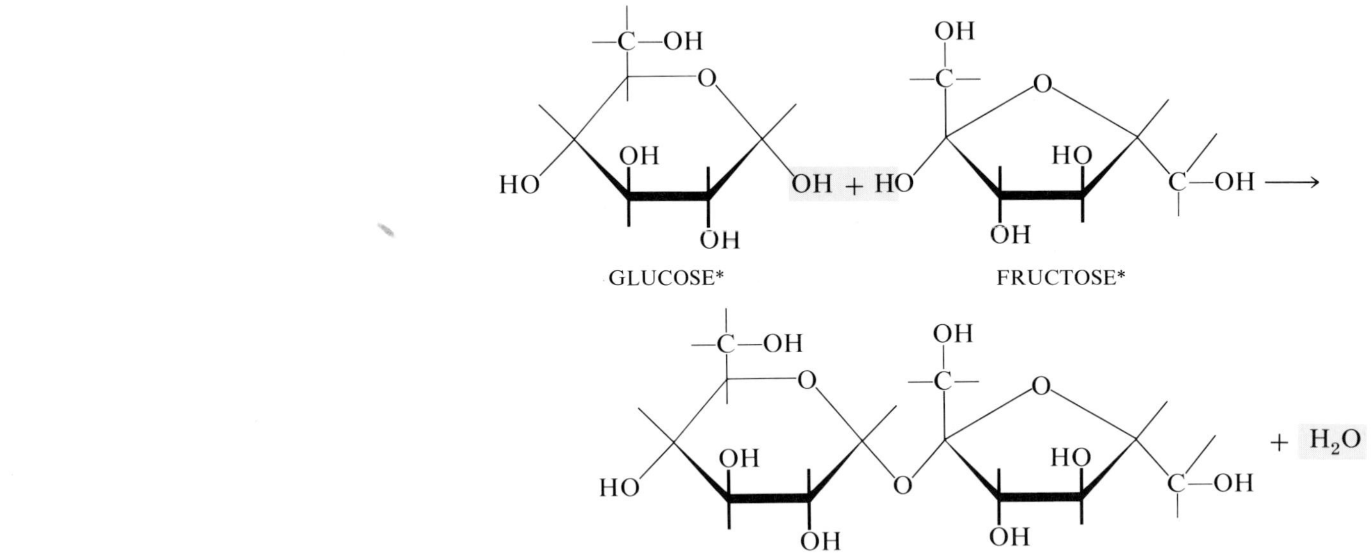

GLUCOSE*   FRUCTOSE*

SUCROSE*   + H_2O

Sugars occur in many plants. The major commercial sources are sugar cane—a large, specialized grass that stores sucrose in the stem—and sugar beet—in which as much as 15 percent of the root is sucrose. In addition, fruits, some vegetables, and honey contain sugars.

Sugars are fairly soluble in water. About 5 moles of sugar dissolve per litre of water, although solubility varies somewhat with the sugar. High solubility in water is readily explained because sugars have many functional groups that can form hydrogen bonds to the oxygen atoms of water.

*These structures are written in a simplified way. Carbon atoms at the corners of the ring and hydrogens on the bonds from carbon are not shown.

Aldehyde sugars are easily oxidized, as in the oxidation reaction that involves cupric hydroxide:

$$R-\overset{\displaystyle O}{\underset{\displaystyle H}{C}} + 2Cu(OH)_2 \longrightarrow R-\overset{\displaystyle O}{\underset{\displaystyle OH}{C}} + Cu_2O(s) + 2H_2O$$

The reaction is more complicated than the equation indicates. The $Cu(OH)_2$ is not very soluble in the basic solution used, so tartaric acid is added to form a complex ion. The $Cu_2O(s)$ is a red solid that precipitates from solution because the Cu^+ ion does not form such a complex ion. The reaction is characteristic and is used as a qualitative test for simple aldehydes and aldehyde sugars.

An important metabolic reaction of disaccharides is the reverse of sucrose formation: the *hydrolysis* of sucrose to produce glucose and fructose. Water in the presence of $H^+(aq)$ ions reacts with sucrose to produce glucose and fructose. The term hydrolysis means "reaction with water."

CELLULOSE AND STARCH

An important part of woody plants is **cellulose.** This substance occurs in cell walls and makes up part of the structural material of stems and trunks. Cotton and flax are almost pure cellulose. Chemically, cellulose is a **polysaccharide,** a *polymer* made by the joining (successive reaction) of many glucose molecules yielding a high-molecular-weight (approximately 600,000) structure. Cellulose is not basically different from the polymers discussed in Chapter 19:

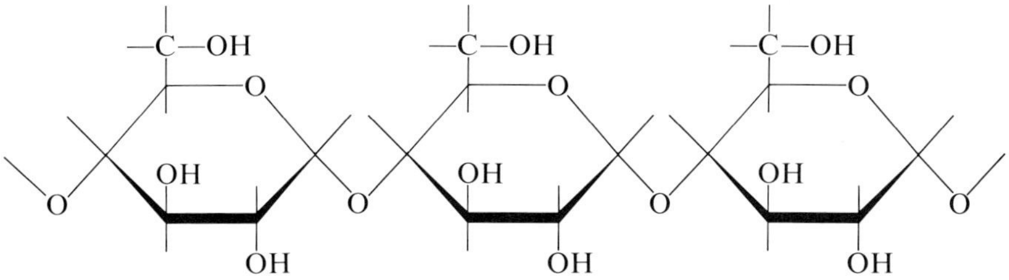

Starch is a mixture of glucose polymers, some of which are water-soluble. The soluble portion consists of comparatively short chains (molecular weight ~4,000). The portion of low solubility consists of much longer chains with frequent branching.

EXERCISE 20-3

The monomer unit in starch and cellulose has the empirical formula $C_6H_{10}O_5$. The units are about 5.0×10^{-10} metre long. Approximately how many units occur, and how long are the molecules of cellulose and soluble starch?

FATS AND OILS

Fats, as well as animal and plant oils, are esters. Actually, they are triple esters of glycerol (1,2,3-propanetriol, or $C_3H_8O_3$). When carboxylic acids similar to those discussed in Chapter 19 react with the —OH group of glycerol, a fat is formed:

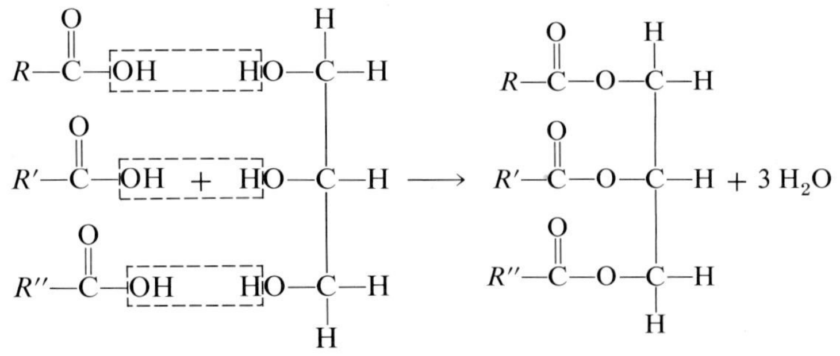

CARBOXYLIC ACIDS   GLYCEROL   A FAT   WATER

The R, R', and R'' represent groups of hydrocarbons. In natural fats, the acids usually have between 12 and 20 carbon atoms, C_{16} or C_{18} acids being most common.

EXERCISE 20-4

Write the formula of glycerol tributyrate, and then write the formula of the fat made from glycerol and one molecule each of stearic ($C_{17}H_{35}COOH$), palmitic ($C_{15}H_{31}COOH$), and myristic ($C_{13}H_{27}COOH$) acids. How many isomers are possible for the last fat?

Table 20-1

Some common fatty acids

Name and formula
lauric ($C_{11}H_{23}COOH$)
myristic ($C_{13}H_{27}COOH$)
palmitic ($C_{15}H_{31}COOH$)
stearic ($C_{17}H_{35}COOH$)
oleic ($C_{17}H_{33}COOH$)
linoleic ($C_{17}H_{31}COOH$)
linolenic ($C_{17}H_{29}COOH$)

Common fats, such as butter and tallow, and oils, such as olive, palm, and peanut, are mixed esters. Each molecule has three, two, or one kind of acid combined with a single glycerol. Three is most common, two less common, and one least common. There are so many combinations of esters in a given sample that fats and oils do not have sharp melting or boiling points. Melting and boiling occur over a range of temperatures. Table 20-1 shows formulas and names for some common acids used to make fats.

If the organic acid of a natural fat has a large number of double bonds (as do all those found in plants), the substance has a low melting point and is an oil. If hydrogen is added to the double bonds in the structure, the substance becomes a solid, such as margarine and the fats used in the kitchen for "shortening." The process is called *hydrogenation* of vegetable oils. It is achieved by adding H_2 to the oil in the presence of solid nickel or of other catalysts.

Fats make up as much as half the diet of many people. Fats are a good source of energy because when they are completely "burned" in the body, they supply twice as much energy per gram as do proteins and carbohydrates. As many people know, this is often a mixed blessing, particularly when body weight is a problem.

20.2   Molecular structures in biochemistry

The study of molecular structures of biologically active compounds has led to some of the most exciting research in recent years. In Chapter 19, we saw that the chemistry of compounds depends on structure. You have also learned how interactions between molecules can influence the properties of water and how attractions between ions and water can arrange the water molecules in preferred positions around an ion. We shall consider a few of these examples in the pages that follow.

PROTEINS

In Chapter 19, we looked at the composition of **proteins.** Proteins are large, amide-linked polymers of amino acids. However, the long-chain formula shown in Figure 19-13 does not represent all that is known about the structure of proteins. It properly shows the covalent structure, but does not indicate the detailed positions of the atoms in space.

The use of X-ray diffraction and the principles of hydrogen bonding have led us to recognize that the chain in natural proteins is coiled. Such a model is consistent with all available evidence and is generally accepted (see Figure 20-1). The form has a great deal of regularity. Thus, the order must have some energy factor sustaining it. In the protein molecule, the energy is provided by hydrogen bonds. These are represented in Figure 20-1 by dotted lines. When hydrogen bonds are broken by heating or by putting the protein in alcohol, the order disappears and the coiled form loses its shape. Often, the damage cannot be repaired and the coil is permanently deformed. Cooking an egg destroys the coiled form of the protein in the egg. If one compares the properties of a raw and cooked egg, it is not hard to see that detailed molecular structure is very important in biochemistry.

Proteins exist in numerous types and shapes. Current intensive research on a protein called *interferon* may produce a new weapon against virus infections and some cancers.*

ENZYMES

Most biochemical reactions, especially those in the human body, take place at about 37 °C (98.6 °F) and proceed at a rate adequate for the role they play. People live, grow, and reproduce under very mild conditions. Many reactions outside of living organisms will not proceed at a temperature as low as 37 °C. Glucose, starches, and fats are very stable compounds and can remain in contact with oxygen without apparent change. This is

*See Derek C. Burke, "The Status of Interferon," *Scientific American,* vol. 236 (April 1977) pp. 42–50. Also, Jean L. Marx, "Interferon (I): On the Threshold of Clinical Application," *Science* (June 15, 1979) pp. 1183–6; and "Interferon (II): Learning about How it Works," *Science* (June 22, 1979), pp. 1293–5.

Figure 20-1

The coiled form of a protein molecule.

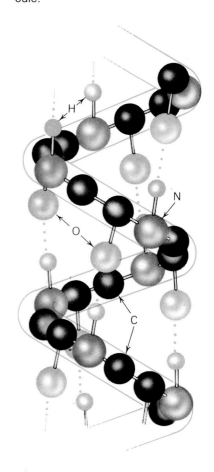

true even though their oxidation to carbon dioxide and water releases large amounts of energy. Catalysts make the reactions proceed by providing new paths with lower activation energies. Because of the catalysts, starches, sugars, and fats can be oxidized in the body at 37 °C to produce energy. Biological catalysts are called **enzymes.** Nearly every step in the breakdown of a complex molecule to a series of smaller ones is catalyzed by a specific enzyme within living cells. Conditions at equilibrium are not affected by enzymes. However, the rates at which reacting substances reach the equilibrium state are affected by enzymes, as with any catalysts.

Enzymes are protein molecules. Although all enzymes are proteins, not all proteins can act as enzymes. The protein molecules of enzymes are very large, with molecular weights on the order of 100,000. In comparison, the substance upon which the enzyme acts, the **substrate,** is made up of small molecules. During the reaction, a small substrate molecule becomes attached to the surface of a large protein molecule (the enzyme). It is on this surface that the reaction occurs. The products of the reaction then dissociate from the enzyme surface. A new substrate molecule is attached to the enzyme and the reaction is repeated. We can write the following sequence:

$$\text{enzyme} + \text{substrate} \longrightarrow \text{enzyme-substrate complex}$$
$$\text{enzyme-substrate complex} \longrightarrow \text{enzyme} + \text{reaction products}$$

Adding these equations and canceling, we have

$$\text{substrate} \longrightarrow \text{reaction products}$$

Despite the large size of an enzyme molecule, there is reason to believe that a reaction can occur at only one or a few places on its surface. These locations are usually referred to as **active centers.**

SPECIFICITY OF ENZYMES

Most enzymes are quite specific for a given substrate. For example, the enzyme urease that catalyzes the reaction

$$O{=}C\underset{\displaystyle NH_2}{\overset{\displaystyle NH_2}{\big<}} + H_2O \underset{}{\overset{\text{urease}}{\rightleftharpoons}} CO_2 + 2NH_3$$

UREA

is specific for urea. If we use urease to try to catalyze the reaction of a very similar molecule, no catalysis is observed. Such specificity suggests that on the surface of the enzyme, there is a special arrangement of atoms. These atoms belong to the amino acids from which the protein is constructed. The arrangement is suitable for attachment of the urea molecule, but not for attachment of other molecules.

Specificity is not always perfect. Sometimes, an enzyme works with any member of a *class* of compounds. Some esterases, enzymes that catalyze the

Table 20-2

Some common proteins

Protein	Number of amino acids	Molecular weight
insulin (hormone)	51	5,700
myoglobin (carries oxygen in muscles)	153	16,900
hemoglobin (transports oxygen in blood stream)	574	64,500
alcohol dehydrogenase (enzyme that metabolizes alcohol)	748	80,000
gamma globulin (antibody)	1,250	150,000
collagen (found in connective tissue, bone, and cartilage)	3,000	300,000

reaction of esters with water, work with numerous esters of similar structure. Usually, in cases of this kind, one of the members of the substrate class reacts faster than the others, so the rates vary from one substrate to another.

ENZYME INHIBITION BY A FALSE SUBSTRATE

Considerable evidence now exists to show that many useful drugs exert their beneficial effects by inhibiting enzyme activity in bacteria. Bacteria such as *staphylococcus* require the simple organic compound *para*-aminobenzoic acid for their growth. Such bacteria can grow and multiply in the human body because sufficient amounts of the compound occur in blood and in body tissue. The control of many diseases caused by staphylococcus and other bacteria was one of the first triumphs of **chemotherapy.*** The first compound found to be an effective drug of this type was sulfanilamide.

PARA-AMINOBENZOIC ACID   SULFANILAMIDE

It seems reasonable that an enzyme that uses the compound *para*-aminobenzoic acid as a substrate might be "deceived" by sulfanilamide. The two compounds are very similar in size and shape and in many chemical properties. To explain the success of sulfanilamide, chemists have proposed that the amide can form an enzyme-substrate complex that blocks the active centers normally occupied by the natural substrate: *para*-aminobenzoic acid.

Usually, fairly high concentrations of such a drug are needed to effectively control an infection because the inhibitor, or false substrate, should occupy as many active centers as possible. Also, the natural substrate probably has a greater affinity for the enzyme. Thus, a high concentration of the false substrate must be used so that the false substrate-enzyme complex will predominate. The bacteria, deprived of a normal metabolic process, cannot grow and multiply. The body's defense mechanisms can now take over and destroy the bacteria.

20.3   DNA and RNA: Genetic messengers

So far, we have discussed the *composition* of living systems, molecular changes in living systems, and certain *structural features* associated with living systems. However, we have not considered one of the most characteristic properties of a living body: its ability to reproduce an organism that is like itself, yet different enough to be a distinct individual.

---

* Chemotherapy is the control and treatment of disease by synthetic drugs. Most of the drugs are organic compounds, often of remarkably simple structure. Sulfanilamide is one example of an organic compound synthesized by chemists for the treatment of bacterial infections.

MARSHALL W. NIRENBERG

Marshall W. Nirenberg is a research biochemist who received his Ph.D. from the University of Michigan in 1957. He is now Chief of the Laboratory of Biochemical Genetics at the National Heart, Lung, and Blood Institute of the National Institutes of Health, in Bethesda, Maryland.

Nirenberg's research has centered on the chemical mechanism by which information carried in the DNA molecule is transferred and used in the synthesis of proteins. He developed a series of experiments to identify the combinations of nucleotides that determine the specific amino acids used in the formation of protein. Previous research by others had suggested that a three-nucleotide "code" existed for each amino acid. Nirenberg was able to "break the code" and identify the sequence that carries information for each of the 20 amino acids. For this work in explaining how genes determine the function of cells, Nirenberg was one of the recipients of the 1968 Nobel Prize for Physiology or Medicine.

Figure 20-2

The double helix of DNA: (a) coiled and (b) partially uncoiled. The double helical structure of DNA consists of two polynucleotide strands twisted around each other.

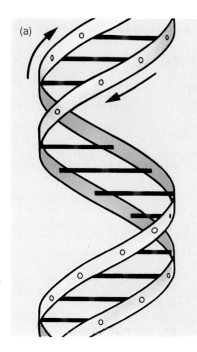
(a)

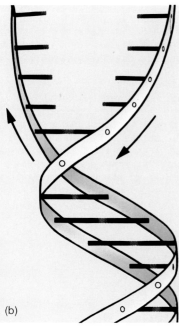
(b)

Growth begins when an egg cell from a female and a sperm cell from a male interact. The "nonliving" chemicals surrounding the now fertilized egg are *organized* and a new *living organism* begins to grow. The organism may be a mouse, an elephant, a red-headed human male, or a dark-haired human female. How are the "directions" for the synthesis of an elephant or a mouse or a boy or a girl carried in the egg and sperm cells? Where is the biological pattern that provides a means for organizing nonliving raw materials into a living organism? This is a profound and exciting question.

A good deal of progress has been made toward answering our question during the last 30 years. We have already mentioned the dramatic synthesis of a virus by Kornberg and Goulian. Let us see if we can use the ideas we have learned to understand what happened in the challenging synthesis of a "living" virus. First, we must learn something about the structure of the *nucleic acids,* called **DNA** (deoxyribonucleic acid) and **RNA** (ribonucleic acid).

DNA: EARLY HISTORY AND SIGNIFICANCE

DNA is the substance in all living cells that transmits hereditary characteristics from one generation to the next. It is the substance that dictates what an organism will be and what its individual cells will produce. It has been called the "blueprint" of life. The *gene,* a common term for units of heredity, consists of such DNA blueprints.* A virus, perhaps the simplest form of living thing, is actually an active DNA thread covered with a nonactive protective sheath, usually of protein.

Scientists approached the study of DNA as they would the study of any other interesting organic chemical—they set out to learn its structural formula. The same principles used to study simpler molecules were used to determine the structure of DNA. However, because of its tremendous polymeric size, the structure of the DNA molecule was much more complex and difficult to determine. Many scientists struggled with this problem. A major breakthrough was made in 1953 when J. D. Watson, Maurice Wilkins, and Francis Crick established with reasonable certainty the now accepted structure of DNA. In 1962, they received a Nobel prize for their work.

THE STRUCTURE OF DNA: NUCLEOTIDES

The most important structural features of DNA are two very long thin polymeric chains twisted around each other in the form of a regular *double helix* (see Figure 20-2). The chains are made up of **nucleotide** units fastened together. The nucleotide units are not very complex. Four different nucleotide units provide the basis for transmitting all necessary hereditary information. A nucleotide unit includes a phosphoric acid molecule to which a sugar molecule, deoxyribose, is linked by loss of water:

---

*DNA was first isolated in impure form in 1869 by the German biochemists Ernst Felix Hoppe-Seyler and Johann Friedrich Miescher. However, its role in heredity was not suggested until 1884 and evidence for that role was not found until 1943.

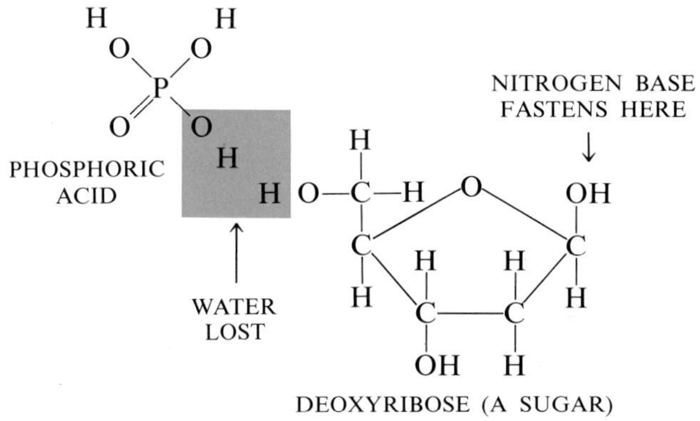

DEOXYRIBOSE (A SUGAR)

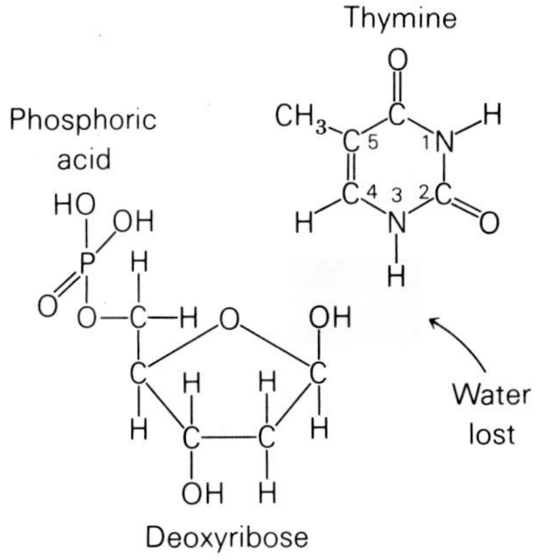

Figure 20-3

This figure shows the formation of a nucleotide of DNA. The complete unit represents one nucleotide.

One of four nitrogen bases is also linked by loss of water to the resulting structure. The base thymine can serve as an illustration (see Figure 20-3). Notice in Figure 20-4 that different nucleotides differ only in the identity of the base linked to the deoxyribose phosphate structure. The phosphate and sugar molecules in each nucleotide are *identical* to those found in every other nucleotide of DNA. The bases attached to these groups differentiate the four distinct nucleotides.

THE SINGLE HELIX

Nucleotides can also link together under the influence of a catalyst by losing one water molecule. A polymeric unit results (see Figure 20-5, on page 570). The resulting polymeric chain, formed by removing a water

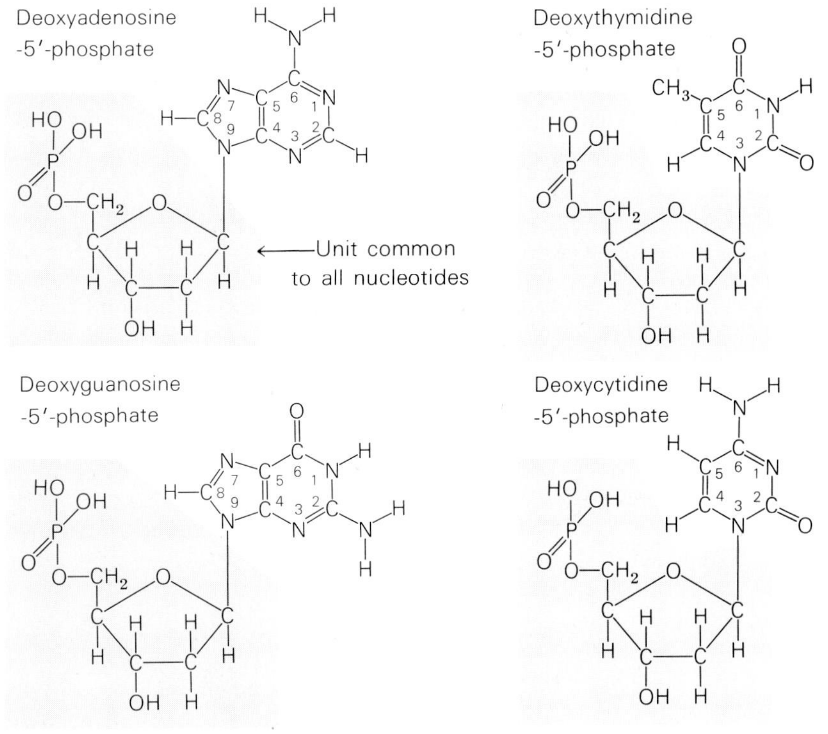

Figure 20-4

The four nucleotide building blocks of DNA.

Figure 20-5

Linkage of nucleotide units in a DNA chain.

molecule from between nucleotide units provides a backbone for the DNA. Notice that each chain is composed of sugar molecules linked together by phosphate groups in a very regular fashion. The backbone of the chain is seen as the shaded portion of Figure 20-6. Attached to each sugar molecule is one of the nitrogen-containing bases adenine, cytosine, guanine, or thymine. The linkage of these bases to the chain is also shown in Figure 20-6. Thymine (abbreviated T) and cytosine (C) are *single, six-membered ring structures containing two nitrogens per ring.* They belong to a chemical family called **pyrimidines.** Adenine (A) and guanine (G) are *double ring structures, each having one five- and one six-membered ring, with each ring containing two nitrogen atoms.* These larger bases belong to a family called **purines.**

The four base molecules identify the four nucleotides used as starting materials in DNA synthesis. The long polymeric chains formed by linking nucleotides assume a helical configuration with different nucleotides making up the chain. That is, different bases are attached to the "backbone." *Hereditary information is transmitted by the order in which the bases are attached to the backbone of deoxyribosephosphate.* The purine and pyrimidine bases are flat, relatively water-insoluble molecules that tend to stack above each other perpendicular to the direction of the helical axis of the backbone (see Figure 20-7, on page 574). The order of the purine and pyrimidine bases along the backbone chain varies greatly from one DNA molecule to another. The order of the bases carries a distinct message.

THE DOUBLE HELIX: TEMPLATE ACTION

Two backbone chains joined by hydrogen bonds between base pairs form a double helix. The convincing experimental data for this structure results from X-ray diffraction studies. An electron micrograph (Figure 20-7, on

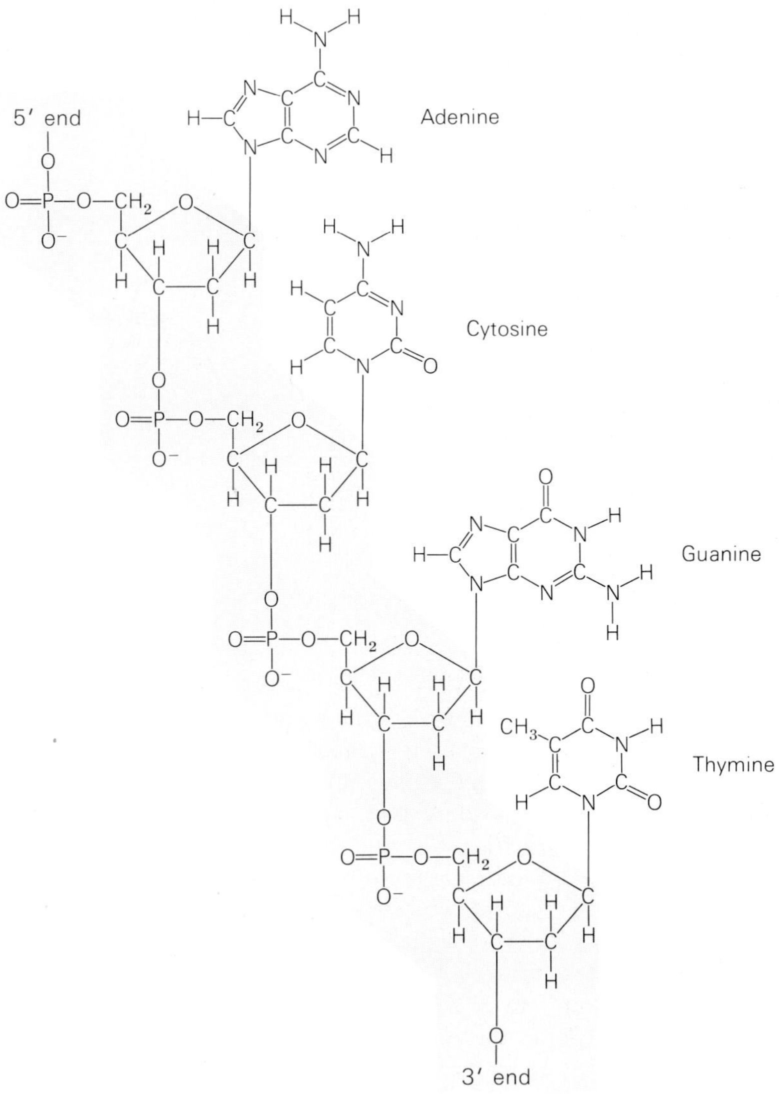

Figure 20-6

A section of the DNA chain.

page 574) provides a visual image that appears to support this idea. Because of the physical sizes of the bases, adenine is always bonded to thymine, and cytosine to guanine (see Figure 20-8, on page 574). In short, A-T linkages and C-G linkages are acceptable but A-C or T-G linkages do not form. Such pairing rules restrict the possible sequence of bases on two intertwined chains. If we have a sequence of ATGTC on one chain, the complementary chain must be TACAG. One polynucleotide chain is a **template,** or mold, for the synthesis of the other.

Genetic information is carried by the sequence of the four bases A, G, C, and T. You may think that having only four nitrogen bases on the DNA molecule limits the amount of information carried. However, the number of possible arrangements is 4^n, where n is the number of nucleotide units, or number of separate base units, in a molecule of DNA. Since each molecule

(continued on page 574)

Cosmetics

People throughout the world have used cosmetics for thousands of years. Today, chemists, physicists, biologists, and health-care scientists all contribute to the research and manufacture of cosmetics. Scientists must test cosmetic products for safety and reliability. Only nontoxic materials can be used in such products. Some of the substances—such as malachite and lead sulfide—used in cosmetics by early peoples are no longer used because of their toxicity.

Cosmetics that are used for skin and hair care usually protect the skin from the drying effects of detergents and the weather. Such products include moisturizers, cleansing creams, lotions, and suntan preparations. These products are basically *emulsions*. An emulsion is a combination of substances that normally do not mix together. Nonpolar molecules such as oils, and polar molecules such as water, are examples of some mutually incompatible substances. Because oils and water must be combined to form most skin and hair care products, the use of chemical agents that will aid in the combination of these substances is very important. Compounds that perform this function are called *emulsifiers*. Emulsifiers permit the combination of substances that do not normally mix together. Glycerol monostearate (a fat) and polyoxyethylene alkyl phenol (a non-ionic detergent) are two of the most commonly used emulsifiers.

$$CH_2OH$$
$$|$$
$$CHOH$$
$$|$$
$$CH_2OH(CH_2)_{16}CH_3$$

GLYCEROL MONOSTEARATE

$$R—\!\!\!\!\bigcirc\!\!\!\!—(OCH_2CH_2)_nOH$$

POLYOXYETHYLENE ALKYL PHENOL

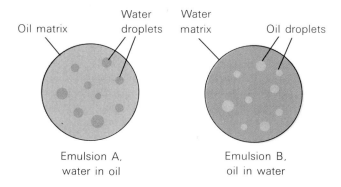

Emulsion A,
water in oil

Emulsion B,
oil in water

Emulsifiers with a low scale number produce emulsions of water in oil, such as A. Emulsifiers with a high scale number produce emulsions of oil in water, such as B.

Glycerol monostearate mixes well with oil and is called *lipophilic,* meaning oil loving. Because of the presence of a long chain that contains many polar oxygen atoms, polyoxyethylene alkyl phenol mixes well with water and is called *hydrophilic.*

To indicate the type of emulsion produced by the use of the two emulsifiers, chemists have developed a scale by which to rate them. The affinity of an emulsifier for an oil or water base is rated on a scale of 1 to 20. The lower the number, the more lipophilic the emulsifier—the higher the number, the more hydrophilic. Glycerol monostearate has a low number and polyoxyethylene alkyl phenol has a high number. Emulsifiers that have a rating between 3 and 6 produce emulsions of water in oil. Those emulsifiers with a rating between 7 and 17 produce emulsions of oil in water. The figure illustrates the differences between the two emulsions.

Cosmetics can be made with either emulsion, depending on the purpose they are to serve. Emulsion A, in which water forms the matrix, allows water to evaporate leaving a film of oil on the surface of the skin. The evaporation of water produces a cooling effect. Emulsion B permits the immediate contact of oil with the skin. This effect is called a warm emulsion because there is no immediate cooling effect due to the evaporation of water.

Other skin care products—moisturizers, cleansing creams, and lotions—replace water that has been lost from the skin. Such products also aid in softening and conditioning the skin. Cleansing creams dissolve away oily materials that hold dirt to the skin. Cold creams are examples of oil-in-water emulsions that are formulated for the quick penetration of the outer skin layers with an accompanying cooling effect.

Some types of cosmetics require the use of pigments. The pigments must not contain any harmful substances. Industrial chemists have been able to produce a variety of pigments, such as yellow, brown, red, and blue, that can be blended into many different shades. To obtain the lusters found in some cosmetics, small amounts of finely divided powders of aluminum, gold, or bronze are added. Usually, the pigments are added to bases that are composed of petroleum jelly, mineral oil, lanolin, or beeswax. These substances protect the surface of the skin.

Toothpastes and other dental-care products are used by all of us. As we are told by many television and radio commercials, toothpastes clean, whiten, and protect our teeth. Chemistry is involved in toothpaste formulation. Modern toothpastes are made by suspending a polishing agent and a variety of minor components such as fluoride, peppermint, and detergent in carriers such as a mixture of glycerol and sorbitol. The polishing agents used to give us "a smile of beauty" are mild abrasives that remove plaque and stains from our teeth. Such compounds include dicalcium phosphate dihydrate ($CaHPO_4 \cdot 2H_2O$), calcium pyrophosphate ($Ca_2P_2O_7$), and various forms of silica (SiO_2). (One form of silica is obtained from diatomaceous earth, which is actually the skeletal remains of very small organisms called diatoms.) As in any cleaning product, a detergent is useful. One such detergent, sodium lauryl sulfate, helps make the toothpaste foam when it is used. In some toothpastes, fluoride is added to minimize tooth decay.

Although suntan preparations have been in use for many years, manufacturers have produced a new product that has the ability to block out some of the harmful rays of the sun. Such a product is known as a sunscreen. By applying a sunscreen, people can protect themselves against overexposure

The use of sunscreens will help protect these people against overexposure to the sun's radiant energy.

to the sun. Overexposure can cause early aging or, in some cases, it may lead to skin cancer. The compounds responsible for protection from the sun are the *para*-aminobenzoates or *para*-aminobenzoic acid (PABA). They screen out the ultraviolet rays of the sun that have wavelengths between 290 and 320 nanometers. These particular wavelengths cause tanning and sunburn. They also aid in the production of vitamin D. The most useful sunscreen products are those that contain derivatives of *para*-aminobenzoic acid (PABA), such as glycerol *para*-aminobenzoate.

GLYCEROL *PARA*-AMINOBENZOATE

In addition to developing new products, the cosmetic industry spends a great deal of time testing the products to make sure that they are not harmful. The Food and Drug Administration also screens the products. Since 1977, cosmetics manufacturers have been required to list on their wrappers the ingredients of their products in order of decreasing concentration. Such lists inform consumers of the products' ingredients and allow the consumers to use them as guides for comparison with other products.

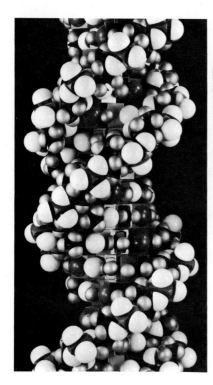

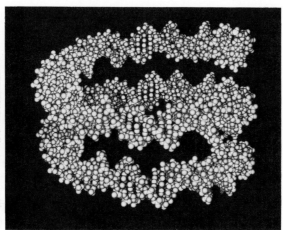

Figure 20-7

Compare a model of a section of a DNA molecule (at left) with a computer-generated model (above). The double helix in the computer model is wound into a coil and scientists believe that this is the form in which DNA is actually packed into chromosomes in the cell nucleus.

Figure 20-8

Hydrogen bonding to link bases of two helical DNA chains.

Thymine   Adenine

Cytosine   Guanine

is very long, many arrangements are possible. DNA molecules have an average molecular weight in excess of 10^6. That means there are at least 1,500 nucleotide units in the chain. Thus, *the possible number of different molecules of DNA is 4^{1500}, more than the number of different molecules of DNA that have existed in all the chromosomes present since the origin of life.* The **genetic code** consists of the sequence of bases on the backbone chain. How do the almost infinitely variable sequences pass on biological information? It is thought that the hydrogen bonds holding the DNA chains together are weak enough to break under certain conditions, permitting the chains to uncoil, as shown in Figure 20-2b on page 568. Each helix can then serve as a template for combining and organizing smaller nucleotide units into the complementary helix for that chain. One chain determines the order of new chains made from it. The nucleotide units are the raw material.*

THE SYNTHESIS OF A VIRUS

We are now in a position to understand the experiments of Kornberg and Goulian described at the beginning of this chapter. The scientists started with four essential, nonliving nucleotides. They then added DNA polymerase, a catalyst that *accelerates* the linking of nucleotide units by removing water. Next, they added some natural DNA from a virus to serve as a template. The result was a strand of DNA consisting of about 6,000 nucleotide units—a strand complementary to that of the original template. Using the complementary molecule, the process was repeated to produce a precise, but *synthetic,* duplicate of the original natural DNA molecule. In effect, the scientists produced the original virus. The result was a virus that could at-

*See the articles in *Scientific American,* Vol. 253 (October 1985). This issue is subtitled "The Molecules of Life."

tack bacteria and kill them as effectively as could the natural virus. Electron photomicrographs of natural and synthetic viral DNA appear at right.

Using the procedure described above, it was possible to assemble a DNA molecule from nonliving nucleotides, using a natural DNA molecule as a pattern. While this was a major accomplishment, it represented only the first step in the long process of unraveling the mysteries surrounding genetic information transfer. Making one DNA molecule like another is a much easier job than directing the synthesis of a complete mouse or of a complete elephant from foodstuffs and water. An intriguing question remains. How does DNA direct the synthesis of a creature 3 centimetres tall rather than one 5 metres tall? We do not know. Nevertheless, some facts and theories have been assembled. The key seems to be ribonucleic acid, or RNA.

RNA

RNA is very closely related to DNA, but it differs from DNA in two important ways. First, the sugar in the chain is ribose,

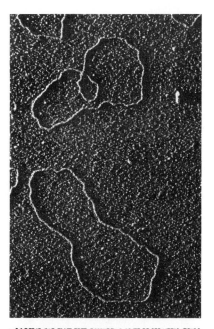

$$\text{RIBOSE} \qquad \text{DEOXYRIBOSE}$$

rather than deoxyribose. *Deoxy-* means "oxygen removed." Second, the thymine on the chain is replaced by uracil:

$$\text{THYMINE} \qquad \text{URACIL}$$

Most RNA exists, *not* as double-helical strands, but as non-hydrogen-bonded, single *polyribonucleotide* strands.

What is the function of RNA, and how does it work? According to molecular biologists, the process can be described in simple terms. DNA directs the synthesis of RNA, which in turn directs the synthesis of all proteins produced by the cell.* Protein structures are the key to life. There are as many as 100,000 different kinds of protein in a single human body, but only 20 amino acids are used in building these proteins. RNA "instructs" the cell as to the order in which to link the amino acids. The order determines the nature of the protein. No two different proteins have the same order of

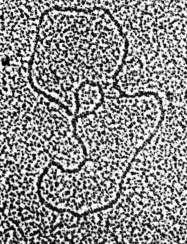

Figure 20-9
Compare natural viral DNA, at top, and synthetic viral DNA, at bottom.

---

*Recent experiments indicate that in certain cancer-producing viruses, genetic information flows "in reverse" from RNA to DNA. It has been hypothesized that the first carriers of genetic information were molecules of RNA. See Manfred Eigen et al., "The Origin of Genetic Information," *Scientific American,* vol. 224 (April 1981) pp. 88–118.

576

linked amino acids. Modern ideas on how RNA functions to direct protein synthesis can best be discussed in a biology course, where more information on biological structures is available. The field of molecular biology is complex and closely related to the fields of medicine, chemistry, and general biology. It is a rapidly developing area of study.

The modern approach to disease is closely related to structural biochemistry. For example, the ideas about template structure developed in an analysis of the operation of DNA and RNA have been used successfully against certain forms of cancer. By putting a fluorine in the 5 position on uracil,

URACIL   5-FLUORO-URACIL

OPPORTUNITIES

Patent law

Patent law, an important and demanding career, is closely related to chemistry. People working in patent law must have good communication skills, well-developed powers of persuasion, and an ability to get along well with others. Graduates who have those attributes as well as expertise in chemistry or chemical engineering will find a career in this area of law intellectually and financially rewarding.

Patent attorneys write patent applications for clients who have invented something they think is new and useful. Patents protect inventors from infringement by others who might use the invention without compensating the inventor for its use. Before writing a patent application, a patent attorney must carefully investigate the invention to see that it is truly original. If the attorney determines that the invention is worthy of being patented, an application is filed with the United States Patent Office. All applications are carefully studied by well-trained patent examiners. If an application is denied, the attorney representing the applicant

must decide whether it is worthwhile to appeal. Appeals are made before the Patent Office Board of Appeals.

Patent attorneys are also needed when two or more inventors claim to have invented the same thing. The Patent Office must resolve such claims. Attorneys file briefs, call witnesses, and present their cases before the Board of Interferences. The activities of patent attorneys before this board are similar to the activities of attorneys in courtrooms. If a patent is finally granted, a patent attorney may arrange for the licensing of the new invention. The financial rewards for the owner of a patent can be very high. Royalties from the use of an invention can amount to millions of dollars.

Patent attorneys are always in demand by companies involved in scientific work. A recent, dramatic case illustrates this point. In 1972, a scientist working for an electric company "built" a new kind of bacterium by using the recombinant DNA techniques described in Section 20.3. This bacterium has the ability to "eat" oil slicks. The company contended that the bacterium was patentable, as is any other invention. The United States Patent

Office disagreed. Patent attorneys for both sides took the case all the way to the United States Supreme Court. On June 16, 1980, the Supreme Court ruled that new life forms are patentable.

Employment opportunities for those trained in law and chemistry are great. Some chemist-attorneys start their own firms and specialize in patent law. A graduate degree in chemistry or chemical engineering and a law degree are essential for this career. For further information, contact the Chemical Manufacturers Association, 2501 M Street, NW, Washington, D.C. 20037.

the molecule is changed just enough to block the use of uracil in cancer growth and destroy enzyme catalysts.

GENE SPLICING

One of the most recent and dramatic advances in biochemical research has been the development of gene splicing. In this process, genetic material—segments of DNA—is transplanted from the cells of one kind of living thing to the cells of another kind of organism. The technique of producing **recombinant DNA** typically involves five basic steps.

First, a *plasmid,* or ring of DNA, is isolated from a one-celled organism such as a bacterium. Then, an appropriate enzyme is added to the plasmid. The enzyme catalyzes a reaction that opens the plasmid. Using the same enzyme, a segment of DNA taken from, for example, a human cell is also isolated. Let us say that the segment from the human cell is a blueprint for the hormone insulin. The insulin gene is inserted in the open space in the bacterial plasmid. The plasmid then closes and is said to be recombined, which accounts for the term *recombinant DNA.* Finally, the new and unique plasmid is reinstated in the bacterial cell (see Figure 20-10).

If the recombined plasmid is functional, it will force the bacterial cell to produce human insulin. That is exactly what researchers achieved in 1978. By 1980, insulin produced by bacteria was being tested in human beings. The substance, which is virtually the only form of treatment for a major form of diabetes, has been in short supply. It is usually obtained from animals, such as pigs and sheep, that produce a substance somewhat different in structure from human insulin. Bacterially produced human insulin is structurally identical to that produced by people, is potentially unlimited in supply, and would produce none of the allergic reactions occasioned by nonhuman, animal insulin.

Gene-splicing techniques have also been used to stimulate bacteria to produce extremely rare human substances, such as HGH, human growth hormone. HGH is made in the pituitary gland. A shortage of this substance causes a condition called dwarfism. Interferon, a substance produced by human cells that combats viral attacks and may be a treatment for cancer, has also been produced by bacteria. Techniques for recombining DNA may

Figure 20-10

Steps in the production of recombinant DNA.

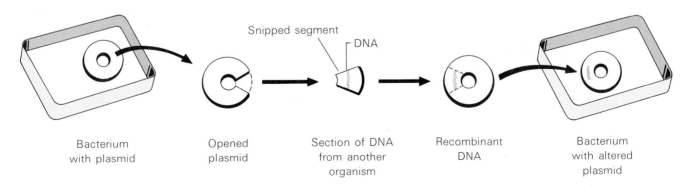

Bacterium with plasmid

Opened plasmid

Snipped segment

DNA

Section of DNA from another organism

Recombinant DNA

Bacterium with altered plasmid

yield a vast number of drugs for the treatment of human diseases.* In addition, genetic research in this area is producing very important basic information about the chemistry of DNA and of genes. For their research in this field, Paul Berg and Walter Gilbert of the United States and Frederick Sanger of Great Britain received the 1980 Nobel prize for Chemistry.

*See W. French Anderson and Elaine G. Diacumakos, "Genetic Engineering in Mammalian Cells," *Scientific American,* vol. 245 (July 1981) pp. 106–121.

Key Terms

active center

biochemistry

carbohydrate

cellulose

chemotherapy

DNA

disaccharide

enzyme

fat

fructose

genetic code

glucose

monosaccharide

nucleotide

polysaccharide

protein

purine

pyrimidine

recombinant DNA

RNA

starch

substrate

sucrose

template

SUMMARY

Living organisms are complex chemical factories employing and synthesizing the same types of molecules studied in Chapter 19. Biologically significant molecules are frequently large polymeric units made by joining simpler units. Starches, cellulose, fats, and proteins are typical of these large polymeric molecules.

DNA, the material of which genes are made, is responsible for passing hereditary information from generation to generation. DNA has a "backbone" of alternating sugar and phosphate units. Attached at each sugar unit is one of four nitrogen bases: thymine, cytosine, adenine, or guanine. Hereditary information is transmitted by the order in which bases are attached to the backbone of the DNA molecule. This system is known as the genetic code. A material called RNA is important in transforming the information on the DNA strand into chains of proteins and into complex biological units. A technique for producing recombinant DNA allows researchers to transfer genetic information between very different organisms. While there is still much to be learned about heredity and life, significant progress is being made.

QUESTIONS AND PROBLEMS

A

1. What laboratory evidence first proved that organic compounds can be made from inorganic compounds.
2. What is biochemistry? What are some of the questions it deals with?
3. (a) Write the cyclic structure for the simple sugar glucose. (b) Write the cyclic structure for the simple sugar fructose. (c) Show how disaccharide sucrose, or table sugar, can be formed by splitting out a molecule of water from one molecule of glucose and one of fructose.
4. What is the empirical formula for a carbohydrate?
5. (a) In what way does starch resemble sucrose in its structure? (b) How does the structure of starch differ from that of sucrose?
6. Cellulose has a polymeric unit derived from glucose of formula $C_6H_{10}O_5$. See page 563. The molecular weight of a cellulose molecule is 600,000. (a) What is the molecular weight of the glucose polymer unit found in cellulose? (b) How many such units exist in a molecule of cellulose? (c) If each polymer unit of the cellulose molecule is 5×10^{-10}

metres long, what is the length of this cellulose molecule if we assume that it is linear?

7. (a) Why do we call fructose a ketone? (b) Why is glucose an aldehyde? To answer these questions, use the structures you drew in answering question 3. (c) Biochemists speak of a reducing sugar and a non-reducing sugar. Which would be a reducing sugar, fructose or glucose? Explain.

8. (a) What is a fat? (b) Why do fats such as butter soften and melt over a wide temperature range while most pure organic compounds have a sharp melting point?

9. What is a protein?

10. What is an enzyme? Why are enzymes important in biology?

11. (a) What is DNA? (b) How does DNA resemble a virus?

12. Why was the work of Kornberg and Goulian important in the field of molecular biology? Explain.

B

13. Use structural formulas to show the straight chain in equilibrium with the cyclic structure of (a) glucose and (b) fructose.

14. What happens when an egg is cooked? Explain in terms of structural change of protein.

15. Why is an enzyme frequently specific for just one process?

16. How can an enzyme inhibitor, or false substrate, serve as a weapon against bacteria or disease?

17. (a) Show through drawing of a structural formula how the nucleotides deoxyadenosine-5'-phosphate, deoxycytidine-5'-phosphate and deoxyguanosine-5'-phosphate combine to make a segment of a single DNA chain. (b) What is the significance of the term "double-helix"?

18. Explain in your own words what occurs in gene splicing.

C

19. Refer to Exercise 20-4 to write the structural formula of the fatty ester glycerol tristearate.

20. Bradykinin, a small peptide hormone, has been successfully synthesized in the laboratory. It is known to have the structure shown below:

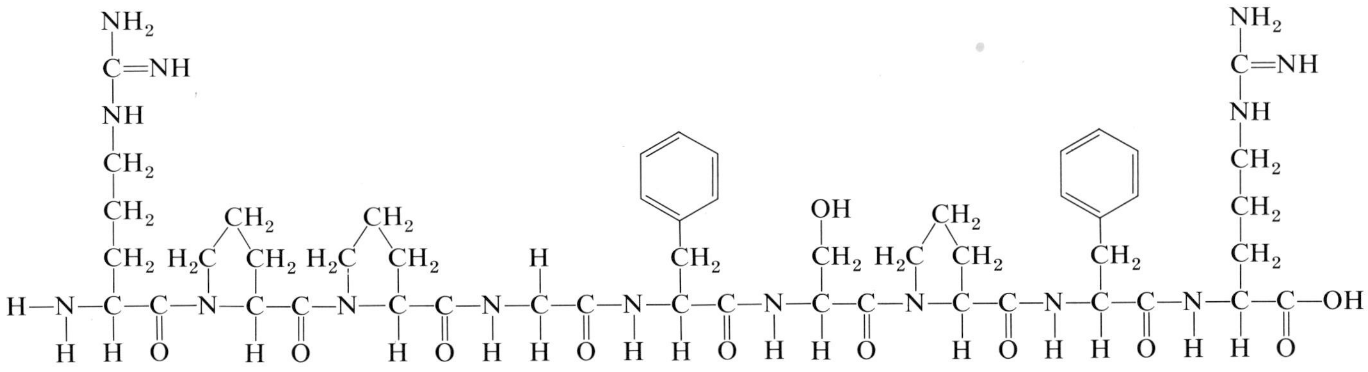

On a separate sheet of paper, draw this structure and use dotted lines to indicate the separate amino acids from which this protein is made. How many amino acid subunits does bradykinin contain?

21. Write the structural backbone of DNA.
22. Show how each of the four nitrogen bases can be put on the DNA chain drawn in answer to question 21.
23. Why does the order of the bases on the DNA chain determine the order of bases on a new chain that uses the old chain as a template?

E X T E N S I O N

The revolution in brain biochemistry

Although people have used drugs for thousands of years, scientists are just beginning to understand how and why drugs work. Recent breakthroughs regarding substances produced in the brain have been quite startling. In the past, many physical conditions that are the results of chemical causes were not properly understood. For example, doctors once thought that pellagra, a disease resulting in skin and digestive disorders, was a form of mental illness. In the 1930s, however, it was found that pellagra was caused by a deficiency of B-complex vitamins in a person's diet.

Sulfa drugs, developed and used extensively since World War II, dramatically revealed the link between a drug's effectiveness and its molecular structure. For example, the structural similarity between the drug *sulfanilamide* and the naturally occurring chemical para-*aminobenzoic acid* (PABA) is believed to be responsible for the drug's effectiveness against certain bacteria. (The molecular structures of the two substances are shown on page 567.) The bacteria normally use PABA to make a life-supporting chemical called folic acid. When sulfanilamide is present, it *replaces* the PABA in the bacteria's biochemical processes. Replacement occurs, presumably, because of the drug's resemblance to the naturally occurring chemical. However, unlike PABA, sulfanilamide *cannot* be used to make folic acid. Thus, the bacteria die.

Another example of how structural similarities between molecules cause drugs to be effective is demonstrated by *antihistamines*. Histamine, formed from the amino acid histidine, is a powerful pharmacological agent that affects the vascular system, smooth muscles, and exocrine glands. The release of histamine within the body can cause severe headaches, skin rashes, and respiratory problems. These are among the symptoms of various allergic reactions such as hay fever and asthma. Antihistamines are drugs that block the actions of histamine and thus reduce the intensity of allergic reactions. The blocking property is thought to be due to the structural similarity between histamine and antihistamine. The antihistamine molecules apparently compete with histamine to occupy receptor sites on *effector* cells. (Such cells transmit "messages," such as the allergic reactions, to muscles and glands.) Histamine is thus excluded or greatly reduced from occupying these receptor sites.

Sometimes, the function of a natural compound can be improved by a slight structural change. For example, the drug cocaine is derived from the leaves of the coca plant. Cocaine was first used by doctors in Europe and North America as a local anesthetic for eye operations. When heat-sterilized, however, the cocaine molecule undergoes changes that interfere with its anesthetic properties. In addition, cocaine was found to have detrimental side effects. Toward the end of the last century, chemists began to synthesize compounds similar to cocaine that would not be heat sensitive. The most successful modification was obtained in 1909 and named *procaine,* although it is better known by its trademark Novocain. Widely used today as a dental anesthetic, Novocain is heat stable, easily synthesized, nonaddictive, and limited in side effects. This synthetic cocaine substitute does *not* exist in nature. Nevertheless, it is a more practical painkiller than the natural compound.

The proof that physical and psychological disturbances involve the same biochemical system came in the early 1950s with the discovery of the first tranquilizers. Tranquilizers such as Thorazine, also known as chlor-promazine, allowed thousands of patients to leave mental hospitals and resume relatively normal lives. However, the mechanisms by which such drugs did their jobs remained a mystery. For example, why did tranquilizing drugs control disordered thoughts and hallucinations in some schizophrenics but not in others? Such questions led researchers to look more closely at the biochemistry of the human brain.

The brain's message system The brain is composed of tens of billions of nerve cells called *neurons*. These elongated, fingerlike cells transport messages, in the form of electrical impulses, to other neurons. In this way, messages eventually reach those areas of the brain that allow us to perceive pain. The neurons communicate with each other across junctions, or gaps, in the nerve pathways. The gaps are called *synapses*. The communication between neurons is achieved by the release of chemicals called *neurotransmitters*. The neurotransmitters are able to move across the synapse. On the other side of the gap, the chemicals attach themselves to the neighboring cell and trigger a new electrical impulse in the adjoining neuron. In this way, the message is transmitted from neuron to neuron.

To date, biochemists have found more than 30 neurotransmitters and suspect that dozens more will be discovered. Each has a unique molecular structure that enables it, or a chemical with a similar structure, to work like a key in a lock. Such a neurotransmitter or a similar structure fits into those sites, or receptors, on the nerves that are specifically designed to receive the chemical.

Too much or too little of a neurotransmitter may cause bodily malfunctions. For example, low levels of the neurotransmitter serotonin may be linked to insomnia and severe depression. A deficiency of the neurotransmitter dopamine has been related to the rigidity and tremor of Parkinson's disease,* while an excess of dopamine may be involved in schizophrenia.† A deficiency of the neurotransmitters acetyl-

choline and serotonin may also contribute to schizophrenia. The low levels of dopamine and serotonin may be caused by an increase in an enzyme called monoamine oxidase, which destroys those neurotransmitters. Antidepressant drugs such as Thorazine act by blocking that enzyme, allowing those neurotransmitters to remain at a normal level.

Opiate drugs and pain Morphine is derived from the opium poppy and is used as an *analgesic,* or painkiller. Studies of the action of morphine and its derivatives have shown that they somehow interfere with the transmission of pain messages to the brain. Recent work with opiates tends to support the concept of specific receptor sites for specific neurotransmitters. This is the lock-and-key concept discussed above. A particular neurotransmitter becomes the "key" that can fit into the "lock" presented by the receptor's structure.

The evidence supporting such a concept can be divided into several categories. First, all opiate painkillers exhibit basic similarities in molecular architecture. For example, morphine, codeine, and heroin have very similar molecular structures. Heroin is obtained synthetically from morphine by replacing the —OH groups of morphine with acetate groups.

Second, most opiates exist in at least two *optical isomers*. Optical isomers are even more closely related than structural isomers. They differ from each other in much the same way as your right hand differs from your left hand. They are *mirror images* of each other but cannot be superimposed upon each other in space. Optical isomers have identical melting and boiling points and can be distinguished from each other only by the direction in which a solution of the isomer affects a beam of polarized light. Usually, only the *levorotatory* isomer, which rotates the plane of polarization of such light to the *left,* can relieve pain or produce the other actions associated with opiates. Optical isomers support the model of a highly specific receptor that can distinguish the right- or left-"handedness" of the opiate molecule.

Third, opiate *agonists*—the name given to opiates that can produce analgesic and euphoric effects—can be transformed by slight molecular modifications into *antagonists*. Antagonists are substances

*Parkinson's disease is known to involve a loss of cells in an area of the brain called the basal ganglia, which is responsible for balancing the activity of opposing muscles. To bend your arm, for example, one muscle must contract while the opposing muscle relaxes. Dopamine transmits messages from muscle cells to the brain. This neurotransmitter is deficient in the basal ganglia of people with Parkinson's disease.
†See Leslie L. Iversen, "The Chemistry of the Brain," *Scientific American,* vol. 241 (September 1979) pp. 134–149.

that specifically block the painkilling and euphoric actions of agonists through very small doses. For example, *nalorphine,* a potent antagonist that blocks all the analgesic and euphoric effects of morphine, is structurally similar to morphine, as shown below. A —CH_3 group on the nitrogen of morphine is replaced by a —CH_2—CH=CH_2 group on the nitrogen of nalorphine. Small changes in structure can convert an agonist into an antagonist for that substance. A person near death from morphine poisoning can be revived almost instantaneously by receiving a much smaller amount of nalorphine than the amount of morphine present. Such a rapid effect implies a common site of action for morphine and nalorphine molecules. It seems as if antagonists such as nalorphine can occupy the *same* opiate receptor sites as agonists such as morphine. In this way, they block access to the opiate agonists.

Morphine
(agonist)

Nalorphine
(antagonist of
morphine)

The kind of evidence discussed above convinced researchers that specific opiate receptors must exist in the brain and possibly in other tissues. The identification of such receptors remained elusive until 1973, when three groups of researchers announced almost simultaneously the discovery of specific receptors for opiates. Those receptors were found in the regions of the mammalian brain and spinal cord involved in the perception and integration of pain and emotions.* The existence of the receptors

---

*The researchers were Solomon H. Snyder and Candace B. Pert at the Johns Hopkins University School of Medicine, Eric J. Simon at the New York University School of Medicine, and Lars Terenius at the University of Uppsala, Sweden. See Solomon H. Snyder, "Opiate Receptors and Internal Opiates," *Scientific American,* vol. 236 (March 1977) pp. 44–56.

strongly suggested the presence of natural, morphine-like substances in the brain. Perhaps the body somehow produced its own internal narcotics. If so, such substances might be neurotransmitters that act at the receptor sites.

Natural brain opiates   Acting on the premise that the body does indeed produce its own opiates, researchers isolated two naturally occuring peptides* from the brains of pigs in 1975. The researchers responsible for this discovery were John Hughes and Hans W. Kosterlitz, of the University of Aberdeen in Scotland.

Hughes and Kosterlitz found that the two peptides bonded tightly to the opiate receptors. They named the peptides *enkephalins* (from the Greek for "in the head"). Each of the enkephalins is a pentapeptide: a molecule containing five amino acids. They are identical except for the terminal amino acid. Much evidence now indicates that enkephalins

---

*As discussed in Section 19.6, peptides are large molecules made by linking various amino acid molecules, in which the carboxyl group (COOH) of one amino acid and the amino group (NH_2) of another are joined by the splitting out of a water molecule.

Figure 20-11

A possible mechanism for enkephalin inhibition in the spinal cord and brain.

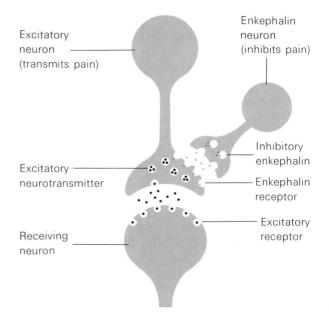

Excitatory
neuron
(transmits pain)

Enkephalin
neuron
(inhibits pain)

Excitatory
neurotransmitter

Inhibitory
enkephalin

Enkephalin
receptor

Excitatory
receptor

Receiving
neuron

are neurotransmitters that mediate pain and emotional behavior. Figure 20-11 shows a possible mechanism for enkephalin inhibition. The human brain may naturally produce *varying* amounts of enkephalins in different individuals accounting, in some cases, for variations in response to pain stimuli.

The discovery of enkephalins has suggested how they may play an important role in opiate addiction. Under normal conditions, opiate receptors are exposed to a certain level of enkephalins. When taken into the body, an opiate, such as morphine or heroin, seems to bind to unoccupied enkephalin receptors. Thus, they mimic the pain-suppressing effects of the enkephalin system. This eliminates the need for the painkilling effects of the system. Repeated doses of morphine or heroin "fill" the receptors so they cannot accept enkephalins. Thus, the enkephalins stop serving as neurotransmitters. The receiving cells are now exposed only to morphine or heroin, so that these cells can tolerate more opiate to make up for the enkephalins no longer present. When administration of morphine or heroin is stopped, the opiate receptors are not occupied by either the foreign or natural opiate. Some time is needed before production of enkephalins is resumed after the morphine or heroin is stopped. The temporary lack of opiate initiates a sequence of physical and mental events that together are called withdrawal symptoms.

Other morphine-like neuropeptides called *endorphins* (meaning "the morphine within") have also been isolated from the brain. Endorphin is the term now used to describe any normally-occurring opiate-like substance in the brain. Beta-endorphin is a chain of 31 amino acids that has 50 times the painkilling effect of morphine when injected directly into the brains of experimental animals.* Alpha-endorphin is a chain of 16 amino acids that is both a tranquilizer and a painkiller. Researchers are convinced that endorphins may help explain many psychological and behavioral phenomena. For example, during World War II, army medics were astonished by some soldiers who lost limbs yet did not complain of pain. It is now thought that the wounded men produced extra enkephalins or endorphins that dulled the pain.

Recent experiments have suggested that certain procedures used to treat chronic pain—such as acupuncture, direct electrical stimulation of the brain, and hypnosis—may work by releasing enkephalins or endorphins in the brain and spinal cord. This hypothesis is based on the finding that such procedures do not work when the antagonist *naloxone* is administered. Naloxone blocks the binding of morphine to the opiate receptors. Naloxone also has been found to block the action of endorphins in the brain of some acutely disturbed schizophrenics, temporarily alleviating their auditory hallucinations. The structural formula of naloxone shows that it is similar to morphine and the previously discussed antagonist nalorphine. The only structural difference between the two antagonists is an additional oxygen atom in naloxone.

Researchers are now attempting to develop nonaddictive compounds that mimic the effects of natural neuropeptides. One possibility for developing nonaddictive analgesics is the use of mixed agonist-antagonist compounds derived from endorphins. Designing such drugs for specific effects is no easy project.

Many problems must be solved before nonaddictive synthetic substitutes are of practical use. One problem is that natural neuropeptides seem to perform multiple roles within the brain and central nervous system. Another problem is that there may be several different types of receptors for a given transmitter substance. In addition, researchers have found that two different neuropeptides may serve as neurotransmitters in a single cell at the same time. Thus, each cell may be able to get two or more messages at a single synapse. However, the functional significance of such multiple systems is not yet known. Although the study of transmitter systems and natural brain opiates has provided important clues to various chemical mechanisms, it is clear that the most exciting discoveries about how the brain works lie ahead.

---

* A newly discovered neuropeptide called *dynorphin,* found in the pituitary gland, is an extremely potent analgesic. It is about 50 times stronger than beta-endorphin, 200 times stronger than morphine, and 700 times stronger than leucine enkephalin.

According to the coordination theory, all six groups in . . . complex radicals . . . are bound to the central atom. The six groups are arranged around the central atom in the positions . . . of the corners of an octahedron.

ALFRED WERNER

21

THE TRANSITION ELEMENTS

Objectives

After completing this chapter, you should be able to:

● Locate the transition elements in the periodic table.

● Relate the oxidation states of transition elements in several compounds to their electron configurations.

● Describe the bonding in various types of complex ions.

● Give the coordination numbers of transition metal ions in various complex ions and molecules and give the geometric shapes of such complexes.

● Explain the amphoteric behavior of certain transition metal hydroxides.

Alloys containing transition metals make this aircraft and other very high-performance aircraft possible.

In Chapter 9, you read that elements range from metals to nonmetals across the periodic table from left to right. In Chapter 11, you read that a large family of metals appears in the fourth horizontal row of the table as the 3d electron levels in the atoms of each element fill up. Similar groups appear in the fifth and sixth rows. Recall from Chapter 9 that metals arising from the filling of the 3d, 4d, and 5d electron levels are called **transition elements.** This chapter discusses these useful and interesting elements.

All transition elements are metals. Most exhibit several oxidation states, and they form a great variety of complex ions. Many of their ions are deeply colored. When transition elements and their compounds are involved in reactions, dramatic color changes often accompany the chemical change.

21.1 Identification and electron arrangement

The transition elements are identified in Figure 21-1, on page 486. There is some disagreement among chemists as to which elements should be included as transition elements. For our purposes, we include elements 21 through 29 as well as those directly beneath them in the periodic table.*

*Sometimes Zn, Cd, and Hg are included in the transition elements, but a careful review of their chemistry shows clearly that they belong in the typical element group.

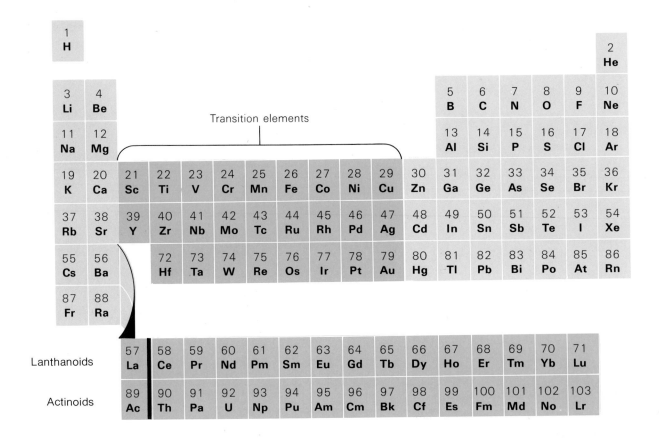

1 H																	2 He
3 Li	4 Be											5 B	6 C	7 N	8 O	9 F	10 Ne
11 Na	12 Mg											13 Al	14 Si	15 P	16 S	17 Cl	18 Ar
19 K	20 Ca	21 Sc	22 Ti	23 V	24 Cr	25 Mn	26 Fe	27 Co	28 Ni	29 Cu	30 Zn	31 Ga	32 Ge	33 As	34 Se	35 Br	36 Kr
37 Rb	38 Sr	39 Y	40 Zr	41 Nb	42 Mo	43 Tc	44 Ru	45 Rh	46 Pd	47 Ag	48 Cd	49 In	50 Sn	51 Sb	52 Te	53 I	54 Xe
55 Cs	56 Ba		72 Hf	73 Ta	74 W	75 Re	76 Os	77 Ir	78 Pt	79 Au	80 Hg	81 Tl	82 Pb	83 Bi	84 Po	85 At	86 Rn
87 Fr	88 Ra																

Transition elements

Lanthanoids	57 La	58 Ce	59 Pr	60 Nd	61 Pm	62 Sm	63 Eu	64 Gd	65 Tb	66 Dy	67 Ho	68 Er	69 Tm	70 Yb	71 Lu
Actinoids	89 Ac	90 Th	91 Pa	92 U	93 Np	94 Pu	95 Am	96 Cm	97 Bk	98 Cf	99 Es	100 Fm	101 Md	102 No	103 Lr

Figure 21-1

The periodic table, showing the transition elements.

Across the first row of the transition region are the elements scandium (Sc), titanium (Ti), vanadium (V), chromium (Cr), manganese (Mn), iron (Fe), cobalt (Co), nickel (Ni), and copper (Cu). On the left is the scandium column, which includes yttrium (Y), lanthanum (La), and actinium (Ac). The elements that follow lanthanum (atomic number 57) in the periodic table—elements 58 through 71—are placed in a separate group known as the rare earths, or **lanthanoids.** They are frequently called inner transition elements. Elements 90 through 103, following actinium (atomic number 89) in the table, are also grouped separately and are known as the rare earths, or **actinoids.** They too are classed as inner transition elements, and they are placed under the lanthanoids in the periodic table.

Two questions come to mind immediately about the transition elements. First, why are they grouped and considered together? Clearly, they are not a single, vertically placed chemical family. Second, what is special about their properties? These questions are closely related because the answers to both depend on the electron configurations of the atoms of such elements.

ELECTRON CONFIGURATION

What is the electron arrangement of the transition elements? We will start by studying calcium. Calcium (atomic number 20) comes just before scandium in the periodic table. The 20 electrons in a calcium atom are distributed as shown in the following diagram:

Calcium nucleus	$1s$	$2s$	$2p$	$3s$	$3p$	$4s$	$3d$	$4p$

Scandium (atomic number 21) follows calcium and is the first transition element. It has one more electron than calcium. Where does that electron go? The twenty-first electron goes into a $3d$ orbital, the next higher level of energy, as the diagram above and Figure 21-2 suggest (see page 588).

Five $3d$ orbitals are available: all have more or less the same energy *for an isolated atom.* If we place two electrons in each of the five orbitals, a total of ten electrons can be accommodated before it is necessary to move to the slightly higher $4p$ energy level. Thus, scandium, as well as the nine elements following it in the table, can be generated by placing additional electrons in $3d$ orbitals. Not until we come to gallium (atomic number 31) is the $4p$ level needed.

EXERCISE 21-1

Using Figure 21-2, decide which orbital will be used after the five $4d$ orbitals have been filled. To what element does this correspond?

With the help of Figure 21-2, you should now be able to predict the electron configurations for most transition elements. You will not be able to predict all details because there are some exceptions to the previously stated rules. The exceptions result from the special stabilities that result when a set of orbitals is either completely filled or half-filled. Fourth-row transition

Table 21-1

Electron configurations of fourth-row transition elements

Element	Symbol	Atomic number	Electron configuration						
scandium	Sc	21	$1s^2$	$2s^2$	$2p^6$	$3s^2$	$3p^6$*	$3d^1\,4s^2$
titanium	Ti	22						$3d^2\,4s^2$
vanadium	V	23						$3d^3\,4s^2$
chromium	Cr	24						$3d^5\,4s^1$
manganese	Mn	25						$3d^5\,4s^2$
iron	Fe	26						$3d^6\,4s^2$
cobalt	Co	27						$3d^7\,4s^2$
nickel	Ni	28						$3d^8\,4s^2$
copper	Cu	29						$3d^{10}4s^1$

*Each fourth-row transition element has these levels filled.

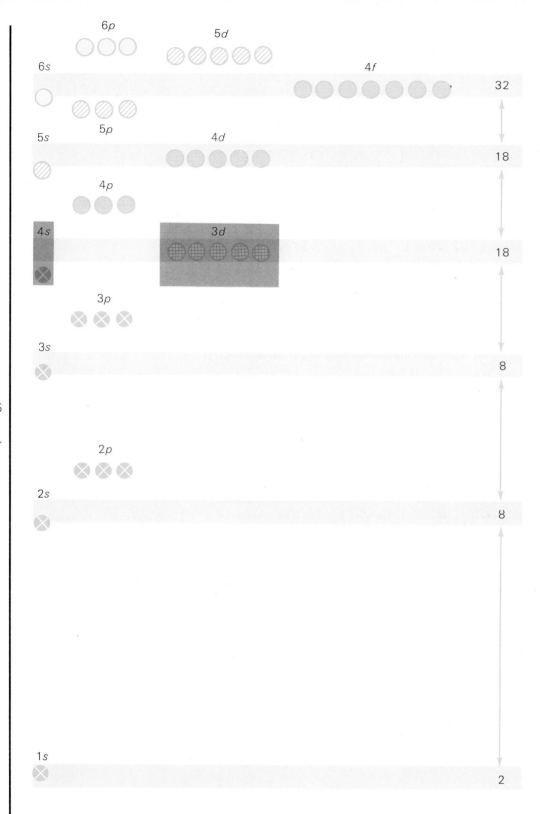

Figure 21-2
The energy-level scheme of a many-electron atom.

elements actually have the set of electron configurations shown in Table 21-1. Notice that chromium (atomic number 24) and copper (atomic number 29) interrupt the continuous build-up of 3d electrons. In the case of chromium, the whole atom has lower energy if one of the 4s electrons moves into the 3d set. Such a move produces an especially stable half-filled set of 3d orbitals and a half-filled 4s orbital. In the case of copper, the atom has lower energy if the 3d set is completely filled by ten electrons and the 4s orbital is half-filled. The arrangement of nine electrons in the 3d orbital and two electrons in the 4s orbital yields a copper atom of very slightly higher energy and, therefore, of somewhat lower stability.

EXERCISE 21-2

Construct an electron configuration table similar to Table 21-1 for the *fifth-row* transition elements, from yttrium (atomic number 39) through silver (atomic number 47). In elements 41 through 45, one of the 5s electrons moves to a 4d orbital, which results in greater stability. In element 46, two 5s electrons behave in the same way. It is clear that energy differences between 5s and 4d levels are very small.

The transition elements are frequently described as that group of elements in which the d levels are being filled with electrons. In elements 21 through 29, electrons are being placed in 3d orbitals. For elements 39 through 47, electrons are filling 4d orbitals. For elements 57 and 72 through 79, electrons are going into the 5d levels. Elements 104 through 106 have electrons in the 6d level.

The inner transition elements are those elements in which electrons are being placed in the f levels. Placing electrons in the 4f orbitals means there will be 14 inner transition elements in any one row. Because differences in electron structure are deep inside the atoms, these 14 elements have almost identical chemical properties. The inner transition elements now include elements 58 through 71 and 90 through 103. As you know, these elements are the lanthanoids and actinoids.

GENERAL PROPERTIES

What properties are shown by the transition elements? What kinds of compounds do they form? How can their properties be interpreted in terms of the electron arrangements of their individual atoms? When clean, all transition elements appear typically metallic. They are shiny, lustrous solids. They are good conductors of electricity and heat. Some transition elements—such as copper, silver, and gold—are among our oldest recognized metals. They are excellent conductors of heat and electricity.

Transition elements show a wide range of chemical reactivity. Some are extremely unreactive. Gold and platinum, for example, can remain exposed to air or water for long periods without any change. Others, such as iron, can be polished to a metallic luster, but slowly corrode from exposure to air and water. Still others, such as scandium and lanthanum, are extremely reactive.

The variety of transition metal compounds is illustrated by the various chromium compounds found on most chemical stockroom shelves. Next to a green powder labeled chromic or chromium(III) oxide (Cr_2O_3), there may well be a bottle of reddish powder, chromium(VI) oxide (CrO_3). Other simple compounds might include black chromous or chromium(II) oxide (CrO), the flaky violet crystalline solid chromic or chromium(III) chloride ($CrCl_3$), and possibly some red chromium(II) acetate, $Cr(CH_3COO)_2$. On other shelves, you might find potassium chromate (K_2CrO_4)—a bright yellow powder—next to a bottle of orange potassium dichromate ($K_2Cr_2O_7$). Chromium seems to exhibit several different *oxidation numbers* in these compounds. It has oxidation numbers of $+2$ in CrO and $Cr(CH_3COO)_2$, $+3$ in Cr_2O_3 and $CrCl_3$, and $+6$ in CrO_3, K_2CrO_4, and $K_2Cr_2O_7$.

EXERCISE 21-3

Find the oxidation number of chromium in each of the following: (a) $Cr_2O_7^{--}$ (b) CrO_4^{--} (c) $Cr(OH)_3$ (d) CrO_2Cl_2.

Along with the simple chromium compounds listed above, you might also find more complex substances. Next to $CrCl_3$, for example, you might find a series of brightly colored solids, such as yellow $CrCl_3 \cdot 6NH_3$, purple-red $CrCl_3 \cdot 5NH_3$, green or violet $CrCl_3 \cdot 4NH_3$, and violet $CrCl_2 \cdot 3NH_3$. The dots in these formulas indicate that several NH_3 molecules are bound to the $CrCl_3$ molecule. The electrically neutral NH_3 molecules have no effect on the oxidation number of the chromium in these compounds. It is always $+3$. Looking further, we might find other complex compounds, such as bright yellow $Na_3Cr(CN)_6$ and violet $KCr(SO_4)_2 \cdot 12H_2O$, in which chromium also exhibits an oxidation number of $+3$. Our chemical stockroom investigation reveals several things about the transition element chromium (atomic number 24). It forms both simple compounds (Cr_2O_3) and complex compounds ($CrCl_3 \cdot 6NH_3$). Many of them are brightly colored, chemically stable solids. Chromium exhibits several different oxidation states in its compounds, such as $+2$, $+3$, and $+6$. Similar behavior is exhibited by many other transition elements.

Is there any regularity in the kinds of compounds formed by the fourth-row transition elements? Table 21-2 lists some of these compounds.

EXERCISE 21-4

Consult a recent handbook of chemistry to find at least one additional compound for each oxidation number given for each element in Table 21-2.

The information given in Table 21-2 reveals several regularities of transition elements:

1. Several oxidation states are possible for most fourth-row transition elements.

Table 21-2

Typical oxidation numbers of fourth-row transition elements

Symbol	Representative compounds					Common oxidation numbers*							Number of valence $3d$ and $4s$ electrons
Sc		Sc_2O_3						**+3**					3
Ti	TiO	Ti_2O_3	TiO_2				+2	+3	**+4**				4
V	VO	V_2O_3	VO_2	V_2O_5			+2	+3	+4	**+5**			5
Cr	CrO	Cr_2O_3		CrO_3			+2	**+3**			+6		6
Mn	MnO	Mn_2O_3	MnO_2	K_2MnO_4	$KMnO_4$		**+2**	+3	**+4**		+6	+7	7
Fe	FeO	Fe_2O_3		K_2FeO_4			+2	**+3**			+6		8
Co	CoO	Co_2O_3					+2	+3					9
Ni	NiO	(Ni_2O_3)	NiO_2				+2	(+3)	**+4**				10
Cu	Cu_2O	CuO				+1	**+2**						11

*The most common oxidation numbers are in bold type. Parentheses indicate uncertainty. For example, Ni in Ni_2O_3 may be one Ni^{++} ion and one Ni^{++++} ion instead of two Ni^{+++} ions.

2. When the same element shows several different oxidation numbers, the numbers often differ from each other by one unit. In the case of vanadium, for example, the numbers form a continuous series from +2 to +3 to +4 to +5.

3. The maximum oxidation state first increases and then decreases as we look across the list of fourth-row transition elements. The maxima are +3 for scandium, +4 for titanium, +5 for vanadium, +6 for chromium, and +7 for manganese. The +7 represents the highest value observed for fourth-row transition elements.

How can we explain such observations? The answer to this question depends on several factors. As already indicated, the combining capacity of an atom depends in part on how many of its electrons are used in bonding to other atoms. The unique feature of transition elements is that they have several electrons in their outermost d and s valence orbitals. It is thus possible for an element such as vanadium to form a series of compounds in which from two to five electrons are either lost to, or shared with, other elements. The oxides VO and V_2O_3, for example, contain V^{++} and V^{+++} ions respectively. While more energy is required to form V^{+++} ions than is needed to form V^{++} ions, this condition is offset by another factor. Because of their greater charge, V^{+++} ions have greater attraction for O^{--} ions than do V^{++} ions. The energy of interaction, or lattice energy, compensates for the extra ionization energy. Thus, both VO and V_2O_3 are stable compounds.

Note from Table 21-2 that the maximum oxidation number of transition elements never exceeds the total number of s and d valence electrons. The higher oxidation states become increasingly more difficult to form as we proceed along a row. We also note that the ionization energies of the d and s electrons increase with atomic number across a given row.

It is characteristic of transition elements that their hydroxides have low basic strength. In general, the hydroxides of transition elements are insoluble. Because of this, they contribute relatively few OH^- ions to solution and appear as very weak bases when compared to hydroxides of such elements as potassium and calcium. Chromium(III) hydroxide even has weakly acidic properties because of the $+3$ charge on the chromium ion.

We have seen that the combining capacity varies from one transition element to another. Now, we shall discuss another regularity.

21.2 Complex ions

The remaining regularity observed for transition elements is that they form a great variety of complex ions. A **complex ion** is one in which other molecules or ions are bonded to the central transition element ion, forming a complex unit. The previously mentioned series of chromium compounds, $CrCl_3 \cdot 6NH_3$, $CrCl_3 \cdot 5NH_3$, $CrCl_3 \cdot 4NH_3$, and $CrCl_3 \cdot 3NH_3$, contain complex ions. Experiments with such compounds reveal some intriguing facts. When 1 mole of each of the above four compounds is dissolved in water, a clear solution is formed. The addition of aqueous silver nitrate should precipitate any chloride ions as AgCl from each solution. The equation is

$$1Ag^+ + 1Cl^- \rightleftharpoons 1AgCl(s)$$

The results of such studies are summarized below. They indicate that not all chloride always precipitates.

Compound	Moles of Cl^- precipitated	Moles of Cl^- not precipitated
$CrCl_3 \cdot 6NH_3$	3 of 3	0
$CrCl_3 \cdot 5NH_3$	2 of 3	1 of 3
$CrCl_3 \cdot 4NH_3$	1 of 3	2 of 3
$CrCl_3 \cdot 3NH_3$	0	3 of 3

Evidently, there are two ways in which chlorine is bonded in the above compounds: one that allows Cl^- to be precipitated by Ag^+, and one that does not. In $CrCl_3 \cdot 6NH_3$, all the chloride can be precipitated; in $CrCl_3 \cdot 3NH_3$, none can be precipitated. Other data also indicate different types of bonding. For example, $CrCl_3 \cdot 6NH_3$ forms a highly conductive solution. On the other hand, $CrCl_3 \cdot 3NH_3$ solution conducts very poorly. The explanation of this behavior was provided in the early 1900s by Alfred Werner (1866–1919), who suggested that complex compounds of Cr^{+++} can be accounted for by assuming that each chromium is bonded to six neighbors. In $CrCl_3 \cdot 6NH_3$, the cation consists of a central Cr^{+++} ion surrounded by six NH_3 molecules at the corners of an octahedron. The three chlorine atoms exist as anions (Cl^-). In $CrCl_3 \cdot 5NH_3$, the cation consists of the central chromium surrounded by the five NH_3 and one of the chlorines; the other two chlorines are anions. In $CrCl_3 \cdot 4NH_3$, the chromium is bonded to four NH_3 and two Cl atoms, leaving one chloride anion. In $CrCl_3 \cdot 3NH_3$, all three Cl atoms and all three NH_3 molecules are bonded to the central chromium. The formulas can be written $[Cr(NH_3)_6]Cl_3$, $[Cr(NH_3)_5Cl]Cl_2$, $[Cr(NH_3)_4Cl_2]Cl$, and $[Cr(NH_3)_3Cl_3]$, respectively.

Table 21-3

Transition metal ions in various gemstones

Trace metal	Color*	Name
Corundum:		
Cr^{+++}	brilliant red	ruby
Mn^{+++}	deep purple	amethyst
Fe^{+++}	yellow to reddish yellow	topez or yellow sapphire
V^{+++} or Co^{+++}	rich blue	blue sapphire
Beryl:		
Cr^{+++}	green	emerald
Fe^{+++}	pale blue to light greenish blue	aqua-marine

*The colors of many familiar gemstones are due to trace amounts of transition metal ions in the minerals corundum (Al_2O_3) and beryl $(Be_3Al_2Si_6O_{18})$. For example, emeralds are a rare variety of beryl with Cr^{+++} substituted for Al^{+++} in trace amounts.

GEOMETRY OF COMPLEX IONS

The way in which atoms or molecules are arranged in space around a central atom has a great influence on the properties of the substance. What kinds of arrangements are found in complex ions? What shapes do the complex ions show? Can we find any regularity in the transition elements that will enable us to predict what complex ions will form?

First, let us introduce a concept useful in giving spatial descriptions: *the* **coordination number** *is the number of near neighbors (ions or molecules) that an atom has.* For example, in the complex ion $CrCl_3 \cdot 6NH_3$, structural studies show that each Cr^{+++} *ion* is surrounded by six NH_3 *molecules* arranged at the corners of a regular octahedron, as shown in Figure 21-3. The electron pair on each NH_3 points in toward the Cr^{+++} ion. We say that chromium has a coordination number of 6 in $CrCl_3 \cdot 6NH_3$.

Some of you have probably seen humidity indicators that have a piece of colored cloth or paper as the moisture-sensitive unit. When the humidity is low, the color of the cloth or paper is blue. When the humidity is high, the color is pink or red. Many different, ingenious, and even humorous devices, purportedly to forecast the weather, have been developed to utilize commercially this change in color. The colored material absorbed by the paper or cloth is cobalt(II) chloride solution ($CoCl_2$). Some dissolved NaCl may also be present in it. The blue color is attributed to a complex ion of cobalt, presumably the well-known blue ion $[CoCl_4]^{--}$. In this complex ion, four chloride ions are arranged tetrahedrally around a central cobalt(II) cation, Co^{++} (see Figure 21-4). The coordination number of cobalt in this complex is 4. The pink color is due to the complex ion $[Co(H_2O)_6]^{++}$, in which six water molecules surround the central Co^{++} ion. The negative ends of the water molecules (electron pairs) point inward (Figure 21-5). The Co^{++} ion has a coordination number of 6 in this species. The basis for the color change is found in the nature of the complex ions in the solution:

$$[CoCl_4]^{--} + 6H_2O \rightleftharpoons [Co(H_2O)_6]^{++} + 4Cl^-$$

TETRAHEDRAL   OCTAHEDRAL
BLUE   PINK
COORDINATION   COORDINATION
NUMBER = 4   NUMBER = 6

It should be clear to us from Le Chatelier's principle that when water evaporates into the room, the equilibrium will shift to produce the blue chlorocomplex. On the other hand, if water is picked up from very moist air, the equilibrium shifts to produce the pink complex. Cobalt(II) shows a coordination number of 4 for chloride and 6 for water under the conditions used in the humidity indicator. The same $CoCl_2$ solution is sometimes used as an invisible ink since the pink color cannot be seen in dilute solution or when on a moist piece of paper, but a blue color develops if the paper is dried.

EXERCISE 21-5

> What is the coordination number of a metal atom in (a) a body-centered cubic structure, (b) a cubic close-packed structure, and (c) a hexagonal close-packed structure? (Refer to pages 343–344.)

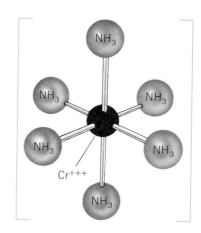

Figure 21-3

A structural model of the cation $[Cr(NH_3)_6]^{+++}$ in $CrCl_3 \cdot 6NH_3$.

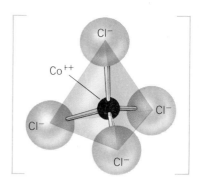

Figure 21-4

The cobalt complex $[CoCl_4]^{--}$ is tetrahedral and colors its solutions blue. The coordination number of the cobalt in this complex is 4.

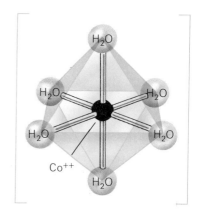

Figure 21-5

The cobalt complex $[Co(H_2O)_6]^{++}$ is octahedral and colors its solutions red. The coordination number of the cobalt in this complex is 6.

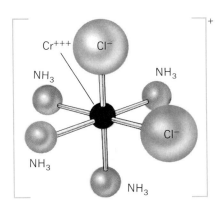

Figure 21-6

The violet-hued *cis*-isomer of $[Cr(NH_3)_4Cl_2]^+$.

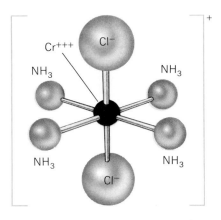

Figure 21-7

The green-hued *trans*-isomer of $[Cr(NH_3)_4Cl_2]^+$.

Water and NH_3 show rather striking similarities in their ability to coordinate around a metal cation. We have already mentioned the complex ions $[Cr(NH_3)_6]^+$ and $[Co(H_2O)_6]^{++}$. As you might expect, we also find $[Cr(NH_3)_5H_2O]^+$. One of the ammonia molecules in $[Cr(NH_3)_6]^{+++}$ has been replaced by a water molecule. Additional replacements of this type produce $[Cr(NH_3)_4(H_2O)_2]^{+++}$, and so on.

Not only water molecules but many anions, such as chloride ions, can replace ammonia molecules. For example, the purple compound with formula $CrCl_3 \cdot 5NH_3$ mentioned earlier has a structure in which one of the three chloride ions of $CrCl_3$ is attached to the chromium. One of the six NH_3 molecules of yellow $CrCl_3 \cdot 6NH_3$ is replaced. Notice that this complex ion, $[Cr(NH_3)_5Cl]^{++}$, would have a charge of $+2$, since one charge on Cr^{+++} is neutralized by Cl^-.

In the compound $CrCl_3 \cdot 4NH_3$, two species of distinctly different color are reported, one violet and the other green. Can we justify this observation? The octahedral models shown in Figure 21-6 and 21-7 indicate that two distinct structures should be possible. As we see, the two chlorine atoms may be on the *same* side of the metal atom (Figure 21-6) or on *opposite* sides of the metal atom (Figure 21-7). The structure in which the two Cl^- groups are located on the same side of the metal atom is called the *cis*-isomer and is violet; the other is called the *trans*-isomer and is green.

The complex ion $[Fe(C_2O_4)_3]^{---}$ is formed when rust stains are bleached out of cloth with oxalic acid solution. The complex ion contains a transition

Figure 21-8

A structural model of $[Fe(C_2O_4)_3]^{---}$.

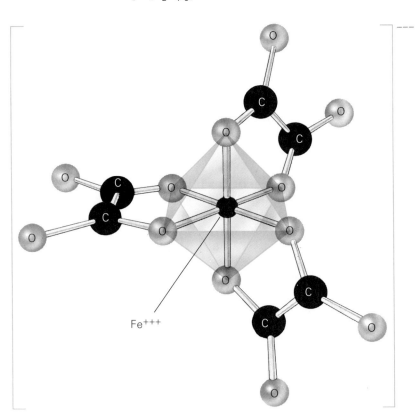

element showing a coordination number of 6. There are only three $[C_2O_4]^{--}$ groups around each iron ion. Figure 21-8 shows the arrangement. Each $[C_2O_4]^{--}$, the oxalate group, uses two of its oxygen atoms to bond with the central iron atom. The number of near neighbors, *as viewed from the iron atom,* is six oxygen atoms at the corners of an octahedron. A *group* such as oxalate, *which can furnish simultaneously two atoms for coordination,* is said to be **bidentate.** This term literally means "double-toothed."

In addition to tetrahedral and octahedral complexes, there are two other arrangements commonly found—the square planar and the linear. In a **square planar complex,** the central atom has four near neighbors at the corners of a square. The coordination number is 4, the same number as in the tetrahedral complexes. An example of a square planar complex is the complex nickel cyanide anion $[Ni(CN)_4]^{--}$.

In a **linear complex,** the coordination number is 2, corresponding to one group on each side of the central atom. An example of such an arrangement is the silver-ammonia complex. This complex generally forms when a very slightly soluble silver salt such as silver chloride dissolves in aqueous ammonia (see Figure 21-9). Another example of a linear complex is $[Ag(CN)_2]^-$, which is formed during the leaching of silver ores with NaCN solution.

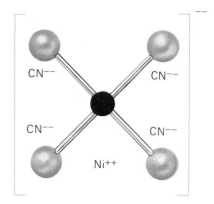

Figure 21-9

A square planar complex, $[Ni(CN)_4]^{--}$.

BONDING IN COMPLEX IONS

What holds the atoms of a complex ion together? There are two possibilities. In some complexes, such as $[AlF_6]^{---}$, the major contribution to bonding comes from the attraction between a positive ion (Al^{+++}) and a negative ion (F^-). The bonding is considered to be ionic. In other complexes, such as $[Fe(CN)_6]^{--}$, there is thought to be substantial sharing of electrons between the central atom and the attached groups. The bonding is considered to be covalent. When there is such sharing, an electron or an electron pair from the attached group spends part of its time in an orbital furnished by the central atom. In either type, as emphasized in Chapter 12, *both* atoms attract the electron.

For transition elements, there are usually empty *d* orbitals to accommodate electrons from attached groups, but any vacant orbital low enough in energy to be populated can be used to form the **coordinate covalent bond** of a complex. The pair of shared electrons in this type of bond is provided by one of the bonded atoms.

As noted in an earlier chapter, we can associate certain structural arrangements with the orbitals used in bond formation. For example, if an *s* and *p* orbital are used in bonding, the geometry of the molecule is linear. If one *s* and 3*p* orbitals are used in bonding, the geometry is tetrahedral. Other geometries associated with electron bonding orbitals are shown below:

Bonding orbitals of central atom	Bonding capacity	Geometric shape of resulting molecule or ion
sp	2	linear
*sp*³	4	tetrahedral
*dsp*²	4	square planar
*d*²*sp*³	6	octahedral

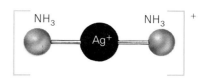

Figure 21-10

A linear complex, $[Ag(NH_3)_2]^+$.

JOHN C. BAILAR, JR.

John Bailar was born in Golden, Colorado, where his father was a professor of chemistry at the Colorado School of Mines. After receiving bachelor and master's of arts degrees from the University of Colorado, Bailar attended the University of Michigan, where he received a doctorate in 1928. Bailar then joined the faculty of the University of Illinois at Urbana. He has remained at that institution as one of the most respected and distinguished members of an outstanding department. Since 1972, Dr. Bailar has been Emeritus Professor of Chemistry.

Bailar is widely recognized for his scientific leadership and his research in metal coordination compounds. His work on the relationship between the structure and chemical reactivity of metal coordination compounds laid the foundation for much of the current work on homogeneous catalysis and for modern studies on the role of metals in biological chemistry.

Bailar has been honored by his students and won national prizes for excellence in teaching. He has also received national and international prizes for his accomplishments in research.

Assuming that the Fe^{+++} ion is spherically symmetrical, what geometric arrangement of six F^- ions around this ion would be predicted by the VSEPR theory of molecular geometry? Draw the structure.

SIGNIFICANCE OF COMPLEX IONS

In addition to their occurrence in solid compounds, complex ions such as those we have mentioned are important for three other reasons:

1. They may determine what species are present in aqueous solutions.
2. Some of them are very important in biological processes.
3. Some of them are very important catalysts.

To understand how a complex can help us determine the species in solution, consider the case of a solution made by dissolving some potassium chrome alum $[KCr(SO_4)_2 \cdot 12H_2O]$ in water. The aqueous solution is distinctly acidic. Apparently, the chromic ion $[Cr(H_2O)_6]^{+++}$ in water solution can act as a weak acid, dissociating to produce a proton (or hydronium ion). Schematically, the dissociation can be represented as the transfer of a proton from one water molecule in the $[Cr(H_2O)_6]^{+++}$ complex to a neighboring H_2O to form a hydronium ion (H_3O^+). Note that removal of a proton from an H_2O bonded to a Cr^{+++} ion leaves an OH^- group at that position. The reaction is reversible and proceeds to equilibrium:

$$[Cr(\overbrace{H_2O)_6}]^{+++} + H_2O \rightleftharpoons [Cr(H_2O)_5OH]^{++} + H_3O^+$$

We see that $[Cr(H_2O)_6]^{+++}$ acts as a proton donor—that is, as an acid. The complex determines the pH of the solution.

AMPHOTERIC COMPLEXES

Our discussion of aqueous solutions in terms of complex ions is easily extended to explain other facts. Chromium hydroxide, $Cr(OH)_3$, dissolves very little in water, but is quite soluble in both acids and very strong bases. A metal hydroxide that is insoluble in water but that dissolves in acidic or in basic solution is said to be **amphoteric.** How does the complexion of chromium hydroxide form and why is it amphoteric? Let us consider both questions.

How is the complex ion of chromium hydroxide, $[Cr(H_2O)_3(OH)_3](s)$, formed? The best starting material is an aqueous solution of $Cr(NO_3)_3$ that contains $[Cr(H_2O)_6]^{+++}$ as the major chromium species. As indicated above, this species ionizes to give $[Cr(H_2O)_5OH]^{++}$. The equation is:

$$[Cr(H_2O)_6]^{+++}(aq) \underset{H_2O}{\rightleftharpoons} [Cr(H_2O)_5(OH)]^{++} + H_3O^+(aq)$$

When NaOH is added to the above equilibrium solution, $OH^-(aq)$ ions remove $H_3O^+(aq)$ ions to give water. The equilibrium is then shifted to the right. Enough NaOH(aq) drives the process to completion. The solution now contains $[Cr(H_2O)_5(OH)]^{++}$ as the major chromium species. The newly formed species $[Cr(H_2O)_5OH]^{++}$ also undergoes ionization. The equation is:

$$[Cr(H_2O)_5(OH)]^{++} \underset{H_2O}{\rightleftharpoons} [Cr(H_2O)_4(OH)_2]^+ + H_3O^+(aq)$$

This ionization process can also be driven to completion if $H_3O^+(aq)$ is removed by addition of NaOH(aq) solution. Finally, the species $[Cr(H_2O)_4(OH)_2]^+$ can undergo ionization as follows:

$$[Cr(H_2O)_4(OH)_2]^+ \underset{H_2O}{\rightleftharpoons} [Cr(H_2O)_3(OH)_3](s) + H_3O^+$$

The green, gelatinous solid, $[Cr(H_2O)_3(OH)_3]$, will be a product. Additional NaOH will drive the last equilibrium to give the insoluble, hydrated chromium hydroxide in quantitative yield.

It is found experimentally that the green, gelatinous $[Cr(H_2O)_3(OH)_3](s)$ will dissolve in either excess NaOH or in excess HCl. The substance is amphoteric. How does each solution process occur?

Let us first consider the dissolving of $[Cr(H_2O)_3(OH)_3](s)$ by excess NaOH solution. Like all the hydrated chromium species considered above, $[Cr(H_2O)_3(OH)_3](s)$ also undergoes a little acid ionization. We can write:

$$[Cr(H_2O)_3(OH)_3](s) \underset{H_2O}{\rightleftharpoons} [Cr(H_2O)_2(OH)_4]^- + H_3O^+$$

Clearly, if a sufficiently high concentration of OH^- is established in the solution by adding NaOH, the above equilibrium can be driven completely to the right and the green, gelatinous, water-insoluble $[Cr(H_2O)_3(OH)_3]$ will dissolve as the species is converted to $[Cr(H_2O)_2(OH)_4]^-$. The final equation for dissolving $[Cr(H_2O)_3(OH)_3](s)$ in NaOH solution is:

$$[Cr(H_2O)_3(OH)_3](s) + Na^+(aq) + OH^-(aq) \longrightarrow$$
$$Na^+(aq) + [Cr(H_2O)_2(OH)_4] + H_2O$$

Now let us focus our attention on the dissolving of $[Cr(H_2O)_3(OH)_3](s)$ by an acid such as aqueous HCl. It is very helpful to consider the process by which $[Cr(H_2O)_3(OH)_3](s)$ is formed. The equation as given earlier is:

$$[Cr(H_2O)_4(OH)_2]^+(aq) \underset{H_2O}{\rightleftharpoons} [Cr(H_2O)_3(OH)_3](s) + H_3O^+(aq)$$

When HCl solution is added, the concentration of $H_3O^+(aq)$ is suddenly increased. Le Chatelier's principle tells us that addition of $H_3O^+(aq)$ will drive the process to the left. If enough HCl is added, the process will be driven to completion and insoluble $[Cr(H_2O)_3(OH)_3](s)$ will be converted quantitatively to soluble $[Cr(H_2O)_4(OH)_2]^{++}$. The overall equation is:

$$[Cr(H_2O)_3(OH)_3](s) + H_3O^+(aq) + Cl^-(aq) \longrightarrow$$
$$[Cr(H_2O)_4(OH)_2]^+(aq) + Cl^-(aq) + H_2O(l)$$

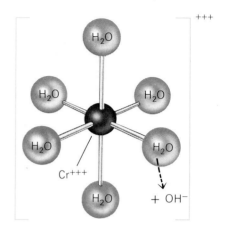

Figure 21-11

A structural model of the cation $[Cr(H_2O)_6]^{+++}$.

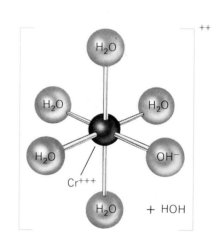

Figure 21-12

A structural model of the cation $[Cr(H_2O)_5OH]^{++}$.

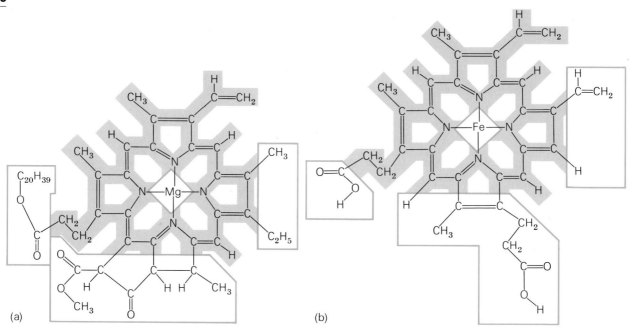

Figure 21-13

The structure of (a) chlorophyll and (b) hemin.

Since insoluble $[Cr(H_2O)_3(OH)_3]$ will dissolve in *either* excess HCl or excess NaOH, we can say that the solid is amphoteric.

COMPLEXES FOUND IN NATURE

Complex ions have important roles in certain life processes. Two such complexes are *hemin,* a part of **hemoglobin**—the pigment in the red cells of the blood—and *chlorophyll,* the green pigment in plants. Hemoglobin contains iron (Fe, atomic number 26). It therefore belongs in a discussion of complex compounds of the transition elements. Chlorophyll, however, is a complex compound of magnesium (Mg, atomic number 12), which is not a transition element.* Nevertheless, considering it here will help us avoid the misconception that only transition elements form complexes.

As extracted from plants, chlorophyll is actually made up of two closely related compounds, chlorophyll *a* and chlorophyll *b*. The two compounds differ slightly in molecular structure. They can be separated because they have different tendencies to be adsorbed on a finely divided solid, such as powdered sugar. The structural formula of chlorophyll *a* is shown in Figure 21-13a. Note the large organic molecule with a magnesium atom in the center. Around the magnesium atom are four near-neighbor nitrogen atoms, each of which is part of a five-member ring.

Figure 21-13b shows the structure of hemin. Note how closely the substance resembles chlorophyll. The portions within the unshaded areas of the illustration identify the differences between the two complexes. Aside from

---

*Chlorophyll was synthesized by Harvard chemist Robert Woodward. Other important syntheses by Woodward and his students include cholesterol, cortisone, vitamin B-12, strychnine, and several antibiotics. At the time of his death in 1979, Woodward was working on the synthesis of erythromycin, an antibiotic used to treat viral pneumonia and throat infections.

the difference in the central metal atom, the differences are all on the periphery of the two molecules.

Hemoglobin in the blood carries oxygen from the lungs to the tissue cells. A weak complex forms between the iron atom in the complex and an oxygen *molecule*. The O_2 is readily released to the cells from the bright-red weak complex (color of arterial blood). When the O_2 is stripped from the iron in the body, the complex becomes purplish red, the color of blood in the veins.

Other groups can bond to hemoglobin in place of O_2. Carbon monoxide molecules can not only replace oxygen in this way, but can bond to hemoglobin more firmly than do O_2 molecules. If we breathe a mixture of CO and O_2 molecules, our red blood cells prefer the CO molecules. The oxygen molecules are thus excluded, and cells die for lack of O_2. If caught in time, carbon monoxide poisoning can be treated by raising the ratio of O_2 to CO in the lungs. This is done by administering fresh air or oxygen. Both of the reactions involved,

$$O_2(g) + \text{hemoglobin} \rightleftharpoons \text{complex 1}$$

$$CO(g) + \text{hemoglobin} \rightleftharpoons \text{complex 2}$$

have tendencies to go to the right. If the concentration of O_2 is much higher than that of CO, O_2 will be used in complex formation instead of CO.

Key Terms

actinoids

amphoteric

bidentate

complex ion

coordinate covalent bond

coordination number

hemoglobin

lanthanoids

linear complex

square planar complex

transition elements

SUMMARY

Elements with atomic numbers 21 through 29 (located in the middle portion of the periodic table) and the elements vertically below them are known as transition elements. In these elements, the *d* orbitals become filled with electrons. The elements in which the *f* orbitals fill with electrons are known as inner transition elements. They include the rare earth, or lanthanoid, elements (atomic numbers 57 through 71) and the actinoid elements (atomic numbers 89 through 103).

Transition elements have metallic properties and high densities. They exhibit several oxidation states and form many deeply colored ions and solid compounds. Most transition elements form complex ions in which other molecules or ions are bonded to the central transition metal ion. The number of molecules or ions surrounding the central ion is known as the coordination number of the ion.

Bonding in complex ions may be ionic, covalent, or a form that lies between the two. In the coordinate covalent bond, the pair of shared electrons is provided by only one of the bonded atoms. Complex ions are important in biological processes and as catalysts. Some transition metal hydroxides, insoluble in water, dissolve in both acidic and basic solutions. Such hydroxides are said to be amphoteric.

QUESTIONS AND PROBLEMS

A

1. Name and give the atomic numbers of the transition elements found in (a) the fourth row of the periodic table and (b) the fifth row of the periodic table.

2. Indicate the position in the periodic table and the atomic numbers of the inner transition elements.

3. Identify (a) a nonreactive transition element and (b) a very reactive transition element.

4. Using the notation of an orbital diagram found in this chapter and elsewhere, write the electronic structure of a titanium atom.

5. When a Ti^{++} ion forms from Ti, the $4s$ electrons are lost. Write the electronic structure of a Ti^{++} ion using the notation of an orbital diagram.

6. Write the electronic structure of Cr using the notation of an orbital diagram. (a) What is the highest oxidation number exhibited by chromium? (b) How many s and d electrons does a chromium atom have? Are the answers to (a) and (b) related? Explain.

7. What is the formula of the oxide of vanadium in which vanadium is in its highest oxidation state? Repeat for the highest oxide of manganese.

8. Write the structural formula showing the geometry of $Co(NH_3)_6Cl_3$.

9. List the names, chemical formulas, and colors of six compounds containing the element chromium.

10. What is the central metal atom in hemin? What is the role of hemoglobin in living things?

B

11. Write the electronic structure for an Nb atom using the notation of an orbital diagram. What is its common valence or oxidation state?

12. Write the electronic structure of a Cr^{+++} ion using the notation of an orbital diagram. (Remember, $4s$ electrons are lost first.) How many unpaired electrons does a Cr^{+++} ion have?

13. Is $KMnO_4$ an oxidizing agent, a reducing agent, or both? Explain your answer.

14. Explain why the species $[Cr(H_2O)_6]^{+++}$ can act as a weak acid.

15. Explain how the precipitate $[Cr(H_2O)_3(OH)_3](s)$ will dissolve upon addition of a solution of (a) a strong base and (b) a strong acid.

16. Explain, in terms of their structure and bonding, why (a) $CrCl_3 \cdot 6NH_3$ forms a highly conducting solution and (b) $CrCl_3 \cdot 3NH_3$ solution conducts very poorly.

17. How many moles of Cl^- ions could be precipitated by Ag^+ ions from (a) a solution containing 1 mole of $CrCl_3 \cdot 6NH_3$ and (b) a solution containing 1 mole of $CrCl_3 \cdot 3NH_3$? Explain your answers.

18. Given the equilibrium

$$[CoCl_4]^{--} + 6H_2O \rightleftharpoons [Co(H_2O)_6]^{++} + 4Cl^-$$

BLUE   PINK

use Le Chatelier's principle to explain why paper coated with cobalt chloride should be blue in dry weather but turn pink in wet weather.

C

19. The ferrous ion, iron(II), forms a complex with six cyanide ions, CN^-. This octahedral complex is called ferrocyanide. The ferric ion, iron(III), forms a complex with six cyanide ions. That octahedral complex is called ferricyanide. Write the structural formulas of the ferrocyanide and ferricyanide complex ions.

20. If $CoCl_2$ is mixed with a solution containing excess NH_3 and some NH_4Cl and then air is bubbled through the solution, the following reaction takes place:

$$2CoCl_2 + 10NH_3 + 2NH_4Cl + \tfrac{1}{2}O_2 \longrightarrow 2CoCl_3 \cdot 6NH_3 + H_2O$$

(a) Draw a structural formula for the yellow compound $CoCl_3 \cdot 6NH_3$. (b) Another product that may be produced in the same reaction is a purple material with formula $CoCl_3 \cdot 5NH_3$. Draw a structural formula for this compound. (c) Two compounds with formula $CoCl_3 \cdot 4NH_3$ can also be isolated from this reaction. Suggest possible structural formulas for these compounds and ways in which you might tell them apart.

21. Why does NH_3 readily form complexes, while an NH_4^+ ion does not? Consider electronic structure as a basis for your answer and write Lewis structures.

22. Chromium(III) oxide (Cr_2O_3) is used as a green pigment and is often made by the reaction between $Na_2Cr_2O_7(s)$ and $NH_4Cl(s)$ to produce $Cr_2O_3(s)$, $NaCl(s)$, $N_2(g)$, and $H_2O(g)$. Write a balanced equation for the reaction and calculate how much pigment can be made from 1.0×10^2 kg of sodium dichromate.

23. Note that the nickel atom in $Ni(CO)_4$ has not lost any of its electrons. Electronically, it is still a nickel atom. What orbitals are available to bind four CO molecules to Ni to get $Ni(CO)_4$? Can you suggest a geometry for $Ni(CO)_4$?

24. How are chlorophyll *a* and hemin molecules structurally similar? How are they different?

25. Nickel carbonyl, $Ni(CO)_4$, boils at $43\,°C$ and uses the sp^3 orbitals of Ni for bonding.* Give reasons to justify the following: (a) it forms a molecular solid; (b) the molecule is tetrahedral; (c) bonding to other molecules is of the van der Waals type; (d) the liquid is a nonconductor of electricity; (e) it is not soluble in water.

26. When ethylene diamine ($H_2NCH_2CH_2NH_2$) is added to a $CoCl_2$ solution containing excess HCl and air is passed through the solution, the compound changes from red to yellow. Under some conditions, a mixture of products is formed. One of the products isolated in large amounts is yellow. Analysis shows that the yellow solid is 17.06 percent Co and 30.80 percent Cl. Two other products are formed in small amounts. One is green; one is purple. Analysis shows that *both* of the latter *products* are *20.65 percent* Co and *37.27 percent* Cl. Presumably, the remainder of the material in all compounds is $H_2NCH_2CH_2NH_2$. (a) Write possible structures for all the compounds. (b) What would happen if extra $AgNO_3$ was added to each solution? Be as specific as possible in your answer. (c) Represent the electronic structure of the cobalt ion in this compound. What is its oxidation state?

27. The yellow compound referred to in question 26 can be separated into two isomers that differ in the way they rotate a plane of polarized light. These are *optical isomers*. Draw the two isomers.

*The 4s electrons of Ni move back to complete the 3d level, and the now empty 4s and three 4p orbitals of Ni are available to bind the four CO molecules.

We recently had to learn constitutional law, and then we had to be economists, and now we have to become chemists and mathematicians. No wonder people long for the good old days.

CONGRESSMAN RICHARDSON PREYER

22

ENVIRONMENTAL CHEMISTRY: SOME UNRESOLVED PROBLEMS

Objectives

After reading this chapter, you should be able to:

- Understand the probable causes and environmental effects of acid rain.

- Define those parts of the earth's atmosphere known as the troposphere and stratosphere and locate their boundaries.

- Understand the role played by stratospheric ozone in shielding the earth's surface from harmful solar radiation.

- Describe the process of scientific modeling and the difficulties associated with modeling complex systems.

Many lakes are now without fish because of acid rain.

The quotation at the beginning of this chapter was prompted by the testimony of research scientists to the United States Congress. The subject being studied at that time was the possible effect on the environment of continued use of certain chemicals in aerosol sprays. Congressman Preyer noted the relevance of chemistry to matters of social concern. He realized that the public must understand the chemical principles involved in health and environmental issues. This chapter will provide some of the historical background and introduce some of the chemistry underlying two contemporary environmental concerns.

Concern for the environment is not a new phenomenon. In 1833, amateur British mathematician William F. Lloyd sketched the scenario for what he called the tragedy of the commons. The idea involved the British "common," a large pasture open to public use. Individual herders received all the profits from the sale of their cattle or milk, while the public common provided the feed. For that reason, it was to each herder's advantage to keep as many cattle on the common as possible. As long as the number of herders and animals remained small, there would be no problem. The common could support them with ease. But as each herd continued to grow, the finite meadow would reach the limit of its carrying capacity. The common would be ruined and could support no one.

In 1968, biologist Garrett Hardin* resurrected Lloyd's scenario, applying it to new concerns over environmental pollution. According to Hardin, the tragedy of the commons reappeared in the pollution that accompanied technological growth. But the modern-day scenario did not involve *removing* something from public property. Rather, it involved *adding* something— sewage, chemicals, radioactive wastes, and so on. Those substances, if released in ever-increasing amounts, might ruin our finite "spaceship earth," just as too many grazing cattle would ruin a finite common pasture.

22.1 The acid rain problem

Probably the most important natural cycle ensuring the continuity of life on earth is the vast circulation of water from the earth's surface into the atmosphere and back again—the so-called hydrologic cycle. The total amount of rain and snow falling on the earth's surface each year is more than a million billion (10^{17}) gallons. Some 80 percent falls over the oceans, while the remaining 20 percent comes down over land areas that include lakes and rivers.

Two hundred years ago, rainfall contained some dissolved CO_2, some dissolved SO_2 from volcanic action, and some other acidic gases. But the actions of human beings had not yet seriously altered the nature of rain. It is increasingly apparent that this situation no longer exists. On April 10, 1974, rainfall during a storm over Pitlochry, Scotland, was found to be as acidic as vinegar. Several storms over Wheeling, West Virginia, in late 1979 brought rain even more acidic than vinegar. While most rain is much less acidic than that, rain has become noticeably more acidic during this century.

In studying **acid rain,** it will be helpful to review briefly the terminology used to define acidity in a quantitative sense. Recall from Chapter 17 that pH equals $-\log [H^+(aq)]$. Solutions with a pH of 7.0 are *neutral*. Those with a pH above 7 are *basic*. Those with a pH below 7 are *acidic*. Recall also that a solution with a pH of 5.0 contains *ten* times as many $H^+(aq)$ ions per litre as a solution with a pH of 6.0. Vinegar, a 5-percent solution of acetic acid, has a pH of about 2.4.

EXERCISE 22-1

Refer to Appendix 2. Given that $pH = -\log [H^+]$, $\log 4 = 0.6$, $\log 10^{-3} = -3$, $\log 10^{-7} = -7$, and $\log 1 = 0$, determine the pH of solutions whose hydrogen ion concentrations, $[H^+]$, are: (a) $1 \times 10^{-7}\ M$ (b) $1 \times 10^{-3}\ M$ (c) $4 \times 10^{-3}\ M$ (d) $1 \times 10^{-4}\ M$.

*Garrett Hardin, "The Tragedy of the Commons," *Science,* vol. 162 (1968) pp. 1243–1248.

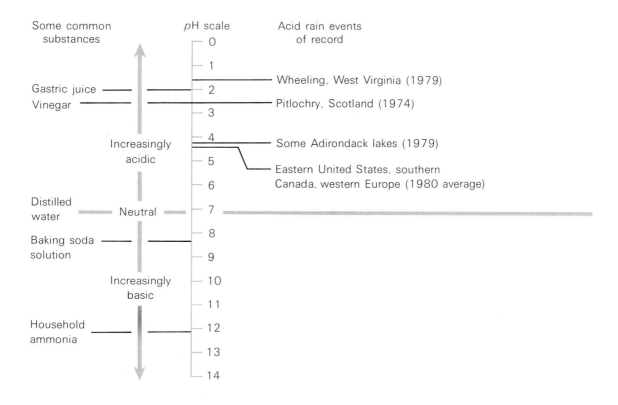

Figure 22-1
Values of pH for acid rain and some common substances.

ACIDITY OF NATURAL PRECIPITATION

What determines the acidity of precipitation? The water vapor in the air comes from the evaporation of water on the earth's surface. Once the water vapor reaches the atmosphere, it condenses on very tiny solid particles and soon reaches equilibrium with atmospheric gases such as nitrogen and oxygen. Neither the $N_2(g)$ nor the $O_2(g)$, which together make up most of the atmosphere, affects the acidity of the water. Atmospheric carbon dioxide, however, although present in small amounts, dissolves in water to form carbonic acid (H_2CO_3). Carbonic acid is known to be a weak acid with a first ionization constant of about 4×10^{-7} in water. Even so, it can increase the hydrogen ion concentration in rainwater sufficiently to create a pH as low as 5.6. The appropriate equation is

$$CO_2(g) + H_2O(l) \rightleftharpoons H_2CO_3(aq) \rightleftharpoons H^+ + HCO_3^-$$

The second ionization of HCO_3^- in carbonic acid is small enough to ignore.

> If the concentration of $H_2CO_3(aq)$ in equilibrium with CO_2 in the air is about 0.01 mole/litre, and K_A for the first ionization of H_2CO_3 is about 4×10^{-7}, show that the pH of the resulting solution would be close to 4.2.

Other natural substances may affect the pH of rainwater. Some evidence suggests that biological processes in the ocean produce reduced sulfur compounds, such as dimethyl sulfide, $(CH_3)_2S$. This compound could oxidize in the atmosphere to form sulfuric acid. Volcanoes such as Mount St. Helens emit gases including sulfur dioxide and hydrogen sulfide, which both dissolve in water to produce acidic solutions. Furthermore, both SO_2 and H_2S react with oxygen to produce SO_3. The latter, in turn, dissolves in water droplets in the air to form sulfuric acid. Thus, the acidity of natural rainfall depends on atmospheric composition.

CAUSES OF ACID PRECIPITATION

As we have noted, dissolved carbon dioxide frequently lowers the pH of normal rainwater to 5.6. *Acid precipitation is defined as rain or snow that produces water with a* pH *value below 5.6.* Acid precipitation has been known for many decades in the vicinity of large cities and industrial plants, but in recent years it has become much more widespread. Large areas of the eastern United States, southern Canada, and western Europe now have yearly average rainfall with pH values close to 4. Recall that the pH scale is logarithmic. Thus, solutions of pH 6, 5, and 4 contain respectively 1, 10, and 100 times as many $H^+(aq)$ ions per litre. A pH of 4 means that the liquid contains about 14 times as much acid as a liquid of pH 5.6.

The increased acidity of rainfall is closely related to increased emissions of nitrogen oxides and sulfur from the combustion of fossil fuels such as oil and coal. According to a recent report by a United States–Canadian research group, the burning of coal and oil for electric power in the United States produced 5.5 million tons of sulfur dioxide in 1950. By 1975, annual production from the same sources had reached 18.6 million tons. SO_2 produced in that manner now represents about two thirds of all SO_2 emissions. More than 40 percent of all nitrogen oxide emissions is thought to come from vehicles, while electric utilities contribute about 30 percent.

Sulfur dioxide, a common air pollutant, is released when sulfur-containing fuels are burned. It is also a byproduct of many smelting operations. In smelting, metal sulfide ores are roasted in air to produce metal oxides and SO_2. We can see, therefore, that human activity accounts for a large proportion of sulfur emissions.

Some simple mathematics reveals the amount of SO_2 that can be released by coal combustion near a big city. In order to generate electricity, the utility company in one major North American city has for many years burned bituminous coal containing 3.5 percent sulfur. (The sulfur content of coal used by European industries and utilities has often been higher.) During one year, the utility in that city burned about 6 million tons of such coal. Since 3.5 percent of 6 million is 210,000, it follows that 210,000 tons of sulfur

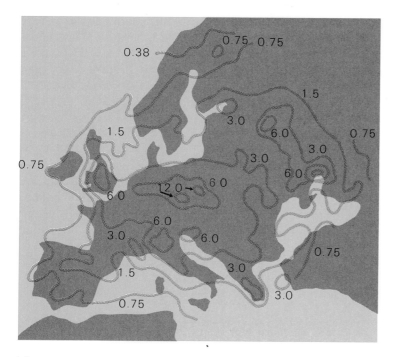

(a)

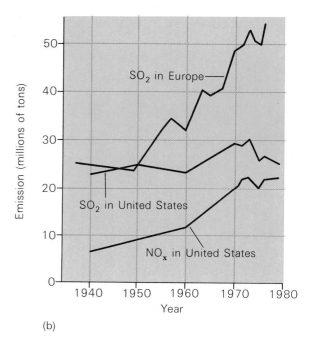

(b)

Figure 22-2

(a) Sulfur falling in Europe as sulfur dioxide (SO_2) in acid rain for the period October 1978 to September 1980. Units are in grams sulfur per square metre per year. Note that the map clearly shows that SO_2 is carried by the wind toward the east. Air currents determine the numbers. Poland and Czechoslovakia have the highest concentrations of SO_2. This map was published in the January 28, 1985 issue of *Chemical and Engineering News*. (b) This graph shows the rise in SO_2 and nitrogen oxide (NO_x) emissions caused by burning fossil fuels and smelting ores. Note the differences in SO_2 levels in the United States and Europe during the late 1970s.

(atomic weight 32) were oxidized to SO_2 (molecular weight 64). Therefore, the utility company in that city alone released 420,000 tons of SO_2 during a single year. That amount accounted for 60 percent of the total SO_2 released in the area that year.

A report to the American Association for the Advancement of Science (AAAS) estimated that yearly worldwide emissions of sulfur dioxide had reached about 80 million tons by the early 1960s. Some 50 to 60 million tons came from the burning of coal and about 11 million tons from crude oil refining. Eleven to 12 million tons came from copper smelting, and up to 4 million tons from lead and zinc smelting.

Since the United States and Canada exchange SO_2 emissions, a joint United States-Canada Commission to study the problem has been created. The United States envoy is A. L. Lewis and the Canadian envoy is W. G. Davis. Mr. Lewis has summed up the problem as follows: "Saying sulfates do not cause acid rain is the same as saying that smoking does not cause lung cancer." But a proposal to try to reduce SO_2 emissions by 1 to 3 million tons per year has not been accepted by either government as of September 1985. Economics is the problem. It is estimated that nearly 50 percent of Canada's SO_2 emissions arise from emissions in the midwestern United States. The problem is even more serious in Europe. See Figure 22-2.

Figure 22-3

Tall smokestacks allow sulfur oxide emissions to spread widely before being deposited, providing time for greater portions to be converted to sulfuric acid. Such airborne acid can fall as acid rain hundreds of miles from the source of pollution. Note the apparent addition of height to the smokestacks in this picture.

In recent years, taller smokestacks have been built in an effort to reduce the local effects of sulfur dioxide discharges. In 1955, only two stacks in the United States were higher than about 180 metres. But almost all stacks being built today are taller than that. By 1975, at least 15 stacks taller than 300 metres had been erected throughout the world. In Britain, the use of tall stacks was one of the important measures employed to reduce the severity of London smog after air pollution disasters there in the 1950s. A copper-nickel smelter in Sudbury, Ontario, has a superstack more than 400 metres tall.

Tall stacks do prevent discharges from falling in the immediate vicinity of the discharger. Pollutants are sent so high into the atmosphere that they are carried thousands of miles by prevailing winds, often across national boundaries. See Figure 22-2(a). However, because the pollutants eventually do fall to earth, local problems simply become regional ones. In addition, the lifting of pollutants such as sulfur dioxide high into the atmosphere has another, more negative effect: it gives the SO_2 more time to be converted into sulfuric acid. Once in the atmosphere, sulfur dioxide is oxidized catalytically to sulfur trioxide (SO_3). Sulfur trioxide combines with water droplets to form sulfuric acid (H_2SO_4).

$$S + O_2 \longrightarrow SO_2$$

$$2SO_2 + O_2 \xrightleftharpoons{\text{catalyst}} 2SO_3$$

$$H_2O + SO_3 \longrightarrow H_2SO_4$$

Tall stacks facilitate more effective conversion of SO_2 into sulfuric acid by prolonging the time SO_2 remains in the atmosphere. Thus, the stacks do not resolve the basic problem of air pollution.

Other major contributors to rainfall acidity are nitrogen oxides produced during the high-temperature combustion of coal and hydrocarbons. The production of nitrogen oxides is a classic illustration of chemical equilibrium and chemical kinetics at work. The formation of $NO(g)$ from $N_2(g)$ and $O_2(g)$ is an endothermic process represented by the equation

$$43.2 \text{ kcal} + N_2(g) + O_2(g) \longrightarrow 2NO(g)$$

Le Chatelier's principle clearly indicates that pressure has no effect on the equilibrium concentration of $NO(g)$ in the atmosphere. The principle also indicates that high temperatures favor $NO(g)$ formation. The high temperatures found in internal combustion engines and in the intense coal and oil fires of power plants lead to the formation of $NO(g)$.

The nitric oxide formed in the reaction described above should decompose when the temperature is lowered. In practice, however, if the temperature is lowered very rapidly as the gases are forced out of the combustion chamber, the *rate* of $NO(g)$ decomposition is very low. Because of this slow decomposition rate, some $NO(g)$ exists at room temperature, even though it would decompose at equilibrium to form $N_2(g)$ and $O_2(g)$. Addition of a catalyst would make the decomposition proceed rapidly at room temperature. That is why late-model automobiles have *catalytic converters* in the exhaust system to decompose $NO(g)$ before the exhaust gases are released into the atmosphere.

Nitric oxide that is not decomposed is present in the air in very low concentration. It is oxidized slowly under the influence of sunlight to produce $NO_2(g)$:

$$2NO(g) + O_2(g) \longrightarrow 2NO_2(g)$$

Nitrogen dioxide (NO_2), an irritating, toxic brown gas, reacts with water droplets in the air to form nitric acid as well as unstable nitrous acid:

$$2NO_2(g) + H_2O(l) \longrightarrow HNO_3(aq) + HNO_2(aq)$$

As a result of this sequence of reactions, some acid rain results from the combustion of sulfur-free coal or hydrocarbons. The United States generates about 25 million tons of nitrogen oxides (NO_x) each year.

As we have seen, the chemistry of acid rain pollution is complex. Currently, this problem is not being resolved, but acid rain reports are appearing more frequently in popular and semitechnical literature.* It is important that control programs be developed now to prevent further damage caused by acid rain, various particulate emissions, and other pollutants in the air.

*See the following articles: "European Concern About Acid Rain is Growing" in *Chemical and Engineering News,* January 28, 1985; "Acid Rain Policy" in *Chemical and Engineering News,* September 23, 1985; "Rain Sulfates in West Linked to Sulfur Emissions" in *Chemical and Engineering News,* September 2, 1985.

OPPORTUNITIES

Toxicology

Toxicology is the study of the adverse effects of chemicals on the human body. The field is one of the fastest-growing branches of the chemical industry. The effects of chemical products on animals, and sometimes on humans, are studied by toxicologists. Toxicologists may be employed by chemical companies or independent testing laboratories. Some experienced toxicologists have set up their own testing laboratories. Increasingly stringent government regulations help independent laboratories flourish, since a great deal of research is required before a new drug can be put on the market.

Toxicologists have varied and interesting jobs. They are particularly concerned with the metabolic reactions of the human body to phar-

maceuticals, cosmetics, detergents, and other chemical products. Toxicologists may be asked to perform simple, one-day tests of specific products, or they may become involved in testing programs that last several years. A toxicologist in a supervisory role may make all the significant decisions in a study, including the type of test animal to be used in the study, the range of dosages, the types of body functions to be monitored, and the monitoring intervals.

Professional toxicologists must have solid undergraduate and graduate training. The investigation of the chemical effects of foreign materials on living systems must be carried out at the cellular-molecular level. To understand reactions at this level, a student must have thorough training in biology and chemistry. A doctorate in chemistry is often required, especially for supervisory

positions. However, there are some assistant and associate positions available for those with master's degrees in toxicology or pharmacology. Further information about this field may be obtained from the Society of Toxicology, in care of Wyeth Laboratories, Box 861, Paoli, Pennsylvania 19301.

(a)

(b)

Figure 22-4

Cleopatra's Needle stood for 3,500 years in Egypt (a) with little change. However, after it was moved to New York City (b), the monument clearly showed the rapid damage caused by atmospheric pollutants.

The damage done by acid rain varies with location. In humid northern forest areas the soil is acidic and thin and there is little natural buffering capacity to neutralize the acid. Water running off from these soils is highly acidic. Streams and lakes become quite acid in a relatively short period of time. In contrast, drier areas with a lot of natural limestone can neutralize acid rain quite effectively. Damaging effects from acid rain in these areas are much less dramatic or nonexistent. Areas of North America where damage would be most severe are shown in Figure 22-5 on page 611.

The effects of such acidity on vegetation, forests, and agricultural soils are just beginning to be understood. Laboratory studies reveal significant reductions in the productivity of plants grown under simulated acid rainfall conditions. As a result, public concern is mounting over the productivity of timber, maple sugar, and fruit and vegetable crops in the northeast. In 1982, the West German Ministry of Food Agriculture and Forestry reported that 8 percent of the country's forested area was damaged by acid rain. By 1983, more detailed studies showed that the damage included 34 percent of the forests. Even more careful studies in 1985 showed that half of the country's tree areas were under some stress. The situation is similar throughout Europe. While the tree damage has not yet reached a critical stage, the problem can not be neglected.

Recent data show that, in North America, acid rain averaging pH 4.5 was primarily confined to southern Canada and the northeastern United States in the mid-1950s. Fifteen years later, acidic rainfall had spread west and south. Now, it seems that few places are untainted by acid rain. In 1975, Carl Schofield and coworkers at Cornell University surveyed 214 Adirondack lakes above 610 metres. They found that over half the lakes had pH values of less than 5. Fish are eliminated if the average pH of precipitation is below 4.3 for a significant length of time. An April 1979 survey noted 170 lakes that did not then support fish. By contrast, only 4 percent of those lakes had a pH below 5 in the 1930s.[*]

In more recent years the effects of acid rain have been seen in our cities. Buildings, statues, monuments, and metal objects have crumbled as the rain became more acidic and corrosive. For example, Figure 22-4 shows changes in Cleopatra's Needle which have occurred since it was moved from Egypt to Central Park in New York City in the 1870's. Acid rain and other air pollutants have caused the deterioration in this 21-metre stonework in a very short period of time, after it had remained relatively unchanged for 3,500 years in Egypt. Many cultural artifacts around the world are being similarly damaged by acid rain and other air pollutants.

Current research studies reflect the realization that acid rain is one of several problems that—unless checked—will affect the entire globe. A knowledge of chemistry is essential to understanding and solving the problems associated with acid precipitation.

22.2   The chlorofluorocarbon problem

June 1974 marked publication of the first scientific paper predicting possible harmful effects from continued use of chlorofluorocarbons as propellents in aerosol sprays. The predictions resulted in four Congressional hearings: two

[*] See George R. Hendrey, "Acid Rain and Gray Snow," *Natural History*, vol. 90 (February 1981) pp. 58–65; and Gene E. Likens et al., "Acid Rain," *Scientific American*, vol. 241 (October 1979) pp. 43–51.

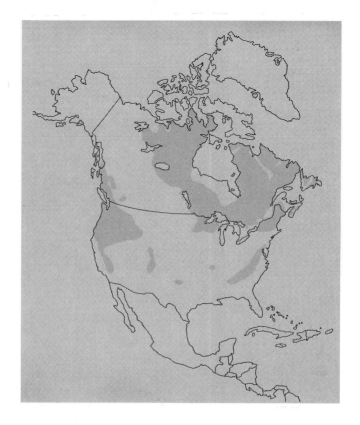

Figure 22-5

The shaded areas represent those parts of North America that are potentially most sensitive to acid rain. The granite bedrock in these regions has little buffering capacity to neutralize acid rain. The extent of damage in such areas depends on their proximity to sulfur dioxide and nitrogen oxide emissions from the combustion of fossil fuels.

in the House of Representatives and two in the Senate. The first hearing prompted the remarks of Congressman Preyer quoted at the beginning of this chapter. In order to understand the nature of these environmental concerns, we must consider some past history.

THE ATMOSPHERE AND STRATOSPHERIC OZONE

One hundred years ago, the temperature of the earth's atmosphere was thought to fall steadily with increasing altitude. The higher one went, the colder it supposedly became. Early in the 1890s, a patient French scientist, Leon Philippe Teisserenc de Bort, began a long series of experiments in which he sent temperature-measuring instruments aloft in kites and balloons. To his surprise, Teisserenc found that somewhere between 8 and 10 kilometres above the earth's surface, the temperature stopped falling, and even showed slight increases. In April 1902, after 236 balloon flights, Teisserenc announced his surprising conclusions.

Teisserenc's discovery defines our present view of the upper atmosphere as that portion above which the temperature stops falling. Although it varies with latitude, its height averages about 10 kilometres. The atmosphere below this point, down to ground level, is known as the **troposphere,** a term suggested by Teisserenc in 1908. Troposphere comes from the Greek word *tropein,* meaning "to turn over." It is an appropriate name because the rising and falling air currents in that area do make it a place of frequent turnover. All our weather—clouds, rain, storms, and so forth—occurs in the troposphere. Except for rockets and supersonic aircraft such as the Concorde, all air travel takes place in the troposphere.

For that portion of the atmosphere above the troposphere, Teisserenc suggested the name **stratosphere,** from the Latin word *stratum,* meaning

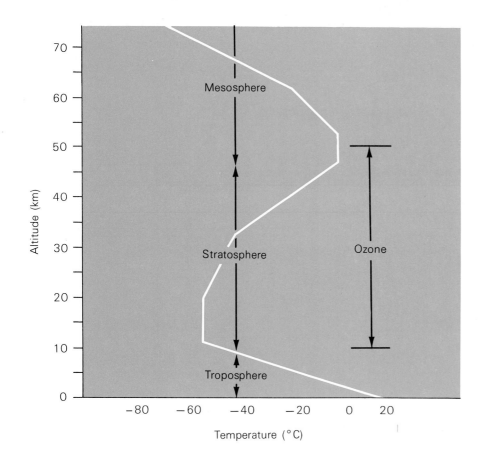

Figure 22-6
Temperature profile of the atmosphere from sea level to an altitude of 60 kilometres.

"flat layer." The name makes sense because the virtual absence of vertical movement of air layers in that region does keep the layers stratified.

Experiments since Teisserenc's time confirm that temperature not only stops falling above a certain altitude, but actually begins to rise in the lower stratosphere. At heights of 40 to 50 kilometres above the earth, the temperature is almost as high as it is on the ground. Figure 22-6 gives a temperature profile of the atmosphere from sea level to an altitude of 60 kilometres. (The atmospheric region above the stratosphere is called the *mesosphere*.)

The reason why temperature rises in the stratosphere in such a surprising way was partially supplied by meteorologist Sydney Chapman in 1930. He proposed that high-energy ultraviolet radiation from the sun is absorbed by certain gaseous molecules in the stratosphere and changes into heat. This heat, according to Chapman, warms the stratosphere and causes the temperature to rise.

The details of Chapman's proposal are contained in the four chemical equations that bear his name. They are called the **Chapman reactions:**

$$uv + O_2 \longrightarrow O + O \qquad (1)$$
$$O + O_2 \longrightarrow O_3 \qquad (2)$$
$$uv + O_3 \longrightarrow O + O_2 \qquad (3)$$
$$O_3 + O \longrightarrow O_2 + O_2 \qquad (4)$$

As these equations indicate, diatomic oxygen molecules are broken apart by ultraviolet radiation, abbreviated uv. The resulting oxygen atoms are extremely reactive. The reactive oxygen atoms can combine with ordinary O_2 molecules to produce a form of oxygen known as **ozone.** Ozone was first reported in the laboratory in 1839 by the Swiss chemist Christian Friedrich Schönbein. As the formula O_3 indicates, this unusual form of oxygen contains three atoms per molecule.

The amount of ozone in the stratosphere is small. Only several of every million gaseous particles are O_3 molecules. As the Chapman reactions indicate, the total amount of ozone in the atmosphere is maintained at a fairly constant level under normal conditions. It is formed by reactions 1 and 2, and destroyed by reactions 3 and 4.

Since ozone absorbs much of the sun's ultraviolet radiation, it is primarily responsible for warming the stratosphere. This is revealed by combining Chapman reactions 2 and 3. After canceling, the net result is absorption of ultraviolet radiation and an increase in atmospheric temperature. In other words, a conversion of radiant energy to molecular kinetic energy of the atmosphere.

EXERCISE 22-3

Add equations 2 and 3 above and show that all chemical species cancel each other out. Verify that the combined effect of the two reactions is absorption of ultraviolet radiation.

The absorption of ultraviolet radiation by stratospheric ozone is of vital importance to life on earth. Large quantities of such radiation are harmful to living organisms. By absorbing much of it, ozone forms a protective layer that reduces the amount of ultraviolet radiation that can reach the earth's surface. The little that does penetrate to ground level is what causes sunburn. Some skin cancers are believed to result from overexposure to ultraviolet radiation.

CHLOROFLUOROCARBONS AND STRATOSPHERIC OZONE

The absorption of harmful solar radiation by stratospheric ozone was first recognized by Chapman in 1930. That same year marked the first synthesis of compounds known as **chlorofluorocarbons.** It was the search for safe refrigerants that led to the development of chlorofluorocarbons, for such compounds do not occur in nature. Before 1930, the gases sulfur dioxide and ammonia, both very reactive and very irritating, were widely used in the cooling systems of commercial and home refrigerators. The synthesis of chlorofluorocarbons was indeed a triumph. Because they are of low reactivity, they pose no direct danger to human health. Chlorofluorocarbons make good refrigerants, and they soon replaced their toxic predecessors, SO_2 and NH_3, in that capacity.

Two widely used chlorofluorocarbons are the compounds $CFCl_3$, known as F-11, and CF_2Cl_2, known as F-12. Both are close relatives of methane (CH_4), the principal constituent of natural gas.

$$
\begin{array}{ccc}
\text{H} & \text{Cl} & \text{Cl} \\
| & | & | \\
\text{H--C--H} & \text{F--C--Cl} & \text{F--C--Cl} \\
| & | & | \\
\text{H} & \text{Cl} & \text{F} \\
\text{METHANE} & \text{F-11} & \text{F-12}
\end{array}
$$

However, there are two important differences between these two compounds and methane. First, methane occurs in nature in huge quantities. It is part of our natural environment. Second, methane can react: the bonds that hold the hydrogen atoms to the central carbon atom are relatively easily broken. Such a reaction occurs, for example, when methane burns, forming CO_2 and water vapor. In contrast, the bonds in F-11 and F-12 are *not* easily broken, which is why neither of these substances reacts. Neither burns and neither dissolves in water. Both are very inert. As already indicated, it is their inert character that makes these compounds so safe as refrigerants.

Aerosol spray cans first came into wide use in the 1940s as efficient dispensers of insecticides. It was found that chlorofluorocarbons, especially F-11, make good spray-can propellants. Such propellants were used to dispense, not only insecticides, but also paint, hair and deodorant sprays, and other household products marketed in spray-can form. F-12, meanwhile, found wide use in air conditioners as well as in refrigerators. By the early 1970s, almost one million tons of chlorofluorocarbons were manufactured annually. In 1974, for example, 500,000 tons of F-12 and 300,000 tons of F-11 were produced worldwide. Some 6 billion spray cans were made that year—half of them in the United States.

In 1970, sensitive measuring instruments began to detect small amounts of both F-11 and F-12 in the lower atmosphere. Only a few of these molecules were present in each billion particles of air, but they were widely distributed over the earth. The American chemist F. Sherwood Rowland wondered what would eventually become of these chlorofluorocarbons. While present in tiny amounts at each spot, their apparently uniform distribution suggested large amounts worldwide. A rough calculation convinced Rowland that the entire amount of F-11 and F-12 produced since 1930 was still present in the troposphere.

Working with the Mexican chemist Mario Molina, Rowland became convinced that all the $CFCl_3$ and CF_2Cl_2 ever made must eventually be released at ground level, either directly from spray cans or when refrigerators and air conditioners leaked or were discarded. Because the compounds do not react, they would simply collect in the lower atmosphere and slowly diffuse upward. Eventually, some of the molecules would rise through stratospheric ozone to reach altitudes of between 25 and 45 kilometres.

Once they reached the ultraviolet-absorbing ozone layer, the chlorofluorocarbons would also be bombarded by high-energy solar radiation. According to the **Rowland–Molina model,** the unreactive molecules would then be broken apart by such radiation into extremely reactive fragments:

$$uv + CFCl_3 \longrightarrow CFCl_2 + Cl \tag{5}$$

$$uv + CF_2Cl_2 \longrightarrow CF_2Cl + Cl \tag{6}$$

$$Cl + O_3 \longrightarrow ClO + O_2 \tag{7}$$

$$O + ClO \longrightarrow Cl + O_2 \tag{8}$$

One of these fragments, atomic chlorine, reacts with ozone, converting it to normal O_2, as shown in equations 7 and 8. Note that the chlorine needed to initiate reaction 7 is produced in reaction 8 and could react with more ozone. Laboratory studies done in England in 1934 and repeated by Rowland and Molina indicate that one chlorine atom can react with and destroy as many as 10,000 ozone molecules. This process, in which the reactant chlorine atom is not consumed, is an example of *catalysis* (see pages 406–410).

The eventual arrival and breakdown in the stratosphere of a significant fraction of the millions of tons of F-11 and F-12 already made would produce large numbers of chlorine atoms, *assuming* the reactions listed above were to occur. If significant ozone loss occurred, more harmful ultraviolet radiation would reach the earth's surface. Concern over this possibility led to a partial ban on use of F-11 as a spray-can propellant in 1979, but the use of freons as refrigerants was still allowed.

What will be the effect of continuing manufacture and use of chlorofluorocarbons for air conditioning and refrigeration? Early predictions made in 1977, using the best experimental and mathematical models available at that time, indicated a decrease in atmospheric ozone of 15 to 18 percent as a result of the release of chlorofluorocarbons into the air. More refined analysis made in 1982 predicted a decrease of 7 to 9 percent in the quantity of ozone. By 1983, the best models predicted a decrease of only 2 to 4 percent. It was noted that some conditions might arise that would result in a slight *increase* in the total ozone content of the atmosphere over the next century!

What do these facts show? Careful analysis of measurements of total ozone (that is, the amount of ozone above a unit area of the earth's surface) shows "no discernable trend in the total column abundance of ozone between 1970 and 1980." This excerpt was taken from the Report of the Committee on Causes and Effects of Changes in Stratospheric Ozone, National Academy of Sciences, 1983.

Research now indicates that ozone is being destroyed in the upper stratosphere as predicted, but the research also indicates that ozone is being generated in the lower stratosphere. The net change to the total amount of ozone will be the difference between these two processes. The change will probably be small. So the problem now appears much less serious than we once thought,* but continued study and modeling of the changes in the atmosphere is essential if we are to be alert to potentially harmful changes in our surroundings.

In the minds of most informed people, it would not be wise to limit the use of chlorofluorocarbons as refrigerants at this time. The benefits of chlorofluorocarbons must be balanced against the potential environmental hazards caused by their use. To date the benefits remain high while the potential environmental problems have sharply fallen. A banning of the use of chlorofluorocarbons in refrigeration could cause serious problems in food preservation and storage. On the other hand, scientists are not certain that serious environmental problems would be produced. The nonessential use of chlorofluorocarbons in spray cans, however, has been banned with little inconvenience to society. Decision-making based on valid scientific research continues to be an important factor in formulating policies of use and disuse.

*See "Ozone Depletion May Be Less Than Thought" in *Chemical and Engineering News,* February 27, 1984.

SUMMARY

The "tragedy of the commons" scenario is analogous to the pollution that has accompanied technological growth on the finite earth. Some unresolved environmental problems include acid rain and the problems associated with chlorofluorocarbons and chlorinated hydrocarbons.

The acidity of rainfall has increased steadily during the nineteenth and twentieth centuries. Current pH values for rain in some areas are close to 4, about 1,000 times more acidic than pure water, with a pH of 7. Acid rain is due to increased emissions of sulfur and nitrogen oxides into the atmosphere. The emissions come from combustion of fossil fuels, crude oil refining, and smelting operations. Acid rain is known to increase the acidity of lakes and streams, resulting in the death of fish as well as of other aquatic life. Forests are also damaged by acid rain.

The chlorofluorocarbons $CFCl_3$ and CF_2Cl_2 have been manufactured in large quantities since 1930 for use as spray-can propellents and refrigerants. Because they are chemically inert, these compounds collect in the lower atmosphere when discarded. These compounds eventually diffuse upward into the stratosphere, according to a model proposed by chemists Rowland and Molina. There, the compounds are broken into reactive fragments by ultraviolet radiation. One such fragment, atomic chlorine, may catalytically destroy ozone. Significant ozone loss in the stratosphere would permit more ultraviolet solar radiation to reach the earth's surface. Ultraviolet radiation is known to damage living cells. No directly observable ozone loss has yet been recorded and the best of our experimental models predict only a possible decrease of 2 to 4 percent in the total ozone content. The problem now appears much less serious than was once thought. Research has added much to our knowledge of the atmospheric system.

A knowledge of chemistry is essential to understanding and solving the problems that can result from release of synthetic chemicals into our finite environment. It is particularly important that we learn to safely dispose of unwanted products of technology.

Key Terms

acid rain

Chapman reactions

chlorofluorocarbons

ozone

Rowland-Molina model

stratosphere

troposphere

QUESTIONS AND PROBLEMS

A

1. Explain, in your own words, what is meant by "the tragedy of the commons".
2. What is the definition of pH? Why is it considered a measure of the acidity of rainfall?
3. Name the principal causes of acid rain.
4. What is the effect of acid rain on freshwater lakes?
5. Define (a) troposphere and (b) stratosphere, using temperature variations as the defining parameter.
6. Why do the compounds F-11 and F-12 collect in the troposphere after they are released into the environment?
7. Explain why chlorofluorocarbons are of environmental importance.

B

8. Which of the following solutions are considered acidic, and which are considered basic, or alkaline: (a) a solution of pH 3, (b) a solution of

*p*H 10, (c) a solution whose [$H^+(aq)$] is 10^{-1} *M*, (d) a solution whose [$H^+(aq)$] is 10^{-12} *M*?

9. List the advantages and dangers of using tall smokestacks for utility and smelting operations.

10. Explain, using appropriate chemical equations, how nitric oxide is produced during high-temperature combustion.

11. How do the Chapman reactions account for stratospheric (a) ozone formation, (b) ozone destruction, and (c) temperature changes? Use and explain the meaning of the appropriate chemical equations needed for each answer.

12. How does stratospheric ozone protect the earth's surface from harmful solar ultraviolet radiation? Include the appropriate chemical equation(s) in your answer.

13. Using appropriate chemical equations, indicate why atomic chlorine is considered to be a catalyst in the destruction of stratospheric ozone.

14. Why is it impossible at present to prove or disprove the Rowland-Molina model?

C

15. What is the *p*H of a solution whose hydrogen ion concentration, [$H^+(aq)$], is: (a) 10^{-1} *M* (b) 10^{-3} *M* (c) 3×10^{-8} *M* (d) 10^{-12} *M*?

16. Given that K_A for the dissociation of H_2CO_3 into $H^+(aq)$ and $HCO_3^-(aq)$ ions is 4.4×10^{-7}, and that the equilibrium concentration of H_2CO_3 is 1.42×10^{-5} *M*, show that the equilibrium [$H^+(aq)$] is 2.5×10^{-6} *M*.

17. Determine the *p*H of the solution of H_2CO_3 described in problem 16. The log of 2.5 is about 0.4.

18. If a city burned 40 million tons of 2-percent (by weight) sulfur-containing coal in a given year, calculate the number of tons of (a) S burned and (b) SO_2 released in that city in that year.

19. Calculate the number of tons of SO_2 released during the roasting of 100 tons of 97-percent-pure ZnS. Indicate all calculations and include the appropriate chemical equation(s).

The exploration of space will inevitably provide a wealth of practical benefits. But the history of science suggests that the most important of these will be unexpected—benefits we are today not wise enough to anticipate.

CARL SAGAN

ASTROCHEMISTRY

Objectives

After reading this chapter, you should be able to:

- Identify and describe the chemical composition of Earth's lithosphere, hydrosphere, and atmosphere.

- Describe the chemical makeup of the moon's surface and interior.

- Describe the chemistries of Venus, Mars, Jupiter, and Saturn and tell how they differ from the chemistry of Earth.

- Describe the chemical makeup of the sun and other stars.

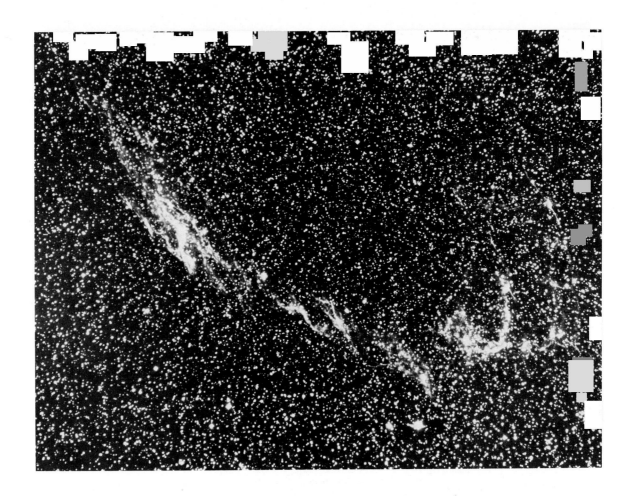

Like our sun, the countless billions of stars are made up mainly of elemental hydrogen.

Your study of chemistry has focused on some of the 109 known elements, as well as a few of the millions of compounds they form. As we look beyond Earth into nearby and deep space, we find evidence of the same elements and some of the same compounds. We also find that such substances apparently behave throughout the visible universe as they do on Earth. No matter how far we look into space, we have yet to discover any elements or compounds that are not known on Earth.

While the same substances seem to exist throughout the known universe, their relative abundances and modes of occurrence differ widely. Elemental nitrogen (N_2) makes up almost 80 percent of Earth's atmosphere. However, it is not present in significant amounts in the atmospheres of the sun and other stars. On the other hand, elemental hydrogen makes up some 95 percent of the stars, including our sun, but is not found as an uncombined element on Earth.

What do we know about the chemistry of Earth, compared to the chemistry of the rest of the universe? This chapter will provide some answers to this intriguing question.

23.1 The chemistry of Earth

Earth materials accessible to direct observation can be divided into three parts. The first part is the **lithosphere,** the outer, solid crust of Earth. The second part is the **hydrosphere,** the liquid portion, made up of Earth's waters. The third part is the **atmosphere,** the gaseous envelope surrounding our planet. First we will consider the lithosphere—the thin, rocky layer of a sphere of material with a radius of about 6,379 kilometres. That is the distance from Earth's surface to its center. Most of this distance is made up of three inner layers.

EARTH'S LAYERS

While we have not directly penetrated the solid lithosphere very deeply, we can construct a rough model of its structure from indirect evidence. Some of the evidence is derived from estimating Earth's mass from gravitational effects and calculating Earth's volume from its diameter. Such computations permit us to calculate the overall average density of Earth, which is about 5.5 grams per cubic centimetre. The average density of surface rocks is only half this value, 2.8 grams per cubic centimetre. The density at the Earth's center may be about 12 grams per cubic centimetre.

Other evidence comes from studies of how earthquake waves pass through Earth. Such studies indicate that Earth is composed of four distinct regions, or layers, of differing density and chemical composition. These layers are the crust, mantle, outer core, and inner core. The crust is made up of all the landmasses and ocean floor. The mantle rocks are a flexible solid that flow very slowly. The outer core is probably a region of constantly churning liquid of mostly iron and nickel. The inner core is probably made up mostly of solid iron and nickel under tremendous pressure. Temperatures in the core are very high, while the pressure there is several million times greater than atmospheric pressure at sea level.

The source of heat for Earth's interior is believed to come from natural radioactivity, with naturally occurring unstable isotopes such as uranium-238 undergoing spontaneous decay. High-energy alpha and beta particles given off in the decay process collide with atoms in the core, transferring much of their energy as heat. It is this heat that is thought to be responsible for the present layered structure of Earth. Because rock is a poor conductor of heat, most of the heat generated by these nuclear reactions was trapped throughout Earth's history. Thus, the interior became hotter and hotter until the iron and nickel and other heavy elements in young Earth melted and sank to the center to form the core. As Earth aged and began to cool, other, less dense materials solidified at different distances from Earth's center because of their various densities and freezing points and thus created the present layered structure.

The mantle layer, probably composed of minerals containing magnesium and iron, comprises almost 70 percent of Earth's total mass. Although the mantle has not been observed directly, geologists believe that over 90 percent of it consists of four elements: magnesium, iron, silicon, and oxygen. These elements are not free but combined in the form of silicates.

The crust makes up less than 0.5 percent of Earth's total mass and ranges in depth from an average of 8 to 32 kilometres. Some 88 elements are found in the crust, most of them in compounds, but just 12 elements make up 99.5

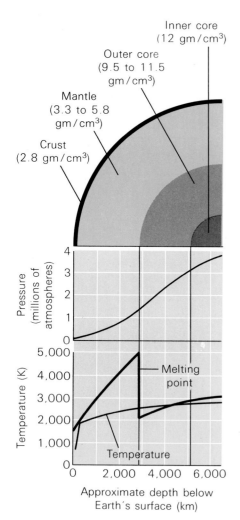

Figure 23-1

Much of this information about Earth's layers has been inferred from the analysis of earthquake waves.

percent of the crust's mass. Of these, oxygen makes up almost half: 49.5 percent. Silicon makes up a little over a quarter: 25.7 percent. The other elements are aluminum, iron, calcium, sodium, potassium, magnesium, hydrogen, titanium, chlorine, and phosphorus. Oxygen and silicon are combined chemically in the common mineral quartz. Chemical combinations of aluminum, silicon, oxygen, and hydrogen make up the bulk of common clay. The metals silver, gold, and platinum are found free in nature because they are of low chemical reactivity.

THE HYDROSPHERE

Less than 30 percent of Earth's surface is land. Over 70 percent is covered with water. More than 97 percent of the water can be found in the world's oceans. All the oceans are connected. Thus, they have roughly the same chemical makeup. Ocean water is called saline because of the various salts dissolved in it. The salinity of seawater averages 35 grams of dissolved salts per kilogram. Because the sea is so vast, it is a huge storehouse of chemicals. The most abundant of the dissolved substances is common salt, sodium chloride. Some 40 million tons of sodium chloride is extracted annually from the ocean.

Seawater is also a major source of bromine, which is extracted from dissolved bromide ions. After sodium, magnesium is the ocean's most abundant metallic element. In addition to metallic elements dissolved in the ocean as ions, certain deposits, which may become future sources of some metals, exist on the ocean floor. Manganese nodules, for example, are found as baseball-sized lumps, scattered over large areas of the ocean bottoms. The nodules are made principally of manganese, iron, and silicon. Metals such as aluminum, sodium, magnesium, titanium, copper, cobalt, and vanadium are also present in smaller quantities.

Table 23-1

Major ionic constituents of seawater

Ionic constituent	Amount (g/kg seawater)	Concentration (M)
chloride (Cl^-)	19.35	0.55
sodium (Na^+)	10.76	0.47
sulfate (SO_4^{--})	2.71	0.028
magnesium (Mg^{++})	1.29	0.054
calcium (Ca^{++})	0.412	0.010
potassium (K^+)	0.40	0.010
carbon dioxide (present in seawater as HCO_3^- and CO_2^{--})	0.106	2.3×10^{-3}
bromide (Br^-)	0.067	8.3×10^{-4}
strontium (Sr^{++})	0.0079	9.1×10^{-5}
boric acid (H_3BO_3)	0.027	4.3×10^{-4}

THE ATMOSPHERE

Table 23-2

Composition of a sample of dry air

Substance	Molecules* (%)
nitrogen (N_2)	78.08
oxygen (O_2)	20.95
argon (Ar)	0.93
carbon dioxide (CO_2)	0.03
neon (Ne)	0.0018
helium (He)	0.00052
krypton (Kr)	0.0001
hydrogen (H_2)	0.00005
xenon (Xe)	0.000008

*Traces of other substances (less than 0.0002 percent of the molecules) are also known to be present.

The atmosphere surrounding our planet forms a blanket in which 99 percent of the air is less than 30 kilometres above sea level. Dry air in Earth's present atmosphere consists principally of nitrogen (N_2) and oxygen (O_2). Nearly four out of every five molecules is nitrogen. Today's atmosphere, however, is believed to differ greatly from the original one, which was probably a combination of water vapor, methane, ammonia, and hydrogen.

How did our present atmosphere develop? The probable answer is that billions of years ago sunlight began to trigger a complex series of chemical reactions in the atmosphere. New substances, such as nitrogen, carbon dioxide, and ozone were formed. Microscopic organisms, such as blue-green algae, used the energy of sunlight to combine carbon dioxide with water to produce food. This photosynthetic process released oxygen as a waste product into the air. Through photosynthesis, the oxygen content in the atmosphere gradually increased and carbon dioxide decreased over millions of years. The various characteristics of our present atmosphere may have stabilized about 600 million years ago.

The pressure of the atmosphere falls off very rapidly with altitude. The atmosphere at Earth's surface exerts pressure due to the entire weight of the atmosphere above it. At higher altitudes, less atmosphere remains above, so less pressure is exerted. At an altitude of about 14 kilometres above Earth, air pressure is only 10 percent of its sea-level value.

The composition of the atmosphere changes with increasing altitude. These are two reasons for this change. One is the faster upward diffusion of lighter molecules, such as helium, over heavier molecules, such as O_2, N_2, and argon. Helium, a trace constituent of the atmosphere at sea level, is a major part of it above 500 kilometres. The second factor in changing atmospheric composition is chemical, and is caused by high-energy electromagnetic radiation from the sun. The ultraviolet radiation splits two-atom oxygen molecules into chemically active single atoms. These atoms attach themselves to oxygen molecules to form three-atom ozone molecules. Ozone then absorbs ultraviolet radiation, which is harmful to living things. A layer of ozone completely surrounds Earth between altitudes of about 10 and 48 kilometres, with the highest concentration between 19 and 29 kilometres.

In addition to ozone, two other atmospheric molecular species are important absorbers of energy. When solar radiation strikes Earth, some of the

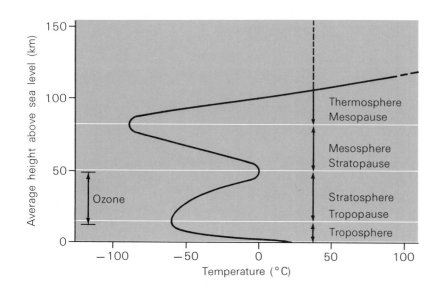

Figure 23-2

Earth's atmosphere is divided vertically into four layers on the basis of temperature.

radiation is reradiated back into space as infrared heat. Much of the infrared is absorbed by atmospheric carbon dioxide and water vapor. Were it not for the small amounts of carbon dioxide and water vapor molecules present in the lower atmosphere, the reradiated heat would be lost to Earth, and temperatures would plummet each night. Instead, much of the solar energy received by our planet is trapped by the atmosphere, keeping nighttime temperatures moderate.

23.2   The chemistry of the moon

The moon is Earth's nearest neighbor in space—only an average 380,000 kilometres away from us. The moon is a dry, airless, barren world, less than one eightieth as massive as Earth. Its diameter is 3,476 kilometres, about one fourth the terrestrial diameter. As a result of its mass and size, the moon's surface gravity is only one sixth that of Earth. Thus, a 180-pound person would weigh only 30 pounds on the moon. Any atmosphere or water that may once have been present on the moon would long ago have escaped into space.

At the end of 1968, Apollo 8 took the first astronauts around the moon and back to Earth. Six months later, on July 20, 1969, the Apollo 11 Lunar Module landed people on the moon for the first time. Astronauts Armstrong and Aldrin brought equipment in order to perform a variety of experiments. The astronauts brought instruments for testing the moon's soil, a camera to take close-up shots of the soil, an aluminum sheet to catch particles given off by the sun, and a seismometer to gather information about moonquakes. The final mission, Apollo 17, took place three and a half years after Apollo 11, in mid-December 1972.

THE LUNAR SURFACE

The Apollo legacy includes 382 kilograms (843 pounds) of moon rock returned to Earth, many kilometres of exposed film, and many rolls of magnetic tape from the lunar seismometers. Before the first moon rocks were collected, we could analyze only two kinds of material from the solar system: material from our own planet and material found in the meteorites that occasionally fall to Earth from outer space. From a study of the Apollo moon rocks, we now know that the moon's surface material is chemically different from both Earth's and from that found in meteorites. However, it more closely resembles Earth's material. (See Figure 23-3, on page 625.)

Lunar and terrestrial rocks are similar in appearance. All moon rocks are igneous, which means they formed by the cooling of molten lava. The dark regions of the moon, known as maria, are low, level areas covered with layers of basalt, a dark volcanic rock similar in composition to some of the lavas that erupt from volcanoes on Earth. The lighter parts of the lunar landscape, called highlands, are higher areas covered with a type of rock called breccia. These rocks are composed of numerous pieces of many kinds of rocks of different sizes, shapes, colors, and compositions.

The most obvious difference between moon rocks and rocks found on Earth is the complete absence of water in the lunar samples. Most rocks found on Earth contain some water chemically combined with some metallic elements. The bluish tint of most copper-containing Earth minerals, for example, is due to the presence of hydrated copper ions with the formula $Cu(H_2O)_4^{++}$. Neither such water of hydration nor free liquid or gaseous water has been found in lunar rocks or anywhere else on the moon's surface.

Another important difference between moon rocks and rocks found on Earth is that the former have never been exposed to appreciable amounts, if any, of either water or oxygen. As a result, an element such as iron, which is always found combined with oxygen on Earth, is found as crystals of pure iron in the lunar samples.

Finally, while moon rocks contain the same chemical elements found on Earth, their proportions are different. Moon rocks contain lower proportions of elements with low melting points, such as sodium (melting point 98 °C) and potassium (melting point 62 °C). Elements with higher melting points, such as calcium (melting point 850 °C), aluminum (melting point 660 °C), and titanium (melting point 1800 °C), are found in much higher proportions in lunar rocks. Titanium, which is only a minor constituent of Earth rocks, was found to make up 10 percent of some lunar samples. Other elements rare on Earth, such as hafnium (melting point 1700 °C) and zirconium (melting point 1869 °C), are also more plentiful in lunar rock samples.

The dating of lunar rocks is based on careful study of the ratio of radioactive to nonradioactive isotopes present in the rock samples. Naturally occurring uranium-238, for example, spontaneously undergoes a series of radioactive transformations, becoming lead-206, a stable form of lead. By knowing both the half-lives of these species and the proportions of uranium-238 to lead-206, it is possible to determine the age of the rock samples. Such studies indicate that the oldest moon rocks collected were formed almost 4.5 billion years ago. The youngest were formed about 3 billion years ago.

THE LUNAR INTERIOR

The seismometers placed on the lunar surface by Apollo astronauts are still recording the tiny vibrations caused by meteorite impacts on the moon's surface and by small moonquakes deep within it. The vibrations provide data from which the moon's inner structure can be inferred. The moon is not uniform inside, as was once thought, but is divided into a series of layers. The outermost layer of the moon is a crust about 65 kilometres thick, probably composed of calcium- and aluminum-rich rocks such as those found by Apollo astronauts in the lunar highlands. Beneath the crust is a thick layer of denser rock, rich in silica, known as the mantle. The mantle extends to a depth of more than 800 kilometres below the lunar surface. But what is beneath this mantle?

The nature of the moon's deep interior is still unknown. It may contain a small iron core at its center, and there is some evidence that the moon may be hot and even partly molten inside. A surprising discovery by the Apollo mission was the identification of regions of high mass concentrations, known as mascons, near or under most maria. The existence of mascons indicates that the entire lunar interior cannot be molten. If it was, the mascons would sink into the interior. Nevertheless, seismic signals made by some meteorite impacts indicate that part of the moon is molten, or at least "plastic," in consistency.

Thus, there is contradictory evidence as to whether the moon's interior is hot or cold. The picture of the moon as Earth-like in some ways, but with different amounts of various elements, does not solve the puzzle of the moon's origin. Today, most scientists believe that Earth and the moon were formed near each other, as a double planet, but the evidence is not conclusive.

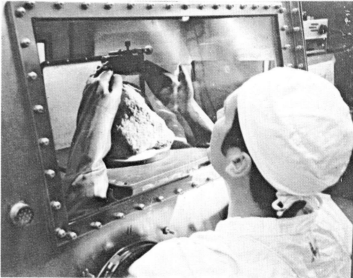

Figure 23-3

At left, an astronaut uses a lunar drill during the Apollo 15 mission, which brought back 173 pounds of lunar rock and soil. At right, a scientist studies a sample of lunar rock.

23.3 The chemistry of Venus

After the sun and the moon, the planet Venus is by far the brightest object in the sky. Its brightness is caused by a dense atmosphere of shimmering yellowish-white clouds that surround the entire planet, obscuring its surface features.

Venus is fascinating and challenging. In some ways, it is Earth's twin. Its mean distance from the sun is 108 million kilometres, nearly three quarters that of Earth's distance. Its year consists of 225 Earth days, and it is almost identical to Earth in size, mass, and density. Although Venus differs from Earth in almost every other way, understanding that planet can help us understand our own.

THE VENUSIAN ATMOSPHERE

Venus comes closer to Earth than does any other planet—as close as 45 million kilometres. However, we still know relatively little about it because of its dense cloud cover. In 1961, by means of radar astronomy from Earth, the clouds were penetrated, and it was determined that Venus rotates once every 243 Earth days (one day on Venus is about 117 days on Earth). However, its rotation is in the *opposite* direction to that of Earth. The sun would rise in the west and set in the east on Venus. This "wrong-way" rotation is an unexplained puzzle.

The first close-up measurements of Venus were radioed to Earth in 1962 by Mariner 2. Data from this first successful interplanetary probe indicated extremely high atmospheric temperatures and pressures. Measurements

Figure 23-4

A full-disc image of cloud-covered Venus, photographed by the Pioneer Venus orbiter from a distance of 59,000 kilometres.

from subsequent space probes indicate a surface pressure 90 times greater than Earth's atmospheric pressure at sea level.

As a result of space probes since 1961, we now have a reasonably accurate picture of the Venusian atmosphere*. It is composed of 97 percent carbon dioxide, 2 percent nitrogen, and 1 percent all other gases combined, such as water vapor, helium, neon, argon, and sulfur dioxide. In contrast, our atmosphere is almost 80 percent nitrogen and less than 0.5 percent carbon dioxide. Scientists believe that the large quantity of relatively heavy CO_2 causes the atmospheric pressure at Venus' surface to be much higher than Earth's. The large amount of carbon dioxide also could account for the fact that Venus' atmosphere is 50 times denser than Earth's.

THE VENUSIAN SURFACE

Before the two United States Pioneer Venus flights in 1978, less than 1 percent of the Venusian surface had been measured. Now, having examined the extensive radar data from the spacecrafts, scientists have identified huge continent-sized features, a mountain higher than Mount Everest, and deep rift valleys on Venus. The data suggest that Venus' terrain is similar to Earth's in having a gently rolling surface with dramatic raised areas. Although not as thick as those of the moon or of Mars, Venus' crust is thicker than Earth's. It apparently consists of a granite-like upper portion, with basalt-like rocks making up a denser lower portion.

The surface temperature of Venus is 480 °C—hot enough to melt lead (melting point 327 °C) and zinc (melting point 420 °C). Because Venus is closer to the sun than Earth, it receives almost twice as much solar energy as we do. However, its cloudy atmosphere reflects much of the solar energy back into space before it can reach the surface. These two competing factors cause the Venusian surface to absorb about the same amount of solar energy as does Earth. Therefore, its temperature *should* be similar to ours. The large difference between its surface temperature and ours is probably caused by a process known as the greenhouse effect. On Venus, the solar radiation that does pass through the atmosphere is absorbed by the planet's surface, which is thus heated. The energy is *re*radiated from the surface as infrared. The infrared radiation in turn is absorbed by Venus' carbon dioxide atmosphere, and so is prevented from escaping into space. Thus, Venus is far hotter than it would be if its atmosphere was transparent to infrared radiation.

Why should the Venusian atmosphere differ so from Earth's? Earth also has an enormous amount of carbon dioxide, but most of it is chemically bound in crustal rocks as calcium carbonate. The process by which certain Earth rocks can react with CO_2 in the air to form carbonates is similar to the reaction of iron with oxygen in the air to form rust. Plant life on Earth has been removing atmospheric CO_2 by photosynthesis for billions of years, thereby enriching our atmosphere with oxygen. The carbon from the CO_2 ends up in deposits of coal and oil.

Scientists believe that the large amount of CO_2 in the Venusian atmosphere is due to that planet's lack of water. On Earth, the oceans provide a huge sink for most of our carbon dioxide. Once dissolved, CO_2 becomes

*See Gerald Schubert and Curt Covey, "The Atmosphere of Venus," *Scientific American,* vol. 245 (July 1981) pp. 66–74.

chemically trapped as limestone, chalk (both $CaCO_3$), and other carbonates. As a result, less than half of 1 percent of Earth's present atmosphere is carbon dioxide, although more than 4,000 times the atmospheric amount is trapped in rocks.

Why is there no water on Venus? We originally thought that water on Earth had formed from the hydrogen and oxygen released from minerals in the hot interior. Could it be that Venus simply did not have such hydrogen- and oxygen-bearing minerals? Many astronomers consider this possibility unlikely. Since Venus and Earth are so alike in size, mass, and density, they probably had very similar makeups originally. Perhaps the high surface temperatures on Venus kept water from liquefying, so that it remained as a gas. As such, it could have been broken apart by intense solar radiation into hydrogen and oxygen, with the lighter, more rapidly moving hydrogen molecules escaping into space. The heavier oxygen molecules might then have combined slowly with surface rocks.

The sequence of events that shaped present-day Venus may never be known. However, radar data from the Pioneer Venus spacecrafts continue to accumulate, and most of the Venusian surface has been mapped. In fact, such maps will probably be more precise than best Earth-made photographic maps of the moon.

23.4 The chemistry of Mars

Seven years after astronauts first landed on the moon, the Viking landers arrived on Mars. The Viking mission to Mars combined two major goals: to study the atmosphere and geology of the entire planet and to analyze its soil and search for life in two specific locations. The landers were designed to remain on the Martian surface and radio information describing their discoveries to Earth.

THE MARTIAN ATMOSPHERE

Mars' atmosphere is thinner and colder than Earth's. Viking's first view of the Martian sky brought a major surprise. Although some scientists expected the Martian sky to be blue, like Earth's, the Viking photographs revealed a creamy, pinkish hue. Such coloration is caused by many fine suspended red dust particles above the surface. Earth's sky is generally not so dusty, except after a major volcanic eruption. The eruptions produce redder-than-normal Earth sunsets, which look something like the color of the Martian sky.

Similarities exist between the weather on Mars and on Earth, even though the thin Martian atmosphere is less than one hundredth as dense as that of Earth and has a surface pressure of only 0.008 atmospheres. On both planets, atmospheric temperature highs occur at about 3:00 P.M. local time. Viking data indicate that surface temperatures on Mars ranged from summertime highs of -31 °C to wintertime lows of -125 °C.

The Martian atmosphere is about 95 percent carbon dioxide. The atmosphere contains small amounts of carbon monoxide; oxygen in the form of O, O_2, and O_3; and water vapor. Viking's instruments also detected between 2 and 3 percent atmospheric nitrogen and between 1 and 2 percent argon. Surface pressure is only 1 percent of Earth's atmospheric pressure.

THE MARTIAN SURFACE

Viking 1 landed on Mars on July 20, 1976. Viking 2 landed some 5,000 kilometres away on September 3, 1976. Chemical analysis of the soil at both sites yield identical results. By weight, the Martian surface is 21 percent silicon, 13 percent iron, 3 percent aluminum, 5 percent magnesium, 4 percent calcium, and 3 percent sulfur. Chlorine, potassium, and titanium are each present in amounts of less than 1 percent. Scientists calculate that oxygen must make up another 42 percent of Martian soil. Thus, about 8 percent of the soil is made up of elements that could not be detected by the Viking sampling methods.

The unusually large amount of iron detected confirms the long-held theory that the red dust of Mars is red iron oxide (Fe_2O_3), similar to rust on Earth. Martian "soil" is much more Earth-like than moonlike. It is believed to be a mixture of iron-rich clay minerals and other compounds, such as iron hydroxide and magnesium sulfate. The surface material contains water and seems to be weathered, in contrast to the dry, unweathered lunar surface.

Among the surprises revealed by photographs of the Martian surface was the indication of volcanic activity on Mars and—most amazing of all—the presence of channels similar to stream beds on Earth. The channels may indicate that many years ago there was running water on the Martian surface. Why is there no detectable amount of liquid water on Mars today? The answer is that at the low Martian temperatures and pressures, water cannot exist as a liquid. Viking's water-vapor mapping team has watched Martian ice turn to vapor each day as the sun heats the surface and then turn back to ice as evening chills the ground.

The Viking team was also able to show that the permanent ice cap around Mars' north pole is made, not of frozen carbon dioxide, as had been suspected, but of frozen water. The minimum temperature on the north polar

Figure 23-5

A boulder-strewn field of red rocks reaches to the horizon, nearly 2 miles from the Viking 2 lander. The salmon color of the sky on Mars is due to dust particles suspended in the atmosphere.

ice cap during summer was found to be $-73\ °C$, a temperature at which carbon dioxide would evaporate. Because Mars' massive north polar cap does not disappear in Martian summer, what remains must be frozen water.

IS THERE LIFE ON MARS?

A major goal of the Viking missions was to determine whether the Martian soil was dead, like the soil of the moon, or teeming with microscopic life, like the soil of Earth. The Viking experiments were designed around two assumptions. First, it was assumed that Martian life, if it existed, would be like Earth life, which is based on the element carbon. Second, it was assumed that where there are large life forms, there are small ones, as on Earth. The small ones are far more abundant on Earth, with large numbers of them in almost every gram of Earth soil. Therefore, it was thought that the best possible way to detect life on Mars would be to search for carbon-based Martian microbes, or similar creatures, living in the soil. Thus, three laboratories on each Viking lander were essentially incubators designed to warm and nourish any life in the Martian soil and to detect the chemical products of the organisms' activity.

The results of Viking's search-for-life experiments are puzzling. While they do not rule out the possibility of life on Mars, the results can all be accounted for by lifeless chemical reactions in the Martian soil. The experiment designed to search for traces of organic chemicals from dead or living organisms found no such traces. Taking all the results together, most scientists concluded that the few lifelike signals emanating from the Viking biology experiments were simply indications of complex soil chemistry. Thus, we still do not know if life exists on Mars, and it may be impossible to determine this solely on the basis of the Viking missions.

23.5 The chemistry of Jupiter

The largest planet in the solar system, giant Jupiter, is 5 times as far from the sun as is Earth. Jupiter revolves around the sun once every 12 Earth years. It is 318 times as massive as Earth and, with its 16 moons, is a miniature solar system in itself. Its 4 largest moons were discovered by Galileo in 1610, when he first looked at Jupiter through his small telescope. Those 4 moons are known as the Galilean satellites. Because the moons change positions relative to Jupiter, Galileo was able to demonstrate that Earth is not the center of the universe: here were 4 objects circling Jupiter, not Earth.

JUPITER'S ATMOSPHERE

Jupiter is huge—so huge that 1,300 Earths could fit inside it. Yet its density is only one fourth that of Earth. In some ways, Jupiter is like a great balloon: if placed in water, it would almost float. All we can see of the giant planet is the gaseous part of the atmosphere, which is only about 1,000 kilometres thick and spectacularly colored. The colors may result from photochemical reactions at the cloud tops of a huge atmosphere that is about 81 percent hydrogen and about 19 percent helium, with traces of methane (CH_4) and ammonia (NH_3). There are also trace amounts of water vapor,

No Life on Mars: A Theory

Why does Mars not seem to have any organic matter? Scientists studied Mars' atmosphere and soil with the Viking landers. They also did experiments on Earth under Mars-like conditions. They now have a theory as to why there probably is no life on Mars.

Scientists have found that three things seem to work together to destroy organic matter on Mars. The three things are oxygen, ultraviolet radiation, and sand. With the help of high levels of ultraviolet radiation, oxygen reacts chemically with any organic matter and quickly breaks it down. This chemical reaction occurs quickly because sand particles act as a catalyst. Support for this theory has come from experiments on Earth that demonstrated that this type of environment prevents the buildup of complicated organic matter. So Mars, like the Moon, is and probably always was a world without life.

Figure 23-6

Jupiter and two of its satellites, Io and Europa, photographed by the Voyager 1 spacecraft from a distance of about 20 million kilometres.

phosphine, acetylene, and ethane. It is the preponderance of hydrogen and helium, the two lightest elements, that accounts for Jupiter's low density.

The outer layer of Jupiter's colorful cloud cover consists of crystals of frozen ammonia sitting atop the hydrogen-helium atmosphere. The temperature there is about −130 °C. The accepted view of this portion of Jupiter's atmosphere is that under the outer cloud layer is a second layer of icy particles of ammonium hydrosulfide (NH_4HS). When made in a dark laboratory, ammonium hydrosulfide is white, like table salt. When sunlight strikes it, the compound turns yellow, then orange, then brown. Such behavior may partially explain much of the color seen on Jupiter.

Calculations based on the measured mass and density of Jupiter indicate that the pressures below Jupiter's cloud tops must be extremely high. Pressures 100,000 times as great as those on Earth's surface are believed to exist some 2,900 kilometres under the cloud tops. The hydrogen and helium in the atmosphere are now in the liquid state. Pressures of several million Earth atmospheres are indicated at depths of about 30,000 kilometres.

The interior of Jupiter is under very high temperatures and pressures. At a depth of about 30,000 kilometres, the temperature is about 11,200 °C and the pressure is about 3 million Earth atmospheres. At about this level hydrogen breaks up into a mixture of protons and electrons. This liquid-metallic phase of hydrogen probably extends nearly all the way to the center of the planet. There is probably a small, dense central core of approximately 28 Earth masses that may be rich in metals and silicates. The temperature at the core is estimated to be about 30,200 °C.

JUPITER'S FEATURES

In 1973 and 1974, Pioneer 10 and Pioneer 11 flew past Jupiter. During the few hours of its closest approach to the planet, Pioneer 10 received a massive and unexpectedly high dose of radiation: a thousand times the lethal amount for a human. As a result of the Pioneer missions, we know that Jupiter has an enormous and powerful magnetic field that is more intense than had been expected.

The planet's gigantic magnetic field may be the largest structure in the solar system. It encompasses all four Galilean satellites and is thought to result from the planet's inner heat stirring up its compressed, metallike liquid hydrogen interior.

In March and July 1979, Voyager 1 and Voyager 2 took a closer look at Jupiter and its many moons. Among other things, these spacecraft found that as Jupiter's magnetic field sweeps past its moon Io, it generates an electric current so intense that it equals the capacity of all Earth's power plants combined. The discovery of active volcanoes on Io that throw sulfur-rich plumes hundreds of kilometres high led to speculation that the volcanoes may be the source of electrically charged sulfur particles detected as far as 7 million kilometres from Jupiter. The particles are heated to temperatures of more than 300 million °C, the highest temperature known in the solar system. They are thought to be heated by interaction with the solar wind, a stream of hot, ionized gases continually blowing away from the sun's outer surface.

The Voyagers also found that Jupiter has a thin ring around it, which is invisible from Earth. The ring could be composed of particles from Io's volcanoes, or it could be debris from comets or meteorite impacts on Jupiter's inner moons.

Another interesting feature of Jupiter is its Great Red Spot in the southern hemisphere, which is large enough to swallow several Earths. Scientists believe that the Red Spot is a permanent, relatively friction-free, stationary whirlpool of swirling gas, driven by the planet's powerful and turbulent winds. The Red Spot is probably able to maintain its shape as long as Jupiter keeps spinning rapidly. The red color may come from countless red phosphorus (P_4) molecules.

Based on new data, scientists now believe that the hurricane description of the Red Spot is inadequate. Storms on Earth are temporary and cyclonic, which means that the winds spiral around and into the center of a low pressure system. The Red Spot, on the other hand, is permanent and anticyclonic, which means that the winds spiral around and out from the center of a high pressure system. Winds blow counterclockwise in the Red Spot. On Earth, cyclonic winds blow clockwise in the southern hemisphere.

Many questions remain about the mechanics of Jupiter's atmosphere, as well as about the planet's interior. Some of the questions may be answered in the late 1980s by a still more sophisticated mission named after Jupiter's first telescopic observer, Galileo. The Galileo flight is expected to shoot a probe deep into Jupiter's atmosphere. Its main spacecraft will orbit the planet, taking extremely close shots of the moons and watching day-to-day atmospheric changes. The atmosphere will be sampled in order to determine whether organic molecules are present.

23.6 The chemistry of the remaining outer planets

In addition to Jupiter, the so-called outer planets include Saturn, Uranus, Neptune, and Pluto. Saturn, Uranus, and Neptune are giant planets, similar in many ways to Jupiter. After its successful visit to Jupiter, Voyager 1 flew

E. MARGARET BURBIDGE

E. Margaret Burbidge was born in England in 1919 and received a doctorate from the University of London in 1943. Dr. Burbidge has been a Research Fellow at the California Institute of Technology, Associate Professor at the University of Chicago's Yerkes Observatory, Research Astronomer at the University of California at San Diego, and Director of the Royal Greenwich Observatory in England.

Burbidge has specialized in stellar evolution, spectroscopic observations of quasars, studies of normal galaxies and radio galaxies, and studies of how various elements may have been formed within stars and through stellar explosions.

Throughout her career, Burbidge has been an active participant in many scientific organizations. She was President of the American Astronomical Society from 1976 to 1978 and in 1980 was elected President of the American Association for the Advancement of Science. In 1978, Dr. Burbidge was elected to the United States National Academy of Sciences.

Professor of Astronomy at the University of California at San Diego since 1964, Dr. Burbidge is also Director of the Center for Astrophysics and Space Sciences at that university.

past Saturn in November 1980 and made some stunning new discoveries.* Voyager 2 took a different trajectory, passing Saturn in August 1981. It is scheduled to continue outward in the solar system for a rendezvous with Uranus in 1986 and Neptune in 1989.

SATURN

Saturn may be the most beautiful object in the solar system. Its series of rings is breathtaking, even when viewed through small telescopes. The path to Saturn was blazed by Pioneer 11. As it flew past Saturn in September 1979, it demonstrated that spacecraft could cross the plane of Saturn's rings without colliding with the particles in the rings.

Voyager 1 cameras picked up a red spot in Saturn's southern hemisphere and another in its northern hemisphere. The spots were reminiscent of Jupiter's Great Red Spot, but smaller in size. Saturn's atmosphere seems at least as violent as Jupiter's. It is a world where lightning bolts a million times more powerful than those on Earth crack through a frigid atmosphere composed primarily of hydrogen and helium, with smaller amounts of methane, ethane, acetylene, ammonia, and phosphine. Saturn has the lowest density of any planet in the solar system: 0.7 grams per cubic centimetre, only 70 percent the density of water.

Saturn's largest moon, Titan, is smothered with a dense atmosphere that makes visual inspection of its surface all but impossible. Voyager 1 data indicate that Titan has a predominantly nitrogen atmosphere. Besides nitrogen, the atmosphere of Titan appears to contain small amounts of argon, methane, ethane, acetylene, hydrogen cyanide, and hydrogen. The thick layers of haze, shown in Figure 23-9, are composed of many organic compounds. Titan is a strange world probably covered by a kilometre-deep muddy ocean of mainly ethane, with a bottom of acetylene mud about 100 to 200 metres thick, and fed by a continual mist of ethane rain and acetylene snow. At Titan's surface temperature of −180 °C and lower, ethane is a liquid and acetylene a solid that settles to the ocean bottom. Organic solids probably coat the ocean bottom and would make the ocean a mucky red color, matching the cloudy and hazy red of the sky.

URANUS, NEPTUNE, AND PLUTO

The other two giant planets beyond Saturn—Uranus and Neptune—also have low density, indicating chemical compositions similar to those of Jupiter and Saturn. Both Neptune and Uranus appear greenish through a telescope. Only hydrogen, helium, and methane have thus far been detected spectroscopically in the atmospheres of both planets. In 1977, it was discovered that Uranus is surrounded by a dark ring system. Uranus is now known to have fifteen moons and Neptune is known to have two.

Pluto, the outermost known planet, deviates from the pattern of the other planets. It has the least circular orbit of any planet. Much of its orbit actually lies inside the orbit of Neptune. The low density of Pluto, less than 1 gm/cc, probably means that it is made up largely of frozen methane, ammonia, and water rather than silicate rocks and heavy metals.

*See Rick Gore, "Riddles of the Rings," *National Geographic,* vol. 160 (July 1981) pp. 3–31.

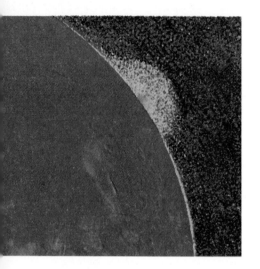

Figure 23-7

This is a special color reconstruction of an erupting volcano on Io, photographed by the Voyager 1 spacecraft from a distance of about half a million kilometres. The Voyager photograph was the first evidence of active volcanism beyond Earth.

Figure 23-8

Saturn, taken by the Voyager 2 spacecraft at a distance of 13.9 million kilometres from the planet.

23.7 The chemistry of stars

The stars we see as points of light in the night sky are huge balls of gas that generate their own heat and light. They are held together by their own gravity. The star we call our sun is so close to us that we can see detail on its surface. Nevertheless, it is just an ordinary star that would appear as a point of light if it were as far from us as the other stars.

OUR SUN

The sun is an average star, hotter than some and cooler than others. As is the case with most stars, some 94 percent of the sun's atmosphere is hydrogen, about 5.9 percent is helium, and a mixture of heavier elements make up the remaining 0.1 percent. The overall composition of the sun's interior is not very different.

The sun's atmosphere is made up of three layers. The **photosphere,** or visible "surface" of the sun, is the lowest atmospheric layer. It is about 550 kilometres thick with a temperature of about 6,000 °C. Above the photosphere is the normally invisible **chromosphere,** which is about 10,000 kilometres thick with a temperature of about 27,800 °C. The outermost layer is the normally invisible **corona.** Gas particles in the corona are moving fast enough to reach temperatures of about 1.7 million degrees Celsius. But the gas particles are so far apart that if a spaceship were flying through the corona and were shielded from the direct rays of the sun, the ship's temperature would hardly rise at all! The outward expansion of the corona is known as the *solar wind,* and the ions and electrons that compose it extend far beyond Earth.

Temperatures on the sun are hottest at its core—about 15 million °C—and they decrease slowly outward toward the sun's visible surface. Within the core, thermonuclear reactions fuse hydrogen into helium to create the sun's energy. The overall equation, in which positrons* are produced is

$$4^1_1\text{H} \rightarrow\ ^4_2\text{He} + 2^0_1e$$

Solar energy is originally produced in the form of deadly gamma radiation. It may take 20,000 years or more for this radiation to reach the sun's visible surface. On their way outward, the gamma rays collide with protons, electrons, and other solar particles. The collisions gradually dissipate the energy of the rays and lower their frequency. By the time the energy leaves the sun, much of it has been changed to infrared, visible light, ultraviolet radiation, X-rays, and radio waves.

Hydrogen fusion occurs only at the sun's core because only there are temperatures and pressures high enough to force protons to collide with sufficient speed and violence to fuse. Some 600 million tons of hydrogen are fused into helium each second within the core. Because the sun contains so much hydrogen, however, this thermonuclear process is expected to continue for at least another 5 billion years.

Galileo was the first to notice black spots on the sun's surface. The number and location of such spots is observed to change periodically. They are

*Positrons are identical in mass to electrons but have a positive electric charge.

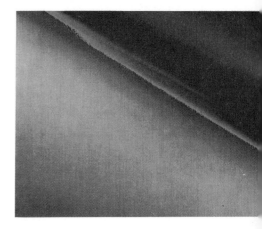

Figure 23-9
Layers of haze covering Saturn's satellite Titan were photographed by Voyager 1 from a distance of 22,000 kilometres. To highlight details of the haze, false colors have been added to this photograph.

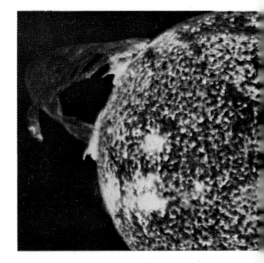

Figure 23-10
One of the most spectacular solar flares ever recorded, spanning more than 588,000 kilometres of the solar surface.

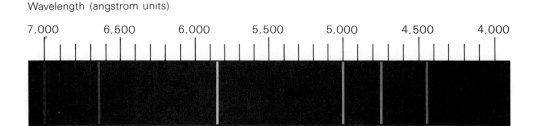

Figure 23-11

Emission spectrum for helium.

now known to be gigantic areas of concentrated magnetic fields, created by motions of the electrically charged particles that make up the gases of the hot solar atmosphere. The sunspots are often larger than Earth. They look darker than the rest of the sun's photosphere, which is about 6,000 °C, only because they are some 1,000 °C cooler. They are cooler because the presence of their magnetic fields tends to block the normal flow of heat from the sun's interior to the solar surface.

OTHER STARS

All known stars appear to have roughly the same chemical composition as our sun. Hydrogen is by far the most abundant element, with helium second. Hydrogen and helium together account for over 95 percent of the mass of a star, while hydrogen alone constitutes from 60 to 80 percent of most stars.

Unlike composition, the *temperatures* of stars vary widely, and so, therefore, do their spectra. If a star's surface temperature lies between 7,500 and 11,000 K, its spectrum shows strong hydrogen lines (see Figure 11-6, on page 272). Hotter stars, with surface temperatures of around 20,000 K, show strong helium spectral lines. Figure 23-9 shows the emission spectrum of helium. On cooler stars, some atoms can combine to form molecules, and the presence of the molecules is revealed by their spectra. The star's hot interior gives off a continuous spectrum, but the atoms and molecules in its cooler atmosphere absorb some of this light. The result is similar to the solar spectrum shown in Figure 11-6. The absorption lines reveal the molecular species present and thus give an indication of the star's surface temperature. Approximate stellar temperatures are indicated by a star's color. Cool stars are red, hotter stars are white, and the hottest visible stars are blue.

23.8 The possibility of extraterrestrial life

Conditions and chemicals similar to those found on Earth may exist elsewhere in the solar system. If they exist in our solar system, they may exist elsewhere in the vast universe. Experiments confirm that the presence of compounds containing carbon, hydrogen, nitrogen, and oxygen can lead to the formation of carbon compounds associated with living systems.

Table 23-3

Some molecules identified in interstellar space

Name	Structure	Name	Structure
acetaldehyde	H₂C(O)–CH₃ (structure)	hydrogen iso-cyanide	H—N≡C
acetonitrile	N≡C—CH₃ (structure)	hydrogen sulfide	H–S–H (structure)
acetylene radical	C≡C—H	hydroxyl	O—H
ammonia	H–NH–H (structure)	isocyanic acid	H–N=C=O (structure)
carbon monosulfide	C=S	methylacetylene	H–CH₂–C≡C—H (structure)
carbon monoxide	C=O	methyl alcohol	O–CH₃ with H (structure)
carbonyl sulfide	O=C=S	methyl amine	H–CH₂–N–H₂ (structure)
cyanoacetylene	H—C≡C—C≡N	methylenimine	N=CH₂ with H (structure)
cyanogen	C≡N	silicon oxide	Si=O
dimethyl ether	H–CH₂–O–CH₂–H (structure)	sulfur monoxide	S=O
formaldehyde	H₂C=O (structure)	thioformaldehyde	S=CH₂ (structure)
formamide	O=C(H)–NH₂ (structure)	vinyl cyanide	H₂C=CH–C≡N (structure)
formic acid	O=C(H)–OH (structure)	water	H–O–H (structure)
hydrogen	H—H		
hydrogen cyanide	H—C≡N		

Recent radio-astronomy data confirm the presence of ammonia and water vapor in interstellar space. Since 1968, about 50 molecules have been identified in outer space. Several of these, such as hydrogen cyanide (HCN), formaldehyde (HCOH), and cyanoacetylene (HC_3N), are known to be key molecules in the chemical reactions leading to the formation of living tissues. Some of the molecules that have been identified in interstellar space are shown in Table 23-3.

As you can see, the study of the chemistry of life on Earth is only part of the broader question of life elsewhere in the universe. If it happened here, it could possibly happen elsewhere. The meteorites and interstellar gas clouds that are now known to carry organic compounds associated with the biochemical precursors of life suggest that living things may not be exclusive to Earth.

Key Terms

atmosphere

chromosphere

corona

hydrosphere

lithosphere

photosphere

SUMMARY

The material that makes up Earth is found in the solid lithosphere, the liquid hydrosphere, and the atmosphere. The bottom of the lithosphere falls within Earth's mantle, one of its layers of differing density and chemical composition. The crust lies above both the mantle and the outer and inner cores below it. The inner core, at Earth's center, is composed primarily of solid iron and nickel. Most of the mantle is made up of magnesium, iron, silicon, and oxygen. Almost half the crust is made up of oxygen, and a little over a quarter of the crust is silicon. The hydrosphere is principally water combined with dissolved salts. The atmosphere is composed largely of nitrogen and oxygen, although significant amounts of helium and ozone are found at higher altitudes.

Lunar and terrestrial surface materials show some similarities and some differences. Moon rocks are similar in composition to some igneous rocks found on Earth. However, there is no water in lunar rocks, and they contain different proportions of the chemical elements found in Earth rocks. Seismological evidence suggests that the moon's interior is layered somewhat like Earth's interior, with a crust, mantle, and core.

Much has been learned about the atmosphere and surface of Venus and Mars as a result of recent space flights. Venus is almost identical to Earth in size, mass, and density, but its atmosphere is 97 percent carbon dioxide. There is no evidence of water on its very hot surface. The Martian atmosphere is about 90 percent carbon dioxide. Its surface material contains large amounts of oxygen, silicon, and iron. There is water on the surface of Mars in the form of ice and water vapor, and there may have been running water on Mars in the past.

Recent space probes have revealed a great deal about the chemistry of the outer planets Jupiter and Saturn. Jupiter's atmosphere is composed largely of hydrogen and helium, with smaller amounts of methane, ammonia, and ammonium hydrosulfide. Jupiter is surrounded by a gigantic magnetic field. Saturn's atmosphere is also composed primarily of hydrogen and helium. It has the lowest density of any planet in the solar system. The next two outer planets, Uranus and Neptune, also have low density. Their chemical composition appears to be similar to that of Jupiter and Saturn. Pluto is probably composed of frozen methane, ammonia, and water.

All known stars appear to have roughly the same chemical composition as our sun. The sun is about 94 percent hydrogen and 6 percent helium. Under the extremely high temperatures and pressures in the sun's core, thermonuclear reactions fuse hydrogen into helium to create the sun's energy. The energy of other stars is produced in a similar fashion, although their temperatures vary widely.

QUESTIONS AND PROBLEMS

A

1. Briefly explain how the layers beneath Earth's solid surface differ in composition, temperature, and pressure.
2. Explain why the composition of Earth's atmosphere changes with increasing altitude.
3. Why might life on Mars not be possible?
4. Explain why astronomers consider the planet Jupiter to be more like a star than a planet.

B

5. How much do you weigh on Earth? How much would you weigh on the moon? Account for the difference.
6. Explain why Earth's moon has no atmosphere.
7. Account for the high surface temperatures on the planet Venus. Would you expect to find liquid water on Venus? Explain.
8. What characteristics of Mars would account for the fact that no *liquid* water can be observed on its surface?
9. What conditions are necessary for a thermonuclear reaction to occur within a star? Describe the overall equation for this reaction.

C

10. Carbon dioxide accounts for 97 percent of the atmosphere of Venus and less than half of 1 percent of the atmosphere of Earth. Yet carbon, in the form of carbonates, is widely distributed throughout Earth's hydrosphere and lithosphere. Account for these differences, using chemical equations where applicable.
11. When hydrogen is subjected to extremely high pressures in Earth laboratories, it exhibits many of the properties of a liquid metal such as mercury. Why should the hydrogen at the bottom of Jupiter's huge atmosphere behave like a liquid metal?
12. Explain how spectrographic analysis can reveal a star's surface temperature.

MEASUREMENT

International system of units (SI)

The metric system is a decimal system—that is, a system based on multiples and subdivisions of 10. The United States monetary system is another example of a decimal system. The International System of Units (SI), a revision of the metric system, was established in 1960. In the SI system, a base unit, or one derived from it, is defined for each kind of measurement. Multiples and subdivisions of a unit are indicated by attaching a prefix to the unit. Some prefixes and conversion factors frequently used in chemistry are listed at the end of this appendix.

LENGTH

The SI base unit of length is the metre, a distance slightly longer than a yard. For example, 1 kilometre = 1,000 metres, 1 centimetre = 0.01 metre, and 1 millimetre = 0.001 metre.

MASS

The SI base unit of mass is the kilogram, but the gram is more frequently used in chemistry. The prefix *kilo-* indicates that 1 kilogram = 1,000 grams.

VOLUME

Volume is derived from length and may be expressed as length × length × length, or length cubed. Thus, the SI base unit of volume is the cubic metre (m^3). The smaller multiple, the cubic centimetre (cm^3), is used in the laboratory.

In measuring the volume of liquids, either the litre or its subdivision the millilitre is used. The prefix *milli-* tells us that 1 litre = 1,000 millilitres. One litre can be pictured as a cube that measures 10 centimetres on each edge. It is slightly larger than a quart. One millilitre is approximately equal to 1 cm^3.

DENSITY

The density of an object is a derived quantity. It is not measured directly but can be calculated from other direct measurements. Density (*D*) is the mass per unit volume of an object: $D =$ mass/volume. Density is expressed in kilograms per cubic metre, grams per cubic centimetre, or, for gases, grams per litre.

Uncertainty in measurement

There is some uncertainty in every measurement. The amount of uncertainty depends not only on your skill in using a measuring instrument but also on the limitations of the instrument itself. For example, a triple-beam centigram balance is accurate to 0.01 gram. Any reading on such a balance could be 0.01 gram higher or 0.01 gram lower than the actual weight. Therefore, the uncertainty in using a triple-beam centigram balance is ±0.01 gram.

Limitations of accuracy and precision contribute to uncertainty. *Precision* is how closely two or more measurements of the same thing agree, when measured with equal care by the same instrument. *Accuracy* is how close a measurement is to an accepted standard.

SIGNIFICANT FIGURES

How many digits should you record when taking a measurement? Being as careful as you can, estimate one more number than the instrument clearly indicates. Thus, the last digit on the right will be uncertain and will occupy the same decimal place as the uncertainty. All other digits are known with a fairly high degree of certainty.

If, for example, you determine the weight of an empty beaker to be 82.64 grams on a centigram balance, you would record 82.64 ± 0.01 grams. All four digits are significant: the first three are known with certainty, and the last is uncertain. According to the plus-or-minus notation, the actual weight could be as high as 82.65, as low as 82.63, or anywhere in between. Note that the last significant figure is in the hundredths place, as is the .01 uncertainty in this example.

EXPONENTIAL NOTATION

Exponential notation is a useful shorthand way to write very large or very small numbers. The number of molecules in a mole is known to be 602,300,000,000,000,000,000,000. Expressing that value as 6.023×10^{23} tells us that the number has *four* significant figures. None of the final 20 zeros in the first representation is significant.

The exponential form of a number consists of two parts. The first part is a value between 1 and 10 and includes all significant figures in the number. The second part, 10 raised to a power (the exponent), locates the decimal point. It indicates how many places we must move the decimal point: to the right if the exponent is positive, to the left if the exponent is negative. The 10^{23} after 6.023 tells us to move the decimal 23 places to the right to get the actual number of molecules in a mole. The mass of an electron, written as 9.11×10^{-28} gram, has a negative exponent. To write the actual mass, we move the decimal 28 places to the left, showing a mass of 0.000 000 000 000 000 000 000 000 000 911 gram.

EXPRESSING UNCERTAINTY IN A SINGLE MEASUREMENT

The number of significant figures used to record a measurement is the simplest expression of uncertainty. For example, a temperature of 10.2 °C is more certain than a temperature of 10 °C. When the uncertainty of the measuring instrument is known, the measurement can be expressed using the plus-or-minus notation. Reporting the temperature as 10.2 ± 0.2 °C is preferable to 10.2 °C or 10 °C, since it indicates how uncertain the reading is.

A third way of expressing uncertainty is to indicate its magnitude as a percentage of the measured quantity. In our example, we can convert the 0.2 uncertainty to a percentage by multiplying it by 100 and dividing the result by the temperature: $0.2 \times 100/10.2 = 2\%$. Thus, we can report the temperature as 10.2 °C $\pm 2\%$.

ACCUMULATION OF ERRORS IN CALCULATED RESULTS

When we calculate with uncertain quantities, our results must be uncertain. When adding or subtracting uncertain quantities, the uncertainty in the result is the sum of the individual uncertainties. When multiplying or dividing uncertain quantities, the uncertainty in the result is the sum of the percentages of uncertainty in the factors.

To better understand these rules, consider an experiment in which a sample of water is heated and the following measurements are made:

Mass of beaker (g)		Temperature of water (°C)	
Empty	Water added	Before	After
83 ± 1	185 ± 1	$10.2 \pm .2$	$15.4 \pm .2$

Suppose we wish to find the number of calories delivered to the water. To find calories, we multiply the mass (grams) of the water by its temperature change. Since both mass and temperature change are found by subtraction, we add the individual uncertainties in each:

$$\text{mass of water} = \frac{\begin{aligned} 185 &\pm 1 \text{ g} \\ - 83 &\pm 1 \text{ g} \end{aligned}}{102 \pm 2 \text{ g}}$$

$$\text{temperature change of water} = \frac{\begin{aligned} 15.4 &\pm .2 \text{ °C} \\ -10.2 &\pm .2 \text{ °C} \end{aligned}}{5.2 \pm .4 \text{ °C}}$$

Multiplying mass by temperature change means we must add percentages of uncertainty. The percentages of uncertainty are $2 \times 100/102 = 2\%$ for the mass of the water and $.4 \times 100/5.2 = 8\%$ for the temperature change. Thus, the calories absorbed by the water are

$$(102 \text{ g} \pm 2\%) \times (5.2 \text{ °C} \pm 8\%) = 530 \text{ cal} \pm 10\%$$

Since 10 percent of 530 is about 50, we can also report our answer as 530 ± 50 calories.

Note that the number of significant figures in the derived answer is the same as the least precise factor. The mass, 102 grams, has three significant figures, but the temperature change, 5.2 °C, has only two. The answer, 530 calories, can thus have only two significant figures. The zero is not significant. We emphasized the number of significant figures by writing the answer exponentially, as 5.3×10^2 calories.

ACCURACY OF EXPERIMENTAL RESULTS

As we have indicated, accuracy is how close an experimental measurement or result is to an accepted standard value. If the accepted value is known, the accuracy of the result can be expressed as a percentage of difference:

$$\% \text{ difference} = \frac{(\text{accepted value} - \text{your value})}{\text{accepted value} \times 100}$$

For example, if it was known that 550 calories were actually delivered to the water in our experiment, the accuracy of the experiment would be

$$\% \text{ difference} = \frac{(550 \text{ cal} - 530 \text{ cal})}{550 \text{ cal} \times 100} = 3.6\%$$

If the accepted value of the result is not known, its accuracy cannot be established with certainty. If the precision, or reproducibility, of a measurement is high, then the accuracy is frequently assumed to be high. This is especially true if the measurements are made using several different experimental techniques.

Selected prefixes used in the metric system

Prefix	Abbreviation	Meaning	Example
mega-	M	10^6	1 megametre (Mn) = 1×10^6 m
kilo-	k	10^3	1 kilometre (km) = 1×10^3 m
deci-	d	10^{-1}	1 decimetre (dm) = 0.1 m
centi-	c	10^{-2}	1 centimetre (cm) = 0.01 m
milli-	m	10^{-3}	1 millimetre (mm) = 0.001 m
micro-	μ	10^{-6}	1 micrometre (μm) = 1×10^{-6} m
nano-	n	10^{-9}	1 nanometre (nm) = 1×10^{-9} m
pico-	p	10^{-12}	1 picometre (pm) = 1×10^{-12} m

Conversion factors

Unit	Abbreviation	Definition or equivalent
angstrom	Å	10^{-8} cm = 10^{-10} m = 0.1 nm
atmosphere	atm	760 torr (or mm Hg) = 101,325 Pa*
torr (mm Hg)	torr	1.316×10^{-3} atm
calorie	cal	4.184 J†
kelvin temperature scale	K	K = °C + 273.16 (freezing point of water = 273.16 K)
unified atomic mass unit	u	1.6605×10^{-24} g

*The pascal (Pa) is the unit of pressure in the internationally accepted SI system. It has the dimensions of newtons/m². The newton (N) is the unit of force in the SI system.
†The joule (J) is the unit of energy in the internationally accepted SI system. It has the dimensions kg × metres²/sec² or mass × velocity², the units of kinetic energy. One joule = 0.24 cal.

HEATS OF FORMATION ($\Delta H_{formation}$) OF VARIOUS SUBSTANCES (25 °C and 1.0 atm)

Positive values of $\Delta H_{formation}$ indicate endothermic reactions. Negative values of $\Delta H_{formation}$ indicate exothermic reactions.

Substance	$\Delta H_{formation}$ (kcal/mole)	Substance	$\Delta H_{formation}$ (kcal/mole)
$AgCl(s)$	−30.4	C_2H_4 (ethylene)(g)	+12.5
$AgI(s)$	−14.9	C_2H_6 (ethane)(g)	−20.2
$AgO(s)$	−7.3	C_3H_4 (methyl acetylene)(g)	+44.3
$AlCl_3(s)$	−166.2	C_3H_6 (propylene)(g)	+4.9
$AlF_3(s)$	−311.0	C_3H_8 (propane)(g)	−24.8
$Al_2O_3(s)$	−399.1	C_4H_{10} (n-butane)(g)	−29.8
$Al_2(SO_4)_3(s)$	−821.0	C_5H_{12} (n-pentane)(l)	−35.0
$Al_2(SO_4)_3 \cdot 6H_2O(s)$	−1,268.1	C_6H_6 (benzene)(l)	+19.8
$AsCl_3(l)$	−80.2	C_6H_{14} (n-hexane)(l)	−40.0
$AsF_3(g)$	−218.3	C_7H_8 (toluene)(l)	+11.9
$AsF_3(l)$	−226.8	C_7H_{16} (n-heptane)(l)	−44.9
$As_2S_3(s)$	−35.0	C_8H_{18} (n-octane)(l)	−49.8
$AuCl(s)$	−8.4	$C_2N_2(g)$	+73.6
$AuCl_3(s)$	−28.3	C_2H_5OH (ethyl alcohol)(l)	−66.4
		$(CH_3)_2O$ (dimethyl ether)(g)	−44.3
$BCl_3(g)$	−94.5	CaC_2 (calcium carbide)(s)	−15.0
$BF_3(g)$	−265.4	$CaCl_2(s)$	−190.0
$B_2H_6(g)$	+7.5	$CaCO_3(s)$	−288.4
$B_5H_9(g)$	+15.0	$CaF_2(s)$	−290.3
$B_2O_3(s)$	−302.0	$CaH_2(s)$	−45.1
$BaCO_3(s)$	−291.3	$CaO(s)$	+151.9
$BaO(s)$	−133.4	$Ca(OH)_2(s)$	−235.8
$BaSO_4(s)$	−350.2	$CaSO_4(s)$	−342.4
$Br(g)$	+26.7	$CaSO_4 \cdot 2H_2O$ (gypsum)(s)	−483.1
		$CdCl_2(s)$	−93.0
$C(g)$	+171.7	$CdSO_4(s)$	−221.4
$CCl_4(l)$	−33.3	$Cl(g)$	+29.0
$CF_4(g)$	−162.5	$Cl_2CO(g)$	−53.3
$CH_4(g)$	−17.9	$CoCl_2(s)$	−75.8
$CHCl_3(g)$	−24.0	$CoF_2(s)$	−157.0
$CHCl_3(l)$	−31.5	$CoF_3(s)$	−185.0
$CH_3Cl(g)$	−19.6	$Co(NO_3)_3(s)$	−100.9
$CH_3NH_2(g)$	−6.7	$Co(OH)_2(s)$	−129.3
CH_3OH (methyl alcohol)(l)	−57.0	$CoSO_4(s)$	−205.5
$CS_2(l)$	+21.0	$CrCl_2(s)$	−94.6
C_2H_2 (acetylene)(g)	+54.2		

(cont.)

Substance	$\Delta H_{formation}$ (kcal/mole)	Substance	$\Delta H_{formation}$ (kcal/mole)
$CrCl_3(s)$	−134.6	$H_3PO_3(aq)$	−232.2
$CrF_3(s)$	−265.2	$H_3PO_4(aq)$	−308.2
$Cr_2O_3(s)$	−269.7	$HgCl_2(s)$	−55.0
$Cr(OH)_3(s)$	−247.1	$Hg_2Cl_2(s)$	−63.3
$CuCl(s)$	−32.5	$HgI_2(s)$	−25.2
$CuCl_2(s)$	−52.3	$HgS(s)$	−12.9
$CuCO_3(s)$	−142.2		
$CuO(s)$	−37.1	$I(g)$	+25.5
$CuS(s)$	−11.6	$ICl(g)$	+4.2
$CuSO_4(s)$	−184.0	$KCl(s)$	−104.2
$CuSO_4 \cdot 5H_2O(s)$	−544.5	$KClO_3(s)$	−93.5
		$KClO_4(s)$	−103.6
$F(g)$	+18.3	$KCN(s)$	−26.9
$FeCl_3(s)$	−96.8	$KF(s)$	−134.5
$Fe_2O_3(s)$	−196.5	$KMnO_4(s)$	−194.4
$Fe_3O_4(s)$	−267.9	$KNO_3(s)$	−117.8
$Fe(OH)_2(s)$	−135.8	$KOH(s)$	−101.8
$Fe(OH)_3(s)$	−197.0	$K_2CO_3(s)$	−273.9
$FeS(s)$	−22.7	$K_2SO_4(s)$	−342.7
$FeSO_4(s)$	−220.5		
		$MgCl_2(s)$	−153.4
$HBr(g)$	−8.7	$Mg(ClO_4)_2(s)$	−140.6
$HCl(g)$	−22.0	$MgCO_3(s)$	−266.0
$HCl(aq)$	−40.0	$MgF_2(s)$	−263.5
$HCN(l)$	+25.2	$Mg(NO_3)_2(s)$	−188.7
$HCN(g)$	+31.2	$MgO(s)$	−143.8
$HCN(aq)$	+25.2	$Mg(OH)_2(s)$	−221.0
$HF(g)$	−64.2	$MgSO_4(s)$	−305.5
$HI(g)$	+6.2	$MnCl_2 \cdot H_2O(s)$	−188.5
$HNO_2(aq)$	−28.4	$MnCO_3(s)$	−213.9
$HNO_3(l)$	−41.4	$MnO_2(s)$	−124.2
$HNO_3(aq)$	−49.4	$MnSO_4(s)$	−254.2
$H_2O(l)$	−68.3		
$H_2O(g)$	−57.8	$N(g)$	+113.0
$H_2O_2(l)$	−44.8	$NH_3(g)$	−11.0
$H_2O_2(aq)$	−45.7	$NH_3(aq)$	−19.3
$(H_2N)_2CO(s)$	−79.6	$NH_4Cl(s)$	−75.4
$(H_2N)_2CO(aq)$	−76.3	$(NH_4)_2SO_4(s)$	−281.9
$H_2S(g)$	−4.8	$NO(g)$	+21.6
$H_2S(aq)$	−9.4	$NO_2(g)$	+8.1
$H_2Se(g)$	+20.5	$NOCl(g)$	+12.6
$H_2SO_3(aq)$	−145.5	$N_2H_4(aq)$	+8.16
$H_2SO_4(l)$	−216.9	$N_2O(g)$	+19.5
$H_3AsO_4(aq)$	−214.8	$N_2O_4(g)$	+2.3
$H_3BO_3(aq)$	−255.2	$NaCl(s)$	−98.0
$H_3BO_3(s)$	−260.2	$NaH(s)$	−13.7

Substance	$\Delta H_{formation}$ (kcal/mole)	Substance	$\Delta H_{formation}$ (kcal/mole)
$NaOH(s)$	−102.0	$SF_6(g)$	−262.0
$NaNO_3(s)$	−111.5	$SO_2(g)$	−70.8
$Na_2CO_3(s)$	−270.3	$SO_3(g)$	−94.4
$Na_2O_2(s)$	−120.6	$SiCl_4(g)$	−145.7
$Na_2S(s)$	−89.2	$SiCl_4(l)$	−153.0
$Na_2SO_4(s)$	−330.9	$SiF_4(g)$	−370.0
$Na_2SO_4 \cdot 10H_2O(s)$	−1,033.5	$SiH_4(g)$	−14.8
$NiCl_2(s)$	−75.5	$SiO_2(s)$	−205.4
$NiO(s)$	−213.0	$SnCl_2(s)$	−83.6
$Ni(OH)_2(s)$	−128.6	$SnCl_4(l)$	−130.3
		$SrCl_2(s)$	−198.0
$O(g)$	+59.2	$SrCO_3(s)$	−291.9
		$SrO(s)$	−141.1
$P(g)$	+75.2	$SrSO_4(s)$	−345.3
$PBr_3(g)$	−35.9		
$PCl_3(g)$	−73.2	$TiCl_3(s)$	−165.0
$PCl_5(g)$	−95.4	$TiCl_4(l)$	−179.3
$PH_3(g)$	+2.2		
$POCl_3(g)$	−141.5	$V_2O_3(s)$	−290.0
$P_4(g)$	+13.1	$V_2O_5(s)$	−373.0
$P_4O_{10}(s)$	−720.0		
$PbCl_2(s)$	−85.8	$ZnCl_2(s)$	−99.4
$PbCO_3(s)$	−167.3	$ZnCO_3(s)$	−194.2
$PbCrO_4(s)$	−225.2	$ZnO(s)$	−83.2
$PbO(s)$	−52.4	$ZnS(s)$	−48.5
$PbO_2(s)$	−66.1	$ZnSO_4(s)$	−233.9
$S(g)$	+53.2	$ZnSO_4 \cdot 6H_2O(s)$	−663.3

MOSELEY'S EXPERIMENT: The relationship between atomic number and the frequency of lines in X-ray spectra

If a beam of cathode rays (a stream of electrons) is allowed to strike a metal target in a vacuum tube, X-rays or light of very short wavelength is produced. A device of this type that generates X-rays, is known as an X-ray tube, as shown below. In 1913, Henry G.J. Moseley studied the nature of the X-radiation given off by X-ray tubes containing different target elements. For example, he measured the frequency of the X-radiation when elemental cobalt was the target. The same measurements were made using nickel, copper, zinc, and so on, as targets. He then found that a plot of the square root of the frequency against the atomic number of the target produced a straight line, as shown below.

The result of Moseley's experiment was of great significance since it clearly identified the atomic number mentioned by Bohr and Rutherford as a fundamental characteristic of the atom. The relationship also was important in showing the atomic numbers of undiscovered elements. For example, if in the graph the element of atomic number 25 was still undiscovered, a blank space would appear in the Moseley plot at this atomic number. If chemists knew the atomic number of a missing element, they could guess its properties and devise procedures for its separation. Furthermore, the expected characteristic X-ray line (predicted by the plot) would provide an easy and certain method for identifying the new element when it was found.

It is also possible to derive the Moseley relationship using Planck's equation $E = hv$, the Bohr model of the atom, and the quantum theory model of the atom. You will not study this derivation until you take a college chemistry course.

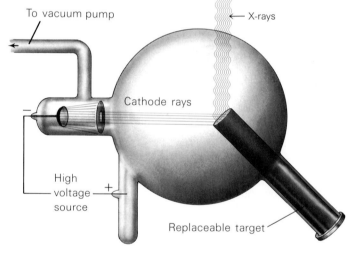

A schematic drawing of an X-ray tube.

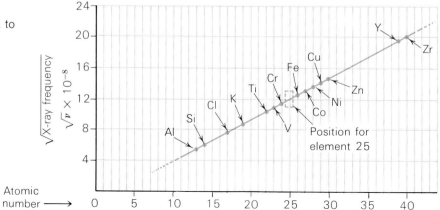

The relationship of X-ray spectra to atomic number.

NAMES, FORMULAS, AND CHARGES OF SOME COMMON IONS

Note that, in ionic compounds, the relative number of positive and negative ions is such that the sum of their electric charges is zero.

Positive ions (cations)

Name	Symbol
ammonium	NH_4^+
cesium	Ce^+
copper(I),* cuprous	Cu^+
hydrogen, hydronium	H^+, H_3O^+
lithium	Li^+
potassium	K^+
silver	Ag^+
sodium	Na^+
barium	Ba^{++}
calcium	Ca^{++}
chromium(II), chromous	Cr^{++}
cobalt	Co^{++}
copper(II), cupric	Cu^{++}
iron(II),* ferrous	Fe^{++}
lead	Pb^{++}
magnesium	Mg^{++}
manganese(II), manganous	Mn^{++}
mercury(I),* mercurous	Hg_2^{++}
mercury(II), mercuric	Hg^{++}
strontium	Sr^{++}
tin(II),* stannous	Sn^{++}
zinc	Zn^{++}
aluminum	Al^{+++}
chromium(III), chromic	Cr^{+++}
iron(III), ferric	Fe^{+++}
tin(IV), stannic	Sn^{++++}

*Aqueous solutions are readily oxidized by air.

Negative ions (anions)

Name	Symbol
acetate	CH_3COO^-
bromide	Br^-
hydrogen carbonate ion, bicarbonate	HCO_3^-
chlorate	ClO_3^-
chloride	Cl^-
chlorite	ClO_2^-
fluoride	F^-
hydroxide	OH^-
hypochlorite	ClO^-
iodide	I^-
nitrate	NO_3^-
nitrite	NO_2^-
hydrogen oxalate ion, bioxalate	$HC_2O_4^-$
perchlorate	ClO_4^-
permanganate	MnO_4^-
dihydrogen phosphate	$H_2PO_4^-$
hydrogen sulfate ion, bisulfate	HSO_4^-
hydrogen sulfide ion, bisulfide	HS^-
hydrogen sulfite ion, bisulfite	HSO_3^-
carbonate	CO_3^{--}
chromate	CrO_4^{--}
dichromate	$Cr_2O_7^{--}$
oxalate	$C_2O_4^{--}$
oxide	O^{--}
monohydrogen phosphate	HPO_4^{--}
sulfate	SO_4^{--}
sulfide	S^{--}
sulfite	SO_3^{--}
phosphate	PO_4^{---}

Appendix 5

RELATIVE STRENGTHS OF ACIDS IN AQUEOUS SOLUTION AT ROOM TEMPERATURE

The equation for the ionization is $HB(aq) \rightleftharpoons H^+(aq) + B^-(aq)$. Since all ions and molecules in water solution are aquated, the (aq) is assumed in the notation. We then write $K_A = [H^+][B^-]/[HB]$.

Acid	Strength	Reaction	K_A
perchloric acid	very strong	$HClO_4 \longrightarrow H^+ + ClO_4^-$	very large
hydriodic acid		$HI \longrightarrow H^+ + I^-$	very large
hydrobromic acid		$HBr \longrightarrow H^+ + Br^-$	very large
hydrochloric acid		$HCl \longrightarrow H^+ + Cl^-$	very large
nitric acid		$HNO_3 \longrightarrow H^+ + NO_3^-$	very large
sulfuric acid	very strong	$H_2SO_4 \longrightarrow H^+ + HSO_4^-$	very large
hydronium ion		$H_3O^+ \longrightarrow H^+ + H_2O$	1
oxalic acid		$HOOCCOOH \longrightarrow H^+ + HOOCCOO^-$	5.4×10^{-2}
sulfurous acid ($SO_2 + H_2O$)		$H_2SO_3 \longrightarrow H^+ + HSO_3^-$	1.7×10^{-2}
hydrogen sulfate ion	strong	$HSO_4^- \longrightarrow H^+ + SO_4^{--}$	1.3×10^{-2}
phosphoric acid		$H_3PO_4 \longrightarrow H^+ + H_2PO_4^-$	7.1×10^{-3}
ferric ion		$Fe(H_2O)_6^{+++} \longrightarrow H^+ + Fe(H_2O)_5(OH)^{++}$	6.0×10^{-3}
hydrogen telluride		$H_2Te \longrightarrow H^+ + HTe^-$	2.3×10^{-3}
hydrofluoric acid	weak	$HF \longrightarrow H^+ + F^-$	6.7×10^{-4}
nitrous acid		$HNO_2 \longrightarrow H^+ + NO_2^-$	5.1×10^{-4}
hydrogen selenide		$H_2Se \longrightarrow H^+ + HSe^-$	1.7×10^{-4}
chromic ion		$Cr(H_2O)_6^{+++} \longrightarrow H^+ + Cr(H_2O)_5(OH)^{++}$	1.5×10^{-4}
benzoic acid		$C_6H_5COOH \longrightarrow H^+ + C_6H_5COO^-$	6.6×10^{-5}
hydrogen oxalate ion		$HOOCCOO^- \longrightarrow H^+ + OOCCOO^{--}$	5.4×10^{-5}
acetic acid	weak	$CH_3COOH \longrightarrow H^+ + CH_3COO^-$	1.8×10^{-5}
aluminum ion		$Al(H_2O)_6^{+++} \longrightarrow H^+ + Al(H_2O)_5(OH)^{++}$	1.4×10^{-5}
carbonic acid ($CO_2 + H_2O$)		$H_2CO_3 \longrightarrow H^+ + HCO_3^-$	4.4×10^{-7}
hydrogen sulfide		$H_2S \longrightarrow H^+ + HS^-$	1.0×10^{-7}
dihydrogen phosphate ion		$H_2PO_4^- \longrightarrow H^+ + HPO_4^{--}$	6.3×10^{-8}
hydrogen sulfite ion		$HSO_3^- \longrightarrow H^+ + SO_3^{--}$	6.2×10^{-8}
ammonium ion	weak	$NH_4^+ \longrightarrow H^+ + NH_3$	5.7×10^{-10}
hydrogen carbonate ion		$HCO_3^- \longrightarrow H^+ + CO_3^{--}$	4.7×10^{-11}
hydrogen telluride ion		$HTe^- \longrightarrow H^+ + Te^{--}$	1.0×10^{-11}
hydrogen peroxide	very weak	$H_2O_2 \longrightarrow H^+ + HO_2^-$	2.4×10^{-12}
monohydrogen phosphate ion		$HPO_4^{--} \longrightarrow H^+ + PO_4^{---}$	4.4×10^{-13}
hydrogen sulfide ion		$HS^- \longrightarrow H^+ + S^{--}$	1.2×10^{-15}
water		$H_2O \longrightarrow H^+ + OH^-$	$1.8 \times 10^{-16*}$
hydroxide ion		$OH^- \longrightarrow H^+ + O^{--}$	$< 10^{-36}$
ammonia	very weak	$NH_3 \longrightarrow H^+ + NH_2^-$	very small

*The acid equilibrium constant, K_A, for water is $\dfrac{K_w}{[H_2O]} = \dfrac{1.00 \times 10^{-14}}{55.6}$.

Appendix 6

STANDARD REDUCTION POTENTIALS FOR HALF-REACTIONS:
Ionic concentrations = 1 molar in water, temperature = 25 °C

Strength as oxidizing agent	Half-reaction	E^0 (volts)	Strength as reducing agent
very strong oxidizing agents	$F_2(g) + 2e^- \longrightarrow 2F^-$	+2.87	very weak reducing agents
	$H_2O_2 + 2H^+ + 2e^- \longrightarrow 2H_2O$	+1.77	
	$MnO_4^- + 8H^+ + 5e^- \longrightarrow Mn^{++} + 4H_2O$	+1.52	
	$Cl_2(g) + 2e^- \longrightarrow 2Cl^-$	+1.36	
	$Cr_2O_7^{--} + 14H^+ + 6e^- \longrightarrow 2Cr^{+++} + 7H_2O$	+1.33	increasing strength as reducing agent
	$MnO_2(s) + 4H^+ + 2e^- \longrightarrow Mn^{++} + 2H_2O$	+1.28	
	$\frac{1}{2}O_2(g) + 2H^+ + 2e^- \longrightarrow H_2O$	+1.23	
	$Br_2(l) + 2e^- \longrightarrow 2Br^-$	+1.06	
	$NO_3^- + 4H^+ + 3e^- \longrightarrow NO(g) + 2H_2O$	+0.96	
	$\frac{1}{2}O_2(g) + 2H^+(10^{-7}\,M) + 2e^- \longrightarrow H_2O$	+0.82	
	$Ag^+ + e^- \longrightarrow Ag(s)$	+0.80	
	$NO_3^- + 2H^+ + e^- \longrightarrow NO_2(g) + H_2O$	+0.78	
	$Fe^{+++} + e^- \longrightarrow Fe^{++}$	+0.77	
	$O_2(g) + 2H^+ + 2e^- \longrightarrow H_2O_2$	+0.68	
	$I_2(s) + 2e^- \longrightarrow 2I^-$	+0.53	
	$Cu^{++} + 2e^- \longrightarrow Cu(s)$	+0.34	
increasing strength as oxidizing agent	$SO_4^{--} + 4H^+ + 2e^- \longrightarrow SO_2(g) + 2H_2O$	+0.17	
	$Sn^{++++} + 2e^- \longrightarrow Sn^{++}$	+0.15	
	$S + 2H^+ + 2e^- \longrightarrow H_2S(g)$	+0.14	
	$2H^+ + 2e^- \longrightarrow H_2(g)$	0.00	
	$Pb^{++} + 2e^- \longrightarrow Pb(s)$	−0.13	
	$Sn^{++} + 2e^- \longrightarrow Sn(s)$	−0.14	
	$Ni^{++} + 2e^- \longrightarrow Ni(s)$	−0.25	
weak oxidizing agents	$2H^+(10^{-7}\,M) + 2e^- \longrightarrow H_2(g)$	−0.41	
	$Fe^{++} + 2e^- \longrightarrow Fe(s)$	−0.44	
	$Cr^{+++} + 3e^- \longrightarrow Cr(s)$	−0.74	
	$Zn^{++} + 2e^- \longrightarrow Zn(s)$	−0.76	
	$2H_2O + 2e^- \longrightarrow 2OH^- + H_2(g)$	−0.83	
	$Mn^{++} + 2e^- \longrightarrow Mn(s)$	−1.18	
	$Al^{+++} + 3e^- \longrightarrow Al(s)$	−1.66	
	$Mg^{++} + 2e^- \longrightarrow Mg(s)$	−2.37	
	$Na^+ + e^- \longrightarrow Na(s)$	−2.71	
	$Ca^{++} + 2e^- \longrightarrow Ca(s)$	−2.87	
	$Ba^{++} + 2e^- \longrightarrow Ba(s)$	−2.90	
very weak oxidizing agents	$Cs^+ + e^- \longrightarrow Cs(s)$	−2.92	very strong reducing agents
	$K^+ + e^- \longrightarrow K(s)$	−2.92	
	$Li^+ + e^- \longrightarrow Li(s)$	−3.00	

EXACT MASSES OF SOME ATOMIC PARTICLES AND ISOTOPES OF ATOMS

(For additional data see *Handbook of Chemistry and Physics*)

Particle	Mass (u)
proton	1.007276
neutron	1.008665
electron	0.0005486
positron	0.0005486
deuteron	2.0130
alpha particle	4.00150

Isotope of Atom		Precise mass of isotope (u)	Isotope of Atom		Precise mass of isotope (u)
^{1}H	(99.985)*	1.007825	^{22}Ne	(8.82)	21.99138
^{2}H	(0.015)	2.01410	^{23}Na	(100)	22.9898
^{4}He	(100)	4.00260	^{24}Mg	(78.70)	23.98504
^{7}Li	(92.58)	7.01601	^{25}Mg	(10.13)	24.98584
^{6}Li	(7.42)	6.01513	^{26}Mg	(11.17)	25.98259
^{9}Be	(100)	9.01	^{27}Al	(100)	26.98153
^{10}B	(19.6)	10.01294	^{28}Si	(92.21)	27.97693
^{11}B	(80.4)	11.00931	^{29}Si	(4.70)	28.97649
^{12}C	(98.89)	12.0000000†	^{30}Si	(3.09)	29.97376
^{13}C	(1.11)	13.00335	^{31}P	(100)	30.97376
^{14}N	(99.63)	14.00307	^{32}S	(95.0)	31.97207
^{15}N	(0.37)	15.00011	^{33}S	(0.76)	32.97146
^{16}O	(99.759)	15.99491	^{34}S	(4.22)	33.96786
^{17}O	(0.037)	16.99914	^{35}Cl	(75.53)	34.96885
^{18}O	(0.204)	17.99916	^{37}Cl	(24.47)	36.96590
^{19}F	(100)	18.99840	^{40}Ar	(99.60)	39.96238
^{20}Ne	(90.92)	19.99244	^{36}Ar	(0.337)	35.96755
^{21}Ne	(0.257)	20.99395			

*Percentage of isotope in the natural element.
†This is the standard by international agreement.

ELECTROLYSIS AND THE PARTICULATE NATURE OF ELECTRIC CHARGE

In the early 1830s, the distinguished British physicist Michael Faraday (1791–1867) did some experiments in electrolysis. The experiments indicated that electric current travels in integral units or packages, now called electrons. (For example, an atom can carry one, two, or three packages of charge but *cannot* carry 1.5872 or 1.4301 packages.) Faraday's work focused attention on the close relationship between matter and electric charge. An experiment similar to one performed by Faraday is illustrated in the figure. An electric current is passed through three electrolysis cells. The two ammeters show the same reading, indicating that the current entering the cell at the right is identical to that leaving the cell at the left. Thus, the electric circuit guarantees that the same amount of electricity passes through each of the three cells. Note also that two of the cells contain a molten compound that carries the electric current, while the third cell contains mercuric nitrate in water solution. This suggests that the process by which current is carried in the cells containing molten material is the same process by which the current is carried through the solution.

In the first cell (containing mercuric nitrate), the net reaction is the production of metallic mercury and gaseous oxygen. After current has passed through the cell for a definite time, the mass of the mercury produced is 6.03 grams. In the second cell, metallic sodium and gaseous chlorine are formed.[*] The same current that produced 6.03 grams of metallic mercury produces 1.38 grams of molten sodium. The third cell produces metallic aluminum

[*] In practice, calcium chloride must be added to such a cell to lower the melting point of the salt and, even then, the temperature must be high (600 °C). This is the commercial method for manufacturing metallic sodium.

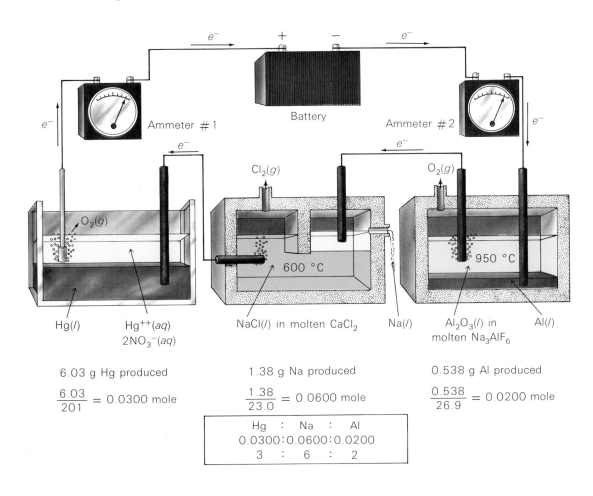

Hg(*l*)   Hg^{++}(*aq*)
              2NO$_3^-$(*aq*)

NaCl(*l*) in molten CaCl$_2$ Na(*l*)

Al$_2$O$_3$(*l*) in   Al(*l*)
molten Na$_3$AlF$_6$

6.03 g Hg produced

$\dfrac{6.03}{201} = 0.0300$ mole

1.38 g Na produced

$\dfrac{1.38}{23.0} = 0.0600$ mole

0.538 g Al produced

$\dfrac{0.538}{26.9} = 0.0200$ mole

Hg	:	Na	:	Al
0.0300	:	0.0600	:	0.0200
3	:	6	:	2

and gaseous oxygen.* But at the temperature used, the oxygen attacks the carbon electrode to form CO_2 gas. The same current that produced 6.03 grams of mercury produces 0.538 grams of aluminum. The amount of oxygen produced in the third cell can be determined from the amount of CO_2 or other products formed. The total oxygen produced is equivalent to the amount produced in the first cell. The conversion of these masses to moles reveals an interesting pattern: the volume or number of moles of chlorine produced is twice the volume or number of moles of oxygen generated. In summary, the same amount of electricity produces 6.03 grams

*This is the basis for the commercial manufacture of aluminum. Another salt is added as solvent to lower the melting point. A mixture of Al_2O_3 and Na_3AlF_6 (cryolite) can be electrolyzed at 950 °C.

of mercury, 1.38 grams of sodium, and 0.538 grams of aluminum.

How are the masses related? If each mass is divided by the appropriate atomic weight, the numbers of moles of each kind of atom are obtained, as shown in the figure. Thus, a given quantity of electric current deposited six atoms of sodium for every three atoms of mercury and two atoms of aluminum. Apparently, the current is "counting" atoms. Our results are quite clear: *a certain amount of electricity will deposit a fixed number of atoms, or some simple multiple of this number, regardless of which element is chosen.* Stated another way, twice as much current is required to deposit 1 mole of mercury as is needed to deposit 1 mole of sodium. Three times as much current is required to deposit 1 mole of aluminum as is required to deposit 1 mole of sodium.

GLOSSARY

This glossary defines many important chemistry terms. Most of the entries appear in boldface type where they are defined in the text and in the list of key terms at the end of each chapter. The page numbers following the definitions refer to the pages on which the terms are defined. More than one page number indicates that the term is discussed in depth in more than one chapter.

absolute temperature scale a temperature scale based on absolute zero, the lowest temperature that can exist; also known as the kelvin scale; 0 kelvin corresponds to -273.1 °C (page 114)

accuracy the closeness of a measured quantity to its accepted value (page 23)

acid a substance that can serve as a proton donor; gives hydrogen ions in water solution (page 459)

acid-base indicator a dye that changes color in solution when the $H^+(aq)$ ion concentration changes across a given specific range (page 464)

acid-base reaction a reaction involving the transfer of a proton from an acid to a base (page 475)

acid equilibrium constant (K_A) the equilibrium constant for the reaction of an acid with water; a strong acid has a large K_A; a weak acid has a small K_A (page 465).

acid rain rain or snow with a pH below 5.6 (page 605)

acid solution a solution in which the concentration of $H^+(aq)$ ions is greater than the concentration of $OH^-(aq)$ ions (page 459)

actinides the elements (atomic numbers 90 through 103) that follow actinium in the periodic table (page 586)

activated complex the unstable arrangement of atoms that forms during the collision of reacting species and from which either products or reactants can be obtained (page 404)

activation energy the minimum energy necessary to get reacting molecules into the activated state so that reaction can proceed (page 403)

active center the place on the surface of an enzyme or catalyst molecule at which reaction can occur (page 566)

addition polymerization the process in which a polymer is formed by reactions between monomer units without the elimination of atoms (page 550)

addition reaction a reaction in which atoms or groups of atoms add to a multiple bond (page 538)

alcohol an organic compound of general formula R—OH where R is a hydrocarbon or related group (page 533)

aldehyde a compound of general formula R—$\overset{\displaystyle O}{\overset{\displaystyle \|}{C}}$—H (page 542)

alkali metal family the elements lithium, sodium, potassium, rubidium, cesium, and francium, located in the first vertical column of the periodic table (page 219)

alkaline earth family the elements beryllium, magnesium, calcium, strontium, barium, and radium, located in the second vertical column of the periodic table (page 244)

alkane a saturated hydrocarbon with the empirical formula C_nH_{2n+2} (page 536)

alpha ray or particle a helium nucleus with a charge of $+2$ and a mass of 4 emitted by certain radioactive elements (page 194)

amide a compound of general formula R—$\overset{\displaystyle O}{\overset{\displaystyle \|}{}}NH_2$, where R is a hydrocarbon (page 549)

amine an organic compound related to ammonia (NH_3) in which one or more of the hydrogen atoms is replaced by an organic group; a primary amine has one H atom replaced, a secondary amine has two H atoms replaced, and a tertiary amine has all three H atoms replaced (page 546)

amphoteric a material which can accept or give up a proton or a metal hydroxide that is insoluble in water but soluble in acidic or basic solution (page 596)

amplitude the vertical distance between the crest and trough of a wave (page 266)

angular molecule a molecule with a bent shape, such as water (page 317)

anion a negatively charged particle (ion), which moves toward the positive electrode, or anode, in an electrochemical cell (page 500)

anode the electrode at which oxidation occurs in an electrochemical cell (page 500)

argon core the stable population of 18 electrons existing in the atoms of all elements beyond argon in the periodic table (page 241)

aromatic compound a compound, such as benzene, containing a ring or rings of carbon atoms with delocalized electrons (page 540)

atom the smallest particle of an element which can participate (or exchange in a chemical reaction; matter is composed of atoms in various combinations (page 63)

atomic number the number of protons in the nucleus of an atom of a particular element (page 191)

atomic weight the relative weight of an atom of a chemical element, based on the weight of an atom of carbon, taken as 12; formerly, atomic weights were based on the weight of an atom of oxygen, taken as 16 (page 72)

Avogadro's hypothesis the proposal made by Amedeo Avogadro in 1811 that equal volumes of different gases under like conditions of temperature and pressure contain the same number of molecules (page 42)

Avogadro's number the number of molecules per mole, numerically equal to 6.02×10^{23} (page 46)

barometer an apparatus for measuring atmospheric pressure (page 107)

base a proton acceptor substance or a substance that generates OH^- in water solution (page 459)

basic solution a water solution in which the concentration of $OH^-(aq)$ ions is greater than the concentration of $H^+(aq)$ ions (page 459)

battery a combination of two or more electrical cells; source of electrical voltage (page 165)

benzene ring a structure of formula C_6H_6 with the six carbon atoms in a flat hexagonal ring and one hydrogen atom attached to each carbon atom (page 539)

beta ray or particle a very fast-moving electron emitted by the atomic nucleus of certain radioactive elements (page 194)

bidentate a molecule that can bind to another atom or ion through two covalent bonds involving two atoms of the bidentate molecule (page 595)

binary compound a compound containing two kinds of atoms and, thus, two elements (page 71)

binding energy the energy released when an atomic nucleus is formed from its component nuclear particles (page 381)

biochemistry the study of chemical processes that occur in living systems (page 560)

body-centered cubic arrangement a structure whose unit cell consists of a simple cube with an atom at each corner and an atom inside the cube at its center (page 343)

boiling point the temperature at which the vapor pressure of a liquid is equal to the pressure of the atmosphere above the liquid (page 145)

Brønsted-Lowry theory a concept of acids and bases as proton donors and acceptors, and of an acid-base reaction as the formation, through proton transfer, of another acid and another base (page 475)

calorie a unit of energy; 1 calorie is the quantity of heat energy required to raise the temperature of 1 gram of water 1 °C; 1 calorie is equivalent to 4.184 joules (page 96)

calorimetry the process of measuring heat transfer (frequently reaction heats) by using a calorimeter (page 364)

carbohydrate a substance with the empirical formula CH_2O; all sugars and starches, for example, are carbohydrates (page 561)

carbonyl group the functional group of atoms $-C\!\!\!\overset{\displaystyle O}{\big\|}-$, found in organic aldehydes, ketones, and acids (page 545)

carboxyl group the functional group of atoms $-C\!\!\!\overset{\displaystyle O}{\big\|}-O-H$ found in organic acids (page 546)

carboxylic acids the family of acids, such as formic acid, acetic acid, oxalic acid, and so on, that contain the carboxyl group (page 546)

catalysis the process of altering the rate of a reaction through the use of a catalyst (page 407)

catalyst a substance or combination of substances that alters the rate of a chemical reaction without being used up itself (page 407)

cathode the electrode at which reduction occurs in an electrochemical cell (page 500)

cation a positively charged particle (or ion), which moves toward the negative electrode, or cathode, in an electrochemical cell (page 500)

cellulose an important part of woody plants such as cotton and flax; chemically, cellulose is a polysaccharide (page 563)

chain reaction a reaction in which the material or energy that starts the reaction is also one of the products which can keep the reaction going (page 379)

Charles's law a relation stating that the volume of a fixed amount of gas at constant pressure varies directly with the absolute temperature; mathematically, $V = kT$, where V = volume, T = absolute temperature of the gas, and k = a constant (page 118)

chemical change a change that produces new substances with new properties; chemical changes are usually accompanied by observable energy changes (page 62)

chemical equation an abbreviated and formalistic representation of a chemical process (page 65)

chemical families vertical groups of elements in the periodic table; members of a given chemical family have similar chemical properties (page 216)

chemical kinetics the study of reaction rates (page 390)

chemotherapy the control and treatment of disease by drugs (page 567)

chlorofluorocarbons synthetic compounds of chlorine, fluoride, and carbon, used as refrigerants (page 613)

chromosphere a jagged, spiky layer of the sun's atmosphere located between the photosphere and the corona, visible from Earth only during a solar eclipse (page 633)

closed-end manometer *see* manometer

close-packed arrangement a geometric packing of spheres which gives a continuing structure of minimum volume; each atom in such an arrangement has 12 nearest neighbor atoms; structural differences between hexagonal and cubic close-packed patterns occur in the arrangement of atoms in the second shell of neighbor atoms; many metals display this close packing of atoms (page 344)

coefficient the number in front of each formula in a chemical equation (page 85)

collision theory a model for a chemical process based on collisions between molecules or molecular fragments (page 393)

complex ion an ion in which molecules or other ions are bonded to a central atom or ion to form a complex unit (page 592)

compound a substance that contains more than one kind of atom in chemical combination (page 68)

concentration (of a solution) the quantity (moles or grams) of a substance in a unit volume or mass of the solute (or solution) such as moles per litre or moles per 1,000 grams of solvent (pages 151, 390)

conceptual definition a definition based on a conceptual model for the system or process (page 472)

condensation polymerization the process by which a polymer is formed from monomer units by elimination of a single molecule, such as water, between functional groups on the monomers (page 550)

coordinate covalent bond a bond in which both shared electrons are provided by one of the bonded atoms (page 595)

coordination number the number of near neighbor ions or molecules surrounding an atom or ion (page 593)

corona the outermost portion of the sun's atmosphere, invisible from Earth except during solar eclipses and extending tens of millions of kilometres into space (page 633)

covalent bond the bond formed when two atoms share a pair of electrons (page 248)

critical mass the amount of radioactive material required to sustain a nuclear chain reaction (page 379)

crystallization the process of forming a crystalline solid from a solution or from molten material (page 148)

cyclic hydrocarbon a saturated hydrocarbon with the carbon atoms linked in rings, with the empirical formula C_nH_{2n} (page 533)

delocalized electron a mobile bonding electron that can move easily between more than two atoms (page 338)

density the mass of a unit volume of a substance (page 104)

deuterium an isotope of hydrogen of atomic number 1 and mass number 2 (page 192)

diatomic (molecule) a molecule made up of two atoms (pages 63, 217)

dimensional analysis *see* unit factor method

disaccharide a sugar containing two simple sugar units per molecule (page 562)

distillation the process of evaporating a substance and condensing the resulting vapor in a separate vessel (page 148)

divalent used to describe atoms that can form two covalent bonds with other atoms (page 306)

DNA abbreviation for deoxyribonucleic acid, the substance in all living cells that transmits hereditary characteristics from one generation to the next (page 568)

double bond the bond formed by sharing four electrons (two electron pairs) between two atoms (page 324)

E° **value** the voltage of a half-cell against a standard hydrogen electrode when all reactants and products are at unit activity or concentration (page 505)

electric charge the numerical difference between the total number of electrons and protons in a given atom or molecule; electric charge is a fundamental property of matter;

there are two kinds of electric charge: positive and negative (page 166)

electric current the movement of electric charge through a medium (page 166)

electrically neutral used to describe an object that has no net electric charge (page 167)

electric dipole a molecule that is electrically positive at one end and electrically negative at the other but has no net electric charge (page 313)

electrochemical, or voltaic, cell a device in which a spontaneous oxidation-reduction reaction produces electrical energy (page 511)

electrode a material that provides a medium for passage of an electric current from a solid to a liquid solution conductor or vice versa; oxidation and reduction reactions occur at the solid-liquid interface (page 500)

electrolysis the process by which an electric current is passed through a liquid phase from solid or liquid electrodes; chemical change occurs at the electrode surfaces (page 177)

electrolytic cell a cell for carrying out electrolysis (page 511)

electromagnetic radiation a form of energy, such as light, X-rays, or radio waves, that travels through a vacuum at the speed of light (page 267)

electrometer a device for detecting and measuring electric charge (page 165)

electron acceptor an atom or ion that will accept electrons (page 495)

electron affinity the energy change associated with the process $A(g) + e^- \longrightarrow A^-(g)$ where A is an atom or ion (page 303)

electron configuration a representation of the number of electrons occupying each occupied orbital of a particular atom (page 282)

element a substance made up of only one kind of atom (page 68)

empirical formula a formula that gives only the relative number of atoms of each element in a molecule, sometimes called the simplest formula; the simplest ratio of each of the atoms in a chemical compound (pages 247, 529)

endothermic a process that absorbs energy; endothermic reactions have positive ΔH values (pages 97, 363)

energy level a specific energy condition characterizing an electron in an atom, molecule, metal, or ion; frequently associated, in general, with the region of space occupied by the electron (page 271)

energy level diagram a diagram representing in graphical form the energy levels in an atom, molecule, ion, or metal (page 278)

enthalpy the heat content of a system at constant pressure, symbolized by the letter H (page 362)

entropy a thermodynamic function related to the disorder of a system (page 441)

enzyme a protein that acts as a biological catalyst (page 566)

equilibrium the state of balance attained for a reversible process occurring in a closed system; the two opposing processes occur at the same time and at the same rate (page 140)

equilibrium constant (K) a number that imposes a condition on reactant and product concentrations in an equilibrium system; for the system at equilibrium represented by the reaction $aA + bB \longrightarrow cC + dD$, the equilibrium constant is $K = \dfrac{[C]^c [D]^d}{[A]^a [B]^b}$ (page 432)

equilibrium vapor pressure the pressure exerted by a vapor in equilibrium with its pure liquid phase (page 140)

equivalence point the point in a titration where one equivalent of reagent A has been added to one equivalent of reagent B; the "end point" of a titration (page 462)

ester a compound of general formula $R-\overset{\displaystyle O}{\overset{\|}{C}}-O-R'$ (page 548)

ether an organic compound in which two hydrocarbon groups are linked by an oxygen atom to give a substance of general formula $R-O-R'$, where R and R' are hydrocarbon groups (page 546)

exothermic a process that releases energy; exothermic reactions have negative ΔH values (pages 97, 363)

experiment a carefully controlled sequence of operations and observations (page 17)

fat an organic compound of general formula

$$R-\overset{\displaystyle O}{\overset{\|}{C}}-O-CH_2$$
$$R'-\overset{\displaystyle O}{\overset{\|}{C}}-O-CH$$
$$R''-\overset{\displaystyle O}{\overset{\|}{C}}-O-CH_2$$

where R, R', and R'' represent hydrocarbon groups (page 564)

free radical a molecule species with one or more unpaired electrons; usually unstable (page 306)

frequency the number of wave crests or troughs that pass a fixed point in a given unit of time; designated by the symbol v, frequently is expressed in the unit 1/sec, or sec^{-1} (page 266)

fructose a simple sugar that is a ketone because the C=O unit is on an internal carbon atom (page 562)

functional group any group of atoms in an organic molecule that undergoes changes during reactions (page 533)

fundamental property a property that is generally observed but for which diligent search has failed to yield a useful model, for example, the attraction of opposite electric charges (page 168)

gamma ray or particle very short wavelength electromagnetic radiation, similar to an X-ray but of higher energy, emitted by certain radioactive elements (page 194)

gas pressure the force exerted by a gas on a given area of its container wall surface (page 20)

genetic code the sequence of nitrogen bases on a DNA polymer (page 574)

glucose a simple sugar that is an aldehyde because its C=O unit is on the end carbon atom (page 561)

ground state the lowest energy level available to an electron (page 271)

half-life the time required for half of the atomic nuclei of a radioactive element to decay (page 198)

half-reaction an equation written to represent the oxidation "half" or the reduction "half" of an oxidation-reduction reaction (page 493)

halide ion the negative ions F$^-$, Cl$^-$, Br$^-$, and I$^-$, formed by halogen atoms during chemical reactions and existing in some solids and solutions (page 224)

halogen family the elements fluorine, chlorine, bromine, iodine, and astatine, located in the seventh vertical column of the periodic table, which have one less electron than each of their noble gas neighbors (page 222)

heat of reaction (ΔH) the *difference* in heat content between products and reactants in a chemical reaction, symbolized by ΔH (page 362)

heats of formation (ΔH_f) the heats of reaction involved in the formation of one mole of product from its elements in their standard, or normal, states at 25 °C and 1 atmosphere (page 365)

helium core the two electrons in the filled 1s orbital plus the nucleus (page 241)

hemoglobin a complex ion, which participates in oxygen transport in the blood, found in the pigment of red blood cells (page 597)

Hess' Law of Constant Heat Summation *see* Law of Additivity of Heats of Reaction

heterogeneous system a system containing two or more distinct phases (page 394)

homogeneous system a system in which all components are in the same phase (page 393)

hybrid orbital an orbital made by combining two or more simple atomic orbitals using appropriate mathematical methods (page 316)

hydration the process in which water molecules add to a molecule or ion (page 347)

hydrocarbon an organic compound containing only carbon and hydrogen (page 533)

hydrogen bond a bond in which a protonic hydrogen on molecule A is attracted to a free electron pair of an electro-negative atom, such as F, O, or N bound in molecule B; hydrogen bonds are about 10 times as strong as van der Waals interactions and about one-tenth as strong as a normal covalent bond (page 349)

hydrogen-ion transfer the transfer of a hydrogen ion from an acidic molecule A to a basic molecule B (page 475)

hydronium ion the name given to the hydrated hydrogen ion; usually represented by the formula H$_3$O$^+$ or abbreviated H$^+$(aq) (page 477)

hydrosphere the liquid portion of Earth including all of the oceans (page 620)

hydroxyl radical the HO· unit with an unpaired electron (page 306)

ideal gas a gas that exhibits the behavior described by the ideal gas law; real gases can approximate this behavior (page 126)

ideal gas constant the numerical quantity represented by the symbol R in the ideal gas law (page 123)

ideal gas law a single expression relating the pressure (P), volume (V), absolute temperature (T), ideal gas constant (R), and number of moles (n) of a gas sample; $PV = nRT$ (page 123)

inert an element that does not form any stable compound (page 212)

infrared a form of electromagnetic radiation with somewhat longer wavelength than visible red light (page 270)

interference, or diffraction, colors the colors displayed by oil films, soap bubbles, and diffraction jewelry that arise from the interference of light waves (page 266)

ion an electrically charged atom or group of atoms (page 169)

ionic bond a chemical bond resulting from attraction of oppositely charged ions (page 246)

ionic character pertaining to a covalent bond in which the bonding electrons move closer to one of the two bonded atoms than to the other, thus moving toward an ionic linkage (page 313)

ionic compound a solid chemical compound composed of oppositely charged ions held together by electrostatic attraction (pages 170, 246)

ionization constant an equilibrium constant for the ionization of a weak acid in water (page 457)

ionization energy the energy required to remove an electron from an isolated gaseous atom; the second ionization energy is the energy required to remove a second electron from a gaseous ion of unit positive charge (page 251)

isomer one of two or more different compounds with identical molecular formulas and different molecular configurations (page 325)

isotope one of two or more kinds of atoms with the same atomic number and different atomic masses (page 192)

kelvin scale *see* absolute temperature

ketone a compound of general formula, $R—C—R'$, where R and R' are hydrocarbon groups (page 544)

kilocalorie a unit of heat energy equal to 1,000 calories (page 360)

kinetic energy the energy possessed by a moving body; $KE = \frac{1}{2}mv^2$ (pages 106, 401)

kinetic-molecular theory of gases a theory that explains the properties of gases in terms of rapidly moving molecules (page 106)

lanthanoides the elements (atomic numbers 58 through 71) that follow lanthanum in the periodic table (page 586)

lattice geometric model showing the regular arrangement of ions in ionic solids (page 221)

law a proven relationship between observations and/or concepts (page 34)

Law of Additivity of Heats of Reaction a relation stating that, when a reaction can be expressed as the algebraic sum of two or more reactions, its heat of reaction is the algebraic sum of the heats of the individual reactions that were added together; also known as Hess' Law of Constant Heat Summation (page 365)

Law of Combining Gas Volumes a relation stating that the volumes of gases used and produced in a chemical reaction can be expressed in ratios of small whole numbers when the volumes are measured at constant temperature and pressure (pages 104, 164)

Law of Conservation of Energy energy is not created or destroyed, but may be exchanged (page 97)

Law of Conservation of Mass mass is always conserved during a physical or chemical process (pages 84, 163)

Law of Definite, or Constant, Composition a relation stating that every sample of a given pure compound is always made up of the same elements in the same proportions by mass (page 163)

Law of Mass Action a relation stating that, when equilibrium exists, there is a simple relationship between the concentrations of products and the concentration of reactants (page 432)

Law of Multiple Proportions a relation stating that when two elements combine to form more than one chemical compound, the masses of one element that combine with a fixed mass of the other element are in the ratio of small, whole numbers (page 164)

Le Chatelier's principle the generalization stating that, if an equilibrium system is subjected to a change, processes occur that tend to partially counteract the imposed change (page 426)

Lewis structure an abbreviated representation of atomic structure in which an element's symbol represents the nucleus and all electrons except the valence electrons and dots around the symbol represent valence electrons (page 243)

limiting reagent a chemical substance that is consumed in a chemical reaction and limits the process (page 87)

linear molecule the geometrical arrangement in which the three or more bonded atoms are arranged in a straight line (page 319)

lithosphere the solid portion of Earth (page 620)

litmus a dye used as an acid-base indicator; it is red in solutions with excess $H^+(aq)$ and blue in solutions with excess $OH^-(aq)$ (page 464)

malleable having the property, usually displayed by metals, of being shaped by beating with hammers, presses, or rollers (page 219)

manometer an instrument for measuring gas pressure; an open-end manometer measures the difference between a gas sample and atmospheric pressure; a properly functioning closed-end manometer gives the absolute pressure of a gas sample (page 108)

mass a measure of the quantity of matter in a body (page 44)

mass number the number of protons plus the number of neutrons in the nucleus of an atom (page 192)

mass spectrometer a device used to determine the absolute mass of an ion using electrostatic and magnetic fields (page 184)

melting point the temperature at which a pure solid changes to a liquid under a given pressure, usually 1 atmosphere (pages 17, 138)

metallic bonds the bonds found in metals (page 341)

metallic state the condition of a substance that exhibits such typical metallic properties as luster, ductility, malleability, and good electrical and heat conductivity (page 339)

microwave radiation a form of electromagnetic radiation with wavelengths somewhat shorter than those of radio waves but longer than those of infrared radiation (page 270)

model a familiar, well-understood system that can be used to explain another system that is poorly understood (page 19)

molar heat of melting or fusion the amount of energy needed to convert 1 mole of solid to a liquid at its melting point; also known as the molar heat of fusion (page 146)

molar heat of vaporization the energy required to vaporize one mole of a pure liquid at its boiling point (page 139)

molarity a term used to define the concentration of a solution given as the number of moles of solute per litre of solution (M) (page 152)

molar mass the mass of 1 mole of a pure substance (page 46)

molar volume the volume occupied by 1 mole of a substance under a given set of conditions; for a gas it is numerically equal to 22.4 litres at STP and to 24.5 litres at 25 °C and 1 atmosphere pressure (pages 51, 104)

mole a mass unit defined as that mass of material containing 6.02×10^{23} molecules, atoms, or other units; one mole of a substance is 6.02×10^{23} molecules or a quantity equal to the molecular weight of the substance in grams (page 46)

molecular formula a shorthand method of using chemical symbols and numerical subscripts to represent the composition of a substance; it denotes the type and number of each kind of atom composing one molecule (pages 65, 530)

molecular weight the weight of a molecule on a scale in which a carbon-12 atom is assigned a mass of 12.0000 (page 44)

molecule the smallest particle into which an element or compound can be divided without changing its properties: a molecule of an element consists of one or more like atoms; a molecule of a compound consists of two or more different atoms (page 39)

momentum a physical quantity equal to the mass of a particle (m) multiplied by its velocity (v) (page 121)

monatomic (molecule) any molecule made up of only one atom (page 64)

monomer the small, individual molecular units that may be joined to form a polymer (page 550)

monosaccharide a sugar with a single, simple sugar unit in each molecule; glucose ($C_6H_{12}O_6$) is a monosaccharide (page 562)

neon core the filled $1s$, $2s$, and $2p$ orbitals plus the nucleus (page 241)

net ionic equation an equation showing only those ions that actively participate in a chemical change; spectator ions are not included in such an equation (page 172)

network solid a solid whose structural units are held together by strong covalent bonds, so that the entire sample can be considered one gigantic molecule; solids, such as diamond, that are characterized by high melting and boiling points (page 336)

neutral solution a solution in which $[H^+(aq)]$ equals $[OH^+(aq)]$ (page 458)

neutron an atomic particle with no electric charge and a mass approximately equal to that of one proton (page 184)

noble gas family the elements helium, neon, argon, krypton, xenon, and radon (page 217)

nuclear equation an equation that represents a nuclear reaction; atomic number and mass number must balance in such an equation (page 195)

nuclear fission the splitting of a heavy atomic nucleus into two lighter nuclei of slightly lighter total mass with the release of energy (page 377)

nuclear fusion the nuclear reaction in which light nuclear particles combine to form heavier atoms with the concurrent release of energy (page 380).

nuclear reactor a device in which the controlled fission (or fusion, hopefully) of nuclei produces new nuclei and energy (page 379)

nucleon a nuclear particle such as a proton or a neutron (page 192)

nucleotide the unit in a DNA or RNA chain in which a phosphoric acid molecule is linked to the sugars ribose or deoxyribose and one of four nitrogen bases is then linked to the sugar molecule; different nucleotides differ only in the identity of the nitrogen base (page 568)

open-end manometer *see* manometer

operational definition a definition based on observable properties or characteristics (page 472)

orbital the region of space around an atomic nucleus where an electron of given energy is most likely to be found associated with an energy level (page 276)

organic chemistry the chemistry of carbon compounds; nearly all compounds containing carbon are called organic compounds (page 525)

oxidation a process involving the loss of electrons, or an increase in the numerical value of a substance's oxidation number (page 486)

oxidation number number assigned to each atom involved in an oxidation-reduction reaction to aid in balancing the equations (page 487)

oxidizing agent a substance that causes the oxidation of another species; in the process, the oxidizing agent is reduced (page 486)

oxygen family the vertical column of elements in the periodic table that includes oxygen, sulfur, selenium, and tellurium (page 246)

ozone a reactive form of oxygen containing three oxygen atoms per molecule, with the formula O_3 (page 613)

partial pressure the pressure each gas in a gaseous mixture would exert if it alone were present; the total pressure of such a mixture is the sum of the partial pressures of each gas present (page 111)

periodic table an arrangement of the chemical elements that emphasizes the regular, periodic recurrences of similar properties (page 212)

phase change the process by which a substance changes from one state to another (page 17)

photosphere the outermost portion, or surface, of the sun, as visible from Earth (page 633)

pH a symbol for the degree of acidity or alkalinity of a solution, defined as $-\log_{10}[H^+(aq)]$; neutral water has a pH of 7, acidic solutions have a pH less than 7, and alkaline solutions have a pH greater than 7 (page 465)

Planck's constant of action the proportionality constant that relates the energy and frequency of electromagnetic radiation; represented by the letter h (E-$h\nu$) (page 271)

polar *see* ionic character

polar molecule a molecule that has electric dipoles; the molecule is positive at one end and negative at the other end but is electrically neutral overall (page 313)

polyester a polymer made from monomer units containing alcohol and carboxylic acid functional groups; the linkages formed between the monomers are ester links (page 552)

polymer a giant molecule made by joining many a small molecular units, called monomers, by the process of polymerization (page 550)

polymerization the process by which polymers are formed; addition polymers are formed by reaction of monomer units without elimination of atoms; condensation polymers are produced by reactions in which a simple molecule, such as water, is eliminated between functional groups (page 550)

polysaccharide a polymer made by joining many simple sugar molecules, such as glucose (the monomer units), producing a structure with a high molecular weight of about 600,000 (page 563)

positive electrode *see* electrode

potential energy energy of position, or stored energy (page 372)

potential energy diagram a graphic representation of a chemical reaction, showing the level of energy of the system as the process proceeds (page 403)

precipitation the formation of a solid precipitate from a solution (page 172)

precision the reproducibility of measured data (page 23)

primary alcohol an alcohol in which the —OH functional group is attached to an end carbon atom (page 543)

principal quantum numbers the numbers that identify the major energy levels of an atom; values are 1, 2, 3, 4, etc. in whole numbers (page 275)

promoting the electron moving an electron to a higher energy level (page 308)

protein a large, amide-linked

polymer made up of amino acid monomer units (page 565)

proton transfer *see* Brønsted-Lowry theory and hydrogen-ion transfer (page 475)

pure substance a material of constant and reproducible composition and properties that contains a single kind of molecule or atom (page 147)

purine a double-ring structure of one five- and one six-membered ring, each ring containing two nitrogen atoms; examples include adenine and guanine (page 570)

pyramidal molecule molecule with atoms arranged in the form of a pyramid; can be triangular, square, etc. (page 318)

pyrimidine a single, six-membered ring structure containing two nitrogen atoms per ring; examples include thymine and cytosine (page 570)

qualitative used to describe qualities as opposed to quantities (page 36)

quantitative used to describe quantities that can be measured and expressed in numerical terms (page 34)

quantum tiny units of energy (page 270)

quantum mechanics a form of mechanics describing atoms in terms of the wave nature of their electrons and with the "quantization" of electron energy states (page 275)

quantum theory a theory of atomic structure stating that an electron in an atom can have only certain discrete energy levels (page 271)

radioactivity the spontaneous breakdown of unstable atomic nuclei; particles and energy are released (page 194)

radio wave a form of electromagnetic radiation with very long wavelength and low frequency (page 270)

rate-determining step the slowest step in a reaction mechanism; it determines the overall reaction rate (page 396)

reaction mechanism the series of steps by which a reaction probably takes place; the several simpler steps must add up algebraically to the overall reaction equation (page 395)

reagent a substance that takes part in a particular chemical reaction (page 87)

recombinant DNA DNA that has undergone a process in which segments of DNA are transplanted from the DNA strands of one organism into the DNA strands of another organism (page 577)

reducing agent a substance that causes the reduction of another substance and, in the process, the reducing agent is oxidized (page 487)

reduction a process involving the gain of electrons, or a decrease in the numerical value of the oxidation number (page 486)

reduction potential (E° value) the electromotive force as measured against a standard hydrogen electrode for the process in which a substance at unit concentration (activity) gains one or more electrons to generate a new substance at unit concentration (activity) (page 495)

RNA abbreviation for ribonucleic acid; the substance in all living cells that directs the synthesis of cellular proteins (page 568)

saturated hydrocarbon a hydrocarbon with the empirical formula C_nH_{2n+2}, containing single bonds between carbon atoms (page 533)

saturated solution a solution in equilibrium, at a given temperature, with excess solid, thus containing all the dissolved solid it can hold in equilibrium at that temperature (page 154)

secondary alcohol an alcohol in which the —OH functional group is attached to a carbon atom that is bound to two hydrocarbon groups, R and R', and one hydrogen atom

$$R-\overset{\overset{\displaystyle H}{\overset{\displaystyle |}{\overset{\displaystyle O}{|}}}}{\underset{\underset{\displaystyle H}{|}}{C}}-R' \text{ (page 543)}$$

significant figures all certain figures in a measured quantity plus one uncertain figure (page 23)

solubility the concentration that results when a fixed amount of liquid has dissolved all the solid it can hold at equilibrium at a given temperature (page 154)

solubility product constant the equilibrium constant representing the reaction in which a slightly soluble solid dissolves in water; represented by K_{sp}, it is the product of the concentrations of the ions formed in a saturated solution, each concentration raised to the appropriate power (page 434)

solute one of the two or more components, or pure substances, mixed to form a solution; for solutions in which a pure solid is dissolved in a pure liquid, the solid is called the solute (page 151)

solution a homogeneous (alike throughout) mixture; properties of solutions depend on the relative amounts of the pure substances in the mixture (page 147)

solution composition the kinds and relative amounts of the components, or pure substances, mixed to form the solution (page 151)

solvent one of the components, or pure substances, mixed to form a solution; the liquid component in which the solute is dissolved (page 151)

spectator ions ions that are present at the beginning as well as at the end of a chemical process, but that play no active role in the reaction (page 172)

spectrometer an instrument used to measure spectra of various kinds; one kind of spectrometer can separate a light beam into its various colors (page 268)

standard state specified concentrations used as guidelines for measurements; the standard state for gases and pure substances is 25 °C and 1 atmosphere pressure; a 1 *M* solution is the standard state for ions; rigorously, the standard state is unit activity of ions or molecules in a solution (page 505)

standard temperature and pressure (STP) 0 °C and 1 atmosphere pressure (page 50)

starch a mixture of glucose polymers, some of which are water-soluble; starch is a polysaccharide (page 563)

stratosphere the portion of Earth's atmosphere located above the troposphere (page 611)

strong electrolyte a substance that undergoes extensive ionization when dissolved in a solvent (page 456)

structural formula a formula that indicates the kind, number, arrangement, and types of bonds between atoms in a molecule (pages 66, 530)

sublimation a phase change directly from a solid to a gas (page 18)

substitution reaction a process in which a given structural unit such as a hydrogen atom in a molecule is *replaced* by another unit, such as a halogen atom (page 538)

substrate the substance upon which an enzyme acts (page 566)

sucrose the chemical name for common table sugar, with the molecular formula $C_{12}H_{22}O_{11}$; an example of the class of compounds called carbohydrates (page 560)

template a mold, or pattern, from which similar patterns can be made; for example, one polynucleotide chain of DNA serves as a template for the synthesis of a similar new DNA chain (page 571)

tetravalent used to describe atoms with a bonding capacity of four (page 308)

theory an explanation of observed natural events using a model to explain the cause of such events; a theory is tested by appeal to experiment (page 34)

titration process the progressive addition of a base to an acid or an acid to a base; applicable also to addition of any reactant A to any other reactant B (page 463)

total pressure *see* partial pressure

transition elements horizontally placed groups of metallic elements in the periodic table formed by the filling of the 3*d*, 4*d*, and 5*d* energy levels with electrons; elements 21 through 29, 39 through 47, 57 through 79, and 89 through 103 (page 585)

transmutation reaction a nuclear reaction in which the identity of a nucleus is changed by al-tering the number of protons and/or neutrons in the nucleus (page 195)

triple bond the bond arising from the sharing of three pairs of electrons between two atoms, usually represented by three dashes between atoms, such as N≡N (page 325)

trivalent used to describe atoms that have a bonding capacity of 3 (page 307)

troposphere the lowest portion of Earth's atmosphere, extending from ground level to an altitude of about 10 kilometres (page 611)

ultraviolet a form of electromagnetic radiation just beyond the violet end of the visible spectrum (page 270)

uncertainty principle the principle stating that there is uncertainty in the precision with which the instantaneous position and momentum of an atomic particle can be specified (page 273)

unit conversion factor a fraction, numerically equal to 1, of the same quantity expressed in different units; for example, 2.54 centimetres per inch, 3 feet per yard, or 60 seconds per minute (page 47)

unit factor method (dimensional analysis) the inclusion and canceling of units in all numerical calculations (page 47)

univalent used to describe atoms with a bonding capacity of 1 (page 304)

unsaturated hydrocarbon a hydrocarbon containing one or more double or triple bond (page 534)

vacuum an enclosed space in which there is very close to zero pressure (page 110)

valence electrons those electrons of an atom assumed to participate in the normal bonding processes (page 243)

valence-shell electron-pair repulsion (VSEPR) model a theory that provides a method for predicting and accounting for molecular architecture; it predicts that electron pairs (free and bonding) around an atom will be situated as far apart as possible (page 316)

van der Waals forces weak intermolecular forces thought to arise because of the temporary attraction of one molecule's electrons for a neighboring molecule's nucleus (page 332)

visible spectrum the pattern of colors formed by passing white light through a prism or grating (page 268)

voltage a measure of the driving force for moving charge through a system (page 500)

voltmeter an instrument for measuring voltage (page 501)

wavelength the distance between two corresponding points on successive waves; represented by the letter λ; large wavelengths are expressed in metres, small wavelengths in nanometres (10^{-9} metre) (page 266)

wave mechanics *see* quantum mechanics

weak electrolyte a substance that dissolves in water to form mostly neutral molecules and few ions (page 456)

weight the force with which a material object is attracted to the earth or any other field of gravity (page 44)

X-ray a form of electromagnetic radiation with very short wavelength and very high frequency (page 270)

INDEX

WORKING ATOMIC WEIGHTS*

Element	Atomic number	Atomic weight	Element	Atomic number	Atomic weight	Element	Atomic number	Atomic weight
actinium (Ac)	89	(227)†	curium (Cm)	96	(247)	lead (Pb)	82	207.2
aluminum (Al)	13	27.0	dysprosium (Dy)	66	162.5	lithium (Li)	3	6.94
americium (Am)	95	(243)	einsteinium (Es)	99	(254)	lutetium (Lu)	71	175.0
antimony (Sb)	51	121.8	erbium (Er)	68	167.3	magnesium (Mg)	12	24.3
argon (Ar)	18	39.9	europium (Eu)	63	152.0	manganese (Mn)	25	54.9
arsenic (As)	33	74.9	fermium (Fm)	100	(257)	mendelevium (Md)	101	(258)
astatine (At)	85	(210)	fluorine (F)	9	19.0	mercury (Hg)	80	200.6
barium (Ba)	56	137.3	francium (Fr)	87	(223)	molybdenum (Mo)	42	95.9
berkelium (Bk)	97	(247)	gadolinium (Gd)	64	157.2	neodymium (Nd)	60	144.2
beryllium (Be)	4	9.01	gallium (Ga)	31	69.7	neon (Ne)	10	20.2
bismuth (Bi)	83	209.0	germanium (Ge)	32	72.6	neptunium (Np)	93	(237)
boron (B)	5	10.8	gold (Au)	79	197.0	nickel (Ni)	28	58.7
bromine (Br)	35	79.9	hafnium (Hf)	72	178.5	niobium (Nb)	41	92.9
cadmium (Cd)	48	112.4	helium (He)	2	4.00	nitrogen (N)	7	14.01
calcium (Ca)	20	40.1	holmium (Ho)	67	164.9	nobelium (No)	102	(259)
californium (Cf)	98	(251)	hydrogen (H)	1	1.008	osmium (Os)	76	190.2
carbon (C)	6	12.01	indium (In)	49	114.8	oxygen (O)	8	16.00
cerium (Ce)	58	140.1	iodine (I)	53	126.9	palladium (Pd)	46	106.4
cesium (Cs)	55	132.9	iridium (Ir)	77	192.2	phosphorus (P)	15	31.0
chlorine (Cl)	17	35.5	iron (Fe)	26	55.8	platinum (Pt)	78	195.1
chromium (Cr)	24	52.0	krypton (Kr)	36	83.3	plutonium (Pu)	94	(244)
cobalt (Co)	27	58.9	lanthanum (La)	57	138.9	polonium (Po)	84	(209)
copper (Cu)	29	63.5	lawrencium (Lr)	103	(260)	potassium (K)	19	39.1

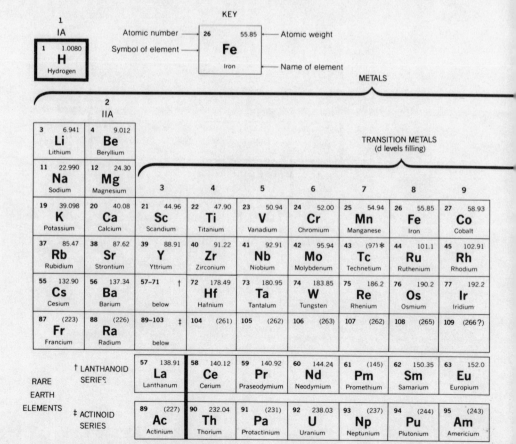

PERIODIC TABLE